Recent Advances and New Challenges in Coupled Systems and Networks: Theory, Modelling, and Applications

Recent Advances and New Challenges in Coupled Systems and Networks: Theory, Modelling, and Applications

Guest Editors

Sundeep Singh
Weizhong Dai

MDPI

Basel • Beijing • Wuhan • Barcelona • Belgrade • Novi Sad • Cluj • Manchester

Guest Editors

Sundeep Singh
Faculty of Sustainable Design Engineering
University of Prince Edward Island
Charlottetown, PE
Canada

Weizhong Dai
College of Engineering and Science
Louisiana Tech University
Ruston, LA
USA

Editorial Office
MDPI AG
Grosspeteranlage 5
4052 Basel, Switzerland

This is a reprint of the Special Issue, published open access by the journal *Mathematical and Computational Applications* (ISSN 2297-8747), freely accessible at: https://www.mdpi.com/journal/mca/special_issues/36OZIA04R7.

For citation purposes, cite each article independently as indicated on the article page online and as indicated below:

Lastname, A.A.; Lastname, B.B. Article Title. *Journal Name* **Year**, *Volume Number*, Page Range.

ISBN 978-3-7258-4415-9 (Hbk)
ISBN 978-3-7258-4416-6 (PDF)
https://doi.org/10.3390/books978-3-7258-4416-6

Contents

About the Editors

Sundeep Singh

Sundeep Singh is an Assistant Professor in the Faculty of Sustainable Design Engineering at the University of Prince Edward Island, Charlottetown, Canada. He received his Master of Technology in Mechanical Engineering (2012) at the National Institute of Technology Hamirpur, India, and his Ph.D. in Mechanical Engineering at the Indian Institute of Technology Ropar, India. He has more than 3 years of postdoctoral research experience in several Canadian universities, such as McMaster University, University of Calgary, and Wilfrid Laurier University. He has co-authored 60+ publications in peer-reviewed journals and conferences and has chaired four special symposiums at international conferences. He is also serving as a Guest Editor and Reviewer for several journals and conferences and has reviewed more than 100 research papers to date. He has over 5 years of teaching experience and has received multiple teaching excellence awards. His current research expertise is highly inter-disciplinary and includes vast areas in the field of mechanical and biomedical engineering, such as thermo-fluids, computational fluid dynamics, minimally invasive thermal therapies, and bio-heat transfer.

Weizhong Dai

Weizhong Dai is currently a McDermott International Professor of Mathematics at the College of Engineering and Science at Louisiana Tech University. He received his B.S. from Huaqiao University in 1983, his M.S. from Xiamen University in 1986, and completed his Ph.D. at the University of Iowa in 1994. His research is interdisciplinary, spanning both science and engineering. It covers numerical methods—such as finite difference, finite element, finite volume, and artificial neural networks—for solving partial differential equations, including fractional-order PDEs. His work also includes numerical heat transfer, micro/nano heat transfer, thermal deformation, bio-heat transfer, quantum physics, computational fluid dynamics, and numerical simulations to study the bio-effects of electromagnetic waves. Additionally, his research explores hydrogen–metal reactions for hydrogen storage, DNA sequencing, and micro-fabrication techniques like X-ray lithography, laser chemical vapor deposition, and melt crystallization. He has authored more than 200 scientific papers in peer-reviewed journals and conferences, receiving more than 3700 citations (H-index 32).

Mathematical and Computational Applications

MDPI

Editorial

Recent Advances and New Challenges in Coupled Systems and Networks: Theory, Modelling, and Applications (Special Issue in Honor of Professor Roderick Melnik)

Sundeep Singh [1,*] and Weizhong Dai [2]

1 Faculty of Sustainable Design Engineering (FSDE), University of Prince Edward Island, Charlottetown, PE C1A 4P3, Canada
2 College of Engineering and Science, Louisiana Tech University, 201 Mayfield Ave., Ruston, LA 71272, USA; dai@coes.latech.edu
* Correspondence: sunsingh@upei.ca

Coupled systems and networks are ubiquitous across all branches of science and engineering, while mathematical and computational models play a fundamental role in their studies. This Special Issue of *Mathematical and Computational Applications* is in honor of Professor Roderick Melnik, who has pioneered many advanced models, allowing for significant breakthroughs in such studies, and has made influential contributions to this field over the past few decades. His professional activities during this time, covering three continents, exemplify ceaseless service to the interdisciplinary scientific community, noted for remarkable achievements and contributions across a wide range of topics.

1. Introduction

Diverse as they may appear, all of Professor Roderick Melnik's research activities stem from his boundless curiosity about our world, its multiscale nature, and the spatio-temporal relationships among the systems, processes, and phenomena within it.

Roderick has always been fascinated by the breathtaking capabilities of the human brain. With some 86 billion neurons forming around 100 trillion connections among themselves, and supported by approximately 85 billion nonneuronal cells, the brain is a remarkable organ that typically fits within an average volume of around 1300 cubic centimetres. But how does this complex structure interact with our vast universe, which is estimated to contain 2 trillion galaxies and 10^{24} stars? Albert Einstein famously said, "There are only two ways to live your life. One is as though nothing is a miracle. The other is as though everything is a miracle". Roderick has always chosen the latter perspective. He has dedicated his career to advancing our understanding of complex systems, processes, and phenomena, as well as their dynamic interactions with the environment. As a distinguished expert in applied and computational mathematics, he also possesses extensive knowledge and expertise across various other disciplines. His groundbreaking contributions are truly interdisciplinary, covering fields such as computational and network sciences, engineering and informatics, and natural and social sciences.

Professor Melnik has held the Tier I Canada Research Chair in Mathematical Modelling for the past two decades. He is an inspiring educator who instils a passion for the power of mathematical models in his students and young scholars. His active engagement with graduate students, postgraduate students, and postdoctoral fellows has earned him a reputation as an enthusiastic and motivational teacher. He has been a founding director of several research institutes, including the MS2Discovery Interdisciplinary Research Institute for Mathematical and Statistical Modelling in Scientific Discovery, Innovation

 https://doi.org/10.3390/mca30030051

and Sustainability. Additionally, he has been a driving force behind international interdisciplinary initiatives, such as the Applied Mathematics, Modelling and Computational Science (AMMCS) framework, Banff International Research Station workshops, and Fields Institute research programs. These initiatives have fostered interdisciplinary activities, and have played a significant role in producing exceptional graduates who make a profound impact in the fields of interdisciplinary science, mathematics, engineering, and industry.

Born to a family of teachers, books about scientists and mathematicians appeared early on the table of Dr. Melnik. His father was a mathematics and physics teacher, while his mother was a linguist, teaching English and Ukrainian. One of Dr. Melnik's favourite books during high school was "Norbert Wiener: A Life in Cybernetics", which significantly influenced his career choice. In 1989, he earned his PhD in Computational Mathematics from Kyiv State University. He began his academic career as an Assistant Professor at the National Technical University KPI, where he was promoted to Associate Professor in 1991, before moving to the University of South Australia in Adelaide.

Dr. Melnik's initial interest in coupled complex systems, particularly in smart materials and structures, as well as in information theory and control, dates back to the early years of his career. After a brief stint at a university in Queensland, his family relocated to Sydney, where he worked as a senior mathematician at the Commonwealth Scientific and Industrial Research Organization, based at the Macquarie University Campus. During this time, he expanded his research interests and significantly contributed to various applied fields, including applied mathematics in electronic and mechanical engineering, climate modelling, and information and network sciences. His work also involved developing innovative models and computational methodologies to describe complex physical processes, such as phase transformations in complex materials with memory, and applications in the life sciences.

At 37, Dr. Melnik became a full professor at the University of Southern Denmark, where he served as the Head of Mathematical Modelling at the Mads Clausen Institute. There, he led multidisciplinary teams of researchers, and actively collaborated with colleagues in both the European Union and the United States. During this time, Dr. Melnik and his teams made significant contributions to coupled multiscale problems and developed innovative models for nanoscience and nanotechnology. After moving to North America, he held full professorships in the USA and Canada, actively maintaining collaborations with his European colleagues. His exceptional achievements have earned him numerous distinctions and honors across Denmark, Spain, Italy, and the United Kingdom, in addition to his Tier I Canada Research Chair role. In North America, Dr. Melnik has made remarkable contributions across a wide range of fields where mathematical, statistical, and computational sciences intersect with natural sciences and engineering, as well as with the humanities and social sciences, particularly in the areas of cognitive systems, networks, and neuroscience.

Professor Roderick Melnik has an extensive body of work, including publications in books, book chapters, and over 500 journal and conference articles. Summarizing his contributions in a brief introduction is a daunting task. However, it is essential to highlight Dr. Melnik's remarkable ability to initiate and advance research in various fields, making significant contributions to each one. He is an endless source of inspiration and innovative ideas. Several key themes consistently persist throughout Professor Melnik's scientific career. He has a deep and lasting interest in complex systems and networks, particularly how they couple and interact with their environment.

The pervasiveness of coupled systems and networks in all branches of science, engineering, human society, and the world surrounding us leads to many forefront challenges. These systems and their networks span a wide range of scales, from the nanoscale to planetary and cosmic scales, often exhibiting complex spatiotemporal interactions. Below, we highlight some representative contributions of Prof. Melnik within this rapidly evolving field. Following this, we provide an overview of the current Special Issue, and summarize the contributions from several esteemed research groups around the globe.

2. Physical and Engineering Coupled Systems and Their Networks

One of the areas of coupled phenomena that attracted the attention of Dr. Melnik at an early age was thermoelasticity, which is based on the principles of thermodynamics, elasticity, and heat conduction. Coupled thermoelastic phenomena are crucial in modern technology, from semiconductor engineering [1] to manufacturing and materials science. His contributions to this area are vast and insightful. Among others, they include essential works with his collaborators on the theory and applications of thermoelasticity with thermal relaxation, also known as hyperbolic thermoelasticity. Working on these problems, he expanded his interest to extended thermodynamics where non-equilibrium states and processes, particularly those involving fast and steep changes, may occur.

In describing the time evolution of a statistical ensemble, relaxation-time approximations and their generalizations play a fundamental role by connecting the distributions of the variables of interest at different stages of their temporal evolution [2]. A hierarchy of these approximations was explored by Dr. Melnik, which uncovered essential features of nonlocal and nonequilibrium phenomena in semiconductors.

Complex nonlinear dynamics, including those exhibited during various phase transformations, have been an ongoing topic in Dr. Melnik's works. Some such topics and, in particular, their applications, have their origins in smart materials and structures. Among them is a class of materials with memory, such as shape memory alloys. In a quite general setting, many related problems can be formulated as dynamic problems of nonlinear thermoelasticity, as presented for the first time in Dr. Melnik's research.

Further generalizations and systematic methodologies for constructing computational models for coupled multiscale dynamic problems were developed in [3]. The early works by Dr. Melnik continue to stand out today. There is a growing recognition within the scientific and engineering communities that many real-world problems are inherently interconnected. This interconnectedness or coupling can arise from various sources, such as the interaction of different physical fields. Examples of such coupled problems include electroelasticity, thermoelasticity, viscoelasticity, and magnetoelasticity. Additionally, analyzing different states of matter together leads to other coupled issues such as fluid–structure interactions, aeroelasticity, and various phase transitions. These problems are often referred to as multi-physics problems. However, this term does not encompass many other significant coupled issues where factors such as chemical and biological fields play a crucial role. For instance, in climate modeling, it is essential to consider the physical, chemical, biochemical, geological, and many other characteristics of the Earth and its surroundings, and how they interact with one another. Scientific fields and engineering disciplines such as biomechanics, climate modeling, geomechanics, and nanotechnology share a common thread: advancements in these areas are closely linked to our ability to effectively solve coupled problems [3].

In the context of smart materials, such as shape memory alloys (SMA), it is essential to use an extension of the standard Ginzburg–Landau theory by incorporating the full coupling between mechanical and thermal fields, leading to non-isothermal Ginzburg–Landau models, as shown in works of Dr. Melnik and his collaborators. Dealing with phase transitions in these materials requires the development of advanced approaches, such as those based on the phase-field theory, allowing them to handle sharp interfaces. Furthermore, at the numerical level, efficient procedures need to be developed, which were put in the framework of isogeometric analysis by Dr. Melnik and his team. Expanding the phase-field approach, they also developed new models and numerical procedures for SMA nanostructures with enhanced properties and other areas covering sensor and actuator technology, biomedical applications, and aerospace.

The field of smart materials and structures is abounded in coupled systems and networks. Another group from this field where Dr. Melnik made significant contributions is the development of mathematical models for piezoelectricity. Piezoelectric materials, ranging from crystals and ceramics to polymers and biological materials, have a wide range of applications, including sensor and actuator technology and energy harvesting. Additionally, piezoelectricity extends beyond the physical and engineering fields, as it is a dynamic phenomenon also observed in biological systems. Dr. Melnik was the first to rigorously prove the well-posedness of a large class of piezoelectricity models in the dynamic case. He also worked and contributed extensively to interactions of coupled piezoelectric systems with various media, such as acoustic and fluid, as well as accounting for the coupling to other physical fields such as thermal, focusing on the dynamic case.

Prof. Melnik's quest for a better understanding of coupled systems and networks at the smallest scales, such as nano, has continued with his outstanding contributions in many different areas of nanoscience and nanotechnology. One such area is low-dimensional nanostructures such as quantum wells, nanorods, quantum dots, and superlattices where, with his doctoral students and colleagues, they made several important discoveries, including predicting barrier localization in modulated nanowires and quantum confinement phenomena in nanowire superlattices. These tiny objects exhibit unique properties and functionalities unseen in the bulk case. Leveraging these unique properties and potential for advanced technologies, they have diverse applications across various fields, including energy storage, catalysis, sensors, optoelectronics, and biomedicine, to name just a few.

In advancing the mathematical theory of these nanostructures, Dr. Melnik's team developed a systematic methodology for incorporating deformation effects, nonlinear strain theory, as well as higher-order nonlinear electromechanical effects. They also proposed an innovative solution to a long-standing problem in the field. One of the prominent features of this research was accounting systematically for the coupling of physical fields that was also carried out to the general multiband case, and revealing new attributes of complex dynamics, as well as new phenomena and effects, particularly in the context of nanowire superlattices. Part of the research in this direction was linked to their studies of SMA nanostructures mentioned earlier.

Two-dimensional materials such as graphene and other layered materials have a special place among low-dimensional nanostructures. With its exceptional properties, including high strength, flexibility, and conductivity, the range of graphene applications continues to grow. A special focus of Dr. Melnik's team in this context was strain engineering for graphene nanoribbons, where they provided new critical insight into the properties of these objects by using state-of-the-art first-principle ab initio calculations.

The other critical class of problems addressed in Dr. Melnik's research was related to spintronics, where we attempt to control or manipulate electron spin for advanced technological applications, including high-density data storage, sensor and actuator technology, as well as emerging technologies pertinent to alternative types of computing such as quantum or neuromorphic. Much of the research in this direction, carried out by Dr. Melnik's group, focused on spintronics in quantum dots and the applications of geometric phases (such as the Berry phase), as well as the Feynman disentangling technique (e.g., [4]). While advancing a general theory of spin–orbit coupling, critical new insight was also obtained for other classes of low-dimensional nanostructures, such as nanowire superlattices and graphene nanoribbons, as well for the theory and potential applications of quantum computing, quantum information processing, and next-generation solid-state quantum devices. The next generation of nanoelectronic devices is poised to benefit significantly from this development by reducing their power consumption and increasing their memory and processing capabilities.

3. Biological and Ecological Coupled Systems and Their Networks

As the importance of coupled biological systems continues to grow, the role of modelling and analysis in this field becomes critical. In many cases, information has to be transferred between scales to develop predictive models of complex behaviour, such as biocell dynamics, tissue evolution, or the progression of neurodegenerative diseases. To extract meaningful information from large datasets or complex networks, adequate mathematical approaches and computational tools, including multiscale techniques, are essential. Dr. Melnik has significantly contributed to these areas, with several breakthrough discoveries made due to the concerted interdisciplinary efforts of his team. Among the diverse areas he contributed to, we mention the analysis of complex nonlinear cell dynamics and the dynamics of molecules that carry the genetic instructions for the development and functioning of an organism, such as DNA. He and his team looked deeper into coupled processes and phenomena within biocells, revealing their importance and evaluating the role of microtubules, contributions of nonlocal effects, and exploring auxeticity in biosystems. Some of this research led them to uncover the role of cell structures and organelles, and to our better understanding of piezoelectricity and flexoelectricity in biological cells.

The research along these directions is just one example that discoveries can often be made at the interface between different disciplines, and such interfaces are never rigid. For example, the research in coupled physical and biological systems substantially overlap through their applications in bioengineering and biomedicine, representing other areas where we can find significant contributions from Dr. Melnik and his team. Some of them include the analysis of coupled effects in biological tissues, while others analyze the emergence of carbon nanotubes and their arrays as promising materials for various biomedical applications, ranging from tissue engineering to bioimaging.

Significant contributions were also made to nucleic architectonics, where the design and construction of complex structures and devices are carried out using nucleic acids as building blocks. Some of the key contributions were made to the development of RNA nanostructures, where Dr. Melnik and his team discovered the new phenomenon of self-stabilization and revealed their new properties through studies by molecular dynamics and atomistic-to-continuum modelling approaches.

Moreover, new nonlinear and nonlocal models were developed to better describe the vibrations of piezoelectric nanowire resonators. Such resonators hold a high potential in detecting tiny bio-particles such as DNA, RNA, proteins, viruses, and bacteria. Many of the models developed in Dr. Melnik's group and described here were designed with data-driven environments in mind.

The intrinsic coupling between mechanical and thermal fields is essential in describing processes and phenomena connected with ultrashort pulsed lasers with their applications in diverse fields, from material processing to biomedicine. Amplified by full thermo–electro–mechanical interactions, it also plays a critical role in thermal ablation of biological tissues where heat relaxation time effects should also be considered [5]. When developing effective therapies that utilize radiofrequency, thermal ablation, and emerging technologies, it is crucial to consider the various domain heterogeneities involved. Additionally, distinct characteristics, such as the non-Fourier heat transfer, play a significant role in the thermal ablation of biological tissues, the laser ablation of brain tissues, and cardiac ablation. These topics were thoroughly examined in the studies of Dr. Melnik's team.

A significant part of Dr. Melnik's group research over recent years has been devoted to studies of the brain, brain networks, and neurodegenerative diseases, where new multiscale approaches have been developed by his group (e.g., [6]). Some of the problems at the forefront of the current knowledge in this field have been confronted by Dr. Melnik's group with their developed nonlocal models (e.g., [7]), while others required in-depth analyses of complex non-Markovian dynamics and the application of nonequilibrium approaches.

Social and biosocial interactions and complex dynamics of knowledge creation have been the areas of interest for Dr. Melnik's research for a long time. Together with his collaborators in Europe, they developed and analyzed the processes of meaningful social interactions and the emergence of collective knowledge through knowledge networks (e.g., [8]), as well as underlying key mechanisms of self-organized criticality in these social processes that led to several important discoveries pertinent to the intrinsic features of social dynamics. This research has underpinned the critical importance of coupled biosocial dynamics in the refined modelling of infection transmissions, including the recent pandemic. Furthermore, modelling human dynamics, including human crowds, has also led to valuable findings by Dr. Melnik's group. The applications of the latter work span diverse fields, ranging from efficient evacuation procedures in emergencies to our better understanding of complex social and urban systems.

Developing new ecological models, including the areas of population dynamics and pattern formations, has been another area of Dr. Melnik's contributions over the years. He and his team discovered that for addressing many problems in this area, accounting for nonlocal interactions is essential for an adequate description of the underlying processes where, in some cases, we observe the emergence of nonequilibrium dynamics. His group has also pioneered the models that allow for the integration of psychological effects and collective decision-making in cooperative behaviour.

4. Intelligent Systems and AI Technologies in Data-Driven Applications

As we mentioned, data-driven applications, coupled systems, and networks have been a driving force for much of Dr. Melnik's research described in the previous sections, including the directions pertinent to complex dynamics and the emergent behaviour. Once human factors are incorporated in the modelling framework for control, as occurred for the first time in [9], a realm of intelligent systems emerge. Such systems are known as technologically advanced devices, computers, or machines that perceive and respond to the world around them through gathering, analyzing, and responding to data, potentially learning and adapting based on experience, mimicking humans. Artificial Intelligence (AI) is a technology behind this development that enables such devices, computers, or machines to simulate human learning, comprehension, problem solving, decision-making, creativity, and autonomy. Using predictive modelling, generative AI is putting this development at a new level, while underlying technologies on which it is built remain machine learning (ML) and deep learning (DL). These two latter technologies have been developed by Dr. Melnik's group in various contexts, ranging from shape memory graphene nanoribbons to bio-inspired robot propulsion.

When AI is integrated with model-based design, reduced-order models can be particularly useful. Conversely, AI technologies can also enhance the development of reduced-order models. However, creating reduced-order models that effectively simplify complex, high-fidelity models while accurately capturing their behavior is a challenging task. Dr. Melnik's group has addressed this issue in various data-driven applications ranging from materials science to biomimetic systems.

Bayesian methods serve as fundamental tools in many AI learning algorithms, helping to extract important information from relatively small datasets and manage missing data effectively. One notable Bayesian technique is Approximate Bayesian Computation, which allows for likelihood-free inference in AI applications. The development of this method, along with other Bayesian approaches, played a crucial role in Dr. Melnik's group's research, contributing to our improved understanding of neurodegenerative diseases such as Alzheimer's and Parkinson's.

The original paper on coupling control with human factors in complex systems was published in *Engineering Applications of Artificial Intelligence* [9]. It used intelligent transportation systems as a prototype example. The theoretical foundations of this control approach can be found in earlier papers (e.g., [10] and others).

The intersection of network science and AI provides powerful tools for understanding and managing complex systems. AI utilizes network structures to analyze data, make predictions, and automate tasks. In its turn, network science offers the framework for modeling and comprehending these intricate relationships. Dr. Melnik and his collaborators on this topic contributed significantly to this field, especially to higher-order networks revealing new intrinsic features of human connectomes and brain hubs (e.g., [11]).

5. Other Applications of Coupled Systems and Networks

There are many other fields pertinent to the research on coupled systems and networks where one can find remarkable contributions from Prof. Melnik. In this section, we collected only some of them.

The development of novel multiscale models and accounting for size effects, which can be captured with nonlocal models, have been one of the highly visible threads in his research. At smaller scales, particularly in nanotechnological applications, such effects can be of great consequence. At large scales, e.g., in climate modelling, their influence can be crucial, requiring far more detailed analyses than currently given in the literature. Both of these scenarios were examined in Dr. Melnik's research.

Coupled systems and their networks are commonly encountered in manufacturing processes and materials science. Complex materials, such as polymers and composites, necessitate the development of efficient methods to enhance our understanding of the relationship between their structure and properties, even at the initiation of the computational process. In addressing these challenges, the results obtained by Dr. Melnik and his collaborators have significant implications, including in drug design and delivery, a field where he has made contributions in various contexts. Furthermore, when considering the dynamics of these complex materials, relaxation effects may become crucial. A significant portion of the research in this field involved the development and application of advanced computational methods, including molecular dynamics (MD) and first-principles studies. Dr. Melnik's group also utilized these methodologies to uncover the properties of complex materials that are important to geosciences and seismology.

Green and alternative energy is a field of ongoing research where Dr. Melnik and his interdisciplinary team have made significant contributions. They developed innovative models for photovoltaic systems that convert sunlight into electricity using semiconductor materials. Additionally, they designed and analyzed energy systems to enhance efficiency. Dr. Melnik's contributions also extend to other renewable energy sources, such as hydrogen production, which offers clean and sustainable energy solutions. Furthermore, he has worked on models to improve the efficiency of electric vehicles by accounting for stick-skid phenomena, enabling better control of their dynamics.

The development of novel models for energy harvesting and environmentally friendly, sustainable technologies continues to be an important avenue of research for Dr. Melnik's team. This development also included research into photosynthetic systems and light-harvesting complexes, as well as lead-free piezocomposites, with detailed analyses of the roles of polycrystallinity, nanoadditives, and anisotropy, nonlinear, and nonlocal effects (e.g., [12]). Based on this research, they proposed several innovative multiscale designs of nanoengineered matrices for lead-free piezocomposites, including those where improved performance can be achieved via controlling anisotropy and auxeticity. Auxeticity is a phenomenon observed in single molecules, crystals, or specific macroscopic structures, and it has been actively explored across various fields [13], leading to an expanding range of applications. Formally, auxetic materials have negative Poisson's ratios, with a counter-intuitive property of expanding rather than contracting perpendicular to an applied stretch. Another compelling property that needs to be considered when modelling lead-free piezoelectric composites is related to flexoelectricity. It is a universal electromechanical coupling effect in which electric polarization occurs in a material due to a gradient of mechanical strain or, conversely, mechanical stress occurs due to a gradient of the electric field. Dr. Melnik's team made essential contributions to the flexoelectric enhancement of green composites, which were further expanded to the development of fully coupled models for such composites, accounting for the influence of thermo–electro–mechanical interactions.

We also note that Dr. Melnik's team made many cutting-edge contributions to developing more efficient industrial processes and innovative technologies in other core parts of the energy sector. Among their notable achievements is their work on altering the wetting state of materials, a promising area of research with potential applications across different fields of engineering and science.

Nonperturbative approaches are becoming increasingly important in various fields of physics and related disciplines. These include equilibrium and nonequilibrium dynamics, quantum mechanics, particle physics, nanoscience, nanotechnology, and quantum gravity. For example, the nonequilibrium Green's function (NEGF) method is a well-known nonperturbative approach that is widely utilized in condensed matter physics and nanoscience. Developing nonperturbative methods and computational algorithms is vital for effectively describing various complex systems. The significance of nonperturbative effects is well-known, particularly in systems such as spin glasses. Such effects often emerge in critical phenomena, where the renormalization group (RG) serves as the most widely used general method. This is the area where Dr. Melnik and his collaborators from the EU made significant advancements. It is crucial to develop nonperturbative approaches, including functional renormalization, Monte Carlo (MC) methods, and Monte Carlo RG (MCRG) methods, on which they have successfully worked.

Notable models in this field, such as the Wetterich and Polchinski equations, are recognized as nonperturbative or exact RG equations. However, these equations can often only be solved approximately. Thus, nonperturbative simulation methods, such as various MC and MCRG algorithms, play a crucial role in studying systems of interacting particles and critical phenomena, as they offer a reliable foundation for comparing and verifying results. Dr. Melnik's team proposed a new and more efficient method for solving the Wetterich equation than existing counterparts. The Wetterich equation is a fundamental concept in the Functional Renormalization Group used to analyze the flow of effective actions in quantum field theories. It has applications in various fields, including statistical physics and high-energy particle physics, especially quantum chromodynamics and quantum gravity, particularly within the framework of the asymptotic safety scenario.

The development and applications of stochastic mathematical models have been a significant part of Dr. Melnik's research portfolio. Some of them we have already mentioned in the context of modelling biosocial systems, while others are more related to business, economics, and finance. Stochastic processes are often used as mathematical models to represent systems and phenomena that appear to change randomly. These models incorporate elements of randomness and uncertainty, allowing for analyzing systems with unpredictable outcomes. By doing so, they enable the estimation of probability distributions for potential results. One of the key contributions of Dr. Melnik's team in this area, pertinent to finance, was related to first passage time problems and jump-diffusion stochastic processes. The first passage refers to the time it takes for a stochastic process to reach a specific state or boundary. This concept is vital and has applications across various fields, including finance, engineering, and biology, as it aids in analyzing and predicting system behaviour.

Materials science explores the properties and behaviour of complex multiscale materials, presenting numerous challenging problems in network science. In its turn, network science offers tools and frameworks for analyzing complex systems, including those encountered in materials science. By combining these two fields, we can gain a deeper understanding of material structures and properties, which includes the ability to predict synthesis routes and optimize material behaviour. Some of the contributions of Dr. Melnik to these fields have already been mentioned, including his works on materials with memory, polymers, and nanostructured materials. Among others, we should also mention here his impressive contributions, made in collaboration with other scientists, on intermetallic compounds, carbon allotropes such as diamond, graphene and graphyne, polymorphic systems and perovskites, as well as special types of nanostructures such as ZnO and diamond nanowires. Diamond nanowires have a variety of potential applications, including quantum technologies, sensing, and biomedical devices. ZnO nanowires also possess a wide range of properties and applications, such as semiconducting, piezoelectric, and photocatalytic capabilities. Beyond these uses, ZnO nanostructures can also be utilized in solar cells, energy storage devices, and energy harvesting devices. Additionally, they play a role in environmental remediation by facilitating the photocatalytic degradation of pollutants in both water and air. Some of this research required the development of new computational methods to handle the problems at hand, ranging from innovative MD procedures and first-principles calculations to efficient combinations of nonequilibrium Green's function and DFT. It is also important to highlight that Dr. Melnik's research in these areas faced significant challenges related to complex phase transformations, requiring determining phase stability limits, as well as to complex interfaces and defects, including those induced by dislocation nucleation.

Gyroscopic systems are utilized in a variety of applications, ranging from everyday devices to advanced technologies in space, such as spacecraft and satellites that maintain their pointing direction. They are crucial in astronomy as well, helping to ensure the orientation and stability of telescopes such as Hubble. This stability is essential for making precise observations of the universe and advancing our understanding of cosmology. Dr. Melnik's team has also made significant contributions to this field.

Plasmonics, electrochemical systems, and the applications of 2D materials remain crucial directions of Dr. Melnik's research, where he continues working with his students and colleagues, discovering new properties of coupled systems in these fields and developing efficient methods for their studies.

The list of areas Professor Roderick Melnik has contributed to is extensive, and we cannot exhaust it here. However, it is important to highlight that while there are many distinguished researchers in mathematical modelling, few have made significant contributions across such a diverse range of fields as he has. One might question how one individual could impact so many disciplines. A key aspect of this is his research's integration through a groundbreaking paradigm shift towards the study of coupled complex systems and networks that he initiated at the beginning of his career. However, there is also another key aspect. Those who know Dr. Melnik well often attribute his success to his broad interdisciplinary knowledge, deep insights, and warm and humble personality. He treats everyone with respect, whether they are distinguished professors or students. It is, therefore, not surprising that he has maintained long and productive collaborations with his former students, postdoctoral fellows, and researchers from various disciplines.

6. Current Issue

The first article of this Special Issue, authored by Wang and Zhai [14], introduces fast explicit numerical schemes for solving the extended Fisher–Kolmogorov (EFK) equation, which is an important model for phase transitions and bistable phenomena. Their approach combines the integrating factor Runge–Kutta method and the Fourier spectral method to obtain a high-precision numerical method for the EFK model. Further, three numerical experiments were conducted to investigate the performance and applicability of their reported method, focusing on convergence analysis and the behavior of diminishing energy. Cui et al. [15] presented efficient finite difference schemes for solving one- and two-dimensional wave equations with fractional damping. The authors proved the unconditional stability and convergence of the proposed difference schemes by testing them on several numerical examples. Yao and Weng [16] introduced a numerical method based on an operator splitting collocation scheme for solving the nonlinear Schrödinger equation. The equation is split into linear and nonlinear parts utilizing a second-order operator splitting method combined with the barycentric Lagrange interpolation collocation method for spatial discretization and the Crank–Nicolson scheme for temporal discretization. The presented numerical experiments validated the accuracy, mass, and energy conservation of the reported method. In another study featured in this Special Issue, Ivanov and Shelyag [17] developed stable periodic solutions for a simple form delay differential equation with nonlinear negative periodic feedback. The theoretical derivations and exact analytical calculations of the considered equation are further verified using extensive numerical simulations. De la Sen [18] investigates the loss of the Sturm–Liouville property of time-varying second-order differential equations in the presence of delayed dynamics. The time-delay systems naturally arise in several real-world scenarios, including biological processes such as epidemic modelling, ecology and medical systems, and the sunflower equation. Furthermore, they play an important role in diffusion phenomena, war and peace problems, and missile/anti-missile strategies. Such delay models are also very relevant in control theory and applications. Wang et al. [19] introduced a conservative and compact finite difference scheme aimed at preserving both the mass change rate and energy when solving the sixth-order Boussinesq equation with surface tension. Through the theoretical analysis, the authors demonstrate that the proposed scheme achieves fourth-order accuracy in spatial discretization and second-order accuracy in temporal discretization. The solvability, convergence, and stability of the scheme are rigorously proven using the discrete energy method. Furthermore, a series of numerical experiments are conducted to highlight the effectiveness and reliability of the scheme for long-time simulations. Torokhti and Soto-Quiros [20] proposed a method for the best constructive approximation of a nonlinear operator within probability spaces. This approach is highly adaptable, and can be extended to several applications, such as the development of efficient algorithms for signal reconstruction and denoising. Moreover, the proposed approach proves valuable in modeling and estimating dynamic systems with stochastic inputs, enhancing predictive accuracy and enabling more effective adaptive control strategies.

Barua and Povitsky [21] developed a comprehensive model to simulate the reactor-scale flow dynamics in chemical vapor deposition, a common industrial process that incorporates a complex combination of fluid flow, chemical reactions, and surface deposition. The model couples the Finite-volume and the Direct Simulation Monte Carlo methods to simulate the complex interactions around the micro- and nano-scale fibers. In another study, Mahoney and Povitsky [22] introduced the Method of Fundamental Solutions (MFS) to model the chemical vapor infiltration with a surface chemical reaction. Their approach computed the dynamics of pore size and location to evaluate the quality of the material obtained by chemical vapor infiltration. The deposition rates are calculated from the fibers' surface concentrations obtained by MFS, and the geometry is updated at each time step for modelling the pore filling over time. The developed model fidelity and integrity are evaluated through comparison with known analytical results for simple geometry as well as published experimental data. In [23], Povitsky introduced a meshless method of fundamental solutions to model particle sedimentation near corrugated (rough) surfaces. The proposed method is able to accurately predict the particle velocity and trajectory within the Stokes flow regime as they move along the corrugated surface under the action of gravity. The developed method holds significant potential for applications in biological systems and microfluidic devices.

Dutta and Layton [24] developed a computational model for simulating the epithelial transport of electrolytes and water along the nephrons in a male rat kidney. The developed model is able to predict the renal transport of Ca^{2+} and Mg^{2+} as well as other electrolytes and water under different physiological conditions. The developed models can serve as a valuable tool for in silico studies on kidney adaptation to alterations in Ca^{2+} and Mg^{2+} homeostasis in different conditions, such as pregnancy, diabetes, and chronic kidney disease. Bianchi et al. [25] reported a network-based study of the dynamics of $A\beta$ and τ proteins in Alzheimer's disease. Four network-based models are presented to capture the spatial and temporal dynamics of $A\beta$ and τ proteins in the brain, which were then directly compared with the clinical data to enhance understanding of disease pathology. In another study, Drapaca [26] introduced a mathematical model to investigate the impact of Alzheimer's drug donepezil hydrochloride on neuronal viscoelasticity and action potentials. The model couples the controlled drug release, transport to the brain, and its effect on action potential. Findings suggest that combining various drug release modalities and dosages could enhance the therapeutic efficacy of donepezil. Jabeen and Ilie [27] introduced a finite-difference strategy for approximating second-order parametric sensitivities for stochastic discrete models of biochemically reacting systems. The proposed strategy employs adaptive tau-leaping schemes and couples perturbed and nominal processes to enhance sensitivity estimation efficiency. The advantages of the proposed technique are demonstrated through applications to various biochemically significant system models. In another study featured in this Special Issue, Sacco et al. [28] reported a reduced-order model for cell volume homeostasis to simulate aqueous humor production in the human eye. The model computes the cell volume dynamics and fluid velocity by coupling intracellular reactions, transmembrane solute exchange, and Starling's Law. The extension of proposed computational framework could serve as a virtual laboratory to further test in vivo experiments and machine learning-based data analysis, assisting in the prevention and treatment of ocular diseases, such as glaucoma.

Hojati et al. [29] proposed a spatio-temporal indexing data structure for distributed networks, DSTree, that operates in the data layer using a distributed InterPlanetary File System (IPFS) network. This proposed framework enables peer-to-peer (P2P) data sharing and queuing without the need for any third-party central entities. The study highlighted that, for time-series data, such as storing sensor data, DSTree improves spatio-temporal query performance by approximately 40% for small and medium datasets. However, challenges remain, such as 20% slower insertion speed and limited semantic query capabilities. The study concluded by highlighting the need for future research efforts from GIScience and related fields to enhance decentralized applications and standardize geographic data sharing on IPFS networks. In another study, Omar et al. [30] analyzed and optimized a propagation model for malware in multi-cloud environments (MCEs) incorporating the effects of Brownian motion. The study explores the MCEs' network dynamics using six-state virtual machines (VMs), deterministic and stochastic dynamics models, and four popular multi-cloud environments: media, healthcare, finance, and educational servers. The analysis covers solution existence, uniqueness, equilibrium, and stability for both modeling approaches. After that, the two developed modeling approaches' solution existence, uniqueness, equilibrium, and stability are carefully investigated. Using an optimal control strategy, the models are tested to maintain VM security levels and minimize malware propagation. The reported results showed the effectiveness of the control methods in managing malware infections.

In conclusion, this Special Issue highlights novel findings of the state-of-the-art computational models that integrate multiscale, multiphysics, and data-driven methodologies, along with artificial intelligence and machine learning techniques. We believe that this Special Issue will provide the readers with ample knowledge and lead to the generation of novel ideas for future advancement, pushing the boundaries of knowledge in mathematical and computational sciences and their diverse applications including natural, social, and engineering systems.

Finally, we extend our sincere gratitude to all the authors who contributed high-quality articles [14–30] to this Special Issue, as well as to the referees whose insights and expertise were invaluable.

Author Contributions: All authors equally contributed to the editorial and approved the submitted version. All authors have read and agreed to the published version of the manuscript.

Acknowledgments: The guest editors would like to express their gratitude to all contributing authors for their outstanding articles, as well as all the reviewers for their insightful comments and valuable support to this Special Issue of Mathematical and Computational Applications.

Conflicts of Interest: The authors declare no conflicts of interest.

References

1. Sherief, H.; Anwar, M.N.; El-Latief, A.A.; Fayik, M.; Tawfik, A.M. A fully coupled system of generalized thermoelastic theory for semiconductor medium. *Sci. Rep.* **2024**, *14*, 13876. [CrossRef] [PubMed]
2. Wilk, G.; Wlodarczyk, Z. Beyond the relaxation time approximation. *Eur. Phys. J. A* **2021**, *57*, 221. [CrossRef]
3. Melnik, R.; Roberts, A.J. Computational Models for Multiscale Coupled Dynamic Problems. *Future Gener. Comput. Syst.* **2004**, *20*, 453–464. [CrossRef]
4. Prabhakar, S.; Melnik, R.; Inomata, A. Geometric spin manipulation in semiconductor quantum dots. *Appl. Phys. Lett.* **2014**, *104*, 142411. [CrossRef]
5. Singh, S.; Melnik, R. Coupled thermo-electro-mechanical models for thermal ablation of biological tissues and heat relaxation time effects. *Phys. Med. Biol.* **2019**, *64*, 245008. [CrossRef]
6. Shaheen, H.; Melnik, R. Brain network dynamics and multiscale modelling of neurodegenerative disorders: A Review. *IEEE Access* **2025**, *13*, 33074–33100. [CrossRef]
7. Pal, S.; Melnik, R. Nonlocal models in biology and life sciences. *Phys. Life Rev.* **2025**, *53*, 24–75. [CrossRef] [PubMed]
8. Dankulov, M.; Melnik, R.; Tadic, B. The dynamics of meaningful social interactions and the emergence of collective knowledge. *Sci. Rep.* **2015**, *5*, 12197. [CrossRef]
9. Melnik, R.V.N. Coupling Control and Human Factors in Mathematical Models of Complex Systems. *Eng. Appl. Artif. Intell.* **2009**, *22*, 351362. [CrossRef]
10. Melnik, R. Markov Chain Network Training and Conservation Law Approximations. *Appl. Math. Comput.* **2008**, *199*, 315–333.
11. Andjelkovic, M.; Tadic, B.; Melnik, R. The topology of higher-order complexes associated with brain hubs in human connectomes. *Sci. Rep.* **2020**, *10*, 17320. [CrossRef] [PubMed]
12. Krishnaswamy, J.; Rodriguez-Tembleque, L.; Melnik, R.; Buroni, F.; Saez, A. Size-dependent electro-elastic enhancement in geometrically anisotropic lead-free piezocomposites. *Int. J. Mech. Sci.* **2020**, *182*, 105745. [CrossRef]
13. Nash, R.J.; Li, Y. On-demand auxeticity and co-existing pre-tension induced compression stage in a sandwich design with kinematically constrained 3D suture tiles. *Nat. Commun.* **2024**, *15*, 6994. [CrossRef]

14. Wang, Y.; Zhai, S. Explicit Integrating Factor Runge–Kutta Method for the Extended Fisher–Kolmogorov Equation. *Math. Comput. Appl.* **2023**, *28*, 110. [CrossRef]
15. Cui, M.; Ji, C.-C.; Dai, W. A Finite Difference Method for Solving the Wave Equation with Fractional Damping. *Math. Comput. Appl.* **2024**, *29*, 2. [CrossRef]
16. Yao, M.; Weng, Z. A Numerical Method Based on Operator Splitting Collocation Scheme for Nonlinear Schrödinger Equation. *Math. Comput. Appl.* **2024**, *29*, 6. [CrossRef]
17. Ivanov, A.; Shelyag, S. Periodic Solutions in a Simple Delay Differential Equation. *Math. Comput. Appl.* **2024**, *29*, 36. [CrossRef]
18. De la Sen, M. Loss of the Sturm–Liouville Property of Time-Varying Second-Order Differential Equations in the Presence of Delayed Dynamics. *Math. Comput. Appl.* **2024**, *29*, 89. [CrossRef]
19. Wang, X.; Dai, W.; Biswas, A. A Conservative and Compact Finite Difference Scheme for the Sixth-Order Boussinesq Equation with Surface Tension. *Math. Comput. Appl.* **2024**, *29*, 112. [CrossRef]
20. Torokhti, A.; Soto-Quiros, P. Modeling of Nonlinear Systems: Method of Optimal Injections. *Math. Comput. Appl.* **2025**, *30*, 26. [CrossRef]
21. Barua, H.; Povitsky, A. Continuum and Molecular Modeling of Chemical Vapor Deposition at Nano-Scale Fibrous Substrates. *Math. Comput. Appl.* **2023**, *28*, 112. [CrossRef]
22. Mahoney, P.; Povitsky, A. Modeling of Chemical Vapor Infiltration for Fiber-Reinforced Silicon Carbide Composites Using Meshless Method of Fundamental Solutions. *Math. Comput. Appl.* **2024**, *29*, 27. [CrossRef]
23. Povitsky, A. Modeling of Sedimentation of Particles near Corrugated Surfaces by the Meshless Method of Fundamental Solutions. *Math. Comput. Appl.* **2024**, *29*, 90. [CrossRef]
24. Dutta, P.; Layton, A.T. Magnesium and Calcium Transport along the Male Rat Kidney: Effect of Diuretics. *Math. Comput. Appl.* **2024**, *29*, 13. [CrossRef]
25. Bianchi, S.; Landi, G.; Marella, C.; Tesi, M.C.; Testa, C.; on behalf of the Alzheimer's Disease Neuroimaging Initiative. A Network-Based Study of the Dynamics of $A\beta$ and τ Proteins in Alzheimer's Disease. *Math. Comput. Appl.* **2024**, *29*, 113. [CrossRef]
26. Drapaca, C.S. A Mathematical Study of Effects of Alzheimer's Drug Donepezil Hydrochloride on Neuronal Viscoelasticity and Action Potentials. *Math. Comput. Appl.* **2024**, *29*, 117. [CrossRef]
27. Jabeen, F.; Ilie, S. Efficient Finite-Difference Estimation of Second-Order Parametric Sensitivities for Stochastic Discrete Biochemical Systems. *Math. Comput. Appl.* **2024**, *29*, 120. [CrossRef]
28. Sacco, R.; Chiaravalli, G.; Guidoboni, G.; Layton, A.; Antman, G.; Shalem, K.W.; Verticchio, A.; Siesky, B.; Harris, A. Reduced-Order Model for Cell Volume Homeostasis: Application to Aqueous Humor Production. *Math. Comput. Appl.* **2025**, *30*, 13. [CrossRef]
29. Hojati, M.; Roberts, S.; Robertson, C. DSTree: A Spatio-Temporal Indexing Data Structure for Distributed Networks. *Math. Comput. Appl.* **2024**, *29*, 42. [CrossRef]
30. Omar, O.A.M.; Ahmed, H.M.; Nofal, T.A.; Darwish, A.; Ahmed, A.M.S. Analysis and Optimal Control of Propagation Model for Malware in Multi-Cloud Environments with Impact of Brownian Motion Process. *Math. Comput. Appl.* **2025**, *30*, 8. [CrossRef]

Mathematical and Computational Applications

MDPI

Article

Explicit Integrating Factor Runge–Kutta Method for the Extended Fisher–Kolmogorov Equation

Yanan Wang and Shuying Zhai *

Fujian Province University Key Laboratory of Computation Science, School of Mathematical Sciences, Huaqiao University, Quanzhou 362021, China; wynmath@yeah.net
* Correspondence: zsy@hqu.edu.cn

Abstract: The extended Fisher–Kolmogorov (EFK) equation is an important model for phase transitions and bistable phenomena. This paper presents some fast explicit numerical schemes based on the integrating factor Runge–Kutta method and the Fourier spectral method to solve the EFK equation. The discrete global convergence of these new schemes is analyzed rigorously. Three numerical examples are presented to verify the theoretical analysis and the efficiency of the proposed schemes.

Keywords: extended Fisher–Kolmogorov; integrating factor Runge–Kutta; error estimate

1. Introduction

The Fisher–Kolmogorov (FK) equation was first proposed by Fisher [1] and Kolmogorov [2] in 1937 to describe the interaction between the spread and adaptation of biological populations. By adding a stabilizing fourth-order derivative term to the FK equation, Collent [3] and Saarloos [4] proposed the extended FK (EFK) equation, which is a very important mathematical and physical model and has been widely used in many physics and engineering applications. In this paper, we consider the following EFK model with periodic boundary conditions:

$$\begin{cases} \partial_t v = -\kappa\Delta^2 v + \Delta v + g(v), & (\mathbf{x},t) \in \Omega \times (0,T], \\ v(\mathbf{x},0) = v_0(\mathbf{x}), & \mathbf{x} \in \Omega, \end{cases} \tag{1}$$

where $\Omega \in [a,b]^d, (d = 1,2)$ is a bounded area and κ is a positive constant. The function $g(v) = -G'(v)$ and $G(v) = \frac{1}{4}(v^2-1)^2$ is a double-well potential. When $\kappa = 0$ in (1), the EFK model reduces to the classical FK model. We assume that the function $g(v)$ exhibits Lipschitz continuity with respect to Ω, where the Lipschitz constant L is defined as follows

$$\max_{v\in\mathbf{R}} |g'(v)| \leq L.$$

The EFK model (1) can be viewed as an L^2 gradient flow associated with the following energy

$$E(v) = \int_\Omega \left(\frac{\kappa}{2}|\Delta v|^2 + \frac{1}{2}|\nabla v|^2 + G(v) \right) dx,$$

which is diminishing in time, i.e., $\frac{d}{dt}E(v) \leq 0$.

All the above properties are determined by the inherent nature of the physical model. Thus, in order to avoid non-physical effects in the simulations over a long period of time, it is highly desirable to design a structure-preserving numerical scheme. There have been some excellent results in numerical research on the EFK model, such as the spline configuration method [5], nonlinear/linear finite difference schemes [6], and the local boundary integral method [7]. However, these works do not consider the physical properties of the EFK model. Recently, Sun et al. [8,9] proposed two convex splitting variable step BDF2/BDF3 schemes for the EFK model, and the proposed schemes preserve the modified discrete energy dissipation law.

The development of high-precision numerical schemes for PDEs of gradient-flow-type has attracted much attention. In this direction, it is worth mentioning the integrating factor Runge–Kutta (IFRK) method [10]. The IFRK method has demonstrated remarkable advantages for equations with stiff linear terms. The exponential function in this method provides the exact solution of the linear part, so the stiffness does not restrict the step size and the solution to the nonlinear part may be approximated explicitly. As a result, the IFRK method is often explicit [11,12]. In recent years, the IFRK method has attracted widespread attention for solving partial differential equations. Ju et al. [13] proposed the MBP-preserving IFRK method for semilinear parabolic equations. Li et al. [14] proposed a class of unconditional MBP-preserving IFRK schemes for the conservative Allen–Cahn equation. Zhang et al. [15,16] developed a class of high-order structure-preserving IFRK methods for the Allen–Cahn equation. Then, they further [17] proposed and analyzed a series of temporal up to fourth-order unconditionally structure-preserving single-step methods to solve the Allen–Cahn equation, and introduced parametric IFRK (pIFRK) schemes which can be used to construct higher-order parametric single-step methods. To the authors' best knowledge, there are very few works in the literature on high-precision numerical methods for the EFK model. The objective of this paper is to develop a class of efficient, high-order-accurate schemes for the EFK model based on the explicit IFRK method coupled with non-decreasing abscissas (eIFRK+) [13,18].

The rest of the article is arranged as follows: In Section 2, a class of eIFRK+ schemes is proposed for solving the EFK model, and the corresponding theoretical analysis is given in Section 3. Numerical experiments are presented to test the performance of the proposed numerical schemes in Section 4, and some concluding remarks are given in Section 5.

2. eIFRK+ Fourier-Spectral Schemes for EFK Model

Define a periodic spatial grid $\cap_h = \Big\{(x_i, y_j) = \Big(a + ih, a + jh\Big),\ 0 \le i, j \le N-1)\Big\}$ with $h = \frac{b-a}{N}$ (N is even). All of the 2D periodic grid functions defined on $\cap_h$ are denoted by $\mathcal{M}_h$.

The discrete Fourier transform $\tilde{\phi} = Q\phi$ and the corresponding inverse transform $\phi = Q^{-1}\tilde{\phi}$ are defined by [19]

$$\begin{aligned} \tilde{\phi}_{pq} &= \frac{h^2}{(b-a)^2}\sum_{i=0}^{N-1}\sum_{j=0}^{N-1}\phi_{ij}\exp\Big(-\mathrm{i}\frac{2p\pi(x_i-a)}{b-a} - \mathrm{i}\frac{2q\pi(y_j-a)}{b-a}\Big), \\ & p, q = 0, \pm 1, \pm 2, \cdots, \frac{N}{2}, \end{aligned} \tag{2}$$

$$\begin{aligned} \phi_{ij} &= \sum_{p=-N/2}^{N/2}\sum_{q=-N/2}^{N/2}\tilde{\phi}_{pq}(t)\exp\Big(\mathrm{i}\frac{2p\pi(x_i-a)}{b-a} + \mathrm{i}\frac{2q\pi(y_j-a)}{b-a}\Big), \\ & \tilde{\phi}_{-\frac{N}{2},q} = \tilde{\phi}_{\frac{N}{2},q},\ \tilde{\phi}_{p,-\frac{N}{2}} = \tilde{\phi}_{p,\frac{N}{2}}, \end{aligned} \tag{3}$$

respectively. We further define the operators $\hat{D}_x$ and $\hat{D}_y$ on $\widehat{\mathcal{M}_h} = \{Q\phi | \phi \in \mathcal{M}_h\}$ as

$$(\hat{D}_x\tilde{\phi})_{pq} = \frac{2p\pi \mathrm{i}}{b-a}\tilde{\phi}_{pq}, \qquad (\hat{D}_y\tilde{\phi})_{pq} = \frac{2q\pi \mathrm{i}}{b-a}\tilde{\phi}_{pq},$$

and then, the Laplace operator can be approximated by

$$\Delta_N = Q^{-1}(\hat{D}_x^2 + \hat{D}_y^2)Q.$$

Next, we give some fully discrete eIFRK+ schemes for solving the EFK equation. Let $\tau = \frac{T}{M}$ and define the time node as $t_k = k\tau$ ($k = 0, 1, \cdots M$). According to the definitions of discrete spatial operators presented in Section 2, the EFK model (1) is reduced to the following nonlinear ordinary differential equations (ODEs) in time

$$\frac{d}{dt}v(t) = \mathcal{L}_N v(t) + g(v(t)), \tag{4}$$

where $\mathcal{L}_N = -\kappa\Delta_N^2 + \Delta_N$ and $v(t) = \{v_{ij}(t)\} \in \mathcal{M}_h$. Let $v(t) = e^{t\mathcal{L}_N}u(t)$, then Equation (4) can be transformed into

$$\frac{d}{dt}u(t) = e^{-t\mathcal{L}_N} g\left(e^{t\mathcal{L}_N}u(t)\right), \tag{5}$$

which can be solved by the following s-stage classical RK scheme [18]

$$\begin{cases} u^{(0)} = u^k, \\ u^{(i)} = \sum\limits_{j=0}^{i-1}[\eta_{ij}u^{(j)} + \tau\gamma_{ij}e^{-t_k^{(j)}\mathcal{L}_N}g(e^{t_k^{(j)}\mathcal{L}_N}u^{(j)})],\ 1 \le i \le n, \\ u^{k+1} = u^{(n)}, \end{cases} \tag{6}$$

where $t_k^{(j)} = t_k + d_j\tau$, d_j is the abscissa at each j-th layer. $\eta_{ij} \ge 0$ and $\sum\limits_{j=0}^{i-1}\eta_{ij} = 1$, γ_{ij} is a real constant.

Since $v^{(j)} = e^{t_k^{(j)}\mathcal{L}_N}u^{(j)}$, then (6) becomes

$$\begin{cases} v^{(0)} = v^k, \\ v^{(i)} = \sum\limits_{j=0}^{i-1} e^{(d_i-d_j)\tau\mathcal{L}_N}[\eta_{ij}v^{(j)} + \tau\gamma_{ij}g(v^{(j)})], 1 \le i \le n, \\ v^{k+1} = v^{(n)}, \end{cases} \tag{7}$$

where $d_i - d_j \ge 0$, $d_0 = 0$ and $d_n = 1$. By a simple derivation, the above scheme can be rewritten in the following equivalent form [13]:

$$\begin{cases} v^{(0)} = v^k, \\ v^{(i)} = e^{d_i\tau\mathcal{L}_N}v^k + \tau\sum\limits_{j=0}^{i-1}\varepsilon_{ij}e^{(d_i-d_j)\tau\mathcal{L}_N}g(v^{(j)}), 1 \le i \le n, \\ v^{k+1} = v^{(n)}, \end{cases} \tag{8}$$

where $\sum\limits_{j=0}^{i-1}\varepsilon_{ij} \le 1$, $\gamma_{ij} = \varepsilon_{ij} - \sum\limits_{s=j+1}^{i-1}\eta_{is}\varepsilon_{sj}$.

Some specific cases of system (7) are presented, which are denoted by eIFRK+(n,q), where q denotes the time accuracy:

eIFRK+(1,1): $v^{k+1} = e^{\tau\mathcal{L}_N}[v^k + \tau g(v^k)].$ (9)

eIFRK+(2,2):
$$\begin{cases} v^{(1)} = e^{\tau\mathcal{L}_N}[v^k + \tau g(v^k)], \\ v^{k+1} = \frac{1}{2}e^{\tau\mathcal{L}_N}v^k + \frac{1}{2}[v^{(1)} + \tau g(v^{(1)})]. \end{cases} \tag{10}$$

eIFRK+(3,3):
$$\begin{cases} v^{(1)} = \frac{1}{2}e^{\frac{2\tau}{3}\mathcal{L}_N}v^k + \frac{1}{2}e^{\frac{2\tau}{3}\mathcal{L}_N}[v^k + \frac{4\tau}{3}g(v^k)], \\ v^{(2)} = \frac{2}{3}e^{\frac{2\tau}{3}\mathcal{L}_N}v^k + \frac{1}{3}[v^{(1)} + \frac{4\tau}{3}g(v^{(1)})], \\ v^{k+1} = \frac{59}{128}e^{\tau\mathcal{L}_N}v^k + \frac{15}{128}e^{\tau\mathcal{L}_N}[v^k + \frac{4\tau}{3}g(v^k)] + \frac{27}{64}e^{\frac{\tau}{3}\mathcal{L}_N}[v^{(2)} + \frac{4\tau}{3}g(v^{(2)})]. \end{cases} \tag{11}$$

$$\textbf{eIFRK+(4,4):}\quad \begin{cases} v^{(1)}=e^{\frac{\tau}{2}\mathcal{L}_N}[v^k+\frac{\tau}{2}g(v^k)], \\ v^{(2)}=\frac{1}{2}e^{\frac{\tau}{2}\mathcal{L}_N}[v^k-\frac{\tau}{2}g(v^k)]+\frac{1}{2}[v^{(1)}+\tau g(v^{(1)})], \\ v^{(3)}=\frac{1}{9}e^{\tau\mathcal{L}_N}[v^k-\tau g(v^k)]+\frac{2}{9}e^{\frac{\tau}{2}\mathcal{L}_N}[v^{(1)}-\frac{3\tau}{2}g(v^{(1)})] \\ \qquad +\frac{2}{3}e^{\frac{\tau}{2}\mathcal{L}_N}[v^{(2)}+\frac{3\tau}{2}g(v^{(2)})], \\ v^{k+1}=\frac{1}{3}e^{\frac{\tau}{2}\mathcal{L}_N}[v^{(1)}+\frac{\tau}{2}g(v^{(1)})]+\frac{1}{3}e^{\frac{\tau}{2}\mathcal{L}_N}v^{(2)}+\frac{1}{3}[v^{(3)}+\frac{\tau}{2}g(v^{(3)})]. \end{cases} \tag{12}$$

3. Discrete Error Estimate

In this section, the discrete error estimation of the scheme (8) is discussed. We first give one lemma that helps to prove the discrete error estimation for the scheme.

Lemma 1 ([20]). *If a matrix A is negative semi-definite, then we have $\|e^{tA}\|_2 \le 1$ for $t > 0$, where $\|\cdot\|_2$ denotes the standard vector or matrix L^2-norm.*

Theorem 1. *(Error estimate.) Assume that $v_0 \in H^p_{per}(\Omega)$={$v \in H^p_{per}(\Omega)$ andvs.is $\Omega-$periodic} with $p > 4$ and the solution to the EFK model belongs to $C^{q+1}(0,T;H^p_{per}(\Omega))$. Under the conditions of Lemma 1, then the numerical solutions $\{\mathbf{V}^k\} \in \mathcal{M}_h$ generated by the eIFRK+ schemes (8) with $\mathbf{V}^0 = I_h v_0$ satisfy the following error estimate*

$$\|\mathbf{V}^k - I_h v(t_k)\|_2 \le \tilde{c}(h^p + \tau^q). \tag{13}$$

where $\tilde{c}$ is a positive constant depending on τ and h. I_h: $H^p_{per}(\Omega) \to \mathcal{M}_h$ is a sample operator as $I_h(v(x,y)) = \{v(x_i,y_j)\}$.

Proof. Let $\mathbf{Z}(t) = \mathbf{V}(t) - I_h v(x,y,t)$ with $\mathbf{Z}(0) = 0$; the difference between Equations (1) and (4) leads to

$$\frac{d}{dt}\mathbf{Z}(t) = \mathcal{L}_N\mathbf{Z}(t) + g(\mathbf{V}(t)) - g(I_h(v)) + R_N(v), \tag{14}$$

where the truncated error $R_N(v) = \mathcal{L}(I_h v) - \mathcal{L}_N(I_h v)$ satisfies$\|R_N\|_2 \le ch^p$.

Multiplying both sides of (14) by $e^{-t\mathcal{L}_N}$, then we have

$$\frac{d}{dt}\left(e^{-t\mathcal{L}_N}\mathbf{Z}(t)\right) = e^{-t\mathcal{L}_N}(g(\mathbf{V}(t)) - g(I_h(v)) + R_N(v)). \tag{15}$$

Using the eIFRK+ scheme (8), we have the following fully discrete scheme for the above equation

$$\begin{cases} \mathbf{Z}^{(i)} = e^{d_i\tau\mathcal{L}_N}\mathbf{Z}^k + \tau\sum\limits_{j=0}^{i-1}\varepsilon_{ij}e^{(d_i-d_j)\tau\mathcal{L}_N}\left(g(\mathbf{V}^j) - g(I_h v^{(j)}) + R_N(v^{(j)})\right), 1 \le i \le n-1, \\ \mathbf{Z}^{k+1} = e^{\tau\mathcal{L}_N}\mathbf{Z}^k + \tau\sum\limits_{i=0}^{n-1}\rho_i e^{(1-d_i)\tau\mathcal{L}_N}\left(g(\mathbf{V}^j) - g(I_h v^{(j)}) + R_N(v^{(j)})\right) + \xi^k, \end{cases} \tag{16}$$

where $\sum\limits_{j=0}^{i-1}\varepsilon_{ij} \le 1$, $\sum\limits_{i=0}^{n-1}\varepsilon_{ni} = \sum\limits_{i=0}^{n-1}\rho_i = 1$, ξ^k is the truncation error satisfying $\|\xi^k\|_2 \le c'\tau^{q+1}$.

By the Lipschitz condition for $g(v)$, we have

$$\|g(\mathbf{V}^j) - g(I_h v^{(j)})\|_2 \le L\|\mathbf{V}^j - I_h v^{(j)}\|_2 = L\|\mathbf{Z}^{(j)}\|_2.$$

Combination of the above estimates leads to

$$\begin{aligned}\|Z^{(i)}\|_2 &\le \|e^{d_i\tau\mathcal{L}_N}Z^k\|_2 + \tau\|\sum_{j=0}^{i-1}\varepsilon_{ij}e^{(d_i-d_j)\tau\mathcal{L}_N}(g(\mathbf{V}^j)-g(I_hv^{(j)}))\|_2\\ &+\tau\|\sum_{j=0}^{i-1}\varepsilon_{ij}e^{(d_i-d_j)\tau\mathcal{L}_N}R_N(v^{(j)})\|_2\\ &\le \|Z^k\|_2 + L\tau\sum_{j=0}^{i-1}\|Z^{(j)}\|_2 + ch^p\tau.\end{aligned}$$

By using the Gronwall's inequality, we further have

$$\|\mathbf{Z}^{(i)}\|_2 \le e^{L(i+1)\tau}(\|\mathbf{Z}^k\|_2 + ch^p\tau). \tag{17}$$

Similar to the estimation of $\|\mathbf{Z}^{k+1}\|_2$, we have

$$\|\mathbf{Z}^{k+1}\|_2 \le \|\mathbf{Z}^k\|_2 + L\tau\sum_{i=0}^{n-1}\|\mathbf{Z}^{(i)}\|_2 + ch^p\tau + \|\xi^k\|_2. \tag{18}$$

Combining (17) with (18), we have

$$\begin{aligned}\|\mathbf{Z}^{k+1}\|_2 &\le \|\mathbf{Z}^k\|_2 + L\tau\sum_{i=0}^{n-1}e^{L(i+1)\tau}(\|\mathbf{Z}^k\|_2 + ch^p\tau) + ch^p\tau + \|\xi^k\|_2\\ &\le \|\mathbf{Z}^k\|_2 + L\tau n e^{Ln\tau}\|\mathbf{Z}^k\|_2 + c''h^p\tau + c'\tau^{q+1}.\end{aligned} \tag{19}$$

Summing the above inequality from $k=0$ to m and using Gronwall's inequality, we obtain the desired result

$$\|\mathbf{Z}^{m+1}\|\|_2 \le e^{LnTe^{LT}}(c''h^pT + c'\tau^qT) \le \tilde{c}(h^p+\tau^q).$$

□

4. Numerical Experiments

We examine the accuracy and efficiency of the proposed method through three numerical experiments, including convergence and diminishing energy.

Example 1. *The purpose of the first numerical experiment is to verify the role of parameter κ. Consider the 1D EFK model in $\Omega = [-4,4]$ with the initial condition*

$$v_0(x) = -\sin(\pi x).$$

We choose $N=256$ and $\tau=0.01$. Figure 1 shows the numerical results using scheme eIFRK+(4,4) at different times $t=0, 0.05, 0.1, 0.15, 0.2$. It is observed that the dynamic evolution of $\kappa=0$ is almost identical to that of $\kappa=10^{-4}$. However, when $\kappa=0.1$, the solution rapidly evolves from the initial state to zeros, which illustrates that the coefficient κ in the EFK model is a stable parameter.

Example 2. *To show the temporal accuracy and convergence of the proposed schemes, we consider the 2D EFK model in $\Omega = [-16,16]^2$ with $\kappa=0.01$ and the initial condition*

$$\begin{aligned}v_0(x,y) &= 0.1 - 0.2\cos\left(\frac{\pi(x-12)}{16}\right)\sin\left(\frac{\pi(y-12)}{16}\right)\\ &+0.1\cos^2\left(\frac{\pi(x+10)}{32}\right)\sin^2\left(\frac{\pi(y+3)}{32}\right) - 0.2\sin^2\left(\frac{\pi x}{8}\right)\cos\left(\frac{\pi(y-6)}{8}\right).\end{aligned}$$

We adopt the uniform 1024×1024 *spatial mesh, which is sufficiently fine so that the errors caused by the spatial approximation can be ignored. The error between two different time steps* τ *and* $\frac{\tau}{2}$ *is calculated. The computational results are presented in Table 1. One may see that the numerical orders of time accuracy are close to the optimal order.*

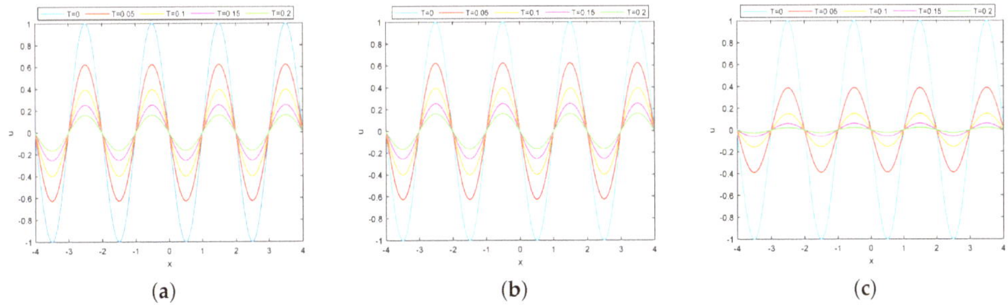

Figure 1. Numerical solutions with different κ values for Example 1. (**a**) $\kappa = 0$; (**b**) $\kappa = 10^{-4}$; (**c**) $\kappa = 0.1$.

Table 1. Temporal errors and convergence orders at $T = 1$ with $\kappa = 0.01$ and $N = 1024$ for Example 2.

eIFRK+	M	Err_{max}	Rate	Err_{L^2}	Rate
(1,1)	16	3.9212×10^{-3}	-	1.9491×10^{-2}	-
	32	1.9980×10^{-3}	0.9728	9.8542	0.9840
	64	1.0086×10^{-3}	0.9862	4.9540×10^{-3}	0.9921
	128	5.0671×10^{-4}	0.9931	2.4837×10^{-3}	0.9961
(2,2)	16	1.2908×10^{-4}	-	6.5886×10^{-4}	-
	32	3.3107×10^{-5}	1.9630	1.6909×10^{-4}	1.9622
	64	8.3832×10^{-6}	1.9816	4.2826×10^{-5}	1.9812
	128	2.1092×10^{-6}	1.9908	1.0776×10^{-5}	1.9906
(3,3)	16	1.4809×10^{-6}	-	5.0559×10^{-6}	-
	32	1.8992×10^{-7}	2.9631	6.5053×10^{-7}	2.9583
	64	2.4046×10^{-8}	2.9815	8.2502×10^{-8}	2.9791
	128	3.0251×10^{-9}	2.9907	1.0388×10^{-8}	2.9896
(4,4)	16	3.1132×10^{-8}	-	1.6382×10^{-7}	-
	32	1.9696×10^{-9}	3.9824	1.0359×10^{-8}	3.9830
	64	1.2395×10^{-10}	3.9901	6.5125×10^{-10}	3.9915
	128	7.7799×10^{-12}	3.9939	4.0823×10^{-11}	3.9958

Example 3. *Consider the EFK model in* $\Omega = [-1,1]^2$ *with the initial condition*

$$v_0(x,y) = -1 + \tanh\left(\frac{0.4 - \sqrt{x^2+y^2}}{\varepsilon\sqrt{2}}\right) - \tanh\left(\frac{0.3 - \sqrt{x^2+y^2}}{\varepsilon\sqrt{2}}\right).$$

We choose $\varepsilon = \frac{5h}{\sqrt{2}tanh^{-1}(0.9)}$, $N = 256$, $\tau = 0.01$ *and different* κ, *Figure 2 shows the evolutions of the snapshots and the energies of the numerical solutions obtained by the eIFRK+(4,4) scheme. It can be seen that the evolution speed of the numerical solution is largely clipped with the increase in* κ, *and the energy obtained from the energy image gradually reaches a steady state with decreasing time.*

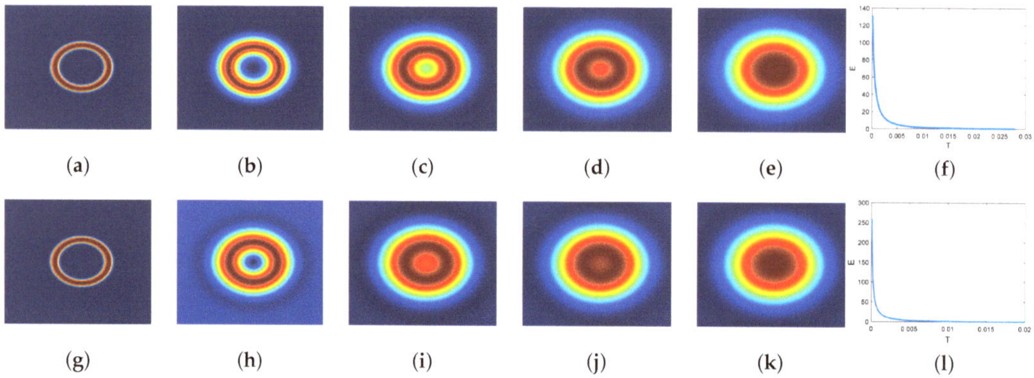

Figure 2. The numerical solution u at different times for Example 3. Top: $\kappa = 10^{-4}$. Bottom: $\kappa = 0.004$. (a) $T = 0$; (b) $T = 0.004$; (c) $T = 0.012$; (d) $T = 0.02$; (e) $T = 0.028$; (f) $E(v)$; (g) $T = 0$; (h) $T = 0.004$; (i) $T = 0.012$; (j) $T = 0.016$; (k) $T = 0.02$; (l) $E(v)$.

5. Conclusions

Based on the explicit IFRK method coupled with nondecreasing abscissas, we have obtained a class of fast and effective numerical schemes for the EFK model. The optimal error estimates of the fully discrete schemes have been analyzed. Three numerical experiments were carried out to test the accuracy and applicability of the proposed schemes.

Author Contributions: Y.W.: Conceptualization, Methodology, Software, Writing—original draft. S.Z.: Methodology, Writing—review & editing. All authors have read and agreed to the published version of the manuscript.

Funding: This research received no external funding.

Data Availability Statement: Data will be made available on reasonable request.

Acknowledgments: The authors would like to thank the editor and referees for their valuable comments and suggestions which helped us to improve the results of this paper.

Conflicts of Interest: The authors declare no conflict of interest.

References

1. Fisher, R.A. The wave of advance of advantageous genes. *Ann. Eugen.* **1937**, *7*, 355–369. [CrossRef]
2. Kolmogorov, A.N.; Petrovskii, I.G.; Piskunov, N.S. A study of the diffusion equation with increase in the quantity of matter, and its application to a biological problem. *Mosc. Univ. Math. Bull.* **1937**, *1*, 1–25.
3. Coullet, P.; Elphick, C.; Repaux, D. Nature of spatial chaos. *Phys. Rev. Lett.* **1987**, *58*, 431. [CrossRef]
4. Saarloos, W.V. Dynamical velocity selection: Marginal stability. *Phys. Rev. Lett.* **1987**, *58*, 2571. [CrossRef]
5. Danumjaya, P.; Pani, A.K. Orthogonal cubic spline collocation method for the extended Fisher–Kolmogorov equation. *J. Comput. Appl. Math.* **2005**, *174*, 101–117. [CrossRef]
6. Li, S.G.; Xu, D.; Zhang, J.; Sun, C.J. A new three-level fourth-order compact finite difference scheme for the extended Fisher–Kolmogorov equation. *Appl. Numer. Math.* **2022**, *178*, 41–51. [CrossRef]
7. Ilati, M.; Dehghan, M. Direct local boundary integral equation method for numerical solution of extended Fisher–Kolmogorov equation. *Eng. Comput.* **2018**, *34*, 203–213. [CrossRef]
8. Sun, Q.H.; Ji, B.Q.; Zhang, L. A convex splitting BDF2 method with variable time-steps for the extended Fisher–Kolmogorov equation. *Comput. Math. Appl.* **2022**, *114*, 73–82. [CrossRef]
9. Sun, Q.H.; Wang, J.D.; Zhang, L.M. A new third-order energy stable technique and error estimate for the extended Fisher–Kolmogorov equation. *Comput. Math. Appl.* **2023**, *142*, 198–207. [CrossRef]
10. Isherwood, L.; Grant, Z.J.; Gottlieb, S. Strong stability preserving integrating factor Runge–Kutta methods. *SIAM J. Numer. Anal.* **2018**, *58*, 3276–3307. [CrossRef]
11. Zhai, S.Y.; Weng, Z.F.; Yang, Y.F. A high order operator splitting method based on spectral deferred correction for the nonlocal viscous Cahn–Hilliard equation. *J. Comput. Phys.* **2021**, *446*, 110632. [CrossRef]

12. Zhai, S.Y.; Weng, Z.F.; Feng, X.L.; He, Y.N. Stability and error estimate of the operator splitting method for the phase field crystal equation. *J. Sci. Comput.* **2021**, *86*, 8. [CrossRef]
13. Ju, L.L.; Li, X.; Qiao, Z.H.; Yang, J. Maximum bound principle preserving integrating factor Runge–Kutta methods for semilinear parabolic equations. *J. Comput. Phys.* **2021**, *439*, 110405. [CrossRef]
14. Li, J.W.; Li, X.; Ju, L.L.; Feng, X.L. Stabilized integrating factor Runge–Kutta method and unconditional preservation of maximum bound principle. *SIAM J. Sci. Comput.* **2021**, *43*, A1780–A1802. [CrossRef]
15. Zhang, H.; Yan, J.Y.; Qian, X.; Chen, X.W.; Song, S.H. Explicit third-order unconditionally structure-preserving schemes for conservative Allen–Cahn equations. *J. Sci. Comput.* **2022**, *90*, 1–29. [CrossRef]
16. Zhang, H.; Yan, J.Y.; Qian, X.; Song, S.H. Numerical analysis and applications of explicit high order maximum principle preserving integrating factor Runge–Kutta schemes for Allen–Cahn equation. *Appl. Numer. Math.* **2021**, *161*, 372–390. [CrossRef]
17. Zhang, H.; Yan, J.Y.; Qian, X.; Song, S.H. Up to fourth-order unconditionally structure-preserving parametric single-step methods for semilinear parabolic equations. *Comput. Methods Appl. Mech. Eng.* **2022**, *393*, 114817. [CrossRef]
18. Shu, C.W.; Osher, S. Efficient implementation of essentially non-oscillatory shock-capturing schemes. *J. Comput. Phys.* **1988**, *77*, 439–471. [CrossRef]
19. Shen, J.; Tang, T.; Wang, L.L. *Spectral Methods: Algorithms, Analysis and Applications*; Springer Science: London, UK; New York, NY, USA, 2011; Volume 41, pp. 1–45.
20. Nan, C.X.; Song, H.L. The high-order maximum-principle-preserving integrating factor Runge–Kutta methods for nonlocal Allen–Cahn equation. *J. Comput. Phys.* **2022**, *456*, 111028. [CrossRef]

Mathematical and Computational Applications

MDPI

Article

Continuum and Molecular Modeling of Chemical Vapor Deposition at Nano-Scale Fibrous Substrates

Himel Barua † and Alex Povitsky *

Department of Mechanical Engineering, College of Engineering, University of Akron, Akron, OH 44325, USA; baruah@ornl.gov

* Correspondence: povitsky@uakron.edu

† Present affiliation: The U.S. Department of Energy (DOE), Oak Ridge National Laboratory (ORNL), 1 Bethel Valley Rd 5200, Oak Ridge, TN 37830, USA.

Abstract: Chemical vapor deposition (CVD) is a common industrial process that incorporates a complex combination of fluid flow, chemical reactions, and surface deposition. Understanding CVD processes requires rigorous and costly experimentation involving multiple spatial scales, from meters to nanometers. The numerical modeling of deposition over macro-scale substrates has been conducted in the literature and results show compliance with experimental data. For smaller-scale substrates, where the corresponding Knudsen number is larger than zero, continuum modeling does not provide accurate results, which calls for the implementation of molecular-level modeling techniques. In the current study, the finite-volume method (FVM) and Direct Simulation Monte Carlo (DSMC) method were combined to model the reactor-scale flow with CVD around micro- and nano-scale fibers. CVD at fibers with round cross-sections was modeled in the reactor, where fibers were oriented perpendicularly with respect to the feedstock gas flow. The DSMC method was applied to modeling flow around the matrix of nano-scale circular individual fibers. Results show that for smaller diameters of individual fibers with the same filling ratio, the residence time of gas particles inside the fibrous media reduces, and, consequently, the amount of material surface deposition decreases. The sticking coefficient on the fibers' surface plays an important role; for instance, increasing the sticking coefficient from 20% to 80% will double the deposition rate.

Keywords: chemical vapor deposition; low-pressure reactor; direct simulation Monte Carlo; computational fluid dynamics; carbon deposition

1. Introduction

Chemical vapor deposition (CVD) is a material deposition process with a vapor-to-solid chemical reaction at a rigid surface. The CVD is generally used in the industrial processes of manufacturing multi-layer materials to create a thin film over a substrate surface [1]. Fluid flow through and around bundles of fibers needs to be modeled to obtain the deposition rate at the fibers' surface. The set-ups of deposition substrate surfaces vary as the surfaces could be smooth or rough, and represent isolated fibers or the collective surface of bundles of fibers [2]. Depending on the shape and size of the substrate, the CVD process involves complex no-slip or slip fluid flow phenomena along with material deposition, which makes the fluid–solid system quite complicated to model and to understand. The scope of the study is to model the CVD of carbon at the fibers' surface (Figure 1) that is qualitatively similar to filtration and physical vapor deposition processes in which solid particles deposit at the fiber surface.

Though for macro-length scales, the traditional Newtonian and continuum description of fluid mechanics with no-slip boundary conditions can resolve the gas-flow phenomena coupled with CVD process, for a smaller-length scale the major discrepancy [3] occurs between continuum approximation and real molecular flows, which causes substantial variation in the CVD rate. To understand the validity of continuum approximation, a well-established non-dimensional group known as the Knudsen number (Kn) is used. The Kn represents the degree of rarefaction of gas by the ratio of the gas molecular mean free path over the geometric length scale of a substrate:

$$Kn = \frac{\lambda}{L}, \quad (1)$$

where λ is the gas molecular mean free path and L is the length scale of the system, such as the diameter of an individual fiber or of a bundle of fibers.

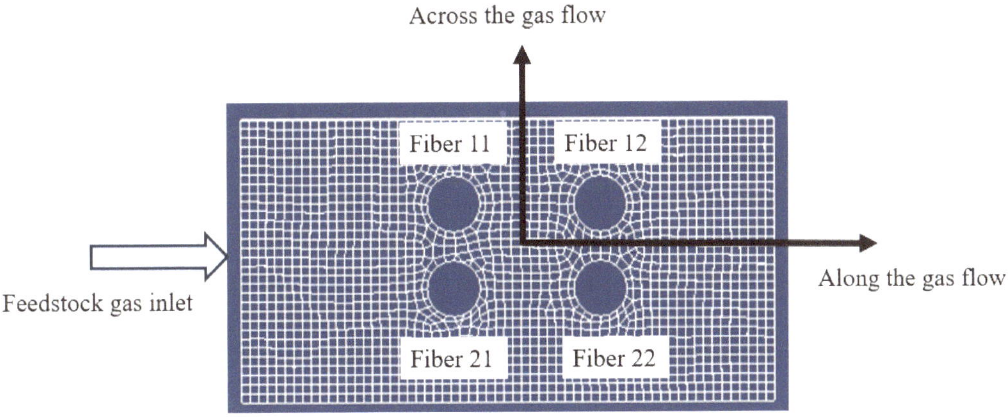

Figure 1. Schematic of DSMC modeling for regular fibers with DSMC cell set-up. For fibers' subscripts, the first number denotes the row of fibers, and the second number denotes the column of fibers. For instance, "Fiber21" means fiber belonging to the 2nd row and 1st column in the bundle.

In general, gas flows encounter compressibility and viscous heating. In the current case, $Ma << 0.3$, and therefore, compressibility and aerodynamic (viscous) heating are not substantial. The Mach number is the ratio of reactor flow velocity and the speed of sound. The speed of sound is evaluated at the feedstock gas temperature of 816 K in a heated reactor [4].

For a macro-scale continuum flow ($Kn < 0.01$), solving Navier–Stokes equations with no-slip boundary conditions produces solutions coinciding fairly with experiments. For smaller linear scales typical for crossflow-oriented long fibers of micron-scale diameters ($0.01 < Kn < 0.1$), the above approach becomes less accurate, and for larger Knudsen numbers ($Kn > 0.1$) the no slip approach is inadequate.

In the vapor deposition process, the gas rarefaction effect creates a thin flow layer near the rigid surface, named the Knudsen layer, which is of the linear scale of a single molecular mean free path. The Knudsen layer cannot be analyzed by using Navier–Stokes equations and requires solution of the Boltzmann equation for the local velocity distribution of molecules. For $Kn < 0.1$, the Knudsen layer covers less than 10% of the channel height for internal flows. For $Kn < 0.1$, the Knudsen layer can be neglected by extrapolating continuous bulk gas flow up to the rigid surface, which results in a finite slip velocity at the wall. This flow regime is called the partial slip flow regime.

For larger Kn numbers, the above continuum approach is no longer accurate and a solution of flow field in the Knudsen layer (Figure 2) is required, which can be achieved using molecular Direct Simulation Monte Carlo (DSMC). In the current research, a methodology was developed to approach problems with a large range of linear flow scales, from the reactor scale to individual fibers of a sub-micron diameter. Two well-established methods, namely, the continuum Finite Volume Method (FVM) for larger linear scales and the molecular Direct Simulation Monte Carlo (DSMC) for flow near fibers, were combined and one-way coupled.

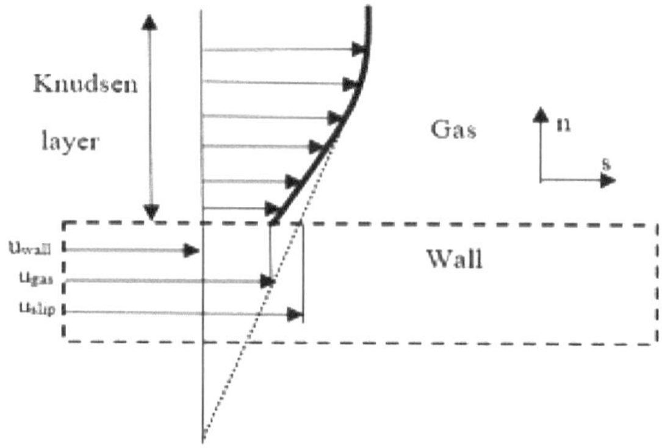

Figure 2. Knudsen layer and slip velocity at wall, where n and s are normal and tangential directions with respect to solid wall.

To approach the problem of determining the slip velocity in the current study, the fluid motion was considered at a molecular scale at least for a part of the domain instead of considering the fluid as a continuous medium. On a molecular scale, the methods that have been used by researchers are categorized as molecular dynamics (MD) and DSMC. When using MD, every computational particle represents an individual fluid molecule. This approach is more applicable for liquids at the nanoscale, where molecules always interact with their neighbors through potential inter-molecular forces. The DSMC allows a fraction of needed computing time compared to MD, as DSMC particles represent billions of physical gas molecules which interact with each other only during their collision. Statistical sampling is used to obtain fluid dynamics properties from DSMC results.

A considerable amount of research has been performed to model CVD process. While CVD modeling over a flat plate has been considered in numerous papers, for example, refs. [5,6], CVD over more complex and multi-scale geometries has drawn much less attention and has been limited to near-atmospheric-pressure reactors [7]. Most of the published approaches have modeled the kinetics of chemical reactions, while the link between the CVD modeling of coating over a single fiber surface has been investigated [8], in which an adaptive finite element method was used to capture the thickness of the deposited material. Ref. [9] conducted an experiment and continuous CFD modeling to investigate the growth process of a multi-walled carbon nanotube by CVD. In Ref [9], the authors used FVM implemented in FLUENT and studied the flow properties. Most of the modeling of carbon deposition on fibers [10–12] is limited to either flow modeling around a single fiber [13] or a reactor bulk flow modeling [14,15]. In the current research, the effects of multiple fibers, which are forming arrays of fibers, have been investigated to discover their critical role in spatial and temporal deposition profiles and the final thickness of the deposited layer of material.

DSMC has been implemented in prior studies [16,17]. Ref. [18] combined FVM with DSMC to model a chemically neutral flow around a fibrous medium. Refs. [19,20] extended DSMC to model adsorption and desorption in porous media. Ref. [21] adopted total collision energy and a quantum kinetic model [22] to implement chemical reactions in a DSMC method. The current research implemented an extension of the DSMC method to solve carbon deposition over wall surfaces. Instead of the direct modeling of volumetric and surface reactions, the proposed approach was implemented by taking the combination of absorbing and reflecting boundary conditions at the substrate wall for DSMC particles embedded in fluid flow. The sticking coefficient was adopted to account for the probability of the DSMC particle undergoing deposition. In general, the magnitude of the sticking coefficient is a function of the temperature, surface coverage, and structural details of the surface. This approach reduces the complexity of DSMC algorithm by simplifying the surface chemical kinetics model and still obtains the local rate of adsorption of particles over the wall surface.

In the CVD process, the surface reaction is the reason particles are deposited over the surface, in which the time available for the reaction is important. The residence time indicates how long it would take for simulated particles to move from the inlet to the outlet of the DSMC domain. One of the major objectives of the current study is to characterize the residence time of particles inside the fibrous medium for different fibers' set-ups and for the range of sticking coefficients at the fiber surface.

In a broader sense, several important and widespread technologies, including Microelectromechanical systems (MEMS), have relevant broad fluid flow scales ranging from meters to nanometers in the frame of a single set-up [23]. The common major challenge in the modeling of fluid flow in these processes is the range of fluid flow scales that requires the combined use of continuum and molecular approaches [24].

The study is composed as follows. In Section 2, the DSMC code is developed and validated using a well-known benchmark named Couette flow. The degree of gas rarefaction is represented by the Knudsen number influencing the flow velocity profile. In Section 2.1, the effect of rarefaction of CVD gas is briefly presented. The velocity profile for a range of Knudsen numbers is validated in Section 2.2 by comparing it to prior Lattice Boltzmann Method (LBM) results [25]. In Section 3, the FVM results of flowfield computations for the entire CVD reactor are presented, and are used as the inlet velocity for the DSMC domain. In Sections 3.2–3.5, the developed in-house DSMC code is used to model the particle deposition rate and residence time of particles around the fibers' substrates (Figure 1). For the DSMC simulations presented in Sections 3.3 and 3.4, infinitely long fibers with circular cross-sections in a regular orientation perpendicular to gas flow are computed for the range of Knudsen numbers, with varying distances between two consecutive fibers, by (i) changing the volume fraction occupied by fibers and (ii) keeping the volume fraction constant but varying the fibers' diameter. For both cases, the residence time of flow particles were calculated, and represents the amount of time the CVD feedstock gas molecules stay inside the fibrous media. In Section 3.5, the effect of rarefaction on the CVD is studied.

2. DSMC and Deposition at Rigid Surface

2.1. Development of DSMC and Continuum Models for Gas Flow in Proximity of Rigid Boundary

In the frame of the partial slip flow approach, $0.01 < Kn < 0.1$, the continuum approximation is still valid apart from the Knudsen layer [26]; see Figure 2. To account for the rarefaction feature at the wall, special boundary conditions are considered that are associated with the velocity slip, named Maxwell's boundary condition [27,28].

The first-order slip velocity boundary condition for a stationary wall developed by Maxwell [24,26] was used in the current study:

$$u_s = \lambda \frac{2-\sigma_u}{\sigma_u} \frac{\partial u}{\partial n}, \tag{2}$$

where u_s is the slip velocity, u is the gas velocity in the Knudsen layer next to the wall, λ is the mean free path, and $\frac{\partial u}{\partial n}$ represents the slope of the velocity normal to the wall. The σ_u term in Equation (2) represents the tangential accommodation coefficient. Momentum and energy transfer between gas molecules and the surface require the specification of interactions between impinging gas molecules and the surface. From a microscopic point of view, this approach is quite complicated and requires detailed knowledge of scattering kernels [24], while from a macroscopic point of view, the accommodation coefficient, σ_u, is accurate enough to describe the gas–wall interactions. For the limit case, $\sigma_u = 0$, gas molecules reflect from the surface specularly and pre-collision and post-collision velocities of gas particles are symmetric with respect to the normal direction to the surface, n. For larger values of the accommodation coefficient σ_u, reflections of gas molecules over the wall become more random, which represents diffusive reflection.

For flows with $Kn > 0.1$, constitutive laws that define stress tensor and heat flux vectors break down [29], which requires higher-order corrections to the constitutive laws, resulting in Burnett or Woods equations [24] that are challenging to solve. In the current research, the DSMC methodology was implemented to compute slip velocity and to calculate the residence time of CVD gas molecules near a representative set of fibers. The DSMC approach comprises five major steps [3,17]:

1. Particle initialization;
2. Computational displacement of simulated particles after time interval Δt;
3. Indexing and cross referencing of simulated particles as they move between DSMC cells;
4. Computing of particles' new velocities after inter-particle collisions and collisions between simulated particles and wall surface;
5. Sampling simulated DSMC particles to obtain macroscopic properties.

The first step of DSMC modeling is initialization of the DSMC particles, which defines the properties of the fluid, the DSMC cell size, the number of DSMC cells and the number of real gas molecules represented by DSMC particles. Simulated particles in DSMC are initialized with random velocity and a random initial position. To evaluate the number of simulated particles using DSMC, let us consider a domain of volume Vc, in which molecular number density is n. The total number of real molecules in the domain is equal to nVc.

The total number of simulated particles $N = \frac{nV_c}{F_n}$, where F_n is the number of real molecules represented by each simulated particle. In the current study, each simulated particle represented 10^{12} gas molecules. The assumption behind DSMC is that molecules have Maxwell–Boltzmann distribution everywhere in the domain and the collision between molecules are elastic in the DSMC domain. The assigned initial position and initial velocity of each DSMC particle were generated by a random number generator.

For each DSMC particle, equations of motion were solved to obtain a new location of particles at the end of the current, n-th, time step.

$$\frac{dr_i}{dt} = V_i, i = 1, \ldots, N, \tag{3}$$

where V_i is the speed of the i-th DSMC particle

During the movement stage of DSMC particles, no collisions of particles are assumed. The position of particles is updated by Equation (4):

$$r_i^{n+1} = r_i^n + \Delta t * V_i^n \tag{4}$$

After moving DSMC particles over a time step Δt, particles are sorted and indexed. Indexing serves the purpose of figuring out which particles are located in which cell. Indexing is necessary because particles might move to the next cell. The new cell locations of the molecules are indexed so that the intermolecular collisions and flow field sampling can be held accurately.

Indexing is also necessary to pick up the collision partner. In each cell, colliding particles are randomly selected. For each pair, pre-collisional velocities are replaced by post-collisional velocities for use in Equations (3) and (4) in the next time step. The proper choice of collision partner and choosing the right number of collisions are crucial to the consistency of DSMC. In the current study, the no-time-counter (NTC) method [30] was used to model the collision. This method uses a multistep collision model. At each time moment, n, only particles which reside in the same DSMC cell participate in collisions.

Every collision is considered a random event with some probability of collision in every cell. The probability of collision between particles is chosen as follows:

$$P_{coll}[i,j] = \left(\frac{1}{2V_c}\right)(F_N N)(N-1)(\sigma_T C_r)_{max}\Delta t, \tag{5}$$

where V_c is the cell volume, $(\sigma_T C_r)_{max}$ is the maximum product of the collision cross section and the relative speed of all possible particle pairs in the cell, Δt is the time step and F_N is the number of real molecules or atoms that each DSMC particle represents. Particle i is chosen randomly from all particles in the cell and j is chosen from the same cell to ensure near-neighbor collision. The collision pair would be selected if the condition below is satisfied:

$$\frac{(\sigma_T C_r)_{ij}}{(\sigma_T C_r)_{max}} > R_f, \tag{6}$$

where R_f is a random number uniformly chosen at interval [0, 1].

In the third stage of DSMC methodology, boundary conditions are implemented. In the current study, the model had inlet, outlet, and symmetrical boundaries (see Figure 1), which is also referred to as a permeable boundary condition. For the inlet, the most probable average speed is taken from the FVM solution for flow velocity inside the reactor, as outlined in Section 3.1. The inlet boundary is located 4 mm upstream from the DSMC domain (see Section 3 and Figure 7). When DSMC particles exit through the upper and lower boundaries of the DSMC domain (see Figure 1), a new DSMC particle is introduced at the opposite boundary at the exit point.

To model the deposition at the substrate surface, a novel approach was implemented by modifying the Maxwell boundary condition of diffuse scattering. A reactive coefficient or sticking coefficient [31] is introduced, which represents the probability of molecules colliding with the substrate to undergo a surface reaction process, which in CVD modeling is deposition. When a simulated particle reaches the substrate surface, a number η is generated by the random number generator. This number is compared with the sticking coefficient. If the value of η is lower than the sticking coefficient, the particle would deposit on the wall. If the value is higher than the sticking coefficient, the particle would reflect onto the fluid domain following the cosine rule. The higher the value of the sticking coefficient, the larger the number of deposited particles would be.

If the DSMC particle is reflected by the surface, the particle post-reflection velocity (v_x, v_y, v_z) is assigned as follows:

$$\begin{aligned} v_x &= \sqrt{-2\left(\tfrac{k}{m}\right)T_w \log(\alpha_1)\cos(2\pi\alpha_2)}, \\ v_y &= \sqrt{-2\left(\tfrac{k}{m}\right)T_w \log(\alpha_2)}, \\ v_z &= \sqrt{-2\left(\tfrac{k}{m}\right)T_w \alpha \sin(2\pi\alpha_2)} \end{aligned} \tag{7}$$

where α_1 or α_2 are the random numbers taken from 0 to 1. In Section 3.3, the effect of the magnitude of the sticking coefficient η on the deposition rate will be explored.

2.2. Validation of DSMC Method and Computer Code

The developed DSMC code was used to model the Couette flow for validation purposes (see Figure 3). The Couette flow is the viscous flow between parallel infinite plates, where the upper plate is moving with uniform speed and the bottom plate is fixed. Couette flow is a shear flow driven by the upper plate movement in the flow direction. DSMC results were validated against the Lattice Boltzmann method (LBM) results developed by Refs. [25,32,33] for Couette flow with no-slip and partial-slip boundary conditions.

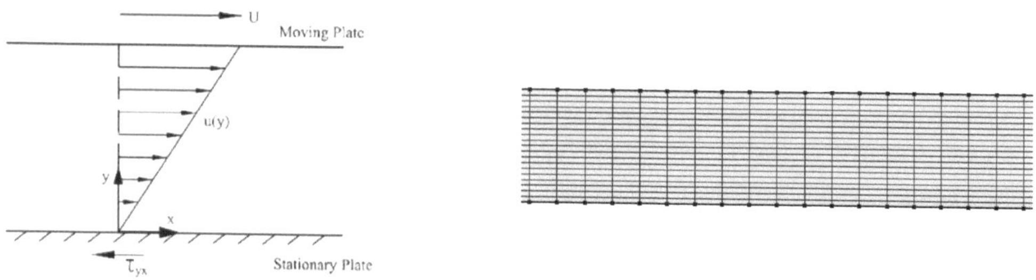

(a) (b)

Figure 3. Validation of DSMC code: (**a**) Couette flow with no-slip boundary conditions and (**b**) DSMC domain with cells.

For validation purposes, the distance between two plates is 1 μm where the fluid domain is filled with the Argon gas [21,34]. At a standard temperature and pressure, the average molecular diameter is 3.66×10^{-10} m, and the molecular mean free path is 65 nm. The plate is moving at the speed of 150 m/s (Ma~0.5) along the positive x direction.

To model the Couette flow, a DSMC simulation domain was created (see Figure 3b). The number of cells along the flow direction (horizontal) was 20 and the number of cells along the height was 20. The length-to-height ratio of each cell was 5. The simulation time step was chosen in such a way that each DSMC particle would stay for at least three time steps in a single cell during the simulation. Here, the time step was chosen as 2.96×10^{-11} s and the number of time steps was 1000. If $Kn = 0.001$, the flow could be assumed to be a continuum. The pressure was 4.89×10^{-6} Pa and the number density of gas was 1.19×10^{27} (see Table 1). The value of $Kn = 0.01$ is a transitional Knudsen number, beyond which the gas flow starts to deviate from continuum approximation. In Table 1, the number density of gas and the corresponding pressure are listed for all the Knudsen numbers.

Table 1. DSMC parameters for Couette flow for various Knudsen numbers.

Knudsen Number, Kn	Length (m)	Mean Free Path (m)	Pressure, Pa	Number Density, n
0.001	1×10^{-6}	1×10^{-9}	4.89×10^{6}	1.19×10^{27}
0.01	1×10^{-6}	1×10^{-8}	4.89×10^{5}	1.19×10^{26}
0.1	1×10^{-6}	1×10^{-7}	4.89×10^{4}	1.19×10^{25}
1	1×10^{-6}	1×10^{-6}	4.89×10^{3}	1.19×10^{24}
10	1×10^{-6}	1×10^{-5}	4.89×10^{2}	1.19×10^{23}

To study the effect of the Knudsen number on the velocity profile, the range of Knudsen numbers was considered and for each case the velocity profile, $u(y)$, was computed by the DSMC code. The Knudsen number in the domain can be controlled by the mean free path, pressure, and number density of gas. The velocity profile and slip velocity are shown in Figures 4 and 5, while the comparison to prior LBM results is shown in Figure 6.

As shown in Figure 4a,b, the Couette flow velocity profile was obtained for the range of Knudsen numbers corresponding to a different degree of rarefaction. The flow with $Kn > 0.01$ represents partial slip flow, in which the velocity at the lower wall is small but not equal to zero. When increasing of the value of the Knudsen number, flow became more rarefied and near the bottom rigid surface the fluid velocity started to increase. In Figure 4a, for $Kn = 0.001$, the flow velocity at the lower wall was 0.05 times the upper-plate velocity, while for $Kn = 0.1$, the ratio increased to 0.12. In Figure 4b, for $Kn = 1$, the wall fluid velocity was 0.28 times the upper plate velocity and for $Kn = 10$, where the flow was quite rarefied, the wall fluid velocity was 0.32 times the upper plate velocity.

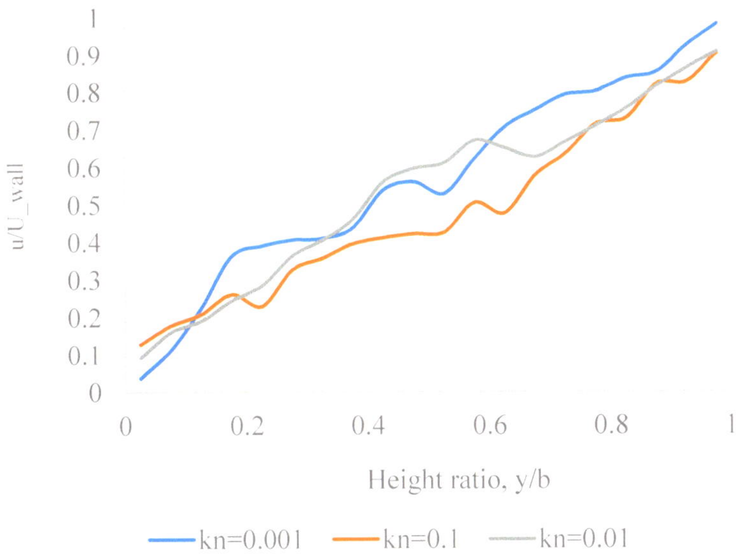

(a)

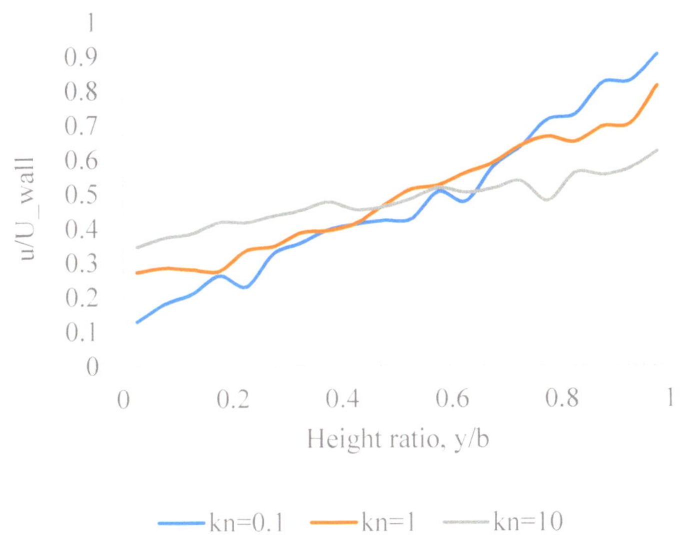

(b)

Figure 4. Couette flow: velocity profile for Knudsen number ranging (**a**) from 0.01 to 0.1 and (**b**) from 0.1 to 10. Velocity profile, u/U_{wall}, is presented as a function of coordinate y normalized by the distance b between plates. The number of simulated particles is equal to 20 per DSMC cell.

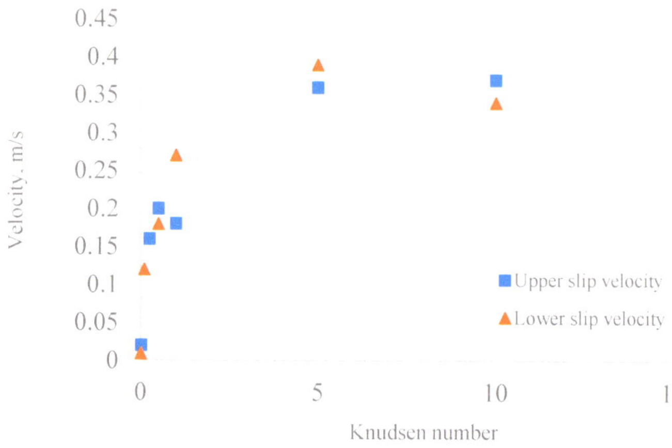

Figure 5. Slip velocity for Couette flow at lower and upper plates as a function of Knudsen number.

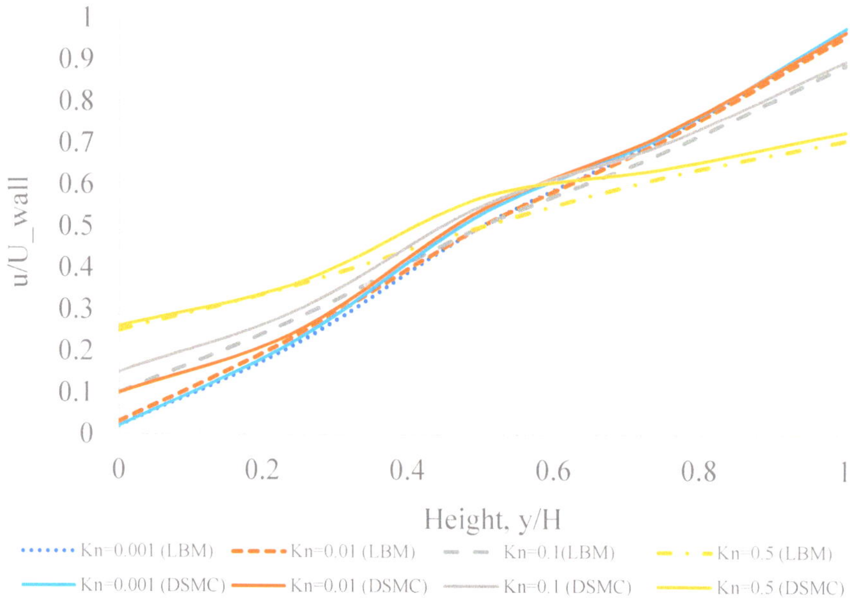

Figure 6. Validation of DSCM code by comparison with LBM results [25,32,33].

Figure 5 reveals the magnitude of slip velocity with the increase in the Knudsen number. It shows that the slip velocity magnitude was quite considerable and approached ~ 30% of upper-wall speed. In this flow regime, the molecular mean free path was either comparable to or larger than the distance between plates, which reduced the probability of intermolecular collisions and increased the probability of molecule-to-wall collisions. Due to higher slip velocity at both walls, the slip velocity profile flattened for $Kn > 0.5$ compared to lower Knudsen number cases.

Figure 6 shows that the DSMC results and results obtained by LBM for the same value of Knudsen number are within the acceptable error margin.

The convergence study for increasing the number of DSMC particles is presented in Appendix A.

3. Model of Deposition of Carbon Particles at Fibers

In this section, structured sets of fibers with a round cross-section (see Figure 1) in the CVD reactor are considered for non-zero Knudsen numbers. The present model does not account for the surface diffusion or re-emission of deposited species that can be accounted for in future research.

3.1. The Reactor Flow Field

The reactor flow field (Figure 7) was obtained by the FVM model for a CVD reactor with a bundle of fibers located at the center of reactor [4]. In Figure 7, the continuous CFD model is described in detail. Pressure, temperature, and velocity were calculated by the FVM at the middle of reactor, 4 mm upstream of fiber bundles, and used as the inlet boundary condition for the DSMC domain. By using the FVM, the concentration of species in feedstock gas flow in reactor was obtained in a prior study of the authors [4]. The DSMC domain encompasses the substrate (fibers), where the major feedstock gas species, which is responsible for surface CVD reaction, is C_2H_2.

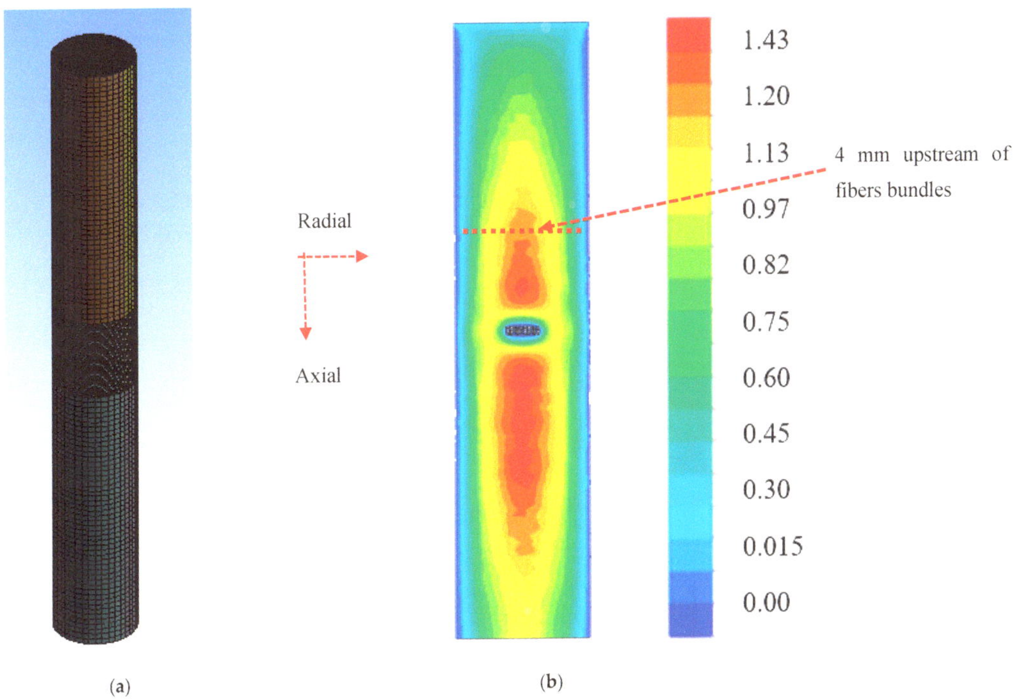

Figure 7. Chemical vapor deposition reactor [4]: (**a**) FVM mesh for CVD reactor and (**b**) isolines of velocity magnitude, m/s, along the vertical plane of the CVD reactor. The inlet boundary of the DSMC domain is shown.

The inlet velocity of the reactor was 1.28 m/s, with an inlet feedstock gas temperature of 25 °C. The static pressure inside the reactor was 1600 Pa and the corresponding Knudsen number of 0.002 for mean free path is $\lambda = 1.87 \times 10^{-5}$m, which represents the continuum flow field. The goal of FVM simulations is to capture the feedstock gas flow velocity upstream of the bundle of fibers. The velocity was captured 4 mm upstream of the fiber bundles at the reactor centerline to be used for DSMC modeling (see Figure 7). To justify the location of the upstream boundary, a parametric study was conducted in which velocity was captured at distances of 1, 2, 3, 4, 5, 6, 7, 8, 9 and 10 mm upstream of the fiber bundles. With the range of 1 mm to 3 mm upstream of the fiber bundles, velocity is lower than at locations above 3 mm due to the formation of a stagnation zone over the fibers' surfaces. Beyond 4 mm upstream of the fibers' bundle, the feedstock gas velocity is no longer affected by the presence of fibers. Thus, the cross-section at 4 mm upstream of the fibers' bundle was chosen to export the CVD gas velocity from the FVM domain to the DSMC domain.

3.2. Setup of Fibers

In Figure 1, the fibers' set up is shown. The diameter, d, of fibers was 1 μm unless specified otherwise in subsequent sections, and the molecular mean free path of the gas was 2.5×10^{-7} m, which corresponds to $Kn = 0.25$. Recall that the range $Kn > 0.1$ corresponded to the rarefied zone computed by DSMC. The DSMC domain length was 10 μm (10d) while the domain height was 6 μm (6d).

The size of the fibers' domain was much smaller compared to the length of the reactor. The end effect of the fibers can be neglected because the fiber length was much larger than the fibers' diameter. The side surfaces of the DSMC domain were taken as symmetrical boundaries to account for the fact that the bundle structure was self-repeating.

In DSMC modeling (see Section 2.1), the first step is to populate the domain with simulated particles, where each simulated particle represents billions of real molecules. The inlet velocity of the DSMC domain was imported from the FVM computations and was equal to 0.97 m/s for current computations (see Figure 7). This velocity was added to the molecular browning motion velocity of DSMC particles. The number density of gas was 1×10^{25}, and the number of DSMC cells was 1800. Each simulated particle represented 10^{12} real gas molecules.

At the fibers' surface, a combined diffuse and reflective boundary condition was considered, where the probability of reflection is p and the probability of sticking to the fiber surface is 1-p (see Section 2).

Figure 8 shows that in the interior of the array of fibers in a bundle, the fluid velocity magnitude was smaller, ~0.33 m/s, compared to the inlet velocity of ~0.97 m/s. Between two fibers, the velocity increased due to the partial blockage of the flow area following the conservation of mass.

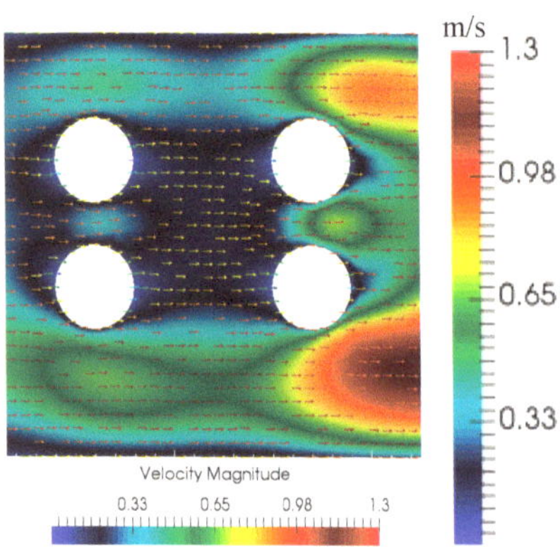

Figure 8. Velocity field around fibers in bundle computed by DSMC.

Figure 9 shows the intensity with which simulated particles are deposited at the fibers' surface. Particles, which flow from the inlet of the DSMC domain, start to deposit at an individual fiber surface after particles reach it. Observation of Figure 8; Figure 9 shows that the first column of the fiber reached a steady state in terms of deposition rate more rapidly compared to the second-column fiber. While after 4 μs the number of deposited particles did not change with respect of time (Figure 9), for the second-column fiber it took ~8 μs (Figure 9) to reach the nearly steady-state phase of deposition. The steady-state number of deposited particles was different by almost 50% between the first and the second-column fibers (Table 2). This is explained by the fact that the second-column fibers were exposed to low-speed flow between fibers while the first-column fibers were exposed to the larger inlet velocity.

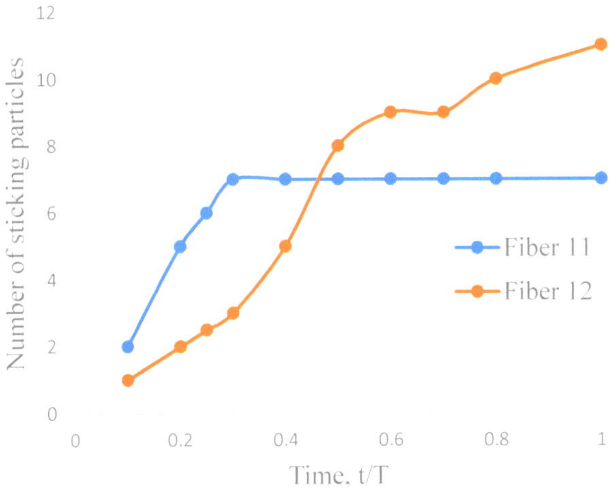

Figure 9. Particle deposition intensity over time at fiber surface—Fiber 11 and Fiber 12—for 50% sticking coefficient. The normalized time, t/T, is used, where T = 10 μs.

Table 2. Number of deposited particles over fiber surfaces.

Fiber	Number of Simulated Particles	Mass of Deposited Layer, Nanogram
Fiber 11	~7	290
Fiber 12	~11	457
Fiber 21	~7	290
Fiber 22	~11	457

Table 2 shows how carbon deposited over the fiber surfaces for the regular orientation of fibers (Figure 1), with Kn = 0.25. The molar mass of C_2H_2 was 26.04 g/mol, after subtracting two atoms of H_2; as hydrogen does not participate in surface deposition, the molecular mass of remaining C_2H_2 was 4.16×10^{-23} kg. As each DSMC particle consisted of 10^{12} molecules, the mass of each DSMC particle would be 4.16×10^{-11} kg.

A comparison to experiments for low Knudsen numbers is available in a prior study of the authors ([4]; see Figure 4), in which the carbon deposition rate obtained by CFD simulations was compared to the experiments [35]. The experiments describe deposition on a bundle of fibers, with the diameter of an individual fiber being 10 μm. The reactor and feedstock gas conditions were the same as in the present study. The fibers' sample was tied together and folded four times. The sample therefore had the form of an "eight"; see Ref. [35], page 30, Figure 3.2. The present study helps to design future experiments using the same reactor and CVD process in which individual fibers of micron-scale diameters are separated and form a structured matrix.

3.3. Effect of Sticking Coefficient on Deposition Rate

In the previous section, simulations were conducted with the sticking coefficient equal to 50%. To understand the role of the value of the sticking coefficient, a parametric study is conducted in the current section for a range of sticking coefficients, η, namely, 20%, 40%, 50%, 60% and 80%. With a higher value of the sticking coefficient, particles would have a higher probability of sticking to the fiber surface, while a lower probability of sticking repels more particles away from the fiber surface. The simulations conducted in the current section quantify the trend.

Figures 10 and 11 show how the particle deposition rate varies with respect to the value of the sticking coefficient and with respect to the position of the fiber in the bundle. Front-column fibers, which face the flow stream first, have a lower deposition rate compared to the deposition at the surface of second-column fibers. A higher sticking coefficient provides a higher deposition rate while a lower sticking coefficient provides a lower deposition rate. This dependence is monotonic and non-linear, as detailed in Figure 12.

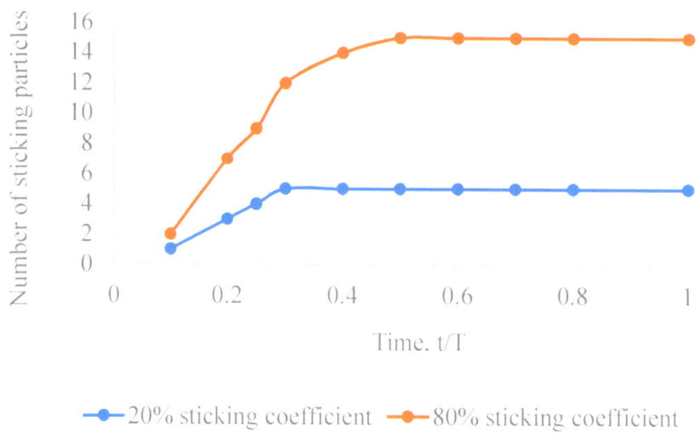

Figure 10. Carbon particle deposition on Fiber 11 for 20% sticking coefficient and 80% sticking coefficient (Kn = 0.25).

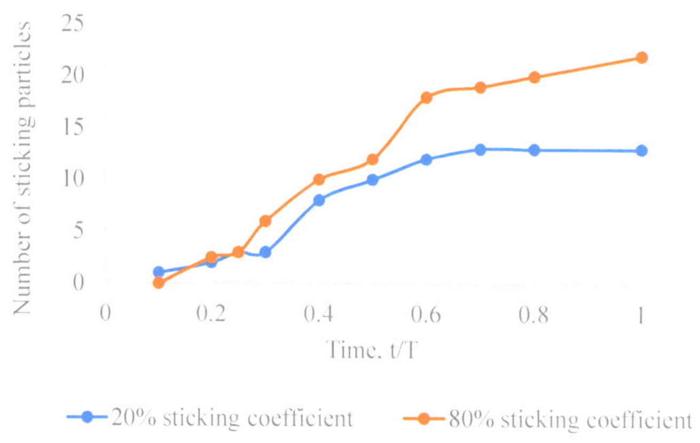

Figure 11. Carbon particle deposition on Fiber 12 (first row, second column fiber) for 20% sticking coefficient and 80% sticking coefficient (Kn = 0.25).

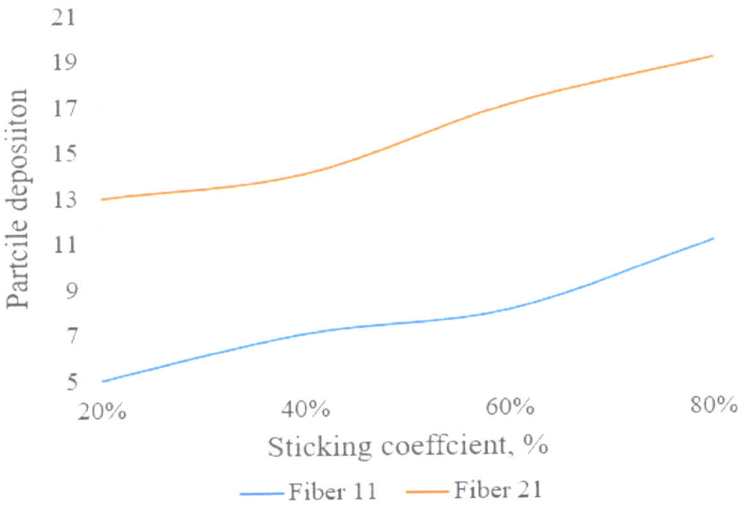

Figure 12. Particle deposition as a function of sticking coefficient at fiber surface.

In the present study, the surface reaction rate was estimated based on the species flux to the surface times the so-called sticking probability (or sticking coefficient). With this simplification, the only necessary input to the DSMC simulations was the sticking probability for each DSMC particle.

To understand the role of the value of sticking coefficient, a parametric study was conducted for a wide range of values of sticking coefficient, from 20% to 80%.

A few authors [22,31] have suggested incorporating reaction rate and activation energy to obtain the sticking coefficient in a more precise, physically accurate way. This will be addressed in the future development of the model proposed in the current study. Recent references have stressed the importance of the use of sticking coefficient. As per Ref. [36], "we adopt a commonly simplified approach—the sticking coefficient (SC) method—to predict the deposition rate when the process is considered to be surface deposition limited". as per Ref. [37], "Central in the idea of the line-of-sight models is the (effective) sticking coefficient which is defined as the ratio of the number of molecules of species that become part of film with a rate, to the total number of impinging molecules, per area per unit of time.... For Monte Carlo methods, the molecules reaching the substrate are adsorbed according to the sticking coefficient S and reflected diffusively with a probability $(1 - S)$".

3.4. *Effect of Distance between the Fibers*

The distance between the centers of fibers plays a vital role in the deposition magnitude and profile. It will be shown in this section that for the same volume fraction of fibers in a bundle, the closer the distance between fibers, the larger the particles' deposition.

Figures 13 and 14 show that, for the same volume fraction of fibers, changing the distance between the fibers affects the deposition pattern and magnitude. A numerical study was conducted in which the volume fraction of fibers was kept constant while changing the diameter and distance between the fibers. Two cases were considered in the study. For case 1, the diameter of fibers was 0.25 µm and the distance (between the circumference) between the fibers was 0.5 µm. For case 2, the diameter of fibers was 0.5 µm and the distance between the fibers was 1 µm. For both cases, the volume fraction was 20%. The flow Knudsen number was 0.25. As the diameters of fibers were chosen according to volume fraction of fibers, the mean free path of gas was changed to keep the Knudsen number constant. For case 1, the mean free path of C_2H_2 was 6.25×10^{-8} m, while for case 2, it was 1.25×10^{-7} m.

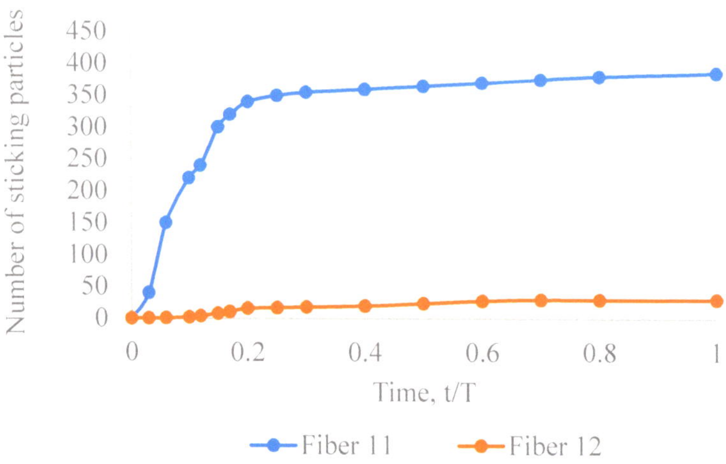

Figure 13. Particle deposition rate for 0.25 μm diameter fiber at 0.5 μm distance between fibers (case 1) for Fiber 11 and Fiber 12 (Kn = 0.25). The volume fraction of fibers was 20% and sticking coefficient was 20%.

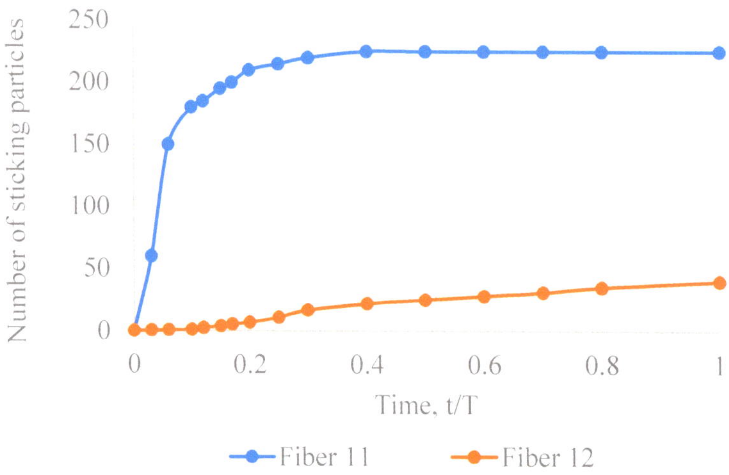

Figure 14. Particle deposition rate for 0.5 μm diameter fiber at 1 μm distance between fibers (case 2). The volume fraction and sticking coefficient were the same as in Figure 13.

For case 1, by time moment t/T = 0.25~360, simulated particles were deposited at the fibers' surface (see Figure 13), while for case 2 with the same 20% volume fraction, at t/T = 0.40, 200, simulated particles were deposited. It shows that the smaller distance between the fibers and smaller diameter of fibers causes ~1.63 times larger deposition for front-column fibers. For the second-column fibers, the magnitude of deposition becomes reversed. Due to the higher distance, fibers of 0.5 μm diameter and 1 μm distance have 1.5 times larger deposition than a 0.25 μm diameter and 0.5 μm distance. Figures 13 and 14 show that, compared with front fibers, a rear fiber accumulates particles at a slower rate over a longer time interval.

3.5. Effect of Rarefaction

Rarefaction in gas affects the deposition profile at fibers' surface. Rarefaction happens either due to lower pressure or due to a smaller diameter of fibers. With a higher Knudsen number (Equation (1)) caused by a smaller fiber diameter with the same volume fraction of fibers in a bundle, gas particles must travel longer to collide due to a larger molecular mean free path. Also, the slip velocity increases with the Knudsen number. With a higher slip velocity, flow resistance is lower due to a smaller velocity gradient normal to a rigid surface and, consequently, reduced skin friction. To quantify the residence time of particles in the vicinity of fibers, an area of influence (Figure 15a) was considered around the bundle of fibers. The length and height of the area of influence were 6d, in microns (square-shaped), respectively. The diameter, d of fibers was 0.25 microns and the distance between the fibers was 0.50 microns (case 1).

For the no-slip condition (Figure 15b), at the surface of fibers all particles which did not stick to the fibers' surface left the domain within 2 microseconds (0.2×10 µs). For $Kn = 0.3$ (Figure 15c), where the gas flow partially slipped at the fibers' surface, the residence time was smaller due to lower friction resistance at the fibers' surface. Particles, which did not stick to the fiber surface left the domain within 0.07×10 µs = 0.7 µs. For $Kn = 0.5$ (Figure 15d), the residence time decreased to 0.06×10 µs = 0.6 µs. With a lower residence time, the available time for particles to participate in surface deposition decreases and the amount of deposition is lower compared to no-slip conditions.

Figure 16 shows that with the increase in the Knudsen number, the residence time decreased nearly linearly for $Kn > 0.1$. The residence time was defined as such that 90% of particles had left the domain. For smaller Knudsen numbers including the continuum zone ($Kn = 0.001$), the residence time was close to 2 µs (see also Figure 15b). In terms of physical significance, with a smaller fiber size, the residence time would be smaller due to the slip velocity at wall, which would offer less time for surface reactions. With less time available for surface reaction, and the smaller surface area available for deposition, the CVD process, which is a relatively slow process itself, would produce less solid deposition in the substrate surface.

In Figure 17, 0 degrees and 180 degrees represent equatorial points, and 90 degree represents the stagnation point. The asymmetric deposition with respect to the centerline of each fiber is caused by the asymmetry of the set-up, where each fiber has neighboring fiber on the one side and domain boundary on the other side (see Figure 1).

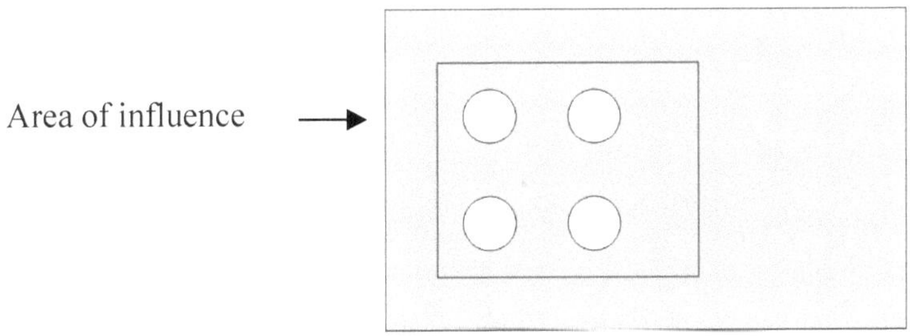

(a)

Figure 15. *Cont.*

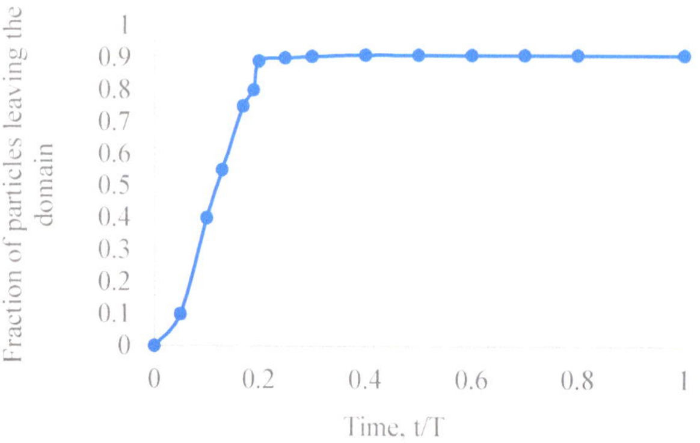

(**b**)

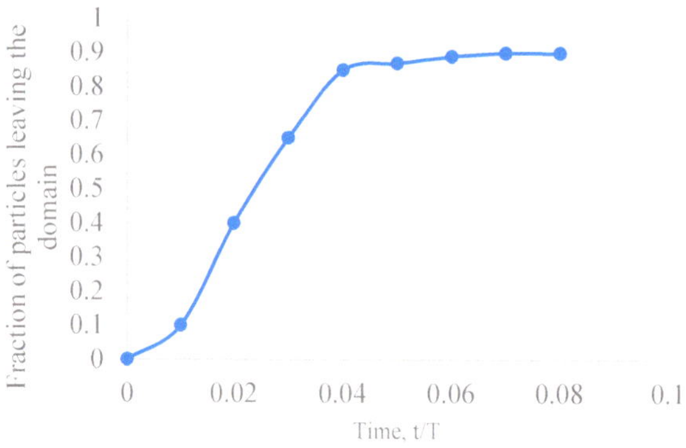

(**c**)

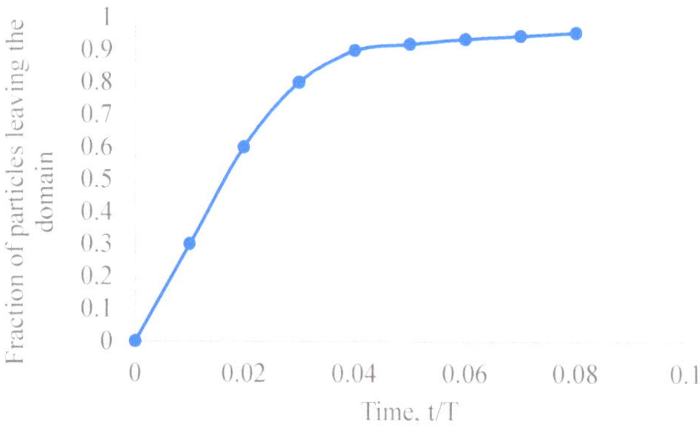

(**d**)

Figure 15. The fraction of particles leaving the domain in flow across the set of fibers for sticking coefficient equal to 50% for the range of Knudsen numbers: (**a**) area of influence around the bundle of fibers; (**b**) no-slip condition; (**c**) Kn = 0.3 and (**d**) Kn = 0.5. The considered time interval T = 10 μs.

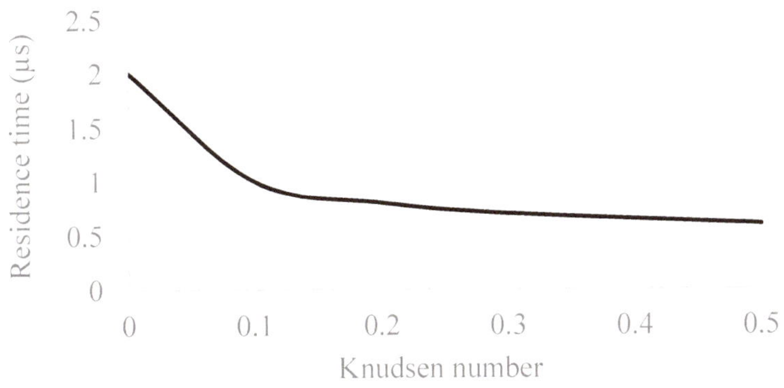

Figure 16. Residence time vs. Knudsen number.

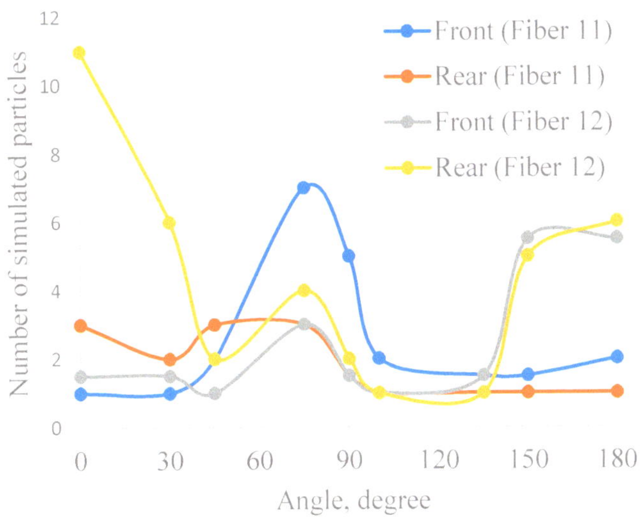

Figure 17. Particle deposition profile at Fiber 11 and Fiber 12 for Kn = 0.3.

Figure 17 shows the deposition profile along the fiber circumference modeled by DSMC method. In the front side of fiber with respect to flow (for angles between 45° and 135°), deposition is larger compared to the rear side for Fiber 11. For Fiber 12, both front and rear semi-circumferences have larger deposition of particles compared to Fiber 11. Due to the low-speed wake downstream of Fiber 11 (see Figure 8), CVD gas has a larger residence time near the second fiber (Fiber 12) that allows larger and more uniform deposition compared to the first-column fibers.

4. Conclusions

The goal of this study was to adopt Direct Simulation Monte Carlo (DSMC) to evaluate the CVD rate for fibers of nano-scale diameter by using simulated C_2H_2 particles. For micro- and nano- scale fibers with a non-zero Knudsen number, the conventional finite volume method (FVM) is not applicable, while molecular methods like DSMC are very computationally intensive to model flowfield for an entire industrial reactor. This study couples FVM with DSMC, where the bulk flow in a reactor and its chemical composition were computed by FVM and flow near the fibers is modeled by DSMC.

With a reduction in fiber diameter, the corresponding Knudsen number increases and velocity at the fibers' surface starts to deviate from no-slip to slip velocity mode. Due to slip velocity at the fibers' surface, the residence time of particles representing reactive molecules within the fibrous medium decreases that eventually decreases the amount of surface deposition at the fibers' surface. The deposition magnitude appears to depend strongly upon the sticking coefficient. For the 80% sticking coefficient, the deposition rate is nearly doubled compared to that for the 20% sticking coefficient. Fibers at the front show a higher deposition rate compared to fibers at the second column due to longer contact with feedstock gas compared to the second-column fibers.

The proposed methodology will be extended to unstructured set-ups of fibers with multiple diameters in future research.

Author Contributions: Conceptualization, A.P. and H.B.; methodology, H.B. and A.P.; software, H.B.; validation, H.B.; investigation, A.P.; data curation, H.B.; writing—original draft preparation, H.B.; writing—review and editing, A.P.; visualization, H.B.; supervision, A.P.; project administration, A.P. All authors have read and agreed to the published version of the manuscript.

Funding: This research received no external funding.

Data Availability Statement: The data presented in this study are available on request from the corresponding author. The data are not publicly available due to privacy.

Conflicts of Interest: The authors declare no conflict of interest.

Appendix A Convergence Study

A total of 40,000 simulated DSMC particles were chosen for the whole domain that corresponds to 100 particles per cell. This is five times larger than the previous simulation case depicted in Figure 4. With the increased number of particles, the velocity profile has a much smaller amplitude of oscillations, as depicted in Figure A1.

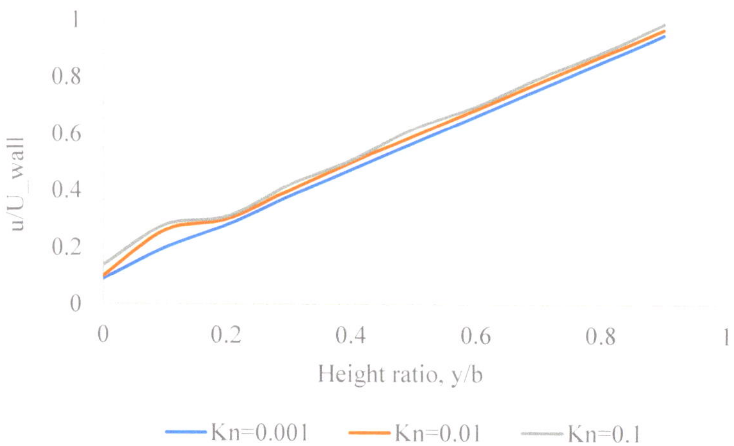

Figure A1. Normalized velocity profile for Couette flow.

References

1. Mittal, M.; Sardar, S.; Jana, A. Nanofabrication Techniques for Semiconductor Chemical Sensors. In *Handbook of Nanomaterials for Sensing Applications*; Elsevier: Amsterdam, The Netherlands, 2021; pp. 119–137.
2. "Vapor Deposition". Available online: https://corrosion-doctors.org/MetalCoatings/VaporDep.htm (accessed on 25 October 2023).
3. Bird, G.A. Monte Carlo Simulation of Gas Flows. *Annu. Rev. Fluid Mech.* **1978**, *10*, 11–31. [CrossRef]
4. Barua, H.; Povitsky, A. Numerical Model of Carbon Chemical Vapor Deposition at Internal Surfaces. *Vacuum* **2020**, *175*, 109234. [CrossRef]
5. Kleijn, C.R.; Hoogendoorn, C.J. A study of 2-and 3-D transport phenomena in horizontal chemical vapor deposition reactors. *Chem. Eng. Sci.* **1991**, *46*, 321–334. [CrossRef]
6. Ouazzani, J.; Chiu, K.-C.; Rosenberger, F. On the 2D modelling of horizontal CVD reactors and its limitations. *J. Cryst. Growth* **1988**, *91*, 497–508. [CrossRef]
7. Cheng, T.S.; Hsiao, M.C. Numerical investigations of geometric effects on flow and thermal fields in a horizontal CVD reactor. *J. Cryst. Growth* **2008**, *310*, 3097–3106. [CrossRef]
8. Adjerid, S.; Flaherty, J.E.; Hudson, J.B.; Shephard, M.S. Modeling and the Adaptive Solution of CVD Fiber-Coating Process. *Comput. Methods Appl. Mech. Eng.* **1999**, *172*, 293–308. [CrossRef]
9. Li, M.; Xu, Z.; Li, Z.; Chen, Y.; Guo, J.; Huo, H.; Zhou, H.; Huangfu, H.; Cao, Z.; Wang, H. An Experimental and CFD Study on Gas Flow Field Distribution in the Growth Process of Multi-Walled Carbon Nanotube Arrays by Thermal Chemical Vapor Deposition. *Cryst. Res. Technol.* **2016**, *51*, 702–707. [CrossRef]
10. Pierson, H.O.; Lieberman, M.L. The Chemical Vapor Deposition of Carbon on Carbon Fibers. *Carbon* **1975**, *13*, 159–166. [CrossRef]
11. Li, A.; Norinaga, K.; Deutschmann, O.; Chemistry, P. Modeling and simulation of materials synthesis: Chemical vapor deposition and infiltration of pyrolytic carbon. *Compos. Sci. Technol.* **2008**, *68*, 1097–1104. [CrossRef]
12. Mishra, P.; Verma, N. A CFD Study on a Vertical Chemical Vapor Deposition Reactor for Growing Carbon Nanofibers. *Chem. Eng. Res. Des.* **2012**, *90*, 2293–2301. [CrossRef]
13. Allu, S.; Simunovic, S.; Lee, T.-S.; Witting, P.; Intl, H. *Development of Mathematical Model and Simulation Tool for the High-Capacity Production of Carbon Fiber CRADA Final Report for CRADA Number NFE-17-06446*; Oak Ridge National Lab. (ORNL): Oak Ridge, TN, USA, 2019.
14. Yu, K.; Hayman, C.C.; Manjunath, S.; Fan, W.; Martin, I.T.; Martin, H.B.; Sankaran, R.M. A Combined CFD Modeling and Experimental Study of Pyrolytic Carbon Deposition. *Diam. Relat. Mater.* **2016**, *70*, 173–178. [CrossRef]
15. Endo, H.; Kuwana, K.; Saito, K.; Qian, D.; Andrews, R.; Grulke, E.A. CFD Prediction of Carbon Nanotube Production Rate in a CVD Reactor. *Chem. Phys. Lett.* **2004**, *387*, 307–311. [CrossRef]
16. Meiburg, E. Comparison of the Molecular Dynamics Method and the Direct Simulation Monte Carlo Technique for Flows around Simple Geometries. *Phys. Fluids* **1986**, *29*, 3107. [CrossRef]
17. Klothakis, A. DSMC Simulations of Rarefied Hypersonic Flows. Master's Thesis, School of Production Engineering and Management, Technical University of Crete, Chania, Greece, 2015.
18. Zhao, S.; Povitsky, A. Coupled Continuum and Molecular Model of Flow through Fibrous Filter. *Phys. Fluids* **2013**, *25*, 112002. [CrossRef]
19. Dreyer, J.A.H.; Riefler, N.; Pesch, G.R.; Kara Mehmedovic, M.; Fritsching, U.; Teoh, W.Y.; Madler, L. Simulation of Gas Diffusion in Highly Porous Nanostructures by Direct Simulation Monte Carlo. *Chem. Eng. Sci.* **2014**, *105*, 69–76. [CrossRef]
20. Pesch, G.R.; Riefler, N.; Fritsching, U.; Ciacchi, L.C.; Madler, L. Gas-Solid Catalytic Reactions with an Extended DSMC Model. *AIChE J.* **2015**, *61*, 2092–2103. [CrossRef]
21. Bird, G.A. *The DSMC Method*, Version 1.2. 2013. Available online: https://www.amazon.com/DSMC-Method-G-Bird/dp/1492112909 (accessed on 25 October 2023).
22. Bird, G.A. The Q-K Model for Gas-Phase Chemical Reaction Rates. *Phys. Fluids* **2011**, *23*, 106101. [CrossRef]
23. What Is MEMS Technology? A MEMS Clearinghouse®and Information Portal for the MEMS and Nanotechnology Community. Available online: https://www.memsnet.org/mems/what_is.html (accessed on 25 October 2023).
24. Karniadakis, G.; Beşkök, A.; Aluru, N.R. *Microflows and Nanoflows: Fundamentals and Simulation*; Springer: Berlin/Heidelberg, Germany, 2005.
25. Shirani, E.; Jafari, S. Application of LBM in Simulation of Flow in Simple Micro-Geometries and Micro Porous Media. *Afr. Phys. Rev.* **2007**, *1*, 0003.
26. Chambre, P.A.; Schaaf, S.A. *Flow of Rarefied Gases*; Princeton Univ. Press: Princeton, NJ, USA, 2017; ISBN 9780691654904.
27. Kennard, E.H. *The Kinetic Theory of Gases. With an Introduction to Statistical Mechanics*, 1st ed.; McGraw-Hill Publishing Co., Ltd.: London, UK, 1938.
28. Colin, S. Gas Microflows in the Slip Flow Regime: A Critical Review on Convective Heat Transfer. *J. Heat Transfer.* **2012**, *134*, 020908. [CrossRef]
29. Chapman, S.; Cowling, T.G. *The Mathematical Theory of Non-Uniform Gases*; Cambridge University Press: Cambridge, UK, 1953.
30. Oran, E.S.; Oh, C.K.; Cybyk, B.Z. Direct simulation Monte Carlo: Recent advances and applications. *Annu. Rev. Fluid Mech.* **1998**, *30*, 403. [CrossRef]

31. Akiyama, Y.; Matsumura, S.; Imaishi, N. Shape of film grown on microsize trenches and holes by chemical vapor deposition: 3-dimensional Monte Carlo simulation. *Jpn. J. Appl. Phys.* **1995**, *34*, 6171–6177. [CrossRef]
32. Lockerby, D.A.; Reese, J.M.; Gallis, M.A. A Wall-Function Approach to Incorporating Knudsen-Layer Effects in Gas Micro Flow Simulations. *AIP Conf. Proc.* **2005**, *762*, 731.
33. Niu, X.D.; Shu, C.; Chew, Y.T. A Lattice Boltzmann BGK Model for Simulation of Micro Flows. *Europhys. Lett.* **2004**, *67*, 600–606. [CrossRef]
34. Garcia, A. *Numerical Methods for Physics (Python)*, Second, Revised (Python) Edition. Available online: https://www.amazon.com/Numerical-Methods-Physics-Python-Alejandro/dp/1548865494 (accessed on 25 October 2023).
35. Balachandran, R. Gas Phase and Surface Modeling of Chemical Vapor Deposition of Pyrolytic Carbon on Silicon Carbide Fibers Using a Pure Methane Precursor. Master Thesis, University of Akron, Akron, OH, USA, 2011. Available online: http://rave.ohiolink.edu/etdc/view?acc_num=akron1303834116 (accessed on 25 October 2023).
36. Guan, K.; Gao, Y.; Zeng, Q.; Luan, X.; Zhang, Y.; Cheng, L.; Wu, J.; Lu, Z. Numerical modeling of SiC by low-pressure chemical vapor deposition from methyl trichlorosilane. *Chin. J. Chem. Eng.* **2020**, *28*, 1733–1743. [CrossRef]
37. Cheimarios, N.; Kokkoris, G.; Boudouvis, A.G. Multiscale Modeling in Chemical Vapor Deposition Processes: Models and Methodologies. *Arch. Comput. Methods Eng.* **2021**, *28*, 637–672. [CrossRef]

Mathematical and Computational Applications

MDPI

Article

A Finite Difference Method for Solving the Wave Equation with Fractional Damping

Manruo Cui [1], Cui-Cui Ji [1,*] and Weizhong Dai [2]

[1] School of Mathematics and Statistics, Qingdao University, Qingdao 266071, China; 2021020248@qdu.edu.cn
[2] Mathematics & Statistics, College of Engineering & Science, Louisiana Tech University, Ruston, LA 71272, USA; dai@coes.latech.edu
* Correspondence: cuicuiahuan@163.com

Abstract: In this paper, we develop a finite difference method for solving the wave equation with fractional damping in 1D and 2D cases, where the fractional damping is given based on the Caputo fractional derivative. Firstly, based on the weighted method, we propose a new numerical approximation for the Caputo fractional derivative and apply it for the 1D case to obtain a time-stepping method. We then develop an alternating direction implicit (ADI) scheme for the 2D case. Using the discrete energy method, we prove that the proposed difference schemes are unconditionally stable and convergent in both 1D and 2D cases. Finally, several numerical examples are given to verify the theoretical results.

Keywords: finite difference method; fractional damping; wave equation; stability; convergence

1. Introduction

Recently, the study of damping effects has been a hot research topic because it appears in a variety of the dynamic processes of complex systems, including electromagnetic shunt [1], extensible beams [2], swelling porous elastic [3], vibration [4], and so on [5–13]. In this paper, we consider the initial value problem for the wave equation with fractional damping as follows:

$$\lambda_1 \frac{\partial^2 u(\mathbf{x},t)}{\partial t^2} + \lambda_2 {}_0^C\mathcal{D}_t^\alpha u(\mathbf{x},t) = D\Delta u(\mathbf{x},t) + f(\mathbf{x},t), \quad \mathbf{x} \in \Omega,\ t \in (0,T], \tag{1}$$

$$u(\mathbf{x},0) = \phi(\mathbf{x}), \quad u_t(\mathbf{x},0) = \psi(\mathbf{x}), \quad \mathbf{x} \in \bar{\Omega}, \tag{2}$$

$$u(\mathbf{x},t)|_{\mathbf{x}\in\partial\Omega} = \varphi(\mathbf{x},t), \quad t \in [0,T], \tag{3}$$

where $\Omega = (a,b) \subset \mathbf{R}^1$ or $\Omega = (a,b) \times (c,d) \subset \mathbf{R}^2$, $\partial\Omega$ is the boundary of Ω, Δ is the Laplacian operator, $D(>0)$ is the diffusion coefficient, $f(\mathbf{x},t)$, $\psi(\mathbf{x},t)$, $\phi(\mathbf{x})$ and $\varphi(\mathbf{x})$ are given functions, and the operator ${}_0^C\mathcal{D}_t^\alpha$ denotes the Caputo fractional derivative defined by

$${}_0^C\mathcal{D}_t^\alpha u(\mathbf{x},t) = \frac{1}{\Gamma(2-\alpha)} \int_0^t \frac{\partial^2 u(\mathbf{x},s)}{\partial s^2} \frac{ds}{(t-s)^{\alpha-1}}, \quad 1 < \alpha < 2. \tag{4}$$

In an overdamped bistable system, the governing model (1)–(3) presents a double-resonance pattern [14]. The fractional derivative in Equation (1) is referred as a damping term, and can account for the memory effects in the system. The case $\alpha = \frac{3}{2}$ in Equation (1) is known as the Lokshin model, which appears in the study of the propagation of the air inside a duct when taking into account viscothermal losses [15,16]. The case $\alpha = 1$ in Equation (1) is known as the damped Klein–Gordon model, arising in relativistic quantum mechanics [17]. The case $\alpha = 2$ in Equation (1) is known as the classical wave equation. Compared with the above commonly discussed models, Equation (1) is more physically flexible due to the term ${}_0^C\mathcal{D}_t^\alpha u(\mathbf{x},t)$ and the parameter λ. Moreover, the inclusion of the time fractional derivative allows for the simulation of the anomalous diffusion and other non-classical transport phenomena.

Although some linear and special fractional partial differential Equations (FPDEs) can be solved using analytical methods, the analytical solutions of the most generalized FPDEs (e.g., multi-term FPDEs) cannot be worked out. Consequently, the study of efficient numerical algorithms for solving the FPDEs plays a critical role in scientific computing and engineering applications. At present, many numerical methods have been developed for solving the FPDEs, such as finite difference methods [18–25], finite element methods [26–30], and spectral methods [31,32].

To the best of our knowledge, only a few numerical methods for the wave equation with fractional damping in Equations (1)–(3) have been considered. Saffarian and Mohebbi [33] developed and analyzed an ADI spectral element method for solving the time fractional damped nonlinear Klein–Gordon equation, in which the equation includes the second-order derivative and the Caputo fractional derivative of order $\alpha \in (0,1)$ in the temporal direction. The WSGL scheme was used to discretize the Caputo derivative based on the relation between the Reimann–Liouville and the Caputo derivative in [33]. Wu et al. [34] used the reduced-order method to develop a Crank–Nicolson compact difference scheme of the time-fractional damped plate vibration equation, in which the equation includes the second-order derivative and the Caputo fractional derivative of order $\alpha \in (1,2)$ in the temporal direction. Based on an efficient sum-of-exponentials (SOE) approximation for the kernel function $t^{-\beta-1}$ with $\beta \in (0,1)$, Lyu et al. [35] presented a fast linearized finite difference scheme for the 1D nonlinear multi-term time-fractional wave equation, where all fractional orders are in $(1,2)$.

In this paper, we aim to propose a new finite difference method for the one- and two-dimensional wave equations with fractional damping in Equations (1)–(3). Note that a direct application of the classic $L1$ approximation to the Caputo derivative in Equation (1) at $t = t_n$ leads to a numerical scheme with only first-order temporal convergence accuracy (see [36]). The difficulty of obtaining a high-order algorithm for Equations (1)–(3) in time lies in how to design a numerical approximation with high accuracy for the following problem:

$$\lambda_1 u''(t) + \lambda_2 {}_0^C\mathcal{D}_t^{\alpha} u(t) = f(t), \quad t \in [0,T]. \tag{5}$$

Therefore, we first consider the difference scheme with high accuracy for Equation (5) in this study. Specifically, we consider the Caputo fractional derivative ${}_0^C\mathcal{D}_t^{\alpha} u(t)$ and the second-order derivative $u''(t)$ as a whole in the numerical discretization, and then take its weighted average at $t = t_{n-1}$ and $t = t_{n+1}$. After careful analysis, a (3–α)-order numerical approximation for Equation (5) is then obtained. This is one key contribution of this paper. In addition, we apply this technique of time discretization to derive numerical algorithms for the wave equation with fractional damping in Equations (1)–(3) in both 1D and 2D cases. We then use the discrete energy method to prove the unconditional stability and convergence of the proposed difference schemes. This is another novelty of our work. For the 2D case, we add some small perturbations to obtain an alternating direction implicit (ADI) scheme in order to reduce the huge computational work and storage. Finally, numerical examples are provided to verify the theoretical analysis, and to show some new diffusion-wave phenomena based on fractional damping.

2. Numerical Discretization for a Fractionally Damped System

In this section, we derive a difference scheme for solving the following fractionally damped system with initial conditions

$$\lambda_1 u''(t) + \lambda_2 {}_0^C\mathcal{D}_t^{\alpha} u(t) = f(t), \ \ t \in (0,T], \tag{6}$$

$$u(0) = \phi, \quad u'(0) = \psi, \tag{7}$$

where the fractional order is $\alpha \in (1,2)$, and $f(t)$ is a given source term. Here, $u'(t)$ and $u''(t)$ denote the first-order and second-order derivative with respect to t, respectively.

Take a uniform partition of the computing interval $[0, T]$, i.e., $t_n = n\tau$, $0 \le n \le N$ with the step size $\tau = \frac{T}{N}$, where N is a given positive integer. The numerical accuracy reaches no more than first order if we directly consider the difference scheme of Equation (6) at $t = t_n$. In order to obtain a stable difference scheme with high accuracy for Equation (6), we average the results of Equation (6) at $t = t_{n-1}$ and $t = t_{n+1}$, i.e.,

$$\lambda_1 \frac{u''(t_{n-1}) + u''(t_{n+1})}{2} + \lambda_2 \frac{{}_0^C\mathcal{D}_t^\alpha u(t_{n-1}) + {}_0^C\mathcal{D}_t^\alpha u(t_{n+1})}{2} = \frac{f(t_{n-1}) + f(t_{n+1})}{2}. \tag{8}$$

For simplicity, we define grid functions $U^n = u(t_n)$, $n \ge 0$. Furthermore, we introduce some difference quotient operators

$$\Delta_\tau U^n = \frac{U^{n+1} - U^{n-1}}{2\tau}, \quad \delta_t U^{n-\frac{1}{2}} = \frac{U^n - U^{n-1}}{\tau}, \quad \delta_t^2 U^n = \frac{\delta_t U^{n+\frac{1}{2}} - \delta_t U^{n-\frac{1}{2}}}{\tau}.$$

Firstly, we can obtain a second-order approximation for the weighted average of $u''(t)$ at $t = t_{n-1}$ and $t = t_{n+1}$ by using the Taylor expansion, i.e.,

$$\frac{u''(t_{n-1}) + u''(t_{n+1})}{2} = \delta_t^2 U^n + O(\tau^2). \tag{9}$$

Next, we adopt the reduced-order method to construct our numerical formula for the Caputo derivative ${}_0^C\mathcal{D}_t^\alpha u(t)$. Let $\beta = \alpha - 1$, $v(t) = u'(t)$. We transform ${}_0^C\mathcal{D}_t^\alpha u(t)$ into its substitute as ${}_0^C\mathcal{D}_t^\beta v(t)$, i.e.,

$${}_0^C\mathcal{D}_t^\beta v(t) = \frac{1}{\Gamma(1-\beta)} \int_0^t \frac{v'(s)ds}{(t-s)^\beta}.$$

The following lemma states the truncation error of the $L1$-type fractional formula.

Lemma 1 (see [37]). *Suppose $g(t) \in C^2[0, t_n]$. Then*

$$\begin{aligned}
&\left| \int_0^{t_n} g'(t) \frac{dt}{(t_n - t)^\beta} - \frac{1}{\tau} \left[\tilde{a}_0 g(t_n) - \sum_{k=1}^{n-1} (\tilde{a}_{n-k-1} - \tilde{a}_{n-k}) g(t_k) - \tilde{a}_{n-1} g(t_0) \right] \right| \\
&\le \frac{1}{1-\beta} \left[\frac{1-\beta}{12} + \frac{2^{2-\beta}}{2-\beta} - (1 + 2^{-\beta}) \right] \max_{0 \le t \le t_n} |g''(t)| \tau^{2-\beta},
\end{aligned} \tag{10}$$

where $0 < \beta < 1$, and

$$\tilde{a}_l = \int_{t_l}^{t_{l+1}} \frac{dt}{t^\beta} = \frac{1}{1-\beta} [(t_{l+1})^{1-\beta} - (t_l)^{1-\beta}] = \frac{\tau^{1-\beta}}{1-\beta} [(l+1)^{1-\beta} - l^{1-\beta}],\ l \ge 0.$$

We assume that the solution $u(t)$ of Equation (6) is smooth. Accordingly, $v(t)$ is smooth. By Lemma 1, we have the following discrete formula at grid point $t = t_n$, i.e.,

$${}_0^C\mathcal{D}_t^\beta v(t_n) = \frac{\tau^{-\beta}}{\Gamma(2-\beta)} \left[a_0^{(\beta)} v(t_n) - \sum_{k=1}^{n-1} (a_{n-k-1}^{(\beta)} - a_{n-k}^{(\beta)}) v(t_k) - a_{n-1}^{(\beta)} v(t_0) \right] + (r_1)^n, \tag{11}$$

where $a_l^{(\beta)} = (l+1)^{1-\beta} - l^{1-\beta}$, $l \ge 0$, and there exists a constant c_1 which satisfies $|(r_1)^n| \le c_1 \tau^{2-\beta}$.

Noticing the relations $\alpha = \beta + 1$ and $u'(t) = v(t)$, we then calculate the weighted average of ${}_0^C\mathcal{D}_t^\alpha u(t_{n+1})$ and ${}_0^C\mathcal{D}_t^\alpha u(t_{n-1})$ to yield

$$\begin{aligned}
\frac{{}_0^C\mathcal{D}_t^\alpha u(t_{n-1}) + {}_0^C\mathcal{D}_t^\alpha u(t_{n+1})}{2} &= \frac{{}_0^C\mathcal{D}_t^\beta v(t_{n-1}) + {}_0^C\mathcal{D}_t^\beta v(t_{n+1})}{2} \\
&= \frac{\tau^{-\beta}}{\Gamma(2-\beta)}\Bigg[a_0^{(\beta)}\frac{v(t_{n-1}) + v(t_{n+1})}{2} - \sum_{k=1}^{n-1}(a_{n-k-1}^{(\beta)} - a_{n-k}^{(\beta)})\frac{v(t_{k-1}) + v(t_{k+1})}{2} \\
&\quad - a_{n-1}^{(\beta)}\frac{v(t_0) + v(t_1)}{2} + a_n^{(\beta)}\frac{v(t_1) - v(t_0)}{2}\Bigg] + \frac{(r_1)^{n-1} + (r_1)^{n+1}}{2} \\
&= \frac{\tau^{1-\alpha}}{\Gamma(3-\alpha)}\Bigg[b_0^{(\alpha)}\frac{u'(t_{n-1}) + u'(t_{n+1})}{2} - \sum_{k=1}^{n-1}(b_{n-k-1}^{(\alpha)} - b_{n-k}^{(\alpha)})\frac{u'(t_{k-1}) + u'(t_{k+1})}{2} \\
&\quad - b_{n-1}^{(\alpha)}\frac{u'(t_0) + u'(t_1)}{2} + b_n^{(\alpha)}\frac{u'(t_1) - u'(t_0)}{2}\Bigg] + \frac{(r_1)^{n-1} + (r_1)^{n+1}}{2},
\end{aligned} \tag{12}$$

where $b_l^{(\alpha)} = (l+1)^{2-\alpha} - l^{2-\alpha}$, $l \geq 0$. Based on the Taylor expansion, the averaging operators in Equation (12) can be approximated using

$$\frac{1}{2}[u'(t_{k-1}) + u'(t_{k+1})] = \Delta_\tau U^k + (r_2)^k, \quad 1 \leq k \leq n, \tag{13}$$

and

$$\frac{1}{2}[u'(t_k) + u'(t_{k-1})] = \delta_t U^{k-\frac{1}{2}} + (r_3)^k, \quad 1 \leq k \leq n, \tag{14}$$

where

$$(r_2)^k = -\frac{\tau^2}{4}\int_0^1 [u'''(t_k - s\tau) + u'''(t_k + s\tau)](1-s)^2 ds, \tag{15}$$

and

$$(r_3)^k = -\frac{\tau^2}{16}\int_0^1 [u'''(t_{k-\frac{1}{2}} - \frac{\tau}{2}s) + u'''(t_{k-\frac{1}{2}} + \frac{\tau}{2}s)](1-s)^2 ds. \tag{16}$$

In Equation (12), we transform $\frac{u'(t_1)-u'(t_0)}{2}$ into the form $\frac{u'(t_1)+u'(t_0)}{2} - u'(t_0)$, and then substitute the approximations in Equations (13) and (14) into Equation (12). This gives

$$\begin{aligned}
\frac{{}_0^C\mathcal{D}_t^\alpha u(t_{n-1}) + {}_0^C\mathcal{D}_t^\alpha u(t_{n+1})}{2} =& \frac{\tau^{1-\alpha}}{\Gamma(3-\alpha)}\Big[b_0^{(\alpha)}\Delta_\tau U^n - \sum_{k=1}^{n-1}(b_{n-k-1}^{(\alpha)} - b_{n-k}^{(\alpha)})\Delta_\tau U^k \\
& - b_{n-1}^{(\alpha)}\delta_t U^{\frac{1}{2}} + b_n^{(\alpha)}(\delta_t U^{\frac{1}{2}} - \psi)\Big] + (r_4)^n,
\end{aligned} \tag{17}$$

where $(r_4)^n$ denotes

$$\begin{aligned}
(r_4)^n =& \frac{\tau^{1-\alpha}}{\Gamma(3-\alpha)}\Big[b_0^{(\alpha)}(r_2)^n - \sum_{k=1}^{n-1}(b_{n-k-1}^{(\alpha)} - b_{n-k}^{(\alpha)})(r_2)^k + (b_n^{(\alpha)} - b_{n-1}^{(\alpha)})(r_3)^1\Big] \\
& + \frac{(r_1)^{n-1} + (r_1)^{n+1}}{2} = O(\tau^{3-\alpha}).
\end{aligned} \tag{18}$$

For simplicity, we define the operator $\delta_\tau^\alpha(U^n, \psi)$ as

$$\delta_\tau^\alpha(U^n, \psi) = \frac{\tau^{1-\alpha}}{\Gamma(3-\alpha)}\Big[b_0^{(\alpha)}\Delta_\tau U^n - \sum_{k=1}^{n-1}(b_{n-k-1}^{(\alpha)} - b_{n-k}^{(\alpha)})\Delta_\tau U^k - b_{n-1}^{(\alpha)}\delta_t U^{\frac{1}{2}} + b_n^{(\alpha)}(\delta_t U^{\frac{1}{2}} - \psi)\Big]. \tag{19}$$

Hence, based on the above numerical formulas in Equations (9) and (17), we obtain a numerical discretization for Equation (6) from the second level to the N-th level as follows:

$$\lambda_1\delta_t^2 U^n + \lambda_2\delta_\tau^\alpha(U^n, \psi) = \frac{f(t_{n-1}) + f(t_{n+1})}{2} + R^n, \quad 1 \leq n \leq N, \tag{20}$$

where $R^n = O(\tau^{3-\alpha})$. We now consider the numerical discretization at the first level. Since ${}_0^C\mathcal{D}_t^{\alpha} u(t_0) = 0$, Equation (6) at $t = t_0$ reads

$$\lambda_1 u''(t_0) = f(t_0). \tag{21}$$

We apply the Taylor expansion to obtain

$$\frac{2\lambda_1}{\tau}\left[\delta_t U^{\frac{1}{2}} - \psi\right] = f(t_0) + R^0, \tag{22}$$

where $R^n = O(\tau)$.

Let u^n be the approximate solution of $u(t_n)$. We drop the truncation error R^n in Equation (20) and the truncation error R^0 in Equation (22) and obtain a difference scheme for Equation (6) as

$$\begin{cases} \lambda_1 \delta_t^2 u^n + \lambda_2 \delta_\tau^{\alpha}(u^n, \psi) = \frac{f(t_{n-1})+f(t_{n+1})}{2}, & n \geq 1, \\ \frac{2\lambda_1}{\tau}\left[\delta_t u^{\frac{1}{2}} - \psi\right] = f(t_0) \end{cases} \tag{23}$$

with $u^0 = \phi$.

To verify the accuracy of the above scheme, we present an example here.

Example 1. *We take* $\lambda_1 = \lambda_2 = 1$, $T = 1$, $f(t) = 12t^2 + \frac{24}{\Gamma(5-\alpha)} t^{4-\alpha}$ *in Equation* (6).

The exact solution of Example 1 is $u(t) = t^4$. In our computation, we use the difference scheme (23) to solve this problem. The numerical error was computed by

$$E_\infty(\tau) = \max_{0 \leq n \leq N} |u(t_n) - u^n|,$$

and the corresponding convergence order was calculated by Order= $\log_2(\frac{E_\infty(\tau)}{E_\infty(\frac{\tau}{2})})$.

We chose $\alpha = 1.3, 1.5, 1.7$, respectively, to test the convergence accuracy. Table 1 shows the numerical errors and the corresponding convergence accuracy for each fixed α. As expected, our scheme (23) generates the numerical approximation accuracy with $\mathcal{O}(\tau^{3-\alpha})$.

Table 1. Numerical approximation error and convergence accuracy in Example 1.

α	τ	$E_\infty(\tau)$	Order
1.3	1/2000	7.0435×10^{-7}	-
	1/4000	2.4095×10^{-7}	1.5476
	1/8000	8.0200×10^{-8}	1.5871
	1/16,000	2.6193×10^{-8}	1.6144
	1/32,000	8.4323×10^{-9}	1.6352
	1/64,000	2.6448×10^{-9}	1.6728
1.5	1/2000	6.4252×10^{-6}	-
	1/4000	2.3176×10^{-6}	1.4711
	1/8000	8.3087×10^{-7}	1.4799
	1/16,000	2.9662×10^{-7}	1.4860
	1/32,000	1.0558×10^{-7}	1.4903
	1/64,000	3.7468×10^{-8}	1.4946
1.7	1/2000	3.9280×10^{-5}	-
	1/4000	1.6027×10^{-5}	1.2933
	1/8000	6.5273×10^{-6}	1.2959
	1/16,000	2.6555×10^{-6}	1.2975
	1/32,000	1.0796×10^{-6}	1.2985
	1/64,000	4.3871×10^{-7}	1.2992

3. 1D Wave Equation with Fractional Damping

In this section, we propose a finite difference method for solving Equations (1)–(3) in a one-dimensional case, i.e.,

$$\lambda_1 \frac{\partial^2 u(x,t)}{\partial t^2} + \lambda_2 {}_0^C\mathcal{D}_t^\alpha u(x,t) = D\frac{\partial^2 u(x,t)}{\partial x^2} + f(x,t), \quad x \in (a,b), \quad t \in (0,T], \tag{24}$$

$$u(a,t) = \varphi_l(t),\ u(b,t) = \varphi_r(t), \quad t \in (0,T], \tag{25}$$

$$u(x,0) = \phi(x), \quad \frac{\partial u(x,0)}{\partial t} = \psi(x), \quad x \in [a,b]. \tag{26}$$

We then analyze the unconditional stability and convergence of the present difference method.

3.1. Derivation of the Finite Difference Method

We first consider the spatial semi-discretization for Equation (24). To this end, we design a set of the grid points as $\Omega_h = \{x_i \mid x_i = a + ih,\ 0 \le i \le M\}$, where the step size $h = (b-a)/M$ and M is a positive integer. Furthermore, we define a space as

$$\mathcal{U}_h = \{u = (u_0, u_1, \dots, u_M) \text{ is a grid function on } \Omega_h \text{ and } u_0 = u_M = 0\}.$$

For simplicity, we introduce some difference operators as

$$\delta_x u_{i-\frac{1}{2}} = \frac{1}{h}(u_i - u_{i-1}), \quad \delta_x^2 u_i = \frac{1}{h}(\delta_x u_{i+\frac{1}{2}} - \delta_x u_{i-\frac{1}{2}}), \quad \forall u \in \mathcal{U}_h.$$

Suppose the solution $u(x,t) \in C_{x,t}^{(4,3)}([a,b] \times [0,T])$ of Equations (24)–(26). Here, we employ a simple second-order difference scheme for Equation (24) in spatial discretization. For simplicity, we denote $U_i(t) = u(x_i,t)$, $0 \le i \le M$. Thus, at the point (x_i,t), we obtain a spatial semi-discretization scheme for Equation (24) as follows:

$$\lambda_1 \frac{\partial^2 U_i(t)}{\partial t^2} + \lambda_2 {}_0^C\mathcal{D}_t^\alpha U_i(t) = D\delta_x^2 U_i(t) + f(x_i,t) + (R_x)_i(t), \quad 1 \le i \le M-1, \tag{27}$$

where $(R_x)_i(t) = O(h^2)$.

Next, we apply the difference method in Equation (23) to derive the temporal discretization for Equation (24). For $n = 1, \cdots, N$, we average Equation (27) at $t = t_{n-1}$ and t_{n+1} to obtain

$$\begin{aligned}
&\frac{\lambda_1}{2}\Big(\frac{\partial^2 U_i(t_{n-1})}{\partial t^2} + \frac{\partial^2 U_i(t_{n+1})}{\partial t^2}\Big) + \frac{\lambda_2}{2}\Big({}_0^C\mathcal{D}_t^\alpha U_i(t_{n-1}) + {}_0^C\mathcal{D}_t^\alpha U_i(t_{n+1})\Big) \\
&= \frac{D}{2}\Big(\delta_x^2 U_i(t_{n-1}) + \delta_x^2 U_i(t_{n+1})\Big) + \frac{f(x_i,t_{n-1}) + f(x_i,t_{n+1})}{2} + \frac{(R_x)_i(t_{n-1}) + (R_x)_i(t_{n+1})}{2}.
\end{aligned} \tag{28}$$

For simplicity, we denote the grid points $U_i^n = u(x_i,t_n)$, $0 \le i \le M$, $0 \le n \le N$ and introduce a symbol $U^{\bar{n}} = \frac{1}{2}(U^{n-1} + U^{n+1})$. Applying Equations (9) and (17) to Equation (28), we have

$$\lambda_1 \delta_t^2 U_i^n + \lambda_2 \delta_\tau^\alpha(U_i^n, \psi_i) = D\delta_x^2 U_i^{\bar{n}} + f_i^{\bar{n}} + R_i^n, \quad 1 \le i \le M-1,\ 1 \le n \le N, \tag{29}$$

where the truncation error R_i^n satisfies

$$|R_i^n| \le c_2(\tau^{3-\alpha} + h^2), \quad 1 \le i \le M-1, \quad 1 \le n \le N \tag{30}$$

with c_2 being a positive constant.

We consider Equation (27) at $t = t_0$, and then transform the result into the following equation:

$$\lambda_1 \frac{\partial^2 U_i(t_0)}{\partial t^2} + \lambda_2 {}_0^C\mathcal{D}_t^\alpha U_i(t_0) = D\delta_x^2 U_i^{\frac{1}{2}} + f_i^0 - \frac{D}{2}(\delta_x^2 U_i(t_1) - \delta_x^2 U_i(t_0)) + (R_x)_i(t_0), \quad (31)$$

where $U_i^{n-\frac{1}{2}} = \frac{1}{2}(U_i^n + U_i^{n-1})$. We notice ${}_0^C\mathcal{D}_t^\alpha U_i(t_0) = 0$ and apply the Taylor expansion to the first term in Equation (31). This gives

$$\frac{2\lambda_1}{\tau}\left[\delta_t U_i^{\frac{1}{2}} - \psi_i\right] = D\delta_x^2 U_i^{\frac{1}{2}} + f_i^0 + R_i^0, \quad (32)$$

where the truncation error R_i^0 satisfies

$$|R_i^0| \leq c_3(\tau + h^2), \quad 1 \leq i \leq M-1, \quad (33)$$

with c_3 being a positive constant.

Using the boundary conditions in Equation (25)

$$U_0^n = \varphi_l(t_n), \quad U_M^n = \varphi_r(t_n), \quad 1 \leq n \leq N, \quad (34)$$

and the initial condition in Equation (26)

$$U_i^0 = \phi(x_i), \quad 0 \leq i \leq M, \quad (35)$$

dropping the small term R_i^n in Equation (29) and the small term R_i^0 in Equation (32), and replacing U_i^n with its numerical approximation u_i^n, we obtain a difference scheme for Equations (24)–(26) as

$$\lambda_1 \delta_t^2 u_i^n + \lambda_2 \delta_\tau^\alpha(u_i^n, \psi_i) = D\delta_x^2 u_i^{\bar{n}} + f_i^{\bar{n}}, \quad 1 \leq i \leq M-1,\ 1 \leq n \leq N, \quad (36)$$

$$\frac{2\lambda_1}{\tau}\left[\delta_t u_i^{\frac{1}{2}} - \psi_i\right] = D\delta_x^2 u_i^{\frac{1}{2}} + f_i^0, \quad 1 \leq i \leq M-1, \quad (37)$$

$$u_0^n = \varphi_l(t_n),\ u_M^n = \varphi_r(t_n), \quad 1 \leq n \leq N, \quad (38)$$

$$u_i^0 = \phi(x_i), \quad 0 \leq i \leq M-1, \quad (39)$$

where the truncation error is $\mathcal{O}(\tau + h^2)$ when $n = 1$, and $\mathcal{O}(\tau^{3-\alpha} + h^2)$ when $n = 2, \cdots N$.

3.2. Analysis of the Difference Scheme

We now analyze the optimal error estimate of the difference scheme (36)–(39). To this end, we first define some discrete inner products and norms. For $u, v \in \mathcal{U}_h$, we introduce the following inner products

$$(u, v) = h\sum_{i=1}^{M-1} u_i v_i, \quad \langle \delta_x u, \delta_x v \rangle = h\sum_{i=1}^{M} (\delta_x u_{i-\frac{1}{2}})\delta_x v_{i-\frac{1}{2}},$$

and the corresponding norms:

$$\| u \| = \sqrt{(u, u)}, \quad |u|_1 = \sqrt{\langle \delta_x u, \delta_x u \rangle}.$$

From the definition of $\{b_n^{(\alpha)}\}_{n=0}^\infty$ in Equation (12), we readily obtain that the properties $b_n^{(\alpha)} > 0, b_{n-1}^{(\alpha)} > b_n^{(\alpha)}$, $n \geq 1$. The following lemma gives an estimate of $\delta_\tau^\alpha(\Delta_\tau v^n, \psi)$ defined in Equation (19), which plays a crucial role in the analysis.

Lemma 1. *Let $\delta_\tau^\alpha(\Delta_\tau v^n, \psi)$ be defined in Equation (19). Then, we have*

$$
\begin{aligned}
&(\delta_\tau^\alpha(\Delta_\tau v^n, \psi))\Delta_\tau v^n \\
&\geq \frac{1}{\tau}\left[\frac{\tau^{2-\alpha}}{2\Gamma(3-\alpha)}\sum_{k=1}^{n} b_{n-k}^{(\alpha)}(\Delta_\tau v^k)^2 - \frac{\tau^{2-\alpha}}{2\Gamma(3-\alpha)}\sum_{k=1}^{n-1} b_{n-k-1}^{(\alpha)}(\Delta_\tau v^k)^2\right] \\
&+\frac{b_{n-1}^{(\alpha)}\tau^{1-\alpha}}{4\Gamma(3-\alpha)}(\Delta_\tau v^n)^2 - \frac{\tau^{1-\alpha}}{\Gamma(3-\alpha)}\left[(b_{n-1}^{(\alpha)} - b_n^{(\alpha)})(\delta_t v^{\frac{1}{2}})^2 + b_n^{(\alpha)}\psi^2\right].
\end{aligned}
\tag{40}
$$

Proof. We re-write $\delta_\tau^\alpha(\Delta_\tau v^n, \psi)$ as the following form

$$
\delta_\tau^\alpha(\Delta_\tau v^n, \psi) = \frac{\tau^{1-\alpha}}{\Gamma(3-\alpha)}\left[b_0^{(\alpha)}\Delta_\tau v^n - \sum_{k=1}^{n-1}(b_{n-k-1}^{(\alpha)} - b_{n-k}^{(\alpha)})\Delta_\tau v^k - (b_{n-1}^{(\alpha)} - b_n^{(\alpha)})\delta_t v^{\frac{1}{2}} - b_n^{(\alpha)}\psi\right]. \tag{41}
$$

Multiplying Equation (41) by $\Delta_\tau v^n$ and using the Cauchy–Schwarz inequality, we have

$$
\begin{aligned}
&(\delta_\tau^\alpha(\Delta_\tau v^n, \psi))\Delta_\tau v^n \\
&= \frac{\tau^{1-\alpha}}{\Gamma(3-\alpha)}\Big[b_0^{(\alpha)}(\Delta_\tau v^n)^2 - \sum_{k=1}^{n-1}(b_{n-k-1}^{(\alpha)} - b_{n-k}^{(\alpha)})(\Delta_\tau v^k)\cdot\Delta_\tau v^n \\
&\quad - (b_{n-1}^{(\alpha)} - b_n^{(\alpha)})\cdot(\delta_t v^{\frac{1}{2}})\cdot\Delta_\tau v^n - b_n^{(\alpha)}\cdot\psi\cdot\Delta_\tau v^n\Big] \\
&\geq \frac{\tau^{1-\alpha}}{\Gamma(3-\alpha)}\Big[b_0^{(\alpha)}(\Delta_\tau v^n)^2 - \sum_{k=1}^{n-1}(b_{n-k-1}^{(\alpha)} - b_{n-k}^{(\alpha)})\frac{(\Delta_\tau v^k)^2 + (\Delta_\tau v^n)^2}{2} \\
&\quad - (b_{n-1}^{(\alpha)} - b_n^{(\alpha)})(\frac{1}{4}(\Delta_\tau v^n)^2 + (\delta_t v^{\frac{1}{2}})^2) - b_n^{(\alpha)}(\frac{1}{4}(\Delta_\tau v^n)^2 + \psi^2)\Big] \\
&\geq \frac{1}{\tau}\left[\frac{\tau^{2-\alpha}}{2\Gamma(3-\alpha)}\sum_{k=1}^{n} b_{n-k}^{(\alpha)}(\Delta_\tau v^k)^2 - \frac{\tau^{2-\alpha}}{2\Gamma(3-\alpha)}\sum_{k=1}^{n-1} b_{n-k-1}^{(\alpha)}(\Delta_\tau v^k)^2\right] \\
&\quad + \frac{b_{n-1}^{(\alpha)}\tau^{1-\alpha}}{4\Gamma(3-\alpha)}(\Delta_\tau v^n)^2 - \frac{\tau^{1-\alpha}}{\Gamma(3-\alpha)}\left[(b_{n-1}^{(\alpha)} - b_n^{(\alpha)})(\delta_t v^{\frac{1}{2}})^2 + b_n^{(\alpha)}\psi^2\right].
\end{aligned}
\tag{42}
$$

Hence, we have completed the proof. □

We now derive a prior estimate of the scheme (36)–(39).

Lemma 2. *Suppose that $\{v_i^n\}$ is the solution of the scheme*

$$
\lambda_1\delta_t^2 v_i^n + \lambda_2\delta_\tau^\alpha(v_i^n, \tilde{\psi}_i) = D\delta_x^2 v_i^{\bar{n}} + g_i^{\bar{n}}, \quad 1 \leq i \leq M-1,\ 1 \leq n \leq N, \tag{43}
$$

$$
\frac{2\lambda_1}{\tau}\left[\delta_t v_i^{\frac{1}{2}} - \tilde{\psi}_i\right] = D\delta_x^2 v_i^{\frac{1}{2}} + g_i^0, \quad 1 \leq i \leq M-1, \tag{44}
$$

$$
v_0^n = 0,\ v_M^n = 0, \quad 1 \leq n \leq N, \tag{45}
$$

$$
v_i^0 = \tilde{\phi}_i, \quad 0 \leq i \leq M-1. \tag{46}
$$

Then, it holds that

$$
|v^n|_1^2 \leq C(D, \lambda_1, \lambda_2, T, \alpha)(|v^0|_1^2 + \|\tilde{\psi}\|^2 + \tau^2\|g^0\|^2 + \tau\sum_{m=1}^{n}\|g^{\bar{m}}\|^2). \tag{47}
$$

Proof. We take an inner product of Equation (44) with $\delta_t v^{\frac{1}{2}}$ to obtain

$$
\frac{2\lambda_1}{\tau}(\delta_t v^{\frac{1}{2}} - \tilde{\psi}, \delta_t v^{\frac{1}{2}}) = D(\delta_x^2 v^{\frac{1}{2}}, \delta_t v^{\frac{1}{2}}) + (g^0, \delta_t v^{\frac{1}{2}}). \tag{48}
$$

Multiplying both sides of Equation (48) by τ, we obtain

$$2\lambda_1\|\delta_t v^{\frac{1}{2}}\|^2 = D\tau(\delta_x^2 v^{\frac{1}{2}}, \delta_t v^{\frac{1}{2}}) + 2\lambda_1(\tilde{\psi}, \delta_t v^{\frac{1}{2}}) + \tau(g^0, \delta_t v^{\frac{1}{2}}). \tag{49}$$

Applying the summation by parts to the first term on the right-hand side of Equation (49) leads to

$$D\tau(\delta_x^2 v^{\frac{1}{2}}, \delta_t v^{\frac{1}{2}}) = -\frac{D}{2}(|v^1|_1^2 + |v^0|_1^2) + D|v^0|_1^2. \tag{50}$$

Using the Cauchy–Schwarz inequality to the last two terms on the right-hand side of Equation (49) gives

$$2\lambda_1(\tilde{\psi}, \delta_t v^{\frac{1}{2}}) \le \frac{\lambda_1}{2}\|\delta_t v^{\frac{1}{2}}\|^2 + 2\lambda_1\|\tilde{\psi}\|^2, \tag{51}$$

and

$$\tau(g^0, \delta_t v^{\frac{1}{2}}) \le \frac{\lambda_1}{2}\|\delta_t v^{\frac{1}{2}}\|^2 + \frac{\tau^2}{2\lambda_1}\|g^0\|^2. \tag{52}$$

Inserting Equations (50)–(52) into Equation (49) leads to

$$\lambda_1\|\delta_t v^{\frac{1}{2}}\|^2 + \frac{D}{2}(|v^1|_1^2 + |v^0|_1^2) \le D|v^0|_1^2 + 2\lambda_1\|\tilde{\psi}\|^2 + \frac{\tau^2}{2\lambda_1}\|g^0\|^2. \tag{53}$$

We take an inner product of Equation (43) with $\Delta_\tau v^n$ as

$$\lambda_1(\delta_t^2 v^n, \Delta_\tau v^n) + \lambda_2(\delta_\tau^\alpha(v^n, \tilde{\psi}), \Delta_\tau v^n) = D(\delta_x^2 v^{\bar{n}}, \Delta_\tau v^n) + (g^{\bar{n}}, \Delta_\tau v^n). \tag{54}$$

After careful analysis, we obtain

$$\lambda_1(\delta_t^2 v^n, \Delta_\tau v^n) = \frac{\lambda_1}{2\tau}\Big(\| \delta_t v^{n+\frac{1}{2}} \|^2 - \| \delta_t v^{n-\frac{1}{2}} \|^2\Big). \tag{55}$$

Applying the summation by parts to the first term on the right-hand side of Equation (54) and noticing the boundary conditions in Equation (45), we have

$$D(\delta_x^2 v^{\bar{n}}, \Delta_\tau v^n) = -\frac{D}{4\tau}(|v^{n+1}|_1^2 - |v^{n-1}|_1^2). \tag{56}$$

We denote $E^n = \frac{\lambda_2\tau^{2-\alpha}}{2\Gamma(3-\alpha)}\sum\limits_{k=1}^{n} b_{n-k}^{(\alpha)}\|\Delta_\tau v^k\|^2$. By Lemma 1, we obtain

$$\begin{aligned}\lambda_2(\delta_\tau^\alpha(v^n, \tilde{\psi}), \Delta_\tau v^n) \ge & \frac{1}{\tau}(E^n - E^{n-1}) + \frac{\lambda_2 b_{n-1}^{(\alpha)}\tau^{1-\alpha}}{4\Gamma(3-\alpha)}\|\Delta_\tau v^n\|^2 \\ & - \frac{\lambda_2\tau^{1-\alpha}}{\Gamma(3-\alpha)}\Big[(b_{n-1}^{(\alpha)} - b_n^{(\alpha)})\|\delta_t v^{\frac{1}{2}}\|^2 + b_n^{(\alpha)}\|\tilde{\psi}\|^2\Big].\end{aligned} \tag{57}$$

We further apply the Cauchy–Schwarz inequality to the last term on the right-hand side of Equation (54). This gives

$$(g^{\bar{n}}, \Delta_\tau v^n) \le \frac{\lambda_2 b_{n-1}^{(\alpha)}\tau^{1-\alpha}}{4\Gamma(3-\alpha)}\|\Delta_\tau v^n\|^2 + \frac{\Gamma(2-\alpha)t_n^{\alpha-1}}{\lambda_2}\|\bar{g}\|^2. \tag{58}$$

For simplicity, we denote $F^n = E^n + \frac{\lambda_1}{2}\|\delta_t v^{n+\frac{1}{2}}\|^2 + \frac{D}{4}(|v^{n+1}|_1^2 + |v^n|_1^2)$. Substituting Equations (55)–(58) into Equation (54) leads to

$$\frac{1}{\tau}(F^n - F^{n-1}) \le \frac{\lambda_2\tau^{1-\alpha}}{\Gamma(3-\alpha)}\Big[(b_{n-1}^{(\alpha)} - b_n^{(\alpha)})\|\delta_t v^{\frac{1}{2}}\|^2 + b_n^{(\alpha)}\|\tilde{\psi}\|^2\Big] + \frac{\Gamma(2-\alpha)t_n^{\alpha-1}}{\lambda_2}\|g^{\bar{n}}\|^2. \tag{59}$$

Multiplying Equation (59) by τ, we obtain

$$F^n \le F^{n-1} + \frac{\lambda_2 \tau^{2-\alpha}}{\Gamma(3-\alpha)}\left[(b_{n-1}^{(\alpha)} - b_n^{(\alpha)})\|\delta_t v^{\frac{1}{2}}\|^2 + b_n^{(\alpha)}\|\tilde{\psi}\|^2\right] + \frac{\Gamma(2-\alpha)t_n^{\alpha-1}}{\lambda_2}\tau\|g^{\bar{n}}\|^2. \quad (60)$$

Replacing n with m in Equation (60) and then summing up for m from 1 to n on both sides of the result, we have

$$\begin{aligned} F^n \le & F^0 + \frac{\lambda_2 \tau^{2-\alpha}}{\Gamma(3-\alpha)}\left[\sum_{m=1}^{n}(b_{m-1}^{(\alpha)} - b_m^{(\alpha)})\|\delta_t v^{\frac{1}{2}}\|^2 + \sum_{m=1}^{n} b_m^{(\alpha)}\|\tilde{\psi}\|^2\right] \\ & + \frac{\Gamma(2-\alpha)}{\lambda_2}\tau\sum_{m=1}^{n} t_m^{\alpha-1}\|g^{\bar{m}}\|^2. \end{aligned} \quad (61)$$

Using the following relations

$$\sum_{m=1}^{n}(b_{m-1}^{(\alpha)} - b_m^{(\alpha)}) = b_0^{(\alpha)} - b_n^{(\alpha)} < b_0^{(\alpha)} = 1, \quad (62)$$

$$\sum_{m=1}^{n} b_m^{(\alpha)} = \sum_{m=1}^{n}[(m+1)^{2-\alpha} - m^{2-\alpha}] = (n+1)^{2-\alpha} - 1 < (n+1)^{2-\alpha}, \quad (63)$$

and $t_m^{\alpha-1} \le t_n^{\alpha-1}$, $m = 1, 2, \cdots, n$ for Equation (61), we obtain

$$F^n \le F^0 + \frac{\lambda_2 \tau^{2-\alpha}}{\Gamma(3-\alpha)}\|\delta_t v^{\frac{1}{2}}\|^2 + \frac{\lambda_2 t_{n+1}^{2-\alpha}}{\Gamma(3-\alpha)}\|\tilde{\psi}\|^2 + \frac{\Gamma(2-\alpha)t_n^{\alpha-1}}{\lambda_2}\tau\sum_{m=1}^{n}\|g^{\bar{m}}\|^2. \quad (64)$$

Applying Equation (53) to estimate F^0 and $\|\delta_t v^{\frac{1}{2}}\|^2$, we have

$$\begin{aligned} F^n \le & \left(\frac{D}{2} + \frac{\lambda_2 D \tau^{2-\alpha}}{\lambda_1 \Gamma(3-\alpha)}\right)|v^0|_1^2 + \left(\lambda_1 + \frac{2\lambda_2 \tau^{2-\alpha}}{\Gamma(3-\alpha)} + \frac{\lambda_2 t_{n+1}^{2-\alpha}}{\Gamma(3-\alpha)}\right)\|\tilde{\psi}\|^2 \\ & + \frac{\lambda_2 \tau^{4-\alpha}}{2\lambda_1^2 \Gamma(3-\alpha)}\|g^0\|^2 + \frac{\Gamma(2-\alpha)t_n^{\alpha-1}}{\lambda_2}\tau\sum_{m=1}^{n}\|g^{\bar{m}}\|^2. \end{aligned} \quad (65)$$

Hence, we have completed the proof. □

Based on Lemma 2, we have the following theorem for stability of the scheme (36)–(39).

Theorem 3 (Stability). *Assume that* $\{(u^1)_i^n\}$ *and* $\{(u^2)_i^n\}$ *are two numerical solutions obtained based on the scheme* (36)–(39) *with same initial and boundary conditions but different source terms* $(f^{(1)})_i^n$ *and* $(f^{(2)})_i^n$. *Let* $u_i^n = (u^{(1)})_i^n - (u^{(2)})_i^n$ *and* $f_i^n = (f^{(1)})_i^n - (f^{(2)})_i^n$. *Then, it holds that*

$$\|u^n\|_\infty^2 \le C(D, \lambda_1, \lambda_2, T, \alpha)(\tau^2\|f^0\|^2 + \tau\sum_{m=1}^{n}\|f^{\bar{m}}\|^2). \quad (66)$$

This implies that the scheme (36)–(39) *is unconditionally stable.*

We now give the error estimate of the difference scheme (36)–(39). Let $e_i^n = U_i^n - u_i^n$, $0 \le i \le M$, $0 \le n \le N$. Subtracting Equations (36)–(39) from Equations (29), (32), (34) and (35), we obtain error equations as follows:

$$\lambda_1\delta_t^2 e_i^n + \lambda_2\delta_\tau^\alpha(e_i^n,0) = D\delta_x^2 e_i^{\bar{n}} + R_i^n,\quad 1\le i\le M-1,\, 1\le n\le N, \tag{67}$$

$$\frac{2\lambda_1}{\tau}\delta_t e_i^{\frac{1}{2}} = D\delta_x^2 e_i^{\frac{1}{2}} + R_i^0,\quad 1\le i\le M-1, \tag{68}$$

$$e_0^n = 0,\, e_M^n = 0,\quad 1\le n\le N, \tag{69}$$

$$e_i^0 = 0,\quad 0\le i\le M-1. \tag{70}$$

Theorem 4 (Convergence). *Suppose that the solution $u(x,t)\in C_{x,t}^{(4,3)}([a,b]\times[0,T])$ of 1D wave equation with fractional damping (24)–(26), and $\{u_i^n \mid 0\le i\le M,\ 0\le n\le N\}$ is the solution of the difference scheme (36)–(39). Then, the following optimal error estimate holds*

$$\|U^n - u^n\|_\infty \le C(\tau^{3-\alpha}+h^2),\quad 1\le n\le N, \tag{71}$$

which implies that the numerical solution is convergent to the analytical solution with the error $O(\tau^{3-\alpha}+h^2)$.

Proof. We take an inner product of Equation (68) with $\delta_t e^{\frac{1}{2}}$ and Equation (67) with $\Delta_\tau e^n$, respectively. This gives

$$\frac{2\lambda_1}{\tau}(\delta_t e^{\frac{1}{2}},\delta_t e^{\frac{1}{2}}) = D(\delta_x^2 e^{\frac{1}{2}},\delta_t e^{\frac{1}{2}}) + (R^0,\delta_t e^{\frac{1}{2}}), \tag{72}$$

and

$$\lambda_1(\delta_t^2 e^n,\Delta_\tau e^n) + \lambda_2(\delta_\tau^\alpha(e^n,0),\Delta_\tau e^n) = D(\delta_x^2 e^{\bar{n}},\Delta_\tau e^n) + (R^n,\Delta_\tau e^n). \tag{73}$$

Adopting the analysis strategy in Lemma 2, we have

$$\|e^n\|_\infty^2 \le C(D,\lambda_1,\lambda_2,T,\alpha)(\tau^2\|R^0\|^2 + \tau\sum_{m=1}^{n}\|R^m\|^2). \tag{74}$$

Using the estimates of the truncation errors in Equations (30) and (33) for Equation (74), we have completed the proof. □

4. 2D Wave Equation with Fractional Damping

4.1. Derivation of the Difference Scheme

We derive an alternating direction implicit (ADI) scheme for solving Equations (1)–(3) in two dimensions

$$\lambda_1\frac{\partial^2 u(x,y,t)}{\partial t^2} + \lambda_2\,{}_0^C\mathcal{D}_t^\alpha u(x,y,t) = D\Delta u + f(x,y,t),\quad (x,y,t)\in\Omega\times(0,T], \tag{75}$$

$$u(x,y,t) = \varphi(x,y,t),\quad (x,y,t)\in\partial\Omega\times(0,T], \tag{76}$$

$$u(x,y,0) = \phi(x,y),\quad \frac{\partial u(x,y,0)}{\partial t} = \psi(x,y),\quad (x,y)\in\bar{\Omega}, \tag{77}$$

where $\Omega=(a,b)\times(c,d)$. We take two positive integers M_1 and M_2, and design a set of the grid points as $\bar{\Omega}_h = \{(x_i,y_j)\mid x_i = a+ih_1,\ y_j = c+jh_2,\ 0\le i\le M_1,\ 0\le j\le M_2\}$ with step sizes $h_1=(b-a)/M_1$ and $h_2=(d-c)/M_2$, and $\partial\Omega_h = \bar{\Omega}_h\cap\partial\Omega$. For $v=\{v_{i,j}\mid 0\le i\le M_1, 0\le j\le M_2\}$, we denote

$$\delta_x v_{i-\frac{1}{2},j} = \frac{1}{h_1}(v_{i,j}-v_{i-1,j}),\quad \delta_x^2 v_{i,j} = \frac{1}{h_1}(\delta_x v_{i+\frac{1}{2},j} - \delta_x v_{i-\frac{1}{2},j}),$$

$$\delta_y v_{i,j-\frac{1}{2}} = \frac{1}{h_2}(v_{i,j}-v_{i,j-1}),\quad \delta_y^2 v_{i,j} = \frac{1}{h_2}(\delta_y v_{i,j+\frac{1}{2}} - \delta_y v_{i,j-\frac{1}{2}}).$$

We further denote the grid function as

$$U_{i,j}^{n} = u(x_i, y_j, t_n), \quad (x_i, y_j) \in \bar{\Omega}_h, \quad 0 \le n \le N.$$

Suppose the solution $u(\mathbf{x}, t) \in C_{\mathbf{x},t}^{(4,3)}(\bar{\Omega} \times [0, T])$ of Equations (75)–(77) with $\mathbf{x} = (x, y)$. Following the derivation of the difference scheme in one-dimensional case, we construct

$$\lambda_1 \delta_t^2 U_{i,j}^{n} + \lambda_2 \delta_\tau^\alpha (U_{i,j}^{n}, \psi_{i,j}) = D(\delta_x^2 U_{i,j}^{\bar{n}} + \delta_y^2 U_{i,j}^{\bar{n}}) + f_{i,j}^{\bar{n}} + O(\tau^{3-\alpha} + h_1^2 + h_2^2), \ (x_i, y_j) \in \Omega_h, \ 1 \le n \le N, \quad (78)$$

$$\frac{2\lambda_1}{\tau}\left[\delta_t U_{i,j}^{\frac{1}{2}} - \psi_{i,j}\right] = D(\delta_x^2 U_{i,j}^{\frac{1}{2}} + \delta_y^2 U_{i,j}^{\frac{1}{2}}) + f_{i,j}^{0} + O(\tau + h_1^2 + h_2^2), \quad (x_i, y_j) \in \Omega_h. \quad (79)$$

We define $\omega = \lambda_1 + \lambda_2 \frac{\tau^{2-\alpha}}{2\Gamma(3-\alpha)}$. Adding small perturbations $\frac{D^2\tau^3}{4\omega}\Delta_\tau \delta_x^2 \delta_y^2 U_{i,j}^{n}$ and $\frac{D^2\tau^3}{8\lambda_1}\delta_t \delta_x^2 \delta_y^2 U_{i,j}^{\frac{1}{2}}$ to both sides of Equations (78) and (79), respectively, we have

$$\lambda_1 \delta_t^2 U_{i,j}^{n} + \lambda_2 \delta_\tau^\alpha (U_{i,j}^{n}, \psi_{i,j}) + \frac{D^2\tau^3}{4\omega}\Delta_\tau \delta_x^2 \delta_y^2 U_{i,j}^{n} = D(\delta_x^2 U_{i,j}^{\bar{n}} + \delta_y^2 U_{i,j}^{\bar{n}}) + f_{i,j}^{\bar{n}} + R_{i,j}^{n}, \quad (x_i, y_j) \in \Omega_h, \ 1 \le n \le N, \quad (80)$$

$$\frac{2\lambda_1}{\tau}\left[\delta_t U_{i,j}^{\frac{1}{2}} - \psi_{i,j}\right] + \frac{D^2\tau^3}{8\lambda_1}\delta_t \delta_x^2 \delta_y^2 U_{i,j}^{\frac{1}{2}} = D(\delta_x^2 U_{i,j}^{\frac{1}{2}} + \delta_y^2 U_{i,j}^{\frac{1}{2}}) + f_{i,j}^{0} + R_{i,j}^{0}, \quad (x_i, y_j) \in \Omega_h, \quad (81)$$

where there exists a positive constant c_4 satisfying

$$|R_{i,j}^{n}| \le c_4(\tau^{3-\alpha} + h_1^2 + h_2^2), \quad |R_{i,j}^{0}| \le c_4(\tau + h_1^2 + h_2^2). \quad (82)$$

Using the boundary condition in Equation (76)

$$U_{i,j}^{n} = \varphi(x_i, y_j, t_n), \quad (x_i, y_j) \in \partial\Omega_h, \quad 1 \le n \le N, \quad (83)$$

and the initial condition in Equation (77)

$$U_{i,j}^{0} = \phi(x_i, y_j), \quad (x_i, y_j) \in \bar{\Omega}_h, \quad (84)$$

and dropping the small term $R_{i,j}^{n}$ in Equation (80) and the small term $R_{i,j}^{0}$ in Equation (81), and then replacing $U_{i,j}^{n}$ with its numerical approximation $u_{i,j}^{n}$, we obtain a difference scheme as

$$\lambda_1 \delta_t^2 u_{i,j}^{n} + \lambda_2 \delta_\tau^\alpha (u_{i,j}^{n}, \psi_{i,j}) + \frac{D^2\tau^3}{4\omega}\Delta_\tau \delta_x^2 \delta_y^2 u_{i,j}^{n} = D(\delta_x^2 u_{i,j}^{\bar{n}} + \delta_y^2 u_{i,j}^{\bar{n}}) + f_{i,j}^{\bar{n}}, \quad (x_i, y_j) \in \Omega_h, \ 1 \le n \le N, \quad (85)$$

$$\frac{2\lambda_1}{\tau}\left[\delta_t u_{i,j}^{\frac{1}{2}} - \psi_{i,j}\right] + \frac{D^2\tau^3}{8\lambda_1}\delta_t \delta_x^2 \delta_y^2 u_{i,j}^{\frac{1}{2}} = D(\delta_x^2 u_{i,j}^{\frac{1}{2}} + \delta_y^2 u_{i,j}^{\frac{1}{2}}) + f_{i,j}^{0}, \quad (x_i, y_j) \in \Omega_h, \quad (86)$$

$$u_{i,j}^{n} = \varphi(x_i, y_j, t_n), \quad (x_i, y_j) \in \partial\Omega_h, \quad 1 \le n \le N, \quad (87)$$

$$u_{i,j}^{0} = \phi(x_i, y_j), \quad (x_i, y_j) \in \bar{\Omega}_h. \quad (88)$$

We multiply Equations (85) and (86) by τ^2, and then factorize the results. This gives

$$\left(\sqrt{w}\mathcal{I} - \frac{D\tau^2}{2\sqrt{w}}\delta_x^2\right)\left(\sqrt{w}\mathcal{I} - \frac{D\tau^2}{2\sqrt{w}}\delta_y^2\right)u_{i,j}^{n+1} = F_{i,j}^{n}, \ (x_i, y_j) \in \Omega_h, \ 1 \le n \le N, \quad (89)$$

$$\left(\sqrt{2\lambda_1}\mathcal{I} - \frac{D\tau^2}{2\sqrt{2\lambda_1}}\delta_x^2\right)\left(\sqrt{2\lambda_1}\mathcal{I} - \frac{D\tau^2}{2\sqrt{2\lambda_1}}\delta_y^2\right)u_{i,j}^{1} = F_{i,j}^{0}, \ (x_i, y_j) \in \Omega_h, \quad (90)$$

where

$$F^n = 2\lambda_1 u_{i,j}^n + \Big(\frac{\lambda_2\tau^{2-\alpha}}{2\Gamma(3-\alpha)} - \lambda_1\Big)u_{i,j}^{n-1} + \frac{D\tau^2}{2}(\delta_x^2 u_{i,j}^{n-1} + \delta_y^2 u_{i,j}^{n-1}) + \tau^2 f_{i,j}^{\bar{n}}$$
$$+ \frac{\lambda_2\tau^{3-\alpha}}{\Gamma(3-\alpha)}\Big[\sum_{k=1}^{n-1}(b_{n-k-1}^{(\alpha)} - b_{n-k-1}^{(\alpha)})\Delta_\tau u_{i,j}^k + b_{n-1}^{(\alpha)}\delta_t u_{i,j}^{\frac{1}{2}} - b_n^{(\alpha)}(\delta_t u_{i,j}^{\frac{1}{2}} - \psi_{i,j})\Big]; \quad (91)$$

$$F^0 = 2\lambda_1 u_{i,j}^0 + 2\lambda_1\tau\psi_{i,j} + \frac{D\tau^2}{2}(\delta_x^2 u_{i,j}^0 + \delta_y^2 u_{i,j}^0) + \tau^2 f_{i,j}^0. \quad (92)$$

We then introduce the intermediate variables $u_{i,j}^*$ and $v_{i,j}^*$ to obtain a D'Yakonov type of ADI scheme:

$$\Big(\sqrt{w}\mathcal{I} - \frac{D\tau^2}{2\sqrt{w}}\delta_x^2\Big)u_{i,j}^* = F_{i,j}^n,\ (x_i, y_j) \in \Omega_h,\ 1 \le n \le N, \quad (93)$$

$$\Big(\sqrt{w}\mathcal{I} - \frac{D\tau^2}{2\sqrt{w}}\delta_y^2\Big)u_{i,j}^{n+1} = u_{i,j}^*,\ (x_i, y_j) \in \Omega_h,\ 1 \le n \le N, \quad (94)$$

$$\Big(\sqrt{2\lambda_1}\mathcal{I} - \frac{D\tau^2}{2\sqrt{2\lambda_1}}\delta_x^2\Big)v_{i,j}^* = F_{i,j}^0,\ (x_i, y_j) \in \Omega_h, \quad (95)$$

$$\Big(\sqrt{2\lambda_1}\mathcal{I} - \frac{D\tau^2}{2\sqrt{2\lambda_1}}\delta_y^2\Big)u_{i,j}^1 = v_{i,j}^*,\ (x_i, y_j) \in \Omega_h. \quad (96)$$

The detailed computation is described as follows:

Firstly, we solve $u_{i,j}^1$ of the scheme (95)–(96) by computing two sets of independent one-dimensional problems. Specially, for fixed $j (1 \le j \le M_2 - 1)$, we obtain $\{v_{i,j}^* \mid (x_i, y_j) \in \Omega_h\}$ by solving the following linear equations

$$\Big(\sqrt{2\lambda_1}\mathcal{I} - \frac{D\tau^2}{2\sqrt{2\lambda_1}}\delta_x^2\Big)v_{i,j}^* = F_{i,j}^0, \quad 1 \le i \le M_1 - 1, \quad (97)$$

$$v_{0,j}^* = \Big(\sqrt{2\lambda_1}\mathcal{I} - \frac{D\tau^2}{2\sqrt{2\lambda_1}}\delta_y^2\Big)u_{0,j}^1,\ v_{M_1,j}^* = \Big(\sqrt{2\lambda_1}\mathcal{I} - \frac{D\tau^2}{2\sqrt{2\lambda_1}}\delta_y^2\Big)u_{M_1,j}^1. \quad (98)$$

For fixed $i (1 \le i \le M_1 - 1)$, we then obtain $u_{i,j}^1$ by solving the following linear equations

$$\Big(\sqrt{2\lambda_1}\mathcal{I} - \frac{D\tau^2}{2\sqrt{2\lambda_1}}\delta_y^2\Big)u_{i,j}^1 = v_{i,j}^*,\ 1 \le j \le M_2 - 1, \quad (99)$$

$$u_{i,0}^1 = \varphi(x_i, y_0, t_1),\ u_{i,M_2}^1 = \varphi(x_i, y_M, t_1). \quad (100)$$

Following the above idea, we can solve $u_{i,j}^n$ of the scheme (93)–(94) for $n \ge 2$ by computing two sets of independent one-dimensional problems.

4.2. Analysis of the Difference Scheme

We now give the optimal error estimate of the difference scheme (85)–(88). To this end, we first define a space as

$$\mathcal{V}_h = \{v \mid v = \{v_{i,j}\} \text{ is a grid function on } \Omega_h \text{ and } v_{i,j} = 0 \text{ if } (x_i, y_j) \in \partial\Omega_h\}.$$

For $u, v \in \mathcal{V}_h$, we introduce some inner products

$$(u, v) = h_1 h_2 \sum_{i=1}^{M_1-1}\sum_{j=1}^{M_2-1} u_{i,j}v_{i,j},$$

$$\langle\delta_x u, \delta_x v\rangle = h_1 h_2 \sum_{i=1}^{M_1}\sum_{j=1}^{M_2-1}(\delta_x u_{i-\frac{1}{2},j})\delta_x v_{i-\frac{1}{2},j},$$

$$\langle \delta_y u, \delta_y v \rangle = h_1 h_2 \sum_{i=1}^{M_1-1} \sum_{j=1}^{M_2} (\delta_x u_{i,j-\frac{1}{2}}) \delta_x v_{i,j-\frac{1}{2}},$$

and the corresponding norms $\|u\| = \sqrt{(u,u)}$ and $|u|_1^2 = \langle \delta_x u, \delta_x u \rangle + \langle \delta_y u, \delta_y u \rangle$.

In the following lemma, we obtain a prior estimate of the scheme (85)–(88).

Lemma 5. *Suppose that $\{v_{i,j}^n\}$ is the solution of the following scheme*

$$\lambda_1 \delta_t^2 v_{i,j}^n + \lambda_2 \delta_\tau^\alpha (v_{i,j}^n, \tilde{\psi}_{i,j}) + \frac{D^2\tau^3}{4\omega} \Delta_\tau \delta_x^2 \delta_y^2 v_{i,j}^n = D(\delta_x^2 v_{i,j}^{\bar{n}} + \delta_y^2 v_{i,j}^{\bar{n}}) + g_{i,j}^{\bar{n}},$$

$$(x_i, y_j) \in \Omega_h,\ 1 \le n \le N, \tag{101}$$

$$\frac{2\lambda_1}{\tau}\left[\delta_t v_{i,j}^{\frac{1}{2}} - \tilde{\psi}_{i,j}\right] + \frac{D^2\tau^3}{8\lambda_1} \delta_t \delta_x^2 \delta_y^2 v_{i,j}^{\frac{1}{2}} = D(\delta_x^2 v_{i,j}^{\frac{1}{2}} + \delta_y^2 v_{i,j}^{\frac{1}{2}}) + g_{i,j}^0,\quad (x_i, y_j) \in \Omega_h, \tag{102}$$

$$v_{i,j}^n = 0,\quad (x_i, y_j) \in \partial\Omega_h,\quad 1 \le n \le N, \tag{103}$$

$$v_{i,j}^0 = \tilde{\varphi}_{i,j},\quad (x_i, y_j) \in \partial\bar{\Omega}_h. \tag{104}$$

Then, it holds that

$$|v^n|_1^2 \le C(D, \lambda_1, \lambda_2, T, \alpha)(|v^0|_1^2 + \|\tilde{\psi}\|^2 + \tau^2\|g^0\|^2 + \tau \sum_{m=1}^{n} \|g^{\bar{m}}\|^2). \tag{105}$$

Proof. We take inner products of Equation (102) with $\delta_t v^{\frac{1}{2}}$ and Equation (101) with $\Delta_\tau v^n$. This gives

$$\frac{2\lambda_1}{\tau}(\delta_t v^{\frac{1}{2}} - \tilde{\psi}, \delta_t v^{\frac{1}{2}}) + \frac{D^2\tau^3}{8\lambda_1}(\delta_t \delta_x^2 \delta_y^2 v^{\frac{1}{2}}, \delta_t v^{\frac{1}{2}}) = D(\delta_x^2 v^{\frac{1}{2}} + \delta_y^2 v^{\frac{1}{2}}, \delta_t v^{\frac{1}{2}}) + (g^0, \delta_t v^{\frac{1}{2}}) \tag{106}$$

and

$$\lambda_1(\delta_t^2 v^n, \Delta_\tau v^n) + \lambda_2(\delta_\tau^\alpha(v^n, \tilde{\psi}), \Delta_\tau v^n) + \frac{D^2\tau^3}{4\omega}(\Delta_\tau \delta_x^2 \delta_y^2 v^n, \Delta_\tau v^n)$$

$$= D(\delta_x^2 v^{\bar{n}} + \delta_y^2 v^{\bar{n}}, \Delta_\tau v^n) + (g^{\bar{n}}, \Delta_\tau v^n). \tag{107}$$

Applying the summation by parts to the second term on the left-hand side of Equation (106) and the third term on the left-hand side of Equation (107), respectively, we have

$$\frac{D^2\tau^3}{8\lambda_1}(\delta_t \delta_x^2 \delta_y^2 v^{\frac{1}{2}}, \delta_t v^{\frac{1}{2}}) \ge 0,\quad \frac{D^2\tau^3}{4\omega}(\Delta_\tau \delta_x^2 \delta_y^2 v^n, \Delta_\tau v^n) \ge 0. \tag{108}$$

Using the same analysis in Lemma 2 to the last terms of Equations (106) and (107), one may obtain the estimate (105). □

Based on Lemma 5, we have the following theorem for stability of the scheme (85)–(88).

Theorem 6 (Stability). *Assume that $\{(u^1)_{i,j}^n\}$ and $\{(u^2)_{i,j}^n\}$ are two numerical solutions obtained based on the difference scheme (85)–(88) with same initial and boundary conditions but different source terms $(f^{(1)})_{i,j}^n$ and $(f^{(2)})_{i,j}^n$. Let $u_{i,j}^n = (u^{(1)})_{i,j}^n - (u^{(2)})_{i,j}^n$ and $f_{i,j}^n = (f^{(1)})_{i,j}^n - (f^{(2)})_{i,j}^n$. Then, it holds that*

$$|u^n|_1^2 \le C(D, \lambda_1, \lambda_2, T, \alpha)(\tau^2\|f^0\|^2 + \tau \sum_{m=1}^{n} \|f^{\bar{m}}\|^2). \tag{109}$$

This implies that the scheme (85)–(88) is unconditionally stable.

We now give the error estimate of the difference scheme (85)–(88). Let $e_{i,j}^n = U_{i,j}^n - u_{i,j}^n$, $(x_i, y_j) \in \Omega_h$, $0 \le n \le N$. Subtracting Equations (85)–(88) from Equations (80), (81), (83) and (84), we obtain error equations as follows:

$$\lambda_1 \delta_t^2 e_{i,j}^n + \lambda_2 \delta_\tau^\alpha (e_{i,j}^n, 0) + \frac{D^2\tau^3}{4\omega} \Delta_\tau \delta_x^2 \delta_y^2 e_{i,j}^n = D(\delta_x^2 e_{i,j}^{\bar{n}} + \delta_y^2 e_{i,j}^{\bar{n}}) + R_{i,j}^n,$$
$$(x_i, y_j) \in \Omega_h,\ 1 \le n \le N, \tag{110}$$
$$\frac{2\lambda_1}{\tau} \delta_t e_{i,j}^{\frac{1}{2}} + \frac{D^2\tau^3}{8\lambda_1} \delta_t \delta_x^2 \delta_y^2 e_{i,j}^{\frac{1}{2}} = D(\delta_x^2 e_{i,j}^{\frac{1}{2}} + \delta_y^2 e_{i,j}^{\frac{1}{2}}) + R_{i,j}^0, \quad (x_i, y_j) \in \Omega_h, \tag{111}$$
$$e_{i,j}^n = 0, \quad (x_i, y_j) \in \partial\Omega_h, \quad 1 \le n \le N, \tag{112}$$
$$e_{i,j}^0 = 0, \quad (x_i, y_j) \in \partial\bar{\Omega}_h. \tag{113}$$

Theorem 7 (Convergence). *Suppose that the solution $u(\mathbf{x}, t) \in C_{\mathbf{x},t}^{(4,3)}(\bar{\Omega} \times [0, T])$ of 2D wave equation with fractional damping (75)–(77), and $\{u_{i,j}^n \mid (x_i, y_j) \in \bar{\Omega}_h,\ 0 \le n \le N\}$ is the solution of the scheme (85)–(88). Then, the following optimal error estimate holds:*

$$|U^n - u^n|_1 \le C(\tau^{3-\alpha} + h_1^2 + h_2^2), \quad 1 \le n \le N, \tag{114}$$

which implies that the numerical solution is convergent to the analytical solution with the error $O(\tau^{3-\alpha} + h_1^2 + h_2^2)$.

Proof. We take an inner product of Equation (111) with $\delta_t e^{\frac{1}{2}}$ and Equation (110) with $\Delta_\tau e^n$, respectively. This gives

$$\frac{2\lambda_1}{\tau}(\delta_t e^{\frac{1}{2}}, \delta_t e^{\frac{1}{2}}) + \frac{D^2\tau^3}{8\lambda_1}(\delta_t \delta_x^2 \delta_y^2 e^{\frac{1}{2}}, \delta_t e^{\frac{1}{2}}) = D(\delta_x^2 e^{\frac{1}{2}} + \delta_y^2 e^{\frac{1}{2}}, \delta_t e^{\frac{1}{2}}) + (R^0, \delta_t e^{\frac{1}{2}}), \tag{115}$$

and

$$\lambda_1(\delta_t^2 e^n, \Delta_\tau e^n) + \lambda_2(\delta_\tau^\alpha(e^n, 0), \Delta_\tau e^n) + \frac{D^2\tau^3}{4\omega}(\Delta_\tau \delta_x^2 \delta_y^2 e^n, \Delta_\tau e^n)$$
$$= D(\delta_x^2 e^{\bar{n}} + \delta_y^2 e^{\bar{n}}, \Delta_\tau e^n) + (R^n, \Delta_\tau e^n). \tag{116}$$

Adopting the analysis strategy in Lemma 5, we have

$$|e^n|_1^2 \le C(D, \lambda_1, \lambda_2, T, \alpha)(\tau^2 \|R^0\|^2 + \tau \sum_{m=1}^{n} \|R^m\|^2). \tag{117}$$

Using the estimates of the truncation errors in Equation (82) for Equation (117), we have completed the proof. □

5. Numerical Examples

In this section, we carry out three numerical examples to test the performance and the applicability of our proposed difference schemes for solving the wave equation with fractional damping (1)–(3) in both 1D and 2D cases. All the numerical experiments are implemented using Matlab R2018b on a desktop with Intel(R) Core(TM) i3-7100U CPU @ 2.40 GHz and 8 GB.

In our computation, we denote τ as the time step size. For the 2D case, we take same step sizes in x and y spatial directions, i.e., $h_1 = h_2 = h$. We denote $E_\infty(h, \tau) = \|U^N - u^N\|_\infty$ as the maximum error, where u^N and U^N are the numerical solution and the analytical solution at the N-th level, respectively.

For the convergence accuracy of the proposed difference schemes, we restrict the time step size and the space step size by $h^2 \approx \tau^{3-\alpha}$. To test the temporal convergence order, we set $h \approx \tau^{\frac{3-\alpha}{2}}$ in order that $E_\infty(h,\tau) \approx E_\infty(\tau)$. The temporal convergence order is obtained using $\text{Rate}_t = \log_2(E_\infty(2\tau)/E_\infty(\tau))$. To obtain the spatial convergence order, we set $\tau \approx h^{\frac{2}{3-\alpha}}$, so that $E_\infty(h,\tau) \approx E_\infty(h)$. The spatial convergence order is obtained using $\text{Rate}_x = \log_2(E_\infty(2h)/E_\infty(h))$.

Example 2. *In Equations (24)–(26), we choose the coefficients $\lambda_1 = 1, \lambda_2 = 1, D = 1$, and the initial values $\phi(x) = 0$ and $\psi(x) = 0$, together with the boundary conditions $\phi_l(t) = \phi_r(t) = 0$. The source term is given by $f(x,t) = \left(12t^2 + \frac{24t^{4-\alpha}}{\Gamma(5-\alpha)} + \pi^2 t^4\right)\sin(\pi x)$.*

The analytical solution of the above problem can be seen to be $u(x,t) = t^4\sin(\pi x)$. In the following simulation, we compute the numerical solution on domain $(x,t) \in [0,1] \times [0,1]$.

Tables 2 and 3 show the errors of numerical approximations and the corresponding convergence orders of the difference scheme (36)–(39) with various fractional orders $\alpha = 1.4$, $\alpha = 1.6$ and $\alpha = 1.8$, respectively. From Tables 2 and 3, one can see that, for all fractional orders considered, the numerical solutions obtained using the difference scheme (36)–(39) agree well with their corresponding analytical solutions. Moreover, it can be seen that the convergence rate of the difference scheme (36)–(39) confirms the accuracy of $\mathcal{O}(\tau^{3-\alpha} + h^2)$, which is consistent with the theoretical error estimate in Theorem 4.

Example 3. *In Equations (75)–(77), we choose coefficients $\lambda_1 = 1, \lambda_2 = 1, D = 1$, and the domain $\Omega = (0,1)^2$ and $T = 1$, and the initial conditions $\phi(x,y) = 0$ and $\psi(x,y) = 0$, and the boundary conditions $\varphi(x,0,t) = t^4\sin(\pi x)$, $\varphi(x,1,t) = -t^4\sin(\pi x)$, $\varphi(0,y,t) = t^4\sin(\pi y)$, $\varphi(1,y,t) = -t^4\sin(\pi y)$. The source term is given by $f(x,y,t) = \left(12t^2 + \frac{24t^{4-\alpha}}{\Gamma(5-\alpha)} + 2\pi^2 t^4\right)\sin(\pi(x+y))$.*

The analytical solution of the above problem can be seen to be $u(x,y,t) = t^4\sin(\pi(x+y))$. In our simulation, we use the difference scheme (78)–(79) and the ADI scheme (85)–(88) to numerically solve Example 3.

Tables 4 and 5 list the errors of the numerical approximations, the convergence orders, and the CPU times took by both schemes for three different fractional orders $\alpha = 1.4$, $\alpha = 1.6$ and $\alpha = 1.8$, respectively. From numerical results, one can see that both schemes effectively solve the governing problem, and the convergence rates of both schemes can achieve the accuracy of $\mathcal{O}(\tau^{3-\alpha} + h^2)$. Furthermore, apparently, the ADI scheme (85)–(88) costs less CPU time than the difference scheme (78)–(79) under the same time and space meshes. Numerical results confirm that the ADI scheme can significantly enhance the computational efficiency as compared with the non-ADI scheme for high-dimensional problems. This is because that the ADI technique factorizes the high-dimensional problem into several sets of independent one-dimensional cases, which can be computed using the Thomas algorithm and hence reduce storage expenses.

Example 4. *We investigate the dynamic of the following 2D wave equation with fractional damping*

$$\begin{cases} \dfrac{\partial^2 u(\mathbf{x},t)}{\partial t^2} + {}_0^C\mathcal{D}_t^\alpha u(\mathbf{x},t) = \Delta u(\mathbf{x},t), \ \mathbf{x} \in \Omega, \ t \in (0,T], \\ u(\mathbf{x},0) = \text{sech}(V1) + \text{sech}(V2), \quad \dfrac{\partial u(\mathbf{x},0)}{\partial t} = \text{sech}(V3), \ \mathbf{x} \in \bar{\Omega}, \\ u(\mathbf{x},t) = 0, \ \mathbf{x} \in \partial\Omega, \ t \in [0,T], \end{cases} \tag{118}$$

where $V1 = (x+2)^2 + y^2$, $V2 = (x-2)^2 + y^2$, $V3 = x^2 + y^2$, and the computational domain $\Omega = (-16,16)^2$.

We use the ADI scheme (85)–(88) to obtain the solution of the problem (118). In our simulation, we divided Ω into meshes with the space step size of $h = 1/8$ and compute the time up to $T = 5$ with the time step size $\tau = 1/20$.

Table 2. Numerical error and temporal convergence accuracy in Example 2 with $h \approx \tau^{\frac{3-\alpha}{2}}$.

α	τ	$E_\infty(\tau)$	Rate_t
1.4	1/2000	3.3961×10^{-6}	-
	1/4000	1.0747×10^{-6}	1.6599
	1/8000	3.4298×10^{-7}	1.6477
	1/16,000	1.1031×10^{-7}	1.6366
1.6	1/2000	1.6207×10^{-5}	-
	1/4000	6.0741×10^{-6}	1.4159
	1/8000	2.2812×10^{-6}	1.4129
	1/16,000	8.5974×10^{-7}	1.4078
1.8	1/2000	8.5707×10^{-5}	-
	1/4000	3.7238×10^{-5}	1.2026
	1/8000	1.6175×10^{-5}	1.2030
	1/16,000	7.0371×10^{-6}	1.2007

Table 3. Numerical error and spatial convergence accuracy in Example 2 with $\tau \approx h^{\frac{2}{3-\alpha}}$.

α	h	$E_\infty(h)$	Rate_x
1.4	1/64	2.0137×10^{-4}	-
	1/128	4.5200×10^{-5}	2.1555
	1/256	1.0417×10^{-5}	2.1174
	1/512	2.4460×10^{-6}	2.0904
	1/1024	5.8316×10^{-7}	2.0685
1.6	1/64	1.7620×10^{-4}	-
	1/128	4.2168×10^{-5}	2.0630
	1/256	1.0291×10^{-5}	2.0348
	1/512	2.5373×10^{-6}	2.0200
	1/1024	6.2953×10^{-7}	2.0110
1.8	1/64	1.9261×10^{-4}	-
	1/128	4.7801×10^{-5}	2.0106
	1/256	1.1916×10^{-5}	2.0041
	1/512	2.9753×10^{-6}	2.0018
	1/1024	7.4337×10^{-7}	2.0009

Table 4. Numerical error, temporal convergence accuracy, and CPU time in Example 3 with $h \approx \tau^{\frac{3-\alpha}{2}}$.

α	τ	ADI Scheme (85)–(88)			Difference Scheme (78)–(79)		
		$E_\infty(\tau)$	Rate_t	CPU(s)	$E_\infty(\tau)$	Rate_t	CPU(s)
1.4	1/100	2.1905×10^{-4}	-	2.5415	2.5769×10^{-4}	-	10.7989
	1/200	7.4343×10^{-5}	1.5590	9.1474	8.3892×10^{-5}	1.6190	235.2666
	1/400	2.4530×10^{-5}	1.5996	84.1986	2.6911×10^{-5}	1.6403	8814.8094
	1/800	8.1856×10^{-6}	1.5834	624.8911	-	-	-
1.6	1/100	6.4844×10^{-4}	-	1.2507	6.8376×10^{-4}	-	1.3788
	1/200	2.4367×10^{-4}	1.4120	4.5492	2.5242×10^{-4}	1.4377	19.2717
	1/400	9.3717×10^{-5}	1.3785	27.1594	9.5907×10^{-5}	1.3961	363.7173
	1/800	3.5293×10^{-5}	1.4089	207.1005	3.5844×10^{-5}	1.4199	10,007.9790
1.8	1/100	1.8465×10^{-3}	-	0.9686	1.9000×10^{-3}	-	1.0571
	1/200	8.1979×10^{-4}	1.1715	2.3394	8.2721×10^{-4}	1.1997	2.6146
	1/400	3.6132×10^{-4}	1.1820	13.4669	3.6318×10^{-4}	1.1876	18.8149
	1/800	1.5601×10^{-4}	1.2116	83.4417	1.5648×10^{-4}	1.2147	281.8330

Table 5. Numerical error, spatial convergence accuracy, and CPU time in Example 3 with $\tau \approx h^{\frac{2}{3-\alpha}}$.

α	h	ADI Scheme (85)–(88)			Difference Scheme (78)–(79)		
		$E_\infty(h)$	Rate$_x$	CPU(s)	$E_\infty(h)$	Rate$_x$	CPU(s)
1.4	1/16	1.2811×10^{-3}	-	0.2432	1.7000×10^{-3}	-	0.2673
	1/32	3.3922×10^{-4}	1.9171	1.0039	4.0663×10^{-4}	2.0638	2.4947
	1/64	8.6549×10^{-5}	1.9706	6.9007	9.8228×10^{-5}	2.0495	134.1365
	1/128	2.1895×10^{-5}	1.9829	90.4336	2.3946×10^{-5}	2.0364	14,028.9227
1.6	1/16	1.5517×10^{-3}	-	0.5325	1.7000×10^{-3}	-	0.5844
	1/32	3.9889×10^{-4}	1.9598	2.6282	4.1654×10^{-4}	2.0290	4.2683
	1/64	1.0014×10^{-4}	1.9940	27.9728	1.0257×10^{-4}	2.0219	279.9205
	1/128	2.5086×10^{-5}	1.9971	466.9703	2.5422×10^{-5}	2.0125	32,390.5134
1.8	1/16	1.8236×10^{-3}	-	0.8777	1.9000×10^{-3}	-	0.9089
	1/32	4.6245×10^{-4}	1.9794	8.7622	4.6530×10^{-4}	2.0298	9.2940
	1/64	1.1576×10^{-4}	1.9982	197.7385	1.1605×10^{-4}	2.0034	797.0400
	1/128	2.8946×10^{-5}	1.9997	4326.2059	2.8975×10^{-5}	2.0019	105,978.3261

Figure 1 displays numerical solutions obtained from the ADI scheme (85)–(88) at different moments $t = 0$(s), $t = 2.5$(s) and $t = 5$(s) with various fractional orders $\alpha = 1$, $\alpha = 1.2$, $\alpha = 1.5$, and $\alpha = 1.8$. The graphs of initial values for all considered fractional orders are showed to be two spike type in first column of Figure 1. It can be seen that, with time increases, the amplitude of solutions decreases for each fractional order. Moreover, the larger the fractional order α is, the smaller the amplitude of solutions becomes. Numerical results indicate that the fractional order α could play an important role in regulating dynamic of damping.

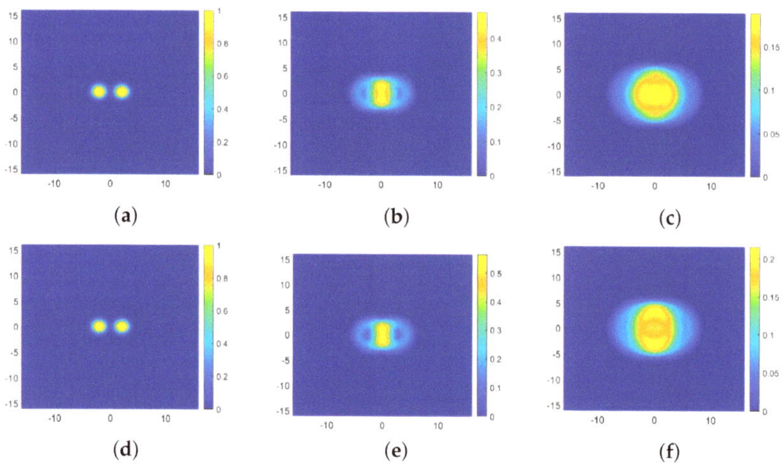

Figure 1. *Cont.*

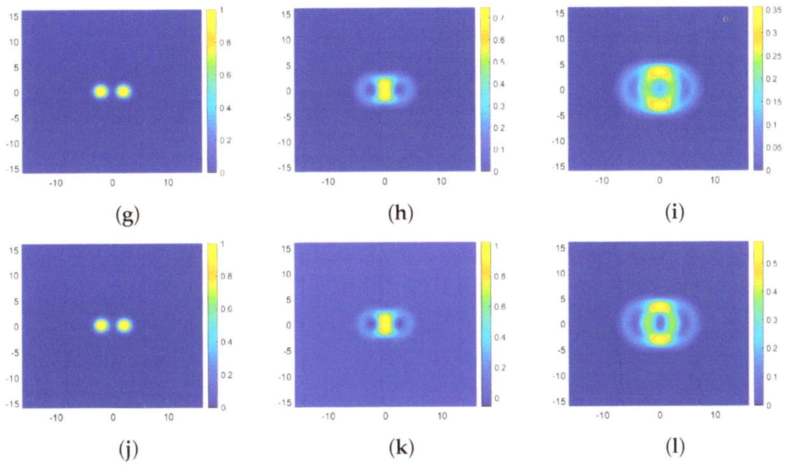

Figure 1. Effect of damping for different derivatives on peak value at different times. (**a**) $\alpha = 1,\ t = 0$, (**b**) $\alpha = 1,\ t = 2.5$, (**c**) $\alpha = 1,\ t = 5$, (**d**) $\alpha = 1.2,\ t = 0$, (**e**) $\alpha = 1.2,\ t = 2.5$, (**f**) $\alpha = 1.2,\ t = 5$, (**g**) $\alpha = 1.5,\ t = 0$, (**h**) $\alpha = 1.5,\ t = 2.5$, (**i**) $\alpha = 1.5,\ t = 5$, (**j**) $\alpha = 1.8,\ t = 0$, (**k**) $\alpha = 1.8,\ t = 2.5$, (**l**) $\alpha = 1.8,\ t = 5$.

6. Conclusions

We have developed efficient difference schemes for solving the one- and two-dimensional wave equations with fractional damping. In addition, we have proved the unconditional stability and convergence of the proposed difference schemes. The accuracy of the present difference schemes and the application of the wave equation with fractional damping have been tested in several numerical examples. Numerical simulations show that the values of fractional orders of the Caputo derivative in the damping model have a significant effect on the wave process.

It should be pointed out that the time discretization technique used in this paper is based on the uniform time mesh, which requires strong regularity assumptions. Recent works [18,25,38,39] have discussed the lack of the smoothness near the initial time of the solution of the time-fractional PDEs. Future study will consider the numerical method in a graded time mesh in order to overcome initial time singularity and provide a theoretical basis for the subsequent numerical scheme.

Author Contributions: Conceptualization, M.C., C.-C.J. and W.D.; methodology, M.C., C.-C.J. and W.D.; software, M.C.; validation, M.C. and C.-C.J.; formal analysis, M.C. and C.-C.J.; investigation, M.C.; data curation, M.C.; writing—original draft preparation, M.C.; writing—review and editing, C.-C.J. and W.D.; visualization, M.C.; supervision, M.C. and C.-C.J.; project administration, M.C. and C.-C.J. All authors have read and agreed to the published version of the manuscript.

Funding: Cui-Cui Ji was partially supported by National Natural Science Foundation of China under grant number 12001307 and Natural Science Foundation of Shandong Province under grant number ZR2020QA033, ZR2021MA072.

Data Availability Statement: Data will be made available on request due to privacy.

Acknowledgments: We would like to express our sincere gratitude to the anonymous reviewers for their valuable comments and suggestions which greatly enhance the quality of this manuscript.

Conflicts of Interest: The authors declare no conflict of interest.

References

1. Ma, H.Y.; Yan, B. Nonlinear damping and mass effects of electromagnetic shunt damping for enhanced nonlinear vibration isolation. *Mech. Syst. Signal. Process.* **2021**, *146*, 107010. [CrossRef]

2. Yayla, S.; Cardozo, C.L.; Silva, M.A.J.; Narciso, V. Dynamics of a Cauchy problem related to extensible beams under nonlocal and localized damping effects. *J. Math. Anal. Appl.* **2021**, *494*, 124620. [CrossRef]
3. Choucha, A.; Boulaaras, S.M.; Abdalla, M. Exponential stability of swelling porous elastic with a viscoelastic damping and distributed delay term. *J. Funct. Space* **2021**, *2021*, 5581634. [CrossRef]
4. Chen, G.W. Infinitely many nontrivial periodic solutions for damped vibration problems with asymptotically linear terms. *Appl. Math. Comput.* **2014**, *245*, 438–446. [CrossRef]
5. Liu, C.C.; Meng, F.J.; Zhong, C.K. A note on the global attractor for the weakly damped wave equation. *Appl. Math. Lett.* **2015**, *41*, 12–16. [CrossRef]
6. de Andrade, B.; Tuan, N.H. A non-autonomous damped wave equation with a nonlinear memory term. *Appl. Math. Opt.* **2022**, *85*, 36. [CrossRef]
7. Wang, M.L.; Zhang, J.L.; Li, E.Q.; Xin, X.F. The generalized Cole-Hopf transformation to a general variable coefficient Burgers equation with linear damping term. *Appl. Math. Lett.* **2020**, *105*, 106299. [CrossRef]
8. Yavuz, M.; Sulaiman, T.A.; Usta, F.; Bulut, H. Analysis and numerical computations of the fractional regularized long-wave equation with damping term. *Math. Method. Appl. Sci.* **2021**, *44*, 7538–7555. [CrossRef]
9. Lian, W.; Radulescu, V.D.; Xu, R.Z.; Yang, Y.B.; Zhao, N. Global well-posedness for a class of fourth-order nonlinear strongly damped wave equations. *Adv. Calc. Var.* **2021**, *14*, 589–611. [CrossRef]
10. Ji, B.Q.; Zhang, L.M. A dissipative finite difference Fourier pseudo-spectral method for the Klein-Gordon-Schrödinger equations with damping mechanism. *Appl. Math. Comput.* **2020**, *376*, 125148. [CrossRef]
11. Kirane, M.; Tatar, N.-E. Exponential growth for a fractionally damped wave equation. *Z. Anal. Anwend.* **2003**, *22*, 167–177. [CrossRef]
12. Tatar, N.-E. A blow up result for a fractionally damped wave equation. *Nodea-Nonlinear. Diff.* **2005**, *12*, 215–226. [CrossRef]
13. Charãoa, R.C.; da Luz, C.R.; Ikehata, R. Sharp decay rates for wave equations with a fractional damping via new method in the Fourier space. *J. Math. Anal. Appl.* **2013**, *408*, 247–255. [CrossRef]
14. Yan, Z.; Wang, W.; Liu, X.B. Analysis of a quintic system with fractional damping in the presence of vibrational resonance. *Appl. Math. Comput.* **2018**, *321*, 780–793. [CrossRef]
15. Lokshin, A.A. Wave equation with singular delayed time (in Russian). *Dokl. Akad. Nauk. SSSR* **1978**, *240*, 43–46.
16. Lokshin, A.A.; Rok, V.E. Fundamental solutions of the wave equation with delayed time (in Russian). *Dokl. Akad. Nauk. SSSR* **1978**, *239*, 1305–1308.
17. Wang, Q.F.; Cheng, D.Z. Numerical solution of damped nonlinear Klein-Gordon equations using variational method and finite element approach. *Appl. Math. Comput.* **2005**, *162*, 381–401. [CrossRef]
18. Stynes, M.; O'Riordan, E.; Gracia, J.L. Error analysis of a finite difference method on graded meshes for a time-fractional diffusion equation. *SIAM J. Numer. Anal.* **2017**, *55*, 1057–1079. [CrossRef]
19. Fu, H.F.; Wang, H. A preconditioned fast finite difference method for space-time fractional partial differential equations. *Fract. Calc. Appl. Anal.* **2017**, *20*, 88–116. [CrossRef]
20. Liu, Z.G.; Cheng, A.J.; Li, X.L. A second order Crank–Nicolson scheme for fractional Cattaneo equation based on new fractional derivative. *Appl. Math. Comput.* **2017**, *311*, 361–374. [CrossRef]
21. Ji, C.-C.; Dai, W.; Sun, Z.-Z. Numerical schemes for solving the time-fractional dual-phase-lagging heat conduction model in a double-layered nanoscale thin film. *J. Sci. Comput.* **2019**, *81*, 1767–1800. [CrossRef]
22. Wang, Y.-M.; Zheng, Z.-Y. A second-order L2-1_σ Crank-Nicolson difference method for two-dimensional time-fractional wave equations with variable coefficients. *Comput. Math. Appl.* **2022**, *118*, 183–207. [CrossRef]
23. Arshad, S.; Bu, W.P.; Huang, J.F.; Tang, Y.F.; Zhao, Y. Finite difference method for time-space linear and nonlinear fractional diffusion equations. *Int. J. Comput. Math.* **2018**, *95*, 202–217. [CrossRef]
24. Ji, C.-C.; Dai, W. Numerical algorithm with fourth-order spatial accuracy for solving the time-fractional dual-phase-lagging nanoscale heat conduction equation. *Numer. Math. Theor. Meth. Appl.* **2023**, *16*, 511–540.
25. Ji, C.-C.; Qu, W.Z.; Jiang, M.S. Numerical method for solving the fractional evolutionary model of bi-flux diffusion processes. *Int. J. Comput. Math.* **2023**, *100*, 880–900. [CrossRef]
26. Du, N.; Wang, H. A fast finite element method for space-fractional dispersion equations on bounded domains in R^2. *SIAM J. Sci. Comput.* **2015**, *37*, A1614–A1635. [CrossRef]
27. Dehghan, M.; Abbaszadeh, M. A finite difference/finite element technique with error estimate for space fractional tempered diffusion-wave equation. *Comput. Math. Appl.* **2018**, *75*, 2903–2914. [CrossRef]
28. Li, C.P.; Chen, A. Numerical methods for fractional partial differential equations. *Int. J. Comput. Math.* **2018**, *95*, 1048–1099. [CrossRef]
29. Li, M.; Huang, C.M.; Ming, W.Y. Mixed finite-element method for multi-term time-fractional diffusion and diffusion-wave equations. *Comput. Appl. Math.* **2018**, *37*, 2309–2334. [CrossRef]
30. Sun, J.; Nie, D.X.; Deng, W.H. A Reduced Finite Element Formulation for Space Fractional Partial Differential Equation. *E. Asian J. Appl. Math.* **2019**, *8*, 678–696. [CrossRef]
31. Liu, Y.Q.; Sun, H.G.; Yin, X.L.; Feng, L.B. Fully discrete spectral method for solving a novel multi-term time-fractional mixed diffusion and diffusion-wave equation. *Z. Angew. Math. Phys.* **2020**, *1*, 71. [CrossRef]

32. Yang, Y.; Chen, Y.P.; Huang, Y.Q.; Wei, H.Y. Spectral collocation method for the time-fractional diffusion-wave equation and convergence analysis. *Comput. Math. Appl.* **2017**, *73*, 1218–1232.
33. Saffarian, M.; Mohebbi, A. Numerical solution of two and three dimensional time fractional damped nonlinear Klein–Gordon equation using ADI spectral element method. *Appl. Math. Comput.* **2021**, *405*, 126182. [CrossRef]
34. Wu, C.; Wei, C.; Yin, Z.; Zhu, A. A Crank-Nicolson Compact Difference Method for Time-Fractional Damped Plate Vibration Equations. *Axioms* **2022**, *11*, 535. [CrossRef]
35. Lyu, P.; Liang, Y.X.; Wang, Z.B. A fast linearized finite difference method for the nonlinear multi-term time-fractional wave equation. *Appl. Numer. Math.* **2022**, *151*, 448–471. [CrossRef]
36. Wang, T.; Jiang, Z.; Zhu, A.; Yin, Z. A Mixed Finite Volume Element Method for Time-Fractional Damping Beam Vibration Problem. *Fractal Fract.* **2022**, *6*, 523. [CrossRef]
37. Sun, Z.Z.; Wu, X.N. A fully discrete difference scheme for a diffusion-wave system. *Appl. Numer. Math.* **2006**, *56*, 193–209. [CrossRef]
38. Lyu, P.; Vong, S. A symmetric fractional-order reduction method for direct nonuniform approximations of semilinear diffusion-wave equations. *J. Sci. Comput.* **2022**, *93*, 34. [CrossRef]
39. Liao, H.-L.; Liu, N.; Lyu, P. Discrete gradient structure of a second-order variable-step method for nonlinear integer-differential models. *SIAM J. Numer. Anal.* **2023**, *61*, 2157–2181. [CrossRef]

Mathematical and Computational Applications

MDPI

Article

A Numerical Method Based on Operator Splitting Collocation Scheme for Nonlinear Schrödinger Equation

Mengli Yao and Zhifeng Weng *

Fujian Province University Key Laboratory of Computation Science, School of Mathematical Sciences, Huaqiao University, Quanzhou 362021, China; yaomengli163@163.com
* Correspondence: zfwmath@163.com

Abstract: In this paper, a second-order operator splitting method combined with the barycentric Lagrange interpolation collocation method is proposed for the nonlinear Schrödinger equation. The equation is split into linear and nonlinear parts: the linear part is solved by the barycentric Lagrange interpolation collocation method in space combined with the Crank–Nicolson scheme in time; the nonlinear part is solved analytically due to the availability of a closed-form solution, which avoids solving the nonlinear algebraic equation. Moreover, the consistency of the fully discretized scheme for the linear subproblem and error estimates of the operator splitting scheme are provided. The proposed numerical scheme is of spectral accuracy in space and of second-order accuracy in time, which greatly improves the computational efficiency. Numerical experiments are presented to confirm the accuracy, mass and energy conservation of the proposed method.

Keywords: nonlinear Schrödinger equation; operator splitting collocation method; barycentric Lagrange interpolation; consistency analysis; convergence analysis

1. Introduction

In 1926, the famous Schrödinger equation was proposed by the Austrian physicist Schrödinger [1]. It's a fundamental equation in the field of quantum mechanics. In recent years, the Schrödinger equation has been studied in different fields of research, such as atomic, molecular, nuclear physics and solid state physics, etc.

In this paper, we consider the following nonlinear Schrödinger (NLS) equation

$$\begin{cases} iu_t + \rho\Delta u + v(\mathbf{x})u + \beta|u|^2u = 0, (\mathbf{x}, \mathrm{t}) \in \Omega \times (0, \mathrm{T}] \\ u(\mathbf{x}, 0) = u_0(\mathbf{x}), \mathbf{x} \in \Omega \\ u(\mathbf{x}, t) = 0, \mathbf{x} \in \partial\Omega, \mathrm{t} \in (0, \mathrm{T}] \end{cases}, \tag{1}$$

where i is the imaginary unit, ρ, β are real-valued constants, $\Omega \subset R^d (d = 1, 2)$ is a bounded area and Δ is the Laplace operator. The function $u_0(\mathbf{x})$ is a given sufficiently smooth function and $v(\mathbf{x})$ is a real-valued potential function. This model reflects the quantum mechanical effects and microscopic system properties and can well describe the state of microscopic particles over time.

In fact, the NLS Equation (1) also conserves both mass and energy with the following mass and energy functions:

$$M(t) = \int_\Omega |u(\mathbf{x}, t)|^2 d\mathbf{x} = M(0) \tag{2}$$

and

$$E(t) = \int_\Omega (\rho|\nabla u(\mathbf{x}, t)|^2 - v(\mathbf{x})|u(\mathbf{x}, t)|^2 - \frac{\beta}{2}|u(\mathbf{x}, t)|^4) d\mathbf{x} = E(0). \tag{3}$$

Recently, many numerical methods have been developed for solving the NLS equation. Gong et al. [2] presented a conservative Fourier pseudo-spectral method for the nonlinear

Schrödinger equation. Cui et al. [3] developed mass- and energy-preserving exponential Runge–Kutta methods for the nonlinear Schrödinger equation. Feng et al. [4] developed the high-order mass- and energy-conserving SAV-Gauss collocation finite element methods for the NLS equation. Wang et al. [5] used the two-grid finite element method for the NLS equation and the given superconvergence analysis of the scheme. Wang et al. [6] studied leapfrog finite element methods for a class of nonlinear Schrödinger equations with damped terms. Hu et al. [7] presented the Newton iterative Crank–Nicolson finite element method for the NLS equation. Chen et al. [8] applied the two-grid finite volume element method for the time-dependent Schrödinger equation. Deng et al. [9] considered a second-order SAV scheme for the nonlinear Schrödinger equation in the whole space with typical generalized nonlinearities and carried out a rigorous error analysis. Su et al. [10] considered the numerical solution of the nonlinear Schrödinger equation with a highly oscillatory potential (NLSE-OP) and rigorously analyzed the error bounds of the splitting schemes for solving the NLSE-OP to a fixed time. Wang et al. [11] proposed finite difference methods for the coupled Gross–Pitaevskii equations in high dimensions and the given error estimates.

There are some advantages, such as no dividing elements, simple formulas, no integrals and easy programming, of the collocation method, which is called the barycentric Lagrange interpolation collocation (BLIC) method. The method has captured the attention of many scholars because of its high accuracy. The numerical stability of the barycentric Lagrange interpolation collocation method with Chebyshev points is very good, and it can also effectively overcome the "Runge" phenomenon. The method has been extended to solve various partial differential equations, such as the sine-Gordon equation [12], the Burgers equation [13], the viscoelastic wave equation [14], the Allen–Cahn equation [15,16], nonlinear convection-diffusion optimal control problems [17] and the fractional telegraph equation [18], among others.

To the best of our knowledge, there are few studies regarding using the barycentric interpolation collocation method combined with the operator splitting method [19–21] for the nonlinear Schrödinger equation. Based on the above work, we focus on the convergence analysis of the proposed scheme. We analyze the fully discretized consistency of the linear subproblem. Moreover, the error estimates of the operator splitting scheme are derived.

The remaining parts of the paper are structured as follows. In Section 2, we present the barycentric Lagrange interpolation collocation method. In Section 3, we present a second-order operator splitting collocation scheme for the NLS equation. The convergence analyses of the proposed method are presented in Section 4. Numerical experiments are conducted in Section 5 to evaluate the accuracy and efficiency of the proposed method, while Section 6 presents some conclusions derived from these experiments.

2. Preliminary

Suppose that $m+1$ distinct interpolation nodes, x_j $(j=0,1,\ldots,m)$, and their corresponding function values, u_j, are provided. Therefore, there exists a unique interpolation polynomial whose degree is not exceeding m, satisfying $q(x_j)=u_j$ $(j=0,1,\ldots,m)$. As we know, $q(x)$ has the Lagrange form as follows,

$$q(x)=\sum_{j=0}^{m} G_j(x)u_j, \tag{4}$$

where $G_j(x)$ represents the Lagrange interpolation basis function and

$$G_j(x)=\frac{\prod\limits_{i=0,i\neq j}^{m}(x-x_i)}{\prod\limits_{i=0,i\neq j}^{m}(x_j-x_i)}, \quad j=0,1,\cdots,m. \tag{5}$$

Suppose that

$$g(x) = (x - x_0)(x - x_1)\dots(x - x_m), \tag{6}$$

and the barycentric weights are defined as follows:

$$w_j = \frac{1}{\prod\limits_{i=0, i\neq j}^{m} (x_j - x_i)}, \quad j = 0, 1, \cdots, m. \tag{7}$$

Then, from Equations (5)–(7), we can obtain

$$G_j(x) = g(x)\frac{w_j}{x - x_j}, \quad j = 0, 1, \cdots, m. \tag{8}$$

By Equations (8) and (4), we have

$$q(x) = g(x)\sum_{j=0}^{m} \frac{w_j}{x - x_j} u_j. \tag{9}$$

If $u = 1$, it has the following form:

$$1 = g(x)\sum_{j=0}^{m} \frac{w_j}{x - x_j}. \tag{10}$$

By Equations (10) and (9), the barycentric interpolation formula of $q(x)$ can be derived:

$$q(x) = \frac{\sum\limits_{j=0}^{m} \frac{w_j}{x - x_j} u_j}{\sum\limits_{j=0}^{m} \frac{w_j}{x - x_j}} = \sum_{j=0}^{m} \frac{\frac{w_j}{x - x_j}}{\sum\limits_{j=0}^{m} \frac{w_j}{x - x_j}} u_j := \sum_{j=0}^{m} \tilde{G}_j(x) u_j. \tag{11}$$

To ensure the numerical stability of the barycentric Lagrange interpolation, we adopt Chebyshev points:

$$x_j = cos(\frac{j}{m}\pi), j = 0, 1, \cdots, m. \tag{12}$$

The v-order derivative of $q(x)$ defined as Equation (11) with respect to x is

$$q^{(v)}(x_i) = \frac{d^v q(x_i)}{dx^v} = \sum_{j=0}^{m} \tilde{G}_j^{(v)}(x_i) u_j = \sum_{j=0}^{m} P_{ij}^{(v)} u_j, v = 1, 2, \cdots, m, \tag{13}$$

where $P_{ij}^{(v)}$ denotes the element of the v-order differentiation matrix $P^{(v)}$.

From Equations (11) and (13), we can obtain [22]

$$\begin{cases} P_{ij}^{(1)} = \tilde{G}_j'(x_i) = \dfrac{w_j/w_i}{x_i - x_j}, j \neq i, \\ P_{ii}^{(1)} = -\sum\limits_{j=0, j\neq i}^{m} P_{ij}^{(1)}, \end{cases} \tag{14}$$

$$\begin{cases} P_{ij}^{(2)} = \tilde{G}_j''(x_i) = -2\dfrac{w_j/w_i}{x_i - x_j}\left(\sum\limits_{k\neq i} \dfrac{w_k/w_i}{x_i - x_k} + \dfrac{1}{x_i - x_j}\right), j \neq i, \\ P_{ii}^{(2)} = -\sum\limits_{j=0, j\neq i}^{m} P_{ij}^{(2)}. \end{cases} \tag{15}$$

Next, we will derive the approximation format for a given function, $u(x,y)$, and its derivative using the barycentric Lagrange interpolation formula.

For $m+1$ distinct nodes, $(x_0,y),(x_1,y),\ldots,(x_m,y)$, the unknown function $u(x,y)$, evaluated at node (x_i,y), can be expressed as follows:

$$u(x_i,y), \quad i=0,1,\ldots,m. \tag{16}$$

From Equations (16) and (11), we can obtain the approximation function of $u(x,y)$:

$$u(x,y)=\sum_{i=0}^{m}\alpha_i(x)u(x_i,y), \quad i=0,1,\ldots,m. \tag{17}$$

Similarly, the expression of $u(x_i,y)$ is as follows:

$$u(x_i,y)=\sum_{j=0}^{n}\beta_j(y)u(x_i,y_j), \quad j=0,1,\ldots,n. \tag{18}$$

From Equations (17) and (18), we can obtain

$$u(x,y)=\sum_{i=0}^{m}\sum_{j=0}^{n}\alpha_i(x)\beta_j(y)u(x_i,y_j), \tag{19}$$

and its second-order partial derivative has the following form:

$$u_{xx}(x,y)=\sum_{i=0}^{m}\sum_{j=0}^{n}\alpha_i''(x)\beta_j(y)u(x_i,y_j), \tag{20}$$

$$u_{yy}(x,y)=\sum_{i=0}^{m}\sum_{j=0}^{n}\alpha_i(x)\beta_j''(y)u(x_i,y_j). \tag{21}$$

3. Operator Splitting Collocation Method

In this section, we propose an operator splitting collocation scheme for the NLS equation, which is based on the Strang splitting procedure. Our approach combines the barycentric Lagrange interpolation collocation method for spatial approximation and a second-order Crank–Nicolson scheme for temporal approximation.

First, rewrite Equation (1) as follows,

$$iu_t=L(u)+N(u), \tag{22}$$

where $L(u)=-\rho\Delta u$ and $N(u)=-v(\mathbf{x})u-\beta|u|^2u$.

Then, Equation (1) is split into the linear part,

$$iu_t=L(u)=-\rho\Delta u, \tag{23}$$

and the nonlinear part,

$$iu_t=N(u)=-v(\mathbf{x})u-\beta|u|^2u. \tag{24}$$

Next, for a given time step, τ, the solution of Equation (1) evolves from t to $t+\tau$ via the Strang splitting method [23], which consists of three substeps:

$$u(x,y,t+\tau)=S^B(\frac{\tau}{2})S^A(\tau)S^B(\frac{\tau}{2})u(x,y,t)+O(\tau^3), \tag{25}$$

where S^A and S^B are the exact solution operators of Equation (23) and Equation (24), respectively.

Then, we will provide numerical approximations S_h^A and S_h^B for the exact solution operators S^A and S^B, respectively. Suppose $\Omega = [a,b] \times [c,d]$ and $\Omega_h = \{(x_i, y_j), i = 0 < 1 < \ldots < m, j = 0 < 1 < \ldots < n\}$. Here, x_i and y_j are Chebyshev mesh points.

To solve Equation (23), the barycentric Lagrange interpolation collocation method is applied to discretize the spatial derivative. The semi-discretized scheme in space is obtained based on the barycentric Lagrange interpolation collocation method, as introduced in Section 2.

$$iu_t(x_i, y_j, t) = -\rho \sum_{i=0}^{m} \sum_{j=0}^{n} \Big(\alpha_i''(x)\beta_j(y) + \alpha_i(x)\beta_j''(y)\Big) u(x_i, y_j, t). \tag{26}$$

The matrix form of (26) can be expressed as

$$i(u_h(t))_t = -\rho \Big(P^{(2)} \otimes I_n + I_m \otimes Q^{(2)}\Big) u_h(t), \tag{27}$$

where $u_h(t) = [u_{00}(t), \cdots, u_{0n}(t), u_{10}(t), \cdots, u_{1n}(t), \cdots, u_{m0}(t), \cdots, u_{mn}(t)]$, $u_{ij}(t) = u(x_i, y_j, t)$; $P^{(2)}$ and $Q^{(2)}$ are second-order differentiation matrices on nodes $x_0, x_1, \ldots, x_m$ and $y_0, y_1, \ldots, y_n$, respectively; $\otimes$ represents the Kronecker product of the matrix; I_m and I_n are the identity matrices of $m+1$ and $n+1$ order, respectively. Then, setting $u_h^k = u_h(t_k)$, we can obtain the following fully discretized scheme:

$$i\frac{u_h^{k+1} - u_h^k}{\tau} = -\rho\Big(\frac{1}{2}(P^{(2)} \otimes I_n + I_m \otimes Q^{(2)})u_h^{k+1} + \frac{1}{2}(P^{(2)} \otimes I_n + I_m \otimes Q^{(2)})u_h^k\Big). \tag{28}$$

Therefore, we obtain

$$S_h^A: \quad u_h^{k+1} = (iI + \frac{\rho\tau}{2}W)^{-1}(iI - \frac{\rho\tau}{2}W)u_h^k, \tag{29}$$

where I is the identity matrix of order $(m+1)(n+1)$, $W = P^{(2)} \otimes I_n + I_m \otimes Q^{(2)}$.

Then, Equation (24) can be solved analytically by

$$S_h^B: \quad u_h^{k+1} = e^{i\frac{\tau}{2}(v(x,y)+\beta|u_h^k|^2)} u_h^k. \tag{30}$$

The second-order operator splitting scheme can be derived as follows:

$$\textbf{MI:}\begin{cases} u_h^{(1)} = e^{i\frac{\tau}{2}(v(x,y)+\beta|u_h^k|^2)} u_h^k \\ u_h^{(2)} = (iI + \frac{\rho\tau}{2}W)^{-1}(iI - \frac{\rho\tau}{2}W)u_h^{(1)} \\ u_h^{k+1} = e^{i\frac{\tau}{2}(v(x,y)+\beta|u_h^{(2)}|^2)} u_h^{(2)} \end{cases}. \tag{31}$$

4. Convergence Analysis

Firstly, in this section we analyze the consistency of the semi-discretized scheme (26). Suppose that $q(x,y)$ is the Lagrange interpolation function of $u(x,y)$, satisfying $q(x_i, y_j) = u(x_i, y_j), i = 0 < 1 < \ldots < m, j = 0 < 1 < \ldots < n$.

By defining

$$r(x,y) = u(x,y) - q(x,y), \tag{32}$$

we can obtain the following estimates [18],

Lemma 1. *Suppose* $u(x,y) \in C^{(\tilde{m}+1)}([a,b]\times[c,d])$, *where* $\tilde{m} = \max\{m,n\}$, *we have*

$$\begin{cases} \|r(x,y)\|_\infty \le C_1 \parallel u^{(m+1)} \parallel_\infty \left(\frac{eh_x}{2m}\right)^m + C_2 \parallel u^{(m+1)} \parallel_\infty \left(\frac{eh_y}{2n}\right)^n \\ \|r_{xx}(x,y)\|_\infty \le C_1^{**}\|u^{(m+1)}\|_\infty \left(\frac{eh_x}{2(m-2)}\right)^{m-2} + C_2\|u^{(n+1)}\|_\infty \left(\frac{eh_y}{2n}\right)^n \\ \|r_{yy}(x,y)\|_\infty \le C_1\|u^{(m+1)}\|_\infty \left(\frac{eh_x}{2m}\right)^m + C_2^{**}\|u^{(n+1)}\|_\infty \left(\frac{eh_y}{2(n-2)}\right)^{n-2} \end{cases},$$

where $h_x = \frac{b-a}{2}$, $h_y = \frac{d-c}{2}$, *and e is the natural constant.*

Let $u(x,y,t)$ be the solution of Equation (1) and $u(x_i,y_j,t)$ is the numerical solution of $u(x,y,t)$ discretized by barycentric Lagrange interpolation collocation method; we then have

$$\mathcal{T}u(x_i,y_j,t) = 0 \tag{33}$$

and

$$\lim_{i,j\to\infty} \mathcal{T}u(x_i,y_j,t) = 0, \tag{34}$$

where $\mathcal{T} = i\frac{\partial}{\partial t} + \frac{\partial^2}{\partial x^2} + \frac{\partial^2}{\partial y^2}$.

Based on the above results, we obtain the following theorem.

Theorem 1. *Let* $u(x_i,y_j,t) : \mathcal{T}u(x_i,y_j,t) = 0$, *we have*

$$\|u(x,y,t) - u(x_i,y_j,t)\|_\infty \le C_1^{**} \parallel u^{(m+1)}\|_\infty \left(\frac{eh_x}{2(m-2)}\right)^{m-2} + C_2^{**} \parallel u^{(n+1)}\|_\infty \left(\frac{eh_y}{2(n-2)}\right)^{n-2}.$$

Proof. As

$$\begin{aligned} &\mathcal{T}u(x,y,t) - \mathcal{T}u(x_i,y_j,t) \\ &= iu_t(x,y,t) + u_{xx}(x,y,t) + u_{yy}(x,y,t) - (iu_t(x_i,y_j,t) + u_{xx}(x_i,y_j,t) + u_{yy}(x_i,y_j,t)) \\ &= iu_t(x,y,t) - iu_t(x_i,y_j,t) + u_{xx}(x,y,t) - u_{xx}(x_i,y_j,t) + u_{yy}(x,y,t) - u_{yy}(x_i,y_j,t) \\ &:= A_1 + A_2 + A_3, \end{aligned} \tag{35}$$

where A_1, A_2 and A_3 represent

$$\begin{aligned} A_1 &= iu_t(x,y,t) - iu_t(x_i,y_j,t), \\ A_2 &= u_{xx}(x,y,t) - u_{xx}(x_i,y_j,t), \\ A_3 &= u_{yy}(x,y,t) - u_{yy}(x_i,y_j,t). \end{aligned} \tag{36}$$

For A_1, we obtain

$$\begin{aligned} A_1 &= i(u_t(x,y,t) - u_t(x_i,y_j,t)) \\ &= i(u_t(x,y,t) - u_t(x_i,y,t) + u_t(x_i,y,t) - u_t(x_i,y_j,t)) \\ &= i(r_t(x_i,y,t) + r_t(x_i,y_j,t)). \end{aligned} \tag{37}$$

From Lemma 1, we can derive

$$\begin{aligned} &\|A_1\|_\infty = \|i(r_t(x_i,y,t) + r_t(x_i,y_j,t))\|_\infty \\ &\le \|r_t(x_i,y,t)\|_\infty + \|r_t(x_i,y_j,t)\|_\infty \\ &\le C_1 \parallel u^{(m+1)}\|_\infty \left(\frac{eh_x}{2m}\right)^m + C_2 \parallel u^{(n+1)}\|_\infty \left(\frac{eh_y}{2n}\right)^n. \end{aligned} \tag{38}$$

Similar estimates can be derived as follow:

$$\begin{aligned}\|A_2\|_\infty &= \|r_{xx}(x_i, y, t) + r_{xx}(x_i, y_j, t)\|_\infty \\ &\leq C_1^{**} \parallel u^{(m+1)} \|_\infty \left(\frac{eh_x}{2(m-2)}\right)^{m-2} + C_2 \parallel u^{(n+1)} \|_\infty \left(\frac{eh_y}{2n}\right)^n,\end{aligned} \tag{39}$$

$$\begin{aligned}\|A_3\|_\infty &= \|r_{yy}(x_i, y, t) + r_{yy}(x_i, y_j, t)\|_\infty \\ &\leq C_1 \parallel u^{(m+1)} \|_\infty \left(\frac{eh_x}{2m}\right)^m + C_2^{**} \parallel u^{(n+1)} \|_\infty \left(\frac{eh_y}{2(n-2)}\right)^{n-2}.\end{aligned} \tag{40}$$

From (38)–(40) and (35), the proof is completed. □

Next, we consider consistency analysis of the full-discretized scheme (28).

Theorem 2. *Suppose $u(x,y,t) \in C^{(\tilde{m}+1)}(\Omega) \times C^2(0,T], \Omega = [a,b] \times [c,d]$, where $\tilde{m} = \max\{m,n\}$ and $u(x_i, y_j, t_{k+1})$ is the corresponding numerical solution of $u(x,y,t)$, we have*

$$\|u(x,y,t) - u(x_i, y_j, t_{k+1})\|_\infty \leq C_3 \left(\tau^3 + \parallel u^{(m+1)} \|_\infty \left(\frac{eh_x}{2(m-2)}\right)^{m-2} + \parallel u^{(n+1)} \|_\infty \left(\frac{eh_y}{2(n-2)}\right)^{n-2}\right),$$

where C_3 is a positive constant.

Proof. Let $u(x,y,t_{k+1})$ be the corresponding numerical solution using the Crank–Nicolson scheme for temporal approximation of $u(x,y,t)$; we can obtain

$$i\delta u_t(x, y, t_{k+\frac{1}{2}}) = -\rho \Delta u(x, y, t_{k+\frac{1}{2}}) + R^k, \tag{41}$$

where $\delta u_t(x,y,t_{k+\frac{1}{2}}) = u(x,y,t_{k+1}) - u(x,y,t_k)$, and $R^k = u_t(x,y,t_{k+\frac{1}{2}}) - \delta u_t(x,y,t_{k+\frac{1}{2}})$ is the truncation error in time. Based on the principle of Taylor expansion, we can obtain

$$\left|R^k\right| \leq C_4 \tau^3. \tag{42}$$

Equation (41) is discretized by the BLIC scheme, and supposing that $u(x_i, y_j, t_{k+1})$ is the numerical solution of $u(x,y,t_{k+1})$ based on the BLIC method, it holds that

$$i\delta u_t(x_i, y_j, t_{k+\frac{1}{2}}) = -\rho \Delta u^h(x_i, y_j, t_{k+\frac{1}{2}}) + R^k + \gamma^{i,j}, \tag{43}$$

where $\gamma^{i,j}$ represents the truncation error in space.

Combining Equations (41) and (43), we have

$$i\delta u_t(x, y, t_{k+\frac{1}{2}}) - i\delta u_t(x_i, y_j, t_{k+\frac{1}{2}}) = -\rho \Delta u(x, y, t_{k+\frac{1}{2}}) + \rho \Delta u^h(x_i, y_j, t_{k+\frac{1}{2}}) - \gamma^{i,j}. \tag{44}$$

If we use a similar technique in Theorem 1, we can derive

$$\left|\gamma^{i,j}\right| \leq C_1^{**} \parallel u^{(m+1)} \|_\infty \left(\frac{eh_x}{2(m-2)}\right)^{m-2} + C_2^{**} \parallel u^{(n+1)} \|_\infty \left(\frac{eh_y}{2(n-2)}\right)^{n-2}. \tag{45}$$

Combining (42) and (45), the proof is completed. □

We will now analyze the error results of the operator splitting scheme.

Define a grid function space on Ω_h,

$$\mathcal{W}^h = \left\{U \middle| U = \{U_{ij} \mid 0 \leq i \leq m, 0 \leq j \leq n\}\right\},$$

and a mapping $I^h : \tilde{H}(\Omega) \longrightarrow \mathcal{W}_h$ by

$$I^h(u) = U,$$

where $\tilde{H}(\Omega) = \{u \in H^2(\Omega) \mid u(\mathbf{x}, t) = 0\}$.

Lemma 2. *Supposing that $U \in \mathcal{W}^h$, $\|W\|_\infty$ is bounded and τ is small enough, we have*

$$\| S_h^A(\tau)U \|_\infty \leq (1+\kappa\tau) \| U \|_\infty .$$

Proof. By Taylor expansion, the approximation of $(iI + \frac{\rho\tau}{2}W)^{-1}$ at the zero matrix is

$$(iI + \frac{\rho\tau}{2}W)^{-1} = -iI + \frac{\rho\tau}{2}W + i\frac{(\rho\tau)^2}{4}W^2 - \frac{(\rho\tau)^3}{8}W^3 + \mathcal{O}(\tau^4). \tag{46}$$

Then, combining with the above formula, we can obtain

$$\| (iI + \frac{\rho\tau}{2}W)^{-1}(iI - \frac{\rho\tau}{2}W)\|_\infty = \| I + i\rho\tau W - \frac{(\rho\tau)^2}{2}W^2 - \frac{(\rho\tau)^4}{16}W^4\|_\infty \leq 1+\kappa\tau. \tag{47}$$

Therefore,

$$\begin{aligned} &\| S_h^A(\tau)U\|_\infty = \| (iI + \frac{\rho\tau}{2}W)^{-1}(iI + \frac{\rho\tau}{2}W)U\|_\infty \\ &\leq \| (iI + \frac{\rho\tau}{2}W)^{-1}(I + \frac{\rho\tau}{2}W)\|_\infty \| U\|_\infty \\ &\leq (1+\kappa\tau) \| U\|_\infty \end{aligned}, \tag{48}$$

where κ is a positive constant independent of τ. □

Lemma 3. *Supposing that $U \in \mathcal{W}^h$, we have*

$$\| S_h^B(\frac{\tau}{2})U \|_\infty = \| U \|_\infty .$$

Proof.

$$\left|S_h^B(\frac{\tau}{2})u_{ij}\right| = \left|e^{i\frac{\tau}{2}(v(x,y)+\beta|u_{ij}|^2)}\right||u_{ij}| = |u_{ij}|. \tag{49}$$

□

By Theorem 2, we can obtain the following result.

Lemma 4. *Supposing that $u_0 \in \tilde{H}(\Omega)$, we have*

$$\| I^h S^A u_0 - S_h^A I^h u_0 \|_\infty \leq C_1^{**} \| u^{(m+1)} \|_\infty \left(\frac{eh_x}{2(m-2)}\right)^{m-2} + C_2^{**} \| u^{(n+1)} \|_\infty \left(\frac{eh_y}{2(n-2)}\right)^{n-2} + C_4\tau^3.$$

For convenience, suppose

$$\eta = C_5\left(\| u^{(m+1)} \|_\infty \left(\frac{eh_x}{2(m-2)}\right)^{m-2} + \| u^{(n+1)} \|_\infty \left(\frac{eh_y}{2(n-2)}\right)^{n-2} + \tau^3\right),$$

where $C_5 = \max\{C_1^{**}, C_2^{**}, C_4\}$.

Theorem 3. *Let $u^{k+1} \in \tilde{H}(\Omega)$ be the exact solution of Equation (1) and $\tilde{u}(x,y,t)$ and U^{k+1} be the exact solution of Equation (25) and the numerical solution at t_{k+1} of Equation (31), respectively. According to Theorem 1 and Lemma 2, we have*

$$\| I^h u^{k+1} - U^{k+1} \|_\infty \leq C\left(\frac{\| u^{(m+1)} \|_\infty}{\tau} \left(\frac{eh_x}{2(m-2)} \right)^{m-2} + \frac{\| u^{(n+1)} \|_\infty}{\tau} \left(\frac{eh_y}{2(n-2)} \right)^{n-2} + \tau^2 \right),$$

where C is a positive constant independent of τ.

Proof. For $k \geq 0$, we obtain

$$\| I^h u^{k+1} - U^{k+1} \|_\infty \leq \| I^h u^{k+1} - I^h \tilde{u}^{k+1} \|_\infty + \| I^h \tilde{u}^{k+1} - U^{k+1} \|_\infty . \tag{50}$$

From [23], we obtain

$$\| I^h \tilde{u}^{k+1} - I^h u^{k+1} \|_\infty \leq C_6 \tau^2. \tag{51}$$

By Lemma 3, we obtain

$$\begin{aligned} &\| I^h \tilde{u}^{k+1} - U^{k+1} \|_\infty \\ &= \| I^h S^B S^A S^B \tilde{u}^k - S_h^B S_h^A S_h^B U^k \|_\infty \\ &\leq \| I^h S^B S^A S^B \tilde{u}^k - S_h^B I^h S^A S^B \tilde{u}^k \|_\infty + \| S_h^B I^h S^A S^B \tilde{u}^k - S_h^B S_h^A S_h^B U^k \|_\infty \\ &\leq \| I^h S^A S^B \tilde{u}^k - S_h^A S_h^B U^k \|_\infty \end{aligned} . \tag{52}$$

From Lemmas 2 and 4, we obtain

$$\begin{aligned} &\| I^h S^A S^B \tilde{u}^k - S_h^A S_h^B U^k \|_\infty \\ &\leq \| I^h S^A S^B \tilde{u}^k - S_h^A I^h S^B \tilde{u}^k \|_\infty + \| S_h^A I^h S^B \tilde{u}^k - S_h^A S_h^B U^k \|_\infty . \\ &\leq \eta + (1 + \kappa\tau) \| I^h S^B \tilde{u}^k - S_h^B U^k \|_\infty \end{aligned} \tag{53}$$

From Lemma 3, we derive

$$\begin{aligned} &\| I^h S^B \tilde{u}^k - S_h^B U^k \|_\infty \\ &\leq \| I^h S^B \tilde{u}^k - S_h^B I^h \tilde{u}^k \|_\infty + \| S_h^B I^h \tilde{u}^k - S_h^B U^k \|_\infty . \\ &\leq \| I^h \tilde{u}^k - U^k \|_\infty \end{aligned} \tag{54}$$

Due to $\| I^h \tilde{u}^0 - U^0 \|_\infty = 0$, and by the Gronwall inequality, we can obtain

$$\begin{aligned} &\| I^h \tilde{u}^{k+1} - U^{k+1} \|_\infty \\ &\leq \eta + (1 + \kappa\tau) \| I^h \tilde{u}^k - U^k \|_\infty \\ &\leq \sum_{j=0}^{k} (1 + \kappa\tau)^j \eta + (1 + \kappa\tau)^{k+1} \| I^h \tilde{u}^0 - U^0 \|_\infty \\ &\leq \frac{(1 + \kappa\tau)^{k+1} - 1}{1 + \kappa\tau - 1} \eta \\ &\leq \frac{e^{\kappa\tau}}{\kappa\tau} \eta \end{aligned} . \tag{55}$$

By the above estimates and the expression of η, we have

$$\begin{aligned} &\| I^h u^{k+1} - U^{k+1} \|_\infty \\ &\leq \frac{e^{\kappa\tau}}{\kappa} \left(C_5 \left(\frac{\| u^{(m+1)} \|_\infty}{\tau} \left(\frac{eh_x}{2(m-2)} \right)^{m-2} + \frac{\| u^{(n+1)} \|_\infty}{\tau} \left(\frac{eh_y}{2(n-2)} \right)^{n-2} + \tau^2 \right) \right), \\ &\leq C \left(\frac{\| u^{(m+1)} \|_\infty}{\tau} \left(\frac{eh_x}{2(m-2)} \right)^{m-2} + \frac{\| u^{(n+1)} \|_\infty}{\tau} \left(\frac{eh_y}{2(n-2)} \right)^{n-2} + \tau^2 \right) \end{aligned} \tag{56}$$

where $C = \max\left\{C_5\frac{e^{\kappa T}}{\kappa}, C_5\frac{e^{\kappa T}}{\kappa} + C_6\right\}$. The proof is completed. □

5. Numerical Experiments

In this section, we will provide some numerical results for the NLS Equation (1) to test the high accuracy and efficiency of our scheme. For convenience, the error notations are given as follows,

$$\mathrm{E}_\infty = \| u_h - u_e \|_\infty, \tag{57}$$

$$\mathrm{E}_r = \frac{\| u_h - u_e \|_\infty}{\| u_e \|_\infty}, \tag{58}$$

where u_h and u_e denote the numerical solution and the exact solution, respectively. $\| \cdot \|_\infty$ is the L^∞ norm. All computations presented in this work were performed on a standard i5 Intel 1.8GHz laptop in MATLAB R2020b.

5.1. Example 1

This example is used to test the accuracy and convergence of our scheme. Considering the following 2D NLS equation on $[0, 2\pi]^2 \times (0, T]$,

$$\begin{cases} iu_t + \frac{1}{2}\Delta u - (1 - \sin^2 x \sin^2 y)u - |u|^2 u = 0 \\ u(x, y, 0) = \sin x \sin y \\ u(0, y, t) = u(2\pi, y, t) = 0 \\ u(x, 0, t) = u(x, 2\pi, t) = 0 \end{cases},$$

where the exact solution is in the following form:

$$u(x, y, t) = e^{-2it} \sin x \sin y.$$

To verify the accuracy and the convergence rate of the operator splitting scheme based on the barycentric Lagrange interpolation collocation method **MI**, the operator splitting scheme based on the barycentric rational interpolation collocation method [24] **MII** and the classical second-order finite difference scheme **SI**, we choose the simulation parameters $n = m$, $\tau = 0.001$ and $T = 1$.

The results are shown in Tables 1–4 and Figures 1–3. Tables 1–3 show the spatial errors of the three schemes. By comparing Tables 1 and 3, it can be seen that the **MI** scheme, based on the barycentric Lagrange interpolation collocation method in space, can achieve a higher accuracy using only 8×8 mesh points. However, for the same accuracy, the **SI** scheme, based on a second-order center difference method in space, requires more than 80×80 mesh points. Furthermore, by comparing Tables 1 and 2, it is easy to see that the **MI** scheme is slightly more efficient than the **MII** scheme. The comparison of the three schemes shows that the barycentric Lagrange interpolation collocation scheme can achieve higher accuracy with fewer points in space. In addition, the CPU time of the **MI** scheme is significantly reduced compared with the **SI** scheme and is similar to the **MII** scheme.

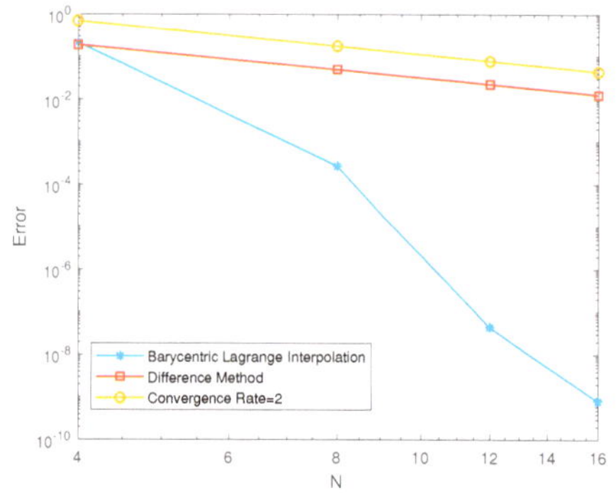

Figure 1. Spatial L^∞ errors at $t = 1$ for NLS equation for Example 1.

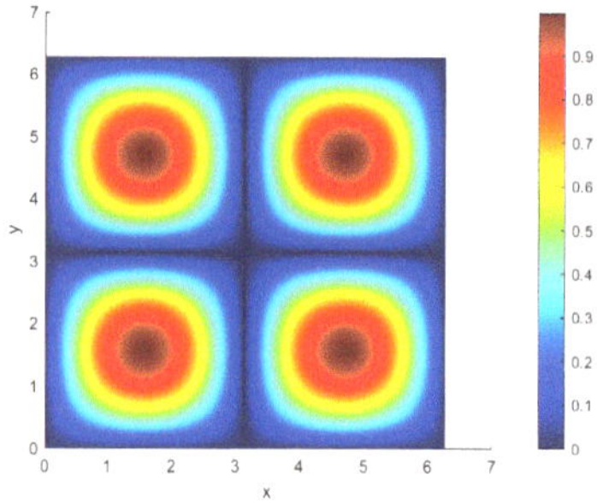

(**a**) The numerical solution diagram

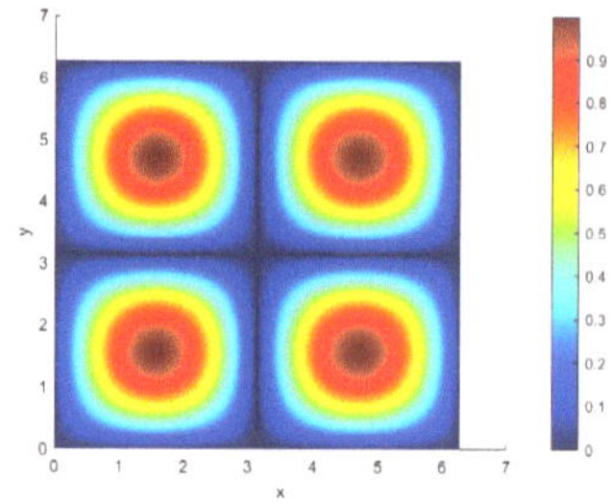

(**b**) The exact solution diagram

Figure 2. The numerical solution and the exact solution diagrams at $T = 1$ for Example 1.

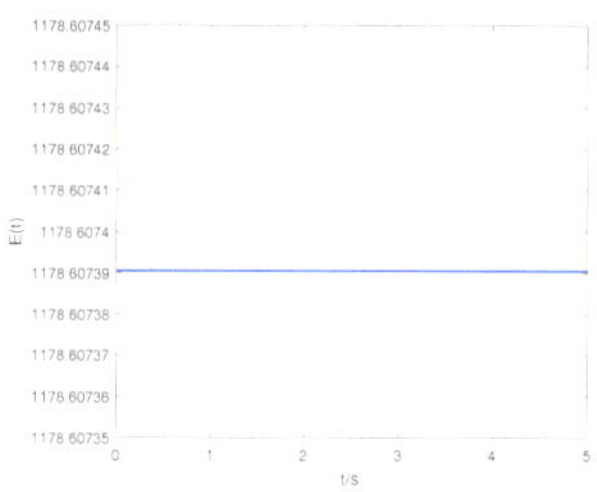

(**a**) Total energy

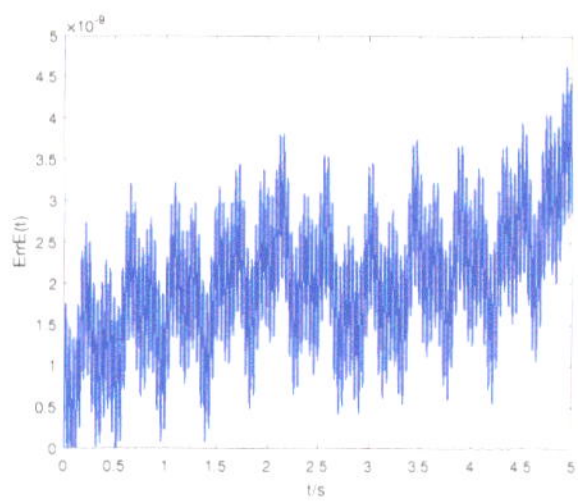

(**b**) Energy relative error

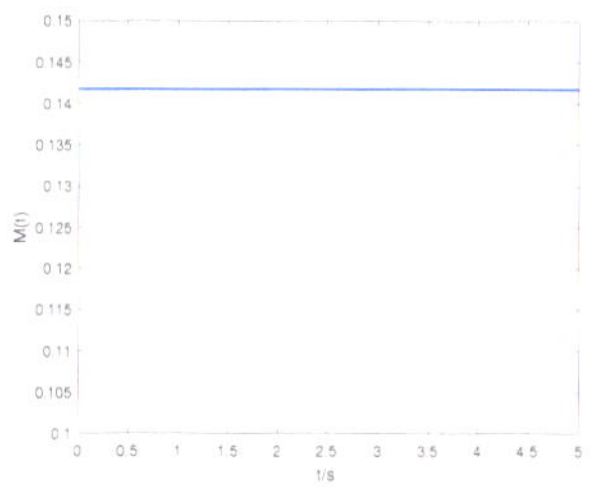

(**c**) Total mass

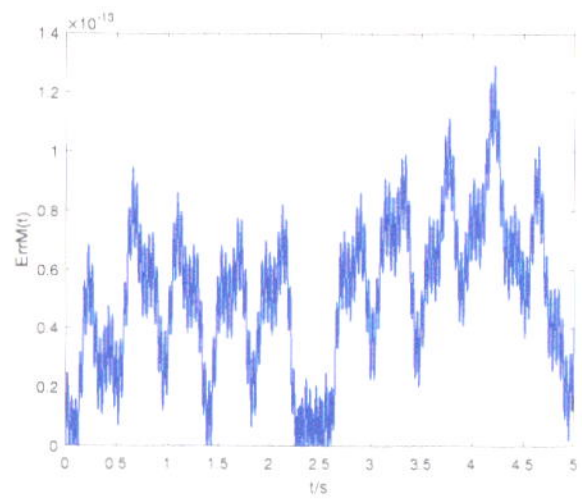

(**d**) Mass relative error

Figure 3. Conservation situation of energy and mass at $T = 1$ for Example 1.

Table 1. Error of barycentric Lagrange interpolation collocation scheme for Example 1.

m	E_∞	E_r	*CPU*
6	8.4010×10^{-3}	8.4010×10^{-3}	0.152 s
8	2.3542×10^{-4}	2.7054×10^{-4}	0.288 s
10	1.9318×10^{-6}	2.0865×10^{-6}	0.694 s
12	5.6755×10^{-8}	5.6755×10^{-8}	1.039 s
16	8.0816×10^{-8}	8.3330×10^{-8}	2.914 s

Table 2. Error of barycentric rational interpolation collocation method for Example 1.

m	E_∞	E_r	*CPU*
6	2.9302×10^{-3}	2.9302×10^{-3}	0.178 s
8	2.7122×10^{-3}	3.1167×10^{-3}	0.289 s
10	5.4257×10^{-6}	5.8602×10^{-6}	0.735 s
12	2.0948×10^{-6}	2.0948×10^{-6}	1.082 s
16	8.3994×10^{-8}	8.6607×10^{-8}	3.066 s

Table 3. Error of difference scheme for Example 1.

m	E_∞	E_r	*CPU*
10	2.9367×10^{-2}	3.2467×10^{-2}	0.202 s
20	8.1977×10^{-3}	8.1977×10^{-3}	3.586 s
40	2.0546×10^{-3}	2.0546×10^{-3}	160.534 s
60	9.1360×10^{-4}	9.1360×10^{-4}	1665.769 s
80	5.1402×10^{-4}	5.1402×10^{-4}	8993.522 s

Table 4. Errors and convergence rate in time for Example 1.

τ	E_∞	*Rate*	E_r	*Rate*
1/8	1.2598×10^{-3}	-	1.2990×10^{-3}	-
1/16	3.1552×10^{-4}	1.9975	3.2533×10^{-4}	1.9975
1/32	7.8913×10^{-5}	1.9994	8.1368×10^{-5}	1.9994
1/64	1.9731×10^{-5}	1.9998	2.0344×10^{-5}	1.9998

If we fix $m = n = 16$ and vary the temporal step, τ, we can obtain errors and temporal convergence rate, as shown in Table 2. It shows that the **MI** scheme based on the barycentric interpolation collocation scheme for the NLS equation has second-order accuracy in time. In addition, the spatial convergence rate is also obtained, and the L^∞ errors at time $T = 1$ for NLS equation are shown in Figure 1.

Choose $m = n = 20$; the numerical solution and the exact solution of the NLS equation are shown in Figure 2a,b. Moreover, it is necessary to verify the conservation of energy and mass. We plot images of mass and energy varying over time. From Figure 3, it is obvious that mass and energy are conserved.

5.2. Example 2

Consider the following 1D problem on $[-10, 40] \times (0, 9]$:

$$\begin{cases} iu_t + \Delta u + 2|u|^2 u = 0, \\ u(-10, t) = u(40, t) = 0, \\ u_0(x) = \operatorname{sech}(x)\exp(2ix) + \operatorname{sech}(x-30)\exp(-i(x-30)). \end{cases}$$

This example shows the collision behavior of two solitary waves. If we provide the initial value condition, a double solitary wave can be generated. When $t = 0$, two solitary waves are separated. The fast wave will catch up with the slow wave over time, and it will then surpass the slow wave after the collision, and there is only one phase change between them. This is consistent with the theory of waves; see Figures 4 and 5.

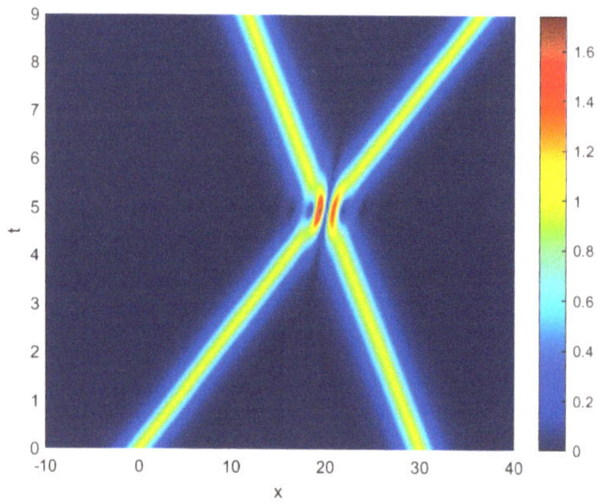

Figure 4. The interaction of two solitary waves without damping at $T = 9$ for Example 2.

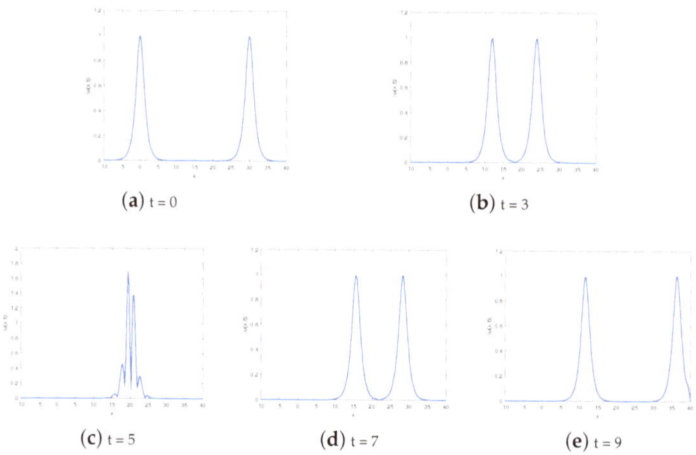

Figure 5. Head-on collisions of two solitary waves without damping for Example 2.

6. Conclusions

In this work, we have proposed an effective operator splitting scheme based on the barycentric Lagrange interpolation collocation method for the nonlinear Schrödinger equation. The convergence analysis is proved theoretically and verified numerically. Numerical examples are presented to show the mass and energy conservation of the proposed scheme. The operator splitting collocation scheme is second-order in time and convergent exponentially in space. The two barycentric interpolation collocation schemes have high accuracy, and the barycentric Lagrange interpolation collocation method is slightly more efficient than the barycentric rational interpolation collocation method. Compared with the finite difference method, the barycentric interpolation collocation method can achieve high accuracy with fewer points. In the future, we plan to extend this method to coupled Schrödinger equations, KdV equations and Klein–Gordon equations, etc.

Author Contributions: M.Y., methodology, conceptualization, design, analysis, software, original draft writing; Z.W., methodology, supervision, conceptualization, editing, draft editing and reviewing. All authors have read and agreed to the published version of the manuscript.

Funding: The research was funded by the Natural Science Foundation of Fujian Province (Nos. 2022J01308, 2021J01306), the NSF of China (No. 11701197), the Program for Innovative Research Teams in Science and Technology in Fujian Province University and the Fuzhou-Xiamen-Quanzhou National Independent Innovation Demonstration Zone Collaborative Innovation Platform under Grant 2022FX5.

Data Availability Statement: No new data were created or analyzed in this study.

Acknowledgments: We thank the anonymous referees for their careful checking and valuable comments which improve the manuscript.

Conflicts of Interest: The authors declare no conflicts of interest.

References

1. Schrödinger, E. The present status of quantum mechanics. *Sci. Nat.* **1935**, *23*, 1–26.
2. Gong, Y.Z.; Wang, Q.; Wang, Y.S.; Cai, J.X. A conservative Fourier pseudo-spectral method for the nonlinear Schrödinger equation. *J. Comput. Phys.* **2017**, *328*, 354–370. [CrossRef]
3. Cui, J.; Xu, Z.Z.; Wang, Y.S.; Jiang, C.L. Mass-and energy-preserving exponential Runge–Kutta methods for the nonlinear Schrödinger equation. *Appl. Math. Lett.* **2020**, *112*, 106770. [CrossRef]
4. Feng, X.B.; Li, B.Y.; Ma, S. High-order mass-and energy-conserving SAV-Gauss collocation finite element methods for the nonlinear Schrödinger equation. *SIAM J. Numer. Anal.* **2021**, *59*, 1566–1591. [CrossRef]
5. Wang, J.J.; Li, M.; Guo, L.J. Superconvergence analysis for nonlinear Schrödinger equation with two-grid finite element method. *Appl. Math. Lett.* **2021**, *122*, 107553. [CrossRef]
6. Wang, L.L.; Li, M. Galerkin finite element method for damped nonlinear Schrödinger equation. *Appl. Numer. Math.* **2022**, *178*, 216–247. [CrossRef]
7. Hu, H.Z.; Li, B.Y.; Zou, J. Optimal convergence of the Newton iterative Crank–Nicolson finite element method for the nonlinear Schrödinger equation. *Comput. Methods Appl. Math.* **2022**, *22*, 91–612. [CrossRef]
8. Chen, C.J.; Lou, Y.Z.; Hu, H.Z. Two-grid finite volume element method for the time-dependent Schrödinger equation. *Comput. Math. Appl.* **2022**, *108*, 185–195. [CrossRef]
9. Deng, B.C.; Shen, J.; Zhuang, Q.Q. Second-order SAV schemes for the nonlinear Schrödinger equation and their error analysis. *J. Sci. Comput.* **2021**, *69*, 88. [CrossRef]
10. Su, C.M.; Zhao, X.F. On time-splitting methods for nonlinear Schrödinger equation with highly oscillatory potential. *Math. Model. Numer. Anal.* **2020**, *54*, 1491–1508. [CrossRef]
11. Wang, T.C.; Zhao, X.F. Optimal L^∞ error estimates of finite difference methods for the coupled Gross–Pitaevskii equations in high dimensions. *Sci. China Math.* **2014**, *57*, 2189–2214. [CrossRef]
12. Li, J.; Qu, J.Z. Barycentric Lagrange interpolation collocation method for solving the Sine-Gordon equation. *Wave Motion* **2023**, *120*, 103159. [CrossRef]
13. Hu, Y.D.; Peng, A.; Chen, L.Q.; Tong, Y.L.; Weng, Z.F. Analysis of the barycentric interpolation collocation scheme for the Burgers equation. *Sci. Asia* **2021**, *47*, 758–765. [CrossRef]
14. Oruç, Ö. Two meshless methods based on local radial basis function and barycentric rational interpolation for solving 2D viscoelastic wave equation. *Comput. Appl. Math.* **2020**, *79*, 3272–3288. [CrossRef]
15. Deng, Y.F.; Weng, Z.F. Barycentric interpolation collocation method based on Crank–Nicolson scheme for the Allen–Cahn equation. *AIMS Math.* **2021**, *6*, 3857–3873. [CrossRef]
16. Deng,Y.F.; Weng, Z.F. Operator splitting scheme based on barycentric Lagrange interpolation collocation method for the Allen–Cahn equation. *J. Appl. Math. Comput.* **2022**, *68*, 3347–3365. [CrossRef]
17. Huang, R.; Weng, Z.F. A numerical method based on barycentric interpolation collocation for nonlinear convection-diffusion optimal control problems. *Netw. Heterog. Media* **2023**, *18*, 562–580. [CrossRef]
18. Yi, S.C.; Yao, L.Q. A steady barycentric Lagrange interpolation method for the 2D higher-order time-fractional telegraph equation with nonlocal boundary condition with error analysis. *Numer. Methods Partial Differ. Equ.* **2019**, *35*, 1694–1716. [CrossRef]
19. Lubich, C. On splitting methods for Schrödinger–Poisson and cubic nonlinear Schrödinger equations. *Math. Comput.* **2008**, *77*, 2141–2153. [CrossRef]
20. Zhai, S.Y.; Wang, D.L.; Weng, Z.F.; Zhao, X. Error analysis and numerical simulations of Strang splitting method for space fractional nonlinear Schrödinger equation. *J. Sci. Comput.* **2019**, *81*, 965–989. [CrossRef]
21. Zhai, S.; Weng, Z.; Zhuang, Q.; Liu, F.; Anh, V. An effective operator splitting method based on spectral deferred correction for the fractional Gray–Scott model. *J. Comput. Appl. Math.* **2023**, *425*, 114959. [CrossRef]
22. Berrut, J.P.; Trefethen, L.N. Barycentric Lagrange interpolation. *SIAM Rev.* **2004**, *46*, 501–507. [CrossRef]

23. Strang, G. On the construction and comparison of difference schemes. *SIAM J. Numer. Anal.* **1968**, *5*, 506–517. [CrossRef]
24. Li, J.; Cheng, Y.L. Linear barycentric rational collocation method for solving heat conduction equation. *Numer. Methods Partial Differ. Equ.* **2021**, *37*, 533–545. [CrossRef]

Mathematical and Computational Applications

MDPI

Article

Magnesium and Calcium Transport along the Male Rat Kidney: Effect of Diuretics

Pritha Dutta [1,*] and Anita T. Layton [1,2,3,4]

1 Department of Applied Mathematics, University of Waterloo, Waterloo, ON N2L 3G1, Canada; anita.layton@uwaterloo.ca
2 Cheriton School of Computer Science, University of Waterloo, Waterloo, ON N2L 3G1, Canada
3 Department of Biology, University of Waterloo, Waterloo, ON N2L 3G1, Canada
4 School of Pharmacy, University of Waterloo, Waterloo, ON N2L 3G1, Canada
* Correspondence: p7dutta@uwaterloo.ca

Abstract: Calcium (Ca^{2+}) and magnesium (Mg^{2+}) are essential for cellular function. The kidneys play an important role in maintaining the homeostasis of these cations. Their reabsorption along the nephron is dependent on distinct trans- and paracellular pathways and is coupled to the transport of other electrolytes. Notably, sodium (Na^+) transport establishes an electrochemical gradient to drive Ca^{2+} and Mg^{2+} reabsorption. Consequently, alterations in renal Na^+ handling, under pathophysiological conditions or pharmacological manipulations, can have major effects on Ca^{2+} and Mg^{2+} transport. One such condition is the administration of diuretics, which are used to treat a large range of clinical conditions, but most commonly for the management of blood pressure and fluid balance. While the pharmacological targets of diuretics typically directly mediate Na^+ transport, they also indirectly affect renal Ca^{2+} and Mg^{2+} handling through alterations in the electrochemical gradient. To investigate renal Ca^{2+} and Mg^2 handling and how those processes are affected by diuretic treatment, we have developed computational models of electrolyte transport along the nephrons. Model simulations indicate that along the proximal tubule and thick ascending limb, the transport of Ca^{2+} and Mg^{2+} occurs in parallel with Na^+, but those processes are dissociated along the distal convoluted tubule. We also simulated the effects of acute administration of loop, thiazide, and K-sparing diuretics. The model predicted significantly increased Ca^{2+} and Mg^{2+} excretions and significantly decreased Ca^{2+} and Mg^{2+} excretions on treatment with loop and K-sparing diuretics, respectively. Treatment with thiazide diuretics significantly decreased Ca^{2+} excretion, but there was no significant alteration in Mg^{2+} excretion. The present models can be used to conduct in silico studies on how the kidney adapts to alterations in Ca^{2+} and Mg^{2+} homeostasis during various physiological and pathophysiological conditions, such as pregnancy, diabetes, and chronic kidney disease.

Keywords: calcium homeostasis; magnesium homeostasis; electrolyte transport; kidney; renal transport

1. Introduction

The divalent cations, Ca^{2+} and Mg^{2+}, are important for various physiological processes. About 99% of the body's Ca^{2+} is stored in bones, where it forms a calcium-phosphate compound called hydroxyapatite [1]. The remaining 1% of body calcium plays an important role in various other physiological processes, such as cell signaling, both skeletal and smooth muscle contraction, and blood clotting [1]. Mg^{2+} plays a pivotal role in energy-demanding metabolic reactions, protein synthesis, ensuring membrane integrity, facilitating nervous tissue conduction, promoting neuromuscular excitability, regulating muscle contraction, influencing hormone secretion, and participating in intermediary metabolism. Nearly 99% of the body's Mg^{2+} is distributed within cells or stored in bone, with only a small fraction in circulation [2]. Tight regulation of the serum Ca^{2+} and Mg^{2+} concentrations is essential since too much or too little Ca^{2+} or Mg^{2+} can have dangerous, potentially fatal consequences. To maintain Ca^{2+} and Mg^{2+} balance, it is crucial to regulate the fluxes of Ca^{2+}

and Mg^{2+} among the primary organs involved in their regulation, namely the intestine, bone, and kidneys.

The kidneys play an important role in maintaining Mg^{2+} and Ca^{2+} homeostasis. The majority, ~60–70% of the filtered Ca^{2+}, is reabsorbed along the proximal tubule through the paracellular pathway [3]. By contrast, paracellular Mg^{2+} permeability in the proximal tubule is very low, and hence only 15–25% of the filtered Mg^{2+} is reabsorbed along this segment [3]. The majority of the filtered Mg^{2+} is reabsorbed along the cortical thick ascending limb (60–70%) paracellularly [3]. The paracellular fractional reabsorption of Ca^{2+} along the thick ascending limb is ~15–25% [3]. The distal convoluted tubule is the final segment that reabsorbs Mg^{2+}; hence, it plays an important role in fine-tuning urinary Mg^{2+} excretion. Approximately 5–10% of the filtered Mg^{2+} is reabsorbed transcellularly along the distal convoluted tubule, mediated by the transient receptor potential melastatin 6/7 (TRPM6/7) heteromeric complex on the apical membrane and the Na^+/Mg^{2+} exchanger on the basolateral membrane [3]. Approximately 5–10% of the filtered Ca^{2+} is reabsorbed transcellularly along the distal convoluted tubule and connecting tubule, mediated by the transient receptor potential vanilloid 5 (TRPV5) on the apical membrane, the Na^+/Ca^{2+} exchanger (NCX1), and plasma membrane Ca^{2+}-ATPase (PMCA) on the basolateral membrane [3]. Finally, ~2–5% of the filtered Mg^{2+} and Ca^{2+} are excreted through urine [3].

Our understanding of Ca^{2+} and Mg^{2+} handling within different segments of the nephron has been greatly advanced through micropuncture and microperfusion studies in rodent nephrons [4,5]. Furthermore, recent genetic studies have expanded our knowledge about the protein mediators of Ca^{2+} and Mg^{2+} transport [6]. Despite these advances, our understanding of the renal handling of these electrolytes remains incomplete. What fraction of the renal reabsorption goes through the transcellular versus paracellular pathway? To what extent is the renal transport in each nephron segment coupled to the transport of other electrolytes, e.g., Na^+, K^+, and Cl^-? To answer these questions, we developed a detailed computational model of epithelial transport of electrolytes and water along the nephrons in a male rat kidney and conducted simulations to predict the renal transport of Ca^{2+} and Mg^{2+} as well as other electrolytes and water under different physiological conditions.

Besides electrolyte and fluid homeostasis, the kidney also plays an essential role in maintaining normal blood pressure. For the management of blood pressure and fluid balance, diuretics are commonly prescribed. Although the pharmacological targets of diuretics directly affect Na^+ transport, they also indirectly affect renal Mg^{2+} and Ca^{2+} reabsorption through changes in the electrochemical gradient. How is renal Ca^{2+} and Mg^{2+} transport affected by the administration of diuretics? To answer this question, we simulate the effect of acute administration of three classes of diuretics—loop; thiazide; and K-sparing diuretics—on renal Mg^{2+} and Ca^{2+} transport and excretion.

2. Materials and Methods

We have previously developed epithelial cell-based computational models of transporter-mediated solute and water transport along the nephron of a rat kidney [7–10], focusing on the renal handling of Na^+, K^+, Ca^{2+}, glucose, and water in physiological and pathophysiological conditions. The superficial nephron model includes the proximal tubule, short descending limb, thick ascending limb, distal convoluted tubule, connecting tubule, and collecting duct segments. Each nephron segment is represented as a tubule lined by a layer of epithelial cells. The model tracks the transport of the following 17 solutes: Na^+, K^+, Cl^-, $HCO_3{}^-$, H_2CO_3, CO_2, NH_3, $NH_4{}^+$, $HPO_4{}^{2-}$, $H_2PO_4{}^-$, H^+, $HCO_2{}^-$, H_2CO_2, urea, glucose, Ca^{2+}, and Mg^{2+}. The segment and cell type determine the type and abundance of transporters found on the apical and basolateral membranes of the cell. Solutes and water may be transported across the epithelium by either moving across the apical and basolateral membranes in the transcellular pathway, mediated by specialized membrane transporters or channels, or via the paracellular pathway between neighboring cells. A schematic diagram for the model nephrons is shown in Figure 1.

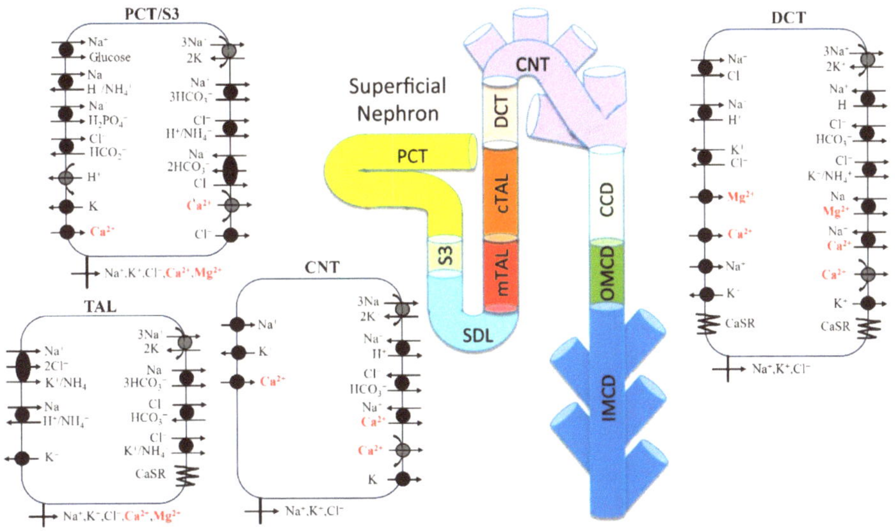

Figure 1. Model diagram of epithelial transport of Ca^{2+} and Mg^{2+} and selected electrolytes along the superficial nephron. Mg^{2+} transport occurs along the proximal tubule (proximal convoluted tubule, PCT, and S3), cortical thick ascending limb (cTAL), and distal convoluted tubule (DCT). Ca^{2+} transport occurs along the proximal tubule (proximal convoluted tubule, PCT, and S3), medullary/cortical thick ascending limb (mTAL/cTAL), distal convoluted tubule (DCT), and connecting tubule (CNT). Only the major Na^+, K^+, Cl^-, Ca^{2+}, and Mg^{2+} transporters are shown. PCT, proximal convoluted tubule; SDL, short descending limb; mTAL, medullary thick ascending limb; limb; CNT, connecting tubule; CCD, cortical collecting duct; OMCD, outer-medullary collecting duct; IMCD, inner-medullary collecting duct.

The model is defined by a large system of coupled differential and algebraic equations that describe mass conservation and determine transmembrane and paracellular fluxes [11]. The model predicts luminal fluid flow, hydrostatic pressure, membrane potential, luminal and cytosolic solute concentrations, transcellular and paracellular fluxes, urine volume, and urinary excretion rates of model solutes.

Below, we summarize a model representation of Mg^{2+} transport along the proximal tubule, cortical thick ascending limb, and distal convoluted tubule. Model parameters that describe Mg^{2+} transport are given in Table 1. The analogous model description and parameter for Ca^{2+} transport can be found in Ref. [10]. Additional model parameters can be found in Ref. [12].

2.1. Mg^{2+} Transport along the Proximal Tubule

The proximal tubule reabsorbs 15–25% of the filtered Mg^{2+} load through the paracellular pathway, which is mediated by claudin-2 and -12 [13] and is driven by the favorable electrochemical gradient established by Na^+/H^+ exchanger 3 (NHE3)-mediated Na^+ reabsorption [5,14,15]. Paracellular electro-diffusive Mg^{2+} flux ($J_{Mg}^{PT,LI}$) is given by

$$J_{Mg}^{PT,LI} = P_{Mg}^{PT,LI} \zeta_{Mg}^{LI} \left(\frac{C_{Mg}^{L} - C_{Mg}^{I} e^{-\zeta_{Mg}^{LI}}}{1 - e^{-\zeta_{Mg}^{LI}}} \right), \tag{1}$$

where the superscripts L and I denote lumen and lateral intercellular space (LIS), respectively; $P_{Mg}^{PT,LI}$ denotes the permeability of Mg^{2+} at the lumen and LIS interface; $\zeta_{Mg}^{LI} = \frac{Z_{Mg}F}{RT}(\psi^{L} - \psi^{I})$; Z_{Mg} is the valence of Mg^{2+} (+2); C_{Mg}^{L} and C_{Mg}^{I} denote Mg^{2+} concentrations in the lumen and LIS, respectively; ψ^{L} and ψ^{I} denote the luminal and LIS membrane potentials, respectively; **RT** = 2.57 J/mmol; and **F** = 96.5 C/mmol represents Faraday's constant.

2.2. Mg^{2+} Transport along the Thick Ascending Limb

The medullary thick ascending limb has negligible Mg^{2+} reabsorption [15–17]. In contrast, the cortical thick ascending limb reabsorbs 60–70% of the filtered Mg^{2+} via the paracellular route. That flux is mediated by claudins 16 and 19 [18], and is driven by the electrochemical gradient established by Na^+-K^+-Cl^- cotransporter 2 (NKCC2)-mediated Na^+ transport [9,14]. Paracellular Mg^{2+} transport in the cortical thick ascending limb $J_{Mg}^{cTAL,LI}$ is represented by an expression analogous to Equation (1), with the superscript "PT" replaced by "cTAL".

2.3. Mg^{2+} Transport along the Distal Convoluted Tubule

Unlike the proximal tubule and cortical thick ascending limb, where Mg^{2+} transport proceeds passively via the paracelluar route, the reabsorptive process along the distal convoluted tubule is active and transcellular and is mediated by the TRPM6 and TRPM7 (TRPM6/7) heteromeric channels, expressed on the apical membranes [19]. Mg^{2+} flux through TRPM6/7 is given by

$$J_{Mg}^{TRPM6/7} = N_{TRPM6/7} \times f_{Mg} \times f_{pH} \times \frac{\Delta\psi^{LC} - E_{Mg}^{LC}}{2F}, \tag{2}$$

f_{Mg} and f_{pH} describe the effects of intracellular Mg^{2+} concentration (C_{Mg}^{C}) and extracellular pH on TRPM6/7 [20], and are given by

$$f_{Mg} = \frac{1}{1 + \left(\frac{C_{Mg}^{C}}{0.51}\right)^{2}}. \tag{3}$$

$$f_{pH} = g_{pH7.4}\left(2 - \frac{1}{1 + \frac{7.4 - pH^{L}}{7.4 - pH_{1/2}}}\right), \tag{4}$$

where $g_{pH7.4}$ denotes the single channel conductance of TRPM6/7 channel at pH 7.4, pH^{L} denotes the luminal fluid pH, and $pH_{1/2}$ denotes the luminal fluid pH for half-maximal conductance.

Mg^{2+} efflux through the basolateral membrane is assumed to be mediated by an Na^+/Mg^{2+} exchanger [19,21], given by

$$J_{Mg}^{NaMgX} = J_{Mg}^{NaMgX,max} \times f_{NaS} \times f_{NaC} \times f_{MgS} \times f_{MgC}, \tag{5}$$

where $J_{Mg}^{NaMgX,max}$ denotes the maximum Mg^{2+} flux through the Na^+/Mg^{2+} exchanger, and the f terms represent regulation by extracellular Na^+, $f_{NaS} = \frac{(c_{Na}^{S})^{2}}{((c_{Na}^{S})^{2} + (K_{M,NaS})^{2})}$, intracellular Na^+, $f_{NaC} = \frac{K_{M,NaC}}{(c_{Na}^{C} + K_{M,NaC})}$, extracellular Mg^{2+}, $f_{MgS} = \frac{K_{M,MgS}}{(c_{Mg}^{S} + K_{M,MgS})}$, and intracellular Mg^{2+}, $f_{MgC} = \frac{(c_{Mg}^{C})^{2}}{((c_{Mg}^{C})^{2} + (K_{M,MgC})^{2})}$. In these expressions, $K_{M,NaS}$, $K_{M,NaC}$, $K_{M,MgS}$, and $K_{M,MgC}$ denote the Michaelis-Menten constants.

2.4. Calcium-Sensing Receptor

The calcium-sensing receptor (CaSR) regulates not only Ca^{2+} and Mg^{2+} reabsorption in the kidneys but other electrolytes and water as well by modifying transporter activities. Ca^{2+} is the primary ligand for activating CaSR. At equimolar concentrations, Mg^{2+} is 1/2 to 2/3 as potent as Ca^{2+} in activating CaSR [22,23]. We model the effect of CaSR on a given parameter v (v may denote paracellular permeability, NKCC2 activity, ROMK activity, or NCC activity; see below) with the following expression:

$$v = v^{*}\left(1 + \alpha_{v,Ca}\left(\frac{(C_{Ca}^{i})^{4}}{(C_{Ca}^{i})^{4} + (EC_{50,Ca})^{4}}\right)\right)\left(1 + \alpha_{v,Mg}\left(\frac{(C_{Mg}^{i})^{4}}{(C_{Mg}^{i})^{4} + (EC_{50,Mg})^{4}}\right)\right), \tag{6}$$

where v^* is the value of v in the absence of the effect of CaSR, c^i_{Ca} and c^i_{Mg} denote the concentration of Ca^{2+} and Mg^{2+} in the luminal (i = L) or interstitial (i = S) fluid, and $EC_{50,\mathrm{Ca}}$ = 1.25 mM and $EC_{50,\mathrm{Mg}}$ = 2.5 mM represent the half-maximal concentrations for Ca^{2+} and Mg^{2+} [24], respectively. CaSR is ubiquitously expressed in the kidney both along the apical and basolateral membranes, with its highest expression being at the basolateral membrane of the cortical thick ascending limb [25]. Hence, we represent v for (i) paracellular permeability, NKCC2 activity, and renal outer-medullary potassium channel (ROMK) in the thick ascending limb, (ii) Na^+-Cl^- cotransporter (NCC) activity in the distal convoluted tubule, (iii) H^+-ATPase flux in outer-medullary collecting duct type A cells, and (iv) water permeability in the inner-medullary collecting duct. The parameters $\alpha_{v,\mathrm{Ca}}$ and $\alpha_{v,\mathrm{Mg}}$ are negative if CaSR has an inhibitory effect on v and positive otherwise. Since the effect of Mg^{2+} on CaSR activation is ~50–66% of that of Ca^{2+}, we set $\alpha_{v,\mathrm{Mg}} = 0.6\alpha_{v,\mathrm{Ca}}$. The values for $\alpha_{v,\mathrm{Ca}}$ and $\alpha_{v,\mathrm{Mg}}$ for each of the segments are given in Table 1.

Table 1. Mg^{2+}-specific parameters for all the segments along the superficial nephron. Values marked (*) are adjusted. PT, proximal tubule; TAL, thick ascending limb; DCT, distal convoluted tubule; CD, collecting duct; OMCD, outer-medullary collecting duct; IMCD, inner-medullary collecting duct.

Parameter	Value
PT	
Tight junction permeability to Mg^{2+} at the lumen-LIS interface ($P^{\mathrm{PT,LI}}_{\mathrm{Mg}}$)	1.1×10^{-5} cm/s [26]
Reflection coefficient of tight junction to Mg^{2+}	0.89 (*)
cTAL	
Tight junction permeability at the lumen-LIS interface in the absence of Mg^{2+} ($P^{\mathrm{TAL,LI*}}_{\mathrm{Mg}}$)	38×10^{-5} cm/s (*)
Maximum half concentration of Ca^{2+} ($EC_{50,\mathrm{Ca}}$)	1.25 mM [24]
Maximum half concentration of Mg^{2+} ($EC_{50,\mathrm{Mg}}$)	2.5 mM [24]
Hill function coefficient, n	4 [24]
Inhibitory coefficient of Ca^{2+} on tight junction permeability ($\alpha_{\mathrm{P,Ca}}$)	−4/7 [27]
Inhibitory coefficient of Mg^{2+} on tight junction permeability ($\alpha_{\mathrm{P,Mg}}$)	−0.34 (*)
Inhibitory coefficient of Ca^{2+} on NKCC2 activity ($\alpha_{\mathrm{NKCC2,Ca}}$)	−0.4 [28]
Inhibitory coefficient of Mg^{2+} on NKCC2 activity ($\alpha_{\mathrm{NKCC2,Mg}}$)	−0.24 (*)
Inhibitory coefficient of Ca^{2+} on ROMK activity ($\alpha_{\mathrm{ROMK,Ca}}$)	−0.8 [29,30]
Inhibitory coefficient of Mg^{2+} on ROMK activity ($\alpha_{\mathrm{ROMK,Mg}}$)	−0.48 (*)
DCT	
TRPM6/7 channel density ($N_{\mathrm{TRPM6/7}}$)	26×10^4 cm^{-2} (*)
Single channel conductance of TRPM6/7 at pH 7.4 ($g_{\mathrm{pH7.4}}$)	56.6 pS [20]
Luminal pH for half-maximal conductance of TRPM6/7 ($\mathrm{pH}_{1/2}$)	5.5 [20]
Maximum Mg^{2+} flux through Na^+/Mg^{2+} exchanger ($J^{\mathrm{NaMgX,max}}_{\mathrm{Mg}}$)	8.6×10^{-9} mmol/cm^2/s (*)
Intracellular Mg^{2+} half-saturation constant ($K_{M,\mathrm{MgC}}$)	3.59 M [19,21]
Extracellular Mg^{2+} half-saturation constant ($K_{M,\mathrm{MgS}}$)	1.3 mM [19,21]
Intracellular Na^+ half-saturation constant ($K_{M,\mathrm{NaC}}$)	12.29 mM [19,21]
Extracellular Na^+ half-saturation constant ($K_{M,\mathrm{NaS}}$)	87.5 mM [19,21]
Excitatory coefficient of Ca^{2+} on NCC activity ($\alpha_{\mathrm{NCC,Ca}}$)	0.5 [31]
Excitatory coefficient of Mg^{2+} on NCC activity ($\alpha_{\mathrm{NCC,Mg}}$)	0.3 (*)
CD	
Ca^{2+} promoting coefficient for apical HATPase activity of type A OMCD cells ($\alpha_{\mathrm{HATP,Ca}}$)	2 [27]
Mg^{2+} promoting coefficient for apical HATPase activity of type A OMCD cells ($\alpha_{\mathrm{HATP,Mg}}$)	1.2 (*)
Ca^{2+} inhibitory coefficient for apical water permeability of IMCD cells ($\alpha_{\mathrm{Pf,Ca}}$)	−3/8 [27]
Mg^{2+} inhibitory coefficient for apical water permeability of IMCD cells ($\alpha_{\mathrm{Pf,Mg}}$)	−0.225 (*)

3. Results

3.1. Baseline Results

Using rat parameters, we computed the model's luminal fluid flow, luminal fluid solute concentrations, cytosolic solute concentrations, membrane potential, and fluxes. Figure 2 shows the predicted segmental delivery, transport, and luminal fluid concentration of Mg^{2+} and Ca^{2+} along the nephron. Results for other electrolytes and water can be found in Ref. [32].

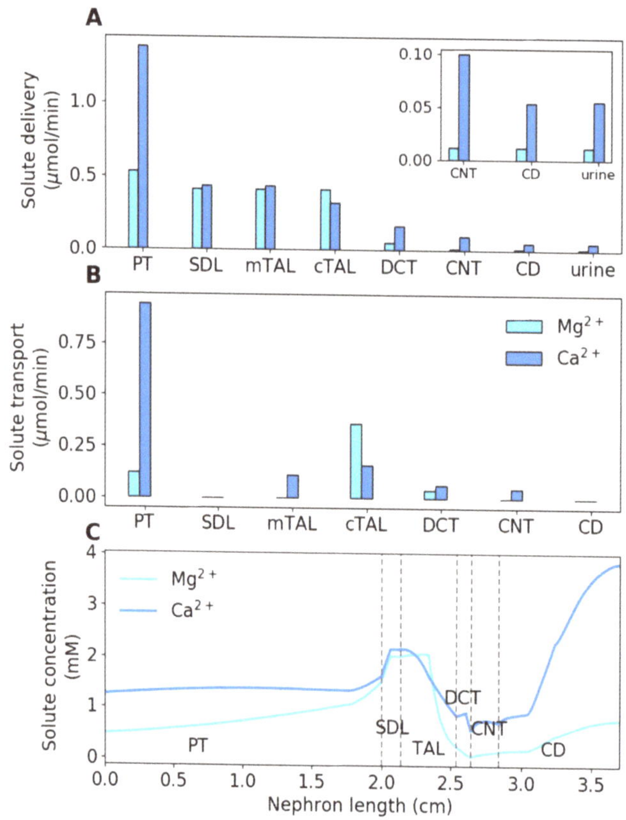

Figure 2. Baseline results. (**A**) Delivery of Mg^{2+} and Ca^{2+} to key nephron segments in male rats, given per kidney. (**B**) Mg^{2+} and Ca^{2+} transepithelial transport along key nephron segments in male rats, given per kidney. (**C**) Luminal Mg^{2+} and Ca^{2+} concentrations along key nephron segments in male rats. PT, proximal tubule; SDL, short descending limb; mTAL, medullary thick ascending limb; limb; cTAL, cortical thick ascending limb; DCT, distal convoluted tubule; CNT, connecting tubule; CD, collecting duct.

The majority of the filtered Ca^{2+} is reabsorbed along the proximal tubule, accounting for 68% of the filtered load; by contrast, Mg^{2+} reabsorption along this segment accounts for only 22% of the filtered load. Since Mg^{2+} reabsorption is low along the proximal tubule, Mg^{2+} concentration increases by ~3-fold [4]. The majority of the overall Mg^{2+} transport occurs downstream along the cortical thick ascending limb (68% of the filtered load), where the lumen-positive membrane potential drives Mg^{2+} reabsorption via the paracellular pathway. The fractional reabsorption of Ca^{2+} along the medullary and cortical thick ascending limb is 20%. The final nephron segment that transports Mg^{2+} is the distal convoluted tubule, where 6.6% of the filtered Mg^{2+} is reabsorbed. The fractional reabsorption of Ca^{2+} along the distal convoluted tubule and connecting tubule is 7.9%. Finally, fractional urinary Mg^{2+} and Ca^{2+} excretions are 3.2% and 3.9%, respectively.

3.2. Effect of Loop Diuretics

Loop diuretics inhibit NKCC2, which is expressed on the apical membrane of the thick ascending limb. We simulated the effect of acute administration of loop diuretics by inhibiting NKCC2 activity by 70%. We assumed that the NKCC2 inhibitor was administered for long enough to significantly impair the kidney's ability to generate an axial osmolality gradient. The cortical interstitial concentrations were assumed to remain unchanged. Since the concentrating mechanism of the outer medulla is significantly impaired following complete NKCC2 inhibition, the interstitial concentrations of Mg^{2+} and Ca^{2+} at the outer-inner medullary boundary are lowered to 0.77 mM (from a baseline value of 0.96 mM) and 2.0 mM (from a baseline value of 2.5 mM), respectively. At the papillary tip, the interstitial concentrations of Mg^{2+} and Ca^{2+} are reduced to 1.0 mM (from 1.54 mM) and 2.62 mM (from 4.0 mM), respectively. For changes in the interstitial concentrations of Na^+, K^+, Cl^-, and urea, refer to [9].

The predicted Mg^{2+} and Ca^{2+} transport along the thick ascending limb and distal tubules and urinary Mg^{2+} and Ca^{2+} excretions following NKCC2 inhibition in male rats are shown in Figure 3. Our NKCC2 inhibition simulations predicted the fractional Mg^{2+} reabsorption along the cortical thick ascending limb to decrease to 57% from the baseline fractional reabsorption of 69% (Figure 3). Administration of furosemide, a loop diuretic, to male mice increased TRPM6 mRNA expression by 30% [33]. Our model simulations predicted a 68% increase in TRPM6/7 channel activity to account for the 240% increase in Mg^{2+} excretion in male rats undergoing furosemide treatment [34]. Fractional Ca^{2+} reabsorption along the thick ascending limb decreased by 17% following NKCC2 inhibition (Figure 3). This resulted in urinary Ca^{2+} excretion increasing to 236% of the baseline excretion value (Figure 3).

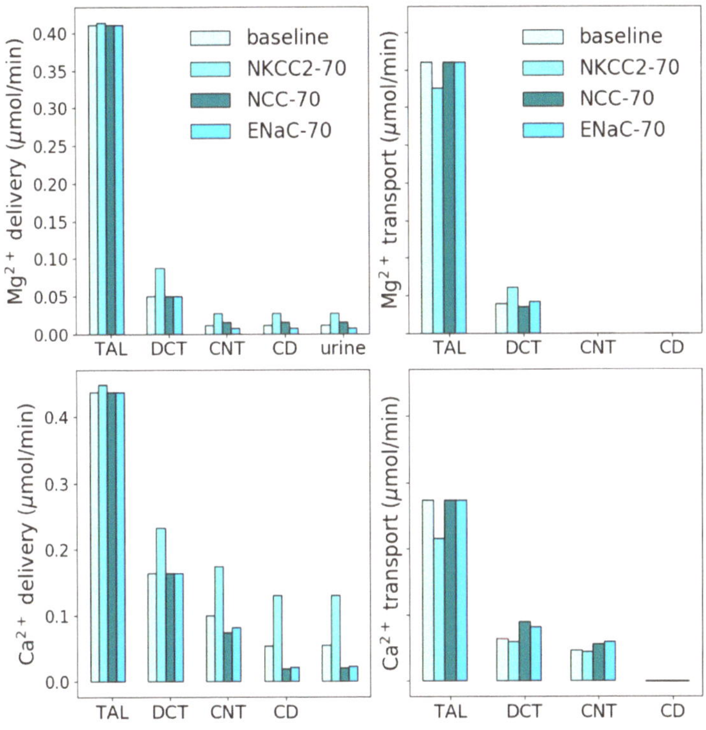

Figure 3. Effects of diuretic treatment. Delivery and transport of Mg^{2+} and Ca^{2+} along key nephron segments in male rats under normal conditions and 70% inhibition of NKCC2, NCC, and ENaC. The values are given per kidney. TAL, thick ascending limb; limb; DCT, distal convoluted tubule; CNT, connecting tubule; CD, collecting duct.

3.3. Thiazide Diuretics

Thiazide diuretics inhibit NCC, which is expressed along the apical membrane of the distal convoluted tubule. We simulated the effect of acute administration of thiazide diuretics by inhibiting NCC activity by 70%. In the NCC inhibition simulations, baseline interstitial concentration profiles were used.

The predicted fractional Mg^{2+} excretion after NCC inhibition increased to 3.1% from the baseline value of 2.8% (Figure 3). This is in agreement with experimental data where male rats treated with bendrofluazide, a thiazide diuretic, did not show any significant change in Mg^{2+} excretion [35]. Acute administration of chlorothiazide to male mice increased TRPV5 mRNA expression by 40–80% and decreased Ca^{2+} excretion by ~60% [36]. Accordingly, we increased TRPV5 activity by 52% following NCC inhibition in our model. This decreased the predicted Ca^{2+} excretion by 63% from the baseline excretion value (Figure 3).

3.4. K-Sparing Diuretics

K-sparing diuretics, such as amiloride, block Na^+ uptake through ENaC, expressed on the apical membrane of the late distal convoluted tubule as well as along the full length of the connecting tubule and collecting ducts. In our model, we simulated the effect of K-sparing diuretics by reducing ENaC activity by 70%.

ENaC inhibition hyperpolarizes the luminal membrane potential and increases K^+, Ca^{2+}, and Mg^{2+} uptake [37–40]. Our model simulations predicted Mg^{2+} reabsorption along the distal convoluted tubule to increase by 8.2% and Ca^{2+} reabsorption along the distal convoluted tubule and connecting tubule to increase by 29% (Figure 3). These increased reabsorptions decreased Mg^{2+} and Ca^{2+} excretions by 31% and 56%, respectively (Figure 3).

3.5. TRPM6/7 Inhibition

Kidney-specific TRPM6 [41] and TRPM7 [42] knock-out mice did not display hypomagnesemia and increased urinary Mg^{2+} excretion. This indicates that in mice, there must be other Mg^{2+} uptake mechanisms in the distal convoluted tubule. In fact, Verschuren et al. [43] reported that fluid shear stress (FSS) stimulated Mg^{2+} uptake in mDCT15 cells, and this uptake was independent of the TRPM6 and TRPM7 channels. However, the pathways or regulatory mechanisms for this FSS-sensitive Mg^{2+} uptake are unclear and hence not included in our present model.

To simulate TRPM6/7 knock-out experiments, we inhibited the TRPM6/7 channel by 100%. Our model predicted fractional Mg^{2+} excretion to increase to 8.8% (from baseline 3.2%), which is significantly above the 2–5% physiological fractional Mg^{2+} excretion. How much should the Mg^{2+} uptake through the FSS-sensitive Mg^{2+} channel be for fractional Mg^{2+} excretion to be within the physiological range when TRPM6/7 is completely inhibited? Our simulations predicted that if Mg^{2+} uptake through the FSS-sensitive channel is at least 60% of the Mg^{2+} uptake through TRPM6/7, then the fractional Mg^{2+} excretion becomes 4.3% (within the physiological range).

4. Discussion

Calcium (Ca^{2+}) and magnesium (Mg^{2+}) are both essential for cellular function. The homeostasis of these cations must be tightly regulated, and that balance is facilitated by intestinal absorption and renal excretion. For Na^+, Cl^-, K^+, Ca^{2+}, and many other major filtered solutes, most of the renal reabsorption occurs along the proximal convoluted tubule (about 1/2 to 2/3 in rats); the same is true for water. Thus, the luminal concentrations of these solutes, including Ca^{2+}, remain close to plasma along the proximal convoluted tubule. The majority of proximal tubule Ca^{2+} reabsorption occurs via a passive paracellular process, driven by Na^+ reabsorption mediated primarily by NHE3 and subsequent water reabsorption. In contrast, only 15–25% of the filtered Mg^{2+} load is reabsorbed along the proximal tubule. As a result, its concentration rises significantly along the proximal tubule. For Mg^{2+}, most of the reabsorption occurs along the cortical thick ascending limb (about 60–70%), while somewhat unexpectedly, essentially none occurs along the medullary thick ascending limb. Most of the remainder of the Mg^{2+} is reabsorbed along the distal convoluted tubule.

What difference does it make for the cortical thick ascending limb and distal convoluted tubule to handle most of the Mg^{2+} transport instead of the proximal tubule, as in the case of Na^{+} and Cl^{-}? Having these distal segments responsible for transporting a substantial fraction of the filtered Mg^{2+} load via the pathways that can be regulated may give the kidney a better ability to regulate Mg^{2+} balance. Recall that the plasma Mg^{2+} level is orders of magnitude lower than Na^{+} or Cl^{-}. Thus, to maintain plasma [Mg^{2+}] within a narrow range, the ability to fine-tune renal Mg^{2+} transport is particularly crucial. Parathyroid hormone, for instance, increases Mg^{2+} reabsorption in both the cortical thick ascending limb and distal convoluted tubule [44]. Transport of Mg^{2+} along these segments can also be regulated by hormones such as calcitonin, vasopressin, glucagon, and β-adrenergic agonists [45]. Coincidentally, some common diuretics also target these segments.

The goal of this study is to better understand the impact of diuretics on renal Ca^{2+} and Mg^{2+} transport. Diuretics are medications that reduce fluid buildup in the body and are often employed in the management of hypertension, edema, and various other conditions influenced by changes in electrolyte transport. In the context of kidney function, diuretics often focus on transport proteins or mechanisms vital for the reabsorption of Na^{+}, Cl^{-}, and water. Considering that renal Ca^{2+} and Mg^{2+} transports are driven primarily by the electrochemical gradients established by tubular NaCl transport processes, potential modifications in the renal handling of Ca^{2+} and Mg^{2+} by the administration of diuretics deserve scrutiny.

Loop diuretics, such as furosemide, induce notable natriuresis by targeting the thick ascending limb of the nephron; specifically, they inhibit NKCC2-dependent transport by competing for the chloride (Cl^{-}) binding site [46]. Given that the inhibition of NKCC2 reduces the lumen-positive transepithelial voltage gradient across the thick ascending limb epithelium [47], it is unsurprising that the transport of Mg^{2+} and Ca^{2+} in this segment is substantially decreased [48,49]. In fact, loop diuretics like furosemide induce hypercalciuria and hypermagnesuria in both experimental animals [50] and human subjects [51]. Thiazide diuretics induce a natriuretic response by inhibiting NCC and blocking NaCl transport in the distal convoluted tubule. A hypocalciuric effect has been reported following thiazide treatment [52]. Consistent with the drug's mode of action, among hypertensive individuals undergoing chronic thiazide treatment, there is a slight decrease in serum Mg^{2+} levels compared to those not taking diuretics [53], although the effect appears to be subtle. K-sparing diuretics function by inhibiting ENaC, a protein expressed in the late distal convoluted tubule, connecting tubule, and collecting duct of the kidney. Research has confirmed that K-sparing diuretics impact the urinary excretion of Ca^{2+} and Mg^{2+} in both human subjects and animals. Hypertensive individuals undergoing K-sparing diuretics treatment exhibit reduced urinary Ca^{2+} excretion [53] and elevated serum Mg^{2+} levels compared to those not receiving treatment [53]. The effect of these three classes of diuretics on urinary Mg^{2+} and Ca^{2+} excretions has been summarized in Figure 4.

The present study considers renal Ca^{2+} and Mg^{2+} transport under normal physiological conditions. The homeostasis of these electrolytes is altered during pregnancy, lactation, and dietary restriction, as well as in diseases such as diabetes and chronic kidney disease. To utilize the present model for in silico studies of how the kidney adapts in terms of Ca^{2+} and Mg^{2+} transport under these physiological and pathophysiological conditions, one can combine the model with computational models of kidney function for a pregnant rat [54,55], a diabetic rat [56–58], and a nephrectomized rat [8,59]. The resulting models may provide insights into altered renal Ca^{2+} and Mg^{2+} transport under these conditions. How do changes in renal Ca^{2+} and Mg^{2+} transport impact whole-body Ca^{2+} and Mg^{2+} homeostasis? A whole-body Ca^{2+} and Mg^{2+} balance model may help answer that question. By incorporating the present model into whole-body electrolyte balance models (e.g., [60–63]), one can obtain an integrative model to study whole-body Ca^{2+} and Mg^{2+} balance.

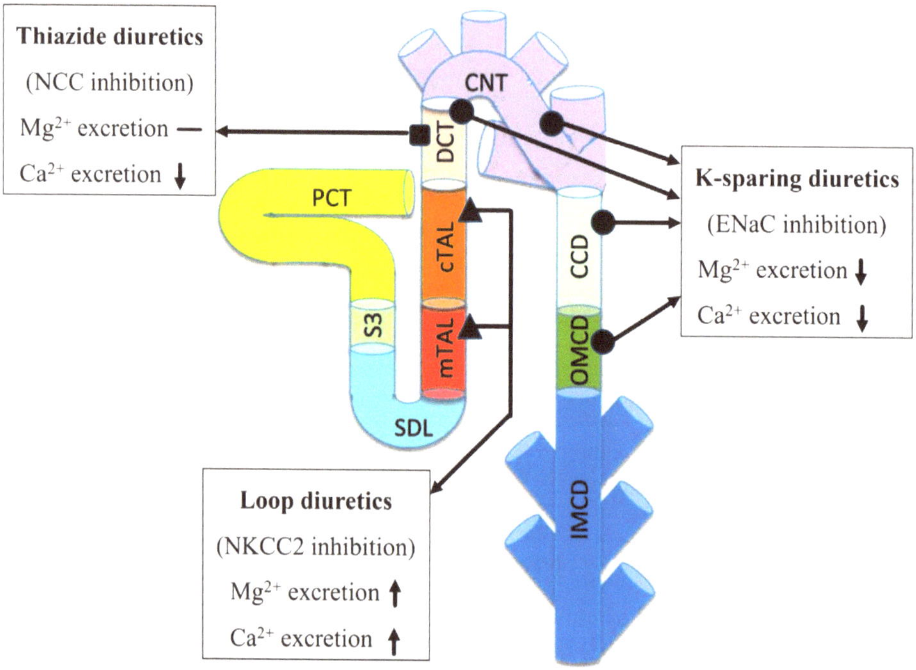

Figure 4. Summary of the effects of diuretic treatment. The effect of loop diuretics, thiazide diuretics, and K-sparing diuretics on Mg^{2+} and Ca^{2+} excretion The upward arrow indicates an increase, the downward arrow indicates a decrease, and the dash indicates no significant change. Notations are analogous to Figure 1.

The present model simulates electrolyte and water transport along a superficial nephron, which constitutes only 2/3 of the total nephron population in a rat kidney. The juxtamedullary nephrons, whose loops of Henle extend into various depths of the inner medulla, comprise the remaining nephron population. These two types of nephrons differ in the single-nephron glomerular filtration rate, transport area, and transporter activities. This study focuses on a superficial nephron model to gain a clearer understanding of segmental Mg^{2+} transport. In future studies, developing a kidney model that incorporates both types of nephrons will enable more accurate predictions of urinary excretion rates.

Another limitation of this model is that it does not include cyclin and CBS domain divalent metal cation transport mediator 2 (CNNM2), expressed on the basolateral membrane, which is also potentially involved in Mg^{2+} transport along the distal convoluted tubule. CNNM2 has been shown to mediate both cellular Mg^{2+} influx and efflux. The role of CNNM2 as an Mg^{2+} transporter has been openly debated (in favor [64]; opposing view [65]). A common view is that CNNM2 is not a Mg^{2+} transporter by itself but is an important protein for regulating Mg^{2+} homeostasis [65,66]. Due to these conflicting views on the role of CNNM2 as a Mg^{2+} transporter, we did not include it in our present model. CNNM2 can be incorporated into our model in the future when more consistent experimental outcomes are available.

Author Contributions: Conceptualization, A.T.L.; Methodology, P.D. and A.T.L.; Software, Validation, Formal Analysis, and Investigation, P.D.; Resources, A.T.L.; Data Curation, P.D.; Writing—Original Draft, P.D. and A.T.L.; Writing—Review and Editing, P.D. and A.T.L.; Visualization, P.D.; Supervision, P.D. and A.T.L.; Funding Acquisition, A.T.L. All authors have read and agreed to the published version of the manuscript.

Funding: This research was funded by the Canada 150 Research Chair program, the National Sciences and Engineering Research Council of Canada (NSERC) Discovery Grant (Grant number: RGPIN-2019-03916), and the Canada Institutes of Health Research (CIHR) Project Grant (Grant number: TNC-174963) (to A.T.L.).

Data Availability Statement: The code used for this study can be accessed at https://github.com/Pritha17/Nephron-Mg_Ca_transport (16 October 2023).

Conflicts of Interest: The authors declare no conflicts of interest.

References

1. Pu, F.; Chen, N.; Xue, S. Calcium intake, calcium homeostasis and health. *Food Sci. Hum. Wellness* **2016**, *5*, 8–16. [CrossRef]
2. Taal, M.W.; Chertow, G.M.; Marsden, P.A.; Skorecki, K.; Alan, S.; Brenner, B.M. *Brenner and Rector's the Kidney E-Book*; Elsevier Health Sciences: Amsterdam, The Netherlands, 2011; ISBN 1-4557-2304-5.
3. Alexander, R.T.; Dimke, H. Effect of diuretics on renal tubular transport of calcium and magnesium. *Am. J. Physiol. Ren. Physiol.* **2017**, *312*, F998–F1015. [CrossRef]
4. Brunette, M.; Vigneault, N.; Carriere, S. Micropuncture study of magnesium transport along the nephron in the young rat. *Am. J. Physiol. Leg. Content* **1974**, *227*, 891–896. [CrossRef] [PubMed]
5. Le Grimellec, C.; Giocondi, M.C.; Philippe, P. Micropuncture study along the proximal convoluted tubule electrolyte reabsorption in first convolutions. *Pflug. Arch.* **1975**, *354*, 133–150. [CrossRef] [PubMed]
6. Curry, J.N.; Yu, A.S.L. Magnesium Handling in the Kidney. *Adv. Chronic Kidney Dis.* **2018**, *25*, 236–243. [CrossRef] [PubMed]
7. Layton, A.T.; Layton, H.E. A Computational Model of Epithelial Solute and Water Transport along a Human Nephron. *PLoS Comput. Biol.* **2018**, *15*, e1006108. [CrossRef] [PubMed]
8. Layton, A.T.; Vallon, V. SGLT2 Inhibition in a Kidney with Reduced Nephron Number: Modeling and Analysis of Solute Transport and Metabolism. *Am. J. Physiol. Ren. Physiol.* **2018**, *314*, F969–F984. [CrossRef] [PubMed]
9. Layton, A.T.; Laghmani, K.; Vallon, V.; Edwards, A. Solute transport and oxygen consumption along the nephrons: Effects of Na^+ transport inhibitors. *Am. J. Physiol. Ren. Physiol.* **2016**, *311*, F1217–F1229. [CrossRef]
10. Hakimi, S.; Dutta, P.; Layton, A.T. Coupling of renal sodium and calcium transport: A modeling analysis of transporter inhibition and sex differences. *Am. J. Physiol. Ren. Physiol.* **2023**, *325*, F536–F551. [CrossRef]
11. Layton, A. A Complete Set of Equations for a Computational Model of Electrolyte and Water Transport along the Nephrons in a Mammalian Kidney. *bioRxiv* **2022**, bioRxiv:509286. [CrossRef]
12. Layton, A.T.; Vallon, V.; Edwards, A. A computational model for simulating solute transport and oxygen consumption along the nephrons. *Am. J. Physiol. Ren. Physiol.* **2016**, *311*, F1378–F1390. [CrossRef] [PubMed]
13. Alexander, R.T.; Dimke, H. Molecular mechanisms underlying paracellular calcium and magnesium reabsorption in the proximal tubule and thick ascending limb. *Ann. N. Y. Acad. Sci.* **2022**, *1518*, 69–83. [CrossRef] [PubMed]
14. Lee, H.Y.; Topham, D.J.; Park, S.Y.; Hollenbaugh, J.; Treanor, J.; Mosmann, T.R.; Jin, X.; Ward, B.M.; Miao, H.; Holden-Wiltse, J. Simulation and prediction of the adaptive immune response to influenza A virus infection. *J. Virol.* **2009**, *83*, 7151–7165. [CrossRef] [PubMed]
15. Wittner, M.; di Stefano, A.; Wangermann, P.; Nitschke, R.; Greger, R.; Bailly, C.; Amiel, C.; Roinel, N.; de Rouffignac, C. Differential effects of ADH on sodium, chloride, potassium, calcium and magnesium transport in cortical and medullary thick ascending limbs of mouse nephron. *Pflug. Arch.* **1988**, *412*, 516–523. [CrossRef] [PubMed]
16. Quamme, G.A. Control of magnesium transport in the thick ascending limb. *Am. J. Physiol. Ren. Physiol.* **1989**, *256*, F197–F210. [CrossRef]
17. Di Stefano, A.; Wittner, M.; Nitschke, R.; Braitsch, R.; Greger, R.; Bailly, C.; Amiel, C.; Roinel, N.; de Rouffignac, C. Effects of parathyroid hormone and calcitonin on Na^+, Cl^-, K^+, Mg^{2+} and Ca^{2+} transport in cortical and medullary thick ascending limbs of mouse kidney. *Pflug. Arch.* **1990**, *417*, 161–167. [CrossRef] [PubMed]
18. De Baaij, J.H.F.; Hoenderop, J.G.J.; Bindels, R.J.M. Magnesium in Man: Implications for Health and Disease. *Physiol. Rev.* **2015**, *95*, 1–46. [CrossRef] [PubMed]
19. Franken, G.A.C.; Adella, A.; Bindels, R.J.M.; de Baaij, J.H.F. Mechanisms coupling sodium and magnesium reabsorption in the distal convoluted tubule of the kidney. *Acta Physiol.* **2021**, *231*, e13528. [CrossRef]
20. Li, M.; Jiang, J.; Yue, L. Functional Characterization of Homo- and Heteromeric Channel Kinases TRPM6 and TRPM7. *J. Gen. Physiol.* **2006**, *127*, 525–537. [CrossRef]
21. Günther, T.; Vormann, J.; Höllriegl, V. Characterization of Na^+-dependent Mg^{2+} efflux from Mg^{2+}-loaded rat erythrocytes. *Biochim. Biophys. Acta BBA Biomembr.* **1990**, *1023*, 455–461. [CrossRef]
22. Ruat, M.; Snowman, A.M.; Hester, L.D.; Snyder, S.H. Cloned and Expressed Rat Ca^{2+}-sensing Receptor: Differential Cooperative Responses to Calcium and Magnesium. *J. Biol. Chem.* **1996**, *271*, 5972–5975. [CrossRef] [PubMed]
23. Bräuner-Osborne, H.; Jensen, A.A.; Sheppard, P.O.; O'Hara, P.; Krogsgaard-Larsen, P. The Agonist-Binding Domain of the Calcium-Sensing Receptor Is Located at the Amino-terminal Domain. *J. Biol. Chem.* **1999**, *274*, 18382–18386. [CrossRef] [PubMed]
24. Brown, E.M.; Chen, C.J. Calcium, magnesium and the control of PTH secretion. *Bone Miner.* **1989**, *5*, 249–257. [CrossRef] [PubMed]

25. Riccardi, D.; Valenti, G. Localization and function of the renal calcium-sensing receptor. *Nat. Rev. Nephrol.* **2016**, *12*, 414–425. [CrossRef] [PubMed]
26. Murayama, Y.; Morel, F.; Le Grimellec, C. Phosphate, calcium and magnesium transfers in proximal tubules and loops of henle, as measured by single nephron microperfusion experiments in the rat. *Pflug. Arch.* **1972**, *333*, 1–16. [CrossRef] [PubMed]
27. Edwards, A. Regulation of calcium reabsorption along the rat nephron: A modeling study. *Am. J. Physiol. Ren. Physiol.* **2015**, *308*, F553–F566. [CrossRef]
28. Toka, H.R.; Al-Romaih, K.; Koshy, J.M.; DiBartolo, S.I.; Kos, C.H.; Quinn, S.J.; Curhan, G.C.; Mount, D.B.; Brown, E.M.; Pollak, M.R. Deficiency of the Calcium-Sensing Receptor in the Kidney Causes Parathyroid Hormone–Independent Hypocalciuria. *J. Am. Soc. Nephrol.* **2012**, *23*, 1879. [CrossRef]
29. Wang, W.H.; Lu, M.; Hebert, S.C. Cytochrome P-450 metabolites mediate extracellular Ca^{2+}-induced inhibition of apical K^+ channels in the TAL. *Am. J. Physiol. Cell Physiol.* **1996**, *271*, C103–C111. [CrossRef]
30. Wang, W.; Lu, M.; Balazy, M.; Hebert, S.C. Phospholipase A2 is involved in mediating the effect of extracellular Ca^{2+} on apical K^+ channels in rat TAL. *Am. J. Physiol. Ren. Physiol.* **1997**, *273*, F421–F429. [CrossRef]
31. Bazúa-Valenti, S.; Rojas-Vega, L.; Castañeda-Bueno, M.; Barrera-Chimal, J.; Bautista, R.; Cervantes-Pérez, L.G.; Vázquez, N.; Plata, C.; Murillo-de-Ozores, A.R.; González-Mariscal, L.; et al. The Calcium-Sensing Receptor Increases Activity of the Renal NCC through the WNK4-SPAK Pathway. *J. Am. Soc. Nephrol.* **2018**, *29*, 1838. [CrossRef]
32. Hu, R.; McDonough, A.A.; Layton, A.T. Functional implications of the sex differences in transporter abundance along the rat nephron: Modeling and analysis. *Am. J. Physiol. Ren. Physiol.* **2019**, *317*, F1462–F1474. [CrossRef] [PubMed]
33. Van Angelen, A.A.; van der Kemp, A.W.; Hoenderop, J.G.; Bindels, R.J. Increased expression of renal TRPM6 compensates for Mg^{2+} wasting during furosemide treatment. *Nephrol. Dial. Transplant. Plus* **2012**, *5*, 535–544. [CrossRef]
34. Devane, J.; Ryan, M. Diuretics and magnesium excretion. *Magnes. Bull.* **1981**, *3*, 122–123.
35. Ryan, M.P.; Devane, J.; Ryan, M.F.; Counihan, T.B. Effects of diuretics on the renal handling of magnesium. *Drugs* **1984**, *28*, 167–181. [CrossRef] [PubMed]
36. Lee, C.-T.; Shang, S.; Lai, L.-W.; Yong, K.-C.; Lien, Y.-H.H. Effect of thiazide on renal gene expression of apical calcium channels and calbindins. *Am. J. Physiol. Ren. Physiol.* **2004**, *287*, F1164–F1170. [CrossRef] [PubMed]
37. Costanzo, L.S. Comparison of calcium and sodium transport in early and late rat distal tubules: Effect of amiloride. *Am. J. Physiol. Ren. Physiol.* **1984**, *246*, F937–F945. [CrossRef]
38. Devane, J.; Ryan, M.P. The effects of amiloride and triamterene on urinary magnesium excretion in conscious saline-loaded rats. *Br. J. Pharmacol.* **1981**, *72*, 285. [CrossRef]
39. Devane, J.; Ryan, M.P. Dose-dependent reduction in renal magnesium clearance by amiloride during frusemide-induced diuresis in rats. *Br. J. Pharmacol.* **1983**, *80*, 421. [CrossRef]
40. Friedman, P.A.; Gesek, F.A. Stimulation of calcium transport by amiloride in mouse distal convoluted tubule cells. *Kidney Int.* **1995**, *48*, 1427–1434. [CrossRef]
41. Chubanov, V.; Ferioli, S.; Wisnowsky, A.; Simmons, D.G.; Leitzinger, C.; Einer, C.; Jonas, W.; Shymkiv, Y.; Bartsch, H.; Braun, A.; et al. Epithelial magnesium transport by TRPM6 is essential for prenatal development and adult survival. *eLife* **2016**, *5*, e20914. [CrossRef]
42. Mittermeier, L.; Demirkhanyan, L.; Stadlbauer, B.; Breit, A.; Recordati, C.; Hilgendorff, A.; Matsushita, M.; Braun, A.; Simmons, D.G.; Zakharian, E.; et al. TRPM7 is the central gatekeeper of intestinal mineral absorption essential for postnatal survival. *Proc. Natl. Acad. Sci. USA* **2019**, *116*, 4706–4715. [CrossRef]
43. Verschuren, E.H.; Hoenderop, J.G.; Peters, D.J.; Arjona, F.J.; Bindels, R.J. Tubular flow activates magnesium transport in the distal convoluted tubule. *FASEB J.* **2019**, *33*, 5034–5044. [CrossRef]
44. Bailly, C.; Roinel, N.; Amiel, C. PTH-like glucagon stimulation of Ca and Mg reabsorption in Henle's loop of the rat. *Am. J. Physiol. Ren. Physiol.* **1984**, *246*, F205–F212. [CrossRef]
45. Quamme, G.A.; de Rouffignac, C. Renal magnesium handling. In *The Kidney: Physiology and Pathophysiology*; Lippincott Williams & Wilkins: Philadelphia, PA, USA, 2000; p. 375.
46. O'Grady, S.M.; Palfrey, H.C.; Field, M. Characteristics and functions of Na-K-Cl cotransport in epithelial tissues. *Am. J. Physiol. Cell Physiol.* **1987**, *253*, C177–C192. [CrossRef]
47. Burg, M.B. Tubular chloride transport and the mode of action of some diuretics. *Kidney Int.* **1976**, *9*, 189–197. [CrossRef] [PubMed]
48. Di Stefano, A.; Roinel, N.; de Rouffignac, C.; Wittner, M. Transepithelial Ca^{2+} and Mg^{2+} Transport in the Cortical Thick Ascending Limb of Henle's Loop of the Mouse Is a Voltage-Dependent Process. *Ren. Physiol. Biochem.* **2008**, *16*, 157–166. [CrossRef]
49. Quamme, G.A. Effect of furosemide on calcium and magnesium transport in the rat nephron. *Am. J. Physiol. Ren. Physiol.* **1981**, *241*, F340–F347. [CrossRef] [PubMed]
50. Duarte, C.G. Effects of ethacrynic acid and furosemide on urinary calcium, phosphate and magnesium. *Metabolism* **1968**, *17*, 867–876. [CrossRef]
51. Leary, W.P.; Reyes, A.J.; Wynne, R.D.; van der Byl, K. Renal Excretory Actions of Furosemide, of Hydrochlorothiazide and of the Vasodilator Flosequinan in Healthy Subjects. *J. Int. Med. Res.* **1990**, *18*, 120–141. [CrossRef] [PubMed]
52. Brickman, A.S.; Massry, S.G.; Coburn, J.W. Changes in Serum and Urinary Calcium during Treatment with Hydrochlorothiazide: Studies on Mechanisms. *J. Clin. Investig.* **1972**, *51*, 945–954. [CrossRef] [PubMed]

53. Kroenke, K.; Wood, D.R.; Hanley, J.F. The Value of Serum Magnesium Determination in Hypertensive Patients Receiving Diuretics. *Arch. Intern. Med.* **1987**, *147*, 1553–1556. [CrossRef]
54. Stadt, M.M.; Layton, A.T. Adaptive changes in single-nephron GFR, tubular morphology, and transport in a pregnant rat nephron: Modeling and analysis. *Am. J. Physiol. Ren. Physiol.* **2022**, *322*, F121–F137. [CrossRef]
55. Stadt, M.M.; West, C.A.; Layton, A.T. Effect of pregnancy and hypertension on kidney function in female rats: Modeling and functional implications. *PLoS ONE* **2023**, *18*, e0279785. [CrossRef]
56. Layton, A.T.; Vallon, V.; Edwards, A. Predicted consequences of diabetes and SGLT inhibition on transport and oxygen consumption along a rat nephron. *Am. J. Physiol. Ren. Physiol.* **2016**, *310*, F1269–F1283. [CrossRef] [PubMed]
57. Hu, R.; Layton, A. A computational model of kidney function in a patient with diabetes. *Int. J. Mol. Sci.* **2021**, *22*, 5819. [CrossRef]
58. Swapnasrita, S.; Carlier, A.; Layton, A.T. Sex-specific computational models of kidney function in patients with diabetes. *Front. Physiol.* **2022**, *13*, 741121. [CrossRef]
59. Layton, A.T.; Edwards, A.; Vallon, V. Renal potassium handling in rats with subtotal nephrectomy: Modeling and Analysis. *Am. J. Physiol. Ren. Physiol.* **2017**, *314*, F643–F657. [CrossRef]
60. Stadt, M.M.; Layton, A.T. Mathematical modeling of calcium homeostasis in female rats: An analysis of sex differences and maternal adaptations. *J. Theor. Biol.* **2023**, *572*, 111583. [CrossRef] [PubMed]
61. Stadt, M.M.; Leete, J.; Devinyak, S.; Layton, A.T. A mathematical model of potassium homeostasis: Effect of feedforward and feedback controls. *PLoS Comput. Biol.* **2022**, *18*, e1010607. [CrossRef] [PubMed]
62. Leete, J.; Layton, A.T. Sex-specific long-term blood pressure regulation: Modeling and analysis. *Comput. Biol. Med.* **2019**, *104*, 139–148. [CrossRef]
63. Ahmed, S.; Layton, A.T. Sex-specific computational models for blood pressure regulation in the rat. *Am. J. Physiol. Ren. Physiol.* **2020**, *318*, F888–F900. [CrossRef] [PubMed]
64. Funato, Y.; Furutani, K.; Kurachi, Y.; Miki, H. CrossTalk proposal: CNNM proteins are Na^+/Mg^{2+} exchangers playing a central role in transepithelial Mg^{2+} (re)absorption. *J. Physiol.* **2018**, *596*, 743. [CrossRef] [PubMed]
65. Arjona, F.J.; de Baaij, J.H. CrossTalk opposing view: CNNM proteins are not Na^+/Mg^{2+} exchangers but Mg^{2+} transport regulators playing a central role in transepithelial Mg^{2+} (re)absorption. *J. Physiol.* **2018**, *596*, 747. [CrossRef] [PubMed]
66. Sponder, G.; Mastrototaro, L.; Kurth, K.; Merolle, L.; Zhang, Z.; Abdulhanan, N.; Smorodchenko, A.; Wolf, K.; Fleig, A.; Penner, R.; et al. Human CNNM2 is not a Mg^{2+} transporter per se. *Pflug. Arch. Eur. J. Physiol.* **2016**, *468*, 1223–1240. [CrossRef]

Mathematical and Computational Applications

MDPI

Article

Modeling of Chemical Vapor Infiltration for Fiber-Reinforced Silicon Carbide Composites Using Meshless Method of Fundamental Solutions

Patrick Mahoney and Alex Povitsky *

Department of Mechanical Engineering, College of Engineering, University of Akron, Akron, OH 44325, USA; pwmahoney79@yahoo.com
* Correspondence: povitsky@uakron.edu

Abstract: In this study, the Method of Fundamental Solutions (MFSs) is adopted to model Chemical Vapor Infiltration (CVI) in a fibrous preform. The preparation of dense fiber-reinforced silicon carbide composites is considered. The reaction flux at the solid surface is equal to the diffusion flux towards the surface. The Robin or third-type boundary condition is implemented into the MFS. From the fibers' surface concentrations obtained by MFS, deposition rates are calculated, and the geometry is updated at each time step, modeling the pore filling over time. The MFS solution is verified by comparing the results to a known analytical solution for a simplified geometry of concentric cylinders with a concentration set at the outer cylinder and a reaction at the inner cylinder. MFS solutions are compared to published experimental data. Porosity transients are obtained by a combination of MFSs with surface deposition to show the relation between the initial and final porosities.

Keywords: chemical vapor infiltration; pores; reinforced silicon carbide (SiC) composites; preforms of fibers; method of fundamental solutions; diffusion equation

1. Introduction

Chemical Vapor Infiltration (CVI) was developed as a method to densify porous graphite bodies [1]. The CVI process has been widely used in the development of composite materials; more than half of all carbon/carbon composites are made by CVI [2]. The composites produced are often highly heat- and wear-resistant, and, as a result, they have been used in the industry [3] as a key enabling technology to reduce fuel consumption and emissions in gas turbine engines [4], heat exchangers and furnaces [5]. CVI is preferable to other methods of ceramic processing because it is a low-stress, low-temperature, and relatively easy-to-control process. There are many different CVI processes used in the industry today to deposit the desired solid product on preform surfaces [6].

At its basic form, CVI involves introducing reactants into a porous preform. The reactants then form precursors, which are deposited on the internal surfaces of the preform, causing the internal surfaces to grow and the solid portion of the preform to become thicker. As fibers grow and the preform becomes increasingly infiltrated, some areas of the geometry are closed off to reactant gas. This leads to voids forming in the geometry and increased porosity when infiltration is completed. Accounting for these voids, which could be detrimental to the quality of manufactured products, is imperative in modeling the CVI.

The modeling of CVI processes is important to predict infiltration, including whether voids may form and their location. The algorithmic goal of this paper is to introduce the Method of Fundamental Solutions (MFSs) to model the CVI of representative nests of fibers representing a portion of an actual preform. The objective of the present study is to implement the Robin or third-type boundary condition to the MFS, which is typically solved for Dirichlet boundary conditions. The main idea of MFSs consists of representing the solution to the problem as a linear combination of the fundamental solutions with respect to source points located outside the domain and particular boundary conditions at collocation points. Then, the initial problem is reduced to the determination of unknown coefficients of the linear combination [7]. As a fundamental solution, it automatically satisfies the governing equation, and only the boundary conditions need to be satisfied [8]. The MFS has substantial advantages over traditional finite-volume and finite-difference mesh-based methods because the MFS does not require cumbersome entire 3-D domain meshing. Instead, MFS requires the placement of fundamental solutions (singularities) only at the boundaries of the considered geometry.

Numerical experiments show that, in contrast to other boundary element methods (BEM), the MFS [9] requires relatively few boundary points and singularities to produce accurate results [10]. The CVI is modelled in the present study by solving the diffusion equation with a surface chemical reaction, using Green's functions for the Laplace operator to obtain the concentration of the reagent throughout the domain. From the concentrations at the fiber surfaces, the fiber growth rate can be calculated, and then the geometry is updated to reflect the calculated fiber growth. During the simulated CVI process, the fiber diameters grow, and the domain is filled with larger-size fibers. When the domain is fully infiltrated by CVI, the MFS solution reaches a steady state. The MFS has not been used to model CVI in prior studies, to the best of the author's knowledge. Past studies were concerned with the use of BEM and related methods for potential theory [11], elasticity and acoustics [12] and multiphase transport in unsaturated swelling porous media [13].

The simplified surface chemical reaction—which is discussed in this paper—is a one-step isothermal irreversible surface reaction involving the transformation of decomposed methyl-trichlorosilane (CH_3SiCl_3, abbreviated as MTS) to solid silicon carbide (SiC) [14–16]. The recent Ref [16] that models the CVI at a macro-scale reaction surface does not predict changes in porosity at the scale of individual fibers.

There are several methods used to transport reactants in a CVI process. These methods are classified by whether they use forced convection or diffusion and whether thermal gradients are imposed [17]. The most widely used CVI is isobaric and isothermal, for which transport occurs entirely by diffusion [6,18]. This type of CVI is modeled in the current study. For an isothermal CVI process, near-surface pores tend to close early in the process, restricting gas flow to the interior surfaces [19]. To counteract this effect, machining and re-infiltration may be necessary to ensure the desired density—this results in a longer production time and more expensive end products. Therefore, the computational prediction of voids and porosity is important for CVI practitioners.

CVI processes are generally slow—taking on the order of days to infiltrate—where fiber growth is on the order of microns per hour. As a result of the time scales of the reaction and the growth rates in the CVI processes, a quasi-steady approach is used in the current study to model infiltration. Most current models are based on the pseudo-steady-state hypothesis [20]—mass and momentum transfer are very rapid compared to changes in pore geometry. This quasi-steady approach, which is used in the present study, involves solving the steady-state Laplacian for concentration at every time step and then updating the geometries based on the calculated growth rates and time step increment.

Due to the complex geometries of the fibrous preforms which are infiltrated, simplified models of the considered processes have been developed. Many of these models are one-dimensional transient pore models. After these simplifications are made, many of these models can be solved analytically. An isothermal CVI reaction was modeled in the frame of the 1-D approach in [21]. A tapered pore was considered in [22]. The analytical solutions to the simplified pore models are limited to the infiltration of a singular round pore [6,23]. In terms of modeling CVI, MFS offers a clear advantage in the broad range of geometries, which can be handled. Using MFS, an arbitrary fiber cross-section is computed in this study.

In Ref [20], the modeling of the CVI process involves understanding the following three steps: (1) the mass transport of the reagent to the fibrous substrate, (2) the deposition kinetics of the solid matrix, and (3) the change in the preform structure that accompanies deposition onto the fibers. The first and second steps are well understood and can be modeled reasonably well. The third step (i.e., the description of the change in the pore structure and surface area) is the most difficult to determine. This step is addressed in the current study.

In Ref [24], the same CVI process is considered in which the MTS dissociates and deposits silicon carbide (SiC) at the solid fiber surface. The Level-Set Method (LSM) [24], based on the solution of the Hamilton–Jacobi level set equation [25], is different from MFS in that it assumes a constant deposition front speed. The advantage of using MFS is that it does not require an assumption about the uniformity of the deposition rate.

The numerical objective of the present study is to implement the MFS derivatives of unknown concentrations of CVI feedstock gas to facilitate the Robin boundary condition at the deposition surface. From the fibers' surface concentration of the reagent obtained by MFS, deposition rates are calculated, and the geometry is updated at each time step; therefore, the pore filling is modeled over time. The MFS solution is verified by comparison to a known analytical solution for a simplified geometry of concentric cylinders with a concentration set at the outer cylinder and a reaction at the inner cylinder. Finally, the MFS solution is compared to published CVI experimental data to demonstrate the practicality of the developed algorithm.

This paper is composed as follows: Section 2 includes governing equations and the MFS methodology, as well as the verification of the MFS methodology by comparing it to an analytical solution for diffusion between cylindrical surfaces and deposition at one of the surfaces; next, Section 3 discusses the obtained computational results and provides a qualitative comparison to the experimental results; finally, Section 4 provides analytical conclusions as well as areas for the further application and development of this study.

2. Theoretical Model and Numerical Methodology

A 3-D fiber preform of straight fibers is approximated by parallel long cylinders of a given initial diameter in contact with feedstock gas (see Figure 1). The one-step isothermal irreversible surface reaction of the transformation of decomposed MTS to solid silicon carbide is assumed. The densification of the set of fibers by CVI is a slow process. The fiber size and the geometry of fibers change because of deposition, where the deposited film thickness is updated after every time step; however, the changes in geometry are small per time step. Instead of modeling the actual transient process, a quasi-steady approach is taken where the change in reagent concentration over each time step is treated by a steady state analysis.

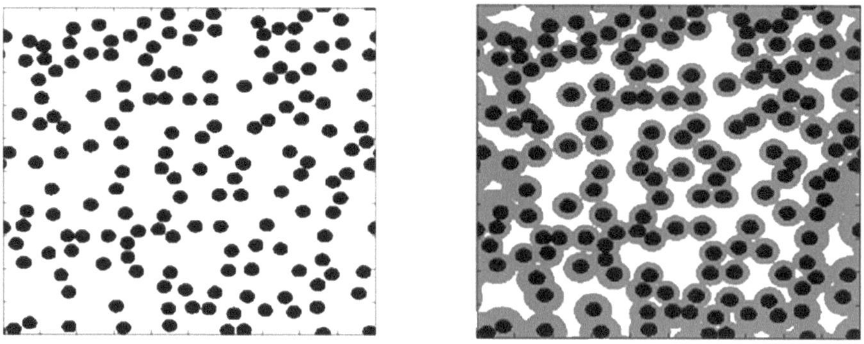

(a) (b)

Figure 1. Infiltration for randomly placed fibers with aligned axis: (**a**) initial geometry of fibers and (**b**) final geometry after the deposition process has concluded. The deposited layer (grey) and fibers (black) are shown.

In CVI modeling, it is necessary to solve the diffusion equation to obtain the concentration of species in the computational domain as well as the surface chemical reaction and growth rate resulting from the deposited material. The distribution of species concentration is described using the following concentration transport equation:

$$\frac{\partial c}{\partial t}+u\frac{\partial c}{\partial x}+v\frac{\partial c}{\partial y}+w\frac{\partial c}{\partial z}=D\left[\frac{\partial^2 c}{\partial x^2}+\frac{\partial^2 c}{\partial y^2}+\frac{\partial^2 c}{\partial z^2}\right], \tag{1}$$

where D is the diffusion coefficient.

Since the concentration update at each time step is obtained by steady-state analysis, the term $\frac{\partial c}{\partial t}$ can be eliminated in the above equation. Also, in this study, geometries involving fibers in cross-section $(x,\ y)$ are considered. These geometries of fibers are assumed to be uniform along the third z-axis, and as a result, the $\frac{\partial c}{\partial z}$ terms can be eliminated in the above equation.

For the considered CVI geometry, the characteristic length, L, is small, and the characteristic velocity, $\overline{u}$, is also small. This results in a very small Peclet number:

$$Pe=\frac{\overline{u}\cdot L}{D}\ll 1 \tag{2}$$

The advective effects are negligible in the CVI process due to the very small Peclet numbers involved, and Equation (1) is further reduced to the following:

$$0=\frac{1}{Pe}\left[\frac{\partial^2 c}{\partial x^2}+\frac{\partial^2 c}{\partial y^2}\right]. \tag{3}$$

Equation (3) is a Laplace operator, which can be solved using Green's functions with boundary singularities. Using Green's function for the Laplace operator in two dimensions, the solution is given by [26]

$$c_j=\frac{1}{2\pi}\sum_{i=1}^{N}F_i\ln\left(r_{i,j}\right), \tag{4}$$

where N is the number of observation points and singularities, F_i represents the strength of the i-th singularity, i represents the index (location) of singularity, j represents the index (location) of the observation point, c represents the concentration at the observation point, and positive $r_{i,j}$ is the distance between the observation point j and location of the singularity, i.

To find the unknown strength of singularity, F, solving a linear system of equations for the number of singularities, $1\le I\le N$, and observation points, $1\le j\le N$, is the same and equal to N. The linear system based on Equation (4) for known concentrations in observation points is given below:

$$\begin{bmatrix} c_1\\ c_2\\ c_3\\ \vdots\\ c_N \end{bmatrix}=\frac{1}{2\pi}\begin{bmatrix} \ln(r_{1,1}) & \ln(r_{1,2}) & \ln(r_{1,3}) & \cdots & \ln(r_{1,N})\\ \ln(r_{2,1}) & \ln(r_{2,2}) & \ln(r_{2,3}) & \ldots & \ln(r_{2,N})\\ \ln(r_{3,1}) & \ln(r_{3,2}) & \ln(r_{3,3}) & \cdots & \ln(r_{3,N})\\ \vdots & \vdots & \vdots & \ddots & \vdots\\ \ln(r_{N,1}) & \ln(r_{N,2}) & \ln(r_{N,3}) & \ldots & \ln(r_{N,N}) \end{bmatrix}\begin{bmatrix} F_1\\ F_2\\ F_3\\ \vdots\\ F_N \end{bmatrix} \tag{5}$$

To avoid the high condition numbers of linear systems that need to be solved, singularities are located outside of the computational domain (submerged). The submergence of singularities for Stokes equations is discussed in [27,28] and the references therein. If the problem set up includes only the Dirichlet (or first-type) boundary condition, the left-hand side of Equation (5) is known, and one should solve the linear system with respect to the unknown vector of strength of singularities, F. Boundary conditions of this type correspond to observation points at domain boundaries, where boundary concentrations are given.

In general, the kinetics of the formation of SiC involves several species and intermediate reactions, while a reduced single-step reaction is used in the present study, see Ref [16] and the references therein, as follows:

$$C_2H_2 + 2SiCl_2 + H_2 \rightarrow 2SiC(solid) + 4HCl \quad (6)$$

The concentration of acetylene (C_2H_2) at the fibers' surface exposed to the above CVI chemical reaction is not known, and can be calculated using the following equation [29], assuming the equality of the reaction flux at the solid surface, $k\, c_{wall}$, to the diffusion flux towards the surface, $D\left(\frac{\partial c}{\partial n}\right)_{wall}$:

$$c_{wall} = \frac{D}{k}\left(\frac{\partial c}{\partial n}\right)_{wall}, \quad (7)$$

where c is the volume molar concentration of C_2H_2 in mol/m^3, c_{wall} is the surface molar concentration of C_2H_2 at the wall in mol/m^2, and k is the reaction rate constant, which is calculated below using the Arrhenius equation [30]:

$$k = A\exp\left(-\frac{E_a}{RT}\right), \quad (8)$$

where A is the pre-exponential factor, E_a is the activation energy, R is the ideal gas constant, and T is the reactor temperature. Since the CVI processes modeled using MFS are isothermal, for a given reaction, k is assumed to be a constant and calculated using (8). The reaction rate is dependent on the temperature only.

Refs. [16,31] reported that the concentration of $SiCl_2$ gas has little effect on SiC growth. The values of parameters in Equation (8), such as activation energy E_a = 355.5 kJ/mol and pre-exponential factor $A = 10^{-3}$ 1/s, are obtained experimentally in Ref. [31] and used in the present study.

The diffusion coefficient, D, of C_2H_2, is computed as the ratio of thermal diffusivity (see [16] for the procedure of its calculation) and the Lewis number. The value of the Lewis number for C_2H_2, $Le_{C2H2} = 0.75$ [16].

Equation (7) represents the Robin, or third-type boundary condition, which is a linear combination of the values of a function and the values of its derivative on the boundary of the domain. MFS is rarely used with this type of boundary condition. One example is the MFS implementation of the partial slip boundary conditions for Stokes equations [27]. To implement this boundary condition for Equation (7), one needs to take the derivative of (4). For instance, the derivative of (4) by x is expressed as follows:

$$\frac{\partial c}{\partial x} = \frac{1}{2\pi}\sum_{i=1}^{N} F_i \frac{\partial\left(\ln\left(r_{i,(x,y)}\right)\right)}{\partial x} = \frac{1}{2\pi}\sum_{i=1}^{N} F_i \frac{1}{r_{i,(x,y)}}\frac{\partial r_{i,(x,y)}}{\partial x} = \frac{1}{2\pi}\sum_{i=1}^{N} F_i \frac{x - x_i}{r_{i,(x,y)}{}^2}, \quad (9)$$

where $r_{i,(x,y)} = \sqrt{(x_i - x)^2 + (y_i - y)^2}$ is the distance between the location of the i th singularity, (x_i, y_i), and an arbitrary point within the computational domain, (x, y).

To derive the above equation, note that the derivative of r by x is $\frac{\partial r_{i,(x,y)}}{\partial x} = \frac{x - x_i}{r_{i,(x,y)}}$.

The concentration derivative normal to the fibers' surface in Equation (7) at observation point j is calculated by taking the directional derivative of (5) and substituting the coordinate of observation points, $x = x_j$, into Equation (9). To calculate the directional derivative, the derivative of the concentration with respect to both x and y Cartesian coordinates is calculated at the following observation points:

$$\begin{bmatrix} \frac{\partial c_1}{\partial x} \\ \frac{\partial c_2}{\partial x} \\ \frac{\partial c_3}{\partial x} \\ \vdots \\ \frac{\partial c_N}{\partial x} \end{bmatrix} = \frac{1}{2\pi} \begin{bmatrix} \frac{r_{x\,(1,1)}}{r^2_{(1,1)}} & \frac{r_{x\,(1,2)}}{r^2_{(1,2)}} & \frac{r_{x\,(1,3)}}{r^2_{(1,3)}} & \cdots & \frac{r_{x\,(1,N)}}{r^2_{(1,N)}} \\ \frac{r_{x\,(2,1)}}{r^2_{(2,1)}} & \frac{r_{x\,(2,2)}}{r^2_{(2,2)}} & \frac{r_{x\,(2,3)}}{r^2_{(2,3)}} & \cdots & \frac{r_{x\,(2,N)}}{r^2_{(2,N)}} \\ \frac{r_{x\,(2,1)}}{r^2_{(3,1)}} & \frac{r_{x\,(3,2)}}{r^2_{(3,2)}} & \frac{r_{x\,(3,3)}}{r^2_{(3,3)}} & \cdots & \frac{r_{x\,(3,N)}}{r^2_{(3,N)}} \\ \vdots & \vdots & \vdots & \ddots & \vdots \\ \frac{r_{x\,(N,1)}}{r^2_{(N,1)}} & \frac{r_{x\,(N,2)}}{r^2_{(N,2)}} & \frac{r_{x\,(N,3)}}{r^2_{(N,3)}} & \cdots & \frac{r_{x\,(N,N)}}{r^2_{(N,N)}} \end{bmatrix} \begin{bmatrix} F_1 \\ F_2 \\ F_3 \\ \vdots \\ F_N \end{bmatrix} \tag{10a}$$

$$\begin{bmatrix} \frac{\partial c_1}{\partial y} \\ \frac{\partial c_2}{\partial y} \\ \frac{\partial c_3}{\partial y} \\ \vdots \\ \frac{\partial c_N}{\partial y} \end{bmatrix} = \frac{1}{2\pi} \begin{bmatrix} \frac{r_{y\,(1,1)}}{r^2_{(1,1)}} & \frac{r_{y\,(1,2)}}{r^2_{(1,2)}} & \frac{r_{y\,(1,3)}}{r^2_{(1,3)}} & \cdots & \frac{r_{y\,(1,N)}}{r^2_{(1,N)}} \\ \frac{r_{y\,(2,1)}}{r^2_{(2,1)}} & \frac{r_{y\,(2,2)}}{r^2_{(2,2)}} & \frac{r_{y\,(2,3)}}{r^2_{(2,3)}} & \cdots & \frac{r_{y\,(2,N)}}{r^2_{(2,N)}} \\ \frac{r_{y\,(3,1)}}{r^2_{(3,1)}} & \frac{r_{y\,(3,2)}}{r^2_{(3,2)}} & \frac{r_{y\,(3,3)}}{r^2_{(3,3)}} & \cdots & \frac{r_{y\,(3,N)}}{r^2_{(3,N)}} \\ \vdots & \vdots & \vdots & \ddots & \vdots \\ \frac{r_{y\,(N,1)}}{r^2_{(N,1)}} & \frac{r_{y\,(N,2)}}{r^2_{(N,2)}} & \frac{r_{y\,(N,3)}}{r^2_{(N,3)}} & \cdots & \frac{r_{y\,(N,N)}}{r^2_{(N,N)}} \end{bmatrix} \begin{bmatrix} F_1 \\ F_2 \\ F_3 \\ \vdots \\ F_N \end{bmatrix}, \tag{10b}$$

where $r_{x\,(j,i)} = x_j - x_i$ and $r_{y\,(j,i)} = y_j - y_i$ are the x and y coordinate differences, in correspondence, between the observation point j and singularity point i in the x and y directions, respectively.

To complete the directional derivative, the dot product of (10) is taken as the unit vector normal to the surface, $[n_x, n_y]$:

$$\frac{\partial c}{\partial n} = \left[\frac{\partial c}{\partial x}, \frac{\partial c}{\partial y}\right] \cdot [n_x, n_y] \tag{11}$$

By rearranging Equation (5) and substituting Equation (10) for an observation point, j, the following equation at the deposition surface is obtained:

$$0 = \frac{1}{2\pi} \begin{bmatrix} \ln(r_{1,j}) - \frac{D}{k}\left(\frac{r_{x\,(1,j)}}{r^2_{(1,j)}} n_x + \frac{r_{y\,(1,j)}}{r^2_{(1,j)}} n_y\right) \\ \ln(r_{2,j}) - \frac{D}{k}\left(\frac{r_{x\,(2,j)}}{r^2_{(2,j)}} n_x + \frac{r_{y\,(2,j)}}{r^2_{(2,j)}} n_y\right) \\ \ln(r_{3,j}) - \frac{D}{k}\left(\frac{r_{x\,(3,j)}}{r^2_{(3,j)}} n_x + \frac{r_{\,,(3,j)}}{r^2_{(3,j)}} n_y\right) \\ \vdots \\ \ln(r_{N,j}) - \frac{D}{k}\left(\frac{r_{x\,(N,j)}}{r^2_{(N,j)}} n_x + \frac{r_{y\,(N,j)}}{r^2_{(N,j)}} n_y\right) \end{bmatrix}^T \begin{bmatrix} F_1 \\ F_2 \\ F_3 \\ \vdots \\ F_N \end{bmatrix}, \tag{12}$$

Boundary conditions (12) are used for observation points located at fiber surfaces at which deposition occurs. For each observation point located at a boundary where deposition occurs, Equation (12) is substituted into system (5) in place of the original row, which corresponds to a given concentration at the boundary.

To construct the domain for a non-structured preform, fibers are placed randomly in the domain, one at a time, until the desired porosity is achieved. The details of porosity computations are presented at the end of this section. It should be noted that—even though the initial porosity of the sample is the same—the actual geometry of the domain could be different due to the random way the fibers are placed. The resulting linear system with respect to the strengths of the singularities, F, is then solved using the backslash operator in MATLAB, solving a linear system of equations [32].

To verify that the MFS methodology solves the Laplace equation correctly for the given boundary conditions, an analytical solution for concentration distribution between the two concentric rigid cylinders is developed. This simplified set up reflects diffusion toward the surface of a single fiber where the chemical reaction occurs. A diagram of this test case is shown in Figure 2. In 1-D cylindrical coordinates, Laplace's equation is written as follows:

$$0 = \frac{1}{r}\frac{\partial}{\partial r}\left(r\frac{\partial c}{\partial r}\right) \tag{13}$$

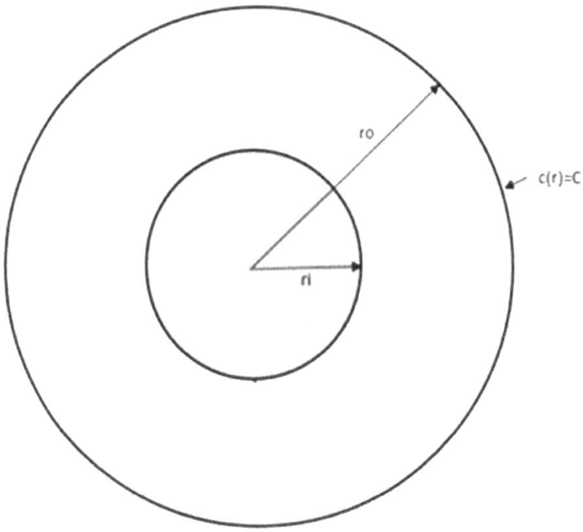

Figure 2. The set up of diffusion between two cylinders with surface reaction at the inner cylinder. Concentration C at the outer cylinder surface is given. The inner cylinder surface reaction has a reaction boundary at equilibrium.

Here, the outer cylinder surface has a known constant concentration, and the inner cylinder, representing an isolated fiber, is a reaction boundary, as described by (7) (see Figure 2). The analytical solution of the above equation with these boundary conditions is given as follows:

$$c(r) = \frac{C}{\frac{D}{k}\frac{1}{r_i} + \ln\left(\frac{r_o}{r_i}\right)}\left(\frac{D}{k}\frac{1}{r_i} + \ln\left(\frac{r}{r_i}\right)\right), \tag{14}$$

where C is the constant concentration at the outer cylinder surface, D is the diffusion coefficient, k is the reaction rate constant, and r_i and r_o represent the inner and outer radius of the cylindrical surfaces, respectively.

The set up of the 2-D MFS solution for the same geometry is shown in Figure 3. The red markers show the location of the singularities, and the blue markers represent observation points where the boundary conditions are set up. The observation points are placed on the inner and outer cylinders while the singularities are placed inside the cylinders at a submergence depth of 0.1 r, where r is the radius of the inner or outer cylinder in correspondence. The number of singular points and collocation points is the same. The analytical solution (14) and the MFS solution calculated using Green's free space function for the Laplace operator with corresponding boundary conditions (5, 12) are shown in Figure 4.

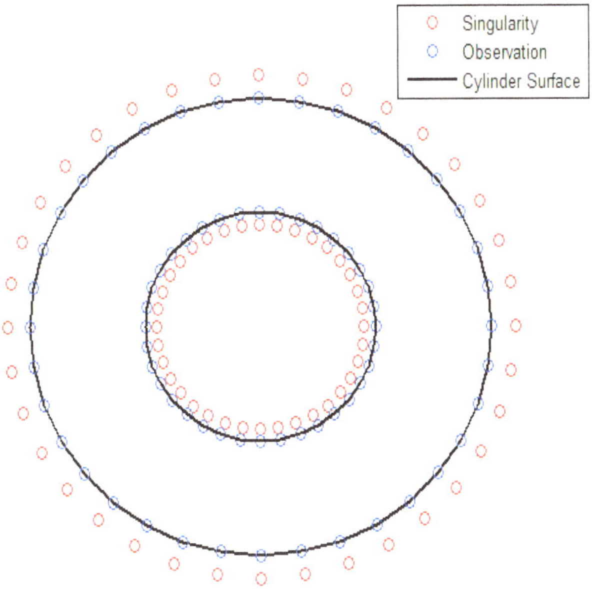

Figure 3. The MFS set up for diffusion between two cylinders is as follows: red markers are submerged singularities, while blue markers are observation points.

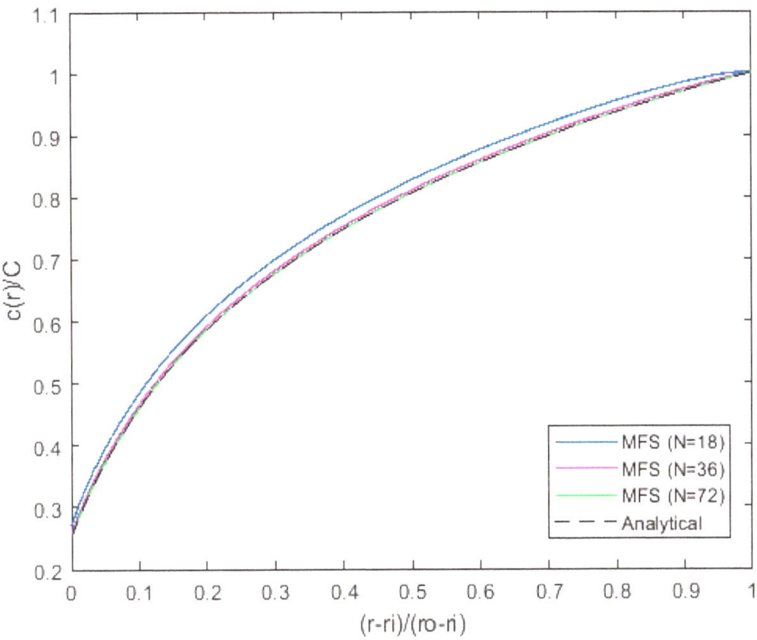

Figure 4. Comparison of the concentration $c(r)$ obtained by MFS with the analytical solution (14).

The MFS computations with N = 18, 36 and 72 Stokeslets per cylinder surface are shown in Figure 4. Each surface has an equal number of observation points and Stokeslets. The obtained $c(r)$ for N = 36 and N = 72 Stokeslets per cylinder practically coincide with each other and with the analytical solution (14). The number of observation points and singularities per fiber, $N_f = 36$, is therefore used in the computations in the next section.

After the steady MFS solution for the concentration of a given geometry of fibers with deposited film has been calculated, the concentration of the reactant at the fiber's surface can then be used to calculate the growth rate [14] of deposited material before proceeding to the next time step. The CVD model [16] considers C_2H_2 as a precursor species of decomposed MTS in the gas phase, where the C_2H_2 concentration determines the SiC film formation rate. After calculating the concentration at the wall using (7), the growth rate of the fiber could then be calculated as follows by multiplying the concentration at the wall, calculating the reaction rate using the Arrhenius Equation (8), and determining the ratio of the molar mass ($M_{reactant}$) to the density of deposited film (ρ_{film}):

$$\delta = 2\,\frac{M_{film}}{\rho_{film}} \cdot k \cdot c_{wall}\,, \tag{15}$$

where δ is the growth rate of the fiber and c_{wall} is the surface molar concentration of C_2H_2 at the wall in mol/m^2.

The density of SiC film is ρ_{SiC} = 3210 kg/m^3 [16]. Coefficient 2 in the above formula appears because, by Equation (6), two molecules of SiC are synthesized by a single C_2H_2 molecule and the reaction rate is calculated by the concentration of C_2H_2 in Equations (7) and (8). The new location of the deposited surface is obtained by adding the deposited layer thickness, $\delta\,\Delta t$, in the normal direction to the fiber.

Void tracking is an important part of modeling CVI. Figure 5 shows a depiction of a void. The area outlined in red is not accessible to reactant gas. The void-forming fiber surfaces do not continue to grow during the rest of the simulation. To determine the location of these voids, the computational domain is meshed. The concentration at boundaries where $C > 0$ is then set to 1 and any mesh point within a fiber is assumed to be a $C = 0$ boundary condition. The Laplace operator (3) for the concentration is then solved using a finite-difference scheme. After the solution has converged, any mesh point with $C > 0$ has access to the reactant and is not in a void, while any mesh point with $C = 0$ is either part of the fiber or in an area surrounded by fibers and has no access to the reactant. Solving (3) using a finite difference scheme is performed entirely for void tracking purposes and is not used to calculate concentrations or growth rates within the computational domain. This is a similar method to that developed in [24].

Figure 5. Example of void (red) in preform of fibers with deposited material (blue).

Once the growth rate is obtained using (15), the singularity and observation points are moved in the normal direction to the local surface, and the MFS solution is repeated at the next time step. This process continues until either the sample has been fully infiltrated (porosity is no longer decreasing) or the specified end time is reached.

The porosity of the samples is calculated by generating 10,000 random points within the computational domain. Whether the generated point is within a fiber or not is determined using the fiber center and observation points. The number of total points in fibers divided by the total number of points is the specimen's porosity. To determine the accuracy of this porosity measurement, a sample with a known 80% porosity is measured 30 times. The average of the measurements is 79.93%, which is close to 80% (known porosity). The porosity minimum and maximum are 79.23% and 80.73%, respectively. The 3-σ range is 78.73–81.27%.

This section introduced the Robin boundary condition to the MFS to solve the diffusion equation with a chemical deposition reaction at the surface. The numerical procedure is coupled with the model of growth of deposited solid film. The validation of this proposed approach by comparing it to an analytical solution was successful.

3. Computational Results and Discussion

To further verify the MFS's ability to model Chemical Vapor Infiltration, comparisons are made to experimental snapshots (Figure 6) and the quantitative data of infiltrated geometry (Table 1).

Table 1. Summary of experimental porosity data from simulations using MFS.

Specimen	Temperature, °C	Run Time, Hours	Initial Computed Porosity	Initial Experimental Porosity	Final Porosity Computed by MFS	Final Experimental Porosity
1	1200	25	0.6378	0.64	0.1445	0.13
2	1200	31	0.6751	0.67	0.1883	0.24
3	1100	30.5	0.6533	0.65	0.1474	0.16
4	1200	66	0.6661	0.67	0.1573	0.20
5	1200	50	0.6318	0.63	0.1273	0.11
6	1200	22	0.7020	0.70	0.2363	0.28

In general, "real-time" simulations of CVI experiments with the same number of fibers are not feasible at this time due to computational limitations and the complex asymmetrical nature of the fiber preforms used. To model a woven fiber preform using MFS, millions of singularities are required. The resulting matrix that would need to be solved to obtain the strength of these singularities would, as a result, be on the order of 10^{12} entries. This would need to be performed multiple times as time progresses throughout the infiltration process. As a result, the set ups of fibers considered using computational techniques, such as unidirectional fiber-reinforced substrates, are simplified versions of the actual experiments [33].

A 3-D woven preform would need to be reduced to the hundreds of 2-D long cylinders, with the N_c of the diameter of 8 microns (see Figure 6) located in a computational domain representing a small section of the actual geometry. The total size S of the 2-D computational domain encompassing N_c cylindrical fibers with porosity f is given by the following expression:

$$S = N_c \pi D^2 / (4(1 - f)) \tag{16}$$

For N_c = 500 fibers used in computations with D = 8 μ and f = 0.7 (specimen #6 in Table 1), the size of the computational domain is S = 83,773 μ^2. When the number of fibers used in MFS computations is doubled or halved, the obtained computational final porosity does not change.

The computational results in Figure 6a are generated by simulating a small section of a woven geometry used in a CVI experiment using the same coordinates of the centers of fibers as those in the experiments depicted in Figure 6b. The obtained value of the diffusion coefficient, as outlined in Section 2, is $D = 0.000528\ \text{m}^2/\text{s}$ at the conditions of experiments in [34], in which the preforms are infiltrated at 1100–1200 °C under atmospheric pressure.

In Ref. [35] the maximum SiC deposition rate is 9.5 nm/min. For MFS simulations, the chosen time step is $\Delta t = 6$ s to ensure the maximum deposition of ~1 nm per time step. This is a small added film thickness compared to ~8 microns diameter fibers (see Figure 6b) that justify the use of an explicit in-time approach to model the film deposition. The initial carbon interphase has a thickness ranging from 75 to 300 nm in experiments [34]. This layer was not accounted for in present MFS computations because the diameter of the fibers is much larger than the thickness of the initial carbon interface layer.

The initial porosity of this sample is 38.81%, and the final porosity after infiltration using MFS is 13.75%. A qualitative comparison of the computational results using MFS (Figure 6a) to the experimental results [35] (Figure 6b) is presented, showing a similar pattern of pores.

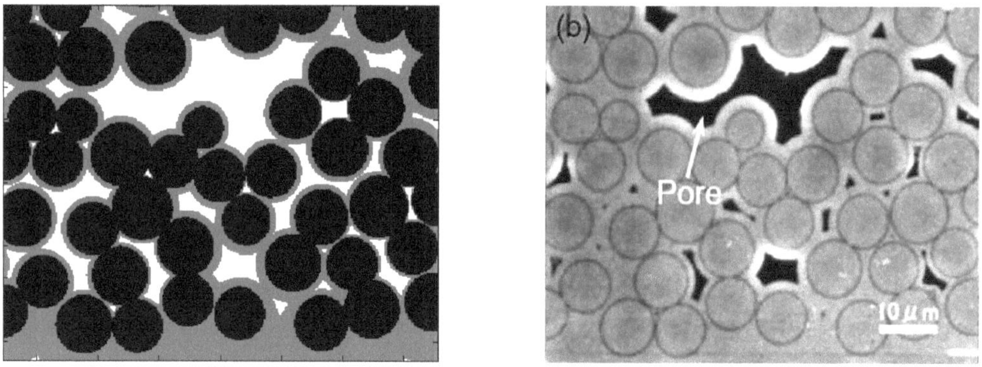

(a) (b)

Figure 6. Comparison of infiltrated geometry (**a**) computed by MFS and (**b**) obtained experimentally, see [35], Figure 4, reprinted with Elsevier's permission.

The results of the experiments [34] and the results of MFS simulations are shown in Table 1 in terms of the dynamics of porosity. The set ups of randomly generated circular fibers at different known porosities are generated by randomly placing fibers in the domain one at a time until the desired porosity is reached. The experimental conditions for specimen 4 in Table 1 [34] are mostly close to the results shown in Figure 6b [34].

The computations for listed specimens differ by initial porosity, run time, and temperature. If specimens differ by temperature, it affects the chemical reaction rate k in Equation (8). The initial porosities in computations are nearly identical to those in the experiments, as they should be. Slight differences between the initial experimental and computational porosities, ~0.005, are due to the computational porosity calculation and computational placement of fibers, as previously discussed.

After the geometry has infiltrated and computed using the MFS, the porosities of the experimental sample and the simulated domain compare well; see the last two columns of Table 1.

The final porosity for low-temperature case 3 is smaller in the experiments and in the MFS simulations as the faster kinetics close the outer surfaces faster at higher temperatures. Comparing case 1 with case 5 and case 2 with case 4 shows that the samples nearly reached terminal porosity after 25–30 h. Processing in similar conditions for double the time did not show a substantial change in the final porosity. To follow discussions in the experimental study [34], decreasing the initial average porosity was very effective at decreasing the final porosity. As a result, the average final porosity decreased in case 1 by decreasing the initial porosity compared to case 6. If the samples were ranked from the least porous to most porous after the MFS-modeled infiltration, the specimens' order would be the same as the experiments (specimen 5, specimen 1, specimen 3, specimen 4, specimen 2, and specimen 6).

This section compared the MFS results to the experimentally obtained pattern of voids and the experimental porosity obtained for six different cases. There is a good correspondence of computational results obtained by the MFS with Robin boundary conditions to experimental data.

4. Conclusions

In the current study, the MFS method was developed and applied to the modeling of Chemical Vapor Infiltration (CVI) with a surface chemical reaction. The MFS method has not been used to solve CVI problems before. The concentrations of the reagent within the preform were obtained using Green's free space functions for the Laplace operator, describing diffusion with boundary conditions corresponding to the CVI reaction at the fiber surfaces. For the quasi-equilibrium at the fiber surfaces considered, the reaction flux at the solid surface was equal to the diffusion flux towards the surface. To account for this condition, the Robin/third-type boundary condition at the fiber surfaces was developed and successfully incorporated with the MFS. These surface concentrations were used to calculate the fiber growth rate due to deposition. Comparisons of analytical solutions for the simplified geometry of two concentric cylinders mimicking a single fiber and MFS solution to the experimental snapshot of the fiber preforms with deposition are presented.

The dynamics of pore size and location were computed using the proposed MFS approach to evaluate the quality of the material obtained by CVI. Porosity transients were obtained for the range of initial porosities to show the relation between the initial and final porosities and the time needed to reach the final porosity.

The Method of Fundamental Solutions appears to be effective in modeling CVI with simplified one-step chemical surface reactions. The method accounts for circumferential variation in the deposition rate along the fiber's surface, the depletion of the chemical reagent at part of the fiber's surface, the merging of fibers with the formation of voids caused by deposition and the non-uniform distribution of the reactant throughout the domain. The present version of MFS was developed to simulate a CVI process that is isobaric and isothermal. The equality of reaction flux at the fiber's surface and the diffusion flux towards the surface was assumed. Extensions of the proposed approach to more complex boundary conditions and the advection–diffusion equation will be addressed in future research.

Author Contributions: Conceptualization, A.P.; Software, P.M.; Validation, P.M.; Investigation, P.M.; Writing—original draft, P.M. and A.P.; Writing—review & editing, A.P.; Project administration, A.P. All authors have read and agreed to the published version of the manuscript.

Funding: This research received no external funding.

Data Availability Statement: Data supporting reported results can be found by contacting the corresponding author.

Conflicts of Interest: The authors declare no conflict of interest.

Nomenclature

Symbols

A	pre-exponential factor (1/s)
C	constant concentration at domain boundaries for analytic solution
$c\ (x, y)$	local concentration (mol/m^3)
c_{wall}	concentration at fiber surface (mol/m^2)
D	diffusion coefficient (m^2/s)

E_a	activation energy (J/mol)
F	singularity strength
f	porosity
i	singularity point index in MFS
j	collocation point index in MFS
k	reaction rate constant (1/s)
M_{film}	molar mass of deposited material (kg/mol)
N	total number of observation points
N_f	number of observation points per fiber surface
N_c	number of cylindrical fibers in MFS computations
n	normal direction
n_x	unit normal x-component
n_y	unit normal y-component
Pe	Peclet number
R	ideal gas constant (J/mol-K)
r	radius (m)
r_i	inner cylinder radius (m)
r_o	outer cylinder radius (m)
$r_{i,j}$	distance from singularity i to observation point j
S	size of computational cross-section of fibers computed by MFS
t	time, s
T	temperature (K or °C)
u, v, w	Cartesian velocity components (m/s)
$\overline{u},$	characteristic velocity to calculate Peclet number (m/s)
x, y, z	Cartesian coordinates
Greek Symbols	
δ	growth rate of the fiber (m/s)
ρ_{film}	density of deposited material (kg/m^3)

References

1. Bickerdike, R.; Brown, A.; Hughes, G.; Ranson, H. The Deposition of Pyrolytic Carbon in the Pores of Bonded and Unbonded Carbon Powders. In Proceedings of the Fifth Conference on Carbon, State College, PA, USA, 19–23 June 1961; Mrosowski, S., Studebaker, M.C., Walters, P.L., Eds.; Pergamon Press: Oxford, UK, 1962; Volume 1, pp. 575–582.
2. Beshmann, T.M.; Sheldon, W.B.; Lowden, R.A.; Stinton, D.P. Vapor-phase Fabrication and Properties of Continuous-filament Ceramic Composites. *Science* **1991**, *253*, 1104–1109. [CrossRef]
3. Fitzer, E.; Manocha, L. Applications of Carbon/carbon Composites. In *Carbon Reinforcements and Carbon/Carbon Composites*; Springer: Berlin/Heidelberg, Germany, 1998; pp. 310–336. [CrossRef]
4. Sankaran, R.; Ramanuj, V.; Cha, C.M.; Liliedahl, D. Level-Set Modeling Simulations of Chemical Vapor Infiltration for Ceramic Matrix Composites Manufacturing. Report by the Oak Ridge National Laboratory; ORNL/TM-2019/1206; CRADA/NFE-16-06412. 2019. Available online: https://info.ornl.gov/sites/publications/Files/Pub127239.pdf (accessed on 27 December 2023).
5. Chalwa, N.; Chawla, K. Chapter 1: Introduction. In *Composite Materials Science and Engineering*, 2nd ed.; Springer: New York, NY, USA, 1998; pp. 3–6. [CrossRef]
6. Naslain, R.; Langlais, F.; Fedou, R. The CVI-processing of Ceramic Matrix Composites. *J. Phys.* **1989**, *50*, 191–207. Available online: https://hal.science/jpa-00229549/document (accessed on 27 December 2023).
7. Askour, O.; Tri, A.; Braikat, B.; Zahrouni, H.; Potier-Ferry, M. Method of Fundamental Solutions and High Order Algorithm to Solve Nonlinear Elastic Problems. *Eng. Anal. Bound. Elem.* **2018**, *89*, 25–35. [CrossRef]
8. Cheng, A.; Hong, Y. An Overview of the Method of Fundamental Solutions—Solvability, Uniqueness, Convergence, and Stability. *Eng. Anal. Bound. Elem.* **2020**, *120*, 118–152. [CrossRef]
9. Liu, Y. *Fast Multipole Boundary Element Method*; Cambridge University Press: Cambridge, UK, 2010.
10. Fairweather, G.; Karageorghis, A. The Method of Fundamental Solutions for Elliptic Boundary Value Problems. *Adv. Comput. Math.* **1998**, *9*, 69–95. [CrossRef]
11. Liu, Y.; Nishimura, N.; Yao, Z. A Fast Multipole Accelerated Method of Fundamental Solutions for Potential Problems. *Eng. Anal. Bound. Elem.* **2005**, *29*, 1016–1024. [CrossRef]
12. Mukherjee, S.; Liu, Y. The Boundary Element Method. *Int. J. Comput. Methods* **2013**, *10*, 1350037. [CrossRef]
13. Zhu, H.; Dhall, A.; Mukherjee, S.; Datta, A.K. A Model for Flow and Deformation in Unsaturated Swelling Porous Media. *Transp. Porous Med.* **2010**, *84*, 335–369. [CrossRef]

14. Roman, Y.G.; Kotte, J.F.A.K.; de Croon, M.H.J.M. Analysis of the Isothermal Forced Flow Chemical Vapour Infiltration Process. Part1: Theoretical Aspects. *J. Eur. Ceram. Soc.* **1995**, *15*, 875–886. [CrossRef]
15. Wei, X.; Cheng, L.; Zhang, L.; Xu, Y.; Zeng, Q. Numerical Simulation of Effect of Methyltrichlorosilane Flux on Isothermal Chemical Vapor Infiltration Process of C/SiC Composites. *J. Am. Ceram. Soc.* **2006**, *89*, 2762–2768. [CrossRef]
16. Ogawa, T.; Fukumoto, K.; Machida, H.; Norinaga, K. CFD Simulation of CVD Reactors in the CH3SiCl3(MTS)/H2 System Using a Two-step MTS Decomposition and One-step SiC growth models. *Heliyon* **2023**, *9*, e15061. [CrossRef]
17. Lackey, W.J.; Starr, T.L. Fabrication of Fiber-reinforced Ceramic Composites by Chemical Vapor Infiltration: Processing, Structure, and Properties. In *Fiber Reinforced Ceramic Composites*; Noyes Publications: Park Ridge, NJ, USA, 1990; pp. 397–450.
18. Fitzer, E.; Gadow, R. Fiber Reinforced Silicon Carbide. *Am. Ceram. Soc. Bull.* **1986**, *65*, 326–335.
19. Chawla, K. Ceramic matrix composites. In *Composite Materials: Science and Engineering*, 3rd ed.; Springer: New York, NY, USA, 2012; pp. 249–292. [CrossRef]
20. Vaidyaraman, S.; Lackey, W.; Agrawal, P.; Starr, T. 1-D Model for Forced Flow-Thermal Gradient Chemical Vapor Infiltration Process for Carbon/carbon Composites. *Carbon* **1996**, *34*, 1123–1133. [CrossRef]
21. Kulik, V.I.; Kulik, A.V.; Ramm, M.S.; Makarov, Y.N. Modeling of SiC-matrix Composite Formation by Isothermal Chemical Vapor Infiltration. *J. Cryst. Growth* **2004**, *266*, 333–339. [CrossRef]
22. Yoshikawa, N.; Evans, J. Modeling of Chemical Vapor Infiltration Rate Considering a Pore Size Distribution. *J. Am. Ceram. Soc.* **2002**, *85*, 1485–1491. [CrossRef]
23. Langhoff, T.A.; Schnack, E. Modeling of Chemical Vapor Infiltration of Pyrolytic Carbon as Moving Boundary Problem. *Chem. Eng. Sci.* **2006**, *63*, 3948–3959. [CrossRef]
24. Jin, S.; Wang, X.; Starr, T.L.; Chen, X. Robust Numerical Simulation of Porosity Evolution in Chemical Vapor Infiltration I: Two Space Dimension. *J. Comput. Phys.* **2000**, *162*, 467–482. [CrossRef]
25. Osher, S.; Sethian, J.A. Fronts Propagating with Curvature-dependent Speed: Algorithms Based on Hamilton-Jacobi Formulations. *J. Comput. Phys.* **1988**, *78*, 12–49. [CrossRef]
26. Duffy, D.G. *Green's Functions with Applications*, 2nd ed.; CRC Press: Boca Raton, FL, USA, 2015.
27. Zhao, S.; Povitsky, A. Coupled Continuum and Molecular Model of Flow through Fibrous Filter. *Phys. Fluids* **2013**, *25*, 112002. [CrossRef]
28. Zhao, S.; Povitsky, A. Three-dimensional Boundary Singularity Method for Partial-slip Flows. *Eng. Anal. Bound. Elem.* **2011**, *35*, 114–122. [CrossRef]
29. Fogler, H.S. *Elements of Chemical Reaction Engineering*; Prentice Hall: Saddle River, NJ, USA, 2006.
30. McNaught, A.D.; Wilkison, A. *IUPAC. Compendium of Chemical Terminology*, 2nd ed.; Scientific Publications: Oxford, UK, 1997.
31. Fukushima, Y.; Sato, N.; Funato, Y.; Sugiura, H.; Hotozuka, K.; Momose, T.; Shimogaki, Y. Multi-scale Analysis and Elementary Reaction Simulation of SiC-CVD Using CH3SiCl3/H2: I. Effect of Reaction Temperature. *ECS J. Solid State Sci. Technol.* **2013**, *2*, 492–497. [CrossRef]
32. *MATLAB*; The MathWorks: Natick, MA, USA, 2022.
33. Zhu, Y.; Schnack, E. Numerical Modeling of Chemical Vapor Infiltration of SiC Composites. *J. Chem.* **2013**, *2013*, 836187. [CrossRef]
34. Igawa, N.; Taguchi, T.; Snead, L.L.; Katoh, Y.; Jitsukawa, S.; Kohyama, A.; McLaughlin, J.C. Optimizing the Fabrication Process for Superior Mechanical Properties in the FCVI SiC Matrix/stoichiometric SiC Fiber Composite System. *J. Nucl. Mater.* **2002**, *307–311*, 1205–1209. [CrossRef]
35. Igawa, N.; Taguchi, T.; Nozawa, T.; Snead, L.L.; Hinoki, T.; McLaughlin, J.C.; Katoh, Y.; Jitsukawa, S.; Kohyama, A. Fabrication of SiC Fiber Reinforced SiC Composite by Chemical Vapor Infiltration for Excellent Mechanical Properties. *J. Phys. Chem. Solids* **2005**, *66*, 551–554. [CrossRef]

Mathematical and Computational Applications

MDPI

Article

Periodic Solutions in a Simple Delay Differential Equation

Anatoli Ivanov [1,†] and Sergiy Shelyag [2,*,†]

1 Department of Mathematics, Pennsylvania State University, Dallas, PA 18612, USA; afi1@psu.edu
2 College of Science and Engineering, Flinders University, Tonsley, Adelaide, SA 5042, Australia
* Correspondence: sergiy.shelyag@flinders.edu.au
† These authors contributed equally to this work.

Abstract: A simple-form scalar differential equation with delay and nonlinear negative periodic feedback is considered. The existence of several types of slowly oscillating periodic solutions is shown with the same and double periods of the feedback coefficient. The periodic solutions are built explicitly in the case with piecewise constant nonlinearities involved. The periodic dynamics are shown to persist under small perturbations of the equation, which make it smooth. The theoretical results are verified through extensive numerical simulations.

Keywords: delay differential equations; periodic negative feedback; slowly oscillating solutions; periodic solutions; piecewise constant nonlinearities; explicit piecewise affine solutions; reduction to interval maps

1. Introduction

The simple-form delay differential equation (DDE),

$$x'(t) = -\mu x(t) + f(x(t-\tau)) \tag{1}$$

represents a broad variety of complex dynamical behaviors that can happen in delay differential models. It is also one of the most studied equations in the field, exhibiting a range of diverse dynamics from the global stability of equilibria to the instability and existence of periodic solutions and chaotic behaviours (in contrast to analogous ordinary differential equations). Basic and fundamental facts about the DDE (1) can be found in many sources; see, e.g., monographs [1–4], among others.

Equation (1) is also extensively used as a mathematical model for a variety of real world phenomena, most notably in biological applications; see, e.g., the monographs [1,4–6] and further references therein. Among the well-known and most studied models of type (1) are the Mackey–Glass physiological equations [5,7], the Wazewska–Lasota blood cell model [8], Nicholson's blowfly equations [9,10], and several others [1,4,6].

Some of the more accurate and adequate mathematical models are explicitly time dependent, which take into account certain intrinsic periodicity factors, such as, e.g., circadian rhythms and seasonal changes. Examples of such differential delay models can be found in publications [5,6,11–14]. Therefore, a natural generalization of Equation (1) is the following DDE:

$$x'(t) = -\mu(t)x(t) + a(t)f(x(t-\tau)), \tag{2}$$

where $\mu(t)$ and $a(t)$ are T-periodic functions, and the nonlinearity $f(x)$ satisfies appropriate feedback conditions. An important theoretical and applied question then arises: whether DDE (2) admits periodic solutions, which are induced by the periodicity in the coefficients μ and a. Therefore, the mathematical problem to address is to derive conditions under which Equation (2) possesses a nontrivial T-periodic solution. This research work is our initial attempt at answering this question for some specific partial cases of the equation under consideration.

This paper deals with the problem of the existence of periodic solutions for the simplest possible scalar differential delay equation of type (2) when $\mu(t) \equiv 0$:

$$x'(t) = a(t)f(x(t-\tau)), \tag{3}$$

where $f(x)$ is a continuous function satisfying the negative feedback assumption $x \cdot f(x) < 0$, $\forall x \neq 0$; the coefficient $a(t) \geq 0$ is a continuous periodic function with a period T satisfying $T > \tau$; and $\tau > 0$ is a delay. The principal problem we are addressing in this work is to derive conditions on the parameters and nonlinearities of Equation (3), which would yield the existence of periodic solutions with the same period T as the coefficient $a(t)$. Though there are some results on the existence of periodic solutions for the DDEs with periodic coefficients, none of them are applicable to the simplest form (Equation (3)), to the best of our assessment and knowledge. In the many cases that have been published in the research literature, generalization from the autonomous case to the periodic one leads to the loss of the steady positive equilibrium. In some other cases, the equilibrium state persists; however, the sufficient conditions for the existence of periodic solutions cannot distinguish between the nontrivial periodic solutions and the steady state. See more details on such relevant results and further references in, e.g., [11–15].

We are stating and dealing with the problem of the existence of nontrivial periodic solutions in simple-form DDE (3). Due to the negative feedback assumption on f, this equation always admits the trivial constant solution $x(t) \equiv 0$. The mathematical problem is to find periodic solutions that are distinct from the trivial one. This is a difficult problem, which has been largely not addressed on the general theoretical level. One way to pave a path for a systematic analysis in this direction is to consider model equations of the form given by Equation (3) with a simple f and a.

We first construct the periodic solutions in an explicit form using piecewise constant functions (both the nonlinearity f and the coefficient a). Then, we modify both functions to be continuous and "close" to the piecewise constant ones by "smoothing" them in a small neighborhood of the discontinuity set (they can also be of the C^∞ type). The periodic dynamics and their stability are shown to persist under such small perturbations.

The rest of this paper is structured as follows. The second section "Preliminaries" contains the basics of DDE (3) and preliminary results from our recent conference papers [16,17], which are necessary for the discussion in subsequent sections. Section 3 deals with the periodic solutions of Equation (3) explicitly constructed for the initially piecewise constant nonlinearity f and the coefficient a. The dynamics on a set of slowly oscillating solutions are reduced to those for induced one-dimensional maps, thus proving both the existence of periodic solutions and the types of their stability. In Section 3, it is shown that the dynamics of the interval maps and the periodic solutions from Section 2 persist when the defining functions f and a of Equation (3) are replaced with continuous or smoothed functions close to them. The arguments of the small regular perturbations of dynamics created by DDE (3) are used, which are behind such "small modifications". The theoretical derivations and exact analytical calculations are verified using extensive numerical simulations.

2. Preliminaries

In this work, we focus on the particular case $\mu = 0$ and $\tau = 1$ when Equation (1) becomes

$$x'(t) = a(t)f(x(t-1)). \tag{4}$$

Note that the case of general delay $\tau > 0$ is easily reduced to the standard consideration $\tau = 1$ through the time scaling $t = \tau s$.

Given an initial function $\varphi(s) \in C([-1,0],\mathbb{R}) := \mathbb{X}$ in the standard phase space $\mathbb{X}$, the corresponding solution $x = x(t,\varphi)$ to Equation (4) is easily found for all $t \geq 0$ through consecutive forward integration (the step method). Let $S^t(\varphi)$ be the shift in $\mathbb{X}$ by time t along the solutions of delay differential Equation (4), that is, $S^t(\varphi) = x(t+s,\varphi), s \in [-1,0]$. It is a straightforward observation that a fixed point φ_0 of the shift by period T ($S^T(\varphi_0) = \varphi_0$) gives rise to a periodic solution $x(t, \varphi_0)$ of DDE (4). Under such an interpretation, the zero solution $x(t) \equiv 0$ is a result of the trivial fixed point $\varphi_0(s) \equiv 0, s \in [-1, 0]$, which always exists for DDE (4), due to the negative feedback assumption. The mathematical problem is to derive conditions for the existence of nontrivial fixed points of the shift operator S^T, which is a principal objective of this work.

A solution is called oscillating if it has an infinite sequence of zeros $\{t_n\} \to \infty, x(t_n) = 0$, $n \in \mathbb{N}$, and $x(t)$ is not an identical zero eventually (on any interval of the form $[t_0, \infty)$). An oscillating solution is called slowly oscillating if the sequence of zeros is such that the distance between consecutive zeros is greater than the delay, i.e., $t_{n+1} - t_n > 1, \forall n \in \mathbb{N}$.

Sufficient conditions for the oscillation of all solutions to Equation (4) are well known; see, e.g., Ref. [18] and further references therein. With the negative feedback assumption $xf(x) < 0, x \neq 0$, and the non-negativity of the coefficient $a(t) \geq 0, a(t) \not\equiv 0$, either $f'(0)$ is sufficiently large, $|f'(0)| \geq f_1 > 0$, and the coefficient a is separated from zero, $a(t) \geq a_1 > 0$, or $f'(0) \neq 0$, and $a_1 > 0$ is large enough (in both cases, the product $|f_1 \cdot a_1|$ must be sufficiently large). These conditions are satisfied for our considerations to follow in the following sections, where the functions f and a are smoothed continuous ones derived from the initial piecewise constant functions (the piecewise constant case can be viewed as a limit of the continuous one).

In this paper, we deal with the slowly oscillating solutions to Equation (4). The slowly oscillating solutions are obtained when the initial functions belong to one of the two sign-definite cones. Consider the following two standard subsets of initial functions:

$$K_- := \{\varphi \in \mathbb{X} \mid \varphi(s) < 0 \ \ \forall s \in [-1, 0]\} \quad \text{and} \quad K_+ := \{\varphi \in \mathbb{X} \mid \varphi(s) > 0 \ \ \forall s \in [-1, 0]\}.$$

Assuming that all solutions to Equation (4) oscillate, it is a straightforward calculation to verify that for arbitrary $\varphi \in K_-$, the corresponding solution $x(t; \varphi)$ has an increasing sequence of zeros $0 < t_1 < t_2 < t_3 < \cdots$ such that $t_{k+1} - t_k > \tau = 1, k \in \mathbb{N}$, and

$$x(t) > 0 \ \ \forall t \in (t_{2k-1}, t_{2k}) \quad \text{and} \quad x(t) < 0 \ \ \forall t \in (t_{2k}, t_{2k+1}). \tag{5}$$

In the case when $\varphi \in K_+, \varphi(0) = h > 0$, the above inequalities on the consecutive intervals (t_{2k-1}, t_{2k}) and (t_{2k-1}, t_{2k}) are reversed into the opposite. Therefore, the solution $x(t; \varphi)$ is slowly oscillating. The consecutive intervals $(t_j, t_{j+1}), j \in \mathbb{N}$, are called semicycles.

Based on well-known facts [18] and the reasoning above, we have the following statement.

Proposition 1. *Suppose that the nonlinearity $f \in C(\mathbb{R}, \mathbb{R})$ satisfies the negative feedback assumption $x \cdot f(x) < 0, \forall x \neq 0$, and $f'(0) < 0$. Let the coefficient $0 < a(t) \in C([0, \infty), \mathbb{R})$ be T-periodic such that $\min\{a(t), t \in [0, T]\} \geq m_0 > 0$. Assume that $|f'(0)| \cdot m_0 > 1/e$. Then, for every initial function $\varphi \in \mathbb{K}_-$ with $\varphi(0) = h < 0$, the corresponding solution $x(t, h), t \geq 0$, is slowly oscillating with the sequence of simple zeros $\{t_n\}$ satisfying the inequalities (5) on the consecutive semicycles. Likewise, for every initial function $\psi \in K_+$ with $\psi(0) = h > 0$, the corresponding solution $x(t, \psi), t \geq 0$, is slowly oscillating, satisfying the inequalities opposite to those in (5).*

Additional related theoretical basics on delay differential equations can be found in, e.g., monographs [2,3].

3. Periodic Solutions

In this section, the periodic solutions of interest are explicitly constructed based on model piecewise constant nonlinearity f and the periodic coefficient a. Let $f(x) = f_0(x) = -\text{sign}(x)$ and define $a = a(t, p_1, p_2, a_1, a_2) := A_0(t)$ as follows:

$$a(t) = A_0(t) = \begin{cases} a_1 > 0, \text{ if } 0 \le t < p_1 \\ a_2 > 0, \text{ if } p_1 \le t < p_1 + p_2 \\ \text{periodic extension outside } [0, p_1 + p_2) \text{ for all } t \in \mathbb{R}. \end{cases} \quad (6)$$

Let an initial function $\varphi(s) \in \mathbb{X}$ be such that either $\varphi(s) < 0$ or $\varphi(s) > 0, \forall s \in [-1, 0]$, and $\varphi(0) = h \neq 0$. Clearly, the corresponding solution $x = x(t, \varphi)$ depends on the value h only and does not depend on the particular values of $\varphi(s) \neq 0$ for $s \in [-1, 0)$. In addition, it is uniquely determined by the parameters p_1, p_2, a_1, and a_2 of the function $a(t)$. The solution x is slowly oscillating and piecewise linear (affine) for $t \geq 0$.

Like in the case of continuous f and a, the corresponding solutions to DDE (4) with $f = f_0$ and $a = a_0$ are slowly oscillating with the consecutive zeros $\{t_n\}$ satisfying the inequalities (5). In this case, we do not need any additional assumptions on f_0 or a_0 as in Proposition 1, as the lower rates of growth or decay of any such solution are bounded away from zero (the rates of growth/decay are also bounded from above, due to the boundedness of f and a). However, since the constructed solutions are slowly oscillating, we assume that, necessarily, the assumption $p_1 + p_2 > 1$ must be in place.

In this section, we follow and expand on ideas and explicit constructions from our recent work [16,17]. In particular, functions f and a are of the same form as in these papers.

3.1. Periodic Solutions with Coefficient Period (Type I)

In this subsection, we explicitly construct piecewise affine periodic solutions to Equation (4), which are of the same period $T > 1$ as the periodic coefficient $a(t)$. Some of the exposition elements are close and based on results in our recent conference paper [17].

We presume that the desired periodic solutions have the shape as shown in Figure 1. That is, the initial value $\varphi(0) = h < 0$ is determined by an initial function $\varphi(s) \in K_-$. The corresponding solution $x = x(t, \varphi), t \geq 0$, has exactly two zeros $\{t_1, t_2\}$ on the initial period $[0, T], T = p_1 + p_2$, with $t_1 < t_2 = t_1 + 2 < T$, for particular assumptions on the parameters p_1, p_2, a_1, and a_2. Such assumptions were verified to be valid through the analytical calculations that follow, and they were confirmed using corresponding numerical simulations.

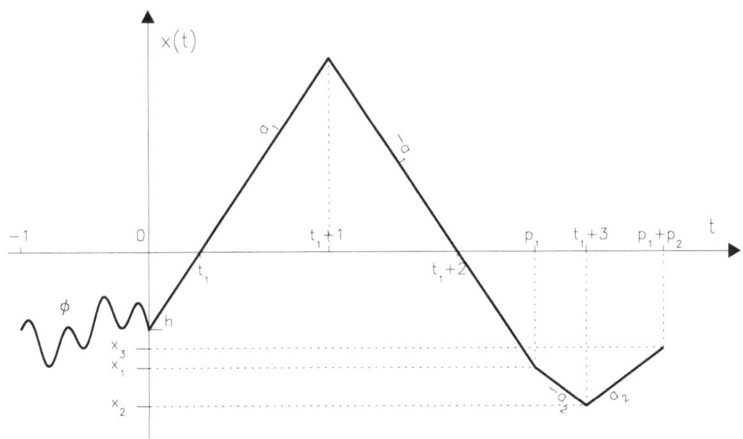

Figure 1. Slowly oscillating piecewise affine solution (Type I).

It is straightforward to find the desired values of the solution $x = x(t,h)$ for any $t \in [0,T]$. We have the following calculations:

$$t_1 = -\frac{h}{a_1}, \quad x_1 = x(p_1) = -h + 2a_1 - a_1 p_1,$$

$$x_2 = x(t_1 + 3) = \left(\frac{a_2}{a_1} - 1\right)h + a_1(2 - p_1) + a_2(p_1 - 3),$$

$$x_3 = x(T) = x(p_1 + p_2) = \left(2\frac{a_2}{a_1} - 1\right)h + a_1(2 - p_1) + a_2[(2p_1 + p_2) - 6] := G(h),$$

where the coefficients of the affine function $G(h) := mh - b$ are given by

$$m = 2\frac{a_2}{a_1} - 1, \quad \text{and} \quad b = a_1(p_1 - 2) + a_2[6 - (2p_1 + p_2)]. \tag{7}$$

It is elementary to deduce that the range for the slope m is the open interval $(-1, \infty)$ with $|m| < 1$ when $0 < a_2 < a_1$ holds, and $m > 1$ when $0 < a_1 < a_2$ is valid.

A fixed point $h_* < 0$ of the map $G(h) = mh - b$ gives rise to a slowly oscillating periodic solution $x = x(t, h_*)$ for Equation (4). Moreover, by the affine type of G and the construction, such a periodic solution is asymptotically stable if $|m| < 1$, which is equivalent to $0 < a_2 < a_1$. The periodic solution is unstable when $m > 1$, which is equivalent to $0 < a_1 < a_2$. The unique fixed point $x = h_*$ is easily found as follows:

$$h_* = b/(m-1) = \frac{a_1[a_1(p_1 - 2) + a_2[6 - (2p_1 + p_2)]]}{2(a_2 - a_1)}. \tag{8}$$

Therefore, we arrive at the following statement.

Theorem 1. *Let the parameters a_1, a_2, p_1, and p_2 be given with the values m and b defined by (7). Then, consider the following:*

(a) In a case when $|m| < 1$ $(0 < a_2 < a_1)$ and $b > 0$ are valid, DDE (4) has an asymptotically stable slowly oscillating T-periodic solution.

(b) In the complimentary case when $m > 1$ $(0 < a_2 < a_1)$ and $b < 0$ are valid, the equation has an unstable slowly oscillating T-periodic solution.

The periodic solutions are generated by the initial function $\varphi(s) \equiv h_*$, $s \in [-1, 0]$, where h_* is given by (8).

Remark 1. *(a) The local stability case $|m| < 1$ is equivalent to the relationship $a_1 > a_2 \geq 0$. At the same time, in order for the fixed point to be negative, one must assume that $b = a_1(p_1 - 2) + a_2[6 - (2p_1 + p_2)] > 0$. The inequality $p_1 > 2$ is valid based on the construction. For arbitrary fixed $p_1 > 2, p_2 > 0$, and $a_2 > 0$, there exists a_1^0 such that the inequalities $|m| < 1$ and $b > 0$ are satisfied for all $a_1 \geq a_1^0$. Thus, the required assumptions are met for all sufficiently large a_1. This case is treated in more detail in our recent paper [17].*

(b) The instability case $m > 1$ is equivalent to $0 < a_1 < a_2$ (note that the case $m < -1$ is not possible for any choices of positive a_1 and a_2). In order to have the fixed point h_* to be negative, one must assume that $b < 0$ is satisfied, that is $a_1(p_1 - 2) + a_2[6 - (2p_1 + p_2)] < 0$. Since $p_1 > 2$, the assumption $p_2 \geq 2$ would imply that $6 - (2p_1 + p_2) < 0$. Therefore, given fixed values of $a_1 > 0, p_1 > 2$, and $p_2 \geq 2$, there exists a_2^0 such that $b < 0$ for all $a_2 \geq a_2^0$. Thus, the required assumptions are met for all sufficiently large a_2.

(c) It is clear that due to the continuous dependence of m and b on the parameters a_1, a_2, p_1, and p_2, the stable or unstable periodic solutions of Theorem 1 exist on an open box in $\mathbb{R}^4$, each side of which is a corresponding open interval about a specific value of each of the four parameters corresponding to a fixed point h_*. These boxes can be multiple, as our numerical insight indicates, and they do not overlap for the stable and unstable solutions due to the respective opposite relationship between a_1 and a_2.

The parametric range for which Theorem 1 guarantees the existence of slowly oscillating periodic solutions, either stable or unstable, is large.

Note that due to the symmetry property of the nonlinearity f (oddness) and the procedure of the construction of the periodic solutions x_{h_*}, under the assumption of Theorem 1, there also exists a periodic solution symmetric to x_{h_*} generated by the initial function $\phi(s) \equiv -h_* > 0$. The two periodic solutions are related by $x_{-h_*}(t) \equiv -x_{h_*}(t)$; thus, they are symmetric to each other.

A large sample of parametric values has been derived for which numerical periodic solutions have been obtained and confirmed to be of the described form. Table 1 is a sample selection of parametric values resulting in asymptotically stable slowly oscillating periodic solutions given by Theorem 1. The values $h_* < 0$ for these periodic solutions are easily found from the explicit formula, Equation (8). They have also been verified to exist numerically.

Table 1. Examples of parametric values for which stable T-periodic solutions of Equation (4) with coefficient a defined by (6) have been demonstrated numerically.

a_1	a_2	p_1	p_2	h_*	T
1	0.25	2.5	1.5	−0.25	4
2	0.5	2.5	2	−1/3	4.5
2	0.25	2.5	1	−4/7	3.5
1	0.5	3	1	−0.5	4
2	1	3	1.5	−0.5	4.5
2.5	0.5	3	4	−0.31	7
3	0.5	3	4.5	−0.45	7.5
4	1	3.5	2	−2	5.5
5	0.5	3	3	−1.94	6
5	1	3	2	−1.88	5
$\sqrt{10}$	$1/\sqrt{5}$	π	$e+1$	−1.0602	$\pi+e+1$

The next table, Table 2, is a sample selection of unstable slowly oscillating periodic solutions implied by Theorem 1. The values h_* for these periodic solutions are found from Equation (8) as well. The unstable periodic solutions cannot be verified numerically as respective initial functions $\varphi \equiv h_* < 0$ actually generate approximately close, nearby solutions (not the exact ones). These solutions, when calculated in forward time, are repelled for large times by the actual unstable ones and are attracted by the stable periodic solutions of the double period $2T$. The latter are studied analytically and numerically and described in detail in the next subsection.

Table 2. Examples of parametric values for which unstable periodic solutions exist for Equation (4) with the coefficient defined by (6). The values h_* are given by (8) with $b < 0$ and $m > 1$.

a_1	a_2	p_1	p_2	h_*	T
0.5	5	5	1	−1.31	6
1	5	4	1	−1.625	5
1	7	4	1	−1.5	5
1	7	3	2	−1.0833	5
1.5	7	2.5	2.5	−1.3295	5
1.5	7	3.5	1.5	−2.0795	5
1.5	7	4	3	−4.3636	7
2	7	3	2	−2.4	5
2	7	3	3	−4.6	6
2	7	3.5	2.5	−4.3	6
2.5	7	2.5	2.5	−1.2614	5
2.5	7	3	2	−1.5682	5
$1/\sqrt{5}$	$3\pi/2$	$2e$	$\pi/3$	−1.382	$2e+\pi/3$

3.2. Periodic Solutions with Double Coefficient Period (Type II)

In this subsection, we explicitly construct piecewise affine periodic solutions to Equation (4), which are of the double period $2T$ of the T-periodic coefficient $a(t)$. Some of the exposition is close and based on the results of our recent conference paper [16].

We presume that the desired periodic solutions have the initial shape shown in Figure 2. That is, the initial value $\varphi(0) = h < 0$ is determined using an initial function $\varphi(s) \in K_-$. The corresponding solution $x = x(t, \varphi), t \geq 0$, has exactly one zero t_1 on the initial periodic interval $[0, T], T = p_1 + p_2$, with $t_1 < p_1$ and $x(t) < 0 \; \forall t \in [0, t_1)$ and $x(t) > 0 \; \forall t \in (t_1, T]$. Such assumptions are verified to be valid through the analytical calculations that follow, as well as confirmed using associated numerical simulations.

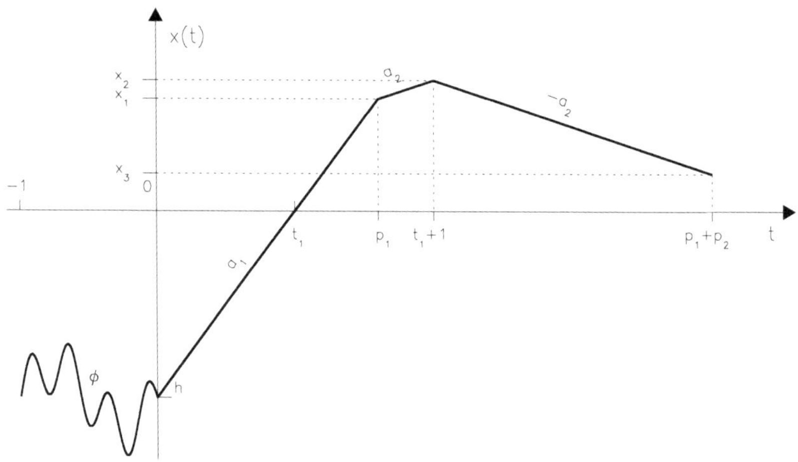

Figure 2. Slowly oscillating piecewise affine solution (Type II) on initial periodic interval.

It is straightforward to calculate the desired values of t_1, x_1, x_2, x_3 as follows:

$$t_1 = -\frac{h}{a_1}, \quad x_1 = x(p_1) = h + a_1 p_1, \quad x_2 = x(t_1 + 1) = \left(1 - \frac{a_2}{a_1}\right)h + p_1(a_1 - a_2) + a_2,$$

$$x_3 = x(p_1 + p_2) = \left(1 - 2\frac{a_2}{a_1}\right)h + a_1 p_1 + a_2(2 - 2p_1 - p_2) := F_1(h) = kh + d,$$

where

$$k = 1 - 2\frac{a_2}{a_1} \quad \text{and} \quad d = a_1 p_1 + a_2(2 - 2p_1 - p_2) \tag{9}$$

are uniquely defined by the parameters p_1, p_2, a_1, and a_2.
Likewise, when analogous calculations $\psi \in K_+$ and $\psi(0) = h > 0$ yield

$$x_3 = x(p_1 + p_2, \psi) = \left(1 - 2\frac{a_2}{a_1}\right) h - a_1 p_1 - a_2(2 - 2p_1 - p_2) := F_2(h) = kh - d,$$

with the same k and d as in the expression for F_1 given by Equation (9), it is easy to see that for arbitrary positive a_1 and a_2, the respective values of k can only belong to the range $(-\infty, 1)$ with $|k| < 1$ if $0 < a_2 < a_1$ and $k < -1$ if $0 < a_1 < a_2$.

We are interested in parametric values p_1, p_2, a_1 and a_2 and initial values $h < 0$ such that $h_1 = F_1(h) > 0$ and $h_2 = F_2(h_1) < 0$. This will be the case when $d > 0$, at least in some vicinity $|h| < \delta_1$, where $h \neq 0$, of the discontinuity point $h = 0$. The through map $F_0 = F_2 \circ F_1$ is given by $F_0(h) = k^2 h + (k - 1)d$. Its only fixed point is found as $h_* = -d/(k + 1)$, and it is negative when $d > 0$ and $|k| < 1$. The fixed point h_* generates an asymptotically stable periodic solution of period $2T$ when the initial function for Equation (4) is chosen as $\varphi(s) \equiv h_*$. The periodic solution can also be described in terms of a cycle of period 2 for an appropriate one-dimensional map F defined by the maps F_1 and F_2 introduced above. Namely, we define the piecewise affine map F for $h \in \mathbb{R} \setminus \{0\}$ as follows:

$$F(h) = \begin{cases} F_1(h) = kh + d, & \text{if} \quad h < 0 \\ F_2(h) = kh - d, & \text{if} \quad h > 0. \end{cases}$$

The map F is discontinuous at $h = 0$ and is symmetric (odd) with respect to $h = 0$, $F(-h) = -F(h)$, $\forall h \neq 0$. Assuming $d > 0$, one can see that the negative feedback condition is satisfied at least locally in a vicinity of $h = 0$. The global dynamics of map F are very simple in this case, as described by the following statement.

Proposition 2. *Suppose that $d > 0$. Then, consider the following:*

(i) *If $0 < a_2 < a_1$ is satisfied, map F has a unique globally attracting cycle of period 2, given by $\{h_*, -h_*\}$, where $h_* = -d/(k+1)$. The two-cycle attracts all the initial values $h \in \mathbb{R} \setminus \{0\}$ in the case when $a_1 > 2a_2$ holds, and all the initial values from the interval $h \in (-d/k, d/k)$ in the case when $2a_2 > a_1 > a_2$ holds;*

(ii) *If $0 < a_1 < a_2$ is satisfied, map F has no cycles of period 2; moreover, any iterative sequence $F^n(h), h \neq 0$, diverges, $\lim_{n\to\infty} |F^n(h)| = \infty$.*

Therefore, by translating the properties of the interval map F to the slowly oscillating solutions of DDE (4), we have the following statement.

Theorem 2. *Let the parameters a_1, a_2, p_1, and p_2 be given with values k and d defined by (9) and $d > 0$. Then, in the case when $|k| < 1\,(0 < a_2 < a_1)$ is valid, DDE (4) has an asymptotically stable slowly oscillating $2T$-periodic solution. The periodic solution is generated by the initial function $\varphi(s) \equiv h_*, s \in [-1, 0]$, where $h_* = -d/(k+1)$. In the case when $k < -1\,(0 < a_2 < a_1)$ is valid, the equation does not have any $2T$-periodic slowly oscillating solutions.*

Table 3 provides sample selections of parametric values for which the stable $2T$-periodic solutions exist. Besides using Formula (8) to calculate respective h_*, their existence was also confirmed numerically.

Table 3. Examplesof parametric values for which stable periodic solutions with the double period of coefficient $a(t)$ exists for Equation (4). The values of h_* are given by Equation (9).

a_1	a_2	p_1	p_2	h_*	$2T$
4	1	0.5	2.5	−0.33	6
4	1	1	3.5	−0.33	9
5	1	1	3.5	−0.938	9
5	1	0.5	3	−0.313	7
6	1	1	3.5	−1.5	9
6	1	1	4.5	−0.9	11
6	1.5	1.0	3.5	−0.5	9
7	1.5	1.5	3.5	−2.39	10
7	2.5	2.5	2.5	−2.92	10
7	2.5	2	3	−1.17	10
$1/\sqrt{5}$	$3\pi/2$	$3\pi/2$	$e/2$	−1.12	$3\pi + e$

3.3. Coexistence of Periodic Solutions

Any slowly oscillating solution considered above in Sections 3.1 and 3.2 is uniquely defined by the sign-definite initial function/values $\varphi \in \mathbb{X}$ such that $\varphi(0) = h \neq 0$ and $\varphi(s) \neq 0 \; \forall s \in [-1, 0]$, and by the parameters p_1, p_2, a_1, and a_2 of the periodic coefficient $a(t)$ (note that $f = f_0(x) = -\text{sign}\,(x)$ is fixed). One solves the initial value problem for $t \geq 0$ as follows:

$$x'(t) = a_0(t)\, f_0(x(t-1)),\; t \geq 0,\; \varphi \in \mathbb{X}, \varphi(0) = h \neq 0, \varphi(s) \neq 0 \; \forall s \in [-1, 0]. \tag{10}$$

Its solution $x_\varphi(t) = x_\varphi(t, h, p_1, p_2, a_1, a_2)$ is easily found to exist for all $t \geq 0$; it is uniquely determined by the set of the parameters p_1, p_2, a_1, and a_2 and the initial value of h at $t_0 = 0$.

Instead of the initial time value $t_0 = 0$, as in the case above, one can consider a similar initial value problem for any $t_0 > 0$. In particular, in the case of the initial time $t_0 = p_1$, one is solving the following initial value problem for $t \geq p_1$:

$$x'(t) = a_0(t)\, f_0(x(t-1)),\ t \geq p_1,\ \psi(p_1) = h \neq 0,\ \psi(s) \neq 0\ \forall s \in [p_1 - 1, p_1]. \tag{11}$$

Similarly, its solution $x_\psi(t) = x_\psi(t, h, p_1, p_2, a_1, a_2)$ is uniquely defined by the set of parameters p_1, p_2, a_1, and a_2 and by the initial value h at $t_0 = p_1$.

Now, due to the T-periodicity of the coefficient $a(t)$, one can easily see that the solutions of the two initial value problems, (10) and (11), are closely related, with an interchange in the set of the parameters and a time shift. Namely, there is just a time shift by p_1 with the parameter interchange in the following sense:

$$x_\varphi(t, h, p_2, p_1, a_2, a_1) = x_\psi(t + p_1, h_1, p_1, p_2, a_1, a_2), \quad \forall t \geq 0, \tag{12}$$

with an appropriate $h_1 = h_1(h)$. This simple observation allows us to conclude the simultaneous existence of periodic solutions of both types I and II, which are described in Theorems 1 and 2. Since the value of the ratio a_1/a_2 with respect to 1 determines the stability of these periodic solutions, their stability types must be opposite to each other. We have the following statement.

Theorem 3. *There is an open set of parameters p_1, p_2, a_1, and a_2 such that DDE (4) has simultaneously an unstable slowly oscillating T-periodic solution of type I and an asymptotically stable $2T$-periodic solution of type II. These two periodic solutions can be paired in the following sense: small perturbations of the unstable solution are repelled by it and attracted by a respective stable solution as $t \to \infty$.*

Such a situation of two coexisting periodic solutions is depicted in Figure 3. The unstable periodic solution $x_u(t)$ of period $T = 4$ is given by Theorem 1 when $p_1 = 3, p_2 = 1$, $a_1 = 1$, and $a_2 = 6$; the respective value of h is $h_* = -0.5$. The stable periodic solution $x_s(t)$ of period $2T = 8$ is given by Theorem 2 for the parameter values $p_1 = 1, p_2 = 3, a_1 = 6$, and $a_2 = 1$ so that the respective values of p_1, p_2 and a_1, a_2 are interchanged for these two solutions. In addition, the solution $x_s(t)$ is shifted by p_1 to the right (graph (b) in Figure 3 shows the solution $x_s(t + p_1)$ for longer time intervals).

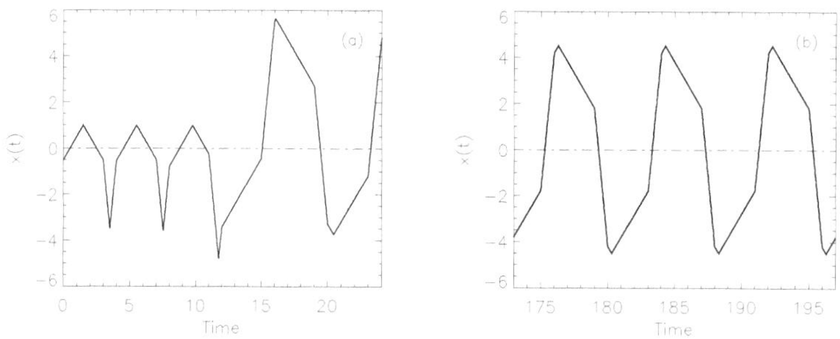

Figure 3. An example of numerical solution with $a_1 = 1, a_2 = 6, p_1 = 3$, and $p_2 = 1$. (**a**)—unstable solution with single period, which is destroyed after two first periods. (**b**)—stable solution with double period. Numerical solution of Equation (4) with $a_1 = 6, a_2 = 1, p_1 = 1$, and $p_2 = 3$, of which its solution has the double period of the T-periodic coefficient and is overplotted as a red dashed line in the right panel.

Tables 4 and 5 provide a sample selection of such dual cases of the coexistence of two types of stable/unstable solutions. Table 5 is a mirror image of Table 4 for the stable periodic solutions with double periods. Namely, each row in Table 5 represents a stable periodic solution of the corresponding unstable solution in the same row number as Table 4.

Table 4. Examples of parametric values for which unstable and stable periodic solutions coexist for DDE (4) with the coefficient defined by (6). The parametric data are shown for the unstable solutions.

a_1	a_2	p_1	p_2	h_*	T
0.5	2.5	3	0.5	−0.09	3.5
0.5	3	5	1	−1.35	6
0.5	5	4	0.5	−0.64	4.5
0.5	7	3	2	−0.52	5
1	6	3	1	−0.5	4
1	7	5	1	−2.67	6
2	7	2	3	−1.4	5
3	7	4	1	−5.62	5
$\pi/6$	e	π	1/2	−0.18	$\pi + 1/2$

Table 5. Examples of parametric values for which stable periodic solutions exist for Equation (4) with the coefficient defined by Equation (6). The parametric data are shown for the stable solutions.

a_1	a_2	p_1	p_2	h_*	$2T$
2.5	0.5	0.5	3	−0.156	7
3	0.5	1	5	−0.3	12
5	0.5	0.5	4	−0.56	9
7	0.5	2	3	−6.19	10
6	1	1	3	−1.8	8
7	1	1	5	−1.17	12
7	2	3	2	−6.3	10
7	3	1	4	−4.375	10
e	$\pi/6$	1/2	π	−1.92	$2\pi + 1$

4. Smoothed Nonlinearities

In this section, we follow a standard procedure of replacing the discontinuous piecewise constant functions f_0 and A_0 with continuous functions f_δ and A_δ close to them for a small $\delta > 0$. This procedure has been used in numerous cases; typical examples can be found in papers [19–22]. In a δ-neighborhood with every discontinuity point for both f_0 and A_0, the jump discontinuity is replaced with an affine function by connecting the two constant levels using a line segment. That is, define $f_\delta(x)$ and $A_\delta(t)$ as follows:

$$f(x) = f_\delta(x) = \begin{cases} +1 \text{ if } x \le -\delta \\ -1 \text{ if } x \ge \delta \\ -(1/\delta)x \text{ if } x \in (-\delta, \delta), \end{cases} \tag{13}$$

and

$$a(t) = A_\delta(t) = \begin{cases} a_2 + \frac{a_1 - a_2}{2\delta}(t + \delta) \text{ if } t \in (-\delta, \delta) \\ a_1 \text{ if } t \in [\delta, p_1 - \delta] \\ a_1 + \frac{a_2 - a_1}{2\delta}[t - (p_1 - \delta)] \text{ if } t \in (p_1 - \delta, p_1 + \delta) \\ a_2 \text{ if } t \in [p_1 + \delta, p_1 + p_2 - \delta] \\ a_2 + \frac{a_1 - a_2}{2\delta}[t - (p_2 - \delta)] \text{ if } t \in (p_1 + p_2 - \delta, p_1 + p_2 + \delta) \\ \text{periodic extension on } \mathbb{R} \text{ outside interval } [0, p_1 + p_2). \end{cases} \tag{14}$$

We shall show next that the values of the piecewise affine slowly oscillating solutions constructed in Section 3 for $t \geq 0$ do not change throughout the entire periodic interval $[0, T]$, except in a small ε-vicinity of the corner points (ε is a multiple of δ).

Indeed, the case of a symmetric corner solution, as it passes through a δ-neighborhood of $x = 0$ for the nonlinearity f_δ, is shown in Figure 4 (which corresponds to, e.g., the type I solution of the first positive semicycle depicted in Figure 1). The δ-neighborhood of $x = 0$ for f corresponds to the $\varepsilon = \delta/a_1$ neighborhood for t at $t = t_1 + 1$. It is immediately clear that the replacement of the corner type solution on the interval $[t_1 + 1 - \varepsilon, t_1 + 1 + \varepsilon]$ with the respective parabolic segment, as a result of the corresponding integration now with a linear nonlinearity f, results in a symmetric parabola with the vertex at $t_1 + 1$ and the same equal values $x(t_1 + 1 - \varepsilon)$ and $x(t_1 + 1 - \varepsilon)$ as for the initial piecewise solution with the discontinuous f_0.

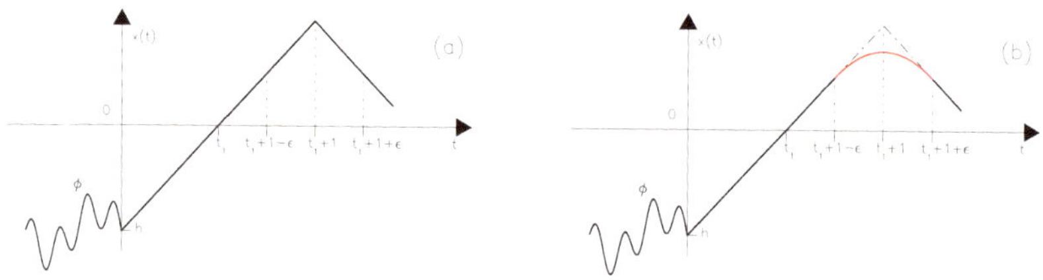

Figure 4. Sketches of (**a**) continuous solution x with corner-type discontinuity for x' at $t_1 + 1$ for functions $A_0(t)$ and $f_0(x)$; (**b**) C^1-smooth solution with parabolic matching on $[t_1 + 1 - \varepsilon, t_1 + 1 + \varepsilon]$, $\varepsilon a_1 = \delta$, for functions $A_\delta(t)$ and $f_\delta(x)$ with small $\delta > 0$.

The general case of a corner-type solution (piecewise affine) is shown in Figure 5, which corresponds to, e.g., point (p_1, x_1) on the graph of a solution of type II (see Figure 2). We shall show now that a smooth parabolic segment solution connecting points $(p_1 - \varepsilon, x(p_1 - \varepsilon))$ and $(p_1 + \varepsilon, x(p_1 + \varepsilon))$ is the precise replacement of the corresponding piecewise affine segment solution that is obtained for the piecewise constant functions f_0 and A_0. The values of the solution x and its derivatives for the piecewise affine solution are given as follows:

$$x(p_1 - \varepsilon) = x_1 - a_1\varepsilon, \quad x'(p_1 - \varepsilon) = a_1, \quad x(p_1 + \varepsilon) = x_1 + a_2\varepsilon, \quad x'(p_1 + \varepsilon) = a_2,$$

By looking for a parabolic connection $y = P(t) = A(t - p_1)^2 + B(t - p_1) + C$ as a solution of DDE (4) on the interval $[p_1 - \varepsilon, p_1 + \varepsilon]$, starting at the point $(p_1 - \varepsilon, x_1 - a_1\varepsilon)$ and satisfying the necessary initial-boundary conditions,

$$x(p_1 - \varepsilon) = x_1 - a_1\varepsilon, \quad x'(p_1 - \varepsilon) = a_1, \quad x'(p_1 + \varepsilon) = a_2,$$

one easily find that the values of the parameters A, B, and C are

$$A = \frac{a_2 - a_1}{4\varepsilon}, \quad B = \frac{a_1 + a_2}{2}, \quad C = \frac{a_2 - a_1}{4}\varepsilon + x_1.$$

Then, by checking the value of the parabolic segment of the solution $x(t) = P(t)$ at $t = p_1 + \varepsilon$, one verifies that

$$\begin{aligned} P(p_1 + \varepsilon) &= A(t - p_1)^2 + B(t - p_1) + C|_{t=p_1+\varepsilon} \\ &= \frac{a_2 - a_1}{4\varepsilon}\varepsilon^2 + \frac{a_1 + a_2}{2}\varepsilon + \frac{a_2 - a_1}{4}\varepsilon + x_1 = x_1 + a_2\varepsilon = x_1 + a_2(t - p_1)|_{t=p_1+\varepsilon}, \end{aligned}$$

which is the same value as the piecewise affine solution at that point.

Thus, the solutions to Equation (4) with continuous $f = f_\delta$ and $a = A_\delta$ are obtained from the solutions when $f = f_0$ and $a = A_0$ by simply cutting out and replacing the small corner segments with respective parabolic segments in their ε-vicinity ($\varepsilon = c \cdot \delta$). An immediate consequence of this fact is that the periodic solutions and their stability persist under such small δ-perturbations of f_0 and a_0 in DDE (4).

Indeed, the easiest way to see this is to consider the new starting point for $t, t = \varepsilon$, with $\varphi(\varepsilon) = H \neq 0$. Then, the corresponding shift map S^T using period T is dynamically equivalent (conjugate) to the respective map constructed in the case of f_0 and a_0. But the latter is equivalent to a composition of respective maps G,F,F_1,F_2, and F_0 derived in Section 3. If one would like to stay with the initial value $H := \varphi(0) \neq 0$ in the case of continuous $f_\delta, a_\delta, \delta > 0$, then a small adjustment in the considerations should be made. Namely, a new $h := H \pm c\varepsilon$ should be introduced, where $c\varepsilon$ is the difference between the two solutions of the differential delay, Equation (4), in the two cases when $\delta = 0$ and small $\delta > 0$. Such a difference is shown in the graph in Figure 4b and in the graph in Figure 5b (in both graphs, it is positive). In general, such a difference is independent of the value x_1 (or equivalent values in other analogous corner situations). Therefore, all the resulting maps G, F, F_0, F_1, and F_2 derived in Section 3 will become finite compositions of the affine maps of the form $h \mapsto \tilde{F}$, where $\tilde{F}(h) = G_*(h + c_1\varepsilon) + c_2\varepsilon$, and $G_*(h)$ is one of the elementary maps considered in Sections 3.1 and 3.2. Therefore, we are in a position to state the following.

Theorem 4. *Suppose that the parametric values a_1, a_2, p_1, and p_2 are such that the assumptions of either one of Theorems 1 or 2 are satisfied. Then, there exists $\delta_0 > 0$ such that for every $0 \leq \delta \leq \delta_0$ the delay differential of Equation (4) with functions $f = f_\delta(x)$ and $a = A_\delta(t)$ has a C^1-smooth periodic solution with the same type of stability as in the respective Theorems 1 or 2. Such a solution converges in uniform metric as $\delta \to +0$ to the respective periodic solution with the corner type discontinuity for x' (when $f = f_0, a = A_0$).*

Note that the smoothing of both functions f_0 and a_0 in a δ-neighborhood of their discontinuity set can be created such that the resulting continuous nonlinearities f_δ and a_δ are continuously differentiable or even of the C^∞ class, and Theorem 4 still remains valid. Such C^1-smoothness is required in some cases when, e.g., one would need to study the Floquet multipliers of the periodic solutions and consider the corresponding linearized equation along the periodic solutions.

Indeed, it is an elementary fact that there are many solutions for the following approximation problem. Find a smooth function $F(t)$ defined on an interval $[t_1, t_2]$ such that it satisfies the given boundary conditions: $F(t_1) = x_1, F'(t_1) = m_1$ and $F(t_2) = x_2, F'(t_1) = m_2$. One of the possible solutions can be suggested as a polynomial of degree 3 or higher. In our case of a C^∞ smooth connection for $f_0(x)$ on the interval $[-\delta, \delta]$, one can use the following function:

$$f^0(x) = e^{\frac{\delta x}{x-\delta}} - 1, \text{ for } x \in [0, \delta], \quad \text{and} \quad f^0(x) = -f^0(x), \text{ for } x \in [-\delta, 0].$$

It is straightforward to verify that such a connecting function is decreasing on the interval $[-\delta, \delta]$, in which it is of class C^∞, with all the matching derivatives at $x = 0, x = -\delta$, and $x = \delta$. An analogous C^∞-connection can be used for the coefficient $a(t)$ at points $0, p_1$, and $T = p_1 + p_2$.

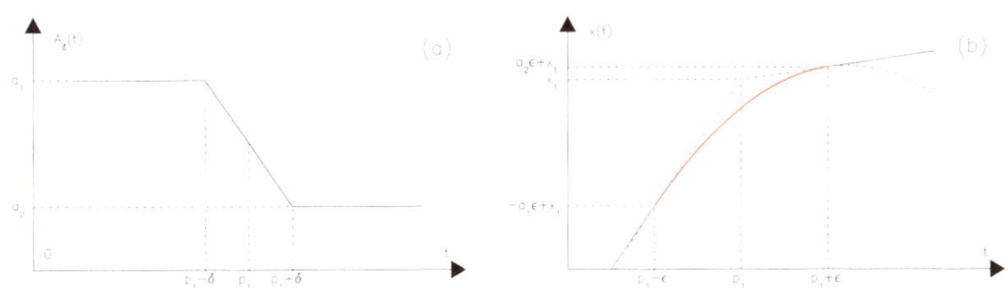

Figure 5. Sketches of (**a**) continuous coefficient $A_\delta(t)$ in δ-vicinity of $t_1 = p_1$; (**b**) C^1-smooth solution with parabolic matching on $[p_1 - \varepsilon, p_1 + \varepsilon]$ for functions $A_\delta(t)$ and $f_\delta(x)$ with small $\delta > 0$.

Author Contributions: Conceptualization, A.I.; Methodology, A.I.; Software, S.S.; Validation, A.I. and S.S.; Formal analysis, A.I. and S.S.; Investigation, A.I. and S.S.; Writing—original draft, A.I. and S.S. All authors have read and agreed to the published version of the manuscript.

Funding: This research was financially supported by Simons Foundation (U.S.) through MATRIX-Simons Collaborative Fund and MATRIX-Simons Travel Grant Scheme, Flinders University (Australia) and Pennsylvania State University (U.S.).

Data Availability Statement: Data are contained within the article.

Acknowledgments: The authors thank the mathematical research institute MATRIX in Australia where part of this research was performed. This work's final version resulted from the collaborative activities of the authors during the workshop "Delay Differential Equations and Their Applications" (https://www.matrix-inst.org.au/events/delay-differential-equations-and-their-applications/ (accessed on 12 March 2024)) held in December 2023. The authors are also grateful for the financial support provided for these research activities by Simons Foundation (USA), Flinders University (Australia), and the Pennsylvania State University (USA). A.I.'s research was also supported in part by the Alexander von Humboldt Stiftung (Germany) during his visit to Justus-Liebig-Universität, Giessen, in June–August 2023.

Conflicts of Interest: The authors declare no conflict of interest.

References

1. Erneux, T. *Applied Delay Differential Equations*; Series: Surveys and Tutorials in the Applied Mathematical Sciences; Springer: Berlin/Heidelberg, Germany, 2009; Volume 3.
2. Diekmann, O.; van Gils, S.; Lunel, S.V.; Walther, H.-O. *Delay Equations: Complex, Functional, and Nonlinear Analysis*; Series: Applied Mathematical Sciences; Springer: Berlin/Heidelberg, Germany, 1995; Volume 110.
3. Hale, J.K.; Lunel, S.M.V. *Introduction to Functional Differential Equations*; Springer Applied Mathematical Sciences: Berlin/Heidelberg, Germany, 1993; Volume 99.
4. Smith, H. *An Introduction to Delay Differential Equations with Applications to the Life Sciences*; Series: Texts in Applied Mathematics; Springer: Berlin/Heidelberg, Germany, 2011; Volume 57.
5. Glass, L.; Mackey, M.C. *From Clocks to Chaos. The Rhythms of Life*; Prinston University Press: Princeton, NJ, USA, 1988.
6. Kuang, Y. *Delay Differential Equations with Applications in Population Dynamics*; Series: Mathematics in Science and Engineering; Academic Press Inc.: Cambridge, MA, USA, 1993; Volume 191, 398p.
7. Mackey, M.C.; Glass, L. Oscillation and chaos in physiological control systems. *Science* **1977**, *197*, 287–289. [CrossRef] [PubMed]
8. Wazewska-Czyzewska, M.; Lasota, A. Mathematical models of the red cell system. *Mat. Stosow.* **1976**, *6*, 25–40. (In Polish)
9. Gurney, W.S.C.; Blythe, S.P.; Nisbet, R.M. Nicholson's blowflies revisited. *Nature* **1980**, *287*, 17–21. [CrossRef]
10. Perez, J.F.; Malta, C.P.; Coutinho, F.A.B. Qualitative Analysis of Oscillations in Isolated Populations of Flies. *J. Theoret. Biol.* **1978**, *71*, 505–514. [CrossRef]
11. Amster, P.; Idels, L. Periodic solutions in general scalar non-autonomous models with delays. *Nonlinear Differ. Equ. Appl.* **2013**, *20*, 1577–1596. [CrossRef]
12. Friedman, H.I.; Wu, J. Periodic solutions of single-species models with periodic delay. *SIAM J. Math. Anal.* **1992**, *23*, 689–701. [CrossRef]
13. Li, Y.; Kuang, Y. Periodic solutions in periodic state-dependent delay equations and population models. *Proc. Am. Math. Soc.* **2001**, *130*, 1345–1353. [CrossRef]
14. Yu, J.; Li, J. Dynamics of interactive wild and sterile mosquitoes with time delay. *J. Biol. Dyn.* **2019**, *13*, 606–620. [CrossRef] [PubMed]
15. Faria, T. Periodic solutions for a non-monotone family of delayed differential equations with applications to Nicholson systems. *J. Differ. Equ.* **2017**, *263*, 509–533. [CrossRef]
16. Ivanov, A.; Shelyag, S. Stable periodic solutions in scalar periodic differential delay equations. *Arch. Math.* **2023**, *59*, 69–76. [CrossRef]
17. Ivanov, A.; Shelyag, S. Explicit periodic solutions in a delay differential equation. *arXiv* **2023**, arXiv:2402.08197.
18. Győry, I.; Ladas, G. *Oscillation Theory of Delay Differential Equations*; Oxford Science Publications, Clarendon Press: Oxford, UK, 1991; 368p.
19. Hale, J.K.; Lin, X.-B. Examples of transverse homoclinic orbits in delay equations. *Nonlinear Anal. Theory Methods Appl.* **1986**, *10*, 693–709. [CrossRef]
20. Ivanov, A.F.; Sharkovsky, A.N. *Oscillations in Singularly Perturbed Delay Equations*; Dynamics Reported (New Series); Springer: Berlin/Heidelberg, Germany, 1991; Volume 1, pp. 165–224.

21. Peters, H. Comportment chaotique d'une equation retardee. *C. R. Acad. Sci. Paris Ser. I Math.* **1980**, *290*, 1119–1122.
22. Walther, H.-O. Homoclinic solution and chaos in $\dot{x}(t) = f(x(t-1))$. *Nonlinear Anal. TMA* **1981**, *5*, 775–788. [CrossRef]

Mathematical and Computational Applications

MDPI

Article

DSTree: A Spatio-Temporal Indexing Data Structure for Distributed Networks

Majid Hojati *, Steven Roberts and Colin Robertson

Department of Geography & Environmental Studies, Wilfrid Laurier University, Waterloo, ON N2L 3C5, Canada; sroberts@wlu.ca (S.R.); crobertson@wlu.ca (C.R.)
* Correspondence: hoja4090@laurier.ca

Abstract: The widespread availability of tools to collect and share spatial data enables us to produce a large amount of geographic information on a daily basis. This enormous production of spatial data requires scalable data management systems. Geospatial architectures have changed from clusters to cloud architectures and more parallel and distributed processing platforms to be able to tackle these challenges. Peer-to-peer (P2P) systems as a backbone of distributed systems have been established in several application areas such as web3, blockchains, and crypto-currencies. Unlike centralized systems, data storage in P2P networks is distributed across network nodes, providing scalability and no single point of failure. However, managing and processing queries on these networks has always been challenging. In this work, we propose a spatio-temporal indexing data structure, DSTree. DSTree does not require additional Distributed Hash Trees (DHTs) to perform multi-dimensional range queries. Inserting a piece of new geographic information updates only a portion of the tree structure and does not impact the entire graph of the data. For example, for time-series data, such as storing sensor data, the DSTree performs around 40% faster in spatio-temporal queries for small and medium datasets. Despite the advantages of our proposed framework, challenges such as 20% slower insertion speed or semantic query capabilities remain. We conclude that more significant research effort from GIScience and related fields in developing decentralized applications is needed. The need for the standardization of different geographic information when sharing data on the IPFS network is one of the requirements.

Keywords: spatio-temporal indexing; temporal topology; query processing; *IPFS*; distributed systems; smart contracts; blockchain

1. Introduction

With the advancement of Internet-based services and sensors, such as the widespread adoption of GPS (Global Positioning System)-based sensors, location-based services, improvements in computational data processing, and satellite imagery, a large amount of novel spatial information is produced daily [1]. The widespread availability of tools to collect and share spatial data enables individuals and small communities to produce their own digital spatial content (e.g., Open Street Maps contributions have tripled between 2012 and 2017 [2]). This enormous production of spatial data requires scalable data management systems [3]. The data infrastructure technology supporting spatial data management and processing has changed from standalone relational database systems to spatial data warehouses that support a variety of data formats and analytical workloads [4] and from centralized infrastructures to decentralized and peer-to-peer (P2P) systems. In addition to the new system architectures, new data structures have been proposed such as, e.g., HDF [5], Data-Cubes [6], Geoparquet [7], and spatial data management and analytic frameworks (e.g., Apache iceberg [8], Digital Earth [9]) have emerged to manage and perform analysis on large-scale, high temporal and spatial resolutions.

Geospatial architectures are another area of technological change developed to handle spatial data management challenges. They have changed from clusters to cloud architectures and more parallel and distributed processing platforms (e.g., Spark [10], Hadoop [11]).

In the realm of network and application architecture, P2P systems have been established in several application areas. In the past decade, P2P systems have been used widely in web3 [12], blockchains [13], and crypto-currencies [14]. The combination of P2P file-sharing systems with blockchains provides scalability, security, immutability, and append-only attributes for sharing information amongst nodes on a network [15]. These attributes make P2P networks suitable for content distribution and service discovery applications [16–19]. However, the main limitation of existing systems is they can only locate data on the network based on a key value using DHT [20]. While DHTs have been used as a main building block for P2P applications, they are seriously deficient in one regard; they only directly support exact match queries and do not allow users to query data by range requests [21–24]. The multi-dimensionality of spatio-temporal data is one of the big challenges in retrieving and querying these data in P2P networks. When querying spatio-temporal data, the ability to perform range queries is required, and it is not currently supported. The rapid increase in spatio-temporal data collection needs a new auxiliary indexing structure. These indexing structures are responsible for tracking the behavior of moving objects through space [25,26]. These indexing methods allow P2P architectures to find and retrieve contents based on the user's filters and address more complex data sharing needs, facilitating data management, query processing, and delivering data to the end-user.

While much work has been carried out towards expediting search in file-sharing P2P systems, issues concerning spatial indexing in P2P systems are significantly more complicated due to cases such as overlaps between spatial objects, avoidance of data scattering, and the complexity of spatial queries [27]. One-dimensional data mainly have been queried using DHTs in P2P networks [28–30]. DHTs are not yet designed for complex spatial queries (e.g., range query, k-nearest neighbor query), and only support the location of data items based on a key value (i.e., equality lookups) [20,31]. For multi-dimensional data, there have been multiple approaches; the first approach includes partitioning data into one-dimensional indexes using space-filling curves or kd-tree-based methods and indexing them using DHT methods (e.g., [32–34]). These methods usually work best with static data, and dynamic content relocation (locality) is dependent on the accuracy of the space-filling curves [34].

Another approach to apply range queries to the P2P networks is to use multiple DHTs in the network (e.g., [35–37]). This method needs a higher level of network and node structure manipulation and is not commonly used in large-scale projects due to interoperability limitations of this approach [38]. A third approach is to construct traditional indexes that have been used in centralized environments and distribute these indexes on the P2P network (e.g., [20,21,39,40]). This approach is implemented by constructing a tree and splitting it into parts and maintaining parts of semi-independent trees at each peer. Prefix Hash Tree (PHT) [21] and, similarly, P-Tree [20] use the same approach by storing a fraction of the overall tree on each peer. In PHT, each node of the tree is labeled with a prefix which is defined recursively. Given a node with the label l, its left and right children are labeled as $l0$ and $l1$, respectively. This pattern constructs a tree structure and enables range queries on a dataset [21].

In this work, we propose a spatio-temporal indexing data structure that works in the data layer and uses a distributed InterPlanetary File System (IPFS) network. Our method is closer to the approach of Ranabhadran et al. [21] and does not require additional DHTs to perform multi-dimensional range queries. The rest of this paper is divided into two main sections. First, we introduce the Distributed Spatio-Temporal Tree (DSTree) as a data structure to perform range spatio-temporal queries on a dataset and we check its features and performance by comparing it to other existing trees. Second, we will look at the integration of DSTree with distributed networks and propose a system architecture to perform queries on the IPFS using DSTree. The originality of this work includes the ability of the spatio-temporal queries on the data shared on IPFS system. This sort of query allows sharing and querying spatio-temporal data on P2P systems without the need for any third-party central entities.

2. Spatio-Temporal Data Indexing Methods

P2P multidimensional query processing refers to the execution of advanced query operators over multidimensional data stored in a distributed system [41]. The retrieved data from a query should be exact and complete. Exactness means the query result should not be approximate. The retrieved data should exactly belong to the query results set. This means that if we run a query on the same dataset on a centralized system, the results should be exactly the same as when we run a query on the distributed system. A basic element of geospatial technology can be defined as three main components including location in space and time, and attribute of that location in space-time [42]. In this work, our focus is to address a methodology to allow users to retrieve a geographic object based on space or time queries from a P2P network.

The spatio-temporal indexing methods in central systems are mainly performed at an abstract level (Figure 1). They are used to improve query performance on the datasets. These indexes usually work as a separate layer on top of the data layer itself. Queries are usually performed against the indexes and then the actual reference to the geographic features are then retrieved from the index. In contrast, our method stores data on the nodes on the network using spatial indexing methods in both abstract and physical storage layers. For example, the purpose of indexing can also be to distribute data that are closer in space, on the nodes that are closer in the network [20] (Figure 1, middle). Through the last couple of decades, many spatio-temporal access methods have been developed. There have been several approaches for spatio-temporal indexing so far [25,26,43–48]. Handling temporal data in GIS ranges from time-stamping GIS layers [49], to more object-oriented approaches such as time-stamping events and processes [44,50]. Another category of spatio-temporal models is trajectory-based access models which track the changes in the geographic object typologies over time [47]. In central systems, indexing models such as Oct-tree [51] are sometimes used to process spatio-temporal queries [47,52,53]. In using Oct-trees for indexing spatial data, each geographic object is considered as a cuboid. The spatial dimension of the data is considered as two dimensions of each cuboid. The third dimension of a cuboid is considered a time interval. There have been some newer approaches to the query process of spatio-temporal data using blockchains. For example, ref. [54] used a block-DAG-based index traversal algorithm to handle spatio-temporal queries on a block-DAG. However, the main issue with the blockchain-based spatial–temporal indexes is the limited data storage capability on the blockchains [55].

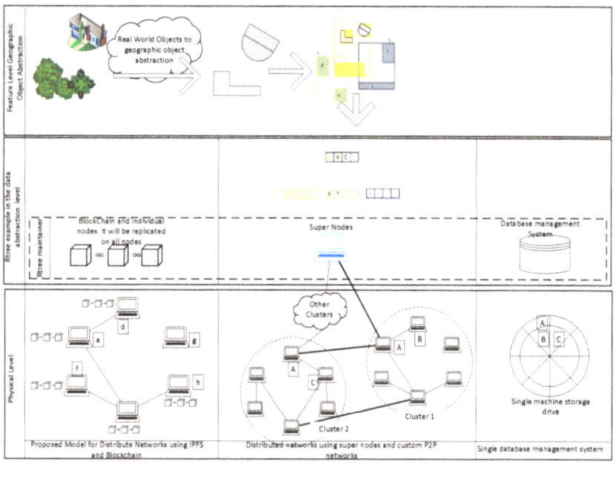

Figure 1. Abstraction levels of geographic information. Spatial indexes cluster units of Geographic Information (GI) at the abstract level and it is used at the storage level in the different architectures.

The Spatial Index maintenance is handled by RDBMS [56] (**Left**) or super nodes (**Middle**) in some studies [20], or using proposed blockchain-based method (**Right**). In the proposed model, each node maintains a spatial index and the latest version of the index is always published on the blockchain. Each node only stores and serves features that they need using IPFS. Rows show different abstraction levels of geographic information from real-world phenomena to machine language.

A decentralized spatial indexing technique must be scalable enough to be able to handle hundreds of thousands of peers and also dynamic enough to deal with peers joining/leaving the system anytime. Another feature of such structures is their ability to preserve the locality and directionality of multidimensional information. Locality implies that multidimensional information is stored in neighboring nodes, while directionality implies that the index structure preserves orientation. The notions of locality and directionality are very important. If an index structure preserves these properties, then searching in the index corresponds to searching in the multidimensional space, which can highly improve query evaluation cost [27]. R-tree-based indexes can efficiently answer various types of multidimensional queries, especially range queries [57]. In addition, a spatio-temporal indexing method is required to support two types of topological relations. The first set of topological relations includes temporal typologies which are based on Allen's temporal algebra covered in [58,59]. These relations include seven typologies that are briefly explained in Table 1. A time interval for Geographic Information (GI) can be defined as the duration in which a GI feature exists with a fixed state. This interval can be as small as a few milliseconds that it takes to collect a GI feature or can be a considerably longer time period, such as geological land classifications or land cover. Defining time intervals and how a GI can be attached to a newer time interval depends on the context of the study. The second type of topological relations which needs to be addressed by a spatio-temporal indexing model is spatial topology.

Each indexing method is optimized for specific types of queries. Our proposed method is more suitable for the storage, query processing, and retrieval of log-based data. These data are produced over the time that different events happen. An example of such data is data that are being collected from a sensor over time or a VGI tool to share images from different locations by users or even open data which are being shared by different government departments, such as crime data that are being shared by the police. Each of these datasets can have different access levels, and geoprivacy levels are being collected over time.

Table 1. Temporal algebra introduced by [58]. X (line border) is time interval of the first GI and Y (dashed border) is the time interval of the second GI.

	X in Relation to Y	Y in Relation to X	Condition
	equal (=)	equal (=)	$X_{t_1} = Y_{t_1}$ & $X_{t_2} = Y_{t_2}$
	meets (X m Y)	is met by (Y mi X)	$X_{t_2} = Y_{t_1}$
	overlaps (X o Y)	is overlapped by (Y oi X)	$Y_{t_1} \leq X_{t_1} \leq Y_{t_2}$
	during (X d Y)	contains (Y di X)	$Y_{t_1} \leq X_{t_1} \leq Y_{t_2}$ & $Y_{t_1} \leq X_{t_2} \leq Y_{t_2}$
	starts (X S Y)	is started by (Y Si X)	$X_{t_1} = Y_{t_1}$
	finishes (X F Y)	is finished by (Y Fi X)	$X_{t_2} = Y_{t_2}$
	precedes (X < Y)	is preceded by (Y > X)	$Y_{t_2} \leq X_{t_2}$

DSTree is a two-level tree structure. The approach of DSTree is close to the work carried out by [60–62]. They constructed a multilevel tree structure to improve the query process of trajectory data. In the method developed by [60], they formed a global space-time subdivision scheme. Sun et al. combined two trees, an R*-Tree and a kd-Tree, to improve the query process on the centralized machines [61]. Tao [62] also used a series of temporal quad-Trees to handle interval queries in centralized systems. In a DSTree, an Interval-Tree [63,64] is used as the top part of the tree, and a quad-Tree [65] is used at the bottom part of the tree.

3. Dstree: A Spatio-Temporal Index

A unit of GI in our model is defined as a spatio-temporal object which can be represented in the form of

$$Gid(GI, GI_{MBR}, GI_{Time\text{-}Interval})$$

where Gid is the identification of the object. GI is geographic location, including longitude and latitude (GI_x, GI_y) along x and y dimensions. GI_{MBR} is Minimum Boundary Rectangle (MBR), which is constructed based on its location. $GI_{Time\text{-}Interval}$ is the temporal interval the GI is valid during. Each GI also has a set of attributes, p, associated with it.

The interval tree is responsible for temporal queries and the quad-Tree is responsible for performing the spatial part of the queries. We define a time interval as a pair of real numbers $[t_1, t_2]$ where $t_1 < t_2$. $[t_1, t_2]$ can be represented as $\{[t_1, t_2] | t \in \mathbb{R} : t_1 \leq t \leq t_2\}$. Different variants of the interval trees are capable of supporting open and half-open intervals. Interval trees are optimized for querying of intervals which overlap with a given interval, but can also be used for point queries. Having the ability to query overlapping intervals allows us to query based on the temporal topology in Table 1. During each time interval, we assume that the GI state does not change.

3.1. Dstree Index

When constructing a DSTree, it is possible to partition data into spatio-temporal chunks and assign a unique ID to each portion of the data. The proposed DSTree indexes are composed of two parts, as shown in Figure 2. The first component (A) is called the interval tree index, and the second component (B) is called the quad index. The interval tree is a binary tree so each node can have only two children. By assigning 1 to the right child and 0 to the left child, a series of IDs will be constructed. The length of digits in the interval tree portion of the index equals the *temporal level* (T). The second section of the DSTree index is a quad index. Each node in a Quad-tree consists of four children. Each child can be assigned an index from 00, 01, 10, and 11 and it can construct the quad index. Figure 3 shows the DSTree index corresponding to quad-tree indexes.

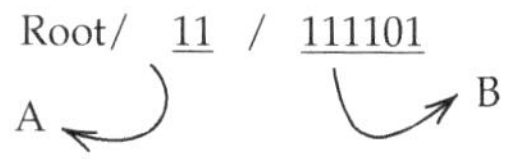

Figure 2. An example of DSTree Index. Each index is constructed of two sub-parts. Part A is the interval tree index. Its length is equal to the *temporal level* (T). Part B is the quad-tree index.

The top part of the DSTree is a regular interval tree. However, the depth of this tree is controlled by a parameter called *temporal level* (T). T is responsible for balancing between the top-level tree and the bottom-level tree. Once the top interval tree is constructed, the GI in the leaves (each leaf includes n GI) is used to construct a quad-tree. As a result, we will have one quad-tree at each leaf of the interval tree at the level of T. Regular quad-trees always have an extent equal to the minimum boundary box of the GI inserted into the tree. However, in a DSTree, all of the quad-trees should have the same extent, e.g., equal to $(-180, -90, 180, 90)$ in geographic coordinates. Having the same spatial extent, the bottom part of DSTree allows us to query data across all of the interval tree leaves. Figure 3 shows an example of a constructed DSTree. Figure 4 shows the points which are used to construct that tree and their relative location and time interval.

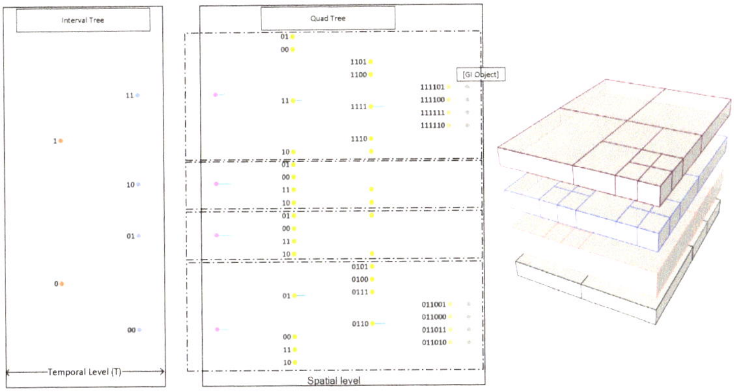

Figure 3. An example of DSTree constructed from a set of sample points. Each DSTree has a temporal (interval tree) component and a spatial (quad-tree) component. The final graph will be a stack of quad-trees on top of each other. Spatial level is the number of levels in the quad-tree.

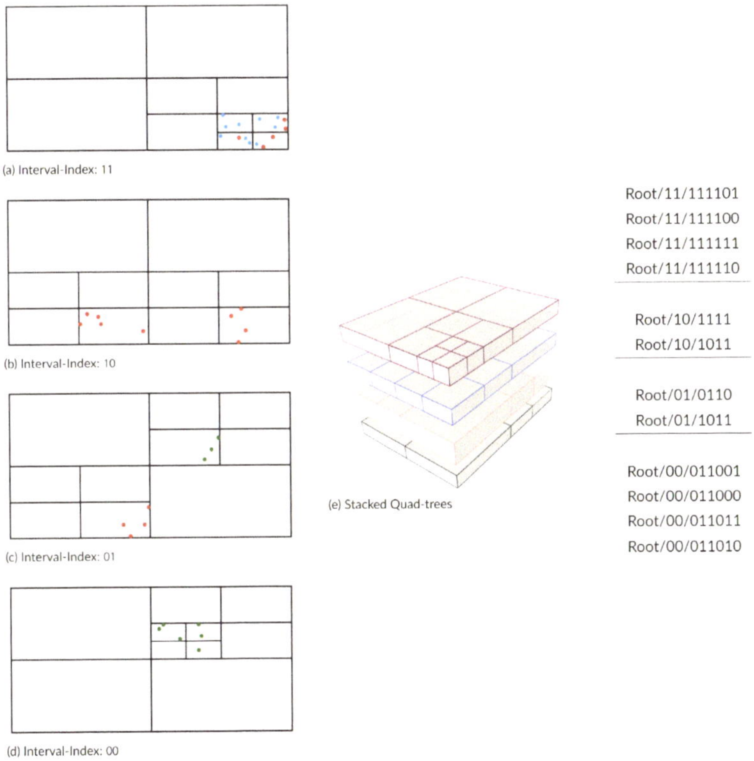

Figure 4. Spatial and temporal location of the points in Figure 3. On the right side, the DSTree-Index related to each section of the graph is listed.

3.2. Insert

In order to insert a GI $Gid(GI_{CID}, GI_{MBR}, GI_{Time\text{-}Interval})$ into a DSTree, a two-step process is required. First, we need to find the proper node on the interval tree part of the DSTree that *Gid* can be added to. Afterward, we will add the *Gid* to the proper quad-tree leaf using GI_{MBR}. To do so, we first obtain the low value of the interval at the root of DSTree. If the root's low value is smaller than $GI_{Time\text{-}Interval}$'s low endpoint, then the new interval goes to the left sub-tree; otherwise, the new node goes to the right sub-tree. We continue the same process until the sub-tree level is equal to the *temporal level* (*T*) parameter of the DSTree. Once the node is selected, if there is already a quad-tree in the node (node is spatial), we insert the *Gid* into the quad-tree using its GI_{MBR} parameter. If the node is empty, we generate a new quad-tree and proceed with adding it. Adding *Gid* to the quad-tree is similar to the regular quad-tree insert (e.g., see [65]). Once all the steps are completed, we update the max value of the ancestors of interval tree portion if needed. Algorithm 1 shows the pseudocode for the process.

Algorithm 1 Insert algorithm for the proposed DSTree

```
function INSERTGIDINTODSTREE(Gid, GI_CID, GI_MBR, GI_Time-Interval)
    current_interval_node ← root_of_DSTree
    current_interval_level ← 0
    interval_level ← DSTree_interval_level
    while current_interval_node is not a leaf ∨current_interval_level ≤ interval_level do
        if current_interval_node.low_value < GI_Time-Interval.low_endpoint then
            current_interval_node ← current_interval_node.left_subtree
        else
            current_interval_node ← current_interval_node.right_subtree
        end if
    end while
    if current_interval_node is spatial then
        INSERTINTOQUADTREE(current_interval_node.quad_tree, Gid, GI_MBR)
    else
        current_interval_node.quad_tree ← NEWQUADTREE
        INSERTINTOQUADTREE(current_interval_node.quad_tree, Gid, GI_MBR)
    end if
    UPDATEANCESTORSMAXVALUE(current_interval_node)
end function
function INSERTINTOQUADTREE(quad_tree, Gid, GI_MBR)    ▷ Perform regular quad-tree
insert    ▷ Implementation details depend on specific quad-tree algorithm    ▷ See
reference [65]
    // Example steps:
    // - Determine quadrant for GI_MBR
    // - Recursively insert Gid into appropriate quadrant of quad_tree
end function
function UPDATEANCESTORSMAXVALUE(node) ▷ Check if max value of ancestors needs
updating    ▷ Traverse up the tree until root, updating max values as needed    ▷
Implementation details depend on DSTree structure
end function
```

Time Complexity of an Insert

The time complexity of inserting a *Gid* into a DSTree involves several steps:

1. Finding the proper node on the interval tree part of the DSTree: The time complexity of this operation depends on the height of the interval tree portion of the DSTree and the number of nodes visited during the traversal. In the worst case, if the interval tree portion is unbalanced, the time complexity can be $O(n)$, where n is the number of intervals in the DSTree.

2. Adding the *Gid* to the proper quad-tree leaf: Once the proper node on the interval tree part is found, adding the *Gid* to the proper quad-tree leaf involves traversing the quad-tree structure. The time complexity of this operation depends on the size and structure of the quad-tree. In general, the time complexity of a quad-tree insertion can be $O(logm)$, where m is the number of objects in the quad-tree.
3. Updating the max value of the ancestors of the interval tree portion: After inserting the *Gid*, the max value of the ancestors of the interval tree portion may need to be updated. The time complexity of this operation depends on the height of the interval tree portion of the DSTree. In the worst case, if the interval tree portion is unbalanced, the time complexity can be $O(n)$, where n is the number of intervals in the DSTree.

Overall, considering all steps, the time complexity of inserting a *Gid* into a DSTree can be approximated as $O(n + logm)$, where n is the number of intervals in the DSTree and m is the number of objects in the quad-tree. However, the actual time complexity may vary depending on the specific implementation details and the characteristics of the DSTree and quad-tree.

3.3. Delete

Deleting GI items from DSTree is a relatively complex process due to the complexities of removing intervals from a regular interval tree. After deleting a *Gid* from the DSTree, if the node containing that *Gid* contains no more objects, that node may be deleted from the tree. This involves promoting a node further from the leaf to the position of the node being deleted, which results in the reconstruction of the top part of the DSTree (for details about deleting items from interval trees, see [64], pp. 348–357). Algorithm 2 shows this process.

Algorithm 2 Delete algorithm for the proposed DSTree

```
function DELETEGIDFROMDSTREE(Gid, GI_CID, GI_MBR, GI_Time-Interval)
    current_node ←QUERYGIDFROMDSTREE(Gid, GI_CID, GI_MBR, GI_Time-Interval)
    if current_node is found then
        // remove the Gid fro current_node
        if current_node is empty after deletion then
            // Delete the node from the tree
            // Reconstruct the DSTree if necessary
        end if
    end if
end function
function QUERYGIDFROMDSTREE(node) ▷ Find the node which contains the current Gid
end function
```

Time Complexity of a Delete

The time complexity of deleting items from a DSTree involves several factors:

1. Finding the node containing the *Gid*: The time complexity of this operation depends on the structure of the DSTree. In the worst case, if the DSTree is unbalanced, the time complexity can be $O(n)$, where n is the number of nodes in the DSTree.
2. Deleting the *Gid* from the node: The time complexity of deleting the *Gid* from the node depends on the data structure used to store intervals in the node. For interval trees, the deletion process can have a time complexity of $O(logi)$, where i is the number of intervals in the node.
3. Deleting the node from the DSTree if it contains no more objects: If the node contains no more objects after deleting the *Gid*, it may need to be deleted from the tree. Deleting a node from a tree can involve restructuring the tree, which can have a time complexity of $O(logn)$, where n is the number of nodes in the tree.

Overall, the time complexity of deleting items from a DSTree can be approximated as $O(nlogi)$, where n is the number of nodes in the DSTree and i is number of intervals.

However, the actual time complexity may vary depending on the specific implementation details and the characteristics of the DSTree.

3.4. Query

A spatio-temporal range search is a query of geographic objects that intersect with a boundary box, $S = (Min_x, Min_y , Max_x, Max_y)$, in two-dimensional space and also is in a temporal topological relation, TP, with a time interval, $I[t_1, t_2]$ [66]. There can be three main variations of queries of a DSTree, which include queries with both temporal interval and spatial extents, queries with only temporal interval, and queries with only spatial extent. Here, we only cover the first variation. The approach to those two variations is similar and explained in more detail in Figure 5. In order to perform a spatio-temporal range on a DSTree, it is required to first query the interval tree portion of the DSTree. If I is in TP relation with the root's interval, we add the root's interval to the candidate node list. If the left child of the root is not empty and the max value of the sub-tree in the left child is greater than I's low value, recur for the left child; otherwise, recur for the right child. Once the candidate nodes are selected, if the selected nodes have a quad-tree as their sub-tree, we perform a quad-tree search for S; otherwise, we only check for the intersection of the selected node with the boundary box, S. Algorithm 3 shows this process.

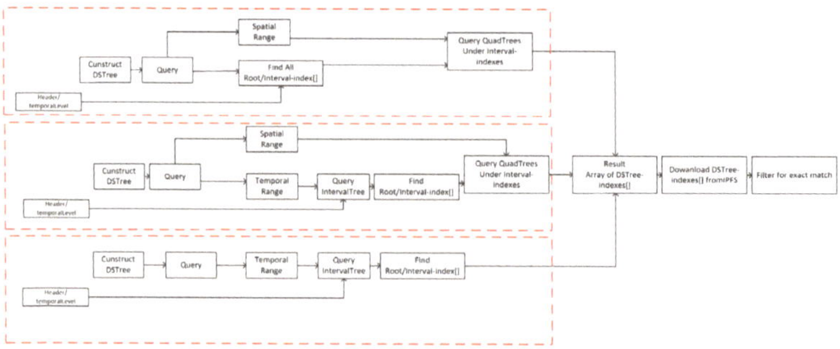

Figure 5. Three main scenarios to process queries using DSTree. Top: When a user requests only a spatial range in which we search all the quad-trees in the DSTree. Middle: When the user queries spatial and temporal ranges together, DSTree first queries interval tree part of the graph and then searches quad-trees that exist at the bottom of those candidate nodes. Bottom: Cases where user only provides a temporal range. As a result, we only search interval tree part of the DSTree and then simply query the root of the quad-tree in each candidate node.

Time Complexity of a Query

The query process has three key operations, as follows:

1. Querying the interval tree portion of the DSTree: This operation involves traversing the interval tree portion of the DSTree recursively. The time complexity of this operation depends on the height of the interval tree portion and the number of nodes visited during the traversal. If the interval tree portion is balanced, the time complexity is $O(logn)$, where n is the number of intervals in the DSTree.
2. Recursing down the left or right child nodes: This operation involves recursively traversing down the left or right child nodes of each candidate node. The time complexity of this operation depends on the structure of the DSTree and the distribution of intervals. In the worst case, if the DSTree is unbalanced, the time complexity can be $O(i)$, where i is the number of intervals in the DSTree.
3. Quad-tree search: If the selected nodes have quad-tree sub-trees, a quad-tree search for S is performed. The time complexity of the quad-tree search depends on the size and structure of the quad-tree and the number of objects in the search area. In general,

the time complexity of a quad-tree search is $O(m + k)$, where m is the number of objects in the search area and k is the number of objects found.

Algorithm 3 Query algorithm for the proposed DSTree

```
function SPATIOTEMPORALRANGESEARCH(DSTree, S, I)
    candidate_nodes ← empty list
    QUERYINTERVALTREE(DSTree.root, I, candidate_nodes)
    if candidate_nodes is not empty then
        for each node in candidate_nodes do
            if isIntervalInRelation(node.interval, I, TP) then
                ADDTORESULT(node.interval)
            end if
            if node.left_child is not empty and node.left_child.max_value > I.low_value
then
                SPATIOTEMPORALRANGESEARCH(node.left_child, S, I)
            else if node.right_child is not empty then
                SPATIOTEMPORALRANGESEARCH(node.right_child, S, I)
            end if
        end for
    end if
    for each node in candidate_nodes do
        if node.has_quad_tree then
            QUADTREESEARCH(node.quad_tree, S)
        else
            if doesNodeIntersect(node, S) then
                ADDTORESULT(node)
            end if
        end if
    end for
end function
function QUERYINTERVALTREE(node, I, candidate_nodes)
    if node is null then
        return
    end if
    if doesNodeIntersect(node.interval, I) then
        candidate_nodes.append(node)
    end if
    QUERYINTERVALTREE(node.left_child, I, candidate_nodes)
    QUERYINTERVALTREE(node.right_child, I, candidate_nodes)
end function
```

Overall, considering all operations, the time complexity of the algorithm can be approximated as $O(logi + m + k)$, where i is the number of intervals in the DSTree, m is the number of objects in the search area, and k is the number of objects found during the search.

3.5. Dstree Construction from Bulk Data

Since DSTree is a two-level tree, the performance of the insert, delete, and construction of the tree depends on both the top level and bottom level of the trees. Construction of the DSTree from scratch for bulk data is a relatively straightforward process. First of all, an interval tree is constructed based on the existing data. Once data are inserted into the hierarchical node structure, the algorithm traverses down from the root of the interval tree until the tree level equals the *temporal level* (T). Once the appropriate level is detected in the interval tree, all of the items under sub-trees (left and right branched) of the selected node are collected into one single node and a quad-tree is constructed in that node.

4. Performance Metrics

The behavior of the proposed tree structure is measured using a number of experiments with real data. These experiments are intended to reflect the conditions of common tasks involved in spatial and temporal queries of the data. Results of multiple models are compared to the method proposed. In the first set of the experiments, data on the occurrence of crime from the Waterloo Regional Police Service (https://www.wrps.on.ca/en/about-us/reports-publications-and-surveys.aspx, accessed on 15 February 2024) are used. These data detail all the police-reported occurrences for the calendar year. The time frame of these data is from 2017 to 2022. The data include occurrence data and time, response time, and geographic coordinates of the occurrence. In this experiment, only the location of each event and the response time of each event are used. In order to have a comparison between DSTree and the other existing models, we have used three other methods to process spatio-temporal queries. In choosing each method, the availability of source code to perform the tests was considered. Oct-tree is one of the methods which is used in the experiments similar to the work carried out by [47,52,53], in which the third dimension of cuboid data is considered as the time interval (source code available at [67]). The second access method used is a regular quad-tree (source code available at [68]). In this method, only a spatial query is performed, and then, to obtain the exact result set, all the results from the spatial query are tested to filter based on the temporal parameters. The third method is an interval tree (source code available at [69]). In this method, in contrast to the quad-tree method, only a temporal query is performed, and once the results are extracted from the tree structure, the spatial filter is applied to them to obtain the exact results. The above experiment is applied to the batches of 50 k, 100 k, 200 k, 400 k, 800 k, and 1 million points. Each test was executed five times then repeated ten times, and the average time of the execution was measured. The spatio-temporal query was constant over the entire experiment. The main query for these different models is defined as follows:

Find the events where their location has an *Intersection* topology relation with an extent equal to

$$Spatial_{Extent} = (\frac{Max_x - Min_x}{4}, \frac{Max_y - Min_y}{4}, 3 \times \frac{Max_x - Min_x}{4}, 3 \times \frac{Max_y - Min_y}{4})$$

and their response time has a topological relation T with a temporal extent of

$$Temporal_{Extent} = [\frac{Max_t - Min_t}{4}, 3 \times \frac{Max_t - Min_t}{4}]$$

where $Max_x, Min_x, Max_y, Min_y$ are the spatial extent of the entire dataset and Max_t, Min_t are temporal extent of the data and T is the temporal topology from Table 1.

In Figure 6, the results of the six temporal queries are discussed. In these six queries, static spatial extent is used. As Figure 6 shows, DSTree shows a good performance for query processing in medium-sized datasets compared to other types of indexing methods. The metrics of the DSTrees in all of the six temporal topologies are close to the quad-tree method. Oct-tree and interval tree also show close performance metrics.

The second experiment is with the number of visited points in order to answer a constant spatio-temporal query. In the experiment, only objects that the index traverses and are checked to answer each query are counted. Figure 7 shows the results of this experiment. In this figure, DSTree has fewer visited points compared to quad-tree. The reason why they have close metrics is that the quad-tree is less computationally heavy compared to DSTree. As Figure 7 shows, the DSTree has visited less points than oct-tree and oct-tree suddenly flattened out after 800k points. This is probably related to the spatial distribution of the points on the area which are clustered and results in a better performance of oct-tree. Further studies are required to address this kind of edge case. However, considering this number of visited points, the time to create an oct-tree is 20% higher, which is discussed in the next section. In addition, since the oct-tree uses functions which involve the third dimension comparison, the query performance does not out-perform DSTree overall.

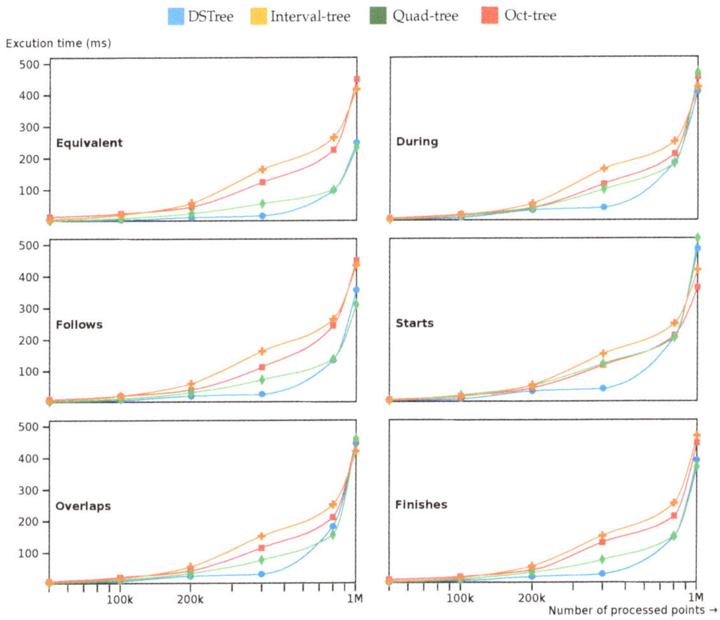

Figure 6. The query processing time for different spatio-temporal access methods. The spatial and temporal extent of each query remained constant. The results are measured for 6 different temporal topologies.

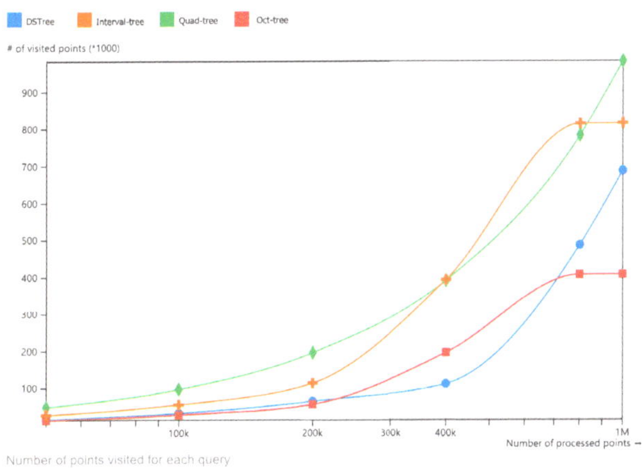

Figure 7. Number of visited points in each model to answer a spatio-temporal query.

5. Ipfs

The first generation of P2P systems, namely file-sharing applications such as BitTorrent, support only keyword lookups and mostly provide no load balancing. The second generation is mainly structured P2P systems supporting basic key-based routing [31]. The InterPlanetary File System (IPFS) is a protocol and a P2P network for storing and sharing data in a distributed file system. IPFS uses content addressing to uniquely identify each file in a global namespace connecting all computing devices [70]. Each file on the IPFS network has a unique hash address which is used as a reference to request it from the network. Any user in the network can serve a file by its content address, and other peers in the network can find and request that content from any node that has it using a distributed hash table (DHT) [71].

At its core, IPFS is built on top of a data structure called InterPlanetary Linked Data (IPLD) [70]. The IPLD model is a set of specifications in support of decentralized data structures for the content-addressable web. Content IDs (CIDs) are hashes generated to allow the user to interact with IPFS in a trustless manner and recover their data. IPLD deals with decoding these hashes so that users can access their data. When new content is added to the IPFS network, that content is separated into several chunks and stored in different blocks. To reconstruct the whole file, a Directed Acyclic Graph (DAG) connects each bit of content together. In a DAG, we can only move from parent nodes to child nodes as each edge is oriented. Hierarchical data in particular are very naturally represented via DAGs.

IPLD creates a series of links to data internally but also allows users to create those links themselves through simple data structures that can be stored on IPFS. This capability allows us to store a DAG graph (in our case, a DSTree) on IPFS. This capability allows users to request a portion of the data from the network without the need to download the entire dataset. For example, a user is able to store a graph, shown in Figure 3, as an IPLD object, as shown in Table 2. IPLD's capability to store and retrieve DAG graphs allows us to store spatio-temporal data as a graph structure, and as a result, we can request them based on the query parameters. DSTree only keeps a CID reference to the actual feature in each tree leaf. So depending on the query parameters, we only need to retrieve a portion of the GI or specific subtree of the DSTree. In an IPLD, each graph node is separated using /. For example, a DSTree leaf can be represented as an IPLD hash as

$$DSTree_{CID}/Interval\text{-}tree_{index}/quad\text{-}tree_{index}/$$

where $DSTree_{CID}$ is the CID of the root of the DSTree, $Interval\text{-}tree_{index}$ is the interval tree portion, and $quad\text{-}tree_{index}$ is the second portion of the DSTree index. Under each leaf, there will be a series of GI objects. Each GI can be stored separately on IPFS, and its own CID, GI_{CID}, can be used as a reference to the object itself. So to access a single feature, we can use an IPFS address similar to $DSTree_{CID}/Interval\text{-}tree_{index}/quad\text{-}tree_{index}/GI_{CID}$. Note that GI_{CID} is generated based on the content of the GI by IPFS, and to have access to it, we need to query it from DSTree.

Table 2. An example of how IPFS stores a DAG graph (based on the graph in Figure 3) and how to request a portion of the graph from the IPFS. *Qmb...R* is $DSTree_{CID}$ which is a hash generated based on the root content of entire DAG graph from the DSTree *Qmb...d* or *Qmb...c* is a GI_{CID}, which is a hash generated based on the content of single *GI* object.

IPFS Address	DAG Result	IPFS Address	DAG Result
Qmb...R Qmb...R/11/111111	Returns entire graph Returns all features under Root/11/111111	Qmb...R/11 Qmb...R/11/111110	Returns Root/11/ Returns all features under Root/11/111110
Qmb...R/10/1011/Qmb...d	Returns a single feature under Root/10/1011/ leaf	Qmb...R/10/1011/Qmb...c	Returns a single feature under Root/10/1011/ leaf
Qmb...R/metadata	Returns entire metadata object	Qmb...R/metadata/key_1	Returns key_1 under metadata

6. Distributed Network Integration

In order to process queries on distributed networks (in our case IPFS), it is required to store data in a DAG format. We use DSTree to construct a DAG graph, and once the tree structure is constructed, an IPLD graph is formed from the DSTree graph. Then, the IPLD object is stored on the IPFS network and a CID of the uploaded contents is used as a root gateway to access the entire tree structure. DSTree in this system acts as the main index structure to perform and answer spatio-temporal queries. Since DSTree is an indexing structure, it does not store the actual GI. It only stores the CID of each individual

GI as a reference to the object itself. This provides a lightweight graph that can be used by each client.

6.1. Data Management

As mentioned earlier, the DSTree only stores a CID reference to the actual GI. The GI can be in any format (e.g., geojson, topojson, or other feature-level standards). In constructing the DSTree, each GI is first uploaded on the IPFS as a regular file or IPLD object. Then, for each GI, we will then have a set of $(GI_{CID}, GI_{MBR}, GI_{Time\text{-}Interval})$. This object is then inserted into a DSTree and the related DAG graph is then constructed. Once all the GI objects are added to the DSTree, the graph structure is converted to an IPLD object and is uploaded on the IPFS and the pair of $(DSTree_{CID}, DSTree_{Metadata})$ will be shared between users.

6.2. Metadata

Metadata are usually defined as data about data [72]. In order to provide interoperability between different systems, it is required to include metadata objects within shared content. In the DSTree, $DSTree_{Metadata}$ is used to store information related to the dataset and can include general spatial metadata objects (e.g., see [73–75]). The following metadata keys (Table 3) are necessary in order to provide minimum interoperability when sharing information using DSTree on IPFS.

Table 3. Necessary metadata objects required for sharing DSTree on IPFS.

Metadata Key	Type	Possible/Example Values	Description
featureType	String	•geojson •topojson •dggsfeature	To identify the decoder for reading GI_{CID}
intervalKeys	Array[[]]	[[1, 10], [5, 15], [28, 20]]	Temporal interval keys in order to decode *interval tree index* values to the exact Intervals
temporalLevel (T)	Integer	1,2,...,n	The interval tree level before converting nodes to quad-tree root nodes
featureTypeProperties	JSON	Variable	Each *featureType* might need specific details and metadata; they can be stored as this object.
quadExtent	Array[]	[−180, −90, 180, 90]	The extent of the quad-trees. The default value is [−180, −90, 180, 90] degrees in geographic coordinates

6.3. Architecture and Implementation

The proposed method to process queries on IPFS networks consists of four main components. Figure 8 shows the flow of the communication between users on a distributed network in order to query, retrieve, and store GI. The start of the data sharing process on IPFS is with a user, *User*1, willing to share a GI, $(GI, GI_{MBR}, GI_{Time\text{-}Interval})$, on the network. Once the user uploads the GI content on IPFS, they use its CID, MBR, and time interval associated to it, $(GI_{CID}, GI_{MBR}, GI_{Time\text{-}Interval})$, to construct a DSTree. In this step, the user is able to keep adding as many GI objects as they want to the DSTree. Once the construction of the DSTree is finished, the necessary metadata are also added to the data structure, and an IPLD object, $DSTree_{IPLD}$, is constructed from DSTree's DAG graph. The $DSTree_{IPLD}$ is then uploaded on the IPFS network and the related IPFS root hash, $DSTree_{CID}$, and its metadata, $DSTree_{Metadata}$, are retrieved from IPFS. Since the $DSTree_{CID}$ is generated by IPFS based on the content of the DAG graph, we need to share this CID with other users to be able to access the index. In order to share the IPFS hash with other users, a smart contract is used. This smart contract is responsible for keeping a history of $DSTree_{CID}$ hashes over time and providing the latest $DSTree_{CID}$ hash to the users at any time. In our example, a simple smart contract using Solidity is developed and deployed on the Ethereum test network. This smart contract is used in a web application in order to provide access to the latest version of $DSTree_{CID}$ hash and its metadata when a user visits the web application. Once the $DSTree_{CID}$ hash is added to the smart contract, it will be available to all the users who connect to this smart contract.

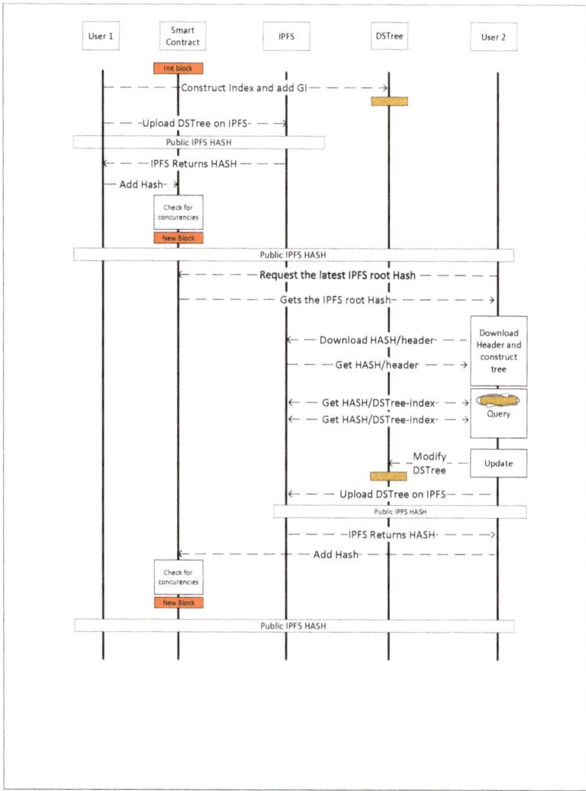

Figure 8. Query process and publishing spatio-temporal data on the IPFS. It shows the process of sharing the DSTree index between users using a blockchain, querying the content from the network by another user, and updating the data on the network.

If a new user, *User*2, accesses the smart contract, then they will be able to fetch metadata and the IPLD DAG graph structure related to the DSTree. Then, they will be able to replicate a version of DSTree on their own local environment, and as a result, they are able to query data from that index. The results of the query, an array of GI_{CID}s, are then requested from the IPFS through the query process explained in Figure 5.

If *User*2 wants to add a new GI, $(GI2, GI2_{MBR}, GI2_{Time\text{-}Interval})$, to the dataset, they first add it to the DSTree and then construct a new IPLD graph and upload the data on IPFS. Since the content of the new DSTree is different from the previous one, a new IPFS hash is generated and the $DSTree2_{CID}$ is returned to the user. In the next step, *User*2 connects to the smart contract and adds $DSTree2_{CID}$ as a new block to the underlying blockchain. At this point, all the users will be able to access the updated data, $DSTree2_{CID}$, throughout the blockchain. The older version of the DSTree, $DSTree_{CID}$, will also remain on the block history of the blockchain and will be accessible too.

7. Discussion

Unlike centralized systems, data storage in P2P networks is distributed across network nodes, providing scalability and no single point of failure. However, managing and processing queries on these networks has always been challenging. The proposed method to share and query spatio-temporal data on distributed networks tackles this issue by tracking and updating a spatio-temporal index between network users. In this approach, a blockchain is responsible for keeping a history of different versions of a DSTree index. Each user can replicate a version of DSTree on their node and run spatio-temporal queries on the index. Since each user performs the queries on their side, the indexing tree should check fewer items and support more topology out of the box. As shown in Figure 7, the number of visited items in the DSTree is generally less than other indexing structures, providing less memory consumption on the client apps. While octree also visits fewer nodes/items during its query process, its tree construction is slower (20%) compared to the other tree indexes. Also, it takes more time to answer the queries (see Figure 6) since the internal intersection functions are three-dimensional. They can also grow quickly if the time intervals are large [47]. In a single interval tree or quad-tree approach, the results of the queries need to be checked for the spatial or temporal topologies accordingly during the post-processing stage. In addition, the DSTree can handle six main temporal topological relations (see Table 1) during the query process without the need to post-process data. Time-wise, the DSTree also performs well on the small to average datasets (see Figure 6). Update, insertion, and deletion of existing data is another requirement for the current data-sharing environments. Due to using an interval tree as the top part of the tree, DSTree is not optimized for deleting items. Inserting a new GI will only update a portion of the tree structure and not impact the entire DAG data graph. However, adding data with large intervals so that the GI temporal interval covers branches from the left to the right side of the interval tree can cause a restructuring of the entire tree. For example, for time-series data such as sensor data or VGI, the DSTree performs better. In our police department example, the newly reported incidents can be added on top of the DSTree, and it does not cause the restructuring of the entire tree.

Conflicts may appear during the update process of the DSTree and publishing the latest version of DSTree by users. To resolve such conflicts, there are several approaches. Conflict resolution between different versions of DSTree can be achieved either on the client side before pushing the latest version on the blockchain or on the smart contract before saving $DSTree_{CID}$. Both of these methods require a mechanism to detect and address conflicts. Because of the size limitations on the smart contracts, we are using a client-side conflict detection approach. In this approach, the DSTree graph is extracted from the latest version available on the blockchain and is compared with the version of the graph on the client side [76]. Supposing the conflicts in the DSTree graph are detected (using tools like [77]), the user will need to resolve the conflict and then publish it on the blockchain. However, this approach needs a trustful user interaction with the network.

The reason for using quad-trees is that since the root boundary box of all the quad-trees is constant, the quad index part of the DSTree index will always point to the same area in the geographic space over different time intervals. This provides a faster access method and the capability to exchange the quad-tree with a Discrete Global Grid System (DGGS). DGGS grid, similar to quad-tree, provides the same index value per each grid cell in the space. It also provides methods to aggregate data on multi-resolution levels [78–80] and also provides built-in data locality and directionality of space [81]. For instance, in sharing police department information over time, the reports can be censored using distributed k-anonymity methods on the P2P networks (e.g., [82]) if the DGGS system is used as feature data storage.

7.1. Data Locality in the IPFS with DSTree

Data locality implies that neighboring multidimensional information is stored in neighboring nodes [31]. In an IPFS network, once each node downloads some particular content, it can be a data provider. Combining data partitioning using DSTree and sharing data on IPFS at the GI level allows users to request only a small portion of the entire dataset. As a result, they can also serve small chunks of a large dataset. This can provide a data locality in the P2P network based on user activities, e.g., on VGI platforms. In our police data sharing example, once the dispatch teams share their location on the network, they have already become a data provider for that shared GI. Once the police officers visit that location, once again, they download that GI, and they will become another data provider for that specific GI on the IPFS network. This approach provides a level of data locality for the nodes close to each other since the nodes in a region usually tend to explore data related to their region. Other examples of use cases of such data locality include sharing geo-tagged information in small communities.

7.2. DSTree Limitations

The proposed DSTree model only supports the spatial topologies that the underlying spatial indexing method supports. In this paper, we only experimented with the intersection topology relation. However, it is possible to perform other topological relations such as overlay, within, and crosses and also perform KNN-based models. All of the temporal typologies are supported except disjoint. The focus of this paper was support for vector-based data structures. However, supporting raster data could be achieved by converting raster data into DGGS-based models or tiling the raster data instead of generating lower-level spatial index using the multi-resolution tiling structure. Table 4 summarizes supported queries and future approaches to support other data models.

Table 4. DSTree capabilities for different spatio-temporal data models.

Data Model	Functions	Notes
Vector	•Temporal typologies on Table 1 •Spatial Intersection •Spatial overlay •Spatial within •Spatial crosses •KNN	KNN is not yet implemented
Raster	•Temporal topology queries •Extent queries •Multi-band data retrieval •Multi-resolution aggregation	This can be achieved using raster tiling methods and constructing a DAG graph or converting data to other models such as DGGS and retrieving them using DGGS DAG graph
TIN	•Extent queries	This can be achieved by converting data to other models such as DGGS and retrieving using DGGS DAG graph

In this work, our focus was on spatio-temporal queries. However semantic queries play an important role in data retrieval. The potential of combining DSTree with models such as STKST-I [83] would allow querying semantic data, along with providing a wider range of temporal topologies.

8. Conclusions

P2P has become very popular decentralized approach for storing and sharing information. The daily amount of spatial data being collected and shared in different sectors with different levels highlights the need for P2P data management, query, and processing. This paper proposes a new spatio-temporal multi-level tree structure, DSTree, which aims to address this problem. DSTree is capable of performing a range of spatio-temporal queries. To integrate this data structure on the IPFS distributed network, a framework that uses blockchain to share the IPFS CID of the index is proposed. Each user is capable of replicating DSTree and querying or updating it. However, this model is not optimized for deletion and is mainly suitable for append-only data over time. In this work, some of the challenges in sharing and querying spatio-temporal data on distributed networks are addressed. Despite the advantages of our proposed framework, challenges remain. We conclude that more significant research effort from GIScience and related fields in developing decentralized applications is needed. The need for the standardization of different feature types and feature type properties when sharing data on the IPFS network is one of the requirements. The possibility of using IPLD objects in sharing GI at the feature level can provide finer access to the information. In addition, it is necessary to address attribute-level query processing, which is not covered in the current work. The use of the smart contract to control access of the users to read and write data to the main chain can also be studied. This access control can even be at the feature level.

Author Contributions: Conceptualization, M.H. and S.R.; methodology, M.H. and C.R.; software, M.H.; writing—original draft preparation, M.H.; writing—review and editing, S.R.; supervision, S.R. and C.R. All authors have read and agreed to the published version of the manuscript.

Funding: This research was funded by Global Water Future grant.

Data Availability Statement: The test dataset for the occurrence of the crime from the Waterloo Regional Police Service is available at (https://www.wrps.on.ca/en/about-us/reports-publications-and-surveys.aspx, accessed on 15 February 2024). Other data used in this paper are available upon request from the corresponding author. The prototype application, screenshots, and the source code for the app and its online version are available on GitHub at https://github.com/am2222/dstree, accessed on 15 February 2024.

Acknowledgments: The authors thank the Global Water Futures research program for funding this work under the Developing Big Data and Decision Support Systems theme.

Conflicts of Interest: The authors declare no conflicts of interest.

Abbreviations

The following abbreviations are used in this manuscript:

IPFS	Interplanetary File System
DSTree	Distributed Spatio-temporal Tree
DHT	Distributed Hash Tree
P2P	Peer-to-Peer
KNN	K-nearest neighbor
DGGS	Discrete global grid system
DAG	Directed acyclic graph
IPLD	InterPlanetary Linked Data
CID	Content ID
GI	Geographic information

References

1. Goodchild, M.F.; Fu, P.; Rich, P. Sharing Geographic Information: An Assessment of the Geospatial One-Stop. *Ann. Assoc. Am. Geogr.* **2007**, *97*, 250–266. [CrossRef]
2. Anderson, J. OpenStreetMap Contributor LifeSpans—Revisiting and Expanding on 2018 Research Paper. Available online: https://www.openstreetmap.org/user/Jennings%20Anderson/diary/398034 (accessed on 15 February 2024).
3. Doulkeridis, C.; Vlachou, A.; Nørvåg, K.; Kotidis, Y.; Vazirgiannis, M. Efficient search based on content similarity over self-organizing P2P networks. *Peer Peer Netw. Appl.* **2010**, *3*, 67–79. [CrossRef]
4. Al-Yadumi, S.; Xion, T.E.; Wei, S.G.W.; Boursier, P. Review on integrating geospatial big datasets and open research issues. *IEEE Access* **2021**, *9*, 10604–10620. [CrossRef]
5. Group, T.H. Hierarchical Data Format (HDF). 2023. Available online: https://www.hdfgroup.org/ (accessed on 15 February 2024).
6. Mahecha, M.D.; Gans, F.; Brandt, G.; Christiansen, R.; Cornell, S.E.; Fomferra, N.; Kraemer, G.; Peters, J.; Bodesheim, P.; Camps-Valls, G.; et al. Earth system data cubes unravel global multivariate dynamics. *Earth Syst. Dyn.* **2020**, *11*, 201–234. [CrossRef]
7. Geoparquet. Available online: https://github.com/opengeospatial (accessed on 15 February 2024).
8. Apache Iceberg—Apache Iceberg—Iceberg.apache.org. 2022. Available online: https://iceberg.apache.org/ (accessed on 15 February 2024).
9. Goodchild, M.F. The future of digital earth. *Ann. GIS* **2012**, *18*, 93–98. [CrossRef]

10. Yu, J.; Wu, J.; Sarwat, M. Geospark: A cluster computing framework for processing large-scale spatial data. In Proceedings of the 23rd SIGSPATIAL International Conference on Advances in Geographic Information Systems, Seattle, WA, USA, 3–6 November 2015; pp. 1–4.
11. Eldawy, A.; Mokbel, M.F. Spatialhadoop: A mapreduce framework for spatial data. In Proceedings of the 2015 IEEE 31st International Conference on Data Engineering, Seoul, Republic of Korea, 13–17 April 2015; IEEE: Piscataway, NJ, USA, 2015; pp. 1352–1363.
12. Bambacht, J.; Pouwelse, J. Web3: A Decentralized Societal Infrastructure for Identity, Trust, Money, and Data. *arXiv* **2022**, arXiv:2203.00398.
13. Nofer, M.; Gomber, P.; Hinz, O.; Schiereck, D. Blockchain. *Bus. Inf. Syst. Eng.* **2017**, *59*, 183–187. [CrossRef]
14. Nakamoto, S. Bitcoin: A Peer-to-Peer Electronic Cash System. Available online: https://www.ussc.gov/sites/default/files/pdf/training/annual-national-training-seminar/2018/Emerging_Tech_Bitcoin_Crypto.pdf (accessed on 15 February 2024).
15. Hojati, M.; Feick, R.; Roberts, S.; Farmer, C.; Robertson, C. Distributed spatial data sharing: A new model for data ownership and access control. *J. Spat. Inf. Sci.* **2023**, *27*. [CrossRef]
16. Djellabi, B.; Amad, M.; Baadache, A. Handfan: A flexible peer-to-peer service discovery system for internet of things applications. *J. King Saud-Univ.-Comput. Inf. Sci.* **2022**, *34*, 7686–7698. [CrossRef]
17. Ye, W.; Khan, A.I.; Kendall, E.A. Distributed network file storage for a serverless (P2P) network. In Proceedings of the 11th IEEE International Conference on Networks, ICON2003, Atlanta, GA, USA, 4–7 November 2003; IEEE: Piscataway, NJ, USA, 2004.
18. Ehiagwina, F.O.; Iromini, N.A.; Olatinwo, I.S.; Raheem, K.; Mustapha, K. A State-of-the-Art Survey of Peer-to-Peer Networks: Research Directions, Applications and Challenges. *J. Eng. Res. Sci.* **2022**, *1*, 19–38. [CrossRef]
19. Achir, M.; Abdelli, A.; Mokdad, L.; Benothman, J. Service discovery and selection in IoT: A survey and a taxonomy. *J. Netw. Comput. Appl.* **2022**, *200*, 103331. [CrossRef]
20. Crainiceanu, A.; Linga, P.; Gehrke, J.; Shanmugasundaram, J. Querying Peer-to-Peer Networks Using P-Trees. In Proceedings of the 7th International Workshop on the Web and Databases: Colocated with ACM SIGMOD/PODS 2004, New York, NY, USA, 17–18 June 2004; pp. 25–30. [CrossRef]
21. Ramabhadran, S.; Ratnasamy, S.; Hellerstein, J.M. Prefix Hash Tree An Indexing Data Structure over Distributed Hash Tables. In Proceedings of the PODC 2004 Conference, St. John's, NL, Canada, 25–28 July 2004.
22. Hassanzadeh-Nazarabadi, Y.; Taheri-Boshrooyeh, S.; Özkasap, Ö. DHT-based Edge and Fog Computing Systems: Infrastructures and Applications. In Proceedings of the IEEE INFOCOM 2022—IEEE Conference on Computer Communications Workshops (INFOCOM WKSHPS), Virtual, 2–5 May 2022; IEEE: Piscataway, NJ, USA, 2022; pp. 1–6.
23. Harren, M.; Hellerstein, J.M.; Huebsch, R.; Loo, B.T.; Shenker, S.; Stoica, I. Complex queries in DHT-based peer-to-peer networks. In *Peer-to-Peer Systems*; Lecture Notes in Computer Science; Springer: Berlin/Heidelberg, Germany, 2002; pp. 242–250.
24. Triantafillou, P.; Pitoura, T. Towards a unifying framework for complex query processing over structured peer-to-peer data networks. In *Databases, Information Systems, and Peer-to-Peer Computing*; Lecture Notes in Computer Science; Springer: Berlin/Heidelberg, Germany, 2004; pp. 169–183.
25. Xia, J.; Yang, C.; Li, Q. Building a spatiotemporal index for earth observation big data. *Int. J. Appl. Earth Obs. Geoinf.* **2018**, *73*, 245–252. [CrossRef]
26. Mokbel, M.F.; Ghanem, T.M.; Aref, W.G. Spatio-temporal access methods. *IEEE Data Eng. Bull.* **2003**, *26*, 40–49.
27. Mondal, A.; Lifu, Y.; Kitsuregawa, M. P2PR-tree: An R-tree-based spatial index for peer-to-peer environments. In *Current Trends in Database Technology—EDBT 2004 Workshops*; Lecture Notes in Computer Science; Springer: Berlin/Heidelberg, Germany, 2004; pp. 516–525.
28. Morton, G.M. A Computer Oriented Geodetic Data Base and a New Technique in File Sequencing. In *International Business Machines*; IBM Ltd.: Ottawa, ON, Canada, 1966.
29. Stocia, I. Chord: A scalable peer-to-peer lookup service for internet applications. *ACM Sigcomm Comput. Commun. Rev.* **2001**, *31*, 149–160. [CrossRef]
30. Sahin, O.D.; Antony, S.; Agrawal, D.; Abbadi, A.E. Probe: Multi-dimensional range queries in p2p networks. In Proceedings of the International Conference on Web Information Systems Engineering, New York, NY, USA, 20–22 November 2005; Springer: Berlin/Heidelberg, Germany, 2005; pp. 332–346.
31. Liang, S. A new peer-to-peer-based interoperable spatial sensor web architecture. *Int. Arch. Photogramm. Remote. Sens. Spat. Inf. Sci.* **2008**, *XXXVII*. Available online: https://citeseerx.ist.psu.edu/document?repid=rep1&type=pdf&doi=3baee6b2370bf3b2cb8f503aeb08cf73c097e77b (accessed on 15 February 2024).
32. Zhang, C.; Krishnamurthy, A.; Wang, R.Y. Skipindex: Towards a Scalable Peer-to-Peer Index Service for High Dimensional Data. Available online: https://www.comp.nus.edu.sg/~cs6203/guidelines/topic3/multi-dimension/skipindex.pdf (accessed on 15 February 2024).
33. Maymounkov, P.; Mazières, D. Kademlia: A Peer-to-Peer Information System Based on the XOR Metric. In Proceedings of the Peer-to-Peer Systems, Cambridge, MA, USA, 7–8 March 2002; pp. 53–65. [CrossRef]
34. Kantere, V.; Skiadopoulos, S.; Sellis, T. Storing and indexing spatial data in P2P systems. *IEEE Trans. Knowl. Data Eng.* **2009**, *21*, 287–300. [CrossRef]

35. Tang, C.; Xu, Z.; Dwarkadas, S. Peer-to-peer information retrieval using self-organizing semantic overlay networks. In Proceedings of the 2003 Conference on Applications, Technologies, Architectures, and Protocols for Computer Communications, Karlsruhe, Germany, 25–29 August 2003; pp. 175–186.
36. Demirbas, M.; Ferhatosmanoglu, H. Peer-to-peer spatial queries in sensor networks. In Proceedings of the Third International Conference on Peer-to-Peer Computing (P2P2003), Linköping, Sweden, 1–3 September 2003; IEEE: Piscataway, NJ, USA, 2003; pp. 32–39.
37. Cai, W.; Zhou, S.; Qian, W.; Xu, L.; Tan, K.L.; Zhou, A. C2: A new overlay network based on can and chord. *Int. J. High Perform. Comput. Netw.* **2005**, *3*, 248–261. [CrossRef]
38. Soro, A.; Lai, C. Range-capable Distributed Hash Tables. In Proceedings of the 3rd ACM Workshop On Geographic Information Retrieval (GIR 2006), Seattle, WA, USA, 10 August 2006.
39. Jagadish, H.; Ooi, B.C.; Vu, Q.H.; Zhang, R.; Zhou, A. VBI-Tree: A Peer-to-Peer Framework for Supporting Multi-Dimensional Indexing Schemes. In Proceedings of the 22nd International Conference on Data Engineering (ICDE'06), Atlanta, GA, USA, 3–7 April 2006; p. 34. [CrossRef]
40. Ganesan, P.; Yang, B.; Garcia-Molina, H. One torus to rule them all: Multi-dimensional queries in p2p systems. In Proceedings of the 7th International Workshop on the Web and Databases: Colocated with ACM SIGMOD/PODS 2004, San Diego, CA, USA, 17–18 June 2004; pp. 19–24.
41. Vlachou, A.; Doulkeridis, C.; Nørvåg, K.; Kotidis, Y. *Peer-to-Peer Query Processing over Multidimensional Data*; Springer: New York, NY, USA, 2012. [CrossRef]
42. Dangermond, J.; Goodchild, M.F. Building geospatial infrastructure. *Geo-Spat. Inf. Sci.* **2019**, *23*, 1–9. [CrossRef]
43. Gebbert, S.; Pebesma, E. A temporal GIS for field based environmental modeling. *Environ. Model. Softw.* **2014**, *53*, 1–12. [CrossRef]
44. Yuan, M. Temporal GIS and spatio-temporal modeling. In Proceedings of the Third International Conference Workshop on Integrating GIS and Environment Modeling, Sante Fe, NM, USA, 21–25 January 1996; Volume 33.
45. Pelekis, N.; Theodoulidis, B.; Kopanakis, I.; Theodoridis, Y. Literature review of spatio-temporal database models. *Knowl. Eng. Rev.* **2004**, *19*, 235–274. [CrossRef]
46. Theodoridis, Y.; Vazirgiannis, M.; Sellis, T. Spatio-temporal indexing for large multimedia applications. In Proceedings of the Third IEEE International Conference on Multimedia Computing and Systems, Hiroshima, Japan, 17–23 June 1996; IEEE: Piscataway, NJ, USA, 1996. [CrossRef]
47. Mahmood, A.R.; Punni, S.; Aref, W.G. Spatio-temporal access methods: A survey (2010–2017). *Geoinformatica* **2019**, *23*, 1–36. [CrossRef]
48. He, Z.; Wu, C.; Liu, G.; Zheng, Z.; Tian, Y. Decomposition Tree: A Spatio-Temporal Indexing Method for Movement Big Data. *Clust. Comput.* **2015**, *18*, 1481–1492. [CrossRef]
49. Armstrong, M.P. Temporality in spatial databases. In *GIS/LIS 88 Proceedings: Accessing the World*; Urban and Regional Information Systems Association: Des Plaines, IL, USA, 1988; pp. 880–889.
50. Peuquet, D.J.; Duan, N. An event-based spatiotemporal data model (ESTDM) for temporal analysis of geographical data. *Int. J. Geogr. Inf. Syst.* **1995**, *9*, 7–24. [CrossRef]
51. Jackins, C.L.; Tanimoto, S.L. Oct-trees and their use in representing three-dimensional objects. *Comput. Graph. Image Process.* **1980**, *14*, 249–270. [CrossRef]
52. Zhang, C.; Zhu, L.; Long, J.; Lin, S.; Yang, Z.; Huang, W. A hybrid index model for efficient spatio-temporal search in HBase. In *Lecture Notes in Computer Science*; Lecture Notes in Computer Science; Springer International Publishing: Cham, Switzerland, 2018; pp. 108–120.
53. Zhao, K.; Chen, L.; Cong, G. Topic Exploration in Spatio-Temporal Document Collections. In Proceedings of the 2016 International Conference on Management of Data, New York, NY, USA, 26 June–1 July 2016.
54. Qu, Q.; Nurgaliev, I.; Muzammal, M.; Jensen, C.S.; Fan, J. On spatio-temporal blockchain query processing. *Future Gener. Comput. Syst.* **2019**, *98*, 208–218. [CrossRef]
55. Zheng, Z.; Xie, S.; Dai, H.N.; Chen, X.; Wang, H. Blockchain challenges and opportunities: A survey. *Int. J. Web Grid Serv.* **2018**, *14*, 352–375. [CrossRef]
56. PostGIS Clustering Data. Available online: https://postgis.net/workshops/postgis-intro/clusterindex.html (accessed on 15 February 2024).
57. Liu, B.; Lee, W.C.; Lee, D.L. Supporting complex multi-dimensional queries in P2P systems. In Proceedings of the 25th IEEE International Conference on Distributed Computing Systems (ICDCS'05), Columbus, OH, USA, 6–10 June 2005; IEEE: Piscataway, NJ, USA, 2005.
58. Allen, J.F. Maintaining knowledge about temporal intervals. In *Readings in Qualitative Reasoning About Physical Systems*; Elsevier: Amsterdam, The Netherlands, 1990; pp. 361–372.
59. Gabbay, D.; Kurucz, A.; Wolter, F.; Zakharyaschev, M. Applied modal logic. In *Many-Dimensional Modal Logics—Theory and Applications*; Studies in logic and the foundations of mathematics; Elsevier: Amsterdam, The Netherlands, 2003; pp. 41–109.
60. Qian, C.; Yi, C.; Cheng, C.; Pu, G.; Wei, X.; Zhang, H. Geosot-based spatiotemporal index of massive trajectory data. *Isprs Int. J. Geo-Inf.* **2019**, *8*, 284. [CrossRef]
61. Sun, Y.; Zhao, T.; Yoon, S.; Lee, Y. A Hybrid Approach Combining R*-Tree and k-d Trees to Improve Linked Open Data Query Performance. *Appl. Sci.* **2021**, *11*, 2405. [CrossRef]

62. Tao, Y.; Papadias, D. MV3R-Tree: A Spatio-Temporal Access Method for Timestamp and Interval Queries. In Proceedings of the 27th International Conference on Very Large Data Bases, San Francisco, CA, USA, 11–14 September 2001; pp. 431–440.
63. de Berg, M.; Cheong, O.; van Kreveld, M.; Overmars, M. *Computational Geometry*, 3rd ed.; Springer: Berlin, Germany, 2008.
64. Cormen, T.H.; Leiserson, C.E.; Rivest, R.L.; Stein, C. *Introduction to Algorithms*, 3rd ed.; The MIT Press: Cambridge, MA, USA, 2009.
65. Finkel, R.A.; Bentley, J.L. Quad trees a data structure for retrieval on composite keys. *Acta Inform.* **1974**, *4*, 1–9. [CrossRef]
66. Erwig, M.; Schneider, M. Developments in spatio-temporal query languages. In Proceedings of the Tenth International Workshop on Database and Expert Systems Applications, DEXA 99, Florence, Italy, 3 September 1999; IEEE: Piscataway, NJ, USA, 1999; pp. 441–449.
67. d3-octree. Available online: https://github.com/vasturiano (accessed on 15 February 2024).
68. quadtree-js. Available online: https://github.com/CorentinTh (accessed on 15 February 2024).
69. flatten-interval-tree. Available online: https://github.com/alexbol99 (accessed on 15 February 2024).
70. Benet, J. IPFS - Content Addressed, Versioned, P2P File System. *arXiv* **1974**, arXiv:1407.3561.
71. Zimmermann, R.; Ku, W.S.; Wang, H. Spatial data query support in peer-to-peer systems. In Proceedings of the 28th Annual International Computer Software and Applications Conference, COMPSAC 2004, Hong Kong, 28–30 September 2004; Volume 2, pp. 82–85. [CrossRef]
72. Coulondre, S.; Libourel, T.; Spéry, L. Metadata And GIS: A Classification of Metadata for GIS. In Proceedings of the International Conference and Exhibition on Geographic Information, Lisbon, Portugal, 7–11 September 1998.
73. Brodeur, J.; Coetzee, S.; Danko, D.; Garcia, S.; Hjelmager, J. Geographic information metadata—An outlook from the international standardization perspective. *ISPRS Int. J. Geoinf.* **2019**, *8*, 280. [CrossRef]
74. Kim, T.J. Metadata for geo-Spatial data sharing: A comparative analysis. *Ann. Reg. Sci.* **1999**, *33*, 171–181. [CrossRef]
75. Bossomaier, T.; Hope, B.A. *Online GIS and Spatial Metadata*, 2nd ed.; CRC Press: London, UK, 2015.
76. Kleppmann, M.; Beresford, A.R. A Conflict-Free Replicated JSON Datatype. *IEEE Trans. Parallel Distrib. Syst.* **2017**, *28*, 2733–2746. [CrossRef]
77. Automerge. Available online: https://github.com/automerge (accessed on 15 February 2024).
78. Li, M.; Stefanakis, E. Geospatial operations of discrete global grid systems—A comparison with traditional GIS. *J. Geovisualization Spat. Anal.* **2020**, *4*, 26. [CrossRef]
79. Robertson, C.; Chaudhuri, C.; Hojati, M.; Roberts, S.A. An integrated environmental analytics system (IDEAS) based on a DGGS. *ISPRS J. Photogramm. Remote Sens.* **2020**, *162*, 214–228. [CrossRef]
80. Hojati, M.; Robertson, C.; Roberts, S.; Chaudhuri, C. GIScience research challenges for realizing discrete global grid systems as a Digital Earth. *Big Earth Data* **2022**, *6*, 358–379. [CrossRef]
81. Sahr, K. Central place indexing: Hierarchical linear indexing systems for mixed-aperture hexagonal discrete global grid systems. *Cartogr. Int. J. Geogr. Inf. Geovisualization* **2019**, *54*, 16–29. [CrossRef]
82. Hojati, M.; Farmer, C.; Feick, R.; Robertson, C. Decentralized geoprivacy: Leveraging social trust on the distributed web. *Geogr. Inf. Syst.* **2021**, *35*, 2540–2566. [CrossRef]
83. Wu, X.; Liu, Y.; Zhao, X.; Chen, J. STKST-I: An Efficient Semantic Trajectory Search by Temporal and Semantic Keywords. *Expert Syst. Appl.* **2023**, *225*, 120064. [CrossRef]

Mathematical and Computational Applications

Article

Loss of the Sturm–Liouville Property of Time-Varying Second-Order Differential Equations in the Presence of Delayed Dynamics

Manuel De la Sen

Institute of Research and Development of Processes, Faculty of Science and Technology, University of the Basque Country (UPV/EHU), 48940 Leioa, Spain; manuel.delasen@ehu.eus

Abstract: This paper considers a nominal undelayed and time-varying second-order Sturm–Liouville differential equation on a finite time interval which is a nominal version of another perturbed differential equation subject to a delay in its dynamics. The nominal delay-free differential equation is a Sturm–Liouville system in the sense that it is subject to prescribed two-point boundary conditions. However, the perturbed differential system is not a Sturm–Liouville system, in general, due to the presence of delayed dynamics. The main objective of the paper is to investigate the loss of the boundary values of the Sturm–Liouville nominal undelayed system in the presence of the delayed dynamics. Such a delayed dynamics is considered a perturbation of the nominal dynamics related to the Sturm–Liouville system with given two-point boundary values. In particular, this loss of the Sturm–Liouville exact tracking of the prescribed two-point boundary values might happen because the nominal boundary values may become lost by the state trajectory solution in the presence of delays, related to the undelayed case, due to the presence of the delayed dynamics. The worst-case error description of the deviation of the two-point boundary values of the current perturbed differential with respect to those of the nominal Sturm–Liouville system is characterized in terms of error norms related to the nominal system. Under sufficiently small deviations of the parameterization of the perturbed system with respect to the nominal one, such a worst-error characterization makes the current perturbed system an approximate Sturm–Liouville system of the nominal undelayed one.

Keywords: ordinary differential equations; Sturm–Liouville system; undelayed system; delayed system; approximate Sturm–Liouville system

1. Introduction

Some of the well-known problems within the so-called Sturm–Liouville class are, for instance, the Bessel's and Legendre's equations and some typical second-order ordinary differential equations with periodic solutions [1–11]. One of the main properties of Sturm–Liouville systems is that their solution trajectory tracks the prescribed two-boundary given points under an infinite countable set of values of a constant real parameter λ which parameterizes the differential equation. Such a set of values is said to be the eigenvalue set of the Sturm–Liouville system [1–3]. In [4], a Sturm–Liouville-type description is developed for the wave equation, while in [5–7], some Sturm–Liouville type systems of interest are described under the formalism of fractional calculus. In particular, in [7], the solvability and the stability of nonlinear impulsive Langevin and Sturm–Liouville equations involving Caputo–Hadamard fractional derivatives and multipoint boundary value conditions are focused on. To unify the two types of equations, we investigate a general nonlinear impulsive coupled implicit system. On the other hand, Sturm–Liouville problems have been studied in the presence of delays. In particular, the uniqueness of the nonlinear inverse problem associated with Sturm–Liouville operators with multiple delays has been investigated in [8], while the multiplicity and the non-existence of positive solutions for a class of impulsive Sturm–Liouville boundary value problems has been discussed in [9]. On the other hand, the optimal state estimation based on deterministic Kalman filtering in both time and the frequency domains is considered in [10]. In particular, the spectral factorization is considered for Kalman filtering, and a class of Riesz-spectral systems is considered in this research. In the study performed in [11], the inverse Sturm–Liouville problem is considered with polynomials of the spectral parameter in the first boundary condition and with entire analytic functions in the second one. More recently, in [12], one proposes a novel approach to inverse Sturm–Liouville eigenvalue problems under Dirichlet boundary conditions. In that way, when a Sturm–Liouville eigenvalue problem with unknown integrable potential interacts with function potentials, a family of perturbation problems is obtained. Also, it is proved in [13] that the Sturm–Liouville problem with mixed boundary conditions can be transformed to a generalized Sturm–Liouville problem endowed with Dirichlet or Neumann boundary conditions. In [14], the eigenvalues of a fractional-type Sturm–Liouville problem are investigated, while a generalization of Picard's Theorem on Sturm–Liouville Equation with Lebesgue integrable function on a finite interval is addressed in [15].

On the other hand, time-delay systems appear in a natural way in the description of many problems of real life like, for instance, in many biological problems (such as epidemic models, ecology and medical models, sunflower equation, etc.), in diffusion problems, in war and peace problems, in missile/anti-missile dynamics, etc. The background literature on this subject has been very rich since several decades ago. See, for instance, [16–25] and the references therein. Delays can typically be either constant or time-varying point delays or distributed delays over intervals of finite or infinite durations. Within the classes of distributed delays, one includes some typical Volterra-type integro-differential equations. Delays can also appear in the internal dynamics, that is, in the states or in the forcing or control functions, that is, in the inputs or outputs. The delays can also be single or multiple and, in the second case, they can be measurable; that is, integer multiples of a basic minimum delay or not then referred to as not measurable. Sets of the above types of delays can appear also in a combined manner in a particular problem. A main characteristic of internal time-delay systems is that they are infinite-dimensional so that they possess infinitely many characteristic zeros. In this context, stability and stabilization criteria and methodologies for time-delay systems are investigated in [16,17]. See also references therein. In particular, in [17], the system's states are assumed to be not measurable. In [18], the effect of time delay on the dynamics of an eco-epidemiological system is investigated. In [19], the existence and the global exponential stability of the periodic solution of a class of Cohen–Grossberg neural networks with time-varying delays is investigated. In [20], the h-manifold stability of neural networks with variable impulsive perturbations and time-varying delays of Cohen–Grossberg-type bidirectional associative memory is investigated. On the other hand, the h-manifold stability for impulsive delayed SIR epidemic models (that is, with Susceptible–Infectious–Recovered subpopulations) have been investigated in [21]. In [22], the dynamics of a Lotka–Volterra competitor–competitor–mutualist system with time-varying delays is investigated and, in [23], the μ analysis and synthesis type for uncertain time-delay systems via Padé approximations is investigated.

It can also be pointed out that the influence of delays is very relevant also in control theory and applications. For instance, the controlled system itself can possess delays through its own nature, which leads to increased stabilization difficulties compared with the cases of absence of delays. Also, the stabilizing controller of an uncontrolled system can be synthesized with designed delays, in order to facilitate the closed-loop stabilization process. In typical problems, there is no precise knowledge of the dynamics or the values of the delays, but only some nominal approximated values, which makes the appropriate resulting stabilization technique still more involved.

The main objective of this paper is to investigate the loss of the boundary values of properties of the Sturm–Liouville nominal undelayed system in the presence of delays in the dynamics, which is considered a perturbation of the nominal dynamics. In particular, this loss of the Sturm–Liouville exact tracking of the prescribed two-point boundary values might happen because the nominal boundary values may become lost in the presence of delays related to the undelayed case, due to the presence of the delayed dynamics, and, respectively, the eventual presence of time-varying parameterizations. See, for instance, refs. [16–25] and, respectively, refs. [26–28], and some of the references therein.

The paper is organized as follows. Section 2 discusses two linear time-varying second-order differential equations associated with the two-boundary values of Sturm–Liouville-type problems. The first one is a Sturm–Liouville system which plays the role of a nominal differential system, while the second one is a current version of the above one which is considered a perturbation caused by an internal constant delay in its dynamics. Both differential equations are equivalently reformulated, for convenience of subsequent analysis, as second-order differential systems of differential equations of first-order. The respective solutions of the undelayed system and the delayed one are calculated analytically in the appropriate closed forms for given finite initial conditions, through the definition and use of "ad hoc" evolution operators. The comparison between such solutions is explicitly addressed. The solution of the nominal undelayed system is calculated under finite-point initial conditions, while that of the delayed one has, in general, a piecewise bounded function of initial conditions on a time interval of length being equal to the delay size value. The dynamics of such a nominal undelayed system is decomposed as a non-unique representation in a time-invariant part plus an incremental time-varying one, whose representative matrix has bounded piecewise continuous entries. Such a decomposition is then useful in the next section of this article, to characterize the boundedness and the asymptotic properties of the evolution operator and the state-trajectory solution, based on the stability, critical stability or instability properties of the time-invariant part of the matrix of dynamics. Sections 3 and 4 are devoted to the characterization of the two-point boundary conditions when the undelayed differential system is a Sturm–Liouville one of uniform boundedness and the asymptotic properties of the evolution operators when the time-invariant part of the undelayed system is stable, critically stable or unstable. The necessary and sufficient condition to keep the prescribed two-point boundary values of the nominal Sturm–Liouville system by the current delayed system is characterized explicitly in Section 3. Such a condition is very difficult to accomplish in general and, therefore, the errors of the two boundary points between those of the nominal Sturm–Liouville system and the current delayed one are characterized in the subsequent section. Section 5 is devoted to characterizing the worst-case errors in terms of error norms of the two-point boundary values between those of the nominal differential system, assumed to be a Sturm–Liouville system, and those of the perturbed current one. The sources of error between the respective two-point boundary values are the norms of the matrix of delayed dynamics and the delay size. The main idea behind the given error approach is to estimate the error norms of the two-point point boundary values related to those of the nominal undelayed system, which is a Sturm–Liouville one. It is seen that, under sufficiently small deviations of the parameterization of the perturbed system with respect to the nominal one, such a worst-error characterization makes the current perturbed system an approximate Sturm–Liouville system of the nominal one. Finally, some concluding remarks end the paper.

Notation

I_n is the n-th identity matrix.

$$\overline{n} = \{1, 2, \cdots, n\}$$

The superscript T denotes transposition.

$$e_1 = (1,0)^T;\ e_2 = (0,1)^T$$

If $f : \boldsymbol{R} \to \boldsymbol{R}^n$ then $\hat{f}_{[\beta\,,\,\alpha]}$ is the strip of $f(t)$ on $[\alpha\,,\,\beta]$, namely, $\hat{f}_{[\beta\,,\,\alpha]} = f(\tau)$ for $\tau \in [\alpha\,,\,\beta]$ and $\hat{f}_{[\beta\,,\,\alpha]} = 0$ for $\tau \notin [\alpha\,,\,\beta]$. In a similar way, one defines $\hat{f}_{(a,b)}$, $\hat{f}_{[a,\,b)}$ and $\hat{f}_{(a\,,\,b]}$.

$M^{\dagger}$ denotes the Moore–Penrose pseudoinverse of $M \in \mathbf{R}^{n\times m}$. In particular, $M^{\dagger} = M^{-1}$ if $n = m$ and M is non-singular.

$\mathrm{R}(M)$ denotes the Range (or Image) subspace of the matrix M.

2. Second-Order Nominal and Delayed Time-Varying Differential Equations

Through this section, we discuss two linear differential equations associated with the two-boundary value Sturm–Liouville-type problems. One of them is considered a nominal version of the Sturm–Liouville problem while the other one is the current version, which is considered a perturbation of the former one. A problem that arises is that the characterization of the conditions under the current differential equation is a Sturm–Liouville system provided that the nominal one is a Sturm–Liouville system. Another problem to be addressed is the characterization of the worst-case errors of the solution deviation of the current differential equation from an achievable nominal Sturm–Liouville system.

Assume that the following linear time-varying differential equation of second-order

$$\left(p_0(t)y_0'(t)\right)' + (q_0(t) + \lambda_0\omega_0(t))y_0(t) = 0 \tag{1}$$

is a nominal version of the perturbed current differential equation with a finite-point time delay $h > 0$:

$$\left(p_0(t)y'(t)\right)' + (q_0(t) + \lambda_0\omega_0(t))y(t) + \left(q_d(t) + \widetilde{\lambda}\omega_d(t)\right)y(t-h) = 0 \tag{2}$$

where $p_0(t) > 0$, $\omega_0(t) > 0$, $q_0(t)$, λ_0 and $\widetilde{\lambda}$ are real numbers, $q_d(t)$ and $\omega_d(t) > 0$ are bounded real functions defined in the real interval $[0,\ b]$ and piecewise-continuous in $(0\ ,\ b)$, and $p_0(t)$ is differentiable in $(0\ ,\ b)$.

Note that Equation (1) can be rewritten, equivalently, as a second-order linear time-varying differential system of two coupled first-order differential equations by defining the second-order vector function $x_0(t) = \left(y_0(t)\ ,\ y_0'(t)\right)^T$ as follows:

$$x_0'(t) = A_0(t)x_0(t);\ t \in [a,b] \tag{3}$$

with $x_0(0) = x_{00}$, where

$$A_0(t) = \begin{bmatrix} 0 & 1 \\ -p_0^{-1}(t)(q_0(t) + \lambda_0\omega_0(t)) & 0 \end{bmatrix} \tag{4}$$

is a piecewise-continuous matrix from the given assumptions on (1), and then (3) and (4) is a second-order piecewise-continuous differential system of first-order differential equations [29], while Equation (2) becomes, in the same way, to be equivalent to the second-order piecewise-continuous delayed differential system of first-order differential equations:

$$x'(t) = A_0(t)x(t) + A_d(t)x(t-h);\ t \in [a,b] \tag{5}$$

with $x(t) = \left(y(t)\ ,y'(t)\right)^T$ subject to a bounded piecewise-continuous function of initial conditions $\varphi : [-h\ ,\ 0] \to \mathbf{R}^2$ with $\varphi(0^+) = x(0^+) = x_0(0^+) = x_{00}$ (note that the piecewise-continuous $\phi(t)$ can have a finite jump at $t = 0$), where

$$A_d(t) = \begin{bmatrix} 0 & 0 \\ -p_0^{-1}(t)\left(q_d(t) + \widetilde{\lambda}\omega_d(t)\right) & 0 \end{bmatrix} \tag{6}$$

The unique solutions of (3), subject to (4), and (5), subject to (6), on $[0\ ,\ \infty)$ for each given initial condition are, respectively:

$$x_0(t) = T_0(t,0)x_{00} \tag{7}$$

$$
\begin{aligned}
x(t) &= \left(\hat{T}(t\,,0)\right)(\varphi) = \left(\hat{T}(t\,,0)\right)\left(\hat{x}_{[0,\,-h]}\right) \\
&= T_0(t,\,0)x_{00} + \int_0^t T_0(t,\tau)A_d(\tau)x(\tau-h)d\tau \\
&= T_0(t,\,0)x_{00} + \int_{-h}^0 T_0(t,\tau)A_d(h+\tau)\varphi(\tau)d\tau + \left(\int_0^{max(t-h,0)} T_0(t,\tau)A_d(\tau)T_0(\tau-h,0)d\tau\right)x_{00} \\
&+ \int_0^{max(t-h,0)} \int_{-h}^0 T_0(t,\tau)A_d(\tau)T_0(\tau-h,\theta)A_d(h+\theta)\varphi(\theta)d\theta d\tau;\ t\in \mathbf{R}_0
\end{aligned} \tag{8}
$$

where $x(t) = \varphi(t)$ for $t \in [-h\,,\,0]$ and $T_0 : [t,\tau] \to \mathbf{R}^{2\times 2}$ and $\hat{T} : [t,\tau] \to \mathbf{R}^{2\times 2}$ are the respective evolution operators of (3) and (5) which satisfy $T_0(t,t) = T(t,t) = I_2$ for $t \geq 0$ and $T_0(t,\tau) = T(t,\tau) = 0$ for $t < \tau$, as well as the respective differential systems (3) and (5), which is

$$T_0'(t) = A_0(t)T_0(t\,,0) \tag{9}$$

$$T'(t) = A_0(t)T(t\,,0) + A_d(t)T(t-h,0) \tag{10}$$

for $t \geq 0$. Note that (7) is the obvious unique solution of the unforced differential system (3) built with the evolution operator $T_0 : [t,\tau] \to \mathbf{R}^{2\times 2}$ which satisfies (3), in fact, the same differential system (3). In the same way, (8) is the unique solution of (5) subject to the evolution operator which satisfies (10), and (5) might be considered indistinctly as an unforced delayed differential system subject to interval initial conditions on $[-h,0]$ or as a forced one whose forcing term is the delayed dynamics. Therefore, the solution can be expressed equivalently as the first expression of (8), via the evolution operator $\hat{T} : [t,\tau] \to \mathbf{R}^{2\times 2}$, or as the subsequent equivalent ones, which are more useful for their evaluation, again via the evolution operator $T_0 : [t,\tau] \to \mathbf{R}^{2\times 2}$ which builds separately the unforced and forced parts of the solution.

Then, the solutions of (3) and (5) for $t \geq 0$ are, respectively,

$$T_0(t\,,\,0) = I_2 + \int_0^t A_0(\tau)T_0(\tau,0)d\tau \tag{11}$$

$$
\begin{aligned}
T(t\,,\,0) &= I_2 + \int_0^t (A_0(\tau)T(\tau,0) + A_d(\tau)T(\tau-h,\,0)\,)d\tau \\
&= T_0(t,0) + \int_0^t \left(A_0(\tau)\widetilde{T}(\tau,0) + A_d(\tau)T(\tau-h,0)\right)d\tau \\
&= T_0(t,0) + \int_0^t T_0(t,\tau)A_d(\tau)T(\tau-h\,,0)d\tau
\end{aligned} \tag{12}
$$

where the error evolution operator between the current and nominal differential systems satisfies the following:

$$\widetilde{T}(t,0) = T(t,0) - T_0(t,0) = \int_0^t \left(A_0(\tau)\widetilde{T}(\tau,0) + A_d(\tau)T(\tau-h,0)\,\right)d\tau \tag{13}$$

for all $t \geq 0$, which is the unique solution of the differential system:

$$
\begin{aligned}
\widetilde{T}'(t) &= A_0(t)\widetilde{T}(t,0) + A_d(t)T(t-h\,,0) \\
&= A_0(t)\widetilde{T}(t,0) + A_d(t)\widetilde{T}(t-h\,,0) + A_dT_0(t-h,\,0)
\end{aligned} \tag{14}
$$

Note that $T(t\,,h) = T_0(t\,,0)$ for $t \in (-\infty\,,\,h]$.

Remark 1. *Since $T(\sigma\,,\sigma) = T_0(\sigma\,,\,\sigma) = I_2$ for any real $\sigma \geq 0$ then (11) and (12) obey the more general expressions:*

$$T_0(t\,,\,\sigma) = I_2 + \int_0^t A_0(\tau)T_0(\tau,\sigma)d\tau \tag{15}$$

$$T(t\,,\,\sigma) = I_2 + \int_\sigma^t (A_0(\tau)T(\tau,\sigma) + A_d(\tau)T(\tau-h,\sigma)\,)d\tau \tag{16}$$

for any real $\sigma \in [0\,,\,t]$.

Remark 2. *Since $A_0(t)$ is not constant, then the evolution operator $T_0(t\,,\tau)$ is not a strongly continuous one-parameter semigroup (commonly referred to also as a C_0-semigroup) from $\mathbf{R}^2_{0+}$ to $L\left(C^2_0(\mathbf{R}_{0+})\right)$ (with $L\left(C^2_0(\mathbf{R}_{0+})\right)$ being the Banach space of bounded continuous real vector functions of dimension two with domain in $C^2_0(\mathbf{R}_{0+})$ endowed with the sup norm) of infinitesimal generator for any $t_0 \in \mathbf{R}_{0+}$ so that it does not hold the property $A_0(t_0)x(t_0) = \lim\limits_{t\to t_0^+} \frac{1}{t}(T(t,t) - I_2)$. It can be seen easily that the operator is not a one-parameter semigroup since it does not fulfil the associative relation [27], $T_0(t+\tau,0) = T_0(t,\,0)T_0(\tau,0) = T_0(\tau,0)T_0(t,\,0)$ for all $t,\tau \in \mathbf{R}_{0+}$ and, in particular, the operator composition is not commutative. In the same way, $T(t,\tau) : \mathbf{R}^2_{0+} \to C^2_0(\mathbf{R}_{0+})$ is not a strongly continuous semigroup either.*

It is now seen that, although the evolution operator $T_0(t,\tau) : \mathbf{R}^2_{0+} \to C^2_0(\mathbf{R}_{0+})$ is not a one-parameter semigroup, it is still a strongly continuous semigroup, since it satisfies the associative property as it is now discussed.

Theorem 1. *The undelayed system has the following properties:*

(i) $T_0(t,\tau) : \mathbf{R}^2_{0+} \to C^2_0(\mathbf{R}_{0+})$ satisfies the associative property $T_0(t,0) = T_0(t,\sigma)T_0(\sigma,0)$ for any $\sigma, t(\geq \sigma) \in \mathbf{R}_{0+}$.

***(ii)** Decompose $A_0(t)$, in a non-unique way, as $A_0(t) = A_{00} + \widetilde{A}_0(t)$ for all $t \in \mathbf{R}_{0+}$, where $A_{00} \in \mathbf{R}^{2\times 2}$ is constant. Then, one has for any $\sigma, t(\geq \sigma) \in \mathbf{R}_{0+}$, which the following equivalent expressions hold:*

$$\begin{aligned} x_0(t) &= T_0(t,\sigma)x_0(\sigma) = T_0(t,0)x_{00} \\ &= e^{A_{00}(t-\sigma)}\left(x_0(t) + \int_0^{t-\sigma} e^{-A_{00}\tau}\widetilde{A}_0(\tau+\sigma)x(\tau+\sigma)d\tau\right) \\ &= e^{A_{00}(t-\sigma)}\left(I_2 + \int_0^{t-\sigma} e^{-A_{00}\tau}\widetilde{A}_0(\tau+\sigma)T_0(\tau+\sigma\,,\sigma)d\tau\right)T_0(\sigma\,,0)x_{00} \\ &= e^{A_{00}(t-\sigma)}\left(I_2 + \int_\sigma^{t} e^{-A_{00}(\tau-\sigma)}\widetilde{A}_0(\tau)T_0(\tau\,,\sigma)d\tau\right)T_0(\sigma\,,0)x_{00} \\ &= e^{A_{00}t}\left(I_2 + \int_0^{t} e^{-A_{00}\tau}\widetilde{A}_0(\tau)T_0(\tau\,,0)d\tau\right)x_{00} \end{aligned} \tag{17}$$

***(iii)** The evolution operator of the solution of (3) satisfies for any $\sigma, t(\geq \sigma) \in \mathbf{R}_{0+}$ and any decomposition $A_0(t) = A_{00} + \widetilde{A}_0(t)$ for all $t \in \mathbf{R}_{0+}$:*

$$T_0(t\,,\sigma) = e^{A_{00}(t-\sigma)}\left(T_0(\sigma\,,0) + \int_0^{t-\sigma} e^{-A_{00}\tau}\widetilde{A}_0(\tau+\sigma)T_0(\tau+\sigma\,,0)d\tau\right) \tag{18}$$

Proof. Note that, since the solution of (3) is unique for each given initial condition, the relations

$$x_0(t) = T_0(t,0)x_{00} = T_0(t,\sigma)x_0(\sigma) = T_0(t,\sigma)T_0(\sigma,0)x_{00} \tag{19}$$

Hold, irrespective of x_{00}. This proves Property (i). On the other hand, if $A_0(t) = A_{00} + \widetilde{A}_0(t)$, then the expressions (17) follow from the solution of (3) on $[0,\,+\infty)$ and Property (i). This proves Property (ii). Property (ii) follows from the fact that (17) hold for any initial condition. □

Now, it is seen that the evolution operator of the delayed system is not a semigroup in the sense that the associative property for composing the solution on a time interval by using intermediate time instants to fix the initial conditions does not work, in general. See also, Remark 2. First, one expresses the solution trajectory of the delayed system $x(t)$ from the previous truncated two-dimensional real vector function $\hat{x}_{[\sigma,\ \sigma-h]}$ through the evolution operator $\hat{T}(t,\tau) : [-h\,,\infty) \times [-h\,,\infty) \to L\left(C^2_0(\mathbf{R})\right)$. For any $\sigma, t(\geq \sigma) \in \mathbf{R}_{0+}$, the following result follows directly from (8) and (17) and (18):

Proposition 1. *The solution trajectory of the delayed system for any given bounded piecewise-continuous function of initial conditions $\varphi : [-h\,,\,0] \to \mathbf{R}^2$, with $\varphi(0^+) = x(0^+) = x_0(0^+) = x_{00}$, is given by the following expressions:*

$$
\begin{aligned}
x(t) &= (\hat{T}(t,\sigma-h))\hat{x}_{[\sigma,\ \sigma-h]} \\
&= T_0(t,\ \sigma)x(\sigma) + \int_0^{h-\sigma} T_0(t,\tau+\sigma)A_d(\tau+\sigma)\varphi(\tau+\sigma-h)d\tau + \int_{h-\sigma}^{t} T_0(t,\tau+\sigma)A_d(\tau+\sigma)x(\tau+\sigma-h)U(\tau+\sigma-h)d\tau \\
&= T_0(t,\ \sigma)\left(\hat{T}(\sigma,0)\right)\hat{x}_{[0,-\hbar]} + \int_0^{h-\sigma} T_0(t,\tau+\sigma)A_d(\tau+\sigma)\varphi(\tau+\sigma-h)d\tau + \left(\int_{h-\sigma}^{t} T_0(t,\tau+\sigma)A_d(\tau+\sigma)\hat{T}(\tau+\sigma-h,0)\right)\hat{x}_{[0,-\hbar]}d\tau \\
&= T_0(t,\ 0)x_{00} + \int_{-h}^{0} T_0(t,\tau)A_d(h+\tau)\varphi(\tau)d\tau + \left(\int_0^{max(t-h),\ 0} T_0(t,\tau)A_d(\tau)T_0(\tau-h,0)d\tau\right)x_{00} \\
&+ \int_0^{max(t-h,0)} \int_{-h}^{0} T_0(t,\tau)A_d(\tau)T_0(\tau-h,\theta)A_d(h+\theta)\varphi(\theta)d\theta d\tau \\
&= e^{A_{00}t}\left(I_2 + \int_0^t e^{-A_{00}\tau}\widetilde{A}_0(\tau)T_0(\tau\ ,0)d\tau\right)x_{00} \\
&+ \left(\int_0^{max(t-h),\ 0} T_0(t,\tau)A_d(\tau)\left[e^{A_{00}(\tau-h)}\left(T_0(\tau-h\ ,0) + \int_0^{\tau-h} e^{-A_{00}\sigma}\widetilde{A}_0(\tau-h+\sigma)T_0(\tau-h+\sigma\ ,0)d\sigma\right)\right]d\tau\right)x_{00} \\
&+ \int_{-h}^{0}\left[e^{A_{00}(t-\tau)}\left(T_0(\tau\ ,0) + \int_0^{t-\tau} e^{-A_{00}\tau}\widetilde{A}_0(\tau+\sigma)T_0(\tau+\sigma\ ,0)d\tau\right)\right]A_d(h+\tau)\varphi(\tau)d\tau \\
&+ \int_0^{max(t-h,0)} \int_{-h}^{0}\left[e^{A_{00}(t-\tau)}\left(T_0(\tau\ ,0) + \int_0^{t-\tau} e^{-A_{00}\sigma}\widetilde{A}_0(\tau+\sigma)T_0(\tau+\sigma\ ,0)d\sigma\right)\right] \\
&\times A_d(\tau)\left[e^{A_{00}(\tau-h-\theta)}\left(T_0(\tau-h\ ,0) + \int_\theta^{\tau-h} e^{-A_{00}\sigma}\widetilde{A}_0(\tau-h+\sigma)T_0(\tau-h+\sigma\ ,0)d\sigma\right)\right]A_d(h+\sigma)\varphi(\sigma)d\sigma d\tau
\end{aligned}
\tag{20}
$$

3. Solutions at the Boundary Points of $[a,b]$ for $+\infty > b > a \geq 0$ and Related Results

This section discusses the loss of the Sturm–Liouville properties associated with a two-point boundary problem in a nominal undelayed system if delayed dynamics is incorporated.

For the undelayed system, the point solutions at the boundary points of the interval of interest are taken from (7) by picking up initial conditions at $t = 0$:

$$x_0(a) = T_0(a\ ,0)x_{00}$$

$$x_0(b) = T_0(b\ ,0)x_{00} = T_0(b\ ,0)T_0^{-1}(a,\ 0)x_0(a) = T_0(b,a)x_0(a)$$

then, by using the associative semigroup property of $T_0(.\,,.)$ proved in Property 1

$$
\begin{aligned}
x(a) = T_0(a,\ 0)x_{00} + \int_{-h}^{0} T_0(a,\tau)A_d(h+\tau)\varphi(\tau)d\tau + \left(\int_0^{max(a-h,0)} T_0(a,\tau)A_d(\tau)T_0(\tau-h,0)d\tau\right)x_{00} \\
+ \int_0^{max(a-h,0)} \int_{-h}^{0} T_0(a,\tau)A_d(\tau)T_0(\tau-h,\theta)A_d(h+\theta)\varphi(\theta)d\theta d\tau
\end{aligned}
\tag{21}
$$

$$
\begin{aligned}
= x_0(a) + \int_{-h}^{0} T_0(a,\tau)A_d(h+\tau)\varphi(\tau)d\tau + \left(\int_0^{max(a-h,0)} T_0(a,\tau)A_d(h+\tau)T_0(\tau,0)d\tau\right)x_{00} \\
+ \int_0^{max(a-h,0)} \int_{-h}^{0} T_0(a,\tau)A_d(h+\tau)T_0(\tau,\theta)A_d(h+\theta)\varphi(\theta)d\theta d\tau
\end{aligned}
\tag{22}
$$

$$
\begin{aligned}
x(b) = T_0(b,\ 0)x_{00} + \int_{-h}^{0} T_0(b,\tau)A_d(h+\tau)\varphi(\tau)d\tau + \left(\int_0^{max(b-h,0)} T_0(b,\tau)A_d(h+\tau)T(\tau,0)d\tau\right)x_{00} \\
+ \int_0^{max(b-h,0)} \int_{-h}^{0} T_0(b,\tau)A_d(h+\tau)T_0(\tau,\theta)A_d(h+\theta)\varphi(\theta)d\theta d\tau
\end{aligned}
\tag{23}
$$

which leads to the equivalent expression:

$$
\begin{aligned}
x(b) = x_0(b) + T_0(b,a)\Big[\int_{-h}^{0} T_0(a,\tau)A_d(h+\tau)\varphi(\tau)d\tau + \left(\int_0^{max(b-h,0)} T_0(a,\tau)A_d(h+\tau)T(\tau,0)d\tau\right)x_{00} \\
+ \int_0^{max(b-h,0)} \int_{-h}^{0} T_0(a,\tau)A_d(h+\tau)T_0(\tau,\theta)A_d(h+\theta)\varphi(\theta)d\theta d\tau\Big] \\
= x_0(b) + T_0(b,a)\Big[\int_{-h}^{0} T_0(a,\tau)A_d(h+\tau)\varphi(\tau)d\tau \\
+ \left(\int_0^{max(a-h,\ 0)} T_0(a,\tau)A_d(h+\tau)T(\tau,0)d\tau + \left(\int_{max(a-h,\ 0)}^{b} T_0(a,\tau)A_d(h+\tau)T(\tau,0)d\tau\right)\right)x_{00} \\
+ \int_0^{max(a-h,0)} \int_{-h}^{0} T_0(a,\tau)A_d(h+\tau)T_0(\tau,\theta)A_d(h+\theta)\varphi(\theta)d\theta d\tau \\
+ \int_{max(a-h,0)}^{max(b-h,0)} \int_{-h}^{0} T_0(a,\tau)A_d(h+\tau)T_0(\tau,\theta)A_d(h+\theta)\varphi(\theta)d\theta d\tau\Big]
\end{aligned}
\tag{24}
$$

Since $T_0(b,a)$ is non-singular, then $x(a) = x_0(a)$ if, and only if,

$$
\begin{aligned}
\int_{-h}^{0} T_0(a,\tau)A_d(h+\tau)\varphi(\tau)d\tau + \left(\int_0^{max(a-h,0)} T_0(a,\tau)A_d(h+\tau)T_0(\tau,0)d\tau\right)x_{00} \\
+ \int_0^{max(a-h,0)} \int_{-h}^{0} T_0(a,\tau)A_d(h+\tau)T_0(\tau,\theta)A_d(h+\theta)\varphi(\theta)d\theta d\tau = 0
\end{aligned}
\tag{25}
$$

and $x(b) = x_0(b)$ from (24), if, and only if,

$$\begin{aligned}\int_{-h}^{0} T_0(b,\tau) \quad & A_d(h+\tau)\varphi(\tau)d\tau \\ & + \left(\int_{0}^{max(a-h,0)} T_0(b,\tau)A_d(\tau)T(\tau-h,0)d\tau + \int_{max(a-h,0)}^{b} T_0(b,\tau)A_d(h+\tau)T(\tau,0)d\tau\right)x_{00} \\ & + \int_{0}^{max(a-h,0)} \int_{-h}^{0} T_0(b,\tau)A_d(h+\tau)T_0(\tau,\theta)A_d(h+\theta)\varphi(\theta)d\theta d\tau \\ & + \int_{max(a-h,0)}^{max(b-h,0)} \int_{-h}^{0} T_0(b,\tau)A_d(h+\tau)T_0(\tau,\theta)A_d(h+\theta)\varphi(\theta)d\theta d\tau = 0\end{aligned} \tag{26}$$

In view of (25) and (26), it follows that $x(a) = x_0(a)$ and $x(b) = x_0(b)$ if, and only if, the constraint (25) holds together with the subsequent constraint which makes $x(b) = x_0(b)$ if $x(a) = x_0(a)$:

$$\begin{aligned}&\left(\int_{max(a-h,\,0)}^{b} T_0(a,\tau)A_d(\tau)T(\tau-h,0)d\tau\right)x_{00} \\ &\quad + \int_{max(a-h,0)}^{max(b-h,0)} \int_{-h}^{0} T_0(a,\tau)A_d(\tau)T_0(\tau-h,\theta)A_d(h+\theta)\varphi(\theta)d\theta d\tau = 0\end{aligned} \tag{27}$$

Since $\varphi\ :\ [-h\ ,\ 0]$ is piecewise continuous, then $x(0^+) = x_{00} = \varphi(0^+)$ if $\phi(t)$ is discontinuous at $t = 0$ and $x(0) = x_{00} = \varphi(0^+)$ if $\phi(t)$ is continuous at $t = 0$.

Define the two-dimensional real matrices of respective orders (4×2) and (4×1):

$$\Omega = \begin{bmatrix}\Omega_1 \\ \Omega_2\end{bmatrix} = \begin{bmatrix}\int_{0}^{max(a-h,\,0)} T_0(a,\tau)A_d(h+\tau)T(\tau,0)d\tau + \int_{max(a-h,\,0)}^{b} T_0(a,\tau)A_d(h+\tau)T(\tau,0)d\tau \\ \int_{max(a-h,\,0)}^{b} T_0(a,\tau)A_d(h+\tau)T(\tau,0)d\tau\end{bmatrix} \tag{28}$$

$$\begin{aligned}\Gamma &= \begin{bmatrix}\Gamma_1 \\ \Gamma_2\end{bmatrix} \\ &= -\begin{bmatrix}\int_{-h}^{0} T_0(a,\tau)A_d(h+\tau)\varphi(\tau)d\tau + \int_{max(a-h,0)}^{max(b-h,0)} \int_{-h}^{0} T_0(b,\tau)A_d(h+\tau)T_0(\tau,\theta)A_d(h+\theta)\varphi(\theta)d\theta d\tau + \int_{0}^{max(a-h,0)} \int_{-h}^{0} T_0(b,\tau)A_d(h+\tau)T_0(\tau,\theta)A_d(h+\theta)\varphi(\theta)d\theta d\tau \\ \int_{max(a-h,0)}^{max(b-h,0)} \int_{-h}^{0} T_0(b,\tau)A_d(h+\tau)T_0(\tau,\theta)A_d(h+\theta)\varphi(\theta)d\theta d\tau\end{bmatrix}\end{aligned} \tag{29}$$

Note that Ω and Γ depend on the piecewise-continuous function of initial conditions $\varphi : [-h\ ,0) \to \boldsymbol{R}^2$. Then, the constraints (25) and (27) can be jointly rewritten in a compact form as $\Omega x_{00} = \Gamma$, which is a compatible algebraic system if, and only if, $rank[\,\Omega] = rank\,[\Omega\ ,\ \Gamma]$ from the Rouché–Capelli theorem from Linear Algebra and, in this case, the whole set of solutions is:

$$x_{00} = \Omega^{\dagger}\Gamma + \left(I_2 - \Omega^{\dagger}\Omega\right)\xi \tag{30}$$

where $\xi \in \boldsymbol{R}^2$ is arbitrary. Thus, the given x_{00} has to be one of the solutions of (30) parameterized by ξ. In other words (30) has to be solvable in ξ for the given x_{00} which holds if, and only if,

$$rank\left[I_2 - \Omega^{\dagger}\Omega\right] = rank\left[I_2 - \Omega^{\dagger}\Omega\ ,\ x_{00} - \Omega^{\dagger}\Gamma\right] \tag{31}$$

Since $x_0(a) = T_0(a,0)x_{00}$ then the Sturm–Liouville system of the undelayed system has to fulfil the boundary condition:

$$x_0(a) = T_0(a,0)\ \left(\Omega^{\dagger}\Gamma + \left(I_2 - \Omega^{\dagger}\Omega\right)\xi\right) \tag{32}$$

The following results have been proved as a direct conclusion of the above discussion:

Theorem 2. *Assume that the differential system is a Sturm–Liouville one on* $[a,\ b] \subset (0\ ,\ \infty)$ *subject to given boundary conditions* $x(a)$ *and* $x(b)$ *if* $A_d = 0$ *and that* $x(0^+) = x_0(0^+) = \varphi(0^+) = x_{00}$. *Assume also that (28) and (29) satisfy the rank condition* $rank[\,\Omega] = rank\,[\Omega\ ,\ \Gamma]$ *for the given piecewise continuous function of initial conditions* $\varphi\ : [-h\ ,0) \to \boldsymbol{R}^2$.

Then, the delayed system is also a Sturm–Liouville system with the same two-point boundary values as the undelayed one if, and only if, $x_0(a) = T_0(a,0)\left(\Omega^{\dagger}\Gamma + (I_2 - \Omega^{\dagger}\Omega)\xi\right)$ *for some* $\xi \in \mathbf{R}^2$.

Theorem 3. *Assume that the undelayed system is a Sturm–Liouville system on* $[a,\ b] \subset (0\ ,\ \infty)$ *subject to given boundary conditions* $x(a)$ *and* $x(b)$*. Then, the following properties hold:*

(i) *If* $rank[\,\Omega] < rank\ [\Omega\ ,\ \Gamma]$ *for the given piecewise continuous function of initial conditions* $\varphi\ : [-h\ ,0) \to \mathbf{R}^2$ *then the delayed system cannot be a Sturm–Liouville system with the same two-point boundary values as the undelayed one; that is,* $x_0(a) = x(a)$ *and* $x_0(b) = x(b)$ *are not fulfilled.*

(ii) *If* $rank[\,\Omega] = rank\ [\Omega\ ,\ \Gamma]$ *for the given piecewise continuous function of initial conditions* $\varphi\ : [-h\ ,0) \to \mathbf{R}^2$ *but there is no* $\xi \in \mathbf{R}^2$ *such that* $x_0(a) = T_0(a,0)\left(\Omega^{\dagger}\Gamma + (I_2 - \Omega^{\dagger}\Omega)\xi\right)$ *then the delayed system is not a Sturm–Liouville one with the same two-point boundary value problems as the undelayed one; that is,* $x(a) = x_0(a)$ *and* $x(b) = x_0(b)$ *are not fulfilled.*

(iii) *If* $rank[\,\Omega] = rank\ [\Omega\ ,\ \Gamma]$ *for the given piecewise continuous function of initial conditions* $\varphi\ : [-h\ ,0) \to \mathbf{R}^2$ *and there is some* $\xi \in \mathbf{R}^2$ *such that* $x_0(a) = T_0(a,0)\left(\Omega^{\dagger}\Gamma + (I_2 - \Omega^{\dagger}\Omega)\xi\right)$ *then the delayed system is a Sturm–Liouville one with the same two-point boundary value problems as the undelayed one; that is,* $x(a) = x_0(a)$ *and* $x(b) = x_0(b)$ *hold.*

(iv) *The delayed system is a Sturm–Liouville one with the same two-point boundary value problems as the undelayed one, that is,* $x(a) = x_0(a)$ *and* $x(b) = x_0(b)$*, for the given piecewise continuous function of initial conditions* $\varphi\ : [-h\ ,0) \to \mathbf{R}^2$*, if, and only if, the following constraints hold together:*

$$rank[\,\Omega] = rank\ [\Omega\ ,\ \Gamma] \tag{33}$$

$$rank\left[\ T_0(a,0)\left(I_2 - \Omega^{\dagger}\Omega\right)\right] = rank\left[T_0(a,0)\left(I_2 - \Omega^{\dagger}\Omega\right)\ ,\ x_0(a) - T_0(a,0)\Omega^{\dagger}\Gamma\right] \tag{34}$$

Outline of Proof. Properties (i)–(iii) follow directly from the previous discussion. Property (iv) is identical to Property (iii), since (34) guarantees the existence of some $\xi \in \mathbf{R}^2$ such that (32) holds for the given $x_0(a)$. □

In the event that $a = 0$, one fixes $x_{00} = x_0(a^+) = x(a^+)$ and (25) holds trivially but, in order that $x(b) = x_0(b)$, (27) has to hold as an individual constraint for $x_{00} = x_0(a)$. Equation (27) can be written compactly as $\Omega_2 x_0(a) = \Gamma_2$, where Ω_2 and Γ_2 are, respectively, 2×2 and 2×1 real matrices defined in the second block matrices of (28) and (29). Then, provided that $rank[\Omega_2] = rank\ [\Omega_2\ ,\ \Gamma_2]$, there should exist some $\varsigma \in \mathbf{R}^2$ such that $x_0(a) = \Omega_2^{\dagger}\Gamma_2 + (I_2 - \Omega_2^{\dagger}\Omega_2)\varsigma$, which would also imply that $x(b) = x_0(b)$. As a result, we conclude the following relaxed counterpart of Theorem 3:

Theorem 4. *Assume that the undelayed system is a Sturm–Liouville system on* $[a,\ b] \subset [0\ ,\ \infty)$ *with* $a = 0$ *subject to given boundary conditions* $x(0)$ *and* $x(b)$*. Then, the following properties hold:*

(i) *If* $rank[\Omega_2] < rank\ [\Omega_2\ ,\ \Gamma_2]$ *then the delayed system cannot be a Sturm–Liouville system with the same two-point boundary values as the undelayed one; that is,* $x_0(a) = x(a)$ *and* $x_0(b) = x(b)$ *are not fulfilled.*

(ii) *If* $rank[\Omega_2] = rank\ [\Omega_2\ ,\ \Gamma_2]$ *but there is no* $\varsigma \in \mathbf{R}^2$ *such that* $x_0(a) = T_0(a,0)\left(\Omega^{\dagger}\Gamma + (I_2 - \Omega^{\dagger}\Omega)\xi\right)$ *then the delayed system is not a Sturm–Liouville one with the same two-point boundary value problems as the undelayed one; that is,* $x(a) = x_0(a)$ *and* $x(b) = x_0(b)$ *are not fulfilled.*

(iii) *If* $rank[\Omega_2] = rank\ [\Omega_2\ ,\ \Gamma_2]$ *and there is some* $\varsigma \in \mathbf{R}^2$ *such that* $x_0(a) = \Omega_2^{\dagger}\Gamma_2 + (I_2 - \Omega_2^{\dagger}\Omega_2)\varsigma$ *then the delayed system is a Sturm–Liouville one with the same two-point boundary value problems as the undelayed one; that is,* $x(a) = x_0(a) = x_{00}$ *and* $x(b) = x_0(b)$ *hold.*

(iv) *The delayed system is a Sturm–Liouville one with the same two-point boundary value problems as the undelayed one; that is,* $x(a) = x_0(a) = x_{00}$ *and* $x(b) = x_0(b)$ *if, and only if, the following constraints hold together:*

$$rank[\Omega_2] = rank\,[\Omega_2\,,\,\Gamma_2] \tag{35}$$

$$rank\Big[\,I_2 - \Omega_2^{\dagger}\Omega_2\Big] = rank\Big[I_2 - \Omega_2^{\dagger}\Omega_2\,,\;x_0(a) - \Omega_2^{\dagger}\Gamma_2\Big] \tag{36}$$

4. Boundedness and Asymptotic Properties of the Evolution Operators

One understands from (18) that the evolution operator of the free of delayed dynamics system, that is, the undelayed system with $A_d(t) \equiv 0$, is

$$T_0(t\,,0) = e^{A_{00}t}\left(I_2 + \int_0^t e^{-A_{00}\tau}\widetilde{A}_0(\tau)T_0(\tau\,,0)d\tau\right); t \in \mathbf{R}_{0+} \tag{37}$$

$$T_0(t\,,\tau) = e^{A_{00}(t-\tau)}\left(I_2 + \int_0^{t-\tau} e^{-A_{00}\tau}\widetilde{A}_0(\tau+\sigma)T_0(\tau+\sigma\,,0)d\tau\right); \tau, t(\geq \tau) \in \mathbf{R}_{0+} \tag{38}$$

Note that the system referred to as the undelayed system is the one free of delayed dynamics; that is, $A_d = 0$ in (5). Also, one understands from (5) that if $h = 0$, the so-called delay-free system of matrix of dynamics being $A_0(t) = A_0(t) + A_d(t)$ which has an evolution operator similar to (18) by replacing $\widetilde{A}_0(t) \to \widetilde{A}_0(t) + A_d(t)$, resulting as

$$T_{00}(t\,,0) = e^{A_{00}t}\left(I_2 + \int_0^t e^{-A_{00}\tau}\left(\widetilde{A}_0(\tau) + A_d(\tau)\right)T_{00}(\tau\,,0)d\tau\right); t \in \mathbf{R}_{0+} \tag{39}$$

$$T_{00}(t\,,\tau) = e^{A_{00}(t-\tau)}\left(I_2 + \int_0^{t-\tau} e^{-A_{00}\sigma}\left(\widetilde{A}_0(\tau+\sigma) + A_d(\tau+\sigma)\right)T_{00}(\tau+\sigma\,,0)d\sigma\right); \tau, t(\geq \tau) \in \mathbf{R}_{0+} \tag{40}$$

And (12), combined with (37) and (38), yields that the evolution operator of the delayed system can be equivalently expressed as

$$\begin{aligned} T(t,0) &= e^{A_{00}t}\left(I_2 + \int_0^t e^{-A_{00}\tau}\widetilde{A}_0(\tau)T_0(\tau\,,0)d\tau\right) + \\ &+ \int_0^t \left(e^{A_{00}(t-\tau)}\left(I_2 + \int_0^{t-\tau} e^{-A_{00}\tau}\widetilde{A}_0(\tau+\sigma)T_0(\tau+\sigma\,,0)d\tau\right)\right)A_d(\tau)T(\tau-h\,,0)d\tau;\; t \in \mathbf{R}_{0+} \end{aligned} \tag{41}$$

Now, some upper-bounding growing laws of the evolution operators with time are calculated in order to obtain approximate Sturm–Liouville boundary conditions of the delayed system, provided that the undelayed one is a Sturm–Liouville system.

It turns out that there exist non-unique real constants $K \geq 1$ (which are norm-dependent) and ρ associated with the dominant eigenvalue of A_{00} such that $\left\|e^{A_{00}t}\right\| \leq Ke^{\rho t}$. In particular, if ρ_{00} is the stability abscissa of A_{00}, that is, the largest of the real parts of its eigenvalues, then $\rho \geq \rho_{00}$ if the dominant eigenvalues are single and $\rho > \rho_{00}$ otherwise [24,25]. It is well-known that if A_{00} is a stability matrix (i.e., all is eigenvalues have a negative real part) then $\rho < 0$. If it is unstable, i.e., at least one of its eigenvalues has a positive real part then $\rho > 0$. If it is critically stable, i.e., at least one of its eigenvalues has null real part and no eigenvalue is unstable, then $\rho = 0$. See, for instance, [26–32]. Then, from (17), by assuming $\sup\limits_{0\leq\tau\leq t}\left\|\widetilde{A}_0(\tau)\right\| \leq \widetilde{a}_0$, one has the following:

(a) if $\rho < 0$, then

$$\|T_0(t,0)\| \leq K\left(1 + |\rho|^{-1}\widetilde{a}_0 \sup_{0\leq\tau\leq t}\|T_0(\tau,\,0)\|\right);\; t \in \mathbf{R}_{0+} \tag{42}$$

thus,

$$\sup_{0\leq\tau\leq t}\|T_0(\tau,0)\| \leq K\left(1 + |\rho|^{-1}\widetilde{a}_0 \sup_{0\leq\sigma\leq\tau}\|T_0(\sigma,\,0)\|\right) \leq K\left(1 + |\rho|^{-1}\widetilde{a}_0 \sup_{0\leq\tau\leq ?t}\|T_0(\tau,\,0)\|\right);\; t,\;\tau(\leq t) \in \mathbf{R}_{0+} \tag{43}$$

and, if $\widetilde{a}_0 < |\rho|/K$, then

$$\sup_{0\leq\tau\leq t} \|T_0(\tau,0)\| \leq \frac{K|\rho|}{|\rho| - K\widetilde{a}_0} \tag{44}$$

(b) For any admissible real ρ, one has from (17) that

$$\|T_0(t,0)\| \leq Ke^{\rho t}\left(1 + \widetilde{a}_0 \int_0^t e^{-\rho\tau} \|T_0(\tau,0)\| d\tau\right);\ t \in \mathbf{R}_{0+} \tag{45}$$

Then,

$$v(t) = e^{-\rho t}\|T_0(t,0)\| \leq K + K\widetilde{a}_0 \int_0^t v(\tau) d\tau;\ t \in \mathbf{R}_{0+} \tag{46}$$

Now, Gronwall's inequality, [2], applied to (46) yields

$$v(t) \leq K + K^2\widetilde{a}_0 \int_0^t e^{\int_\tau^t K\widetilde{a}_0 dr} d\tau = K + K^2\widetilde{a}_0 \int_0^t e^{K\widetilde{a}_0(t-\tau)} d\tau = Ke^{K\widetilde{a}_0 t};\ t \in \mathbf{R}_{0+} \tag{47}$$

Thus, one gets from (46) and (47),

$$\|T_0(t,0)\| \leq Ke^{(\rho + K\widetilde{a}_0)t};\ t \in \mathbf{R}_{0+} \tag{48}$$

so that:

(b1) If $\rho < 0$ and $\widetilde{a}_0 < |\rho|/K$ then $\lim_{t\to+\infty} \|T_0(t,0)\| = 0$ with exponential convergence and also and the bounding condition $\sup_{0\leq\tau\leq t} \|T_0(\tau,0)\| \leq K|\rho|/(|\rho| - K\widetilde{a}_0) < +\infty$ for all $t \in \mathbf{R}_{0+}$ according to (44) can be refined to $\sup_{0\leq\tau\leq t} \|T_0(\tau,0)\| \leq K$ for $t \in \mathbf{R}_{0+}$, according to (48).

(b2) If $\rho = \widetilde{a}_0 = 0$ then $\sup_{0\leq\tau\leq t} \|T_0(\tau,0)\| \leq K$ for all $t \in \mathbf{R}_{0+}$.

Example 1. *Assume that the time-varying component $\widetilde{A}_0(t) = A_0(t) - A_{00}$ of $A_0(t)$ is zero, except on a subset S of $[0,\infty)$ of nonzero finite Lebesgue measure L_S and that $\rho = 0$, that is, the constant part A_{00} of $A_0(t)$ is critically stable. Then, (47) becomes*

$$\|T_0(\tau,0)\| = v(t) \leq K + K^2\widetilde{a}_0 \int_0^t e^{\int_\tau^t K\widetilde{a}_0 dr} d\tau \leq K\left(1 + K\widetilde{a}_0 e^{K\widetilde{a}_0 L_S} t\right);\ t \in \mathbf{R}_{0+} \tag{49}$$

since

$$e^{\int_\tau^t K\|\widetilde{A}_0(r)\|dr} \leq e^{K\int_S \|\widetilde{A}_0(r)\|dr} \leq e^{K\widetilde{a}_0 L_S} < +\infty \tag{50}$$

for all $0 \leq \tau(\leq t),\ \ t \leq +\infty$.

Therefore, Gronwall's inequality does not guarantee the boundedness of the undelayed evolution operator if $A_0(t)$ is constant, except in a set of zero measure, while the constant additive component is critically stable.

Then, one understands from (12) and (48), if $\widetilde{a}_0 < |\rho|/K$ and $\rho < 0$, that

$$\|T(t,0)\| \leq \|T_0(t,0)\| + \int_0^t \|T_0(t,\tau)\|\|A_d(\tau)\|\|T(\tau - h\ ,0)\|d\tau;\ t \in \mathbf{R}_{0+} \tag{51}$$

If $a_d = \sup_{0\leq\tau\leq t} \|A_d(\tau)\|$ then

$$\begin{aligned}\sup_{0\leq\tau\leq t} \|T(\tau,0)\| &\leq \sup_{0\leq\tau\leq t} \|T_0(\tau,0)\| + \sup_{0\leq\tau\leq t-h} \|T(\tau\ ,0)\| \textstyle\int_0^t \|T_0(t,\tau)\|\|A_d(\tau)\|d\tau \\ &\leq \sup_{0\leq\tau\leq t} \|T_0(\tau,0)\| + a_d \sup_{0\leq\tau\leq t} \|T(\tau\ ,0)\| \textstyle\int_0^t \|T_0(t,\tau)\|d\tau;\ t \in \mathbf{R}_{0+}\end{aligned} \tag{52}$$

Now, if $\widetilde{a}_0 < |\rho|/K, \rho < 0, a_d < 1/\left(\int_0^\infty \|T_0(t,\tau)\|d\tau\right)$, so that $\|T_0(t,0)\| \leq Ke^{-(|\rho|-K\widetilde{a}_0)t}$ and $\int_0^t \|T_0(t,\tau)\|d\tau \leq K/(|\rho| - K\widetilde{a}_0)$ from (48) for all $t \in \mathbf{R}_{0+}$, so that the above constraint $a_d < 1/\left(\int_0^\infty \|T_0(t,\tau)\|d\tau\right)$ is guaranteed if $a_d < (|\rho| - K\widetilde{a}_0)/K$, equivalently if $(\widetilde{a}_0 + a_d) < |\rho|/K$, and one understands from (52) that

$$\left(1 - \frac{Ka_d}{|\rho| - K\widetilde{a}_0}\right) \sup_{0\leq\tau\leq t} \|T(\tau,0)\| \leq \sup_{0\leq\tau\leq t} \|T_0(\tau,0)\|;\ t \in \mathbf{R}_{0+} \tag{53}$$

Then, it follows from (52) that

$$\sup_{0\leq\tau\leq t} \|T(\tau,0)\| \leq \frac{|\rho| - K\widetilde{a}_0}{|\rho| - K(\widetilde{a}_0 + a_d)} \sup_{0\leq\tau\leq t} \|T_0(\tau,0)\| = \frac{K\left(|\rho| - K\widetilde{a}_0\right)}{|\rho| - K(\widetilde{a}_0 + a_d)};\ t \in \mathbf{R}_{0+} \tag{54}$$

Then, the subsequent result holds from the above discussion and partial results (51)–(53):

Theorem 5. *Assume that $\rho < 0$. Then, the following properties hold:*
(*i*) *if $\widetilde{a}_0 < |\rho|/K$ then*

$$\sup_{\sigma\leq\tau\leq t} \|T_0(t,\sigma)\| \leq \frac{K|\rho|}{|\rho| - K\widetilde{a}_0};\ \sigma,\ t(\geq \sigma) \in \mathbf{R}_{0+} \tag{55}$$

(*ii*) *if $(\widetilde{a}_0 + a_d) < |\rho|/K$ then $\sup_{\sigma\leq\tau\leq t} \|T_0(\tau,\sigma)\| \leq K;\ \sigma,\ t(\geq \sigma) \in \mathbf{R}_{0+}$, and*

$$\sup_{\sigma\leq\tau\leq t} \|T(\tau,\sigma)\| \leq \frac{|\rho| - K\widetilde{a}_0}{|\rho| - K(\widetilde{a}_0 + a_d)} \sup_{\sigma\leq\tau\leq t} \|T_0(\tau,\sigma)\| = \frac{K\left(|\rho| - K\widetilde{a}_0\right)}{|\rho| - K(\widetilde{a}_0 + a_d)};\ \sigma,\ t(\geq \sigma) \in \mathbf{R}_{0+} \tag{56}$$

$$\lim_{\sigma,t\to\infty} \sup_{t\leq\tau\leq t+\sigma} \|T(t+\sigma,t)\| \leq \frac{|\rho| - K\widetilde{a}_0}{|\rho| - K(\widetilde{a}_0 + a_d)} \lim_{\sigma,t\to\infty}\left(\sup_{\sigma\leq\tau\leq t} \|T_0(t+\sigma,t)\|\right) = 0 \tag{57}$$

It is of interest in the context of Sturm–Liouville systems to investigate the uniform boundedness of non-asymptotically vanishing evolution operators of the undelayed system; that is, when A_{00} is critically stable so that $\rho = 0$. This covers a wide class of situations when defining two-boundary value problems on finite intervals, including periodicity of such boundary values. A typical and elementary example is concerned with second-order linear time-invariant differential equations whose matrix of dynamics is driven by a couple of imaginary complex-conjugate eigenvalues. Sufficiency-type conditions of boundedness of the evolution operators when $\rho = 0$ are now given:

Proposition 2. *Assume that $\rho = 0$ and* $\left\| \begin{pmatrix} \left(\int_0^\infty e^{A_{00}(t-\tau)}\right)_{12} A_{0_{21}}(\tau)d\tau \\ \left(\int_0^\infty e^{A_{00}(t-\tau)}\right)_{22} A_{0_{21}}(\tau)d\tau \end{pmatrix} \right\| < 1$. *Then,* $\|T_0(t,0)\| < +\infty;\ t \in \mathbf{R}_{0+}$.

A stronger sufficient condition which guarantees that $\|T_0(t,0)\| < +\infty;\ t \in \mathbf{R}_{0+}$ is $\widetilde{a}^I_{0_{21}} = \int_0^\infty |A_{0_{21}}(\tau)d\tau| < K^{-1}$.

Proof. Note from (37) that, if $\rho = 0$, then

$$\begin{aligned} \|T_0(t,0)\| &\leq K + \left\|\int_0^t e^{A_{00}(t-\tau)}\widetilde{A}_0(\tau)T_0(\tau,0)d\tau\right\| \\ &\leq K + \left(\left\|\int_0^t e^{A_{00}(t-\tau)}\widetilde{A}_0(\tau)d\tau\right\|\right) \sup_{0\leq\tau\leq t} \|T_0(\tau,0)\|;\ t \in \mathbf{R}_{0+} \end{aligned} \tag{58}$$

which implies that

$$\underset{0\le\tau\le t}{sup}\ \|T_0(\tau\,,0)\| \le K + \left(\left\|\int_0^t e^{A_{00}(t-\tau)}\widetilde{A}_0(\tau)d\tau\right\|\right)\underset{0\le\tau\le t}{sup}\ \|T_0(\tau\,,0)\|;\ t\in \mathbf{R}_{0+} \tag{59}$$

so that, if $1 > \left\|\int_0^\infty e^{A_{00}(t-\tau)}\widetilde{A}_0(\tau)d\tau\right\|$, which is equivalent to $1 > \left\|\begin{pmatrix} \left(\int_0^\infty e^{A_{00}(t-\tau)}\right)_{12}\widetilde{A}_{0_{21}}(\tau)d\tau \\ \left(\int_0^\infty e^{A_{00}(t-\tau)}\right)_{22}\widetilde{A}_{0_{21}}(\tau)d\tau \end{pmatrix}\right\|$

which is guaranteed under the sufficiency-type condition $\widetilde{a}^I_{0_{21}} = \int_0^\infty \left|\widetilde{A}_{0_{21}}(\tau)d\tau\right| < K^{-1}$ (since $\left\|e^{A_{00}(t-\tau)}\right\| \le K;\ \tau,\ t(\ge\tau) \in \mathbf{R}_{0+}$) because of the structure of (4), which makes the entry $A_{0_{21}}(.)$ of $A_0(t)$ being the only one which is eventually non-null for any established decomposition $A_0(t) = A_{00} + \widetilde{A}_0(t)$. Then, one has

$$\underset{0\le\tau\le t}{sup}\ \|T_0(\tau\,,0)\| \le K\left(1 - \left\|\int_0^t e^{A_{00}(t-\tau)}\widetilde{A}_0(\tau)d\tau\right\|\right)^{-1} < +\infty;\ t\in\mathbf{R}_{0+} \tag{60}$$

$$\underset{0\le t<\infty}{sup}\ \|T_0(t\,,0)\| \le K\left(1 - \left\|\int_0^\infty e^{A_{00}(t-\tau)}\widetilde{A}_0(\tau)d\tau\right\|\right)^{-1} < +\infty \tag{61}$$

If, furthermore, $\widetilde{a}^I_{0_{21}} < K$ then $\underset{0\le t<\infty}{sup}\ \|T_0(t\,,0)\| \le K/\left(1 - Ka^I_{0_{21}}\right)^{-1} < +\infty$.

Which proves the result. □

Proposition 3. *Assume that* $\rho = 0$ *and that the following conditions hold:*

$$(1)\ \left\|\begin{pmatrix} \left(\int_0^\infty e^{A_{00}(t-\tau)}\right)_{12}\widetilde{A}_{0_{21}}(\tau)d\tau \\ \left(\int_0^\infty e^{A_{00}(t-\tau)}\right)_{22}\widetilde{A}_{0_{21}}(\tau)d\tau \end{pmatrix}\right\| < 1 \tag{62}$$

(2) $A_d : [-h\,,+\infty) \to \mathbf{R}^{2\times 2}$ is in $\mathbf{L}^1(\,[0\,,+\infty);\mathbf{R}^{2\times 2})$ so that $a^I_d = \int_0^\infty \|A_d(\tau)\|d\tau < +\infty$ fulfils $a^I_d < 1/\left(\underset{0\le t\le\infty}{sup}\ \|T_0(t\,,0)\|\right) < +\infty$ (63)

Then, the following properties hold:

(i) *The constraint (62) is guaranteed if*

$$a^I_d < K^{-1}\left(1 - \left\|\int_0^\infty e^{A_{00}(t-\tau)}\widetilde{A}_0(\tau)d\tau\right\|\right) < +\infty \tag{64}$$

$$\textbf{\textit{(ii)}}\ \underset{0\le t\le\infty}{sup}\ \|T(t\,,\,0)\| \le \frac{\underset{0\le t\le\infty}{sup}\ \|T_0(t\,,\,0)\|}{1 - a^I_d\ \underset{0\le t\le\infty}{sup}\ \|T_0(t\,,\,0)\|} < +\infty \tag{65}$$

(iii) *If (64) holds, then*

$$\underset{0\le t\le\infty}{sup}\ \|T(t\,,\,0)\| \le \frac{K\left(1 - \left\|\int_0^\infty e^{A_{00}(t-\tau)}\widetilde{A}_0(\tau)d\tau\right\|\right)}{1 - a^I_d K\left(1 - \left\|\int_0^\infty e^{A_{00}(t-\tau)}\widetilde{A}_0(\tau)d\tau\right\|\right)} < +\infty \tag{66}$$

Proof. From Proposition 2, the condition 1, Equation (62), implies that $\|T_0(t\,,0)\| < +\infty$. This proves Property (i). From (12) and the condition 2, it follows that (65) holds, which proves Property (ii). If the condition 2 holds under the strongest one (64) then (65) implies (66). Also, (64) and (65) imply (66) and then Property (iii). □

5. Worst-Case Deviation of the Delayed System from a Sturm–Liouville Behavior Compared to the Undelayed Sturm–Liouville Nominal One

Theorem 2 and Theorem 3 establish conditions for the Sturm Liouville two-boundary value conditions to be kept in the delayed system, from the undelayed one. The respective deviations from zero of the left-hand sides of (25) and (26), in terms of error norms, give estimations of the errors in the two-point boundary conditions of the delayed system at $t = a$ and $t = b$, with respect to the undelayed Sturm–Liouville one. Note that if the delayed system loses the Sturm–Liouville property, then the zero right-hand-side values in (25) and (26) are replaced by $(x(a) - x_0(a))$ and $(x(b) - x_0(b))$, respectively, since the error between the current and nominal systems at the boundary points is no longer zero. That concern is now addressed by using the results about the boundedness and the asymptotic behaviors of the evolution operators which have been characterized in the above section, together with the results of Theorem 2, Theorem 3, and Equations (25) and (26) of Section 2.

Now, suppose that

$$\overline{\varphi} = \sup_{-h\le\tau\le 0}\|\varphi(\tau)\|;\ a_{da} = \sup_{0\le\tau\le a}\|A_d(\tau)\|;\ a_{db} = \sup_{0\le\tau\le b}\|A_d(\tau)\| = max\left(a_{da},\ \sup_{a\le\tau\le b}\|A_d(\tau)\|\right) \tag{67}$$

Taking norms in the right-hand sides of (25) and (26), one obtains:

$$\begin{cases} x(a) = x_0(a)\ if\ a = 0 \\ \|x(a) - x_0(a)\| \le a a_{da}\overline{\varphi}\sup_{0\le\tau\le a}\|T_0(a,\tau)\|\left(1 + h a_{da}\sup_{0\le\tau\le a}\|T_0(a,\tau)\|\right)\ if\ a > 0 \end{cases} \tag{68}$$

and

$$\|x(b) - x_0(b)\| \le b a_{db}\overline{\varphi}\sup_{0\le\tau\le b}\|T_0(b,\tau)\|\left(1 + h a_{db}\sup_{0\le\tau\le b}\|T_0(b,\tau)\|\right) \tag{69}$$

Case a: $\rho = 0$ and $\left\|\begin{pmatrix}\left(\int_0^\infty e^{A_{00}(t-\tau)}\right)_{12}\widetilde{A}_{0_{21}}(\tau)d\tau \\ \left(\int_0^\infty e^{A_{00}(t-\tau)}\right)_{22}\widetilde{A}_{0_{21}}(\tau)d\tau\end{pmatrix}\right\| < 1$, which is guaranteed if $K^{-1} > \widetilde{a}^I_{0_{21}} = \int_0^\infty\left|\widetilde{A}_{0_{21}}(\tau)d\tau\right|$ (Proposition 2). Then, one has, from (60), (68) and (69), the following:

$$\begin{cases} x(a) = x_0(a)\ if\ a = 0 \\ \|x(a) - x_0(a)\| \le a a_{da}\overline{\varphi}K\left(1 - \left\|\int_0^a e^{A_{00}(a-\tau)}\widetilde{A}_0(\tau)d\tau\right\|\right)^{-1}\left(1 + h a_{da}K\left(1 - \left\|\int_0^a e^{A_{00}(a-\tau)}\widetilde{A}_0(\tau)d\tau\right\|\right)^{-1}\right)\ if\ a > 0 \end{cases} \tag{70}$$

$$\|x(b) - x_0(b)\| \le b a_{db}\overline{\varphi}K\left(1 - \left\|\int_0^b e^{A_{00}(a-\tau)}\widetilde{A}_0(\tau)d\tau\right\|\right)^{-1}\left(1 + h a_{db}K\left(1 - \left\|\int_0^b e^{A_{00}(a-\tau)}\widetilde{A}_0(\tau)d\tau\right\|\right)^{-1}\right) \tag{71}$$

If, furthermore, the stronger constraint $\widetilde{a}^I_{0_{21}} = \int_0^\infty\left|\widetilde{A}_{0_{21}}(\tau)d\tau\right| < K^{-1}$ holds, then two stronger sufficiency-type conditions than (70) and (71), which in turn guarantee them, are

$$\begin{cases} x(a) = x_0(a)\ if\ a = 0 \\ \|x(a) - x_0(a)\| \le a a_{da}\overline{\varphi}K\left(1 - K\widetilde{a}^I_{0_{21}}\right)^{-1}\left(1 + h a_{da}K\left(1 - K\widetilde{a}^I_{0_{21}}\right)^{-1}\right)\ if\ a > 0 \end{cases} \tag{72}$$

and

$$\|x(b) - x_0(b)\| \le b a_{db}\overline{\varphi}K\left(1 - K\widetilde{a}^I_{0_{21}}\right)^{-1}\left(1 + h a_{db}K\left(1 - K\widetilde{a}^I_{0_{21}}\right)^{-1}\right) \tag{73}$$

Case b: $\rho < 0$ and $\widetilde{a}_0 < |\rho|/K$

One uses (44), (48), (55) and the left-hand sides of (25) and (26). Then,

$$\begin{cases} x(a) = x_0(a)\ if\ a = 0 \\ \|x(a) - x_0(a)\| \le \frac{K|\rho|a_{da}\overline{\varphi}a}{|\rho| - K\widetilde{a}_0}\left(1 + \frac{K|\rho|}{|\rho| - K\widetilde{a}_0}(1 + a_{da}a)\right)\ if\ a > 0 \end{cases} \tag{74}$$

and

$$\|x(b) - x_0(b)\| \leq \frac{K|\rho|a_{db}\overline{\varphi}b}{|\rho| - K\widetilde{a}_0}\left(1 + \frac{K|\rho|}{|\rho| - K\widetilde{a}_0}(1 + a_{db}b)\right) \tag{75}$$

Case c: $\rho \in \boldsymbol{R}$

One uses (48) and the left-hand sides of (25) and (26). Then,

$$\begin{cases} x(a) = x_0(a) \; if \; a = 0 \\ \|x(a) - x_0(a)\| \leq Ke^{(\rho + K\widetilde{a}_0)a}a_{da}\overline{\varphi}a\left(1 + Ke^{(\rho + K\widetilde{a}_0)a}(1 + a_{da}a)\right) \; if \; a > 0 \end{cases} \tag{76}$$

and

$$\|x(b) - x_0(b)\| \leq Ke^{(\rho + K\widetilde{a}_0)b}a_{db}\overline{\varphi}b\left(1 + Ke^{(\rho + K\widetilde{a}_0)b}(1 + a_{db}b)\right) \tag{77}$$

Note that Case c also includes the situation of possible instability of A_{00}; that is, the case of positive stability abscissa. As a result, (76) and (77) apply for any real ρ but they are not very tight for $\rho \leq 0$ compared to (72)–(75).

Note that there exist $\widetilde{\lambda}_0$, $\widetilde{\lambda}_0^I$, λ_{da}, λ_{db}, $\lambda_{\overline{\varphi}} \in \boldsymbol{R}_{0+}$, depending on the given ε, such that

$$\widetilde{a}_0 \leq \widetilde{\lambda}_0\varepsilon; \; \widetilde{a}_{0_{21}}^I \leq \lambda_0^I\varepsilon; \; a_{da} \leq \lambda_{da}\varepsilon; \; a_{db} \leq \lambda_{db}\varepsilon; \; \overline{\varphi} \leq \lambda_{\overline{\varphi}}\varepsilon \tag{78}$$

The worst-case errors at the two-point boundary points are kept below any prescribed arbitrary threshold $\varepsilon > 0$ under the following conditions.

Then,

(1) If $\rho = 0$ and $a > 0$ (Case a of critical stability of A_{00}), then one has from (70) and (71), that

$$a\lambda_{da}\lambda_{\overline{\varphi}}K\left(1 - K\widetilde{\lambda}_0 a\right)^{-1}\left(1 + h\lambda_{da}K\left(1 - K\widetilde{\lambda}_0 a\right)^{-1}\right) \leq 1 \Rightarrow \|x(a) - x_0(a)\| \leq \varepsilon \text{ if } \widetilde{\lambda}_0 < 1/Ka \tag{79}$$

$$b\lambda_{db}\lambda_{\overline{\varphi}}K\left(1 - K\widetilde{\lambda}_0 b\right)^{-1}\left(1 + h\lambda_{db}K\left(1 - K\widetilde{\lambda}_0 b\right)^{-1}\right) \leq 1 \Rightarrow \|x(b) - x_0(b)\| \leq \varepsilon \text{ if } \widetilde{\lambda}_0 < 1/Kb \tag{80}$$

and, from (72) and (73),

$$a\lambda_{da}\lambda_{\overline{\varphi}}K\left(1 - K\widetilde{\lambda}_0^I\right)^{-1}\left(1 + h\lambda_{da}K\left(1 - K\widetilde{\lambda}_0^I\right)^{-1}\right) \leq 1 \Rightarrow \|x(a) - x_0(a)\| \leq \varepsilon \text{ if } \widetilde{\lambda}_0^I < 1/K \tag{81}$$

$$b\lambda_{db}\lambda_{\overline{\varphi}}K\left(1 - K\widetilde{\lambda}_0^I\right)^{-1}\left(1 + h\lambda_{db}K\left(1 - K\widetilde{\lambda}_0^I\right)^{-1}\right) \leq 1 \Rightarrow \|x(b) - x_0(b)\| \leq \varepsilon \text{ if } \widetilde{\lambda}_0^I < 1/K \tag{82}$$

The constraint $\widetilde{a}_{0_{21}}^I = \int_0^\infty \left|\widetilde{A}_{0_{21}}(\tau)d\tau\right| < K^{-1} \leq \lambda_0^I\varepsilon$ can be replaced by the weakest ones:

$$\widetilde{a}_{0_{a21}}^I = \int_0^a \left|\widetilde{A}_{0_{21}}(\tau)d\tau\right| < K^{-1} \leq \lambda_{0a}^I\varepsilon; \; \widetilde{a}_{0_{b21}}^I = \int_0^b \left|\widetilde{A}_{0_{21}}(\tau)d\tau\right| < K^{-1} \leq \lambda_{0b}^I\varepsilon \tag{83}$$

Thus, (81) and (82) might be replaced with

$$a\lambda_{da}\lambda_{\overline{\varphi}}K\left(1 - K\widetilde{\lambda}_{0a}^I\right)^{-1}\left(1 + h\lambda_{da}K\left(1 - K\widetilde{\lambda}_{0a}^I\right)^{-1}\right) \leq 1 \Rightarrow \|x(a) - x_0(a)\| \leq \varepsilon \text{ if } \widetilde{\lambda}_{0a}^I < 1/K \tag{84}$$

$$b\lambda_{db}\lambda_{\overline{\varphi}}K\left(1 - K\widetilde{\lambda}_{0b}^I\right)^{-1}\left(1 + h\lambda_{db}K\left(1 - K\widetilde{\lambda}_{0b}^I\right)^{-1}\right) \leq 1 \Rightarrow \|x(b) - x_0(b)\| \leq \varepsilon \text{ if } \widetilde{\lambda}_{0b}^I < 1/K \tag{85}$$

(2) If $a > 0$, $\rho < 0$ and $\widetilde{a}_0 < |\rho|/K$ (Case b of stability of A_{00}) then one has from (74) and (75) that

$$\frac{K|\rho|\lambda_{da}\lambda_{\overline{\varphi}}}{|\rho| - K\widetilde{\lambda}_0}\left(1 + \frac{K|\rho|}{|\rho| - K\overset{\leftrightarrow}{\lambda}_0}(1 + \lambda_{da}a)\right) \leq 1 \Rightarrow \|x(a) - x_0(a)\| \leq \varepsilon \text{ if } \widetilde{\lambda}_0 < |\rho|/K \tag{86}$$

$$\frac{K|\rho|\lambda_{db}\lambda_{\overline{\varphi}}}{|\rho|-K\widetilde{\lambda}_0}\left(1+\frac{K|\rho|}{|\rho|-K\overset{\leftrightarrow}{\lambda}_0}(1+\lambda_{db}a)\right)\leq 1\Rightarrow \|x(b)-x_0(b)\|\leq\varepsilon \text{ if } \widetilde{\lambda}_0<|\rho|/K \tag{87}$$

(3) If $\rho \in \boldsymbol{R}$ (Case c including the potential stability, critical stability or instability of A_{00}), then one has from (76) and (77) that

$$Ke^{(\rho+K\widetilde{\lambda}_0)a}\lambda_{da}\lambda_{\overline{\varphi}}a\left(1+Ke^{(\rho+K\widetilde{\lambda}_0)a}(1+\lambda_{da}a)\right)\Rightarrow \|x(a)-x_0(a)\|\leq\varepsilon \tag{88}$$

$$Ke^{(\rho+K\widetilde{\lambda}_0)b}\lambda_{db}\lambda_{\overline{\varphi}}b\left(1+Ke^{(\rho+K\widetilde{\lambda}_0)b}(1+\lambda_{db}b)\right)\Rightarrow \|x(b)-x_0(b)\|\leq\varepsilon \tag{89}$$

6. Concluding Remarks

This paper has discussed the relations between two linear time-varying second-order differential equations with two-point boundary values. One of them is undelayed and it is a Sturm–Liouville system on a finite time interval, while the other one is a delayed perturbed current version of the first one. It is proved that the delayed differential system loses, in general, the Sturm–Liouville property, due to the lack of tracking of the prescribed two-point boundary conditions fixed for the undelayed nominal Sturm–Liouville differential system. Both the undelayed nominal equation and the delay-perturbed differential one are described equivalently by second-order differential systems, each of them formed by two first-order differential equations. The reason for this equivalent description is to facilitate a joint calculation of the respective solutions and their first-order time derivatives through time on the interval of interest via the involvement and "ad hoc" use of their respective evolution operators. In this context, the formal results are developed for the equivalent second-order differential systems of first-order equations, based on their respective evolution operators. This allows one to easily visualize how the nominal Sturm–Liouville system is able to track the given admissible two-point boundary conditions, while the delayed perturbed one loses that property, in general.

The admissible two-point boundary conditions given for the nominal differential equation translate in a natural way into four two-point boundary values at the initial and final values on the time interval of interest of the respective solutions and its first time-derivatives with respect to time. The above-mentioned second differential system is seen as a perturbation of the nominal one, since it is subject to a constant delay in its dynamics. This is the reason for the loss of the Sturm–Liouville property, since the four single two-point boundary values cannot be jointly kept, in general, identical for both systems. The undelayed Sturm–Liouville differential system plays the role of a nominal differential system, while the delayed one is viewed as a perturbation of the above one. Finally, the errors of the two-point boundary values between those of the nominal differential system and those of the perturbed current one are evaluated, and mutually compared, in terms of worst-case norm errors for small deviations of the perturbed system related to the nominal one. For that purpose, one takes advantage of the proposed equivalent descriptions of both differential equations via second-order differential systems, via their respective evolution operators. It is foreseen to extend these results in future works to systems with the presence of multiple internal delays in their dynamics and to forced systems with delays in their forcing terms.

Funding: The author would like to thank the Basque Government for funding his research work through Grant IT1555-22. He also thanks MICIU/AEI/10.13039/501100011033 and ERDF/E for partially funding his research work through Grants PID202-123543OB-C21 and PID2021-123543OB-C22.

Data Availability Statement: The original contributions presented in the study are included in the article, further inquiries can be directed to the corresponding author.

Conflicts of Interest: The author declares no conflicts of interest.

References

1. Al-Gwaiz, M.A. *Sturm-Liouville Theory and Its Applications*; Springer: London, UK, 2008.
2. Zetti, A. Sturm-Liouville Theory. In *Mathematical Surveys and Monographs*; American Mathematical Society: Providence, RI, USA, 2005; Volume 121.
3. Keller, H.B. *Numerical Methods for Two-Point Boundary-Value Problems*; Dover Publications Inc.: Minneola, NY, USA, 2018.
4. Daci, A.; Tola, S.; Çepa, N. Application of Sturm-Liouville problem in the wave equation. *Int. Sci. J. Math. Model.* **2023**, *7*, 76–79.
5. Zivlaei, L.G.; Mingarelli, A.B. Existence and uniqueness of some unconventional fractional Sturm-Liouville equations. *Fractal Fract.* **2024**, *8*, 148. [CrossRef]
6. Gulsen, T.; Goktas, S.; Abdeljawad, T.; Gurefe, Y. Sturm-Liouville problem in multiplicative fractional calculus. *AIMS Math.* **2024**, *9*, 22794–22812. [CrossRef]
7. Zhao, K.H.; Liu, J.Q.; Lv, X.J. A unified approach to solvability and stability of multipoint BVPs for Langevin and Sturm–Liouville equations with CH–fractional derivatives and impulses via coincidence theory. *Fractal Fract.* **2024**, *8*, 111. [CrossRef]
8. Gao, C.H.; Du, G.F. Uniqueness of nonlinear inverse problem for Sturm-Liouville operators with multiple delays. *J. Nonlinear Math. Phys.* **2024**, *31*, 13. [CrossRef]
9. Yang, X.; Liu, P.; Wang, W. Multiplicity and non-existence of positive solutions to impulsive Sturm-Liouville boundary value problems. *Bound. Value Probl.* **2024**, *2024*, 51. [CrossRef]
10. Hastir, A.; Mohet, J.; Winkin, J.J. A frequency domain approach to Kalman filtering on Hilbert spaces: Application to Sturm–Liouville systems with pointwise measurement. *Annu. Rev. Control* **2023**, *55*, 379–389. [CrossRef]
11. Bondarenko, N.P.; Chitorkin, E.E. Inverse Sturm-Liouville problem with spectral parameter in the boundary conditions. *Mathematics* **2023**, *11*, 1138. [CrossRef]
12. Zhao, M.; Qi, J.; Chen, X. A new approach to inverse Sturm-Liouville problems based on point interaction. *arXiv* **2024**, arXiv:2407.17223.
13. Liu, C.S.; Chen, Y.W.; Chang, C.W. Precise eigenvalues in the solutions of generalized Sturm–Liouville problems. *Math. Comput. Simul.* **2024**, *217*, 354–373. [CrossRef]
14. Al-Jararha, M.; Al-Refai, M.; Luchko, Y. A self-adjoint fractional Sturm-Liouville problem with the general fractional derivatives. *J. Differ. Equ.* **2024**, *413*, 110–128. [CrossRef]
15. Samudra, D.R.; Indrat, C.R. Fundamental solution of Sturm-Liouville equation. *AIP Conf. Proc.* **2024**, *3029*, 040006.
16. de la Sen, M. Sufficiency-type stability and stabilization criteria for linear time-invariant systems with constant point delays. *Acta Appl. Math.* **2004**, *83*, 235–256. [CrossRef]
17. Delasen, M. On some structures of stabilizing control laws for linear and time-invariant systems with bounded point delays and unmeasurable states. *Int. J. Control* **1994**, *59*, 529–541.
18. Xu, C.J.; Wu, Y.S. The effect of time delay on dynamical behaviour in an ecoepidemiological model. *J. Appl. Math.* **2012**, *2012*, 286961. [CrossRef]
19. Liu, K.Y.; Zhang, Z.Q.; Wang, L.P. Existence and global exponential stability of periodic solution to Cohen-Grossberg BAM neural networks with time-varying delays. *Abstr. Appl. Anal.* **2012**, *2012*, 805846. [CrossRef]
20. Stamov, G.; Stamova, I.; Simeonov, S.; Torlakov, T. On the stability with respect to H-manifolds for Cohen-Grossberg-type bidirectional associative memory neural networks with variable impulsive perturbations and time-varying delays. *Mathematics* **2020**, *8*, 335. [CrossRef]
21. Bohner, M.; Stamov, G.; Stamova, I.; Spirova, C. On h-manifolds stability for impulsive delayed SIR epidemic models. *Appl. Math. Model.* **2023**, *118*, 853–862. [CrossRef]
22. Cao, L.; Halik, A.; Muhammadhaji, A. Further studies on the dynamics of a Lotka-Volterra competitor-competitor-mutualist system with time-varying delays. *Mathematics* **2023**, *11*, 2902. [CrossRef]
23. Pinheiro, R.F.; Colon, D. On the μ-analysis and synthesis for uncertain time-delay systems with Padé approximations. *J. Frankl. Inst.-Eng. Appl. Math.* **2024**, *361*, 106643. [CrossRef]
24. Hua, C.; Zhang, L.; Guan, X. *Robust Control for Nonlinear Time-Delay Systems*; Springer: Singapore, 2018.
25. Zhao, D.Q.; Wang, Y.C.; Xu, L.J.; Wu, H. Adaptive robust control for a class of uncertain neutral systems with time delays and nonlinear uncertainties. *Int. J. Control Autom. Syst.* **2021**, *19*, 1215–1227. [CrossRef]
26. Kumar, V.; Djemal, M. Existence, stability and controllability of piecewise impulsive dynamic systems on arbitrary time domain. *Appl. Math. Model.* **2023**, *117*, 529–548. [CrossRef]
27. Zadeh, L.A.; Desoer, C.A. Linear Systems Theory. In *The State Space Approach*; Mc Graw-Hill: New York, NY, USA, 1963.
28. Chen, C.T. *Linear System Theory and Design*; Oxford University Press: New York, NY, USA, 1999.
29. Feng, G. Controller design and analysis of uncertain piecewise-linear systems. *IEEE Trans. Circuits Syst.-I Fundam. Theory Appl.* **2002**, *49*, 224–232. [CrossRef]
30. Barcena, R.; de la Sen, M.; Sagastabeitia, I. Improving the stability properties of the zeros of sampled data systems with fractional order hold. *IEE Proc.-Control Theory Appl.* **2000**, *147*, 456–464. [CrossRef]

31. Luo, N.; Rodellar, J.; De la Sen, M. Composite semiactive control of a class of seismically excited structures. *J. Frankl. Inst.-Eng. Appl. Math.* **2001**, *338*, 225–240. [CrossRef]
32. Zakian, V. Computation of the abscissa of stability by repeated use of the Routh test. *IEEE Trans. Autom. Control* **1979**, *24*, 604–607. [CrossRef]

Mathematical and Computational Applications

MDPI

Article

Modeling of Sedimentation of Particles near Corrugated Surfaces by the Meshless Method of Fundamental Solutions

Alex Povitsky

Department of Mechanical Engineering, The University of Akron, Akron, OH 44325-3903, USA; povitsky@uakron.edu

Abstract: The velocity and trajectory of particles moving along the corrugated (rough) surface under the action of gravity is obtained by a modified Method of Fundamental Solutions (MFS). This physical situation is found often in biological systems and microfluidic devices. The Stokes equations with no-slip boundary conditions are solved using the Green's function for Stokeslets. In the present study, the velocity of a moving particle under the action of the gravity force is not known and becomes a part of the MFS solution. This requires an adjustment of the matrix of the MFS linear system to include the unknown particle velocity and incorporate in the MFS the balance of hydrodynamic and gravity forces acting on the particle. The study explores the combination of the regularization of Stokeslets and placement of Stokeslets outside the flow domain to ensure the accuracy and stability of computations for particles moving in proximity to the wall. The MFS results are compared to prior published approximate analytical and experimental results to verify the effectiveness of this methodology to predict the trajectory of particles, including their deviation from the vertical trajectory, and select the optimal set of computational parameters. The developed MFS methodology is then applied to the sedimentation of a pair of two spherical particles in proximity to the corrugated wall, in which case, the analytical solution is not available. The MFS results show that particles in the pair deviate from the trajectory of a single particle: the particle located below moves farther away from vertical wall, and the particle located above shifts closer to the wall.

Keywords: boundary singularity method; method of fundamental solutions; Stokes flow; corrugated surface; sedimentation

1. Introduction

The flow field with a low value of Reynolds number, Re << 1, involving motion of a particle or a micro-swimmer in proximity to the corrugated (rough) surface is often found in biological systems and microfluidic devices [1–3]. In the current study, the particle is moving under the action of the gravity force acting in the vertical direction. To ensure that the sum of hydrodynamic Stokes forces (shear stress and pressure) and gravity force exerted on the particle is equal to zero, the particle must have a non-zero component of velocity perpendicular to the corrugated wall. That is, the particle could be either attracted or repelled by the rough wall along which it moves. In addition, the particle undergoes lateral displacement away from its vertical trajectory when corrugations are tilted relative to the gravitational force. Prior experiments (see Ref. [3] and references therein) revealed oscillatory 3D particle dynamics with an overall drift along the surface corrugations.

Interactions between particles and nearby boundaries make possible a variety of processes to achieve optimal sorting, mixing, and focusing of complex particulate suspensions in microfluidic devices [3]. Corrugated surfaces have proved to be a powerful tool to manipulate particle motion for a variety of applications.

The MFS has substantial advantages over traditional finite volume (FV) and finite difference (FD) mesh-based methods in computational fluid dynamics, because it does not require cumbersome 3D domain meshing. Instead, the MFS requires the placement of fundamental solutions (singularities) only near the boundaries of the considered geometry. This feature of the MFS becomes especially critical for problems involving Stokes flows where the geometry of the domain is complex, including the computation of the flow field about arbitrary-shaped particles or ensembles of particles in the vicinity of a rough surface. To model the flow field, the linear MFS system is modified in the present study to account for particles are moving under the action of the gravity force. Using the MFS, an arbitrary distance between particle and wall, arbitrary shapes of wall corrugation and the shape of a particle (or multiple particles), can be handled.

In recent years, boundary element methods (BEMs), including the MFS, have validated and proven their efficiency in numerous fields of computational physics. Numerical experiments have shown that, in contrast to other boundary element methods [4,5], the MFS [6] requires relatively few boundary points and singularities to produce accurate results [7]. The main idea of the MFS consists in representing the solution of the problem as a linear combination of fundamental solutions with respect to source points located outside the domain and particular boundary conditions at collocation points. Then, the initial problem is reduced to the determination of unknown coefficients of the linear combination [8]. As a fundamental solution automatically satisfies the governing equation, only the boundary conditions need to be satisfied [9].

Micro- and nanoscale Stokes flows in channels, near particles, around fibers and their ensembles were considered in prior studies of the author [10–13], while gravity was not accounted for. As opposed to a regular MFS problem set-up explored in the above cited papers, the velocity of a moving particle is unknown in the present study and is a part of the MFS solution. This requires the adjustment of the matrix of the MFS linear system to include the unknown particle velocity and incorporate zero balance of the forces acting on the particle.

In Ref. [14], the no-slip boundary condition is imposed on the inclusions; however, neither traction nor velocity values are known on the particle boundary. The regular BEM formulation requires imposition of a boundary condition on the boundary of the solution domain. For the sake of simplicity, the authors [14] assumed particles are buoyant in the buffer solution, and hence, the only force acting on the particles is the hydrodynamic force, neglecting the gravity force that will drive particles in the present study.

The problem considered is challenging and represents a broader class of Stokes problems involving flow fields between surfaces in proximity. When the sphere and the wall surface are close together, the collocation points located near one surface can adversely affect the integral taken over the Stokeslets at the other nearby surface [15], and therefore, the set-up of the MFS must be carefully selected. The MFS still suffers from issues related to the non-invertible or ill-conditioned matrix problem [16]; therefore, it is important to maintain moderate values of the condition number. For this purpose, a combination of the regularization of singularities and placement of singularities outside of the flow domain will be explored in the present study.

The analytical solution [1,3,17] of steady Stokes equations, to which the MFS results are compared in this study, is based on the first approximation of the flow field by considering the sedimentation of a sphere near a planar wall using a bi-spherical coordinate representation. Such an approach can be used for a single spherical particle but not for particles of non-spherical shape or ensembles of particles in proximity. The linear expansion about the first approximation is considered for small surface roughness amplitude with the effective slip velocity along the planar wall to mimic a corrugated surface via a domain perturbation approach. As opposed to the MFS, the approach is limited to small distances between the particle and wall and small magnitudes of wall corrugations compared to the size of a particle.

In Ref. [17], the authors studied the hydrodynamic coupling between particles and solid, rough boundaries characterized by random surface textures. They compared results by the approximate analytical solution of the Stokes equations (see above) to those obtained by the Boundary Integral Method (BIM). In particular, the analytical predictions for the translational velocity showed good agreement with the numerical solutions. The rotational velocity showed some deviations from the numerical results, which could be, amongst other factors, due to the accuracy of the method, as the latter velocity components were one order in magnitude smaller than the former ones and thus more difficult to resolve. Although the BIM used in Ref. [17] is different from the MFS used in the current study, one can expect that the difference between the analytical solution and the MFS approximation observed in the present study can be caused by underlying assumptions of the analytical method, as well as the number and location of Stokeslets.

The goal of this paper is to extend the MFS to model the flow field of the representative particle-to-corrugated wall combination, where the particle is moving under the action of gravity. This will be done by solving the Stokes equations with no-slip boundary conditions using the Green's function for the Stokeslet (single point force). Comparisons will be made to prior experimental and analytical results to verify the effectiveness of this methodology, including the prediction of lateral displacement of a single particle by a flow field near the corrugation.

Using the developed and validated MFS methodology, the sedimentation of two particles near a corrugated wall is modeled to extend the MFS to cases in which the above-mentioned analytical solution for the Stokes equations is not available. The objective is to show how trajectories of particle ensembles in proximity deviate from the trajectory of a single particle where the first particle in the pair is moving differently from the second particle, which is located on top of the first particle.

In Section 2, the geometric model, governing equations, and modified MFS are introduced to account for the force of gravity. The regularization of singularities and selection of their location are introduced in this section. In Section 3, a comparison of the MFS solution with prior analytical solutions for a particle at a fixed location with respect to a corrugation wall is conducted, and a set of computational parameters is selected to ensure numerical accuracy. Other methods of reduction in the condition number of the MFS, which represents limitations to this numerical framework, are introduced briefly in this section and will be tested in future research. In Section 4, the MFS is extended to a moving particle along the corrugated wall. In this section, the corrugations are located at 45 degrees to the direction of gravity to compare the particle velocity and trajectory to the available experimental and analytical results [3]. In Section 5, the developed MFS is applied to the sedimentation of two spherical particles in proximity to a corrugated surface in a case where the analytical solution is not readily available.

2. The Geometric Model and Numerical Methodology

The considered geometric set-up includes a spherical particle in proximity to a corrugated vertical wall (Figure 1). For the corrugated wall, the roughness is represented as

$$z = d + z_m \sin(2\pi x / w), \tag{1}$$

where x is the vertical direction along the wall, z is the normal direction to the un-corrugated wall into the flow domain, as shown in Figure 1a, d is the distance between the sphere and the smooth wall, z_m is the magnitude of the wall roughness, and w is the wavelength of the periodical wall roughness. The variables are normalized by the radius of the particle. The value of $\frac{x}{w}$ determines the local height of the roughness.

A viscous steady flow with Re < 1, which is caused by a moving particle under the action of gravity at the terminal state, is represented by the Stokes equations, which can be written as follows:

$$\begin{cases} \nabla \cdot \vec{u} = 0 \\ -\nabla p' + \mu \nabla^2 \vec{u} = 0, \end{cases} \tag{2}$$

where $\vec{u}(x,y,z)$ is the velocity vector field, $p'(x,y,z) = p(x,y,z) - \rho g x$ is the modified static pressure that accounts for the gravity term, and μ is the viscosity. The modified static pressure appears from the pressure and gravity forces in the original Stokes equations: $\partial p/\partial x - \rho g = \partial(p - \rho g x)/\partial x$. The stationary (fixed) particle is considered in Sections 2 and 3, while the unsteady (moving) particle is considered in Section 4.

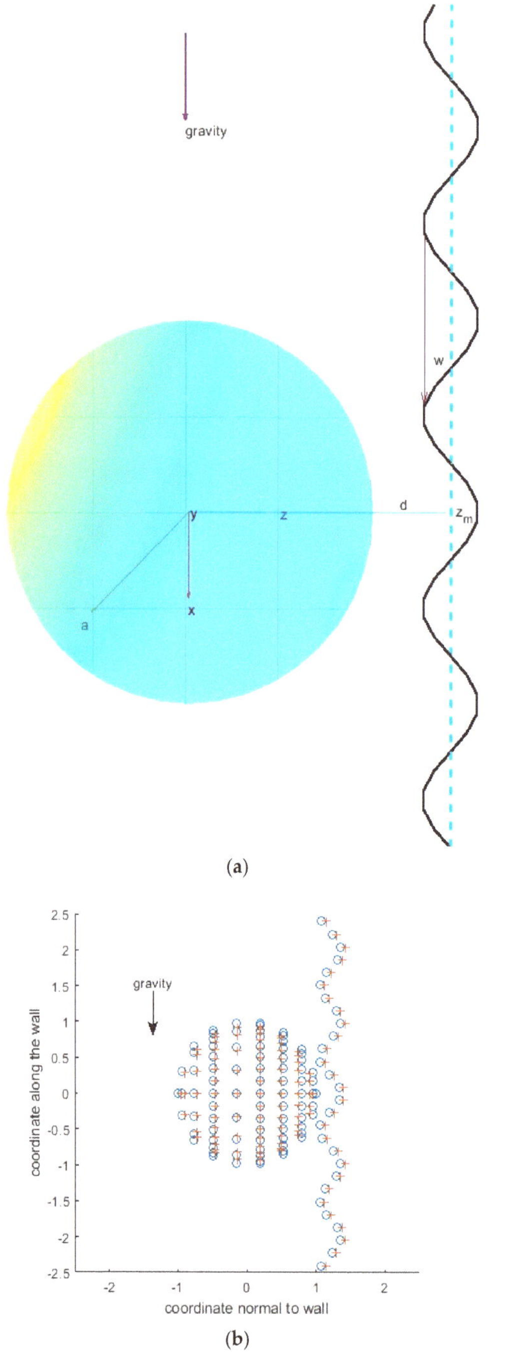

Figure 1. Spherical particle moving under the action of gravity near a corrugated vertical wall: (**a**) domain set-up and (**b**) collocation points (o) and location of Stokeslets (+). Here, coordinate x is vertical along the action of the gravity force, coordinate z is directed normally to the wall, and coordinate y is perpendicular to the figure plane.

In experiments [3] using tweezers, a particle is placed nearby the corrugated surface and then released. Similar to Ref. [3], the flow field around the individual particle is tracked at the limit of low concentration of the particles. The viscosity of the fluid as opposed to the apparent macro-viscosity of a fluid with suspended particles, which will be dependent on the concentration of the particles, is assumed to be constant, as in analytical computations [3].

The boundary conditions are no-slip and no-penetrating at the wall and at the particle(s) surface. Therefore, the fluid velocity at the particle(s) surface is equal to the velocity of the particle.

The MFS involves two steps in its application: first, the substitution of the velocity, generated by the total N_T Stokeslets with an unknown strength into the proper boundary conditions to construct a desired linear algebraic system of equations and the solution of this system; second, the use of the obtained strength of the Stokeslets to compute the velocity and pressure distributions for the entire domain. Three-dimensional velocity of the flow and the pressure induced by a point force (the Stokeslet) are given by [18]

$$u^{(k)} = \frac{1}{8\pi\mu}\left(\frac{F^{(k)}}{\left|\vec{r}\right|} + \frac{F^{(m)}r^{(m)}}{\left|\vec{r}\right|^3}r^{(k)}\right) \tag{3}$$

and

$$p' = \frac{1}{4\pi}\left(\frac{F^{(m)}r^{(m)}}{\left|\vec{r}\right|^3}\right), \tag{4}$$

where $\vec{F}$ is the strength of an individual Stokeslet, $\left|\vec{r}\right|$ is the distance between the location of the Stokeslet and the collocation point, and the superscripts k and m denote the components of the vectors $\vec{F}$ and $\vec{r}$. The Einstein summation rule is used in the above equations.

The collocation points, in which the no-slip and no-penetration boundary conditions should be satisfied, are located at the surface of the particle and at the vertical wall (see Figure 1b). A uniform distribution of collocation points [10] is adopted in this study. This distribution assumes an approximately equal distance between collocation points over the spherical surface of the particle (Figure 1b).

To avoid a high condition number, c, of the linear systems to be solved, singularities (Stokeslets) are located outside of the computational domain, as shown in Figure 1b. The Stokeslets are submerged underneath the wall surface and the sphere surface. This way, the pole of the Green's function moves away from the boundary onto a surface exterior to the flow [18–21]. The submergence of singularities for Stokes equations and the optimum depth of submergence were discussed in [10] and the references therein. The optimum depth of submergence is needed, because the condition number becomes large when the Stokeslets are moved excessively towards the sphere center. In Figure 2, the angle β, which is between the vectors directed from a collocation point to the locations of neighboring Stokeslets, tends to be zero when the Stokeslets are located farther away from the particle surface; therefore, the vectors directed from a collocation point to neighboring Stokeslets become nearly parallel to each other. As a result, the system matrix becomes ill-conditioned, because it contains nearly equal columns originating from the neighboring Stokeslets. A similar situation occurs when the number of Stokeslets and collocation points become large: the vectors directed from a collocation point to neighboring Stokeslets, which are located close to each other, become nearly parallel to each other. It should be mentioned here that the MFS suffers from an ill-conditioned matrix issue noticed by prior studies (see Ref. [22] and references therein). This linear algebra drawback can lead to inaccurate Stokeslets strengths and, consequently, inaccuracies in the induced flow field. It requires the selection of an optimal number of collocation points and Stokeslets that will be addressed in the next section.

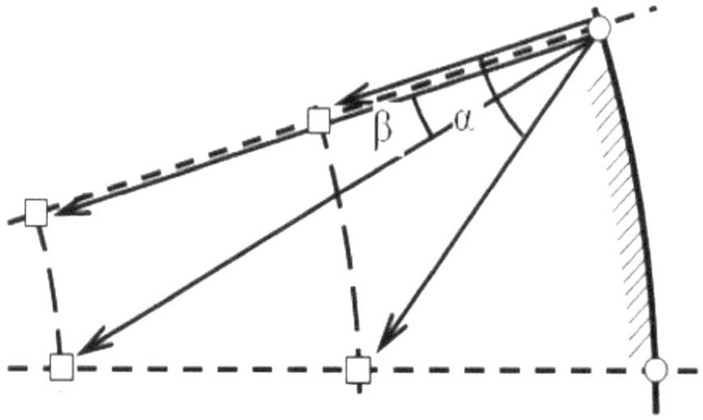

Figure 2. Submergence of Stokeslets: vectors directed from a collocation point to two neighboring Stokeslets. Here, squared and circular symbols refer to the Stokeslets and collocation point positions, respectively, and α and β are the angles between vectors, where the angle β corresponds to a larger distance between collocation points and locations of Stokeslets.

As another way to improve the accuracy of the MFS and reduce the condition number, we can use the Stokeslets' regularization technique presented in [23,24]. This technique was introduced to avoid singularity at the boundaries. The accuracy of the MFS with submerged and/or regularized singularities depends on the placement of singularities and on the value of the regularization parameter. The regularization is introduced using a so-called blob function with its regularization parameter, ε [23,24], replacing the Dirac delta function for Stokeslets, which shows the singularity at the center of the Stokeslet. The spatially smoothed approximation *by the blob function* to the singular 3D Dirac delta function is plot in [24] for a few representative values of ε. The authors [23] showed theoretically (see [23], Equation (21)) the order of discretization errors as a function of the regularization parameter, ε. Also, they presented the error obtained by numerical computations as a function of this parameter. The value of $\varepsilon = \Delta s/4$, where Δs is the distance between Stokeslets, was introduced first in Ref. [24], Section 3.1, by numerical experimentation. The authors (see [24], Section 3.2) stated that the MFS solution is not very sensitive to the coefficient of proportionality between ε and Δs as soon as the coefficient is smaller than one. By more extensive analysis [23], the value of $\varepsilon = \Delta s/4$ corresponds to the area of the minimum of the discretization error (see [23], Figure 2), and therefore, this value is selected in the present study.

The equations for the velocity for the Stokeslets with the regularized singularities are as follows:

$$u^{(k)} = \frac{1}{8\pi\mu}\left(\frac{F^{(k)}\left(\left|\vec{r}\right|^2 + 2\varepsilon^2\right)}{\left|\vec{r}_\varepsilon\right|^3} + \frac{F^{(m)}r^{(m)}}{\left|\vec{r}_\varepsilon\right|^3}r^{(k)}\right), \tag{5}$$

where $\left|\vec{r}_\varepsilon\right|^2 = \left|\vec{r}\right|^2 + \varepsilon^2$.

As will be shown in the next section, the most accurate MFS results are obtained by a combination of the submergence of Stokeslets and their regularization.

The above system (3) or its regularized version (5) can be expressed in the matrix form for an arbitrary 3D Stokes flow as follows:

$$\vec{U} = M\vec{F}, \tag{6}$$

where $\vec{U}$ is a velocity vector at the collocation points (boundary conditions), $\vec{F}$ is a Stokes force vector with $3N_T$ unknowns, and M is the $3N_T \times 3N_T$ matrix of the above system. The Stokes force vector has the structure of $\vec{F} = \{F_1{}^x, F_1{}^y, F_1{}^z, \ldots, F_{NT}{}^x, F_{NT}{}^y, F_{NT}{}^z\}$. Since the pressure is modified (see explanation after Equation (2)) the normal force exerted on the particle includes the buoyance force exerted by the fluid. For the past implementation of the MFS [10,13] for the flow field determined by far field velocity, the velocity vector at the collocation points is set up as equal to zero for no-slip and no-penetrating boundary conditions, while the known non-zero constant velocity is selected at the far field. In the current study, the velocity vector, $\vec{U}$, includes the unknown velocities at the particle surface, u_x, u_y, and u_z. As soon as the particle is non-deformable and non-rotating, its velocity components, u_x, u_y, and u_z, are the same for all collocation points. It is assumed that the particle has reached its terminal velocity.

Ref. [25] and the references therein compared the Stokes force calculated by Faxen, who solved Oseen's equation allowing the sphere to rotate, and by O'Neill, who solved the Stokes equation using bipolar coordinates without rotation. In Ref. [25], in Figure 1, the authors concluded that the obtained Stokes force for non-rotating computations by O'Neill and rotating computations by Faxen were close to each other and close to the experimental results [25] as soon as the parameter determining the proximity of the particle to the wall, e, was between 0.04 and 0.4. In the present study, the value of e is within this range (see Section 3). Therefore, the rotation of particles is neglected in the present study and will be considered in future research.

Note that, if the rotational velocity needs to be accounted for, the no-slip boundary conditions at the particle surface are extended, $u = \Omega \wedge r$, where $u = (u_x, u_y, u_z)$, Ω is the instantaneous rotational velocity of the sphere, and r is the radius vector from the center of the sphere to the sphere surface. The torque, M_Ω, needs to be calculated to obtain Ω [17].

Because of the presence of the three new unknowns, u_x, u_y, and u_z, in the vector $\vec{U}$, the system (6) requires the addition of three equations representing the terminal (sedimentation) state of the particle motion with no acceleration:

$$\sum_{i=1}^{N} F_i^x = W; \sum_{i=1}^{N} F_i^y = 0; \sum_{i=1}^{N} F_i^z = 0, \tag{7}$$

where N is the number of Stokeslets per particle, and W is the weight force directed in the x direction and equal to one in the normalized variables.

The above equations represent the condition that the resultant force is equal to the sum of Stokeslets at the terminal state of the particle motion, where the gravity force is directed along coordinate x. After the addition of three equations (7) and three unknown velocities to the system (6), its size becomes $(3N_T + 3) \times (3N_T + 3)$. The numerical solution of the above linear system (6) is performed using the MATLAB backslash linear algebraic operation, $F = U\backslash M$ [26].

Once the above linear system is solved, the velocity at any point of the flow field can be calculated by substituting the Stokeslets, which are obtained by solving (6), either into (3) or into its regularized version, (5).

3. Computational Parameters of the MFS and Comparison of the Numerical Results with an Approximate Analytical Solution

In this section, the particle is located at a fixed location with respect to corrugation (Figure 1a). The corrugations are horizontal, and the number of corrugations is unlimited. The MFS solution is compared to the prior approximate analytical solution of the flow field around the sedimented particle near the corrugated wall [1,3,17]. The set of MFS computational parameters is selected to ensure numerical accuracy and computational efficiency.

The geometric parameters listed below are normalized by the radius of the particle. The center of particle is located at the origin. The given amplitude of corrugation, z_m, is equal to 0.15, and the corrugation wavelength, w, is equal to one. The distance between the sphere and the corrugation-free wall in the z direction, d, is equal to 0.2, so the minimum distance between the particle's surface and the wall is 0.05, and the distance between the wall and the center of the particle is 1.2. The ratio of distance to the wall to the particle radius is e = 0.2/1 = 0.2. The location of the center of particle with respect to a neutral level of corrugation is determined by the parameter φ:

$$z = d + z_m \sin(2\pi(\varphi + x/w)) \tag{8}$$

For φ at its limit values, 0 or 1, the particle is facing the corrugated wall corresponding to the zero value of the sinus; that is, the distance from particle to wall is $z = d$. The values of φ = 0.25 and φ = 0.75 correspond to the maximum, $d + z_m$, and minimum, $d - z_m$, distance between particle and wall; that is, the maximum and minimum of the *sin* function in Equation (8). At the mid-values of the phase parameter, φ~0.5, the particle is facing a neutral level of corrugation; however, the gravity force is acting toward the corrugation in this case, whereas, for φ at 0 or 1, the gravity force is acting toward the fluid. For the MFS computations presented in Figures 3 and 4, the interval $0 \leq \varphi \leq 1$ is divided into 36 equal cells, from 0 or 1, and the MFS system of Equations (2)–(7) is solved for each value of φ. The corrugations are non-tilted (horizontal) with respect to gravity, and therefore, they do not create a velocity component in the y direction.

The most accurate MFS results are obtained by regularized submerged Stokeslets for a submergence depth for the Stokeslets of 0.05 for the spherical particle and 0.075 for the wall. The approximate distance between collocation points on the sphere is

$$\Delta = \sqrt{\frac{4\pi a^2}{N}}, \tag{9}$$

where a is the radius of the sphere, and N is the number of submerged Stokeslets and the number of collocation points on the sphere surface. The locations of the Stokeslets and collocation points are steady with respect to the center of the particle.

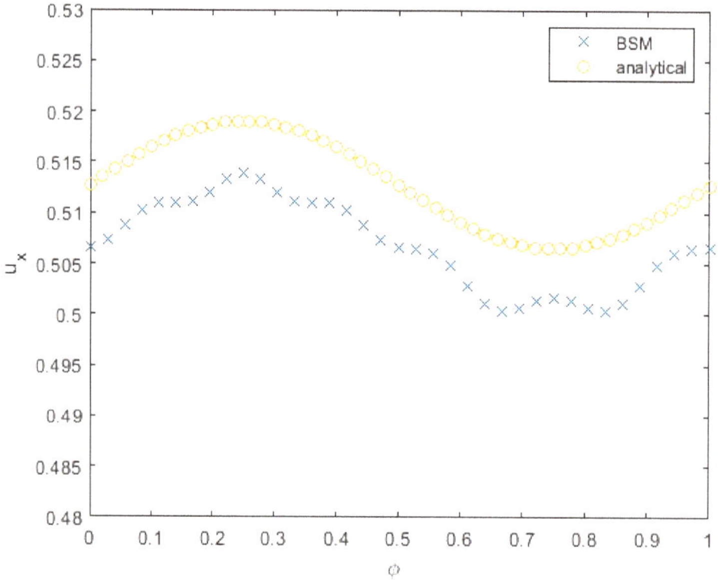

(a)

Figure 3. *Cont.*

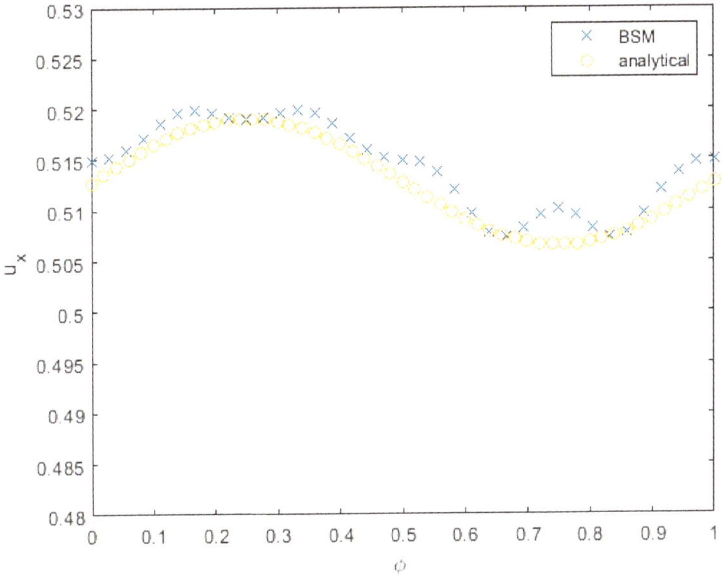

(b)

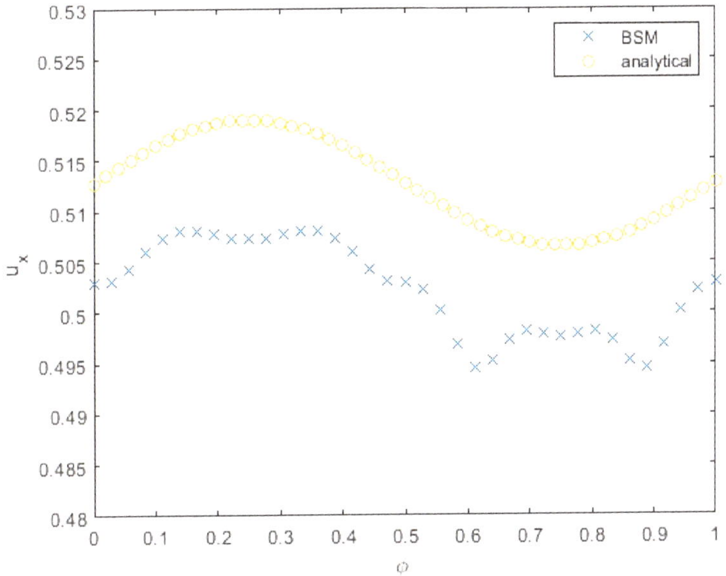

(c)

Figure 3. Velocity component u_x parallel to the un-corrugated wall (along gravity) as a function of the parameter φ (phase of sin-type corrugation, see Equation (8)), and analytical solution (Ref. [3], p. 7) of the Stokes equations near the corrugated wall. The number of Stokeslets and number of collocation points per spherical particle: (**a**) N = 100, (**b**) N = 150, and (**c**) N = 200.

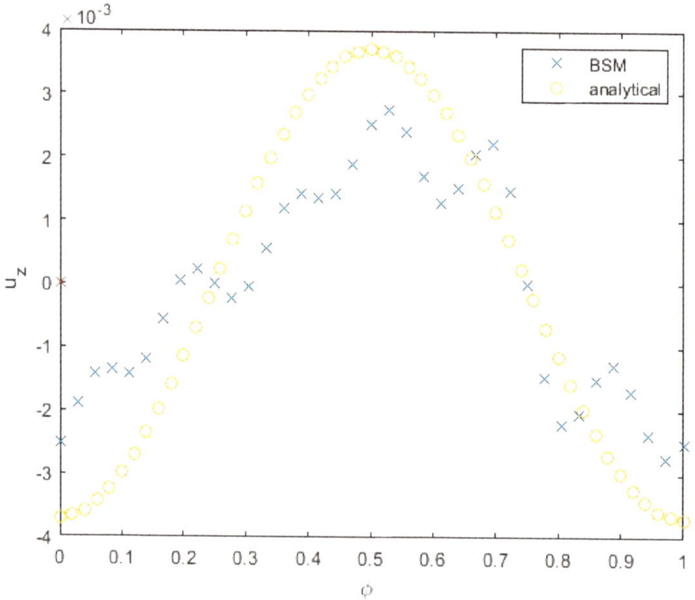

(a)

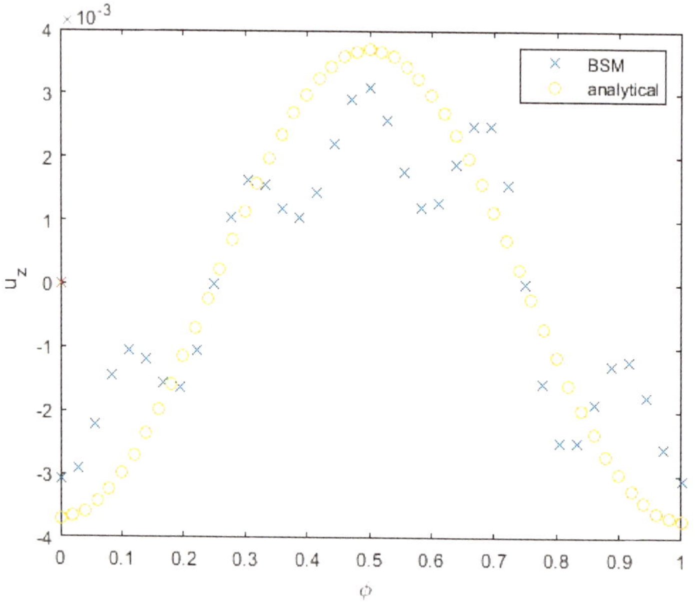

(b)

Figure 4. *Cont.*

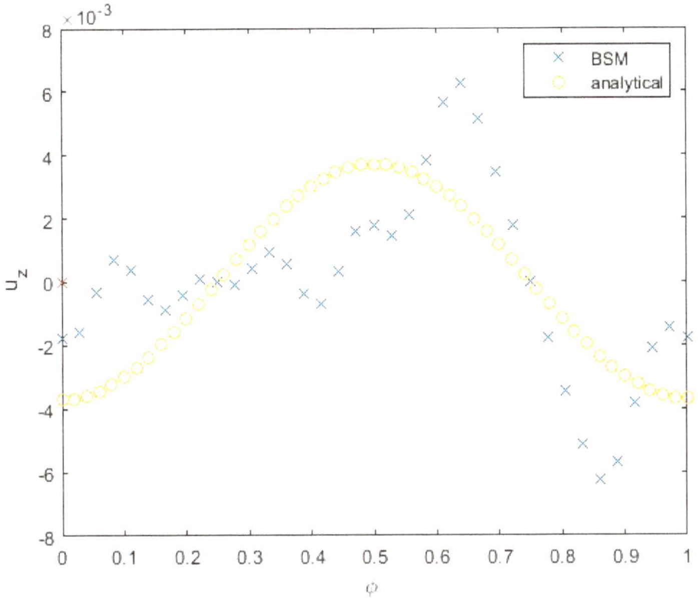

(c)

Figure 4. Velocity normal to the wall for (**a**) N = 100, (**b**) N = 150, and (**c**) N = 200 as a function of the parameter φ. Velocity obtained by the MFS is compared to the analytical solution (Ref. [3], p. 7).

The Stokeslets located along the wall are placed half the distance between the neighboring Stokeslets in the x and z directions, $\frac{\Delta}{2}$, where Δ is defined by Equation (9). The MFS solution with N = 150 for the velocity component parallel to the wall coincides fairly with the analytical solution [1,3] of the Stokes equations near the corrugated wall (Figure 3b). The results in terms of vertical velocity for N = 100 (Figure 3a) deviate from the analytical solution by 2%, whereas the results for N = 200 (Figure 3c) deviate from the analytical solution by 2.4%. The larger deviation from the analytical solution is explained by the insufficient number of Stokeslets for N = 100 and the large condition number of the matrix, M, for N = 200 (see the discussion below). In Table 1, the MFS results are also presented for a smooth vertical wall located at the same distance from the particle for comparison to the u_x results (Figure 3). The velocity for a smooth vertical wall in Table 1 can be compared to that obtained by O'Neill, who solved the Stokes equation using bipolar coordinates without rotation [25]. For this purpose, the velocity is obtained from the expression for the normalized vertical Stokes force depicted in Figure 1, Ref. [25]. The obtained value of the velocity for a particle near the smooth wall is ~0.5, which corresponds fairly well to the MFS results presented in the rightmost column in Table 1.

Table 1. The elapsed computational time of the presented numerical simulations.

Number of Stokeslets per Spherical Particle, N	Elapsed Time for Backslash Operation, $F = U\backslash M$, per Single Value of φ, s	Total Elapsed Time per Single Value of φ, s	Velocity along Gravity Direction, u_x, for Smooth Wall
100	6.16	8.17	0.574
150	20.77	25.25	0.558
200	45.95	54.59	0.549

The velocity component normal to the wall differs from the analytical solution of the Stokes equations developed in Refs. [1,3] for some values of φ (Figure 4), especially those corresponding to the particle facing the depth of the sin profile of roughness—that is, for mid-values of the phase parameter. The magnitude of the normal component of the velocity is $O(10^{-3})$ (Figure 4), while the magnitude of the component of the velocity parallel to the wall is O(1) (Figure 3); therefore, the relative error of the velocity normal to the wall, u_z, is larger than that for the velocity in the direction of gravity, u_x, while the absolute error is larger for the component u_x.

The results obtained by using a 10×10 x-y domain are very close to those obtained by a 15×15 domain, therefore, the results in Figure 4 are presented for a 10×10 domain. In Figure 4, the MFS results are presented for the following numbers of Stokeslets per sphere: N = 100 (Figure 4a), N = 150 (Figure 4b), and N = 200 (Figure 4c). The results for N = 150 are closer to the analytical solution compared to those for N = 100 (compare Figure 4a to Figure 4b). The results for N = 200 show the deviation from the analytic values with the increase in magnitude of the velocity, u_z, for some values of φ. This can be explained by the dramatic increase in the condition number, $c \sim 10^7$, for N = 200. This value of the condition number is much larger than that for N = 100 ($c \sim 650$) and for N = 150 ($c \sim 2300$).

The convergence study with respect to the number of Stokeslets, N, should be conducted until the increase in condition number by an order of magnitude that negatively affects the accuracy of the MFS.

In Table 1, the computational cost (in terms of elapsed computational time) for a single value of the parameter φ is presented for the range of numbers of Stokeslets at the sphere: N = 100, 150, and 200. The computer used is a PC OptiPlex 9020 manufactured by Dell, Round Rock, Texas USA, with a processor Intel(R) Core(TM) i7-4790 CPU @ 3.60 GHz, with 4 Cores. MATLAB reads the internal time at the execution of the toc function and displays the elapsed time since the most recent call to the tic function without an output. The elapsed time is expressed in seconds.

The computational time includes the solution of the linear algebraic system (6), $F = U\backslash M$, using MATLAB's linear algebraic backslash operation. The total elapsed time includes the solution of the linear system by Equations (6) and (7) and the setting up of matrix M by Equations (2.3) and (2.5). The computational cost of the solution of the linear MFS system is $O(N^{\gamma})$, where the exponent γ satisfies $2 < \gamma < 3$. The construction of matrix M before the solution of the linear MFS system is $O(N^2)$, where N^2 is the number of elements of M. By computing a linear regression of the logarithm of the computer time for the backslash operation with log N (see Table 1), we obtain the slope of regression line $\gamma = 2.9$.

The fraction of the total computer time to compute the matrix M drops from 24.6% for N = 100 to 15.8% for N = 200. Assuming the total computer time $\sim O(N^{\delta})$, by a linear regression of the logarithm of the total computer time with log N (see Table 1), we obtain $\delta = 2.74$.

As shown above in this section, the MFS suffers from an ill-conditioned matrix issue associated with finding the Stokeslets' strengths, which can lead to inaccurate Stokeslet strengths and, consequently, inaccuracies in the computed flow field (see Ref. [22] and references therein). The authors [22] used Tikhonov regularization to solve ill-conditioned linear systems of equations. Their algorithm is based on introducing a regularization parameter into the original MFS matrix. This approach could be tested in future research.

Recently, a localized MFS (LMFS) was developed (see [27] and references therein). The LMFS combines the concept of localization and the MFS. The resultant system of linear algebraic equations in the LMFS is sparse and banded and, thus, could reduce the condition number, storage, and computational burden of the MFS. The LMFS will be verified and adopted for the considered set-up in future research.

4. MFS Computations for a Particle Moving near the Corrugations Tilted with Respect to Gravity

In this section, the results of the MFS computations are presented to determine the trajectory of the particle near the corrugated wall with tilted corrugations with respect to gravity. The computational parameters are the same as those obtained in previous sections, unless otherwise stated. The N = 150 Stokeslets per spherical particle will be used in this section, as this number of Stokeslets corresponds to the most accurate solution for the u_x (Figure 3) and u_z (Figure 4) velocity components in the past section. While the locations of Stokeslets and collocation points at the particle are steady with respect to the center of the particle, their locations move with the moving particle.

The author compared the theoretical and experimental results (see Figure 3 in Ref. [3]) against the MFS computations. The derivation of the approximate analytical solution of the Stokes equations for prediction of the roughness-induced velocity is presented in Ref. [3], p. 7. The experimental methodology of imaging of the trajectory of the particle involved in sedimentation near a 3D-printed corrugated surface is presented in Ref. [3], p. 6. In these experiments using tweezers, a particle is placed nearby the corrugated surface and then released. To minimize the effect of the initial release/unsteadiness, the particle is allowed to sediment ~10 times its radius before recording its trajectory [3]. The two 3D-printed corrugations of sinusoidal shape, Equation (1), are tilted at an angle of 45 degrees to the vertical gravity force. The tilted corrugation wavelength λ = 6 mm. The particle's radius is 1.6 mm, and the initial distance between the particle surface and the smooth wall is 0.2 mm.

The length of the domain along the vertical coordinate, x, is $L = 3.5\lambda$, centered at $x = 0$. A particle starts at 0.75λ upstream of the first corrugation, and the trajectory ends at 0.75λ downstream of the second corrugation. Initially, the particle distance to the wall is equal to 0.2 mm, as specified above. The size of the wall surface domain where the collocation points are located—that is, in the y and x directions—is $10a \times 10a$, where a is the particle radius. The computational domain is centered at the current particle location and moves together with the moving particle after each time step. Varying the domain size from $8a \times 8a$ to $15a \times 15a$ does not affect the obtained particle trajectory. Note that there is no need to define the domain in the z direction, within the fluid domain, as the Stokeslets and collocation points are not located in the z direction. A total of 200 equal time steps of length Δt is taken to follow the particle's motion along the vertical x coordinate over two corrugations.

At each time step, the flow field around the particle is assumed to become relaxed to the steady-state Stokes Equation (2). The new values of velocity, u_x, u_y, and u_z, are found by the solution of the linear system (6). After computing the new velocity, the new coordinates of the particle are found by the following relations:

$$x_{new} = x_{old} + u_x\,\Delta t;\; y_{new} = y_{old} + u_y\,\Delta t \text{ and } z_{new} = z_{old} + u_z\,\Delta t, \quad (10)$$

where x_{new}, y_{new}, z_{new} are the new coordinates of the particle after the time step Δt, x_{old}, y_{old}, z_{old} are the old coordinates of the particle, and Δt is the time step.

For each time step, the linear system (6) representing the MFS must be formed and solved [28].

The results presented in Figure 5 compare MFS computations against published experimental and analytical results (see Figure 3B [3]). From Figure 5, the distance in the normal direction from the smooth wall is normalized by the corrugation wavelength, z/λ, and presented as a function of the vertical coordinate along the wall, x/λ. As depicted in Figure 5, the results of the MFS computations are located closer to the experimental data points compared to the results obtained by the analytical method in areas near the minimum of z/λ. For areas near the maximum of z/λ, the analytical and MFS computations are closer to each other compared to the areas of the minimum. From Figure 5, the amplitude of the sin-type oscillations, z/λ, computed by the MFS is smaller than that computed by the analytical method.

The results in Figure 6 compare the lateral displacements obtained by the MFS computations against the published experimental and analytical results (see Figure 3C [3] for sinusoidal corrugation). In Figure 6, the lateral horizontal displacement along the wall, y, is normalized by the corrugation wavelength, y/λ, and presented as a function of the normalized coordinate along the wall, x/λ. The results in Figure 6 show that the obtained MFS maxima of the magnitude of the lateral displacement, in terms of y/λ, are not as prominent as the maxima obtained analytically and experimentally [3], but they still coincide in terms of their locations, x/λ, and distance between the maxima. The correspondence of the locations of the extrema for the analytical and MFS solutions in Figures 5 and 6 is explained by the fact that the difference between the analytical and computed values of the vertical velocity, u_x, is within 1% for N = 150 (Figure 3b).

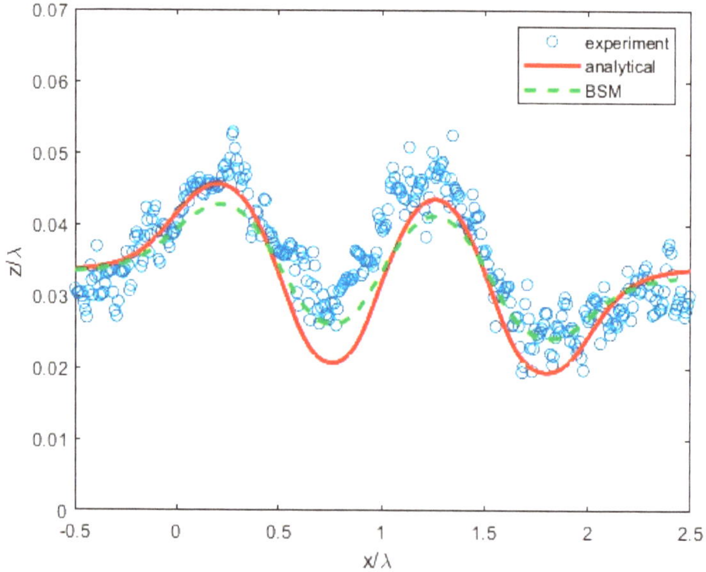

Figure 5. Comparison of the distance to the wall obtained by the MFS to the published analytical and experimental results [3] for particle moving by the action of gravity along tilted corrugations, where x/λ is the normalized distance along the wall and z/λ is the normalized distance normal to the wall.

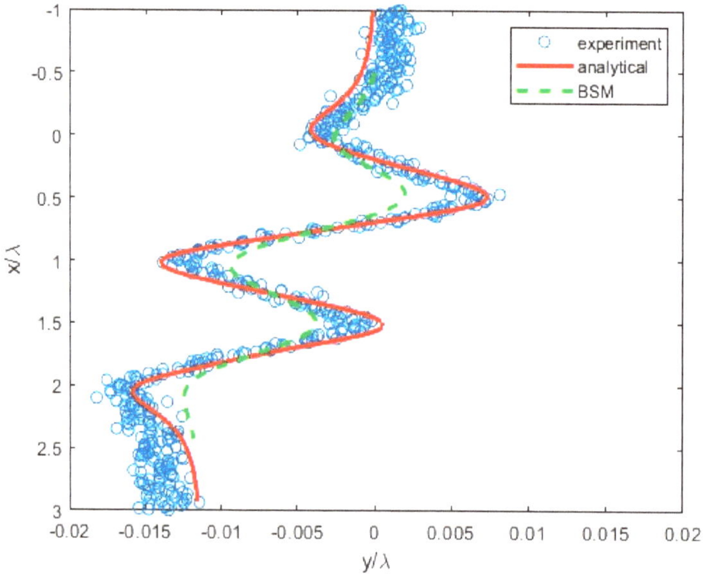

Figure 6. Comparison of the normalized lateral displacement of the particle, y/λ, obtained by the MFS to the published analytical and experimental results [3].

5. MFS Modeling of the Motion of Two Particles near the Corrugated Wall

The discussion in this section aims at monitoring the difference between the motion of a single particle discussed in the previous section and the tandem of particles and investigating the role of the computational parameters, such as the number of Stokeslets per spherical particle.

Each of the two particles is assumed to be non-deformable and non-rotating; therefore, their velocity components, u_x, u_y, and u_z, are the same for all collocation points belonging to the same particle. The rotation of the particles will be considered in the future research, as outlined in Section 2. It is assumed that each particle has reached its individual terminal velocity. The presence of six new unknowns, $u_{x,1}$, $u_{y,1}$, $u_{z,1}$, and $u_{x,2}$, $u_{y,2}$, and $u_{z,2}$, in the vector $\vec{U}$ requires the addition of six equations to the system (6). These equations represent the condition for the resultant force to become equal to zero at the terminal state of each particle motion (see the similar Equation (7) for a single particle). The sum of the Stokeslets strengths in the y and z directions is therefore equal to zero. The particle's weight is equal to the sum of the Stokeslets in the x direction.

The set-up of two particles near the corrugated wall and the distribution of static pressure computed by Equation (4) with the modification for regularized Stokeslets [24] is shown in Figure 7. Initially, the particles were in the same proximity to the vertical wall. The initial distance between the centers of the particles in the vertical direction, x, is equal to $5a$, where a is the radius of the particle. The first particle is initially located at 0.75λ up the first corrugation (the same location as the single particle in the previous section). The second particle is located at 5a up the first particle. Initially, the particles' distance to the wall in the z direction is equal to 0.2 mm, as specified in Section 4. When particles move, each of them appears at a different distance z from the wall, as shown in Figure 8a.

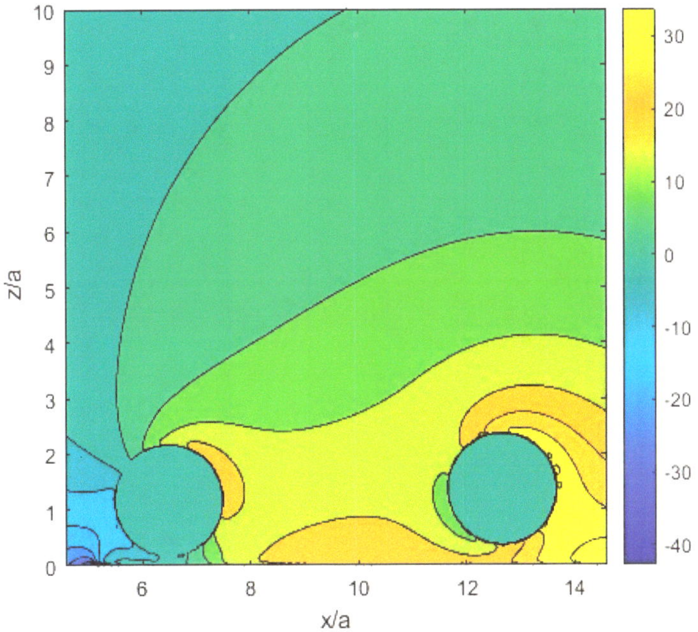

Figure 7. Set-up of two particles located near the corrugated wall with a computed distribution of the static pressure. The coordinate x is centered at the mid-point between particles and extends 5a upstream and downstream.

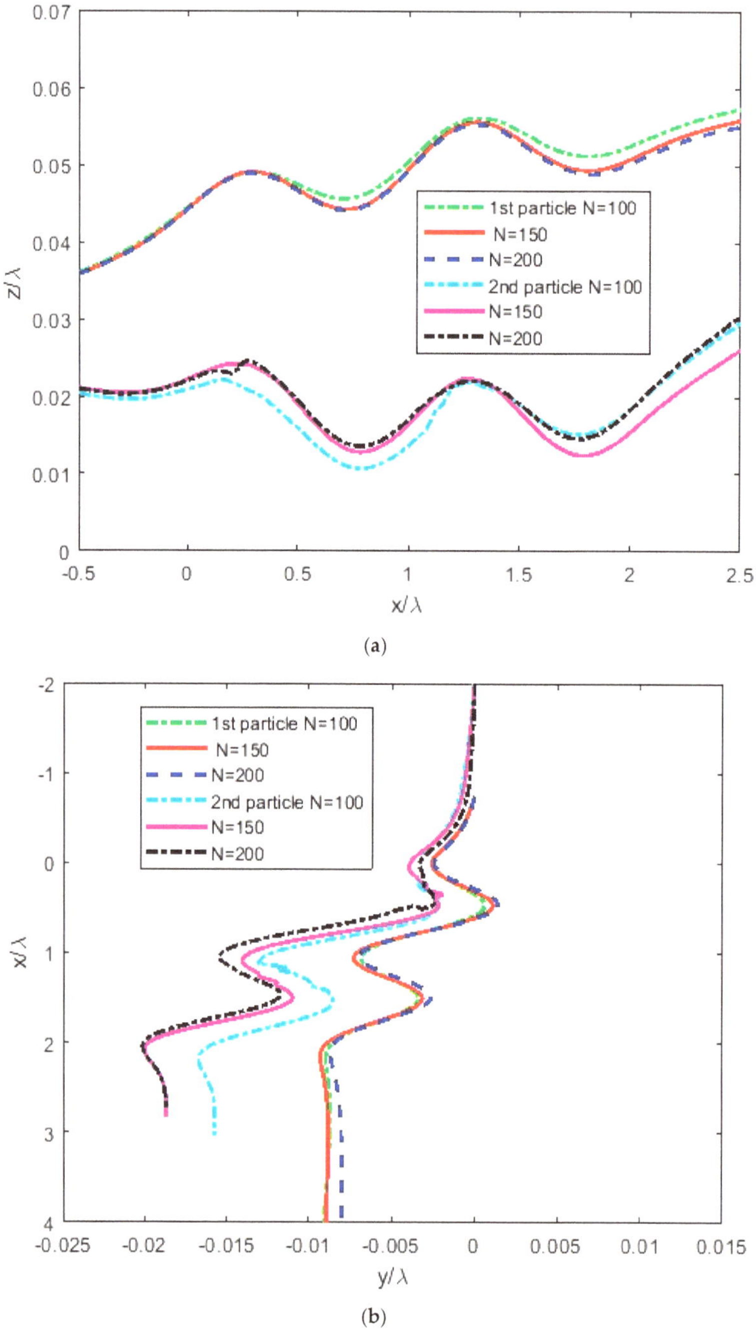

Figure 8. Trajectories of two particles undergoing sedimentation along a rough wall, x/λ, computed by the MFS: (**a**) deviation from the vertical trajectory in terms of the normal direction to the wall, z/λ, and (**b**) deviation in the horizontal direction along the wall, y/λ. The MFS computations are presented for the number of Stokeslets per spherical particle: N = 100, 150, and 200. Particles move along the *x* axis from left to right in (**a**) and up–down in (**b**).

In Figure 8, the trajectories of the two particles moving along the corrugated wall under the action of gravity are depicted. To obtain the trajectories, MFS computations for 200 time steps are conducted for the number of Stokeslets per spherical particle: N = 100, 150, and 200. In Figure 8a, the deviation from the vertical trajectory in terms of the normal direction to the wall, z/λ, is shown. Recall that the initial value of z/λ = 0.2 mm/6 mm = 0.033 for both particles. In Figure 8b, the deviation in the horizontal direction along the wall, y/λ, is depicted.

In Figure 8a, the first (located down the wall) particle is pushed away from the wall by the moving second (located up the wall) particle in its wake, whereas the second particle is moving in closer proximity to the wall compared to its initial distance from the wall. In Figure 8b, the first particle follows a path that is qualitatively like that of a single particle in Figure 6. However, its deviation from the vertical trajectory, y/λ, is less prominent compared to that for a single particle (Figure 6), as its distance from the surface roughness is larger. On the contrary, the deviation of the second particle in the y direction is more than doubled compared to that of a single particle (Figure 6), as the second particle passes closer to the wall roughness.

The vertical, z/λ, displacements are qualitatively similar for a varying number of Stokeslets, from N = 100 to N = 200, as shown in Figure 8a. As shown in Figure 8a, the trajectories converge when the number of Stokeslets per spherical particle increases from N = 100 to N = 200 because of the cancellation of the MFS errors in the computed velocity when the velocity is integrated in time by Equation (10). The horizontal, y/λ, displacement for the second particle shows some oscillations (see Figure 8b). For the smallest number of Stokeslets, N = 100, the deviation of the second particle from the vertical direction is less prominent compared to that for the larger number of Stokeslets, N = 150 and 200.

6. Conclusions

The Method of Fundamental Solutions (MFS) is developed and applied to a particle(s) moving along a corrugated wall under the action of gravity to obtain the particles' local velocity and trajectory. The MFS linear system is modified to compute the unknown particles' velocity and account for the gravity force as a driver for the particles' motion. The combination of regularization and submergence of Stokeslets is implemented to maintain a moderate value of the condition number of the MFS matrix to ensure the accuracy of the solved linear system for the challenging case of closed proximity between the particle and the wall. The obtained trajectory of a particle, including its lateral displacement, does correspond to published experimental data and approximate analytical solutions of the Stokes equations.

Using the obtained set of MFS parameters, the method is extended to the case of two moving particles in proximity to the corrugated wall in a case where an analytical solution for the Stokes equations is not available. The results show that the particles deviate from the trajectory of a single particle, where the forward (down) particle moves farther away from vertical wall in the z direction and the back (up) particle shifts closer to the wall. As a result, the horizontal deviation along the wall of the forward particle, y, is smaller, whereas the horizontal deviation of the back particle is larger compared to that for a single particle.

In future research, the MFS will be extended to more complex situations of sedimentation in which an analytical solution is not available, such as stochastic non-periodical wall roughness and clusters of particles shaped as either multiple spheres or Cassini ovals.

Funding: The author would like to thank the Visitors Program of Max Planck Institute for the Physics of Complex Systems (MPI) in Dresden, Germany, where this study started, to enable his research visit in Fall 2022 during his sabbatical leave from the University of Akron.

Data Availability Statement: The original contributions presented in the study are included in the article, further inquiries can be directed to the corresponding author.

Acknowledgments: The author acknowledges Christina Kurzthaler, Research Group Leader, MPI, for the discussion and for pointing the author to her published analytical results for the Stokes equations for comparison to the presented MFS computations. MATLAB version R2022b is used in the current study.

Conflicts of Interest: The author declares no conflicts of interest.

List of Symbols

a	Radius of spherical particle
c	Condition number of matrix M
d	Distance between the sphere and the smooth wall
$\vec{F}$	Strength of Stokeslet
M	Matrix of MFS system
N	Number of Stokeslets and collocation points per spherical particle
p	Static pressure in the flow field
$p' = p - \rho g x$	Modified static pressure that accounts for gravity term
$\left\|\vec{r}\right\|$	Distance between the location of the Stokeslet and the collocation point
$\vec{U}$	Velocity vector at collocation points
$\vec{u}$	Fluid velocity vector
u_x, u_y and u_z	Particle velocity components
W	Weight force exerted on particle in the x direction
w	Wavelength of periodical wall roughness
x	Up–down coordinate along the wall
$x_{new}, y_{new}, z_{new}$	Coordinates of particle at the end of time interval
$x_{old}, y_{old}, z_{old}$	Coordinates of particle at the beginning of time interval
y	Horizontal coordinate along the wall
z	Normal direction to the wall
z_m	Magnitude of the wall roughness.
β	Power in the estimate of the computational cost of the solution of the linear MFS system, $O(N^{\beta})$
λ	Tilted corrugation wavelength
δ	Power in the estimate of the total elapsed computer time, $O(N^{\delta})$
Δ	Distance between collocation points at sphere surface
Δs	Distance between Stokeslets
Δt	Time step to compute the new location of a particle
ε	Value of the regularization parameter
μ	Viscosity
φ	Phase parameter of wavelength corrugation
k and m	Superscripts that denote components of the vectors $\vec{F}$ and $\vec{r}$

References

1. Kurzthaler, C.; Stone, H.A. Microswimmers near corrugated, periodic surfaces. *Soft Matter* **2021**, *17*, 3322. [CrossRef] [PubMed]
2. Lauga, E.; Powers, T.R. The hydrodynamics of swimming microorganisms. *Rep. Prog. Phys.* **2009**, *72*, 096601. [CrossRef]
3. Chase, D.L.; Kurzthaler, C.; Stone, H.A. Hydrodynamically Induced Helical Particle Drift due to Patterned Surfaces. *Proc. Natl. Acad. Sci. USA* **2022**, *119*, e2202082119. [CrossRef] [PubMed]
4. Yao, Z.; Wang, H. Some benchmark problems and basic ideas on the accuracy of boundary element analysis. *Eng. Anal. Bound. Elem.* **2013**, *37*, 1674–1692. [CrossRef]
5. Mukherjee, S.; Liu, Y. The Boundary element method. *Int. J. Comput. Methods* **2013**, *10*, 1350037. [CrossRef]
6. Liu, Y. *Fast Multipole Boundary Element Method*; Cambridge University Press: Cambridge, MA, USA, 2010.
7. Fairweather, G.; Karageorghis, A. The method of fundamental solutions for elliptic boundary value problems. *Adv. Comput. Math.* **1998**, *9*, 69–95. [CrossRef]
8. Askour, O.; Tri, A.; Braikat, B.; Zahrouni, H.; Potier-Ferry, M. Method of fundamental solutions and high order algorithm to solve nonlinear elastic problems. *Eng. Anal. Bound. Elem.* **2018**, *89*, 25–35. [CrossRef]
9. Cheng, A.; Hong, Y. An overview of the method of fundamental solutions—Solvability, uniqueness, convergence, and stability. *Eng. Anal. Bound. Elem.* **2020**, *120*, 118–152. [CrossRef]
10. Mikhaylenko, M.; Povitsky, A. Optimal Allocation of Boundary Singularities for Stokes Flows about Pairs of Particles. *Eng. Anal. Bound. Elem. (EABE)* **2014**, *41*, 122–138. [CrossRef]

11. Zhao, S.; Povitsky, A. Three-dimensional boundary singularity method for partial slip flows. *Eng. Anal. Bound. Elem.* **2011**, *35*, 114–122. [CrossRef]
12. Zhao, S.; Povitsky, A. Boundary Singularity Method for Partial Slip Flows. *Int. J. Numer. Methods Fluids* **2009**, *61*, 255–274. [CrossRef]
13. Mikhaylenko, M.; Povitsky, A. Combined boundary singularity method and finite volume method with application to viscous deformation of polymer film in synthesis of sub-micron fibers. *Eng. Anal. Bound. Elem.* **2017**, *83*, 265–274. [CrossRef]
14. Topuz, A.; Baranoğlu, B.; Çetin, B. A multi-domain direct boundary element formulation for particulate flow in microchannels. *Eng. Anal. Bound. Elem.* **2021**, *132*, 221–230. [CrossRef]
15. Sun, Q.; Klaseboer, E.; Khoo, B.C.; Chan, D.Y. Boundary regularized integral equation formulation of Stokes flow. *Phys. Fluids* **2015**, *27*, 023102. [CrossRef]
16. Aboelkassem, Y.; Staples, A.E. Stokeslets-meshfree computations and theory for flow in a collapsible microchannel. *Theor. Comput. Fluid Dyn.* **2013**, *27*, 681–700. [CrossRef]
17. Kurzthaler, C.; Zhu, L.; Pahlavan, A.A.; Stone, H.A. Particle motion nearby rough surfaces. *Phys Rev. Fluids* **2020**, *5*, 082101. [CrossRef]
18. Pozrikidis, C. *Boundary Integral and Singularity Methods for Linearized Viscous Flow*; Cambridge University Press: Cambridge, MA, USA, 1992.
19. Gonzalez, O. On stable, complete and singularity-free boundary integral formulations of exterior Stokes flow. *Soc. Ind. Appl. Math. (SIAM) J. Appl. Math.* **2009**, *69*, 933–958. [CrossRef]
20. Hebeker, F.K. A boundary element method for Stokes equations in 3-D exterior domains. In *The Mathematics of Finite Elements and Applications*; Whiteman, J.R., Ed.; Academic Press: London, UK, 1985; pp. 257–263.
21. Koens, L.; Lauga, E. The boundary integral formulation of Stokes flows includes slender-body theory. *J. Fluid Mech.* **2018**, *850*, R1. [CrossRef]
22. Aboelkassem, Y.; Staples, A.E. A three-dimensional model for flow pumping in a microchannel inspired by insect respiration. *Acta Mech.* **2014**, *225*, 493–507. [CrossRef]
23. Cortez, R.; Fauci, L.; Medovikov, A. The method of regularized Stokeslets in three dimensions: Analysis, validation, and application to helical swimming. *Phys. Fluids* **2005**, *17*, 031504. [CrossRef]
24. Cortez, R. The method of regularized Stokeslets. *SIAM J. Sci. Comput.* **2001**, *23*, 1204–1225. [CrossRef]
25. Ambari, A.; Manuel, B.G.; Guyon, E. Effect of a plane wall on a sphere moving parallel to it. *J. Phys. Lett.* **1983**, *44*, 143–146. [CrossRef]
26. Chapra, S.C. *Applied Numerical Methods with MATLAB*, 3rd ed.; McGraw Hill: New York, NY, USA, 2012; p. 588.
27. Qu, W.; Fan, C.-M.; Li, X. Analysis of an augmented moving least squares approximation and the associated localized method of fundamental solutions. *Comput. Math. Appl.* **2020**, *80*, 13–30. [CrossRef]
28. Mitchell, W.H.; Spagnolie, S.E. Sedimentation of spheroidal bodies near walls in viscous fluids: Glancing, reversing, tumbling, and sliding. *J. Fluid Mech.* **2015**, *772*, 600–629. [CrossRef]

Mathematical and Computational Applications

MDPI

Article

A Conservative and Compact Finite Difference Scheme for the Sixth-Order Boussinesq Equation with Surface Tension

Xiaofeng Wang [1,*], Weizhong Dai [2] and Anjan Biswas [3,4,5]

1 School of Mathematics and Statistics, Minnan Normal University, Zhangzhou 363000, China
2 Mathematics & Statistics, College of Engineering & Science, Louisiana Tech University, Ruston, LA 71272, USA; dai@coes.latech.edu
3 Department of Mathematics and Physics, Grambling State University, Grambling, LA 71245, USA; biswas.anjan@gmail.com
4 Department of Applied Sciences, Cross-Border Faculty of Humanities, Economics and Engineering, Dunarea de Jos University of Galati, 111 Domneasca Street, 800201 Galati, Romania
5 Department of Mathematics and Applied Mathematics, Sefako Makgatho Health Sciences University, Pretoria 0204, South Africa
* Correspondence: wxfmeng@163.com

Abstract: In this study, we propose a conservative and compact finite difference scheme designed to preserve both the mass change rate and energy for solving the sixth-order Boussinesq equation with surface tension. Theoretical analysis confirms that the proposed scheme achieves second-order accuracy in temporal discretization and fourth-order accuracy in spatial discretization. The solvability, convergence, and stability of the difference scheme are rigorously established through the application of the discrete energy method. Additionally, a series of numerical experiments are conducted to illustrate the effectiveness and reliability of the conservative scheme for long-time simulations.

Keywords: Boussinesq equation; convergence; conservation; stability

1. Introduction

The numerical solution of partial differential equations (PDEs) is a fundamental component in numerous scientific and engineering applications. In recent years, there has been a growing emphasis on developing high-order finite difference methods to improve both the accuracy and efficiency of solutions for a wide range of PDEs. A particular area of interest is the Boussinesq equation, which was first introduced by Joseph Boussinesq in 1872 to model nonlinear wave propagation in dispersive media [1]. Consequently, this equation has been extensively applied across multiple domains, including the investigation of ion sound waves in plasma, solitary waves in vascular systems, tsunami waves in oceanography and coastal sciences, as well as pressure waves in liquid–gas bubble mixtures, among others [2,3].

This article investigates the sixth-order Boussinesq equation, taking into account the effects of surface tension, as outlined in the following [4]:

$$u_{tt} - k^2 u_{xx} + a_1 u_{xxxx} - a_2 u_{xxtt} - b_1 u_{xxxxxx} + b_2 u_{xxxxtt} + c(u^{2p})_{xx} = 0, \tag{1}$$

with the specified initial conditions

$$u(x,0) = \varphi(x),\ \ u_t(x,0) = \psi(x), \tag{2}$$

where the parameters $a_1 \geq 0$, $a_2 \geq 0$, $b_1 \geq 0$, and $b_2 \geq 0$ are non-negative, while k and c are constants. Furthermore, $p \geq 1$ is a positive integer, and $\varphi(x)$ and $\psi(x)$ are two provided smooth functions.

Solitary waves have attracted considerable interest in various fields of physics and engineering. Biswas [5] successfully derived an exact solitary wave solution for the Korteweg–de Vries equation, which is characterized by power law nonlinearity and time-varying coefficients. The singular solitary wave solution for the Rosenau–KdV equation is presented in [6], while a specific 1-soliton solution for the Rosenau–KdV RLW equation is detailed in [7]. Equation (1) is frequently encountered in numerous mathematical models concerning solitary wave propagation, as emphasized by Lu and Helal [8,9]. Furthermore, Biswas et al. [4] provided a variety of solitary wave, shock wave, and singular solitary wave solutions for the Boussinesq equation (1), incorporating the effects of surface tension. Under the assumption of solitary wave behavior, it is observed that the solution to the sixth-order Boussinesq equation (1) and its derivatives approaches zero as $|x| \to \infty$, as noted by Burde and Yimnet et al. [10,11]

$$\lim_{x\to\pm\infty} u = 0, \quad \lim_{x\to\pm\infty} \frac{\partial^n u}{\partial x^n} = 0, \ n \geq 1.$$

Thus, Equation (1) is linked to the following global conservation law:

$$\frac{d}{dt}\int_{-\infty}^{+\infty} u(x,t)dx = C.$$

This research focuses on the solitary wave solution and aims to develop a conservative linear finite difference scheme for the sixth-order Boussinesq Equation (1). To implement the numerical method effectively, we analyze a sufficiently large yet finite spatial domain, denoted as $[x_l, x_r]$, instead of the unbounded interval $(-\infty, +\infty)$. This ensures that the solution remains sufficiently small at the domain boundaries. Consequently, we consider the following boundary conditions:

$$u(x_l,t) = u(x_r,t) = 0, \ u_x(x_l,t) = u_x(x_r,t) = 0, \ u_{xx}(x_l,t) = u_{xx}(x_r,t) = 0, \tag{3}$$

where $0 \leq t \leq T$. Considering the requirement for continuity, we further assume $\varphi(x_l) = \varphi(x_r) = 0$.

The Boussinesq equation and its variants have been extensively studied through both theoretical and numerical methods. Esfahani and Farah [12] investigated the following nonlinear sixth-order Boussinesq equation in the theoretical context:

$$u_{tt} - u_{xx} \pm u_{xxxx} - u_{xxxxxx} + (u^2)_{xx} = 0, \ x \in R, \ t \geq 0. \tag{4}$$

The authors established the local well-posedness of Equation (4) within the framework of non-homogeneous Sobolev spaces $H^s(\mathbb{R})$, where $s < 0$ and $s \in \mathbb{R}$. Subsequently, Wang and Esfahani [13] conducted an investigation into the sixth-order Boussinesq equation as follows:

$$u_{tt} - u_{xx} \pm u_{xxxx} - u_{xxxxxx} - (|u|^2 u)_{xx} = 0, \ x \in R, \ t \geq 0. \tag{5}$$

They demonstrated that the problem described by Equation (5) is globally well-posed within the Sobolev space $H^s(\mathbb{R})$, where $3/2 < s < 2$ and $s \in \mathbb{R}$.

In a numerical context, Feng et al. [14] proposed a symmetric three-level implicit difference scheme that exhibits second-order accuracy for the following sixth-order Boussinesq equation

$$u_{tt} - u_{xx} - u_{xxxx} - 0.4u_{xxxxxx} - 6(u^2)_{xx} = 0. \tag{6}$$

A linearized stability analysis indicates that the difference scheme, which incorporates a free parameter denoted as θ, exhibits stability for values of θ that are equal to or greater than

1/4. In 2017, Kolkovska and Vucheva [15] introduced a nonlinear second-order difference scheme aimed at addressing the sixth-order Boussinesq problem, as detailed below:

$$\begin{cases} u_{tt} - u_{xx} - \beta_1 u_{xxtt} + \beta_2 u_{xxxx} - \beta_3 u_{xxxxxx} + \alpha(u^2)_{xx} = 0, \\ u(x,0) = \varphi(x), \ \ u_t(0,x) = \psi(x), \\ u \to 0, \ \ u_{xx} \to 0, \ \ u_{xxxx} \to 0, \ \ |x| \to \infty. \end{cases} \tag{7}$$

However, the scheme presented in [15] demonstrates conditional stability, which is dependent on a strict limitation on the subsequent ratio

$$\frac{\tau}{h^2} < \frac{3\sqrt{\beta_1}}{2\sqrt{1+4\beta_2+16\beta_3}}, \ \ \beta_1 \geq 0, \ \ \beta_2 \geq 0, \ \ \beta_3 \geq 0.$$

Moreover, there is a significant lack of theoretical analysis regarding the solvability and convergence properties of the scheme proposed by Kolkovska [15]. Recently, Arslan [16] introduced a novel methodology combining the differential transform method with the finite difference technique to approximate solutions for singularly perturbed ill-posed problems and sixth-order Boussinesq equations. In another study, Zhang et al. [17] introduced a meshless numerical technique called the generalized finite difference method, which has proven effective for solving enhanced Boussinesq-type equations, especially for simulating wave propagation over irregular bottom topographies.

In numerical simulations, compact difference schemes generally require fewer stencils than standard finite difference methods while providing better resolution and maintaining the same order of accuracy. Hou et al. [18] introduced an energy-preserving, high-order compact finite difference scheme specifically designed for two-dimensional nonlinear wave equations. Mohanty et al. [19–22] developed highly accurate compact difference schemes for several equations, including the coupled viscous Burgers equations, the good Boussinesq equation, the time-dependent viscous Burgers–Huxley equation, and a generalized version of fourth-order parabolic partial differential equations.

On the other hand, the principle of conservation of mass and energy is crucial in physics. A numerical scheme that fails to uphold local conservation may produce nonphysical outcomes [23]. Therefore, preserving mass and energy is fundamental in the development of numerical schemes for solving PDEs. Significant research has been conducted on conservative finite difference schemes for various nonlinear wave equations. For instance, Deng et al. [24] proposed energy-preserving finite difference methods for two-dimensional nonlinear wave equations. Bayarassou et al. [25] investigated a high-order conservative linearized finite difference scheme for the regularized long wave (RLW) Korteweg–de Vries equation. Additionally, Nanta et al. [26] presented a wave model integrating the classical Camassa–Holm equation with the BBM–KdV equation, incorporating dual-power law nonlinearities while ensuring energy conservation.

This article aims to present a conservative difference scheme for solving the sixth-order Boussinesq problem as defined in Equations (1)–(3). Additional, it will examine the solvability, convergence, and stability of the proposed scheme.

Theorem 1. *Consider u as the solution to the sixth-order Boussinesq problem defined by Equations* (1)–(3). *Let us denote $u_t = v_{xx}$ and assume that the supplementary boundary conditions are provided by $v_x(x_l,t) = v_x(x_r,t) = 0$ and $v_{xx}(x_l,t) = v_{xx}(x_r,t) = 0$. Under these conditions, it can be concluded that $E(t) = E(0)$ for all $t \in [0,T]$, where*

$$E(t) = k^2\|u\|_{L^2}^2 + a_1\|u_x\|_{L^2}^2 + a_2\|u_t\|_{L^2}^2 + b_1\|u_{xx}\|_{L^2}^2 + b_2\|u_{xt}\|_{L^2}^2 + \|v_x\|_{L^2}^2 - \frac{2c}{2p+1}\int_{x_l}^{x_r} u^{2p+1}dx.$$

Proof. Let $u_t = v_{xx}$. We substitute this expression into Equation (1) and subsequently take the antiderivative twice. This process results in the following equation:

$$v_t = k^2u - a_1u_{xx} + a_2u_{tt} + b_1u_{xxxx} - b_2v_{xxtt} - cu^{2p},$$

we the obtain

$$\begin{aligned}
\frac{dE}{dt} &= 2\int_{x_l}^{x_r}\left(k^2uu_t + a_1u_xu_{xt} + a_2u_tu_{tt} + b_1u_{xx}u_{xxt} + b_2u_{xt}u_{xtt} + v_xv_{xt} - cu^{2p}u_t\right)dx \\
&= 2\int_{x_l}^{x_r}\left[u_t\left(k^2u + a_2u_{tt} - cu^{2p}\right) + \left(a_1u_xu_{xt} + b_1u_{xx}u_{xxt} + b_2u_{xt}u_{xtt} + v_xv_{xt}\right)\right]dx \\
&= 2\int_{x_l}^{x_r}\left[u_t\left(v_t + a_1u_{xx} - b_1u_{xxxx} + b_2u_{xxtt}\right)\right. \\
&\quad \left.+ \left(a_1u_xu_{xt} + b_1u_{xx}u_{xxt} + b_2u_{xt}u_{xtt} + v_xv_{xt}\right)\right]dx \\
&= 2\int_{x_l}^{x_r}(u_tv_t + v_xv_{xt})dx + 2a_1\int_{x_l}^{x_r}(u_tu_{xx} + u_xu_{xt})dx \\
&\quad + 2b_2\int_{x_l}^{x_r}(u_tu_{xxtt} + u_{xt}u_{xtt})dx - 2b_1\int_{x_l}^{x_r}(u_tu_{xxxx} - u_{xx}u_{xxt})dx \\
&\equiv 2(l_1 + a_1l_2 + b_2l_3 - b_1l_4).
\end{aligned} \tag{8}$$

Utilizing the boundary conditions delineated in Equation (3) along with the additional assumptions, we derive the following results

$$l_1 = \int_{x_l}^{x_r}(u_tv_t + v_xv_{xt})dx = \int_{x_l}^{x_r}(v_{xx}v_t + v_xv_{xt})dx = v_tv_x|_{x_l}^{x_r} = 0, \tag{9}$$

$$l_2 = \int_{x_l}^{x_r}(u_tu_{xx} + u_xu_{xt})dx = u_tu_x|_{x_l}^{x_r} = 0, \tag{10}$$

$$l_3 = \int_{x_l}^{x_r}(u_tu_{xxtt} + u_{xt}u_{xtt})dx = u_tu_{xtt}|_{x_l}^{x_r} = v_{xx}u_{xtt}|_{x_l}^{x_r} = 0. \tag{11}$$

Since

$$\begin{aligned}
&\int_{x_l}^{x_r} u_{xx}u_{xxt}dx = \int_{x_l}^{x_r} u_{xx}d(u_{xt}) = u_{xt}u_{xx}|_{x_l}^{x_r} - \int_{x_l}^{x_r} u_{xt}u_{xxx}dx \\
&= -\int_{x_l}^{x_r} u_{xt}u_{xxx}dx = -\int_{x_l}^{x_r} u_{xxx}d(u_t) = -u_tu_{xxx}|_{x_l}^{x_r} + \int_{x_l}^{x_r} u_tu_{xxxx}dx \\
&= -v_{xx}u_{xxx}|_{x_l}^{x_r} + \int_{x_l}^{x_r} u_tu_{xxxx}dx = \int_{x_l}^{x_r} u_tu_{xxxx}dx,
\end{aligned}$$

which yields

$$l_4 = \int_{x_l}^{x_r}(u_tu_{xxxx} - u_{xx}u_{xxt})dx = \int_{x_l}^{x_r} u_tu_{xxxx}dx - \int_{x_l}^{x_r} u_{xx}u_{xxt}dx = 0. \tag{12}$$

Consequently, from Equations (8)–(12), we deduce that $dE/dt = 0$, which indicates that $E(t) = E(0)$ for $t \in [0, T]$. This completes the proof. □

It is essential to emphasize that, if parameters a_1, a_2, b_1, b_2, k, and c and the initial conditions $\varphi(x)$ and $\psi(x)$ satisfy $E(0) \geq 0$, then $E(t)$ can be defined as the energy, and the solution to Equations (1)–(3) complies with the principle of energy conservation, where

$$\begin{aligned}
E(0) &= k^2\|\varphi\|_{L^2}^2 + a_1\|\varphi_x\|_{L^2}^2 + a_2\|\psi\|_{L^2}^2 + b_1\|\varphi_{xx}\|_{L^2}^2 + b_2\|\psi_x\|_{L^2}^2 + \|v_x(x,0)\|_{L^2}^2 \\
&\quad - \frac{2c}{2p+1}\int_{x_l}^{x_r}\varphi^{2p+1}dx.
\end{aligned}$$

The remainder of this article is organized as follows. Section 2 introduces a compact finite difference scheme for solving the sixth-order Boussinesq Equations (1)–(3). Section 3 rigorously establishes the discrete conservation properties of mass change rate and energy. A comprehensive theoretical analysis addressing the scheme's solvability, convergence, and stability is presented in Section 4. Numerical experiments validating the theoretical results are included in Section 5 . Finally, Section 6 provides a concise summary of the findings.

2. Compact Difference Scheme

In this section, we present a compact and conservative difference scheme to solve the sixth-order Boussinesq problem as delineated in Equations (1)–(3). For two positive integers J and N, we define the spatial and temporal discretization parameters as $h = (x_r - x_l)/J$ and $\tau = T/N$, respectively. The uniform grid points are established as $x_j = x_l + jh$ and $t^n = n\tau$, where $j = 0, 1, \ldots, J$ and $n = 0, 1, \ldots, N$. We denote $U_j^n \approx u(x_j, t^n)$ as the approximation of the function u at the point (x_j, t^n) and

$$Z_h^0 = \{U = (U_j) | U_{-2} = U_{-1} = U_0 = U_1 = U_2 = U_{J-2} = U_{J-1} = U_J = U_{J+1} = U_{J+2} = 0\},$$

where $-2 \le j \le J + 2$. The Sobolev space is defined as described in [27]:

$$H_0^k(\Omega) = \left\{ u(x) \in H^k(\Omega) : \frac{\partial^i u}{\partial x^i}\Big|_{\partial\Omega} = 0,\ i = 0, 1, 2, \ldots, k-1 \right\}.$$

For any two mesh functions $U^n, V^n \in Z_h^0$, we define the following difference operators, inner product and norms:

$$\begin{aligned}
&(U_j^n)_x = \frac{1}{h}(U_{j+1}^n - U_j^n),\ (U_j^n)_{\bar{x}} = \frac{1}{h}(U_j^n - U_{j-1}^n),\ (U_j^n)_{\hat{x}} = \frac{1}{2h}(U_{j+1}^n - U_{j-1}^n),\\
&(U_j^n)_t = \frac{1}{\tau}(U_j^{n+1} - U_j^n),\ (U_j^n)_{\bar{t}} = \frac{1}{\tau}(U_j^n - U_j^{n-1}),\ (U_j^n)_{\hat{t}} = \frac{1}{2\tau}(U_j^{n+1} - U_j^{n-1}),\\
&\bar{U}_j^n = \frac{1}{2}(U_j^{n+1} + U_j^{n-1}),\ U_j^{n+\frac{1}{2}} = \frac{1}{2}(U_j^{n+1} + U_j^n),\\
&\langle U^n, V^n \rangle = h\sum_{j=1}^{J-1} U_j^n V_j^n,\ \|U^n\|^2 = \langle U^n, U^n \rangle,\ \|U^n\|_\infty = \max_{1 \le j \le J-1} |U_j^n|.
\end{aligned}$$

By setting

$$w = k^2 u_{xx} - a_1 u_{xxxx} + a_2 u_{xxtt} + b_1 u_{xxxxxx} - b_2 u_{xxxxtt} - c\left(u^{2p}\right)_{xx}, \tag{13}$$

then Equation (1) can be expressed as $w = u_{tt}$. By utilizing the Taylor expansion for the variable x, we obtain

$$\begin{aligned}
w_j^n = {}& k^2\left[(U_j^n)_{x\bar{x}} - \frac{h^2}{12}(\partial_x^4 u)_j^n\right] - a_1\left[(U_j^n)_{x\bar{x}x\bar{x}} - \frac{h^2}{6}(\partial_x^6 u)_j^n\right] + a_2\left[(U_j^n)_{x\bar{x}t\bar{t}} - \frac{h^2}{12}(\partial_x^4\partial_t^2 u)_j^n\right]\\
& + b_1\left[(U_j^n)_{x\bar{x}x\bar{x}x\bar{x}} - \frac{h^2}{4}(\partial_x^8 u)_j^n\right] - b_2\left[(U_j^n)_{x\bar{x}x\bar{x}t\bar{t}} - \frac{h^2}{6}(\partial_x^6\partial_t^2 u)_j^n\right]\\
& - c\left\{[(U_j^n)^{2p}]_{x\bar{x}} - \frac{h^2}{12}(\partial_x^4 u^{2p})_j^n\right\} + O(\tau^2 + h^4).
\end{aligned} \tag{14}$$

Furthermore, Equation (2) is associated with

$$\begin{aligned}
b_1(\partial_x^8 u)_j^n = {}& -k^2(\partial_x^4 u)_j^n + a_1(\partial_x^6 u)_j^n + a_2(\partial_x^4\partial_t^2 u)_j^n + b_2(\partial_x^6\partial_t^2 u)_j^n\\
& + c(\partial_x^4 u^{2p})_j^n + (\partial_x^2 w)_j^n.
\end{aligned} \tag{15}$$

Substituting Equation (15) into Equation (14) yields

$$\begin{aligned} w_j^n &= k^2(U_j^n)_{\bar{x}\hat{x}} - a_1(U_j^n)_{\bar{x}\hat{x}\bar{x}\hat{x}} + a_2(U_j^n)_{\bar{x}\hat{x}\bar{t}\hat{t}} + b_1(U_j^n)_{\bar{x}\hat{x}\bar{x}\hat{x}\bar{x}\hat{x}} - b_2(U_j^n)_{\bar{x}\hat{x}\bar{x}\hat{x}\bar{t}\hat{t}} \\ &- c[(U_j^n)^{2p}]_{\bar{x}\hat{x}} + \frac{k^2h^2}{6}(\partial_x^4 u)_j^n - \frac{a_1h^2}{12}(\partial_x^6 u)_j^n + \frac{a_2h^2}{6}(\partial_x^4\partial_t^2 u)_j^n \\ &- \frac{b_2h^2}{12}(\partial_x^6\partial_t^2 u)_j^n - \frac{ch^2}{6}(\partial_x^4 u^{2p})_j^n - \frac{h^2}{4}(\partial_x^2 w)_j^n. \end{aligned}$$

By applying second-order accuracy in our approximations, we achieve

$$\begin{aligned} &U_j^n = \bar{U}_j^n + O(\tau^2),\ w_j^n = (\partial_t^2 u)_j^n = (U_j^n)_{\bar{t}\hat{t}} + O(\tau^2), \\ &(\partial_x^4 u)_j^n = (U_j^n)_{\bar{x}\hat{x}\bar{x}\hat{x}} + O(h^2),\ (\partial_x^6 u)_j^n = (U_j^n)_{\bar{x}\hat{x}\bar{x}\hat{x}\bar{x}\hat{x}} + O(h^2), \\ &(\partial_x^2\partial_t^2 u)_j^n = (U_j^n)_{\bar{x}\hat{x}\bar{t}\hat{t}} + O(\tau^2 + h^2),\ (\partial_x^4\partial_t^2 u)_j^n = (U_j^n)_{\bar{x}\hat{x}\bar{x}\hat{x}\bar{t}\hat{t}} + O(\tau^2 + h^2), \\ &(\partial_x^6\partial_t^2 u)_j^n = (U_j^n)_{\bar{x}\hat{x}\bar{x}\hat{x}\bar{x}\hat{x}\bar{t}\hat{t}} + O(\tau^2 + h^2),\ (\partial_x^4 u^{2p})_j^n = [(U_j^n)^{2p}]_{\bar{x}\hat{x}\bar{x}\hat{x}} + O(h^2). \end{aligned}$$

Consequently, the proposed compact finite difference scheme can be expressed as follows:

$$\begin{aligned} &(U_j^n)_{\bar{t}\hat{t}} - k^2(\bar{U}_j^n)_{\bar{x}\hat{x}} + \left(a_1 - \frac{k^2h^2}{6}\right)(\bar{U}_j^n)_{\bar{x}\hat{x}\bar{x}\hat{x}} - \left(a_2 - \frac{h^2}{4}\right)(U_j^n)_{\bar{x}\hat{x}\bar{t}\hat{t}} - \left(b_1 - \frac{a_1h^2}{12}\right)(\bar{U}_j^n)_{\bar{x}\hat{x}\bar{x}\hat{x}\bar{x}\hat{x}} \\ &+ \left(b_2 - \frac{a_2h^2}{6}\right)(U_j^n)_{\bar{x}\hat{x}\bar{x}\hat{x}\bar{t}\hat{t}} + \frac{b_2h^2}{12}(U_j^n)_{\bar{x}\hat{x}\bar{x}\hat{x}\bar{x}\hat{x}\bar{t}\hat{t}} + c[(U_j^n)^{2p}]_{\bar{x}\hat{x}} + \frac{ch^2}{6}[(U_j^n)^{2p}]_{\bar{x}\hat{x}\bar{x}\hat{x}} = 0, \end{aligned} \quad (16)$$

$$U_j^0 = \varphi(x_j),\ (U_j^0)_{\hat{t}} = \psi(x_j),\ 0 \le j \le J, \quad (17)$$

$$\begin{aligned} &U_0^n = 0,\ \frac{4}{3}(U_0^n)_{\hat{x}} - \frac{1}{3}(U_0^n)_{\ddot{x}} = 0,\ \frac{4}{3}(U_0^n)_{\bar{x}x} - \frac{1}{3}(U_0^n)_{\hat{x}\check{x}} = 0, \\ &U_J^n = 0,\ \frac{4}{3}(U_J^n)_{\hat{x}} - \frac{1}{3}(U_J^n)_{\ddot{x}} = 0,\ \frac{4}{3}(U_J^n)_{\bar{x}x} - \frac{1}{3}(U_J^n)_{\hat{x}\check{x}} = 0,\ 0 \le n \le N. \end{aligned} \quad (18)$$

It is essential to recognize that, since $U_0^n = 0$, $\frac{4}{3}(U_0^n)_{\hat{x}} - \frac{1}{3}(U_0^n)_{\ddot{x}} = 0$, and $\frac{4}{3}(U_0^n)_{\bar{x}x} - \frac{1}{3}(U_0^n)_{\hat{x}\check{x}} = 0$ hold based on Equation (18), we may proceed with the assumption that $U_{-2}^n = U_{-1}^n = U_1^n = U_2^n = 0$ for the sake of simplicity. Similarly, we can also assume that $U_{J-2}^n = U_{J-1}^n = U_{J+1}^n = U_{J+2}^n = 0$, where $j = -2, -1, J+1$, and $J+2$ denote ghost points under the condition $1 \le n \le N$. Consequently, it can be concluded that $U^n \in Z_h^0$.

In this context, we employ the following methodology to derive U^1:

$$\begin{aligned} &(U_j^0)_{\bar{t}\hat{t}} - k^2(\bar{U}_j^0)_{\bar{x}\hat{x}} + \left(a_1 - \frac{k^2h^2}{6}\right)(\bar{U}_j^0)_{\bar{x}\hat{x}\bar{x}\hat{x}} - \left(a_2 - \frac{h^2}{4}\right)(U_j^0)_{\bar{x}\hat{x}\bar{t}\hat{t}} - \left(b_1 - \frac{a_1h^2}{12}\right)(\bar{U}_j^0)_{\bar{x}\hat{x}\bar{x}\hat{x}\bar{x}\hat{x}} \\ &+ \left(b_2 - \frac{a_2h^2}{6}\right)(U_j^0)_{\bar{x}\hat{x}\bar{x}\hat{x}\bar{t}\hat{t}} + \frac{b_2h^2}{12}(U_j^0)_{\bar{x}\hat{x}\bar{x}\hat{x}\bar{x}\hat{x}\bar{t}\hat{t}} + c[(U_j^0)^{2p}]_{\bar{x}\hat{x}} + \frac{ch^2}{6}[(U_j^0)^{2p}]_{\bar{x}\hat{x}\bar{x}\hat{x}} = 0. \end{aligned} \quad (19)$$

From Equation (17), we obtain

$$\frac{1}{2\tau}(U_j^1 - U_j^{-1}) = (U_j^0)_{\hat{t}} = \psi(x_j) + O(\tau^2),\ 0 \le j \le J;$$

that is,

$$U_j^{-1} = U_j^1 - 2\tau\psi(x_j) + O(\tau^3),\ 0 \le j \le J, \quad (20)$$

which yields

$$(U_j^0)_{\bar{t}t} = \frac{1}{\tau^2}(U_j^1 - 2U_j^0 + U_j^{-1})$$
$$= \frac{2}{\tau^2}\Big[U_j^1 - U_j^0 - \tau\psi(x_j)\Big] = \frac{2}{\tau^2}\Big[U_j^1 - \varphi(x_j) - \tau\psi(x_j)\Big] + O(\tau),$$

where $0 \le j \le J$.

3. Discrete Conservative Laws

In this section, we demonstrate that the proposed methodology, as outlined in Equations (16)–(18), maintains both the rate of mass change and energy conservation.

We will now present a modified version of the previously discussed scheme and define an auxiliary variable $V^n \in Z_h^0$ that fulfills the following conditions:

$$(V_j^{n+1/2})_{\bar{x}\bar{x}} = (U_j^n)_{\bar{t}}, \ j = 0,1,\ldots,J, \ n = 0,1,\ldots,N-1. \tag{21}$$

Lemma 1 ([28–30]). *For any two discrete functions $U, V \in Z_h^0$, it follows that*

$$\langle U_{\bar{x}\bar{x}}, U\rangle = -\langle U_{\bar{x}}, U_{\bar{x}}\rangle = -\|U_{\bar{x}}\|^2,$$
$$\langle U_{\bar{x}\bar{x}}, V\rangle = -\langle U_{\bar{x}}, V_{\bar{x}}\rangle = \langle U, V_{\bar{x}\bar{x}}\rangle.$$

Furthermore, we have

$$\langle U_{\bar{x}\bar{x}\bar{x}\bar{x}}, U\rangle = \|U_{\bar{x}\bar{x}}\|^2.$$

Lemma 2. *For any discrete function $U \in Z_h^0$, it follows that*

$$\langle U_{\bar{x}\bar{x}\bar{x}\bar{x}\bar{x}\bar{x}}, U\rangle = -\langle U_{\bar{x}\bar{x}\bar{x}}, U_{\bar{x}\bar{x}\bar{x}}\rangle = -\|U_{\bar{x}\bar{x}\bar{x}}\|^2.$$

Proof. According to Lemma 1, it can be concluded that

$$\langle U_{\bar{x}\bar{x}\bar{x}\bar{x}\bar{x}\bar{x}}, U\rangle = \langle (U_{\bar{x}\bar{x}\bar{x}\bar{x}})_{\bar{x}\bar{x}}, U\rangle = \langle U_{\bar{x}\bar{x}\bar{x}\bar{x}}, U_{\bar{x}\bar{x}}\rangle$$
$$= \langle (U_{\bar{x}\bar{x}})_{\bar{x}\bar{x}}, U_{\bar{x}\bar{x}}\rangle = -\langle (U_{\bar{x}\bar{x}})_{\bar{x}}, (U_{\bar{x}\bar{x}})_{\bar{x}}\rangle = -\|U_{\bar{x}\bar{x}\bar{x}}\|^2.$$

This completes the proof. □

Lemma 3 ([11,30]). *For any discrete function $U^n \in Z_h^0$, it follows that*

$$\|U^n\| \le C\|U_{\bar{x}}^n\|, \ \|U^n\|_\infty \le C\|U_{\bar{x}}^n\|.$$

Lemma 4 ([11,31]). *For any discrete function $U^n \in Z_h^0$, it follows that*

$$\|U_{\bar{x}}^n\| \le \frac{4}{h^2}\|U^n\|.$$

Theorem 2. *Let $U^n \in Z_h^0$ denote the solution to Equations (16)–(18). We define the discrete mass as $M^n = h\sum_{j=1}^{J-1} U_j^n$. Consequently, the rate of change regarding discrete mass is described by the following equation:*

$$R^n \equiv \frac{h}{\tau}\sum_{j=1}^{J-1}(U_j^{n+1} - U_j^n) = R^{n-1} = \cdots = R^0 = h\sum_{j=1}^{J-1}\psi(x_j) + O(\tau^2 h),$$

where $0 \le n \le N-1$.

Proof. By multiplying Equation (16) by h and subsequently summing over j from 1 to J, we obtain the following result:

$$h\sum_{j=1}^{J-1}(U_j^n)_{\bar{t}\bar{t}} - k^2h\sum_{j=1}^{J-1}(\bar{U}_j^n)_{\bar{x}\bar{x}} + \left(a_1 - \frac{k^2h^2}{6}\right)h\sum_{j=1}^{J-1}(\bar{U}_j^n)_{\bar{x}\bar{x}\bar{x}\bar{x}} - \left(a_2 - \frac{h^2}{4}\right)h\sum_{j=1}^{J-1}(U_j^n)_{\bar{x}\bar{x}\bar{t}\bar{t}}$$
$$-\left(b_1 - \frac{a_1h^2}{12}\right)h\sum_{j=1}^{J-1}(\bar{U}_j^n)_{\bar{x}\bar{x}\bar{x}\bar{x}\bar{x}\bar{x}} + \left(b_2 - \frac{a_2h^2}{6}\right)h\sum_{j=1}^{J-1}(U_j^n)_{\bar{x}\bar{x}\bar{x}\bar{x}\bar{t}\bar{t}} + \frac{b_2h^3}{12}\sum_{j=1}^{J-1}(U_j^n)_{\bar{x}\bar{x}\bar{x}\bar{x}\bar{x}\bar{x}\bar{t}\bar{t}}$$
$$+ch\sum_{j=1}^{J-1}[(U_j^n)^{2p}]_{\bar{x}\bar{x}} + \frac{ch^3}{6}\sum_{j=1}^{J-1}[(U_j^n)^{2p}]_{\bar{x}\bar{x}\bar{x}\bar{x}} = 0. \tag{22}$$

Noting that $U^n \in Z_h^0$, we can deduce

$$p_1 = h\sum_{j=1}^{J-1}(U_j^n)_{\bar{t}\bar{t}} = \frac{h}{\tau^2}\sum_{j=1}^{J-1}(U_j^{n+1} - 2U_j^n + U_j^{n-1}), \tag{23}$$

$$p_2 = h\sum_{j=1}^{J-1}(\bar{U}_j^n)_{\bar{x}\bar{x}} = \frac{1}{h}\sum_{j=1}^{J-1}(\bar{U}_{j+1}^n - 2\bar{U}_j^n + \bar{U}_{j-1}^n) = \frac{1}{h}\sum_{j=1}^{J-1}\left[(\bar{U}_{j+1}^n - \bar{U}_j^n) - (\bar{U}_j^n - \bar{U}_{j-1}^n)\right]$$
$$= \frac{1}{h}\sum_{j=1}^{J-1}(\bar{U}_{j+1}^n - \bar{U}_j^n) - \frac{1}{h}\sum_{j=1}^{J-1}(\bar{U}_j^n - \bar{U}_{j-1}^n) = \frac{1}{h}\left[(\bar{U}_{J+1}^n - \bar{U}_J^n) - (\bar{U}_1^n - \bar{U}_0^n)\right] = 0, \tag{24}$$

$$p_3 = h\sum_{j=1}^{J-1}(\bar{U}_j^n)_{\bar{x}\bar{x}\bar{x}\bar{x}} = \frac{1}{h^3}\sum_{j=1}^{J-1}(\bar{U}_{j+2}^n - 4\bar{U}_{j+1}^n + 6\bar{U}_j^n - 4\bar{U}_{j-1}^n + \bar{U}_{j-2}^n)$$
$$= \frac{1}{h^3}\sum_{j=1}^{J-1}\left[(\bar{U}_{j+2}^n - \bar{U}_{j+1}^n) - 3(\bar{U}_{j+1}^n - \bar{U}_j^n) + 3(\bar{U}_j^n - \bar{U}_{j-1}^n) - (\bar{U}_{j-1}^n - \bar{U}_{j-2}^n)\right]$$
$$= \frac{1}{h^3}\left[(\bar{U}_{J+1}^n - \bar{U}_2^n) - 3(\bar{U}_J^n - \bar{U}_1^n) + 3(\bar{U}_{J-1}^n - \bar{U}_0^n) - (\bar{U}_{J-2}^n - \bar{U}_{-1}^n)\right] = 0, \tag{25}$$

$$p_4 = h\sum_{j=1}^{J-1}(\bar{U}_j^n)_{\bar{x}\bar{x}\bar{x}\bar{x}\bar{x}\bar{x}} = \frac{1}{h^5}\sum_{j=1}^{J-1}(\bar{U}_{j+3}^n - 6\bar{U}_{j+2}^n + 15\bar{U}_{j+1}^n - 20\bar{U}_j^n + 15\bar{U}_{j-1}^n - 6\bar{U}_{j-2}^n + \bar{U}_{j-3}^n)$$
$$= \frac{1}{h^5}\left[(\bar{U}_{J+2}^n - \bar{U}_3^n) - 5(\bar{U}_{J+1}^n - \bar{U}_2^n) + 10(\bar{U}_J^n - \bar{U}_1^n) - 10(\bar{U}_{J-1}^n - \bar{U}_0^n)\right.$$
$$\left.+5(\bar{U}_{J-2}^n - \bar{U}_{-1}^n) - (\bar{U}_{J-3}^n - \bar{U}_{-2}^n)\right] = 0, \tag{26}$$

$$p_5 = h\sum_{j=1}^{J-1}(U_j^n)_{\bar{x}\bar{x}\bar{t}\bar{t}} = \frac{1}{h}\sum_{j=1}^{J-1}\left[(U_{j+1}^n - 2U_j^n + U_{j-1}^n)_{\bar{t}\bar{t}}\right]$$
$$= \frac{1}{h}\sum_{j=1}^{J-1}(U_{j+1}^n - U_j^n)_{\bar{t}\bar{t}} - \frac{1}{h}\sum_{j=1}^{J-1}(U_j^n - U_{j-1}^n)_{\bar{t}\bar{t}}$$
$$= \frac{1}{h}\left[(U_{J+1}^n - U_J^n) - (U_1^n - U_0^n)\right]_{\bar{t}\bar{t}} = 0. \tag{27}$$

Similarly, we obtain

$$p_6 = h\sum_{j=1}^{J-1}[(U_j^n)^{2p}]_{\bar{x}\bar{x}} = 0, \quad p_7 = h\sum_{j=1}^{J-1}[(U_j^n)^{2p}]_{\bar{x}\bar{x}\bar{x}\bar{x}} = 0, \tag{28}$$

$$p_8 = h\sum_{j=1}^{J-1}(U_j^n)_{\bar{x}\bar{x}\bar{x}\bar{x}\bar{t}\bar{t}} = 0, \quad p_9 = h\sum_{j=1}^{J-1}(U_j^n)_{\bar{x}\bar{x}\bar{x}\bar{x}\bar{x}\bar{x}\bar{t}\bar{t}} = 0. \tag{29}$$

Substituting Equations (23)–(29) into Equation (22), we obtain

$$h\sum_{j=1}^{J-1}(U_j^n)_{\bar{t}t}=\frac{h}{\tau^2}\sum_{j=1}^{J-1}(U_j^{n+1}-2U_j^n+U_j^{n-1})=0;$$

that is,

$$\frac{h}{\tau}\sum_{j=1}^{J-1}(U_j^{n+1}-U_j^n)=\frac{h}{\tau}\sum_{j=1}^{J-1}(U_j^n-U_j^{n-1}),$$

which yields $R^n=R^{n-1}=\cdots=R^0$.

Additionally, as indicated by Equation (20), we obtain

$$\begin{aligned}R^0&=\frac{h}{\tau}\sum_{j=1}^{J-1}(U_j^1-U_j^0)=\frac{h}{\tau}\sum_{j=1}^{J-1}(U_j^0-U_j^{-1})\\&=\frac{h}{\tau}\sum_{j=1}^{J-1}[U_j^0-U_j^1+2\tau\psi(x_j)+O(\tau^3)]\\&=-R^0+2h\sum_{j=1}^{J-1}\psi(x_j)+O(\tau^2h),\end{aligned}$$

which yields $R^0=h\sum_{j=1}^{J-1}\psi(x_j)+O(\tau^2h)$. This completes the proof. □

Theorem 3. *Let $U^n\in Z_h^0$ represent the solution to Equations* (16)–(18) *and let $V^n\in Z_h^0$ denote the solution to Equation* (21). *It can be inferred that the discrete energy is conserved, such that $E^n=E^{n-1}=\cdots=E^0$, where*

$$\begin{aligned}E^n\equiv&\|V_{\tilde{x}}^{n+\frac{1}{2}}\|^2+\frac{k^2}{2}\Big(\|U^{n+1}\|^2+\|U^n\|^2\Big)+\frac{1}{2}\Big(a_1-\frac{k^2h^2}{6}\Big)\Big(\|U_{\tilde{x}}^{n+1}\|^2+\|U_{\tilde{x}}^n\|^2\Big)\\&+\Big(a_2-\frac{h^2}{4}\Big)\|U_t^{n+1}\|^2+\frac{1}{2}\Big(b_1-\frac{a_1h^2}{12}\Big)\Big(\|U_{\tilde{x}\tilde{x}}^{n+1}\|^2+\|U_{\tilde{x}\tilde{x}}^n\|^2\Big)\\&+\Big(b_2-\frac{a_2h^2}{6}\Big)\|U_{\tilde{x}t}^{n+1}\|^2-\frac{b_2h^2}{12}\|U_{\tilde{t}t}^{n+1}\|^2\\&-\frac{ch\tau}{6}\sum_{k=0}^{n}\Big\{\sum_{j=1}^{J-1}\Big[(U_{j+1}^k)^{2p}+4(U_j^k)^{2p}+(U_{j-1}^k)^{2p}\Big](U_j^{k+1}+U_j^k)_t\Big\},\end{aligned}$$

where $0\le n\le N-1$.

Proof. By computing the inner product of Equation (16) with $V^{n+\frac{1}{2}}+V^{n-\frac{1}{2}}$, we obtain

$$\begin{aligned}&\langle U_{\bar{t}t}^n,V^{n+\frac{1}{2}}+V^{n-\frac{1}{2}}\rangle-k^2\langle\bar{U}_{\tilde{x}\tilde{x}}^n,V^{n+\frac{1}{2}}+V^{n-\frac{1}{2}}\rangle+\Big(a_1-\frac{k^2h^2}{6}\Big)\langle\bar{U}_{\tilde{x}\tilde{x}\tilde{x}\tilde{x}}^n,V^{n+\frac{1}{2}}+V^{n-\frac{1}{2}}\rangle\\&-\Big(a_2-\frac{h^2}{4}\Big)\langle U_{\tilde{x}\tilde{x}\bar{t}t}^n,V^{n+\frac{1}{2}}+V^{n-\frac{1}{2}}\rangle-\Big(b_1-\frac{a_1h^2}{12}\Big)\langle\bar{U}_{\tilde{x}\tilde{x}\tilde{x}\tilde{x}\tilde{x}\tilde{x}}^n,V^{n+\frac{1}{2}}+V^{n-\frac{1}{2}}\rangle\\&+\Big(b_2-\frac{a_2h^2}{6}\Big)\langle U_{\tilde{x}\tilde{x}\tilde{x}\tilde{x}\bar{t}t}^n,V^{n+\frac{1}{2}}+V^{n-\frac{1}{2}}\rangle+\frac{b_2h^2}{12}\langle U_{\tilde{x}\tilde{x}\tilde{x}\tilde{x}\tilde{x}\tilde{x}\bar{t}t}^n,V^{n+\frac{1}{2}}+V^{n-\frac{1}{2}}\rangle\\&+c\langle(U^n)_{\tilde{x}\tilde{x}}^{2p},V^{n+\frac{1}{2}}+V^{n-\frac{1}{2}}\rangle+\frac{ch^2}{6}\langle(U^n)_{\tilde{x}\tilde{x}\tilde{x}\tilde{x}}^{2p},V^{n+\frac{1}{2}}+V^{n-\frac{1}{2}}\rangle=0.\end{aligned}\tag{30}$$

Utilizing Lemmas 1 and 2, along with Equation (21), a simple computation yields

$$
\begin{aligned}
q_1 &= \langle U^n_{\bar{t}\hat{t}}, V^{n+\frac{1}{2}} + V^{n-\frac{1}{2}} \rangle = \langle (V^{n+\frac{1}{2}}_{\tilde{x}\bar{x}})_{\bar{t}}, V^{n+\frac{1}{2}} + V^{n-\frac{1}{2}} \rangle \\
&= \frac{1}{\tau} \langle V^{n+\frac{1}{2}}_{\tilde{x}\bar{x}} - V^{n-\frac{1}{2}}_{\tilde{x}\bar{x}}, V^{n+\frac{1}{2}} + V^{n-\frac{1}{2}} \rangle = -\frac{1}{\tau} (\| V^{n+\frac{1}{2}}_{\tilde{x}} \|^2 - \| V^{n-\frac{1}{2}}_{\tilde{x}} \|^2), \qquad (31) \\
q_2 &= \langle \bar{U}^n_{\tilde{x}\bar{x}}, V^{n+\frac{1}{2}} + V^{n-\frac{1}{2}} \rangle = \frac{1}{2} \langle U^{n+1} + U^{n-1}, (V^{n+\frac{1}{2}} + V^{n-\frac{1}{2}})_{\tilde{x}\bar{x}} \rangle \\
&= \frac{1}{2} \langle U^{n+1} + U^{n-1}, (U^n + U^{n-1})_{\bar{t}} \rangle = \frac{1}{2\tau} \langle U^{n+1} + U^{n-1}, U^{n+1} - U^{n-1} \rangle \\
&= \frac{1}{2\tau} (\| U^{n+1} \|^2 - \| U^{n-1} \|^2), \qquad (32) \\
q_3 &= \langle \bar{U}^n_{\tilde{x}\bar{x}\tilde{x}\bar{x}}, V^{n+\frac{1}{2}} + V^{n-\frac{1}{2}} \rangle = \frac{1}{2} \langle (U^{n+1} + U^{n-1})_{\tilde{x}\bar{x}}, (V^{n+\frac{1}{2}} + V^{n-\frac{1}{2}})_{\tilde{x}\bar{x}} \rangle \\
&= \frac{1}{2} \langle (U^{n+1} + U^{n-1})_{\tilde{x}\bar{x}}, (U^n + U^{n-1})_{\bar{t}} \rangle = \frac{1}{2\tau} \langle (U^{n+1} + U^{n-1})_{\tilde{x}\bar{x}}, U^{n+1} - U^{n-1} \rangle \\
&= -\frac{1}{2\tau} (\| U^{n+1}_{\tilde{x}} \|^2 - \| U^{n-1}_{\tilde{x}} \|^2), \qquad (33) \\
q_4 &= \langle U^n_{\tilde{x}\bar{x}\bar{t}\hat{t}}, V^{n+\frac{1}{2}} + V^{n-\frac{1}{2}} \rangle = \langle (V^{n+\frac{1}{2}}_{\tilde{x}\bar{x}})_{\tilde{x}\bar{x}\bar{t}}, V^{n+\frac{1}{2}} + V^{n-\frac{1}{2}} \rangle \\
&= \frac{1}{\tau} \langle V^{n+\frac{1}{2}}_{\tilde{x}\bar{x}\tilde{x}\bar{x}} - V^{n-\frac{1}{2}}_{\tilde{x}\bar{x}\tilde{x}\bar{x}}, V^{n+\frac{1}{2}} + V^{n-\frac{1}{2}} \rangle = \frac{1}{\tau} (\| V^{n+\frac{1}{2}}_{\tilde{x}\bar{x}} \|^2 - \| V^{n-\frac{1}{2}}_{\tilde{x}\bar{x}} \|^2) \\
&= \frac{1}{\tau} (\| U^n_{\bar{t}} \|^2 - \| U^{n-1}_{\bar{t}} \|^2) = \frac{1}{\tau} (\| U^{n+1}_{\bar{t}} \|^2 - \| U^n_{\bar{t}} \|^2), \qquad (34) \\
q_5 &= \langle \bar{U}^n_{\tilde{x}\bar{x}\tilde{x}\bar{x}\tilde{x}\bar{x}}, V^{n+\frac{1}{2}} + V^{n-\frac{1}{2}} \rangle = \frac{1}{2} \langle (U^{n+1} + U^{n-1})_{\tilde{x}\bar{x}\tilde{x}\bar{x}}, (V^{n+\frac{1}{2}} + V^{n-\frac{1}{2}})_{\tilde{x}\bar{x}} \rangle \\
&= \frac{1}{2} \langle (U^{n+1} + U^{n-1})_{\tilde{x}\bar{x}\tilde{x}\bar{x}}, (U^n + U^{n-1})_{\bar{t}} \rangle = \frac{1}{2\tau} \langle (U^{n+1} + U^{n-1})_{\tilde{x}\bar{x}\tilde{x}\bar{x}}, U^{n+1} - U^{n-1} \rangle \\
&= \frac{1}{2\tau} (\| U^{n+1}_{\tilde{x}\bar{x}} \|^2 - \| U^{n-1}_{\tilde{x}\bar{x}} \|^2), \qquad (35) \\
q_6 &= \langle U^n_{\tilde{x}\bar{x}\tilde{x}\bar{x}\bar{t}\hat{t}}, V^{n+\frac{1}{2}} + V^{n-\frac{1}{2}} \rangle = \langle (V^{n+\frac{1}{2}}_{\tilde{x}\bar{x}})_{\tilde{x}\bar{x}\tilde{x}\bar{x}\bar{t}}, V^{n+\frac{1}{2}} + V^{n-\frac{1}{2}} \rangle \\
&= \frac{1}{\tau} \langle V^{n+\frac{1}{2}}_{\tilde{x}\bar{x}\tilde{x}\bar{x}\tilde{x}\bar{x}} - V^{n-\frac{1}{2}}_{\tilde{x}\bar{x}\tilde{x}\bar{x}\tilde{x}\bar{x}}, V^{n+\frac{1}{2}} + V^{n-\frac{1}{2}} \rangle = -\frac{1}{\tau} (\| V^{n+\frac{1}{2}}_{\tilde{x}\bar{x}\tilde{x}} \|^2 - \| V^{n-\frac{1}{2}}_{\tilde{x}\bar{x}\tilde{x}} \|^2) \\
&= -\frac{1}{\tau} (\| U^n_{\tilde{x}\bar{t}} \|^2 - \| U^{n-1}_{\tilde{x}\bar{t}} \|^2) = -\frac{1}{\tau} (\| U^{n+1}_{\tilde{x}\bar{t}} \|^2 - \| U^n_{\tilde{x}\bar{t}} \|^2), \qquad (36) \\
q_7 &= \langle U^n_{\tilde{x}\bar{x}\tilde{x}\bar{x}\tilde{x}\bar{x}\bar{t}\hat{t}}, V^{n+\frac{1}{2}} + V^{n-\frac{1}{2}} \rangle = \langle (V^{n+\frac{1}{2}}_{\tilde{x}\bar{x}})_{\tilde{x}\bar{x}\tilde{x}\bar{x}\tilde{x}\bar{x}\bar{t}}, V^{n+\frac{1}{2}} + V^{n-\frac{1}{2}} \rangle \\
&= \frac{1}{\tau} \langle V^{n+\frac{1}{2}}_{\tilde{x}\bar{x}\tilde{x}\bar{x}\tilde{x}\bar{x}\tilde{x}\bar{x}} - V^{n-\frac{1}{2}}_{\tilde{x}\bar{x}\tilde{x}\bar{x}\tilde{x}\bar{x}\tilde{x}\bar{x}}, V^{n+\frac{1}{2}} + V^{n-\frac{1}{2}} \rangle = \frac{1}{\tau} (\| V^{n+\frac{1}{2}}_{\tilde{x}\bar{x}\tilde{x}\bar{x}} \|^2 - \| V^{n-\frac{1}{2}}_{\tilde{x}\bar{x}\tilde{x}\bar{x}} \|^2) \\
&= \frac{1}{\tau} (\| U^n_{\hat{t}\bar{t}} \|^2 - \| U^{n-1}_{\hat{t}\bar{t}} \|^2) = \frac{1}{\tau} (\| U^{n+1}_{\hat{t}\bar{t}} \|^2 - \| U^n_{\hat{t}\bar{t}} \|^2), \qquad (37) \\
q_8 &= \langle (U^n)^{2p}_{\tilde{x}\bar{x}}, V^{n+\frac{1}{2}} + V^{n-\frac{1}{2}} \rangle = \langle (U^n)^{2p}, (V^{n+\frac{1}{2}} + V^{n-\frac{1}{2}})_{\tilde{x}\bar{x}} \rangle \\
&= \langle (U^n)^{2p}, (U^n + U^{n-1})_{\bar{t}} \rangle = \langle (U^n)^{2p}, (U^{n+1} + U^n)_{\bar{t}} \rangle = h \sum_{j=1}^{J-1} \left[(U^n_j)^{2p} (U^{n+1}_j + U^n_j)_{\bar{t}} \right] \\
&= h \sum_{k=0}^{n} \left\{ \sum_{j=1}^{J-1} \left[(U^k_j)^{2p} (U^{k+1}_j + U^k_j)_{\bar{t}} \right] \right\} - h \sum_{k=0}^{n-1} \left\{ \sum_{j=1}^{J-1} \left[(U^k_j)^{2p} (U^{k+1}_j + U^k_j)_{\bar{t}} \right] \right\}, \qquad (38) \\
q_9 &= \langle (U^n)^{2p}_{\tilde{x}\bar{x}\tilde{x}\bar{x}}, V^{n+\frac{1}{2}} + V^{n-\frac{1}{2}} \rangle = \langle (U^n)^{2p}_{\tilde{x}\bar{x}}, (V^{n+\frac{1}{2}} + V^{n-\frac{1}{2}})_{\tilde{x}\bar{x}} \rangle \\
&= \langle (U^n)^{2p}_{\tilde{x}\bar{x}}, (U^n + U^{n-1})_{\bar{t}} \rangle = \langle (U^n)^{2p}_{\tilde{x}\bar{x}}, (U^{n+1} + U^n)_{\bar{t}} \rangle = h \sum_{j=1}^{J-1} \left[(U^n_j)^{2p}_{\tilde{x}\bar{x}} (U^{n+1}_j + U^n_j)_{\bar{t}} \right] \\
&= h \sum_{k=0}^{n} \left\{ \sum_{j=1}^{J-1} \left[(U^k_j)^{2p}_{\tilde{x}\bar{x}} (U^{k+1}_j + U^k_j)_{\bar{t}} \right] \right\} - h \sum_{k=0}^{n-1} \left\{ \sum_{j=1}^{J-1} \left[(U^k_j)^{2p}_{\tilde{x}\bar{x}} (U^{k+1}_j + U^k_j)_{\bar{t}} \right] \right\}. \qquad (39)
\end{aligned}
$$

As a result, following Equations (38) and (39), we arrive at the following conclusions

$$
\begin{aligned}
&c\langle (U^n)^{2p}_{\bar{x}\bar{x}}, V^{n+\frac{1}{2}} + V^{n-\frac{1}{2}}\rangle + \frac{ch^2}{6}\langle (U^n)^{2p}_{\bar{x}\bar{x}\bar{x}\bar{x}}, V^{n+\frac{1}{2}} + V^{n-\frac{1}{2}}\rangle \\
&= ch\sum_{k=0}^{n}\Big\{\sum_{j=1}^{J-1}\Big[(U_j^k)^{2p}(U_j^{k+1}+U_j^k)_{\bar{t}}\Big]\Big\} + \frac{ch^3}{6}\sum_{k=0}^{n}\Big\{\sum_{j=1}^{J-1}\Big[(U_j^k)^{2p}_{\bar{x}\bar{x}}(U_j^{k+1}+U_j^k)_{\bar{t}}\Big]\Big\} \\
&\quad - ch\sum_{k=0}^{n-1}\Big\{\sum_{j=1}^{J-1}\Big[(U_j^k)^{2p}(U_j^{k+1}+U_j^k)_{\bar{t}}\Big]\Big\} - \frac{ch^3}{6}\sum_{k=0}^{n-1}\Big\{\sum_{j=1}^{J-1}\Big[(U_j^k)^{2p}_{\bar{x}\bar{x}}(U_j^{k+1}+U_j^k)_{\bar{t}}\Big]\Big\} \\
&= \frac{ch}{6}\sum_{k=0}^{n}\Big\{\sum_{j=1}^{J-1}\Big[(U_{j+1}^k)^{2p}+4(U_j^k)^{2p}+(U_{j-1}^k)^{2p}\Big](U_j^{k+1}+U_j^k)_{\bar{t}}\Big\} \\
&= \frac{ch}{6}\sum_{k=0}^{n-1}\Big\{\sum_{j=1}^{J-1}\Big[(U_{j+1}^k)^{2p}+4(U_j^k)^{2p}+(U_{j-1}^k)^{2p}\Big](U_j^{k+1}+U_j^k)_{\bar{t}}\Big\}.
\end{aligned}
\tag{40}
$$

By substituting Equations (31)–(37) and (40) into Equation (30), we obtain

$$
\begin{aligned}
&\frac{1}{\tau}\|V_{\bar{x}}^{n+\frac{1}{2}}\|^2 + \frac{k^2}{2\tau}\|U^{n+1}\|^2 + \frac{1}{2\tau}\Big(a_1 - \frac{k^2h^2}{6}\Big)\|U_{\bar{x}}^{n+1}\|^2 + \frac{1}{\tau}\Big(a_2 - \frac{h^2}{4}\Big)\|U_{\bar{t}}^{n+1}\|^2 \\
&+ \frac{1}{2\tau}\Big(b_1 - \frac{a_1h^2}{12}\Big)\|U_{\bar{x}\bar{x}}^{n+1}\|^2 + \frac{1}{\tau}\Big(b_2 - \frac{a_2h^2}{6}\Big)\|U_{\bar{x}\bar{t}}^{n+1}\|^2 - \frac{b_2h^2}{12\tau}\|U_{\bar{t}\bar{t}}^{n+1}\|^2 \\
&- \frac{ch}{6}\sum_{k=0}^{n}\Big\{\sum_{j=1}^{J-1}\Big[(U_{j+1}^k)^{2p}+4(U_j^k)^{2p}+(U_{j-1}^k)^{2p}\Big](U_j^{k+1}+U_j^k)_{\bar{t}}\Big\} \\
&= \frac{1}{\tau}\|V_{\bar{x}}^{n-\frac{1}{2}}\|^2 + \frac{k^2}{2\tau}\|U^{n-1}\|^2 + \frac{1}{2\tau}\Big(a_1 - \frac{k^2h^2}{6}\Big)\|U_{\bar{x}}^{n-1}\|^2 + \frac{1}{\tau}\Big(a_2 - \frac{h^2}{4}\Big)\|U_{\bar{t}}^{n}\|^2 \\
&+ \frac{1}{2\tau}\Big(b_1 - \frac{a_1h^2}{12}\Big)\|U_{\bar{x}\bar{x}}^{n-1}\|^2 + \frac{1}{\tau}\Big(b_2 - \frac{a_2h^2}{6}\Big)\|U_{\bar{x}\bar{t}}^{n}\|^2 - \frac{b_2h^2}{12\tau}\|U_{\bar{t}\bar{t}}^{n}\|^2 \\
&- \frac{ch}{6}\sum_{k=0}^{n-1}\Big\{\sum_{j=1}^{J-1}\Big[(U_{j+1}^k)^{2p}+4(U_j^k)^{2p}+(U_{j-1}^k)^{2p}\Big](U_j^{k+1}+U_j^k)_{\bar{t}}\Big\}.
\end{aligned}
\tag{41}
$$

Multiplying Equation (41) by τ, and subsequently adding the formula

$$
\frac{k^2}{2}\|U^n\|^2 + \frac{1}{2}\Big(a_1 - \frac{k^2h^2}{6}\Big)\|U_{\bar{x}}^n\|^2 + \frac{1}{2}\Big(b_1 - \frac{a_1h^2}{12}\Big)\|U_{\bar{x}\bar{x}}^n\|^2
$$

to both sides, we obtain

$$
\begin{aligned}
&\|V_{\bar{x}}^{n+\frac{1}{2}}\|^2 + \frac{k^2}{2}\Big(\|U^{n+1}\|^2 + \|U^n\|^2\Big) + \frac{1}{2}\Big(a_1 - \frac{k^2h^2}{6}\Big)\Big(\|U_{\bar{x}}^{n+1}\|^2 + \|U_{\bar{x}}^n\|^2\Big) \\
&+ \Big(a_2 - \frac{h^2}{4}\Big)\|U_{\bar{t}}^{n+1}\|^2 + \frac{1}{2}\Big(b_1 - \frac{a_1h^2}{12}\Big)\Big(\|U_{\bar{x}\bar{x}}^{n+1}\|^2 + \|U_{\bar{x}\bar{x}}^n\|^2\Big) + \Big(b_2 - \frac{a_2h^2}{6}\Big)\|U_{\bar{x}\bar{t}}^{n+1}\|^2 \\
&- \frac{b_2h^2}{12}\|U_{\bar{t}\bar{t}}^{n+1}\|^2 - \frac{ch\tau}{6}\sum_{k=0}^{n}\Big\{\sum_{j=1}^{J-1}\Big[(U_{j+1}^k)^{2p}+4(U_j^k)^{2p}+(U_{j-1}^k)^{2p}\Big](U_j^{k+1}+U_j^k)_{\bar{t}}\Big\} \\
&= \|V_{\bar{x}}^{n-\frac{1}{2}}\|^2 + \frac{k^2}{2}\Big(\|U^n\|^2 + \|U^{n-1}\|^2\Big) + \frac{1}{2}\Big(a_1 - \frac{k^2h^2}{6}\Big)\Big(\|U_{\bar{x}}^n\|^2 + \|U_{\bar{x}}^{n-1}\|^2\Big) \\
&+ \Big(a_2 - \frac{h^2}{4}\Big)\|U_{\bar{t}}^n\|^2 + \frac{1}{2}\Big(b_1 - \frac{a_1h^2}{12}\Big)\Big(\|U_{\bar{x}\bar{x}}^n\|^2 + \|U_{\bar{x}\bar{x}}^{n-1}\|^2\Big) + \Big(b_2 - \frac{a_2h^2}{6}\Big)\|U_{\bar{x}\bar{t}}^n\|^2 \\
&- \frac{b_2h^2}{12}\|U_{\bar{t}\bar{t}}^n\|^2 - \frac{ch\tau}{6}\sum_{k=0}^{n-1}\Big\{\sum_{j=1}^{J-1}\Big[(U_{j+1}^k)^{2p}+4(U_j^k)^{2p}+(U_{j-1}^k)^{2p}\Big](U_j^{k+1}+U_j^k)_{\bar{t}}\Big\};
\end{aligned}
\tag{42}
$$

that is, $E^n = E^{n-1} = \cdots = E^0$.

In a similar manner, by computing the inner product of Equation (19) with $V^{\frac{1}{2}} + V^{-\frac{1}{2}}$, we obtain

$$
\begin{aligned}
E^0 &= \|V_{\bar{x}}^{\frac{1}{2}}\|^2 + \frac{k^2}{2}\Big(\|U^1\|^2 + \|U^0\|^2\Big) + \frac{1}{2}\Big(a_1 - \frac{k^2h^2}{6}\Big)\Big(\|U_{\bar{x}}^1\|^2 + \|U_{\bar{x}}^0\|^2\Big) + \Big(a_2 - \frac{h^2}{4}\Big)\|U_{\bar{t}}^1\|^2 \\
&+ \frac{1}{2}\Big(b_1 - \frac{a_1h^2}{12}\Big)\Big(\|U_{\bar{x}\bar{x}}^1\|^2 + \|U_{\bar{x}\bar{x}}^0\|^2\Big) + \Big(b_2 - \frac{a_2h^2}{6}\Big)\|U_{\bar{x}\bar{t}}^1\|^2 - \frac{b_2h^2}{12}\|U_{\bar{t}\bar{t}}^1\|^2 \\
&- \frac{ch\tau}{6}\sum_{j=1}^{J-1}\Big\{\Big[(U_{j+1}^0)^{2p} + 4(U_j^0)^{2p} + (U_{j-1}^0)^{2p}\Big](U_j^1 + U_j^0)_{\bar{t}}\Big\} \\
&= \|V_{\bar{x}}^{-\frac{1}{2}}\|^2 + \frac{k^2}{2}\Big(\|U^0\|^2 + \|U^{-1}\|^2\Big) + \frac{1}{2}\Big(a_1 - \frac{k^2h^2}{6}\Big)\Big(\|U_{\bar{x}}^0\|^2 + \|U_{\bar{x}}^{-1}\|^2\Big) \\
&+ \Big(a_2 - \frac{h^2}{4}\Big)\|U_{\bar{t}}^0\|^2 + \frac{1}{2}\Big(b_1 - \frac{a_1h^2}{12}\Big)\Big(\|U_{\bar{x}\bar{x}}^0\|^2 + \|U_{\bar{x}\bar{x}}^{-1}\|^2\Big) \\
&+ \Big(b_2 - \frac{a_2h^2}{6}\Big)\|U_{\bar{x}\bar{t}}^0\|^2 - \frac{b_2h^2}{12}\|U_{\bar{t}\bar{t}}^0\|^2,
\end{aligned} \tag{43}
$$

where $U_j^{-1} = U_j^1 - 2\tau\psi(x_j), 0 \le j \le J$. This completes the proof. □

Note 1. In Theorem 1, setting

$$\eta(t) = \frac{2c}{2p+1}\int_{x_l}^{x_r} u^{2p+1}dx,$$

we have

$$\eta'(t) = 2c\int_{x_l}^{x_r} u^{2p}u_t dx.$$

By using Simpson's formula, one obtains

$$
\begin{aligned}
\eta'(t^k) &\approx \frac{2ch}{6}\sum_{j=1}^{J-1}\Big[(U_{j+1}^k)^{2p} + 4(U_j^k)^{2p} + (U_{j-1}^k)^{2p}\Big]u_t(x_j, t^k) \\
&\approx \frac{2ch}{6}\sum_{j=1}^{J-1}\Big[(U_{j+1}^k)^{2p} + 4(U_j^k)^{2p} + (U_{j-1}^k)^{2p}\Big]\Big[\frac{1}{2}(U_j^{k+1} + U_j^k)_{\bar{t}}\Big] \\
&\approx \frac{ch}{6}\sum_{j=1}^{J-1}\Big[(U_{j+1}^k)^{2p} + 4(U_j^k)^{2p} + (U_{j-1}^k)^{2p}\Big](U_j^{k+1} + U_j^k)_{\bar{t}}.
\end{aligned}
$$

We further obtain

$$
\begin{aligned}
\eta(t) &\approx \int_0^T \eta'(t^k)dt^k = \frac{ch}{6}\int_0^T\sum_{j=1}^{J-1}\Big[(U_{j+1}^k)^{2p} + 4(U_j^k)^{2p} + (U_{j-1}^k)^{2p}\Big](U_j^{k+1} + U_j^k)_{\bar{t}}dt^k \\
&\approx \frac{ch\tau}{6}\sum_{k=0}^{n}\Big\{\sum_{j=1}^{J-1}\Big[(U_{j+1}^k)^{2p} + 4(U_j^k)^{2p} + (U_{j-1}^k)^{2p}\Big](U_j^{k+1} + U_j^k)_{\bar{t}}\Big\},
\end{aligned}
$$

which is an approximation of the continuous term $\frac{2c}{2p+1}\int_{x_l}^{x_r} u^{2p+1}dx$ in Theorem 1.

Theorem 4. *Let $\varphi(x), \psi(x) \in H_0^2(\Omega)$; then, it follows that the solution U^n to Equations* (16)–(18) *satisfies*

$$\|U^n\| \le C, \ \ \|U_{\bar{x}}^n\| \le C, \ \ \|U^n\|_\infty \le C, \ \ 0 \le n \le N.$$

Proof. We employ mathematical induction to derive the estimate. From Equation (17), it can be deduced that $\|U^0\| \le C$ and $\|U^0_{\tilde{x}}\| \le C$. From Lemma 3, we obtain $\|U^0\|_\infty \le C$. We will proceed with the assumption that

$$\|U^m\| \le C, \ \ \|U^m_{\tilde{x}}\| \le C, \ \ \|U^m\|_\infty \le C,$$

where $m = 1, 2, \cdots, n$. By taking the inner product of Equation (16) with $U^n_{\hat{t}}$, we can express the components of the inner product as follows:

$$\begin{aligned}
&\langle U^n_{\bar{t}t}, U^n_{\hat{t}}\rangle - k^2\langle \bar{U}^n_{\tilde{x}\tilde{x}}, U^n_{\hat{t}}\rangle + \left(a_1 - \frac{k^2h^2}{6}\right)\langle \bar{U}^n_{\tilde{x}\tilde{x}\tilde{x}\tilde{x}}, U^n_{\hat{t}}\rangle - \left(a_2 - \frac{h^2}{4}\right)\langle U^n_{\tilde{x}\tilde{x}\bar{t}t}, U^n_{\hat{t}}\rangle \\
&- \left(b_1 - \frac{a_1h^2}{12}\right)\langle \bar{U}^n_{\tilde{x}\tilde{x}\tilde{x}\tilde{x}\tilde{x}\tilde{x}}, U^n_{\hat{t}}\rangle + \left(b_2 - \frac{a_2h^2}{6}\right)\langle U^n_{\tilde{x}\tilde{x}\tilde{x}\tilde{x}\bar{t}t}, U^n_{\hat{t}}\rangle + \frac{b_2h^2}{12}\langle U^n_{\tilde{x}\tilde{x}\tilde{x}\tilde{x}\tilde{x}\tilde{x}\bar{t}t}, U^n_{\hat{t}}\rangle \\
&= -c\langle (U^n)^{2p}_{\tilde{x}\tilde{x}}, U^n_{\hat{t}}\rangle - \frac{ch^2}{6}\langle (U^n)^{2p}_{\tilde{x}\tilde{x}\tilde{x}\tilde{x}}, U^n_{\hat{t}}\rangle.
\end{aligned} \tag{44}$$

By utilizing Lemmas 1 and 2 in conjunction with the Cauchy–Schwarz inequality [28,32], a simple calculation yields

$$s_1 = \langle U^n_{\bar{t}t}, U^n_{\hat{t}}\rangle = \frac{1}{2\tau}\langle U^{n+1}_{\bar{t}} - U^n_{\bar{t}}, U^{n+1}_{\bar{t}} + U^n_{\bar{t}}\rangle = \frac{1}{2\tau}(\|U^{n+1}_{\bar{t}}\|^2 - \|U^n_{\bar{t}}\|^2), \tag{45}$$

$$s_2 = \langle \bar{U}^n_{\tilde{x}\tilde{x}}, U^n_{\hat{t}}\rangle = \frac{1}{4\tau}\langle U^{n+1}_{\tilde{x}\tilde{x}} + U^{n-1}_{\tilde{x}\tilde{x}}, U^{n+1} - U^{n-1}\rangle = -\frac{1}{4\tau}(\|U^{n+1}_{\tilde{x}}\|^2 - \|U^{n-1}_{\tilde{x}}\|^2), \tag{46}$$

$$s_3 = \langle \bar{U}^n_{\tilde{x}\tilde{x}\tilde{x}\tilde{x}}, U^n_{\hat{t}}\rangle = \frac{1}{4\tau}\langle U^{n+1}_{\tilde{x}\tilde{x}\tilde{x}\tilde{x}} + U^{n-1}_{\tilde{x}\tilde{x}\tilde{x}\tilde{x}}, U^{n+1} - U^{n-1}\rangle = \frac{1}{4\tau}(\|U^{n+1}_{\tilde{x}\tilde{x}}\|^2 - \|U^{n-1}_{\tilde{x}\tilde{x}}\|^2), \tag{47}$$

$$s_4 = \langle U^n_{\tilde{x}\tilde{x}\bar{t}t}, U^n_{\hat{t}}\rangle = -\langle (U^n_{\tilde{x}})_{\bar{t}t}, (U^n_{\tilde{x}})_{\hat{t}}\rangle = -\frac{1}{2\tau}(\|U^{n+1}_{\tilde{x}\bar{t}}\|^2 - \|U^n_{\tilde{x}\bar{t}}\|^2), \tag{48}$$

$$\begin{aligned}
s_5 &= \langle \bar{U}^n_{\tilde{x}\tilde{x}\tilde{x}\tilde{x}\tilde{x}\tilde{x}}, U^n_{\hat{t}}\rangle = \frac{1}{4\tau}\langle U^{n+1}_{\tilde{x}\tilde{x}\tilde{x}\tilde{x}\tilde{x}\tilde{x}} + U^{n-1}_{\tilde{x}\tilde{x}\tilde{x}\tilde{x}\tilde{x}\tilde{x}}, U^{n+1} - U^{n-1}\rangle \\
&= -\frac{1}{4\tau}(\|U^{n+1}_{\tilde{x}\tilde{x}\tilde{x}}\|^2 - \|U^{n-1}_{\tilde{x}\tilde{x}\tilde{x}}\|^2),
\end{aligned} \tag{49}$$

$$s_6 = \langle U^n_{\tilde{x}\tilde{x}\tilde{x}\tilde{x}\bar{t}t}, U^n_{\hat{t}}\rangle = \langle (U^n_{\tilde{x}\tilde{x}})_{\bar{t}t}, (U^n_{\tilde{x}\tilde{x}})_{\hat{t}}\rangle = \frac{1}{2\tau}(\|U^{n+1}_{\tilde{x}\tilde{x}\bar{t}}\|^2 - \|U^n_{\tilde{x}\tilde{x}\bar{t}}\|^2), \tag{50}$$

$$s_7 = \langle U^n_{\tilde{x}\tilde{x}\tilde{x}\tilde{x}\tilde{x}\tilde{x}\bar{t}t}, U^n_{\hat{t}}\rangle = -\langle (U^n_{\tilde{x}\tilde{x}\tilde{x}})_{\bar{t}t}, (U^n_{\tilde{x}\tilde{x}\tilde{x}})_{\hat{t}}\rangle = -\frac{1}{2\tau}(\|U^{n+1}_{\tilde{x}\tilde{x}\tilde{x}\bar{t}}\|^2 - \|U^n_{\tilde{x}\tilde{x}\tilde{x}\bar{t}}\|^2), \tag{51}$$

$$\begin{aligned}
s_8 &= \langle (U^n)^{2p}_{\tilde{x}\tilde{x}}, U^n_{\hat{t}}\rangle = -\langle (U^n)^{2p}_{\tilde{x}}, U^n_{\tilde{x}\hat{t}}\rangle = -h\sum_{j=1}^{J-1}\left[(U^n_j)^{2p}_{\tilde{x}}(U^n_j)_{\tilde{x}\hat{t}}\right] \\
&= -h\sum_{j=1}^{J-1}\left[(U^n_j)^{2p-1}_{\tilde{x}}U^n_{j+1} + (U^n_j)^{2p-1}(U^n_j)_{\tilde{x}}\right](U^n_j)_{\tilde{x}\hat{t}} \\
&\le h\|U^n_{\tilde{x}}\|^{2p-1}_\infty\sum_{j=1}^{J-1}\left[|U^n_{j+1}|\cdot|(U^n_j)_{\tilde{x}\hat{t}}|\right] + h\|U^n\|^{2p-1}_\infty\sum_{j=1}^{J-1}\left[|(U^n_j)_{\tilde{x}}|\cdot|(U^n_j)_{\tilde{x}\hat{t}}|\right] \\
&\le C(\|U^n\|^2 + \|U^n_{\tilde{x}}\|^2 + \|U^n_{\tilde{x}\hat{t}}\|^2).
\end{aligned}$$

Since

$$U^n_{\tilde{x}\hat{t}} = \frac{1}{2}(U^n_{\tilde{x}\bar{t}} + U^{n+1}_{\tilde{x}\bar{t}}), \ \ \|U^n\| \le C\|U^n_{\tilde{x}}\|,$$

then we finally have

$$s_8 = \langle (U^n)^{2p}_{\tilde{x}\tilde{x}}, U^n_{\hat{t}}\rangle \le C(\|U^n_{\tilde{x}}\|^2 + \|U^n_{\tilde{x}\bar{t}}\|^2 + \|U^{n+1}_{\tilde{x}\bar{t}}\|^2). \tag{52}$$

Similarly, we have

$$\begin{aligned}
s_9 &= h^2\langle (U^n)^{2p}_{\bar{x}\bar{x}\bar{x}\bar{x}}, U^n_{\hat{t}}\rangle = h^2\langle (U^n)^{2p}_{\bar{x}\bar{x}}, U^n_{\bar{x}\bar{x}\hat{t}}\rangle = h^3\sum_{j=1}^{J-1}\left[(U^n_j)^{2p}_{\bar{x}\bar{x}}(U^n_j)_{\bar{x}\bar{x}\hat{t}}\right] \\
&= h\sum_{j=1}^{J-1}\left[(U^n_{j+1})^{2p} - 2(U^n_j)^{2p} + (U^n_{j-1})^{2p}\right](U^n_j)_{\bar{x}\bar{x}\hat{t}} \le C(\|U^n\|^2 + \|U^n_{\bar{x}\bar{x}\hat{t}}\|^2) \\
&\le C(\|U^n_{\bar{x}}\|^2 + \|U^n_{\bar{x}\bar{x}\bar{t}}\|^2 + \|U^{n+1}_{\bar{x}\bar{x}\bar{t}}\|^2).
\end{aligned} \tag{53}$$

By substituting Equations (45)–(53) into Equation (44), we obtain

$$\begin{aligned}
&\frac{1}{2}(\|U^{n+1}_{\bar{t}}\|^2 - \|U^n_{\bar{t}}\|^2) + \frac{k^2}{4}(\|U^{n+1}_{\bar{x}}\|^2 - \|U^{n-1}_{\bar{x}}\|^2) + \frac{1}{4}\left(a_1 - \frac{k^2h^2}{6}\right)(\|U^{n+1}_{\bar{x}\bar{x}}\|^2 - \|U^{n-1}_{\bar{x}\bar{x}}\|^2) \\
&+ \frac{1}{2}\left(a_2 - \frac{h^2}{4}\right)(\|U^{n+1}_{\bar{x}\bar{t}}\|^2 - \|U^n_{\bar{x}\bar{t}}\|^2) + \frac{1}{4}\left(b_1 - \frac{a_1h^2}{12}\right)(\|U^{n+1}_{\bar{x}\bar{x}\bar{x}}\|^2 - \|U^{n-1}_{\bar{x}\bar{x}\bar{x}}\|^2) \\
&+ \frac{1}{2}\left(b_2 - \frac{a_2h^2}{6}\right)(\|U^{n+1}_{\bar{x}\bar{x}\bar{t}}\|^2 - \|U^n_{\bar{x}\bar{x}\bar{t}}\|^2) - \frac{b_2h^2}{24}(\|U^{n+1}_{\bar{x}\bar{x}\bar{x}\bar{t}}\|^2 - \|U^n_{\bar{x}\bar{x}\bar{x}\bar{t}}\|^2) \\
&\le C\tau(\|U^n_{\bar{x}}\|^2 + \|U^n_{\bar{x}\bar{t}}\|^2 + \|U^{n+1}_{\bar{x}\bar{t}}\|^2 + \|U^n_{\bar{x}\bar{x}\bar{t}}\|^2 + \|U^{n+1}_{\bar{x}\bar{x}\bar{t}}\|^2).
\end{aligned} \tag{54}$$

Defining

$$\begin{aligned}
B^n &= \frac{1}{2}\|U^{n+1}_{\bar{t}}\|^2 + \frac{k^2}{4}(\|U^{n+1}_{\bar{x}}\|^2 + \|U^n_{\bar{x}}\|^2) + \frac{1}{4}\left(a_1 - \frac{k^2h^2}{6}\right)(\|U^{n+1}_{\bar{x}\bar{x}}\|^2 + \|U^n_{\bar{x}\bar{x}}\|^2) \\
&+ \frac{1}{2}\left(a_2 - \frac{h^2}{4}\right)\|U^{n+1}_{\bar{x}\bar{t}}\|^2 + \frac{1}{4}\left(b_1 - \frac{a_1h^2}{12}\right)(\|U^{n+1}_{\bar{x}\bar{x}\bar{x}}\|^2 + \|U^n_{\bar{x}\bar{x}\bar{x}}\|^2) \\
&+ \frac{1}{2}\left(b_2 - \frac{a_2h^2}{6}\right)\|U^{n+1}_{\bar{x}\bar{x}\bar{t}}\|^2 - \frac{b_2h^2}{24}\|U^{n+1}_{\bar{x}\bar{x}\bar{x}\bar{t}}\|^2,
\end{aligned}$$

then, based on Lemma 4, we can conclude that

$$\begin{aligned}
B^n &\ge \frac{1}{2}\|U^{n+1}_{\bar{t}}\|^2 + \frac{k^2}{4}(\|U^{n+1}_{\bar{x}}\|^2 + \|U^n_{\bar{x}}\|^2) + \frac{a_1}{4}(\|U^{n+1}_{\bar{x}\bar{x}}\|^2 + \|U^n_{\bar{x}\bar{x}}\|^2) + \frac{a_2}{2}\|U^{n+1}_{\bar{x}\bar{t}}\|^2 \\
&+ \frac{b_1}{4}(\|U^{n+1}_{\bar{x}\bar{x}\bar{x}}\|^2 + \|U^n_{\bar{x}\bar{x}\bar{x}}\|^2) + \frac{b_2}{2}\|U^{n+1}_{\bar{x}\bar{x}\bar{t}}\|^2 - \frac{k^2}{6}(\|U^{n+1}_{\bar{x}}\|^2 + \|U^n_{\bar{x}}\|^2) - \frac{1}{2}\|U^{n+1}_{\bar{t}}\|^2 \\
&- \frac{a_1}{12}(\|U^{n+1}_{\bar{x}\bar{x}}\|^2 + \|U^n_{\bar{x}\bar{x}}\|^2) - \frac{a_2}{3}\|U^{n+1}_{\bar{x}\bar{t}}\|^2 - \frac{b_2}{6}\|U^{n+1}_{\bar{x}\bar{x}\bar{t}}\|^2 \\
&= \frac{k^2}{12}(\|U^{n+1}_{\bar{x}}\|^2 + \|U^n_{\bar{x}}\|^2) + \frac{a_1}{6}(\|U^{n+1}_{\bar{x}\bar{x}}\|^2 + \|U^n_{\bar{x}\bar{x}}\|^2) + \frac{a_2}{6}\|U^{n+1}_{\bar{x}\bar{t}}\|^2 \\
&+ \frac{b_1}{4}(\|U^{n+1}_{\bar{x}\bar{x}\bar{x}}\|^2 + \|U^n_{\bar{x}\bar{x}\bar{x}}\|^2) + \frac{b_2}{3}\|U^{n+1}_{\bar{x}\bar{x}\bar{t}}\|^2.
\end{aligned} \tag{55}$$

Consequently, Equation (54) can be reformulated as follows

$$B^n - B^{n-1} \le C\tau(B^n + B^{n-1}),\ \ 1 \le n \le N-1.$$

Assuming that τ is sufficiently small, specifically $\tau \le \frac{k_0-2}{k_0C}$ with $k_0 \ge 2$, we further obtain

$$B^n \le \left(\frac{1+\tau C}{1-\tau C}\right)B^{n-1} \le (1+\tau k_0C)B^{n-1} \le \exp(k_0TC)B^0.$$

This indicates that B^n is bounded, which leads to the conclusions that $\|U^{n+1}_{\bar{x}}\| \le C$ as derived from Equation (55). By applying Lemma 3, we further establish that $\|U^{n+1}\| \le C$, and $\|U^{n+1}\|_\infty \le C$. This completes the proof. □

4. Solvability, Convergence, and Stability

In this section, we examine the solvability, convergence, and stability of the previously discussed scheme.

Theorem 5. *The difference scheme outlined in Equations* (16)–(18) *exhibits a unique solution.*

Proof. From Equation (17), we know that U^0 has a unique solution. To demonstrate the unique solvability of U^1, we will analyze the homogeneous version of Equation (19) as detailed below:

$$\frac{1}{\tau^2}U^1 - \frac{k^2}{2}U^1_{\bar{x}x} + \frac{1}{2}\left(a_1 - \frac{k^2h^2}{6}\right)U^1_{\bar{x}x\bar{x}x} - \frac{1}{\tau^2}\left(a_2 - \frac{h^2}{4}\right)U^1_{\bar{x}x} - \frac{1}{2}\left(b_1 - \frac{a_1h^2}{12}\right)U^1_{\bar{x}x\bar{x}x\bar{x}x} + \frac{1}{\tau^2}\left(b_2 - \frac{a_2h^2}{6}\right)U^1_{\bar{x}x\bar{x}x} + \frac{b_2h^2}{12\tau^2}U^1_{\bar{x}x\bar{x}x\bar{x}x} = 0. \quad (56)$$

By computing the inner product of Equation (56) with U^1, we obtain

$$\|U^1\|^2 + \left(\frac{\tau^2k^2}{2} + a_2 - \frac{h^2}{4}\right)\|U^1_{\bar{x}}\|^2 + \left[\frac{\tau^2}{2}\left(a_1 - \frac{k^2h^2}{6}\right) + \left(b_2 - \frac{a_2h^2}{6}\right)\right]\|U^1_{\bar{x}x}\|^2 + \left[\frac{\tau^2}{2}\left(b_1 - \frac{a_1h^2}{12}\right) - \frac{b_2h^2}{12}\right]\|U^1_{\bar{x}x\bar{x}}\|^2 = 0. \quad (57)$$

Furthermore, from Lemma 4, one obtains

$$\begin{aligned}
&\|U^1\|^2 + \left(\frac{\tau^2k^2}{2} + a_2\right)\|U^1_{\bar{x}}\|^2 + \left(\frac{\tau^2}{2}a_1 + b_2\right)\|U^1_{\bar{x}x}\|^2 + \frac{\tau^2b_1}{2}\|U^1_{\bar{x}x\bar{x}}\|^2 \\
&= \frac{h^2}{4}\|U^1_{\bar{x}}\|^2 + \left(\frac{\tau^2k^2h^2}{12} + \frac{a_2h^2}{6}\right)\|U^1_{\bar{x}x}\|^2 + \left(\frac{a_1\tau^2h^2}{24} + \frac{b_2h^2}{12}\right)\|U^1_{\bar{x}x\bar{x}}\|^2 \\
&\leq \|U^1\|^2 + \left(\frac{\tau^2k^2}{3} + \frac{2a_2}{3}\right)\|U^1_{\bar{x}}\|^2 + \left(\frac{a_1\tau^2}{6} + \frac{b_2}{3}\right)\|U^1_{\bar{x}x}\|^2,
\end{aligned}$$

which yields

$$\left(\frac{\tau^2k^2}{6} + \frac{a_2}{3}\right)\|U^1_{\bar{x}}\|^2 + \left(\frac{a_1\tau^2}{3} + \frac{2b_2}{3}\right)\|U^1_{\bar{x}x}\|^2 + \frac{\tau^2b_1}{2}\|U^1_{\bar{x}x\bar{x}}\|^2 \leq 0. \quad (58)$$

Considering the parameters $a_1 \geq 0$, $a_2 \geq 0$, $b_1 \geq 0$, and $b_2 \geq 0$, we then obtain $\|U^1_{\bar{x}}\| = 0$, $\|U^1_{\bar{x}x}\| = 0$, $\|U^1_{\bar{x}x\bar{x}}\| = 0$. We further obtain from Equation (57) that $\|U^1\| = 0$. Consequently, Equation (19) admits only the trivial solution, indicating that U^1 is uniquely determined.

We then assume that $U^0, U^1, U^2, \ldots, U^n$ are uniquely determined, where $n \leq N-1$. By examining the homogeneous version of Equation (16) for U^{n+1}, we obtain

$$\frac{1}{\tau^2}U^{n+1} - \frac{k^2}{2}U^{n+1}_{\bar{x}x} + \frac{1}{2}\left(a_1 - \frac{k^2h^2}{6}\right)U^{n+1}_{\bar{x}x\bar{x}x} - \frac{1}{\tau^2}\left(a_2 - \frac{h^2}{4}\right)U^{n+1}_{\bar{x}x} - \frac{1}{2}\left(b_1 - \frac{a_1h^2}{12}\right)U^{n+1}_{\bar{x}x\bar{x}x\bar{x}x} + \frac{1}{\tau^2}\left(b_2 - \frac{a_2h^2}{6}\right)U^{n+1}_{\bar{x}x\bar{x}x} + \frac{b_2h^2}{12\tau^2}U^{n+1}_{\bar{x}x\bar{x}x\bar{x}x} = 0. \quad (59)$$

Similarly, by computing the inner product of Equation (59) with U^{n+1}, we obtain

$$\|U^{n+1}\|^2 + \left(\frac{\tau^2k^2}{2} + a_2 - \frac{h^2}{4}\right)\|U^{n+1}_{\bar{x}}\|^2 + \left[\frac{\tau^2}{2}\left(a_1 - \frac{k^2h^2}{6}\right) + \left(b_2 - \frac{a_2h^2}{6}\right)\right]\|U^{n+1}_{\bar{x}x}\|^2 + \left[\frac{\tau^2}{2}\left(b_1 - \frac{a_1h^2}{12}\right) - \frac{b_2h^2}{12}\right]\|U^{n+1}_{\bar{x}x\bar{x}}\|^2 = 0, \quad (60)$$

which yields $\|U^{n+1}\|=0$, $\|U_{\tilde{x}}^{n+1}\|=0$, $\|U_{\tilde{x}\bar{x}}^{n+1}\|=0$, $\|U_{\tilde{x}\bar{x}\tilde{x}}^{n+1}\|=0$. The difference scheme (16)–(18) is demonstrated to possess a unique solution through the application of the method of induction. This completes the proof. □

Theorem 6. *Assuming that $\varphi(x),\psi(x)\in H_0^2(\Omega)$ and $u\in C_{x,t}^{10,4}([x_l,x_r]\times(0,T])$, it can be concluded that the solution U^n to Equations (16)–(18) converges to the solution u^n of the problem defined by Equations (1)–(3). Furthermore, the rate of convergence is characterized by $O(\tau^2+h^4)$ in both the $\|\cdot\|$ and $\|\cdot\|_\infty$ norms. Specifically, this implies that*

$$\|u^n-U^n\|\le O(\tau^2+h^4),\ \ \|u^n-U^n\|_\infty\le O(\tau^2+h^4).$$

Proof. The truncation errors associated with Equations (16)–(18) can be expressed as follows:

$$\begin{aligned}&(u_j^n)_{\bar{t}t}-k^2(\bar{u}_j^n)_{\tilde{x}\bar{x}}+\Big(a_1-\frac{k^2h^2}{6}\Big)(\bar{u}_j^n)_{\tilde{x}\bar{x}\tilde{x}\bar{x}}-\Big(a_2-\frac{h^2}{4}\Big)(u_j^n)_{\tilde{x}\bar{x}\bar{t}t}-\Big(b_1-\frac{a_1h^2}{12}\Big)(\bar{u}_j^n)_{\tilde{x}\bar{x}\tilde{x}\bar{x}\tilde{x}\bar{x}}\\&+\Big(b_2-\frac{a_2h^2}{6}\Big)(u_j^n)_{\tilde{x}\bar{x}\tilde{x}\bar{x}\bar{t}t}+\frac{b_2h^2}{12}(u_j^n)_{\tilde{x}\bar{x}\tilde{x}\bar{x}\tilde{x}\bar{x}\bar{t}t}+c[(u_j^n)^{2p}]_{\tilde{x}\bar{x}}+\frac{ch^2}{6}[(u_j^n)^{2p}]_{\tilde{x}\bar{x}\tilde{x}\bar{x}}=r_j^n,\end{aligned}\tag{61}$$

$$u_j^0=\varphi(x_j),\ \ (u_j^0)_{\hat{t}}=\psi(x_j),\ \ 0\le j\le J,\tag{62}$$

$$\begin{aligned}&u_0^n=0,\ \ \frac{4}{3}(u_0^n)_{\hat{x}}-\frac{1}{3}(u_0^n)_{\ddot{x}}=0,\ \ \frac{4}{3}(u_0^n)_{\tilde{x}\bar{x}}-\frac{1}{3}(u_0^n)_{\hat{x}\hat{x}}=0,\\&u_J^n=0,\ \ \frac{4}{3}(u_J^n)_{\hat{x}}-\frac{1}{3}(u_J^n)_{\ddot{x}}=0,\ \ \frac{4}{3}(u_J^n)_{\tilde{x}\bar{x}}-\frac{1}{3}(u_J^n)_{\hat{x}\hat{x}}=0,\ \ 0\le n\le N.\end{aligned}\tag{63}$$

Through the application of the Taylor expansion, it can be demonstrated that the inequality $|r_j^n|\le O(\tau^2+h^4)$ is valid as both τ and h approach zero for $n\ge 0$. Letting $e_j^n=u_j^n-U_j^n$ and subtracting Equations (16)–(18) from Equations (61)–(63), we obtain

$$\begin{aligned}r_j^n=&(e_j^n)_{\bar{t}t}-k^2(\bar{e}_j^n)_{\tilde{x}\bar{x}}+\Big(a_1-\frac{k^2h^2}{6}\Big)(\bar{e}_j^n)_{\tilde{x}\bar{x}\tilde{x}\bar{x}}-\Big(a_2-\frac{h^2}{4}\Big)(e_j^n)_{\tilde{x}\bar{x}\bar{t}t}-\Big(b_1-\frac{a_1h^2}{12}\Big)(\bar{e}_j^n)_{\tilde{x}\bar{x}\tilde{x}\bar{x}\tilde{x}\bar{x}}\\&+\Big(b_2-\frac{a_2h^2}{6}\Big)(e_j^n)_{\tilde{x}\bar{x}\tilde{x}\bar{x}\bar{t}t}+\frac{b_2h^2}{12}(u_j^e)_{\tilde{x}\bar{x}\tilde{x}\bar{x}\tilde{x}\bar{x}\bar{t}t}+c[(u_j^n)^{2p}-(U_j^n)^{2p}]_{\tilde{x}\bar{x}}\\&+\frac{ch^2}{6}[(u_j^n)^{2p}-(U_j^n)^{2p}]_{\tilde{x}\bar{x}\tilde{x}\bar{x}},\end{aligned}\tag{64}$$

$$e_j^0=0,\ \ (e_j^0)_{\hat{t}}=0,\ \ 0\le j\le J,\tag{65}$$

$$\begin{aligned}&e_0^n=0,\ \ \frac{4}{3}(e_0^n)_{\hat{x}}-\frac{1}{3}(e_0^n)_{\ddot{x}}=0,\ \ \frac{4}{3}(e_0^n)_{\tilde{x}\bar{x}}-\frac{1}{3}(e_0^n)_{\hat{x}\hat{x}}=0,\\&e_J^n=0,\ \ \frac{4}{3}(e_J^n)_{\hat{x}}-\frac{1}{3}(e_J^n)_{\ddot{x}}=0,\ \ \frac{4}{3}(e_J^n)_{\tilde{x}\bar{x}}-\frac{1}{3}(e_J^n)_{\hat{x}\hat{x}}=0,\ \ 0\le n\le N.\end{aligned}\tag{66}$$

By computing the inner product of Equation (64) with $(e_j^n)_{\hat{t}}$, one obtains

$$\begin{aligned}&\frac{1}{2}(\|e_{\bar{t}}^{n+1}\|^2-\|e_{\bar{t}}^{n}\|^2)+\frac{k^2}{4}(\|e_{\tilde{x}}^{n+1}\|^2-\|e_{\tilde{x}}^{n-1}\|^2)+\frac{1}{4}\Big(a_1-\frac{k^2h^2}{6}\Big)(\|e_{\tilde{x}\bar{x}}^{n+1}\|^2-\|e_{\tilde{x}\bar{x}}^{n-1}\|^2)\\&+\frac{1}{2}\Big(a_2-\frac{h^2}{4}\Big)(\|e_{\tilde{x}\bar{t}}^{n+1}\|^2-\|e_{\tilde{x}\bar{t}}^{n}\|^2)+\frac{1}{4}\Big(b_1-\frac{a_1h^2}{12}\Big)(\|e_{\tilde{x}\bar{x}\tilde{x}}^{n+1}\|^2-\|e_{\tilde{x}\bar{x}\tilde{x}}^{n-1}\|^2)\\&+\frac{1}{2}\Big(b_2-\frac{a_2h^2}{6}\Big)(\|e_{\tilde{x}\bar{x}\bar{t}}^{n+1}\|^2-\|e_{\tilde{x}\bar{x}\bar{t}}^{n}\|^2)-\frac{b_2h^2}{24}(\|e_{\tilde{x}\bar{x}\tilde{x}\bar{t}}^{n+1}\|^2-\|e_{\tilde{x}\bar{x}\tilde{x}\bar{t}}^{n}\|^2)\\&=\tau\langle r^n,e_{\hat{t}}^n\rangle-c\tau\langle[(u^n)^{2p}-(U^n)^{2p}]_{\tilde{x}\bar{x}},e_{\hat{t}}^n\rangle-\frac{c\tau h^2}{6}\langle[(u^n)^{2p}-(U^n)^{2p}]_{\tilde{x}\bar{x}\tilde{x}\bar{x}},e_{\hat{t}}^n\rangle.\end{aligned}\tag{67}$$

According to Lemma 1 and the Cauchy–Schwarz inequality [28,32], it can be concluded that

$$\begin{aligned}
&\langle|[(u^n)^{2p}-(U^n)^{2p}]_{\tilde{x}\bar{x}},e^n_{\hat{t}}\rangle|=\langle(u^n)^{2p}-(U^n)^{2p},e^n_{\tilde{x}\bar{x}\hat{t}}\rangle\\
&=h\sum_{j=1}^{J-1}[(u^n_j)^{2p}-(U^n_j)^{2p}](e^n_j)_{\tilde{x}\bar{x}\hat{t}}=h\sum_{j=1}^{J-1}\left\{e^n_j\left[\sum_{k=0}^{2p-1}(u^n_j)^{2p-1-k}(U^n_j)^k\right](e^n_j)_{\tilde{x}\bar{x}\hat{t}}\right\}\\
&\le C(\|e^n_{\tilde{x}}\|^2+\|e^{n+1}_{\tilde{x}\bar{x}\bar{t}}\|^2+\|e^n_{\tilde{x}\bar{x}\bar{t}}\|^2).
\end{aligned}\tag{68}$$

Similarly, we have

$$\begin{aligned}
&h^2\langle[(u^n)^{2p}-(U^n)^{2p}]_{\tilde{x}\bar{x}\tilde{x}\bar{x}},e^n_{\hat{t}}\rangle=h^2\langle[(u^n)^{2p}-(U^n)^{2p}]_{\tilde{x}\bar{x}},e^n_{\tilde{x}\bar{x}\hat{t}}\rangle\\
&=h\sum_{j=1}^{J-1}\left\{\left[(u^n_{j+1})^{2p}-(U^n_{j+1})^{2p}\right]-2\left[(u^n_j)^{2p}-(U^n_j)^{2p}\right]+\left[(u^n_{j-1})^{2p}-(U^n_{j-1})^{2p}\right]\right\}(U^n_j)_{\tilde{x}\bar{x}\hat{t}}\\
&\le C(\|e^n_{\tilde{x}}\|^2+\|e^n_{\tilde{x}\bar{x}\bar{t}}\|^2+\|e^{n+1}_{\tilde{x}\bar{x}\bar{t}}\|^2).
\end{aligned}\tag{69}$$

Furthermore, we obtain

$$\langle r^n,e^n_{\hat{t}}\rangle=\langle r^n,\frac{1}{2}(e^{n+1}_{\bar{t}}+e^n_{\bar{t}})\rangle\le\frac{1}{2}\|r^n\|^2+\frac{1}{4}(\|e^{n+1}_{\bar{t}}\|^2+\|e^n_{\bar{t}}\|^2).\tag{70}$$

Substituting Equations (68)–(70) into Equation (67), we obtain

$$\begin{aligned}
&\frac{1}{2}(\|e^{n+1}_{\bar{t}}\|^2-\|e^n_{\bar{t}}\|^2)+\frac{k^2}{4}(\|e^{n+1}_{\tilde{x}}\|^2-\|e^{n-1}_{\tilde{x}}\|^2)+\frac{1}{4}\left(a_1-\frac{k^2h^2}{6}\right)(\|e^{n+1}_{\tilde{x}\bar{x}}\|^2-\|e^{n-1}_{\tilde{x}\bar{x}}\|^2)\\
&+\frac{1}{2}\left(a_2-\frac{h^2}{4}\right)(\|e^{n+1}_{\tilde{x}\bar{t}}\|^2-\|e^n_{\tilde{x}\bar{t}}\|^2)+\frac{1}{4}\left(b_1-\frac{a_1h^2}{12}\right)(\|e^{n+1}_{\tilde{x}\bar{x}\tilde{x}}\|^2-\|e^{n-1}_{\tilde{x}\bar{x}\tilde{x}}\|^2)\\
&+\frac{1}{2}\left(b_2-\frac{a_2h^2}{6}\right)(\|e^{n+1}_{\tilde{x}\bar{x}\bar{t}}\|^2-\|e^n_{\tilde{x}\bar{x}\bar{t}}\|^2)-\frac{b_2h^2}{24}(\|e^{n+1}_{\tilde{x}\bar{x}\tilde{x}\bar{t}}\|^2-\|e^n_{\tilde{x}\bar{x}\tilde{x}\bar{t}}\|^2)\\
&\le\frac{\tau}{2}\|r^n\|^2+C\tau(\|e^n_{\tilde{x}}\|^2+\|e^{n+1}_{\tilde{x}\bar{x}\bar{t}}\|^2+\|e^n_{\tilde{x}\bar{x}\bar{t}}\|^2+\|e^{n+1}_{\bar{t}}\|^2+\|e^n_{\bar{t}}\|^2).
\end{aligned}\tag{71}$$

Setting

$$\begin{aligned}
D^n=&\frac{1}{2}\|e^{n+1}_{\bar{t}}\|^2+\frac{k^2}{4}(\|e^{n+1}_{\tilde{x}}\|^2+\|e^n_{\tilde{x}}\|^2)+\frac{1}{4}\left(a_1-\frac{k^2h^2}{6}\right)(\|e^{n+1}_{\tilde{x}\bar{x}}\|^2+\|e^n_{\tilde{x}\bar{x}}\|^2)\\
&+\frac{1}{2}\left(a_2-\frac{h^2}{4}\right)\|e^{n+1}_{\tilde{x}\bar{t}}\|^2+\frac{1}{4}\left(b_1-\frac{a_1h^2}{12}\right)(\|e^{n+1}_{\tilde{x}\bar{x}\tilde{x}}\|^2+\|e^n_{\tilde{x}\bar{x}\tilde{x}}\|^2)\\
&+\frac{1}{2}\left(b_2-\frac{a_2h^2}{6}\right)\|e^{n+1}_{\tilde{x}\bar{x}\bar{t}}\|^2-\frac{b_2h^2}{24}\|e^{n+1}_{\tilde{x}\bar{x}\tilde{x}\bar{t}}\|^2,
\end{aligned}\tag{72}$$

then, from Lemma 4, we have

$$\begin{aligned}
D^n\ge&\frac{1}{2}\|e^{n+1}_{\bar{t}}\|^2+\frac{k^2}{4}(\|e^{n+1}_{\tilde{x}}\|^2+\|e^n_{\tilde{x}}\|^2)+\frac{a_1}{4}(\|e^{n+1}_{\tilde{x}\bar{x}}\|^2+\|e^n_{\tilde{x}\bar{x}}\|^2)+\frac{a_2}{2}\|e^{n+1}_{\tilde{x}\bar{t}}\|^2\\
&+\frac{b_1}{4}(\|e^{n+1}_{\tilde{x}\bar{x}\tilde{x}}\|^2+\|e^n_{\tilde{x}\bar{x}\tilde{x}}\|^2)+\frac{b_2}{2}\|e^{n+1}_{\tilde{x}\bar{x}\bar{t}}\|^2-\frac{k^2}{6}(\|e^{n+1}_{\tilde{x}}\|^2+\|e^n_{\tilde{x}}\|^2)-\frac{1}{2}\|e^{n+1}_{\bar{t}}\|^2\\
&-\frac{a_1}{12}(\|e^{n+1}_{\tilde{x}\bar{x}}\|^2+\|e^n_{\tilde{x}\bar{x}}\|^2)-\frac{a_2}{3}\|e^{n+1}_{\tilde{x}\bar{t}}\|^2-\frac{b_2}{6}\|e^{n+1}_{\tilde{x}\bar{x}\bar{t}}\|^2\\
=&\frac{k^2}{12}(\|e^{n+1}_{\tilde{x}}\|^2+\|e^n_{\tilde{x}}\|^2)+\frac{a_1}{6}(\|e^{n+1}_{\tilde{x}\bar{x}}\|^2+\|e^n_{\tilde{x}\bar{x}}\|^2)+\frac{a_2}{6}\|e^{n+1}_{\tilde{x}\bar{t}}\|^2\\
&+\frac{b_1}{4}(\|e^{n+1}_{\tilde{x}\bar{x}\tilde{x}}\|^2+\|e^n_{\tilde{x}\bar{x}\tilde{x}}\|^2)+\frac{b_2}{3}\|e^{n+1}_{\tilde{x}\bar{x}\bar{t}}\|^2.
\end{aligned}\tag{73}$$

Consequently, from Equations (71) and (72), we obtain

$$D^n - D^{n-1} \le \frac{\tau}{2}\|r^n\|^2 + C\tau(D^n + D^{n-1}).$$

That is

$$(1 - C\tau)(D^n - D^{n-1}) \le \frac{\tau}{2}\|r^n\|^2 + 2C\tau D^{n-1}.$$

Therefore, if τ is sufficiently small such that $1 - C\tau = 1/c_0 > 0$, then

$$D^n - D^{n-1} \le \frac{c_0}{2}\tau\|r^n\|^2 + 2Cc_0\tau D^{n-1}. \tag{74}$$

By summing Equation (74) from 1 to n, we obtain

$$D^n \le D^0 + \frac{c_0}{2}\tau\sum_{l=1}^{n}\|r^l\|^2 + 2Cc_0\tau\sum_{l=1}^{n}D^{l-1}.$$

It is essential to note that

$$\tau\sum_{l=1}^{n}\|r^l\|^2 \le n\tau \max_{1\le l\le n}\|r^l\|^2 \le T\cdot O(\tau^2+h^4)^2,\;\; D^0 \le O(\tau^2+h^4)^2.$$

We then obtain from the Discrete Gronwall's inequality [33,34] that $D^n \le O(\tau^2+h^4)^2$. This leads to the conclusion that $\|e_x^n\| \le O(\tau^2+h^4)$ and $\|e_{\bar{x}}^n\| \le O(\tau^2+h^4)$. Ultimately, we utilize Lemma 3 to demonstrate that $\|e^n\| \le O(\tau^2+h^4)$ and $\|e^n\|_\infty \le O(\tau^2+h^4)$. This completes the proof. □

We subsequently obtain the following theorem.

Theorem 7. *Under the conditions specified in Theorem 6, the solution U^n of the Equations* (16)–(18) *exhibits stability. The analysis is conducted with respect to both the $\|\cdot\|$ norm and the $\|\cdot\|_\infty$ norm.*

5. Numerical Experiments

In this section, we conduct a series of numerical experiments aimed at assessing the efficacy and accuracy of the theoretical analysis presented in the preceding sections. Following this, we will quantify the corresponding errors utilizing the designated error norms [31,33,35]:

$$\|e^n\| = \left[h\sum_{j=1}^{J-1}|u_j^n - U_j^n|^2\right]^{\frac{1}{2}},\;\; \|e^n\|_\infty = \max_{1\le j\le J-1}|u_j^n - U_j^n|.$$

Example 1. *Consider the parameters*

$$k=1,\; a_1=0,\; a_2=1,\; b_1=0,\; b_2=0,\; c=-1,\; p=1,$$

which is the following improved Boussinesq equation [2,31]

$$u_{tt} - u_{xx} - u_{xxtt} - (u^2)_{xx} = 0, \tag{75}$$

and the solitary wave solution is expressed as follows

$$u(x,t) = A\,\mathrm{sech}^2\left[\frac{1}{c_0}\sqrt{\frac{A}{6}}\left(x - x_0 - c_0 t\right)\right],\;\; c_0 = \pm\sqrt{1+\frac{2}{3}A}, \tag{76}$$

where A, x_0, and c_0 are provided constants. The initial conditions may be established by evaluating Equation (76) and its derivative for t at the point $t = 0$:

$$\varphi(x) = A\,\mathrm{sech}^2\left[\frac{1}{c_0}\sqrt{\frac{A}{6}}\left(x - x_0\right)\right], \tag{77}$$

$$\psi(x) = 2A\sqrt{\frac{A}{6}}\,\mathrm{sech}^2\left[\frac{1}{c_0}\sqrt{\frac{A}{6}}\left(x - x_0\right)\right]\tanh\left[\frac{1}{c_0}\sqrt{\frac{A}{6}}\left(x - x_0\right)\right]. \tag{78}$$

First, we choose the following parameters:

$$x_0 = 0,\ \ A = 0.5,\ \ c_0 = \sqrt{1 + 2A/3},\ \ x_l = -60,\ \ x_r = 100,\ \ t \in [0,1],$$

and compare our findings with the numerical results presented in [2]. The computational accuracy for both temporal and spatial variables is evaluated at $T = 1$ and is detailed in Tables 1 and 2, where $\tau = 0.001$ in Table 1 and $\tau/h = 0.1$ in Table 2. The data in Tables 1 and 2 indicate that the current difference scheme (16)–(18) significantly outperforms the finite volume element scheme described in [2].

Table 1. Comparative analysis of spatial errors and convergence rates for Example 1.

h	$\|\|e^n\|\|$[2]	Rate	$\|\|e^n\|\|$	Rate	$\|\|e^n\|\|_\infty$[2]	Rate	$\|\|e^n\|\|_\infty$	Rate
0.4	9.05×10^{-4}	–	5.44×10^{-5}	–	4.50×10^{-4}	–	3.15×10^{-5}	–
0.2	2.25×10^{-4}	2.01	3.21×10^{-6}	4.08	1.13×10^{-4}	1.99	2.05×10^{-6}	3.94
0.1	5.62×10^{-5}	2.00	1.90×10^{-7}	4.07	2.81×10^{-5}	2.00	1.19×10^{-7}	4.09

Table 2. Comparison of the temporal errors and convergence rates for Example 1.

h	$\|\|e^n\|\|$[2]	Rate	$\|\|e^n\|\|$	Rate	$\|\|e^n\|\|_\infty$[2]	Rate	$\|\|e^n\|\|_\infty$	Rate
0.04	8.69×10^{-4}	–	2.76×10^{-5}	–	4.32×10^{-4}	–	1.51×10^{-5}	–
0.02	2.19×10^{-4}	1.98	6.58×10^{-6}	2.06	1.10×10^{-4}	1.97	3.63×10^{-6}	2.05
0.01	5.53×10^{-5}	1.99	1.59×10^{-6}	2.04	2.78×10^{-5}	1.99	9.04×10^{-7}	2.00

On the other hand, to further demonstrate the computational efficiency of the proposed compact difference scheme (16)–(18), we compare the numerical errors with the previous studies [3,36,37]. The results are detailed in Table 3, where the error is defined as follows

$$Error = \|e^n\|/\|u^n_{exact}\|.$$

It is evident that, under the same conditions, the computational errors of the proposed difference scheme (16)–(18) are smaller than those reported in other studies.

Table 3. Comparison of the computational errors between the present scheme and the other methods for Example 1.

h	τ	(16)–(18)	Error [3]	Error [36]	Error [37]
0.4	0.04	2.29716×10^{-5}	2.19×10^{-3}	2.07164×10^{-4}	1.22×10^{-5}
0.2	0.02	5.96777×10^{-6}	5.33×10^{-4}	5.27850×10^{-5}	8.41×10^{-6}
0.1	0.01	1.50726×10^{-6}	1.33×10^{-4}	1.33205×10^{-5}	8.11×10^{-6}

The present analysis investigates the solitary wave and its corresponding numerical solutions, as illustrated in Figure 1. The spatial domain is defined as $x \in [-20, 30]$ and $[-40, 60]$, with parameters set at $h = 0.1$ and $\tau = h^2$ at the time instances $T = 0, 5, 10, 15, 20$. Figure 1 indicates that the patterns generated by the current scheme (16)–(18), demonstrate a significant degree of agreement with the solitary wave solutions.

Furthermore, the numerical solutions for the auxiliary variable V^n, as estimated by Equation (21), are visually represented in Figure 2. The graphs displaying the approximate solutions U^n and V^n, obtained from the difference scheme (16)–(18) with the parameters $x \in [-40, 60]$, $h = 0.1$, $\tau = h^2$, $c_0 = \pm\sqrt{1 + 2A/3}$, $T = 10$ are shown in Figure 3. It is observed that, when $c_0 > 0$, the solitary wave propagates to the right, while, for $c_0 < 0$, the solitary wave propagates to the left. These results are consistent with those reported in [31].

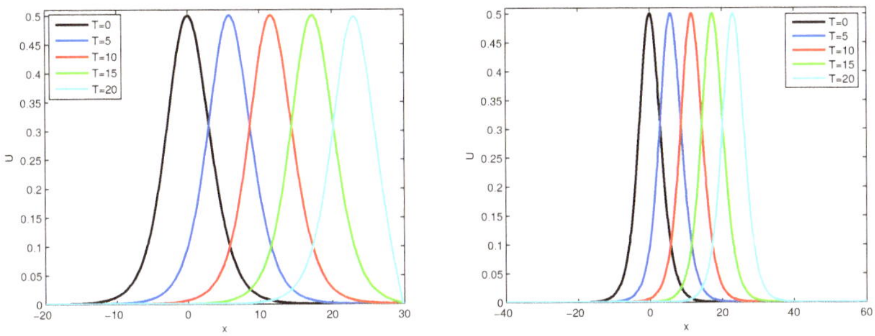

Figure 1. Solitary wave solutions of $u(x,t)$ at $T = 0$ and numerical solutions at $T = 5$, 10, 15, and 20, $x_l = -20$, and $x_r = 30$ (**left**) and $x_l = -40$ and $x_r = 60$ (**right**) for Example 1.

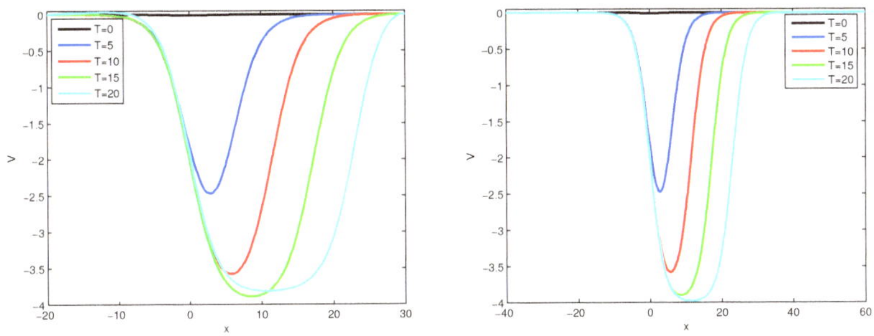

Figure 2. Numerical solutions of the auxiliary variable V at $T = 0$, 5, 10, 15, and 20, $x_l = -20$, and $x_r = 30$ (**left**) and $x_l = -40$ and $x_r = 60$ (**right**) for Example 1.

Finally, we present the discrete mass M^n, the discrete rate of mass change R^n, and the discrete energy E^n at various time intervals, as detailed in Table 4. The parameters are defined with $c_0 > 0$, $x \in [-40, 60]$, $h = 0.1$, and $\tau = h^2$. Analysis of Table 4 indicates that the discrete rate of mass change R^n associated with soliton wave propagation is minimal, indicating that the mass change R^n is conservative. Additionally, it is noted that the energy E^n also exhibits conservative properties.

Table 4. Discrete mass M^n, discrete rate of mass change R^n, and discrete energy E^n at various time intervals for Example 1.

T	M^n	R^n	E^n
0	3.99999998662202	0.00000000770792	2.59519634357874
5	3.99999998684879	0.00000000748522	2.59519634801190
10	3.99999998721201	0.00000000711875	2.59519636602745
15	3.99999998755727	0.00000000676334	2.59519639725431
20	3.99999998788500	0.00000000641617	2.59519644155837

Example 2. *Consider the parameters*

$$k = 1, \ a_1 = \frac{363}{169}, \ a_2 = 2, \ b_1 = \frac{1}{2}, \ b_2 = 0, \ c = \frac{1}{2}, \ p = 1,$$

which is the sixth-order Boussinesq equation [15]

$$u_{tt} - u_{xx} - 2u_{xxtt} + \frac{363}{169}u_{xxxx} - \frac{1}{2}u_{xxxxxx} + \frac{1}{2}(u^2)_{xx} = 0, \tag{79}$$

and the solitary wave solution is provided by

$$u(x,t) = \frac{210}{169}\operatorname{sech}^4\left(\frac{x}{\sqrt{26}} - \sqrt{\frac{97}{4394}}t\right). \tag{80}$$

The initial data are

$$\varphi(x) = \frac{210}{169}\operatorname{sech}^4\left(\frac{x}{\sqrt{26}}\right), \tag{81}$$

$$\psi(x) = \frac{420}{28,561}\sqrt{2522}\operatorname{sech}^4\left(\frac{x}{\sqrt{26}}\right)\tanh\left(\frac{x}{\sqrt{26}}\right). \tag{82}$$

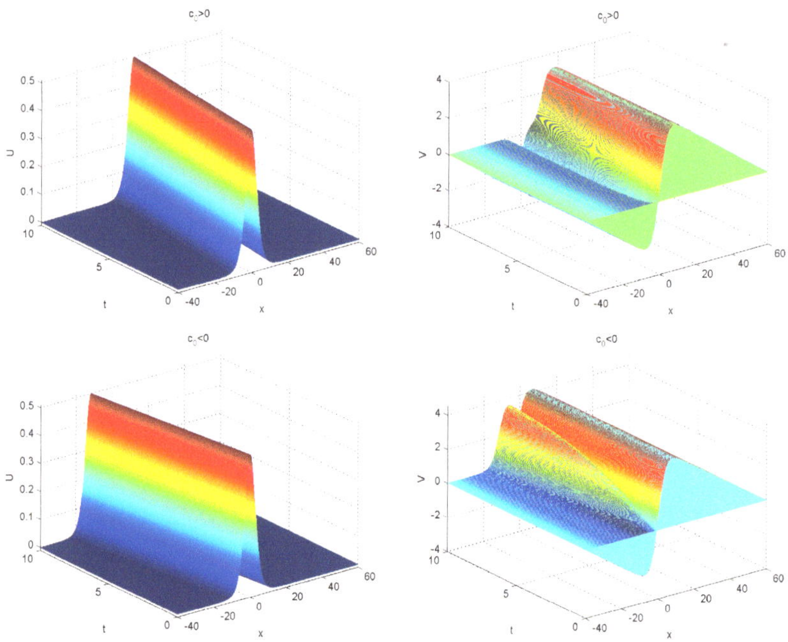

Figure 3. 3D plots of numerical solutions U^n and V^n at $T = 10$ with $x_l = -40$ and $x_r = 60$ for Example 1.

In this example, we first calculate the solution over the spatial domain defined by $[-40, 60] \times [0, 40]$ with a grid size of $h = 0.1$ and a time step $\tau = h^2$. The comparison between the solitary wave and the numerical solutions at the time instances $T = 10, 20,$ and 30 is presented in Figure 4. The results indicate that the errors across all grid points remain below 4.5×10^{-5}.

Figure 5 illustrates the numerical solutions at various time intervals, specifically $T = 0,$ 10, 20, 30, and 40. A detailed analysis of Figure 5 indicates that the wave heights at these specified time points are nearly indistinguishable. Figure 6 presents the graphs of the approximate solutions U^n obtained from the current scheme described by Equations (16)–(18) at times $T = 5, 10, 15$, and 20, with the parameters set to $h = 0.1$, $\tau = h^2$, $x_l = -40$, and $x_r = 60$.

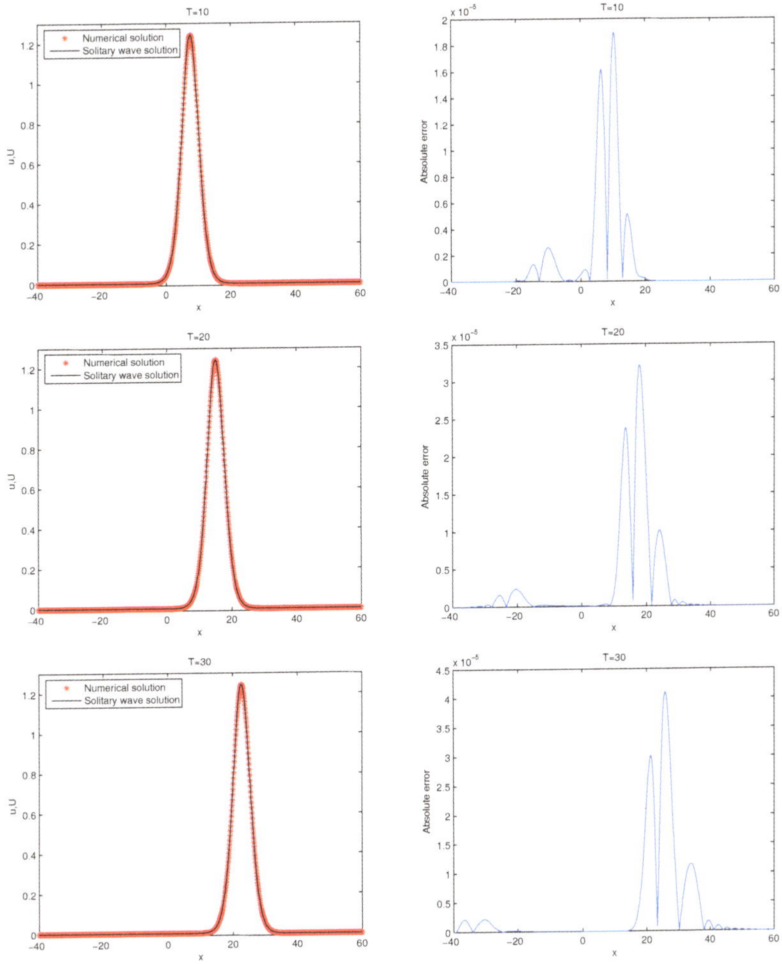

Figure 4. Numerical and solitary wave solutions and absolute error at $T = 10$, 20 and 30 for Example 2.

Additionally, Table 5 presents a summary of the discrete mass M^n, the discrete rate of mass change R^n, and the discrete energy E^n at various time points $T = 0, 5, 10, 15$, and 20. The findings suggest that both the discrete rate of mass change R^n and the discrete energy E^n exhibit conservative behavior in the discrete sense.

Table 5. Discrete mass M^n, discrete rate of mass change R^n, and discrete energy E^n with $h = 0.1$ and $\tau = h^2$ at various time intervals for Example 2.

T	M^n	R^n	E^n
0	8.44807966748734	1.88271050488737	15.35813895730075
5	8.44807966747724	1.88289006817254	15.35813896808486
10	8.44807966744347	1.88285678675310	15.35813901287684
15	8.44807966738941	1.88269314441971	15.35813909109440
20	8.44807966731402	1.88292585889068	15.35813920256866

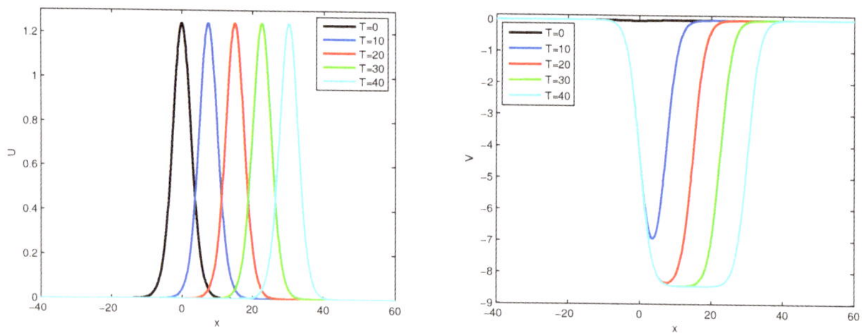

Figure 5. Numerical solutions of U and V with $h = 0.1$ and $\tau = h^2$ at $T = 0, 10, 20, 30,$ and 40, $x_l = -40$, and $x_r = 60$ for Example 2.

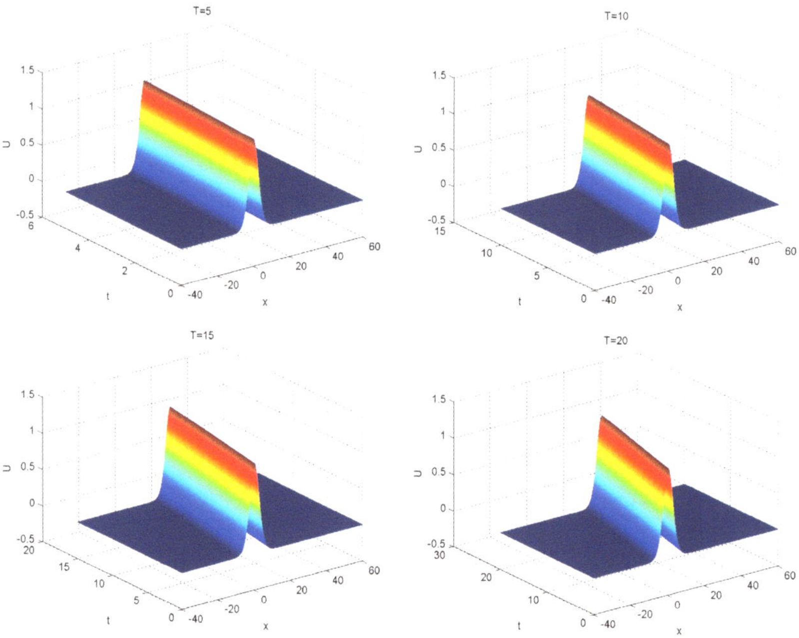

Figure 6. 3D plots of the numerical solutions U^n at $T = 5, 10, 15,$ and 20 with $h = 0.1$, $\tau = h^2$, $x_l = -40$, and $x_r = 60$ for Example 2.

Example 3. *Consider the following sixth-order Boussinesq equation with the effects of surface tension*

$$u_{tt} - k^2 u_{xx} + a_1 u_{xxxx} - a_2 u_{xxtt} - b_1 u_{xxxxxx} + b_2 u_{xxxxtt} + c(u^{2p})_{xx} = 0, \tag{83}$$

where $a_1 \geq 0$, $a_2 \geq 0$, $b_1 \geq 0$, $b_2 \geq 0$, *and* $p \geq 1$. *In their research, Biswas et al. [4] examined the solitary wave solution corresponding to Equation* (83)*; that is,*

$$u(x,t) = A\,\mathrm{sech}^{\frac{2}{2p-1}}[B(x - vt)], \tag{84}$$

where

$$A = \left[\frac{(2p+1)(b_2k^2 - b_1)}{2cb_2}\right]^{\frac{1}{2p-1}}, \quad B = \frac{|2p-1|}{2}\sqrt{\frac{b_2k^2 - b_1}{a_1b_2 - a_2b_1}}, \quad v = \sqrt{\frac{b_1}{b_2}}.$$

To perform numerical simulations, the initial conditions can be determined by evaluating Equation (84) *and its derivative at the time point* $t = 0$*; that is,*

$$\varphi(x) = A\,\mathrm{sech}^{\frac{2}{2p-1}}(Bx), \tag{85}$$

$$\psi(x) = \frac{2}{2p-1} ABv\,\mathrm{sech}^{\frac{2}{2p-1}}(Bx)\tanh(Bx). \tag{86}$$

To demonstrate the efficacy of the present difference scheme, we provide numerical results that pertain to the errors and rates of convergence. The parameters are configured with $h = 0.4$ and $\tau = h^2$ at time $T = 10$, as illustrated in Table 6, where $x_l = -40$ and $x_r = 60$. From Table 6, it is evident that the convergence rate achieved by the current finite difference scheme is approximately 4.0, which is consistent with the theoretical order of convergence specified in Theorem 6.

Table 6. Errors and convergence rates of the current scheme when $h = 0.4$ and $\tau = h^2$ at $T = 10$ for Example 3.

p		h, τ	$h/2, \tau/4$	$h/4, \tau/16$
	$\|\|e\|\|$	$1.530067340 \times 10^{-2}$	$9.709362247 \times 10^{-4}$	$5.777896747 \times 10^{-5}$
$p = 1$	rate	–	3.9780748	4.07076021
	$\|\|e\|\|_\infty$	$1.015003674 \times 10^{-2}$	$6.556924437 \times 10^{-4}$	$4.127486212 \times 10^{-5}$
	rate	–	3.9523219	3.9896839
	$\|\|e\|\|$	$7.862155174 \times 10^{-2}$	$4.719448787 \times 10^{-3}$	$3.012389794 \times 10^{-4}$
$p = 2$	rate	–	4.0582346	3.9696380
	$\|\|e\|\|_\infty$	$7.973541804 \times 10^{-2}$	$4.682164328 \times 10^{-3}$	$3.018001249 \times 10^{-4}$
	rate	–	4.0899732	3.9555103

Figure 7 depicts the numerical solutions and absolute errors at times $T = 10$ and 20 with $h = 0.1$, $\tau = h^2$, $x_l = -40$, $x_r = 60$, and $p = 2$. Furthermore, Table 7 presents additional information regarding the discrete mass M^n, the discrete rate of mass change R^n, and the discrete energy E^n at various time intervals $T = 0$, $T = 5$, $T = 10$, $T = 15$, $T = 20$, $T = 25$, and $T = 30$. It is clear that both the rate of mass change R^n and the energy E^n exhibit conservative behavior in the discrete sense.

Table 7. Discrete mass M^n, discrete rate of mass change R^n, and discrete energy E^n with $h = 0.1$, $\tau = h^2$, and $P = 2$ at various time intervals for Example 3.

T	M^n	R^n	E^n
0	3.02090976925920	1.52260617289068	8.70669460321910
5	3.02090976827031	1.52400439056038	8.70669481288416
10	3.02090976455933	1.52404485849716	8.70669563361528
15	3.02090975829348	1.52179604152168	8.70669698162941
20	3.02090974941932	1.52420689642572	8.70669874989782
25	3.02090973801432	1.52375610223547	8.70670081739405
30	3.02090972410221	1.52240106874032	8.70670306135933

Furthermore, for $p = 1$, Equation (83) exhibits two conserved quantities [4]:

$$I_1 = \frac{\nu}{10}\int_{-\infty}^{\infty}\left(7b_2u_{xxx}^2 - 10a_2u_{xx}^2 + 10u_x^2\right)dx, \tag{87}$$

$$I_2 = \frac{1}{30}\int_{-\infty}^{\infty}\left\{-\left(15b_1 + 2b_2\nu^2\right)u_{xxx}^2 - 15a_2\nu^2u_{xx}^2 + 15\left(c^2 - \nu^2 - k^2\right)u_x^2\right\}dx. \tag{88}$$

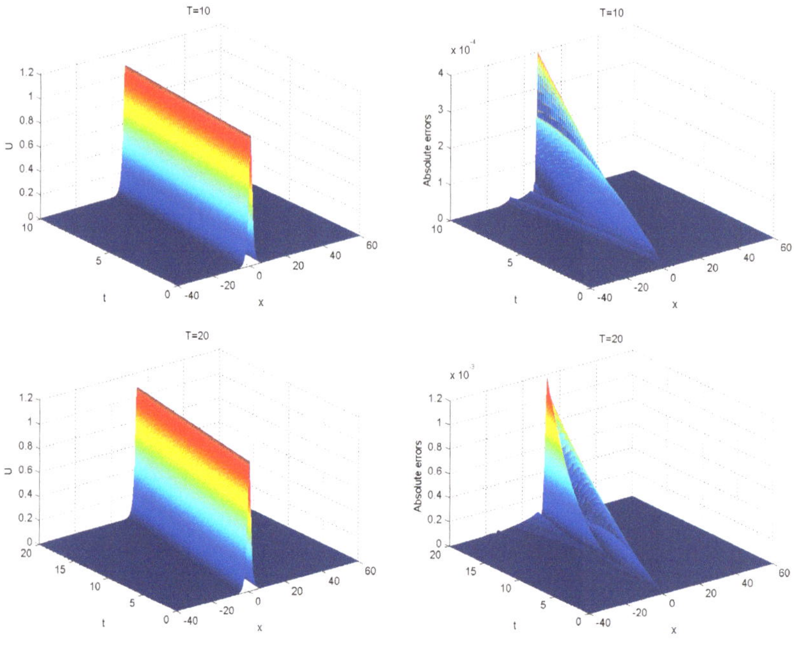

Figure 7. Numerical solutions and absolute errors at $T = 10$ and 20 with $h = 0.1$, $\tau = h^2$, $x_l = -40$, $x_r = 60$, and $p = 2$ for Example 3.

Furthermore, for any arbitrary $p > 1$, Equation (83) possesses two conserved quantities [4]:

$$I_3 = \int_{-\infty}^{\infty}\left(-\frac{1}{3}b_2u_{xxxxt} + \frac{1}{2}a_2u_{xxt} - u_t\right)dx, \tag{89}$$

$$I_4 = \int_{-\infty}^{\infty}\left(-\frac{1}{3}b_2tu_{xxxxt} + \frac{1}{15}b_2u_{xxx} + \frac{1}{2}a_2tu_{xxt} - \frac{1}{6}a_2u_{xx} - tu_t + u\right)dx. \tag{90}$$

We will evaluate our numerical method by employing these invariants and selecting the approximate values as follows:

$$\tilde{I}_1^n = \frac{\nu h}{10}\sum_{j=1}^{J-1}\Big\{\frac{7b_2}{2}[3(U_j^n)_{\hat{x}\hat{x}\hat{x}} - (U_j^n)_{\check{x}\check{x}\check{x}}]^2 - \frac{10a_2}{3}[4(U_j^n)_{\check{x}\hat{x}} - (U_j^n)_{\hat{x}\hat{x}}]^2 + \frac{10}{3}[4(U_j^n)_{\hat{x}} - (U_j^n)_{\check{x}}]^2\Big\},$$

$$\tilde{I}_2^n = \frac{h}{30}\sum_{j=1}^{J-1}\Big\{-\frac{1}{2}(15b_1 + 2b_2\nu^2)[3(U_j^n)_{\hat{x}\hat{x}\hat{x}} - (U_j^n)_{\check{x}\check{x}\check{x}}]^2 - 5a_2\nu^2[4(U_j^n)_{\check{x}\hat{x}} - (U_j^n)_{\hat{x}\hat{x}}]^2$$

$$+ 5(c^2 - \nu^2 - k^2)[4(U_j^n)_{\hat{x}} - (U_j^n)_{\check{x}}]^2\Big\},$$

$$\tilde{I}_3^n = h\sum_{j=1}^{J-1}\Big\{-\frac{b_2}{9}\Big[5(U_j^n)_{\check{x}\hat{x}\check{x}\hat{x}} - 2(U_j^n)_{\hat{x}\hat{x}\hat{x}\hat{x}}\Big]_{\hat{t}} + \frac{a_2}{6}\Big[4(U_j^n)_{\check{x}\hat{x}} - (U_j^n)_{\hat{x}\hat{x}}\Big]_{\hat{t}} - (U_j^n)_{\hat{t}}\Big\},$$

$$\tilde{I}_4^n = h\sum_{j=1}^{J-1}\Big\{-\frac{b_2 t^n}{9}\Big[5(U_j^n)_{\check{x}\hat{x}\check{x}\hat{x}} - 2(U_j^n)_{\hat{x}\hat{x}\hat{x}\hat{x}}\Big]_{\hat{t}} + \frac{b_2}{30}[3(U_j^n)_{\hat{x}\hat{x}\hat{x}} - (U_j^n)_{\check{x}\check{x}\check{x}}]$$

$$+ \frac{a_2 t^n}{6}\Big[4(U_j^n)_{\check{x}\hat{x}} - (U_j^n)_{\hat{x}\hat{x}}\Big]_{\hat{t}} - \frac{a_2}{18}\Big[4(U_j^n)_{\check{x}\hat{x}} - (U_j^n)_{\hat{x}\hat{x}}\Big] - t^n(U_j^n)_{\hat{t}} + U_j^n\Big\}.$$

To demonstrate the conservative properties of the current difference scheme represented by Equations (16)–(18), we have compiled the invariants $\tilde{I}_1^n$, $\tilde{I}_2^n$, $\tilde{I}_3^n$, $\tilde{I}_4^n$, M^n, R^n, and E^n at various temporal intervals, as presented in Tables 8 and 9. The data illustrated in these tables indicate that the conservative quantities $\tilde{I}_j^n$ $(j = 1, 2, 3, 4)$ and R^n and E^n remain approximately constant over a time frame of up to 30 units. Consequently, the compact scheme proposed, as outlined by Equations (16)–(18), demonstrates both efficiency and reliability for simulations conducted over long periods.

Table 8. Conservative quantities $\tilde{I}_1^n$ and $\tilde{I}_2^n$ and discrete quantities M^n, R^n, and E^n with $h = 0.1$, $\tau = h^2$, and $p = 1$ at various time intervals for Example 3.

T	$\tilde{I}_1^n$	$\tilde{I}_2^n$	M^n	R^n	E^n
0	0.77574398822	0.72568182168	2.99999999979	1.06047283470	3.78558199462
5	0.77574415451	0.72568197184	2.99999999869	1.06079683168	3.78558200635
10	0.77574474392	0.72568250450	2.99999999479	1.06080624484	3.78558205484
15	0.77574571092	0.72568337974	2.99999998836	1.06028637400	3.78558213956
20	0.77574703291	0.72568457913	2.99999997943	1.06084514881	3.78558226063
25	0.77574867931	0.72568607763	2.99999996804	1.06074072374	3.78558241692
30	0.77575061250	0.72568784435	2.99999995426	1.06042851618	3.78558260687

Table 9. Conservative quantities $\tilde{I}_3^n$ and $\tilde{I}_4^n$ and discrete quantities M^n, R^n, and E^n with $h = 0.1$, $\tau = h^2$, and $p = 2$ at various time intervals for Example 3.

T	$\tilde{I}_1^n$	$\tilde{I}_2^n$	M^n	R^n	E^n
0	0.00000000879	3.02090976945	3.02090976927	1.52260617358	8.70669460355
5	0.00000001689	3.02090976957	3.02090976906	1.52318643159	8.70669464967
10	0.00000002587	3.02090976979	3.02090976876	1.52365257478	8.70669471985
15	0.00000003535	3.02090977012	3.02090976836	1.52400439292	8.70669481421
20	0.00000004386	3.02090977051	3.02090976788	1.52424172749	8.70669493260
25	0.00000005221	3.02090977097	3.02090976732	1.52436447198	8.70669507504
30	0.00000006009	3.02090977148	3.02090976668	1.52437256971	8.70669524020

6. Conclusions

A conservative and compact fourth-order accurate difference scheme has been formulated for solving the sixth-order Boussinesq equation. A comprehensive analysis of solvability, convergence, and stability has been conducted. Numerical results show that the numerical solution effectively preserves the discrete rates of mass change R^n and energy E^n.

Author Contributions: Conceptualization, X.W. and W.D.; methodology, W.D. and A.B.; validation, X.W.; formal analysis, X.W.; writing—original draft preparation, X.W.; writing—review and editing, X.W. and W.D.; supervision, W.D. and A.B. All authors have read and agreed to the published version of the manuscript.

Funding: X.W. was supported by Fujian Provincial Natural Science Foundation of China (No. 2024J01802).

Data Availability Statement: The data that support the findings of this study are available from the corresponding author, X.W., upon reasonable request.

Acknowledgments: We would like to express our sincere gratitude to the anonymous reviewers for their valuable comments and suggestions, which greatly enhanced the quality of this manuscript.

Conflicts of Interest: The authors declare no conflicts of interest.

References

1. Boussinesq, J. Théorie des ondes et des remous qui se propagent le long d'un canal rectangulaire horizontal, en communiquant au liquide contenu dans ce canal des vitesses sensiblement pareilles de la surface au fond. *J. Math. Pures Appl.* **1872**, *17*, 55–108.
2. Yan, J.; Deng, D.; Lu, F.; Zhang, Z. A new efficient energy-preserving finite volume element scheme for the improved Boussinesq equation. *Appl. Math. Model.* **2020**, *87*, 20–41. [CrossRef]
3. Wang, Q.; Zhang, Z.; Zhang, X.; Zhu, Q. Energy-preserving finite volume element method for the improved Boussinesq equation. *J. Comput. Phys.* **2014**, *270*, 58–69. [CrossRef]
4. Biswas, A.; Vega Guzman, J.; Bansal, A.; Kara, A.H.; Aphane, M.; Yıldırım, Y.; Alshehri, H.M. Solitary waves, shock waves and conservation laws with the surface tension effect in the Boussinesq equation. *Proc. Est. Acad. Sci.* **2023**, *72*, 17–29. [CrossRef]
5. Biswas, A. Solitary wave solution for KdV equation with power-law nonlinearity and time-dependent coefficients. *Nonlinear Dyn.* **2009**, *58*, 345–348. [CrossRef]
6. Razborova, P.; Triki, H.; Biswas, A. Perturbation of dispersive shallow water waves. *Ocean. Eng.* **2013**, *63*, 1–7. [CrossRef]
7. Razborova, P.; Kara, A.H.; Biswas, A. Additional conservation laws for Rosenau-KdV-RLW equation with power law nonlinearity by Lie symmetry. *Nonlinear Dyn.* **2015**, *79*, 743–748. [CrossRef]
8. Lu, F.; Song, Z.; Zhang, Z. A compact fourth-order finite difference scheme for the improved Boussinesq equation with damping terms. *J. Comp. Math.* **2016**, *34*, 462–478.
9. Helal, M.A.; Seadawy, A.R.; Zekry, M. Stability analysis of solutions for the sixth-order nonlinear Boussinesq water wave equations in two-dimensions and its applications. *Chinese J. Phys.* **2017**, *55*, 378–385. [CrossRef]
10. Burde, G.I. Solitary wave solutions of the high-order KdV models for bi-directional water waves. *Commun. Nonlinear. Sci. Numer. Simulat.* **2011**, *16*, 1314–1328. [CrossRef]
11. Yimnet, S.; Wongsaijai, B.; Rojsiraphisal, T.; Poochinapan, K. Numerical implementation for solving the symmetric regularized long wave equation. *Appl. Math. Comput.* **2016**, *273*, 809–825. [CrossRef]
12. Esfahani, A.; Farah, L.G. Local well-posedness for the sixth-order Boussinesq equation. *J. Math. Anal. Appl.* **2012**, *385*, 230–242. [CrossRef]
13. Wang, H.; Esfahani, A. Global rough solutions to the sixth-order Boussinesq equation. *Nonlinear Anal.* **2014**, *102*, 97–104. [CrossRef]
14. Feng, B.; Kawahara, T.; Mitsui, T.; Chan, Y. Solitary-wave propagation and interactions for a sixth-order generalized Boussinesq equation. *Int. J. Math. Math. Sci.* **2005**, *9*, 1435–1448. [CrossRef]
15. Kolkovska, N.; Vucheva, V. Energy preserving finite difference scheme for sixth-order Boussinesq equation. *Procedia Eng.* **2017**, *199*, 1539–1543. [CrossRef]
16. Arslan, D. Approximate solutions of singularly perturbed nonlinear ill-posed and sixth-order Boussinesq equations with hybrid method. *Bitlis Eren Üniversitesi Fen Bilim. Dergisi.* **2019**, *8*, 451–458. [CrossRef]
17. Zhang, T.; Lin, Z.-H.; Huang, G.-Y.; Fan, C.M.; Li, P.W. Solving Boussinesq equations with a meshless finite difference method. *Ocean Engin.* **2020**, *198*, 106957. [CrossRef]
18. Hou, B.; Liang, D. The energy-preserving time high-order AVF compact finite difference scheme for nonlinear wave equations in two dimensions. *Appl. Numer. Math.* **2021**, *170*, 298–320. [CrossRef]
19. Mohanty, R.K.; Dai, W.; Han, F. Compact operator method of accuracy two in time and four in space for the numerical solution of coupled viscous Burgers' equations. *Appl. Math. Comput.* **2015**, *256*, 381–393. [CrossRef]

20. Mohanty, R.K.; Dai, W.; Liu, D. Operator compact method of accuracy two in time and four in space for the solution of time dependent Burgers-Huxley equation. *Numer. Algor.* **2015**, *70*, 591–605. [CrossRef]
21. Kaur, D.; Mohanty, R.K. Highly accurate compact difference scheme for fourth order parabolic equation with Dirichlet and Neumann boundary conditions: Application to good Boussinesq equation. *Appl. Math. Comput.* **2020**, *378*, 125202. [CrossRef]
22. Kaur, D.; Mohanty, R.K. High-order half-step compact numerical approximation for fourth-order parabolic PDEs. *Numer. Algor.* **2024**, *95*, 1127–1153. [CrossRef]
23. Zhou, H.; Sheng, Z.; Yuan, G. A conservative gradient discretization method for parabolic equations. *Adv. Appl. Math. Mech.* **2021**, *13*, 232–260. [CrossRef]
24. Deng, D.; Liang, D. The energy-preserving finite difference methods and their analyses for system of nonlinear wave equations in two dimensions. *Appl. Numer. Math.* **2020**, *151*, 172–198. [CrossRef]
25. Bayarassou, K.; Rouatbi, A.; Omrani, K. Uniform error estimates of fourth-order conservative linearized difference scheme for a mathematical model for long wave. *Int. J. Comput. Math.* **2020**, *97*, 1678–1703. [CrossRef]
26. Nanta, S.; Yimnet, S.; Poochinapan, K.; Wongsaijai, B. On the identification of nonlinear terms in the generalized Camassa-Holm equation involving dual-power law nonlinearities. *Appl. Numer. Math.* **2021**, *160*, 386–421. [CrossRef]
27. Wang, X.; Dai, W. A three-level linear implicit conservative scheme for the Rosenau-KdV-RLW equation. *J. Comput. Appl. Math.* **2018**, *330*, 295–306. [CrossRef]
28. Wongsaijai, B.; Poochinapan, K.; Disyadej, T. A compact finite difference method for solving the General Rosenau-RLW equation. *Int. J. Appl. Math.* **2014**, *44*, 192–199.
29. Mohanty, R.K.; Dai, W.; Han, F. A new high accuracy method for two-dimensional biharmonic equation with nonlinear third derivative terms: Application to Navier-Stokes equations of motion. *Int. J. Comput. Math.* **2015**, *92*, 1574–1590 [CrossRef]
30. Ye, H.; Liu, F.; Anh, V. Compact difference scheme for distributed-order time-fractional diffusion-wave equation on bounded domains. *J. Computat. Phys.* **2015**, *298*, 652–660. [CrossRef]
31. Wongsaijai, B.; Oonariya, C.; Poochinapan, K. Compact structure-preserving algorithm with high accuracy extended to the improved Boussinesq equation. *Math. Comput. Simulat.* **2020**, *178*, 125–150. [CrossRef]
32. Zhou, Y. *Application of Discrete Functional Analysis to the Finite Difference Methods;* International Academic Publishers: Beijing, China, 1990.
33. Wang, X.; Dai, W.; Guo, S. A conservative linear difference scheme for the 2D regularized long-wave equation. *Appl. Math. Comput.* **2019**, *342*, 55–70. [CrossRef]
34. Tamang, N.; Wongsaijai, B.; Mouktonglang, T.; Poochinapan, K. Novel algorithm based on modification of the Galerkin finite element method to general Rosenau-RLW equation in (2+1)-dimensions. *Appl. Numer. Math.* **2020**, *148*, 109–130. [CrossRef]
35. Ucar, Y.; Karaagac, B.; Esen, A. A new approach on numerical solutions of the Improved Boussinesq type equation using quadratic B-spline Galerkin finite element method. *Appl. Math. Comput.* **2015**, *270*, 148–155. [CrossRef]
36. Bratsos, A. A second order numerical scheme for the improved Boussinesq equation. *Phys. Lett. A* **2007**, *370*, 145–147. [CrossRef]
37. Jang, T.S. A new dispersion-relation preserving method for integrating the classical Boussinesq equation. *Commun. Nonlinear Sci. Numer. Simul.* **2017**, *43*, 118–138. [CrossRef]

Mathematical and Computational Applications

MDPI

Article

A Network-Based Study of the Dynamics of $A\beta$ and τ Proteins in Alzheimer's Disease

Stefano Bianchi [1,†], Germana Landi [1,*,†], Camilla Marella [2,†], Maria Carla Tesi [1,†], Claudia Testa [2,†] and on behalf of the Alzheimer's Disease Neuroimaging Initiative [‡]

1 Department of Mathematics, University of Bologna, 40126 Bologna, Italy; stefano.bianchi22@unibo.it (S.B.); mariacarla.tesi@unibo.it (M.C.T.)

2 Department of Physics and Astronomy "Augusto Righi", University of Bologna, 40126 Bologna, Italy; camilla.marella2@unibo.it (C.M.); claudia.testa@unibo.it (C.T.)

* Correspondence: germana.landi@unibo.it

† These authors contributed equally to this work.

‡ Data used in preparation of this article were obtained from the Alzheimer's Disease Neuroimaging Initiative (ADNI) database (adni.loni.usc.edu). As such, the investigators within the ADNI contributed to the design and implementation of ADNI and/or provided data but did not participate in analysis or writing of this report. A complete listing of ADNI investigators can be found at: http://adni.loni.usc.edu/wp-content/uploads/how_to_apply/ADNI_Acknowledgement_List.pdf.

Abstract: Due to the extreme complexity of Alzheimer's disease (AD), the etiology of which is not yet known, and for which there are no known effective treatments, mathematical modeling can be very useful. Indeed, mathematical models, if deemed reliable, can be used to test medical hypotheses that could be difficult to verify directly. In this context, it is important to understand how $A\beta$ and τ proteins, which, in abnormal aggregate conformations, are hallmarks of the disease, interact and spread. We are particularly interested, in this paper, in studying the spreading of misfolded τ. To this end, we present four different mathematical models, all on networks on which the protein evolves. The models differ in both the choice of network and diffusion operator. Through comparison with clinical data on τ concentration, which we carefully obtained with multimodal analysis techniques, we show that some models are more adequate than others to simulate the dynamics of the protein. This type of study may suggest that, when it comes to modeling certain pathologies, the choice of the mathematical setting must be made with great care if comparison with clinical data is considered decisive.

Keywords: Alzheimer's disease; models on graphs; $A\beta$ and τ proteins; medical imaging; numerical simulations

1. Introduction

Alzheimer's disease (AD) is a neurodegenerative disorder characterized by a progressive decline in memory and other cognitive functions, leading inevitably to death. It is an incurable disease affecting more than 50 million people, with figures set to increase significantly in the coming years (World Alzheimer Report 2023). The etiology of the disease is unclear to date, but two proteins are universally recognized as playing a crucial role in the development of the disease: amyloid-beta ($A\beta$) and tau (τ). Both proteins are physiologically present in the brain but, in the presence of the disease, they form abnormal aggregates, in a progressive and irreversible way. Indeed, $A\beta$ plaques and neurofibrillary tangles (NFTs) of τ are pathological hallmarks of AD. Experimental studies suggest non-uniform distributions of pathological proteins in the brain [1]. Both $A\beta$ and τ exhibit characteristic spatiotemporal deposition patterns. NFTs appear first in the entorhinal cortex and then spread to the amygdala, temporal areas, and, finally, throughout the cortex [2]. On the other hand, $A\beta$ plaques form first in the temporal and frontal areas and then spread to other areas of the brain [2,3]. Recent literature suggests that the interplay between the two proteins should be crucial in the development of the disease and must be taken into account for the development of new therapies [4–6]. Clearly, when one is interested in modeling complex pathologies such as AD, comparison with clinical and experimental data is crucial to the reliability of the study. Clinical multimodal neuroimages, if properly processed in terms of harmonization, combination, and a quantitative approach, can give information of local character, providing, for instance, the concentration of proteins in different brain regions [7].

This work has two main objectives. One is to capture such types of non-uniform distributions through appropriate mathematical models. The other purpose is to compare the results from the mathematical models with clinical data in the context of Alzheimer's disease. Comparison with clinical data is a fundamental step in the development of a trustable mathematical model but it is rarely addressed in the mathematical modeling literature. To this aim, we have processed a considerable amount of clinical neuroimages to extract quantitative features, which can be compared with the τ concentration values provided by the mathematical models.

Networks are mathematical tools that provide an ideal setting for comparisons between clinical data and numerical results from models. Complex network properties have been identified with some consistency in all modalities of neuroimaging data and over a range of spatial and time scales. Thus, brain networks allow to describe the information process in neurons and its characteristics of being both locally specialized or segregated and globally distributed or integrated [8]. In particular, small-worldness characterizes an intermediate regime of networks between the two extreme regimes of a regular lattice and a random network, efficiently representing the brain structure [9]. The mathematical modeling of AD can be very useful for integrating clinical and experimental data in a theoretical setting, which could allow, once reliability has been verified, to test various hypotheses otherwise difficult to verify. For an exhaustive overview of existing mathematical models of proteins spreading on networks and of various related challenging questions, we refer to [10]. Here, we wish to emphasize that the dynamics of the two proteins we are interested in differ in several respects, both spatial ($A\beta$ spreads over small distances, τ over large distances) and temporal (the dynamics of $A\beta$ are much faster than those of τ). Differences in spatial dynamics should, in some way, be reflected by the choice of network used for the evolution of the dynamics. This paper is inspired by the paper [11] as far as the idea of using distinct networks for different protein dynamics is concerned; however, its purpose is quite different. Indeed, in [11], the authors were mainly interested in investigating the synergistic interactions of the two proteins $A\beta$ and τ, whose relevance was confirmed there by testing various modeling hypotheses. However, the authors were not interested in any comparison with medical data. Here, we strongly rely on the comparison with medical data to test different modeling hypotheses concerning the diffusion of τ on properly chosen networks, to see if there is an optimal way to deal with the spreading of proteins when it comes to modeling AD. Therefore, the main target of this paper is twofold: 1. to consider different models for the spreading of $A\beta$ and, especially, of τ, developed on networks with a precise biomedical meaning; 2. to have medical data for a comparison as meaningful as possible with the models, and to use that comparison to test the trustability of the model.

The plan of the work is the following: in Section 2, we describe the networks we use for our models. We start from publicly available data and, from that, we construct several different weighted graphs on which we evolve proteins. In Section 3, we describe the models we study; in particular, we present four different models for various possible mechanisms of the spreading of τ. Section 4 is concerned with the procedures we have adopted to obtain reliable medical data. We emphasize that we used a multimodal approach, based on MRI and PET, to determine the concentrations of τ protein in the brain. Finally, in Section 5, we present the results obtained by numerical simulations and we make comparisons with medical data.

2. Brain Networks via Weighted Graphs

The central concept of a brain network constructed to segregate and integrate information processing comes from the advent of the "disconnection syndromes" hypothesis [12], mainly on the basis of clinico-pathological correlations. The conceptualization of the relationship between the structure and function of the brain has led to understand that brain regions participate in many functions, introducing the structure–function framework described by a network of brain regions. Mesulam and colleagues [13] introduced the network approach to understand the localization of complex functions. The connection matrix of the human brain, known as the human "connectome", today represents an indispensable tool for mapping brain structure to functional processes and has a valuable impact on understanding brain diseases.

To model the spreading of different proteins in the brain, we consider several connectomes, each corresponding to a different undirected weighted graph. Data used for the construction of these connectomes have been downloaded from the website https://braingraph.org (accessed on 24 November 2024) and consist of an *averaged* graph of 477 healthy subjects, each with 1015 nodes (the weight calculation mode was set as median and the number of fibers launched was set to 20) [14,15]. We stress that the vertices of such a graph do not have associated coordinates that determine their position in space, since it is precisely an averaged graph. This will have consequences when we talk about "distances", which must be understood in an intrinsic sense and not in the classical Euclidean metric sense.

In the following, we first recall the definition of a weighted graph and how to construct an associated Laplacian, mimicking diffusion on it. Then, we describe the weighted graphs we will use to model the spreading of $A\beta$ and τ proteins. A *graph* is a pair $\mathcal{G} = (V, E)$, where V is a set of vertices and $E \subset V \times V$ is a set of the edges; $\mathcal{G}$ is said to be undirected if $(i, j) \in E$ implies that $(j, i) \in E$. The graph $\mathcal{G}$ is called weighted if a measure $\omega : E \to \mathbb{R}^+$ exists, assigning a unique positive number to each edge; the value $\omega(i, j)$ is referred to as the weight of the edge (i, j). A weighted graph can be represented through the adjacency matrix A whose entries $A_{i,j}$ represent the weights of the edge (i, j). Let $N = |V|$ be the number of vertices in $\mathcal{G}$; then, $A \in \mathbb{R}^{N\times N}$ is defined as

$$A_{i,j} = \begin{cases} w(i,j) & \text{if } (i,j) \in E; \\ 0 & \text{otherwise.} \end{cases}$$

There are several possible definitions for the Laplacian associated with a given graph. Given the adjacency matrix $A \in \mathbb{R}^{N\times N}$ of $\mathcal{G}$, following [16], we define the graph Laplacian L as $L = D - A$ where D is the weighted degree matrix whose jth diagonal element $D_{j,j}$ is given by

$$D_{j,j} = \sum_{i=1}^{N} A_{i,j}.$$

2.1. Structural Connectome

We call *structural connectome* a weighted graph $\mathcal{G} = (V, E)$ extracted from the tractography of diffusion tensor images of 477 healthy subjects of the Human Connectome Project [17] using the Budapest Reference Connectome v3.0 [15]. In this graph, vertices correspond to parcellated regions of gray matter formed by neurons, which share similarities in cytoarchitecture, functional activity, and structural connections to other regions, and edges represent the connectivity between the regions. We used a high-resolution connectome with $N = 1015$, which can be downloaded at the website https://pitgroup.org/connectome/ (accessed on 24 November 2024). It is possible to choose, for the edges of this graph, different weights, based on the mean number of fibers connecting two regions, and on their mean length. Accordingly, we denote by $\mathcal{G}_{\mathrm{NL}}$ the graph with weights given by

$$w_{\mathrm{NL}}(i,j) = \frac{n_{ij}}{\ell_{ij}}, \tag{1}$$

where n_{ij} is the mean number of fibers connecting vertices i and j, and ℓ_{ij} is the mean length of such fibers. Analogously, we denote by $\mathcal{G}_{\mathrm{L}}$ the graph with weights given by

$$w_{\mathrm{L}}(i,j) = \ell_{ij}. \tag{2}$$

Therefore, the two weighted graphs $\mathcal{G}_{\mathrm{NL}}$ and $\mathcal{G}_{\mathrm{L}}$ have the same vertices and edges, but different weights.

2.2. Intrinsic Proximity Connectome

Let $R_{\mathrm{P}} \in \mathbb{R}^+$ be a given positive value; the *intrinsic proximity connectome* is the weighted graph $\mathcal{G}_{\mathrm{P}} = (V, E)$ whose set of vertices V is the same as $\mathcal{G}_{\mathrm{L}}$'s and whose set of edges $E_{\mathrm{P}} \subset E_{\mathrm{L}}$ is the subset of edges of $\mathcal{G}_{\mathrm{L}}$ connected by a fiber with length ℓ less than R_{P}. The weights are given by

$$w_{\mathrm{P}}(i,j) = \begin{cases} e^{-\ell_{ij}^2/\sigma} & \text{if } \ell_{ij} \leq R_{\mathrm{P}} \\ 0 & \text{otherwise,} \end{cases}$$

where $\sigma \in \mathbb{R}^+$ is a fixed parameter. In this way, we connect two vertices only if they are close in an intrinsic sense, as opposed to a "geometric" vicinity measured with some kind of Euclidean distance. Indeed, as already mentioned, the vertices of the graph do not have associated spatial coordinates. The weights we assign are stronger for intrinsically close vertices. For this reason, $\mathcal{G}_{\mathrm{P}}$ is referred to as *intrinsic proximity connectome*. It is appropriate to emphasize here that this proximity connectome is quite different from the one used in [11]. Indeed, here, we use only *intrisic* connections, without adding any new connection, and we give them a weight proportional to the intrinsic distance.

2.3. Cumulative Connectome

The *cumulative connectome* is a weighted graph $\mathcal{G}_{\mathrm{C}} = (V, E)$ with the same V and E as $\mathcal{G}_{\mathrm{NL}}$, but with weights defined as follows. Let $\mathcal{S}_{ij}$ be the set of all paths in the graph $\mathcal{G}_{\mathrm{NL}}$ starting at vertex i and ending at vertex j, i.e.,

$$\mathcal{S}_{ij} = \{p_{ij}^1, p_{ij}^2, \ldots, p_{ij}^{m_{ij}}\}, \tag{3}$$

where $m_{ij} = |\mathcal{S}_{ij}|$ and each path $p_{ij}^k \in \mathcal{S}_{ij}$ consists of a sequence of edges:

$$p_{ij}^k = \{e_{ij}^{k_1}, e_{ij}^{k_2}, \ldots, e_{ij}^{k_r}\}, \quad k = 1, \ldots, m_{ij}.$$

We recall that the graphs $\mathcal{G}_{\mathrm{NL}}$ and $\mathcal{G}_{\mathrm{L}}$ differ only in the weights assigned to the edges; thus, we can define the length of the path p_{ij}^k, considering it as a part of the graph $\mathcal{G}_{\mathrm{L}}$ or the graph $\mathcal{G}_{\mathrm{NL}}$. Since the length of a path is usually defined as the sum of the weights of the edges forming the path, in $\mathcal{G}_{\mathrm{NL}}$, we obtain the length $l_{\mathrm{NL}}(p_{ij}^k)$ defined as

$$l_{\mathrm{NL}}(p_{ij}^k) = \sum_{s \in \{k_1, \ldots, k_r\}} w_{\mathrm{NL}}(e_{ij}^s), \quad k = 1, \ldots, m_{ij},$$

and, in $\mathcal{G}_{\mathrm{L}}$ we obtain the length $l_{\mathrm{L}}(p_{ij}^k)$ given by

$$l_{\mathrm{L}}(p_{ij}^k) = \sum_{s \in \{k_1, \ldots, k_r\}} w_{\mathrm{L}}(e_{ij}^s), \quad k = 1, \ldots, m_{ij}.$$

Finally, let $\mathcal{S}_{ij}^{\delta}$ be the subset of $\mathcal{S}_{ij}$ such that

$$\mathcal{S}_{ij}^{\delta} = \{p_{ij}^{k} \in \mathcal{S}_{ij} \mid l_{\mathrm{L}}(p_{ij}^{k}) \leq \delta\} \tag{4}$$

where $\delta \in \mathbb{R}^{+}$ is a fixed parameter. Two vertices i and j are connected in the cumulative connectome $\mathcal{G}_{\mathrm{C}}$ if there is at least one path of length in $\mathcal{G}_{\mathrm{L}}$ less or equal to δ joining them (we call such a path an admissible path). In the affirmative case, the weight of the edge connecting vertices i and j is given by the sum of the lengths in $\mathcal{G}_{\mathrm{NL}}$ of all admissible paths (hence the name cumulative connectome):

$$w_{\mathrm{C}}(i,j) = \sum_{p_{ij}^{k} \in \mathcal{S}_{ij}^{\delta}} l_{\mathrm{NL}}(p_{ij}^{k}).$$

Summarizing, in the cumulative connectome, two vertices i and j are connected if there is at least one path starting at i and ending at j whose length is smaller than a fixed value δ. Vertices that are not connected in $\mathcal{G}_{\mathrm{NL}}$ are connected in $\mathcal{G}_{\mathrm{C}}$ if there is an admissible path in $\mathcal{G}_{\mathrm{NL}}$ joining them. In this way, we encode in $\mathcal{G}_{\mathrm{C}}$ connections between brain regions over long distances. In other words, we connect two vertices if the corresponding brain regions are joined by axonal paths with a length determined by a δ parameter that we choose. The weights of the connections depend on the number and length of the axonal fibers in the connections.

3. The Models

By choosing appropriate graphs among the ones introduced in the previous section, we set up different models to describe the dynamics of $A\beta$ and τ proteins. The choice of the graphs is guided by the main biological features characterizing the two proteins and their dynamics.

Concerning $A\beta$, it is well known that monomeric $A\beta$ peptides originate physiologically from the cleavage of the transmembrane protein APP (amyloid precursor protein), and are considered non toxic [18]. Then, for a reason not yet clarified, an imbalance between the production and clearance of the protein can occur, giving rise to a process of agglomerations. This leads to the formation of toxic amyloid fibrils [19,20], often referred to as oligomers. These oligomers eventually accumulate in insoluble agglomerates known as senile plaques, nowadays considered non toxic [19]. To describe this process, we adopted a so-called compartmental model, that is, we considered $A\beta$ proteins assuming only three types of conformations: monomers, oligomers (which include all soluble conformations that are not monomers), and plaques (insoluble conformations). Moreover, we wanted to take into account that the $A\beta$ protein diffuses on short distances and its dynamics (production, aggregation, and clearance) are fast [21]. The molar concentration of a protein on the vertices V of a graph are denoted by an N-dimensional vector, where $N = |V|$ is the number of vertices of the graph. Accordingly, the vector-valued functions $\mathbf{u}_j : \mathbb{R} \to \mathbb{R}^N$ describe the molar concentration of $A\beta$ monomers ($j = 1$), oligomers ($j = 2$), and plaques ($j = 3$). We stress that with oligomers we mean aggregates made by two units while with plaques we mean aggregates made by three or more units.

We model $A\beta$ dynamics with the following equations [11]:

$$\begin{cases} \epsilon \dfrac{d\mathbf{u}_1(t)}{dt} = -\gamma_1 L_{\mathrm{P}} \mathbf{u}_1(t) + C_{\mathbf{u}_1} - \alpha \mathbf{u}_1(t) .* \displaystyle\sum_{j=1}^{3} \mathbf{u}_j(t) - \sigma_1 \mathbf{u}_1(t) & \text{(5a)} \\ \epsilon \dfrac{d\mathbf{u}_2(t)}{dt} = -\gamma_2 L_{\mathrm{P}} \mathbf{u}_2(t) + \dfrac{\alpha}{2} \mathbf{u}_1(t) .* \mathbf{u}_1(t) - \alpha \mathbf{u}_2(t) .* \displaystyle\sum_{j=1}^{3} \mathbf{u}_j(t) - \sigma_2 \mathbf{u}_2(t) & \text{(5b)} \\ \epsilon \dfrac{d\mathbf{u}_3(t)}{dt} = \dfrac{\alpha}{2} \displaystyle\sum_{3 \leq j+k < 6} \mathbf{u}_j(t) .* \mathbf{u}_k(t) - \sigma_3 \mathbf{u}_3(t) & \text{(5c)} \end{cases}$$

where $.*$ denotes the element-wise product between vectors. System (5) is endowed with initial conditions at $t = 0$ as follows (at the initial time, the brain is healthy; therefore, only $A\beta$ monomers are present):

$$\begin{cases} \mathbf{u}_1(0) = C_{\mathbf{u}_1}\mathbf{1} \\ \mathbf{u}_2(0) = \mathbf{0} \\ \mathbf{u}_3(0) = \mathbf{0} \end{cases} \tag{6}$$

where $\mathbf{0}$ and $\mathbf{1}$ denote the N-dimensional vectors whose components are all equal to zero and one, respectively. Since t denotes a slow time variable, the ϵ in front of the equations takes into account the fact that the processes described are fast [21]. The first term in Equations (5a) and (5b) models the diffusion of $A\beta$ monomers and oligomers along the network. Since this protein spreads along short distances, we choose the intrinsic proximity connectome to model the network. Indeed, on the intrinsic proximity connectome $\mathcal{G}_P$, two vertices i and j are connected only if the mean length of the axonal fibers connecting the corresponding regions of the brain is sufficiently short; in addition, the shorter the fiber length, the greater the weight of the corresponding edge. Accordingly, here, diffusion is driven by the graph Laplacian L_P. In the first equation, the term $C_{\mathbf{u}_1}$ represents a source term (gain) due to the physiological production of $A\beta$ monomers, the term $-\alpha\mathbf{u}_1(t).*\sum_{j=1}^{3}\mathbf{u}_j(t)$ is a loss due to the aggregation of monomers with other $A\beta$ proteins and the term $-\sigma_1\mathbf{u}_1(t)$ is a loss due to clearance phenomena. In the second equation, the term $\frac{\alpha}{2}\mathbf{u}_1(t).*\mathbf{u}_1(t)$ is a gain in oligomers due to the aggregation of two monomers, while the next two terms are losses analogous to those in the first equation. Finally, in the third equation, the diffusion term is absent since plaques are insoluble (they are too heavy to diffuse); there is a gain term due to aggregation (for a detailed explanation of this term and, in general, on the use of Smoluchowki's equations in this context, see [6]) and a loss term due to clearance. We stress that, in the equations governing the dynamics of $A\beta$ proteins, there is no coupling with the τ protein. Indeed, such a coupling will be present in the equation governing the dynamics of τ.

The protein τ is a physiological microtubule-associated protein: its main function is to stabilize microtubules and to regulate axonal transport. In brains affected by AD, misfolded τ has been found in the form of aggregates called neurofibrillary tangles, which, together with $A\beta$ plaques, are a hallmark of AD [22]. Misfolded τ is toxic for the neuron; indeed, it compromises stabilization, transport, and, in general, interferes with neuronal functions. As already mentioned in the Introduction, in recent years, there has been a consensus in the scientific community to consider a synergistic effect of $A\beta$ and τ when it comes to AD [4,23,24]. More precisely, the "trigger and bullet" hypothesis identifies in toxic $A\beta$ peptides the trigger for τ misfolding: the bullet is the misfolded τ, which, in turn, causes neural damage until, eventually, the neuron's death [25–27]. Concerning the spreading mechanism of misfolded τ, it has been proposed that neuronal damage spreads in the neuronal net through a neuron-to-neuron prion-like propagation mechanism [28–31]. Based on the considerations presented so far, the dynamics of misfolded τ, whose concentration is given by the vector-valued function $\mathbf{w} : \mathbb{R} \to \mathbb{R}^N$, are governed by the following equation [11]:

$$\frac{d\mathbf{w}(t)}{dt} = \gamma_3 K[\mathbf{w}] + C_{\mathbf{w}} \cdot \left(\mathbf{u}_2(t) - U_{\mathbf{w}}\right)^+ + s_{\mathbf{w}}(t) - \sigma_4\mathbf{w}(t) \tag{7}$$

endowed with the initial condition (at the initial time, the brain is healthy; therefore, there is no misfolded τ):

$$\mathbf{w}(0) = \mathbf{0}. \tag{8}$$

Here, we see the coupling between $A\beta$ and τ. Indeed, the second term $C_{\mathbf{w}} \cdot \left(\mathbf{u}_2(t) - U_{\mathbf{w}}\right)^+$ in the equation rules the interaction of the two proteins: if $\mathbf{u}_2$, the toxic form of $A\beta$, is above a given threshold, then it triggers the misfolding of τ. The term $s_{\mathbf{w}}(t)$ represents a source of misfolded τ, typically located in the enthorinal region of the brain [2], while the last term is the loss due to clearance.

The main purpose of this work is to investigate different possible forms for the operator K, appearing as the first term in (7), modeling the spreading of τ. As already mentioned, the spreading of such a protein can take place over long distances [32], possibly following a prion-like type of process [30], on a slow time scale. Several models have appeared in the literature concerning various possible mechanisms of τ spreading on networks [11,33–40]. Here, we considered four different possibilities (corresponding to four different mathematical models) for the operator K, and we made a comparison between the results obtained for each model and clinical data. In this way, we tried to determine if there is one form for K better describing the dynamics we were interested in reproducing. The full system of equations we studied is, therefore, given by

$$\begin{cases} \epsilon \dfrac{d\mathbf{u}_1(t)}{dt} = -\gamma_1 L_{\mathrm{P}} \mathbf{u}_1(t) + C_{\mathbf{u}_1} - \alpha \mathbf{u}_1(t) .* \displaystyle\sum_{j=1}^{3} \mathbf{u}_j(t) - \sigma_1 \mathbf{u}_1(t) & \text{(9a)} \\ \epsilon \dfrac{d\mathbf{u}_2(t)}{dt} = -\gamma_2 L_{\mathrm{P}} \mathbf{u}_2(t) + \dfrac{\alpha}{2} \mathbf{u}_1(t) .* \mathbf{u}_1(t) - \alpha \mathbf{u}_2(t) .* \displaystyle\sum_{j=1}^{3} \mathbf{u}_j(t) - \sigma_2 \mathbf{u}_2(t) & \text{(9b)} \\ \epsilon \dfrac{d\mathbf{u}_3(t)}{dt} = \dfrac{\alpha}{2} \displaystyle\sum_{3 \le j+k < 6} \mathbf{u}_j(t) .* \mathbf{u}_k(t) - \sigma_3 \mathbf{u}_3(t) & \text{(9c)} \\ \dfrac{d\mathbf{w}(t)}{dt} = \gamma_3 K[\mathbf{w}] + C_{\mathbf{w}} \cdot \left(\mathbf{u}_2(t) - U_{\mathbf{w}}\right)^+ + s_{\mathbf{w}}(t) - \sigma_4 \mathbf{w}(t) & \text{(9d)} \end{cases}$$

3.1. Model 1: Diffusion of τ Along the Structural Connectome $\mathcal{G}_{\mathrm{NL}}$

Following previous works in the literature [6,11,33,38], a quite standard choice to model the spreading of τ protein along edges of a connectome consists of choosing the Laplacian L_{NL} associated with the structural connectome $\mathcal{G}_{\mathrm{NL}}$. In this way, spreading corresponds to diffusion and we have

$$K[\mathbf{w}] = -L_{\mathrm{NL}} \mathbf{w}. \tag{10}$$

3.2. Model 2: Diffusion of τ Along the Cumulative Connectome $\mathcal{G}_{\mathrm{C}}$

In the cumulative connectome, two vertices are connected if there exists at least one admissible path between them. Such a path can be formed by several consecutive edges, each with its own weight contributing to the final weight. The corresponding Laplacian L_{C} therefore encodes the structure of the brain with connections spanning long distances. For this reason, we choose

$$K[\mathbf{w}] = -L_{\mathrm{C}} \mathbf{w}. \tag{11}$$

We stress again that all distances mentioned in the paper are to be understood not in a geometric but in an intrinsic sense, i.e., referring to fiber lengths, that is, edges, in the graph.

3.3. Model 3: Spreading of τ via Convolution on $\mathcal{G}_{\mathrm{L}}$

A common way of dealing with dynamics occurring over long distances is to make use of integral operators, typically convolutions with an appropriate kernel taking into account distances. With a notation similar to (4), let $\mathcal{S}_{ij}^{\tilde{\delta}}$ be the subset of $\mathcal{S}_{ij}$ containing all the paths in $\mathcal{G}_{\mathrm{L}}$ from node i to node j with a length less than $\tilde{\delta}$:

$$\mathcal{S}_{ij}^{\tilde{\delta}} = \{p_{ij}^k \in \mathcal{S}_{ij} \mid l_{\mathrm{L}}(p_{ij}^k) \le \tilde{\delta}\} \tag{12}$$

where $\widetilde{\delta} \in \mathbb{R}^+$ is a fixed parameter. We define the signal k_i on each node i of the graph $\mathcal{G}_\mathrm{L}$ as follows:

$$k_i = \sum_{p_{ij}^k \in \mathcal{S}_{ij}^{\widetilde{\delta}}} l_\mathrm{L}(p_{ij}^k). \tag{13}$$

Intuitively, k_i accounts for all the paths between i and each vertex j in $\mathcal{G}_\mathrm{L}$ that is not too far from i and plays the role of the convolution kernel. The convolution product on $\mathcal{G}_\mathrm{L}$ between $\mathbf{k}$ and $\mathbf{w}$ is defined by using the graph Laplacian eigenvectors [41]. Let $\{\mathbf{v}_\ell\}_{\ell=1,\dots,N}$ be a complete set of orthonormal eigenvectors of the Laplacian L_L of $\mathcal{G}_\mathrm{L}$ with eigenvalues $\{\lambda_\ell\}_{\ell=1,\dots,N}$ ordered in nondecreasing order and let $U \in \mathbb{R}^{N\times N}$ be the matrix whose columns correspond to the eigenvectors $\mathbf{v}_\ell$, $\ell = 1,\dots,N$. The graph Fourier transform $\hat{\mathbf{k}}$ of the signal $\mathbf{k}$ on $\mathcal{G}_\mathrm{L}$ is defined as the expansion of $\mathbf{k}$ in terms of the eigenvectors of L_L, i.e.,

$$\hat{\mathbf{k}} = U^*\mathbf{k} \tag{14}$$

where * denotes the conjugate transpose of a matrix. The inverse graph Fourier transform is given by

$$\mathbf{k} = U\hat{\mathbf{k}}. \tag{15}$$

The convolution product on $\mathcal{G}_\mathrm{L}$ between $\mathbf{k}$ and $\mathbf{w}$ is the vector $\mathbf{k} * \mathbf{w}$ such that

$$\mathbf{k} * \mathbf{w} = U(\hat{\mathbf{k}}.*\hat{\mathbf{w}}) \tag{16}$$

where $\hat{\mathbf{w}}$ is the graph Fourier transform of $\mathbf{w}$ defined as in (14). We choose $K[\mathbf{w}]$ as the convolution product of $\mathbf{w}$ with $\mathbf{k}$:

$$K[\mathbf{w}] = \mathbf{k} * \mathbf{w}. \tag{17}$$

3.4. Model 4: Spreading of τ via Convolution on $\mathcal{G}_\mathrm{NL}$

In [36,37], the authors introduce a graph convolution operator as a nonlocal model for the conversion from a healthy protein to a toxic one. Inspired by these works, we use their nonlocal operator to model the prion-like spreading of misfolded τ proteins. Let $M \in \mathbb{R}^{N\times N}$ be the matrix whose elements $m_{i,j}$, $i,j = 1,\dots,N$, are the length of the shortest path from vertex i to vertex j in the graph $\mathcal{G}_\mathrm{NL}$. If two vertices are not connected, $m_{i,j} = 0$. Moreover, let $\widetilde{M} \in \mathbb{R}^{N\times N}$ be the matrix with elements

$$\widetilde{m}_{i,j} = e^{-\eta^2 m_{i,j}^2}, \quad i,j = 1,\dots,N, \tag{18}$$

where η is a positive parameter. Finally, let $C \in \mathbb{R}^{N\times N}$ be the matrix defined as

$$c_{i,j} = \frac{\widetilde{m}_{i,j}}{\sum_{k=1}^N \widetilde{m}_{i,k}}, \quad i,j = 1,\dots,N. \tag{19}$$

Then, the operator $K[\mathbf{w}]$ is chosen as

$$K[\mathbf{w}] = C\mathbf{w}. \tag{20}$$

4. Methods

4.1. Data Availability

Data used in the preparation of this article were obtained from the ADNI Initiative (ADNI) database (https://adni.loni.usc.edu/, accessed on 24 November 2024). As described on the ADNI website, ADNI was launched in 2003 as a public–private partnership, led by Principal Investigator Michael W. Weiner, MD. The primary goal of ADNI has been to test whether serial MRI, positron emission tomography (PET), other biological markers, and clinical and neuropsychological assessments can be combined to measure the progression of mild cognitive impairment (MCI) and early Alzheimer's disease. For up-to-date information, see www.adni-info.org (accessed on 24 November 2024).

4.2. Subjects

The study included a total of 261 participants: 238 Cognitive Normal (CN) subjects (mean age $\pm$ *st.dev.* $= 72.7y \pm 8.7y$; M:F = 86:152), and 23 AD (mean age $\pm$ *st.dev.* $= 73.1y \pm 9.6y$; M:F = 11:12) from the Alzheimer's Disease Neuroimaging Initiative (ADNI). All data were downloaded from ADNI3. The number of included subjects is the consequence of the following inclusion criteria. To quantitatively measure the concentration of τ in the brain regions, we considered subjects having undergone a PET imaging with the radiotracer [^{18}F]-AV1451 and a corresponding accelerated sagittal T1-weighted anatomical image acquired using a 3D magnetization-prepared rapid acquisition gradient echo (MPRAGE) sequence from Siemens 3T MRI scanners to minimize effects of inter-scanner variability. To eliminate the known variability of the results due to the radiopharmaceutical used, we only considered PET images obtained with the same radiopharmaceutical ([^{18}F]-AV151). Moreover, we decided to select MR scan obtained only by 3T scanners produced by a single vendor, again to reduce variability. PET and MRI imaging were chosen to be acquired at no more than 3 months of temporal distance from one another. This substantial limitation was applied to prevent different image modalities from reflecting a different pathological condition of the same subject due to the progression of dementia.

4.3. PET Acquisition and Processing

[^{18}F]-AV1451 imaging acquisition and pre-processing in ADNI3 consisted of the baseline [^{18}F]-AV1451 PET scan (first scan acquired) for each subject, injection of 370 MBq (10 mCi) of tracer, and a 30 min dynamic scan consisting of six 5 min frames 75 min post-injection. Pre-processing consisted of the co-registration of separated frames to minimize patient motion effect, average of the 5 frames to obtain a single 30 min PET scan, re-orienting to a standardized space and normalization of intensity, and, finally, the application of a filter specific for the type of scanner.

4.4. MRI Structural Imaging Acquisition and Processing

The MRI structural images chosen within 3 months to the date of [^{18}F]-AV151 PET, were pre-processed, segmented, and parcellated with FreeSurfer 6.0 (http://surfer.nmr.mgh.harvard.edu/, accessed on 24 November 2024), in order to subdivide the brain volumes into a set of 83 anatomical cortical and subcortical region of interest (ROIs), separated between left and right hemisphere (with the addition of the brain stem) and belonging to six main networks within the brain: the frontal, the parietal, the occipital, the temporal, and the limbic lobes, plus the basal ganglia [42,43]. The subdivision just described was also chosen based on the prior hypothesis that the limbic system is specially affected by the disease. A complete list of the regions, the abbreviation of their names, and the network they belong to can be seen in the Appendix A. Nodes of the brain graphs were put in correspondence with the 83 ROIs, considering the anatomical information of each vertex, to compare τ concentrations relative to the 83 ROIs in the PET images and the concentrations estimated by the mathematical models.

4.5. PET and MRI Co-Registration and Quantification of τ

To calculate the differences in [^{18}F]-AV1451 uptake between AD and CN, we co-registered the PET images of each subject to the corresponding MRI structural image using PETSurfer (https://surfer.nmr.mgh.harvard.edu/fswiki/PetSurfer, accessed on 24 November 2024). Once the PET images were aligned in the same space as the structural image, since the ROIs' parcellization was in the latter space, we could calculate the τ concentration in each of the 83 ROIs. To have an absolute quantification of each subject, we normalized the concentration values to the cerebellum value (i.e., we divided the concentration values of each brain region by the concentration value in the cerebellum).

Indeed, the cerebellum is known to not be affected by the pathological accumulation of τ protein and it is considered a reference region for [^{18}F-AV1451 metabolism].

4.6. Statistical Analysis

Firstly, the normality and homogeneity of variances of the distributions of τ concentrations for each group (AD and CN) were tested using the Shapiro–Wilks and Levene's tests. This procedure was needed to test the hypothesis of whether the distributions of τ concentrations for AD and CN were those of normal random variables. Then, the difference of [^{18}F]-AV1451 PET uptake between AD and CN was calculated by a Mann–Whitney–Wilcoxon test. The null hypothesis (i.e., AD and CN's tau concentration values belong to the same distribution) was rejected when p-values were below 0.05. In these cases, the post hoc test, Tukey's HSD test, was employed since it is considered robust even when data are unbalanced [44]. Post hoc tests control the family-wise error rate by the Benjamini–Hochberg step-up procedure [45]. The ROIs for which the null hypothesis was rejected were considered significantly different between AD and CN. Significant ROIs were ordered from the most significantly different between AD and CN to the less significant until the non-significant ROIs and the corresponding value of [^{18}F]-AV1451 concentration was considered. In Figure 1, we depict the 83 cortical and subcortical nodes of the brain with colors corresponding to different τ concentrations. For the statistical analysis, we used Python libraries.

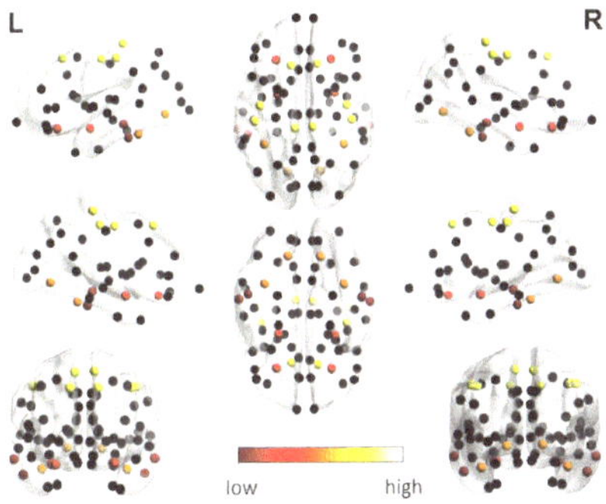

Figure 1. Nodes of the 83 cortical and subcortical ROIs of the brain. The color defines the nodes in which the concentration of τ is significantly greater in AD with respect to CN. The color code defines the increasing order in τ concentration.

5. Results

This section presents the results of the numerical simulations of the four mathematical models corresponding to the different choices for the operator K governing τ spreading. All simulations were conducted using Matlab R2021a on an Intel Core i5 processor with 2.50 GHZ and a Windows operating system. The codes used for the current experiments can be made available upon request to the authors.

5.1. Experimental Setting

The differential system (9) was numerically solved by using the Matlab function `ode45` implementing the Dormand–Prince 5(4) Runge–Kutta method with a variable time step for efficient computation [46]. Since misfolded τ protein has been shown to originate in the entorhinal region of the brain, the source term $s_{\mathbf{w}}(t)$ was chosen as non-null and equal to one only on those vertices belonging to the entorhinal region, i.e., for all $t > 0$, we have

$$(s_{\mathbf{w}}(t))_j = \begin{cases} 1 & j \in \text{entorhinal region} \\ 0 & \text{otherwise} \end{cases} \tag{21}$$

where $(s_{\mathbf{w}}(t))_j$ denotes the jth component of $s_{\mathbf{w}}(t)$. The seeding region for misfolded τ is shown in Figure 2.

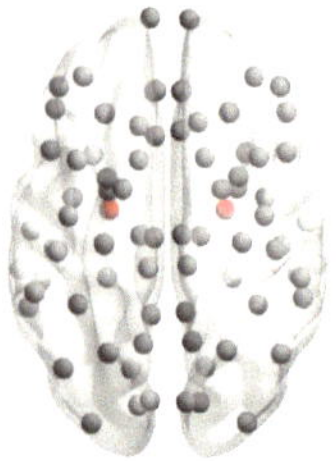

Figure 2. Nodes of the 83 ROIs corresponding to the cortical and subcortical regions of the brain. Colored nodes correspond to the enthorinal regions.

The differential system was integrated from the initial time $t_0 = 0$ to the final time $t_f = 150$. The final time was chosen to correspond to the stabilization of the solutions. In the numerical simulations of the four models, we fixed the values of all parameters of system (9), except γ_3, as reported in Table 1. These values were chosen, on the basis of different tests, as the ones giving the best approach to asymptotic equilibrium in a pathological situation.

Table 1. Fixed parameter values for system (9).

γ_1	γ_2	α	C_{u_1}	C_w	σ_1	σ_2	σ_3	σ_4
0.001	0.001	0.12	0.1	0.5	0.1	0.1	0.1	0.11

Moreover, we fixed, for the four models, with the same criterion as above, the parameters reported in Table 2. The procedure used for identifying, for each model, a good γ_3 value will be described in the following subsection.

Table 2. Fixed parameter values for the four models.

R_p	δ	$\tilde{\delta}$	η
25	30	15	5

5.2. Identification of the Parameter γ_3

The statistical analysis described in Section 4.6 shows that the distribution of ^{18}F-AV1451 uptake is not normal for all ROIs. Subsequently, the non-parametric Mann–Whitney–Wilcoxon Test with a Bonferroni correction for multiple comparisons showed that 29 ROIs significantly differ between AD and CN. Table 3 shows the six ROIs, which are mostly different (p-value < 10^{-5}) and their respective averaged τ concentrations.

Table 3. p-value and τ concentration (mean ± st.dev.) in AD and CN for the six most significant ROIs.

ROI	Significance	AD τ Concentration	CN τ Concentration
Fusiform region	8.3×10^{-9}	1.6 ± 0.3	1.2 ± 0.1
Inferior temporal region	1.6×10^{-8}	1.7 ± 0.5	1.2 ± 0.2
Middle temporal region	2.9×10^{-6}	1.6 ± 0.5	1.2 ± 0.2
Lingual region	3.1×10^{-6}	1.4 ± 0.3	1.1 ± 0.1
Lateral orbitofrontal region	5.4×10^{-6}	1.5 ± 0.4	1.2 ± 0.2
Amygdala	6.5×10^{-6}	1.4 ± 0.3	1.2 ± 0.1

The most significant ROIs for the accumulation of τ in AD with respect to CN are, in decreasing order, the inferior temporal region, the fusiform region, the middle temporal region, the lateral orbitofrontal region, the amygdala, and the lingual region. These six regions belong to three different networks (see Appendix A): the inferior temporal and middle temporal regions belong to the temporal network, the fusiform and the lingual regions are part of the occipital network, and the lateral orbitofrontal region and the amygdala are included in the limbic network. Since the brain is organized in several functional discrete networks, as introduced in Section 4, it is convenient to refer to the networks to which these regions belong rather than to the single ROIs. For this reason, we average the values of τ protein in the significant regions belonging to the same network. Let $w_{\mathrm{T}}^{(*)}$ be the average of the τ values obtained from clinical data in the inferior temporal and middle temporal regions. Analogously, let $w_{\mathrm{O}}^{(*)}$ be the mean τ value in the fusiform and the lingual regions and let $w_{\mathrm{L}}^{(*)}$ be the mean τ value in the latter orbital and the amygdala regions.

To evaluate the results obtained from four models, we also considered the value of the τ protein in the sensorimotor network that is known not to be characterized by the pathological accumulation of τ in AD. A good model for AD progression should, in fact, describe the evolution of τ both in the brain areas affected by the disease and in those that are not influenced. Therefore, let $w_{\mathrm{S}}^{(*)}$ be the average τ value in the paracentral, postcentral, precentral, and superior frontal regions forming the sensorimotor network. The τ mean values obtained from clinical data constitute our benchmark; they are ordered in decreasing order as follows:

$$w_{\mathrm{T}}^{(*)} > w_{\mathrm{L}}^{(*)} > w_{\mathrm{O}}^{(*)} > w_{\mathrm{S}}^{(*)}$$

resulting in the string

$$s^{(*)} = \text{“TLOS”}$$

which, therefore, represents the decreasing order of τ values in clinical data. Brain regions with a higher concentration of misfolded τ are more deteriorated; therefore, $s^{(*)}$ represents a *clinical deterioration pattern* of a brain with AD. We stress that this deterioration pattern considers only the brain regions that are significative for AD, according to our statistical analysis. Mathematical models should reproduce, in the best possible way, this deterioration pattern to be reliable tools for the analysis of AD progression. We remark that our benchmark TLOS is consistent with the literature and with the Braak stages, considering that the temporal and limbic regions are mostly affected, even in the early stages (Stages I and II) and central stages (in particular, the amygdala is involved in Stage IV), while, only in the latest stage (Stage VI), is there a final involvement of the motor and sensory areas, which corresponds to the fact that, in our data, the sensorimotor regions have low tau concentrations values. The parieto-occipital involvement has been demonstrated in more recent literature as being correlated with visuospatial dysfunction, especially in the early stage of the disease [47].

To evaluate the ability of the models to reproduce the deterioration pattern, we considered, for each model j, $j = 1, \dots, 4$, the values of τ protein at the final time t_f. We averaged the values on the vertices belonging to the inferior temporal and middle temporal regions to obtain $w_{\mathrm{T}}^{(j)}$, $j = 1, \dots, 4$. Proceeding similarly, we obtained $w_{\mathrm{L}}^{(j)}$, $w_{\mathrm{O}}^{(j)}$, and $w_{\mathrm{S}}^{(j)}$ as mean τ values on the vertices belonging to the later orbital and the amygdala regions, the fusiform and the lingual regions and the paracentral, postcentral, precentral, and superior frontal regions, respectively. For each model j, we defined the string $s^{(j)}$ formed using the letters T, L, O, and S sorted in decreasing order of τ values $w_{\mathrm{T}}^{(j)}$, $w_{\mathrm{L}}^{(j)}$, $w_{\mathrm{O}}^{(j)}$, and $w_{\mathrm{S}}^{(j)}$. We used the Hamming distance (HD) between the strings $s^{(*)}$ and $s^{(j)}$, $j = 1, \dots, 4$ to measure the distance between the clinical and the models' deterioration patterns. We recall that the Hamming distance between two strings is the number of positions at which the two strings differ [48]. The value of the parameter γ_3 was experimentally determined by trial and error to identify, for each model, the best outcome in terms of Hamming distance. Since, for each model, intervals of values exist for γ_3 giving the same optimal Hamming distance value, we identified the best γ_3 value in such intervals as follows. We define the vector

$$\mathbf{w}^{(*)} = (\tilde{w}_{\mathrm{T}}^{(*)}, \tilde{w}_{\mathrm{L}}^{(*)}, \tilde{w}_{\mathrm{O}}^{(*)}, \tilde{w}_{\mathrm{S}}^{(*)})^T$$

whose components, which are normalized τ values ranging from 0 to 1, are defined as follows:

$$\tilde{w}_{\mathrm{T}}^{(*)} = \frac{w_{\mathrm{T}}^{(*)}}{w_{\mathrm{T}}^{(*)}},\ \tilde{w}_{\mathrm{L}}^{(*)} = \frac{w_{\mathrm{L}}^{(*)}}{w_{\mathrm{T}}^{(*)}},\ \tilde{w}_{\mathrm{O}}^{(*)} = \frac{w_{\mathrm{O}}^{(*)}}{w_{\mathrm{T}}^{(*)}},\ \tilde{w}_{\mathrm{S}}^{(*)} = \frac{w_{\mathrm{S}}^{(*)}}{w_{\mathrm{T}}^{(*)}}.$$

Then, we define the vectors

$$\mathbf{w}^{(j)} = (\tilde{w}_{\mathrm{T}}^{(j)}, \tilde{w}_{\mathrm{L}}^{(j)}, \tilde{w}_{\mathrm{O}}^{(j)}, \tilde{w}_{\mathrm{S}}^{(j)})^T,$$

where

$$\tilde{w}_{\mathrm{T}}^{(j)} = \frac{w_{\mathrm{T}}^{(j)}}{M^{(j)}},\ \tilde{w}_{\mathrm{L}}^{(j)} = \frac{w_{\mathrm{L}}^{(j)}}{M^{(j)}},\ \tilde{w}_{\mathrm{O}}^{(j)} = \frac{w_{\mathrm{O}}^{(j)}}{M^{(j)}},\ \tilde{w}_{\mathrm{S}}^{(j)} = \frac{w_{\mathrm{S}}^{(j)}}{M^{(j)}},$$

and $M^{(j)} = \max\{w_{\mathrm{T}}^{(j)}, w_{\mathrm{L}}^{(j)}, w_{\mathrm{O}}^{(j)}, w_{\mathrm{S}}^{(j)}\}$ for $j = 1, \ldots, 4$. As a measure of the distance between $\mathbf{w}^{(*)}$ and $\mathbf{w}^{(j)}$, we can use the Root Mean Square Error (RMSE) and the Mean Absolute Error (MAE). We recall that, given a computed vector $\mathbf{w} \in \mathbb{R}^4$ of τ values, these error measures are , respectively, defined as

$$\mathrm{RMSE} = \sqrt{\frac{\|\mathbf{w}^{(*)} - \mathbf{w}\|_2^2}{4}},$$

$$\mathrm{MAE} = \frac{\|\mathbf{w}^{(*)} - \mathbf{w}\|_1}{4}$$

where $\|\cdot\|_2$ and $\|\cdot\|_2$ denote the L_2 and L_1 norms. Among all the γ_3 values corresponding to the same optimal HD value, we heuristically determine the value that achieves the best RMSE reduction. This leads to four values for γ_3, one for each model, shown in Table 4. We observe that the identified optimal γ_3 value is the same for Model 1 and Model 2.

Table 4. Identified optimal values for γ_3.

Model 1	Model 2	Model 3	Model 4
0.001	0.001	150	0.05

5.3. Numerical Results

Table 5 shows the clinical and computed deterioration patterns; Table 6 reports the corresponding HD, RMSE, and MAE values.

Table 5. Clinical and models' deterioration patterns.

Clinical Data	Model 1	Model 2	Model 3	Model 4
T	L	T	L	L
L	T	L	S	S
O	O	O	T	O
S	S	S	O	T

Table 6. Numerical results for the four models.

	Model 1	Model 2	Model 3	Model 4
HD	2	0	4	3
RMSE	0.11574	0.11543	0.10931	0.10615
MAE	0.09256	0.09212	0.08911	0.08950

In Figure 3, we show the deterioration pattern in the significative brain regions from clinical data and from the four different models.

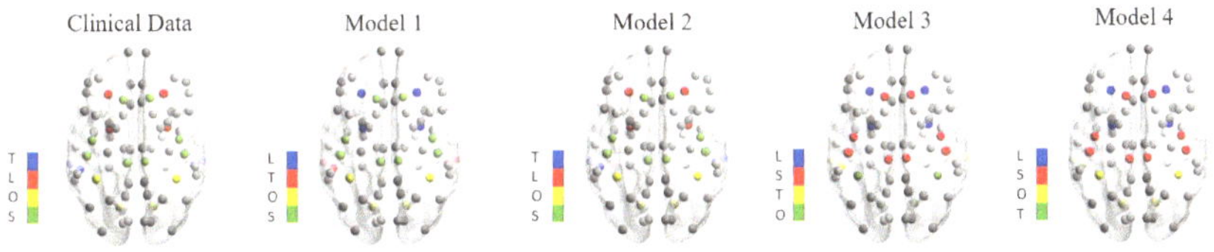

Figure 3. Deterioration pattern in the significative brain regions. From left to right: clinical data, Model 1, Model 2, Model 3, and Model 4. The non-significative regions are displayed in gray. In the color bar, the colors correspond to decreasing τ values in the significative regions.

From the results shown above, it is clear that the only model capable of reproducing the clinical deterioration pattern is Model 2. A diffusion operator along the cumulative connectome seems to be the most appropriate operator for modeling the spreading of τ protein in a brain with AD. Model 1 uses a diffusion operator too, but, comparing the deterioration patterns of Model 1 and Model 2, it is evident that the cumulative connectome, encoding information on brain regions connected along long distances, is a key tool. Even if Model 1 does not exactly reproduce the clinical deterioration pattern, it is able to distinguish between deteriorated and non-deteriorated regions, since the letter S is in the last position in the string $w^{(1)}$, as it should be. We observe from Table 6 that models 3 and 4 give the best results in terms of RMSE and MAE reduction. However, they do not reproduce the deterioration pattern of AD and do not distinguish between deteriorated and non-deteriorated regions (the letter S is in the second position both in $w^{(3)}$ and $w^{(4)}$). For this reason, they do not seem to be good models for the spreading of misfolded τ in presence of AD.

6. Discussion

In this study, we had two main objectives. We were interested in (1) comparing different modeling possibilities concerning the spreading of τ protein in a brain with AD, when the $A\beta$ protein is also present and the synergy between the two was considered; (2) producing clinical data that can be compared with the outputs of the models in order to verify their reliability. We believe that the lack of comparison between the results of a theoretical model and clinical data could be considered a deficiency that should be rectified in order to support the usefulness of mathematical models in AD [49]. To this end, we evolved the two proteins on appropriate networks, created from the medical data of human connectomes. The need for different connectomes was dictated by both the physiology of the brain and the biological characteristics of the proteins themselves. We therefore considered, for the evolution of $A\beta$, which travels only on short distances, an intrinsic proximity connectome, and a standard diffusion Laplacian on it. With regard to τ spreading, which is supposed to possibly travel with a prion-like mechanism and over long distances, we considered four distinct mathematical models on as many networks: a diffusion via Laplacian on a structural connectome and on a cumulative connectome, and a spreading via convolution on two further different connectomes. We emphasize that the so-called intrinsic proximity connectome and the cumulative connectome were introduced in this work for the first time. We identified from public multimodal data, with a careful statistical analysis, six regions relevant when it comes to AD and we evaluated τ concentrations in them, obtaining a degradation pattern that was crucial for us to verify the goodness of the models.

By comparing the results of the simulations with clinical data, we saw that only the model using the cumulative connectome was able to correctly reproduce the clinical degradation pattern, which also includes a network that is not damaged by the disease. We hypothesize that this result is due to the fact that the cumulative connectome encodes information on brain regions connected along long distances. On the basis of these findings, we feel that we can say that the models are not all equivalent, and that the comparison with clinical data is a crucial element to be able to assess their reliability. Of course, the results obtained here relate to the specific case of AD, and it is plausible that, for other diseases, the same models give different performances. As a next step, we are going to apply the same paradigm of analysis, i.e., a close comparison with appropriate clinical data, to the models enriched by taking into account atrophy, to be more realistic and see if something relevant happens. We believe that a highly interdisciplinary study, such as the one reported in the paper, could be a very promising direction in which cutting-edge biomedical mathematical research could be heading.

Author Contributions: All authors have contributed equally to this work. All authors have read and agreed to the published version of the manuscript.

Funding: This research was supported by the AlmaIdea 2022 programme of the Alma Mater Studiorum—Università di Bologna, project n. CUPJ45F21002000001.

Institutional Review Board Statement: As per ADNI protocols, all procedures performed in studies involving human participants were in accordance with the ethical standards of the institutional and/or national research committee and with the 1964 Helsinki declaration and its later amendments or comparable ethical standards. More details can be found at adni.loni.usc.edu. (This article does not contain any studies with human participants performed by any of the authors)

Informed Consent Statement: Authors received the consent for publication from ADNI.

Data Availability Statement: Publicly available datasets were analyzed in this study. These data can be found at the following urls: http://adni.loni.usc.edu and https://braingraph.org (accessed on 24 November 2024); see [14,15]. All data produced by the authors are available upon request from the authors.

Acknowledgments: Data collection and sharing for this project was funded by the Alzheimer's Disease Neuroimaging Initiative (ADNI) (National Institutes of Health Grant U01 AG024904) and DOD ADNI (Department of Defense award number W81XWH-12-2-0012). ADNI is funded by the National Institute on Aging, the National Institute of Biomedical Imaging and Bioengineering, and through generous contributions from the following: AbbVie, Alzheimer's Association; Alzheimer's Drug Discovery Foundation; Araclon Biotech; BioClinica, Inc.; Biogen; Bristol-Myers Squibb Company; CereSpir, Inc.; Cogstate; Eisai Inc.; Elan Pharmaceuticals, Inc.; Eli Lilly and Company; EuroImmun; F. Hoffmann-La Roche Ltd. and its affiliated company Genentech, Inc.; Fujirebio; GE Healthcare; IXICO Ltd.; Janssen Alzheimer Immunotherapy Research & Development, LLC.; Johnson & Johnson Pharmaceutical Research & Development LLC.; Lumosity; Lundbeck; Merck & Co., Inc.; Meso Scale Diagnostics, LLC.; NeuroRx Research; Neurotrack Technologies; Novartis Pharmaceuticals Corporation; Pfizer Inc.; Piramal Imaging; Servier; Takeda Pharmaceutical Company; and Transition Therapeutics. The Canadian Institutes of Health Research is providing funds to support ADNI clinical sites in Canada. Private sector contributions are facilitated by the Foundation for the National Institutes of Health (www.fnih.org). The grantee organization is the Northern California Institute for Research and Education, and the study is coordinated by the Alzheimer's Therapeutic Research Institute at the University of Southern California. ADNI data are disseminated by the Laboratory for Neuro Imaging at the University of Southern California.

Conflicts of Interest: The authors declare no conflicts of interest.

Appendix A

Table A1. List of anatomical cortical and subcortical ROIs identified with FreeSurfer, and the six networks: frontal, occipital, temporal, limbic, sensorimotor network, and basal ganglia.

Region of Interest	Network
Nucleus accumbens	Limbic
Amygdala	Limbic
Bankssts	Temporal
Brain Stem	Limbic
Caudal anterior cingulate gyrus	Limbic
Caudal middle frontal gyrus	Frontal
Caudate	Basal ganglia
Cuneus	Occipital
Entorhinal gyrus	Limbic
Frontal pole	Frontal
Fusiform gyrus	Occipital
Hippocampus	Limbic
Inferior parietal gyrus	Parietal
Inferior temporal gyrus	Temporal
Insula	Limbic
Isthmus cingulate gyrus	Limbic
Lateral occipital gyrus	Occipital
Lateral orbito frontal gyrus	Limbic
Lingual gyrus	Occipital
Medial orbito frontal gyrus	Limbic
Middle temporal gyrus	Temporal
Pallidum	Basal ganglia
Paracentral gyrus	Sensorimotor
Parahippocampal gyrus	Limbic
Pars opercularis	Frontal
Pars orbitalis	Frontal
Pars triangularis	Frontal
Peri calcarine gyrus	Occipital
Postcentral gyrus	Sensorimotor
Posterior cingulate gyrus	Limbic
Precentral gyrus	Sensorimotor
Precuneus	Parietal
Putamen	Basal ganglia
Rostral anterior cingulate gyrus	Limbic
Rostral middle frontal gyrus	Frontal
Superior frontal gyrus	Sensorimotor
Superior parietal gyrus	Parietal
Superior temporal gyrus	Temporal
Supramarginal gyrus	Parietal
Temporal pole	Temporal
Transverse temporal gyrus	Temporal

References

1. Ossenkoppele, R.; Rabinovici, G.D.; Smith, R.; Cho, H.; Schöll, M.; Strandberg, O.; Palmqvist, S.; Mattsson, N.; Janelidze, S.; Santillo, A.; et al. Discriminative accuracy of [18F] flortaucipir positron emission tomography for Alzheimer disease vs. other neurodegenerative disorders. *JAMA* **2018**, *320*, 1151–1162. [CrossRef] [PubMed]
2. Braak, H.; Braak, E. Neuropathological stageing of Alzheimer-related changes. *Acta Neuropathol.* **1991**, *82*, 239–259. [CrossRef]
3. Grothe, M.J.; Barthel, H.; Sepulcre, J.; Dyrba, M.; Sabri, O.; Teipel, S.J.; Initiative, A.D.N.; Initiative, A.D.N. In vivo staging of regional amyloid deposition. *Neurology* **2017**, *89*, 2031–2038. [CrossRef]
4. Busche, M.A.; Hyman, B.T. Synergy between amyloid-β and tau in Alzheimer's disease. *Nat. Neurosci.* **2020**, *23*, 1183–1193. [CrossRef]
5. Thompson, T.B.; Chaggar, P.; Kuhl, E.; Goriely, A.; Initiative, A.D.N. Protein-protein interactions in neurodegenerative diseases: A conspiracy theory. *PLoS Comput. Biol.* **2020**, *16*, e1008267. [CrossRef] [PubMed]

6. Bertsch, M.; Franchi, B.; Raj, A.; Tesi, M.C. Macroscopic modelling of Alzheimer's disease: Difficulties and challenges. *Brain Multiphys.* **2021**, *2*, 100040. [CrossRef]
7. Kim, J.; Jeong, M.; Stiles, W.R.; Choi, H.S. Neuroimaging modalities in Alzheimer's disease: Diagnosis and clinical features. *Int. J. Mol. Sci.* **2022**, *23*, 6079. [CrossRef]
8. Bassett, D.S.; Bullmore, E.T. Human brain networks in health and disease. *Curr. Opin. Neurol.* **2009**, *22*, 340–347. [CrossRef]
9. Sporns, O.; Zwi, J.D. The small world of the cerebral cortex. *Neuroinformatics* **2004**, *2*, 145–162. [CrossRef]
10. Torok, J.; Anand, C.; Verma, P.; Raj, A. Connectome-based biophysics models of Alzheimer's disease diagnosis and prognosis. *Transl. Res.* **2023**, *254*, 13–23. [CrossRef]
11. Bertsch, M.; Franchi, B.; Tesi, M.C.; Tora, V. The role of A β and Tau proteins in Alzheimer's disease: A mathematical model on graphs. *J. Math. Biol.* **2023**, *87*, 49. [CrossRef] [PubMed]
12. Catani, M.; Mesulam, M. What is a disconnection syndrome? *Cortex* **2008**, *44*, 911–913. [CrossRef]
13. Mesulam, M.M. A cortical network for directed attention and unilateral neglect. *Ann. Neurol. Off. J. Am. Neurol. Assoc. Child Neurol. Soc.* **1981**, *10*, 309–325. [CrossRef] [PubMed]
14. Szalkai, B.; Kerepesi, C.; Varga, B.; Grolmusz, V. High-resolution directed human connectomes and the consensus connectome dynamics. *PLoS ONE* **2019**, *14*, e0215473. [CrossRef] [PubMed]
15. Szalkai, B.; Kerepesi, C.; Varga, B.; Grolmusz, V. Parameterizable consensus connectomes from the human connectome project: The budapest reference connectome server v3. 0. *Cogn. Neurodyn.* **2017**, *11*, 113–116. [CrossRef] [PubMed]
16. Grigor'yan, A. *Introduction to Analysis on Graphs*; American Mathematical Society: Providence, RI, USA, 2018; Volume 71.
17. McNab, J.A.; Edlow, B.L.; Witzel, T.; Huang, S.Y.; Bhat, H.; Heberlein, K.; Feiweier, T.; Liu, K.; Keil, B.; Cohen-Adad, J.; et al. The Human Connectome Project and beyond: Initial applications of 300 mT/m gradients. *Neuroimage* **2013**, *80*, 234–245. [CrossRef]
18. Giuffrida, M.L.; Caraci, F.; Pignataro, B.; Cataldo, S.; De Bona, P.; Bruno, V.; Molinaro, G.; Pappalardo, G.; Messina, A.; Palmigiano, A.; et al. β-amyloid monomers are neuroprotective. *J. Neurosci.* **2009**, *29*, 10582–10587. [CrossRef]
19. Haass, C.; Selkoe, D.J. Soluble protein oligomers in neurodegeneration: Lessons from the Alzheimer's amyloid β-peptide. *Nat. Rev. Mol. Cell Biol.* **2007**, *8*, 101–112. [CrossRef]
20. Ono, K.; Condron, M.M.; Teplow, D.B. Structure–neurotoxicity relationships of amyloid β-protein oligomers. *Proc. Natl. Acad. Sci. USA* **2009**, *106*, 14745–14750. [CrossRef]
21. Meyer-Luehmann, M.; Spires-Jones, T.L.; Prada, C.; Garcia-Alloza, M.; De Calignon, A.; Rozkalne, A.; Koenigsknecht-Talboo, J.; Holtzman, D.M.; Bacskai, B.J.; Hyman, B.T. Rapid appearance and local toxicity of amyloid-β plaques in a mouse model of Alzheimer's disease. *Nature* **2008**, *451*, 720–724. [CrossRef]
22. Grundke-Iqbal, I.; Iqbal, K.; Tung, Y.; Quinlan, M.; Wisniewski, H.; Binder, L. Abnormal phosphorylation of the microtubule-associated protein τ (tau) in Alzheimer cytoskeletal pathology. *Alzheimer Dis. Assoc. Disord.* **1987**, *1*, 202. [CrossRef]
23. Ittner, L.M.; Götz, J. Amyloid-β and tau—A toxic pas de deux in Alzheimer's disease. *Nat. Rev. Neurosci.* **2011**, *12*, 67–72. [CrossRef] [PubMed]
24. Ricciarelli, R.; Fedele, E. The amyloid cascade hypothesis in Alzheimer's disease: It's time to change our mind. *Curr. Neuropharmacol.* **2017**, *15*, 926–935. [CrossRef]
25. Bloom, G.S. Amyloid-β and tau: The trigger and bullet in Alzheimer disease pathogenesis. *JAMA Neurol.* **2014**, *71*, 505–508. [CrossRef]
26. Bennett, R.E.; DeVos, S.L.; Dujardin, S.; Corjuc, B.; Gor, R.; Gonzalez, J.; Roe, A.D.; Frosch, M.P.; Pitstick, R.; Carlson, G.A.; et al. Enhanced tau aggregation in the presence of amyloid β. *Am. J. Pathol.* **2017**, *187*, 1601–1612. [CrossRef]
27. Small, S.A.; Duff, K. Linking Aβ and tau in late-onset Alzheimer's disease: A dual pathway hypothesis. *Neuron* **2008**, *60*, 534–542. [CrossRef]
28. Braak, H.; Del Tredici, K. Alzheimer's pathogenesis: Is there neuron-to-neuron propagation? *Acta Neuropathol.* **2011**, *121*, 589–595. [CrossRef] [PubMed]
29. Goedert, M.; Spillantini, M.G. Propagation of Tau aggregates. *Mol. Brain* **2017**, *10*, 18. [CrossRef]
30. Tatarnikova, O.; Orlov, M.; Bobkova, N. Beta-amyloid and tau-protein: Structure, interaction, and prion-like properties. *Biochemistry* **2015**, *80*, 1800–1819. [CrossRef]
31. Dujardin, S.; Hyman, B.T. Tau prion-like propagation: State of the art and current challenges. In *Tau Biology*; Takashima, A., Wolozin, B., Buee, L., Eds.; Springer: Singapore, 2020; pp. 305–325.
32. Ahmed, Z.; Cooper, J.; Murray, T.K.; Garn, K.; McNaughton, E.; Clarke, H.; Parhizkar, S.; Ward, M.A.; Cavallini, A.; Jackson, S.; et al. A novel in vivo model of tau propagation with rapid and progressive neurofibrillary tangle pathology: The pattern of spread is determined by connectivity, not proximity. *Acta Neuropathol.* **2014**, *127*, 667–683. [CrossRef] [PubMed]
33. Raj, A.; Kuceyeski, A.; Weiner, M. A network diffusion model of disease progression in dementia. *Neuron* **2012**, *73*, 1204–1215. [CrossRef] [PubMed]
34. Raj, A.; Tora, V.; Gao, X.; Cho, H.; Choi, J.Y.; Ryu, Y.H.; Lyoo, C.H.; Franchi, B. Combined model of aggregation and network diffusion recapitulates Alzheimer's regional tau-positron emission tomography. *Brain Connect.* **2021**, *11*, 624–638. [CrossRef] [PubMed]
35. Raj, A. Graph models of pathology spread in Alzheimer's disease: An alternative to conventional graph theoretic analysis. *Brain Connect.* **2021**, *11*, 799–814. [CrossRef] [PubMed]

36. Pal, S.; Melnik, R. Nonlocal multiscale interactions in brain neurodegenerative protein dynamics and coupled proteopathic processes. In Proceedings of the 9th Edition of the International Conference on Computational Methods for Coupled Problems in Science and Engineering (Coupled Problems 2021), Online Event, 13–16 June 2021; pp. 14–16.
37. Pal, S.; Melnik, R. Nonlocal models in the analysis of brain neurodegenerative protein dynamics with application to Alzheimer's disease. *Sci. Rep.* **2022**, *12*, 7328. [CrossRef]
38. Fornari, S.; Schäfer, A.; Jucker, M.; Goriely, A.; Kuhl, E. Prion-like spreading of Alzheimer's disease within the brain's connectome. *J. R. Soc. Interface* **2019**, *16*, 20190356. [CrossRef]
39. Fornari, S.; Schäfer, A.; Kuhl, E.; Goriely, A. Spatially-extended nucleation-aggregation-fragmentation models for the dynamics of prion-like neurodegenerative protein-spreading in the brain and its connectome. *J. Theor. Biol.* **2020**, *486*, 110102. [CrossRef]
40. Goriely, A.; Kuhl, E.; Bick, C. Neuronal oscillations on evolving networks: Dynamics, damage, degradation, decline, dementia, and death. *Phys. Rev. Lett.* **2020**, *125*, 128102. [CrossRef]
41. Shuman, D.I.; Narang, S.K.; Frossard, P.; Ortega, A.; Vandergheynst, P. The emerging field of signal processing on graphs: Extending high-dimensional data analysis to networks and other irregular domains. *IEEE Signal Process. Mag.* **2013**, *30*, 83–98. [CrossRef]
42. Evangelisti, S.; Testa, C.; Ferri, L.; Gramegna, L.L.; Manners, D.N.; Rizzo, G.; Remondini, D.; Castellani, G.; Naldi, I.; Bisulli, F.; et al. Brain functional connectivity in sleep-related hypermotor epilepsy. *NeuroImage Clin.* **2018**, *17*, 873–881. [CrossRef]
43. Mesulam, M.M. From sensation to cognition. *Brain A J. Neurol.* **1998**, *121*, 1013–1052. [CrossRef]
44. Brown, A.M. A new software for carrying out one-way ANOVA post hoc tests. *Comput. Methods Programs Biomed.* **2005**, *79*, 89–95. [CrossRef] [PubMed]
45. Benjamini, Y.; Hochberg, Y. Controlling the false discovery rate: A practical and powerful approach to multiple testing. *J. R. Stat. Soc. Ser. B (Methodol.)* **1995**, *57*, 289–300. [CrossRef]
46. Dormand, J.R.; Prince, P.J. A family of embedded Runge–Kutta formulae. *J. Comput. Appl. Math.* **1980**, *6*, 19–26. [CrossRef]
47. Cho, H.; Choi, J.Y.; Lee, S.H.; Lee, J.H.; Choi, Y.C.; Ryu, Y.H.; Lee, M.S.; Lyoo, C.H. Excessive tau accumulation in the parieto-occipital cortex characterizes early-onset Alzheimer's disease. *Neurobiol. Aging* **2017**, *53*, 103–111. [CrossRef] [PubMed]
48. Hutcheson, K. A test for comparing diversities based on Shannon formula. *J. Theor. Biol.* **1970**, *29*, 151–154. [CrossRef]
49. Moravveji, S.; Doyon, N.; Mashreghi, J.; Duchesne, S. A scoping review of mathematical models covering Alzheimer's disease progression. *Front. Neuroinform.* **2024**, *18*, 1281656. [CrossRef] [PubMed]

Mathematical and Computational Applications

MDPI

Article

A Mathematical Study of Effects of Alzheimer's Drug Donepezil Hydrochloride on Neuronal Viscoelasticity and Action Potentials

Corina S. Drapaca

Department of Engineering Science and Mechanics, Pennsylvania State University, University Park, PA 16802, USA; csd12@psu.edu

Abstract: Alzheimer's disease (AD) is a degenerative disorder characterized by progressive cognitive decline and memory loss. The few contemporary therapies may ease symptoms and/or slow down AD progression but cannot cure the disease. The orally administered AD drug donepezil hydrochloride enhances the availability of acetylcholine that supports cholinergic neurotransmission. In this paper, a generalized Hodgkin-Huxley model is proposed that uses Caputo fractional order temporal derivatives to link action potentials and viscoelasticity of cholinergic receptors. The model provides not only structurally dependent action potentials for health and AD but also a possible mechanism of donepezil effect on action potentials: the binding between the acetylcholine and the receptors preserves the structural fitness of these receptors. In addition, a generalized pharmacokinetic model of donepezil transport to the brain is proposed that incorporates controlled release modalities. Caputo fractional order temporal derivatives are used again to model anomalous drug release. Numerical simulations show how controlled release donepezil recovers the structural integrity of the receptors which further brings the abnormal action potentials due to AD to their healthy state. The results suggest that combining various drug release modalities and dosages may improve treatment effectiveness with donepezil.

Keywords: fractional calculus; viscoelasticity; action potentials; Alzheimer's disease; donepezil

1. Introduction

The first case of Alzheimer's disease (AD) was diagnosed in 1906 [1]. More than a century later, there are tens of millions of people with AD worldwide and millions of new cases are confirmed annually [2]. Currently, AD is considered to be a multifactorial degenerative disorder of the neuro-glial-vascular units characterized by progressive cognitive and memory deficits. The few existing therapies may ease symptoms and/or slow down AD progression but cannot cure the disease.

The lack of a cure and effective treatments for AD is because of the complexities of brain's structure and biochemical processes that have hindered the discovery of critical mechanisms governing the AD onset and progression. Pathological factors like the aggregation of toxic extracellular amyloid-β ($A\beta$) plaques, the accumulation of intracellular tau-containing neurofibrillary tangles (NFT) and the rise in neuroinflammation [3], as well as the intricate interlinks among them are presently considered possible causes of AD. For instance, the large loss of cells in the amygdala, hippocampus, and neocortex disrupts the cholinergic neurotransmission that leads to cognitive and functional dysfunctions. The accumulation of NFT causes the loss of neuronal functionality and eventually neuronal apoptosis. The aggregation of toxic $A\beta$ plaques impedes the clearance of brain waste via the glymphatic pathway and the blood-brain barrier which leads to neuroinflammation and neurodegeneration [4,5]. Cerebral chronic inflammation due to prolonged activation of microglia cells and astrocytes contributes not only to neurodegeneration but also to the aggregation of $A\beta$ plaques and NFT [6].

Given the rather futile attempts at treating AD by tackling various aspects of the accumulations of $A\beta$ plaques and NFT [7], recent years have seen renewed efforts in understanding the interplay between neurodegeneration, characterized by abnormal action potentials and damaged neuronal structure, and the buildups of $A\beta$ plaques and NFT that can improve the AD therapies aimed at enhancing the cholinergic neurotransmission [8–17]. One of the few FDA-approved drugs for treating AD that is safe and provides cognitive benefits at all stages of the disease is donepezil hydrochloride [7,18]. Donepezil is a fast-acting and reversible inhibitor of acetylcholinesterase (AChE), an enzyme that becomes abundant during AD progression and may be involved in the formation of the $A\beta$ plaques [7,8]. Specifically, the presence of $A\beta$ plaques increases the amount of one molecular form of AchE, while the accumulation of NFT causes an increase in the levels of all major molecular forms of AChE [8]. AChE disintegrates acetylcholine (ACh), a key neurotransmitter released by neurons involved in neuronal communications through binding to cholinergic ACh receptors [19,20]. Nicotinic and muscarinic receptors are the two main types of cholinergic ACh receptors. Nicotinic receptors are ligand-gated ion channels whose structural conformations are temporarily changed by the ACh binding to allow the fast depolarization of the postsynaptic neuronal membrane. Muscarinic receptors are G-protein coupled receptors that, when binded to ACh, activate intracellular signaling pathways via the G-proteins that cause the opening/closing of neuronal ion channels (hyperpolarization/depolarization of the membrane) and other slower cellular responses. The binding between AChE and donepezil increases temporarily the bioavailability of ACh and, thus, improves cholinergic transmission. Additionally, donepezil and its analogues are involved in the enhancement of cerebral blood flow and the reduction of neuroinflammation and oxidative stress and thus they may inhibit the aggregation of the $A\beta$ plaques [7,21].

Donepezil lessens AD symptoms but, as the disease progresses and negatively impacts the individual functions of the two types of cholinergic ACh receptors and the functional crosstalks between them [22,23], its dosage is increased gradually which can worsen gastrointestinal and heart-related adverse effects of the drug. To reduce side effects, various delivery and release systems for donepezil were proposed in the literature. For example, some donepezil-loaded systems that can reach the brain and provide nontoxic prolonged drug release are mango gum polymeric nanoparticles [24], extracellular vesicles isolated from human plasma [25], solid lipid nanoparticles with or without ApoE [26,27], cholesterol-modified pullulan nanoparticle with polysorbate 80 surface coverage [28], ternary sodium alginate based hydrogels [29], lyotropic liquid crystalline mesophases composed by monoolein/oleic acid/water [30], and sodium alginate microspheres [31]. Although in clinical practice donepezil is administered orally, nasal [32–34], subcutaneous [35] and transdermal [36] administrations have also been investigated and shown promising results.

Two aspects relevant to improving the treatment with donepezil that need more clarification are: (1) finding controlled release modalities for donepezil that provide effective treatments for longer time periods without increasing the drug dosage, and (2) finding mechanisms that link donepezil intake and action potentials. Mathematical modeling can provide insights into brain's processes and mechanisms worthy of further experimental/clinical explorations which ultimately may lead to better therapies for AD. Mathematical models of drug release (Fick, Higuchi, Hopfenberg Peppas, Weibull, etc. models) existing in the pharmacology literature predict cumulative drug release over time and can be easily fitted to experimental data [37,38]. In particular, the Ritger-Peppas model was used in experimental studies of controlled release donepezil to describe both Fickian and non-Fickian temporal release behavior [31]. The transport of a drug through the body can be described mathematically using pharmacokinetic (compartmental) models [39–41]. Specifically, pharmacokinetic models fitted to experimental observations of the transport of orally and nasally administrated donepezil were given in [32]. Lastly, the propagation of action potentials can be described mathematically by the Hodgkin-Huxley model [42] (see also [43]). Mathematical models of action potentials in the presence of AD were proposed in [44–46]. These models are obtained by coupling the Hodgkin-Huxley model and ion concentration dynamics specific to AD via the Nernst equations. Potential linkages among the above-mentioned pharmacokinetic and Hodgkin-Huxley mathematical models have not been explored yet.

In this paper, a mathematical model that couples controlled release of donepezil, transport to the brain of donepezil, and the effect of donepezil on action potentials is proposed to study the two aspects stated earlier. The model's main assumptions are that: (1) AD onset and progression cause structural damages to the ion channels within a neuronal membrane that lead to abnormal action potentials, and (2) in the presence of AD and donepezil intake, the binding between ACh and ligand-gated ion channels temporarily recovers the structural fitness of the channels accompanied by healthy action potentials. These assumptions are supported by [10–12] (assumption 1) and respectively [47,48] (assumption 2). Mathematically, the coupling is realized through fractional order derivatives, versatile integro-differential operators that have been successfully used to model long-range (spatial) interactions and long-term (temporal) memory characteristic to various systems. Specifically, anomalous diffusion (diffusion process characterized by a non-linear power law relationship between the mean squared displacement of a particle and time) [49] and viscoelastic fading memory (distant past deformations have less influence on the present state than those which happened in the more recent past) [50] have been modeled with fractional order derivatives and these modeling techniques are also used in this paper. Thus, a generalized pharmacokinetic model for the transport of donepezil with anomalous release patterns is proposed that combines the pharmacokinetic model in [32] and the controlled release modalities in [31]. Also, a generalized Hodgkin-Huxley model linking the electric activity of neuronal membrane (action potentials) and the viscoelasticity of ion channels is adapted from [51] (the focus in [51] was to study mechanical deformations of a neuron with a healthy electric activity described by the classic Hodgkin-Huxley model exposed to traumatic mechanical events). The viscoelastic behavior of the channels is described by the variable-order fractional Maxwell linear viscoelastic model which was introduced in [52] without a detailed presentation of its properties. The Haar piecewise-constant approximants of the variable fractional orders are used to highlight the creep and relaxation properties of the chosen viscoelastic model. Lastly, using the fact that ACh acts upon the structure of the ligand-gated ion channels to facilitate action potentials and assuming that the concentrations of ACh and donepezil in a brain with AD treated with donepezil are proportional, the coupling of the two generalized models is achieved by taking the variable fractional order describing the viscoelastic fading memory of the ion channels to be dependent on the concentration of donepezil. Numerical simulations show the following: (1) in the brain, an anomalous release donepezil decays slower than a donepezil solution available on the market, (2) abnormal action potentials correspond to longer viscoelastic fading memory of the ion channels, and (3) donepezil intake temporarily recovers healthy action potentials which agrees with clinical observations [7].

Thus, the original contributions of this paper are:

1. a pharmacokinetic model showing that anomalous release modalities of donepezil can slow down the decay of the drug in the brain which may provide effective treatments for AD with smaller drug dosages;
2. an electromechanical model of ion channels showing that: (1) a shorter viscoelastic fading memory modulated by the ACh binding is associated with a properly functioning neuron, (2) a longer viscoelastic memory corresponding to ion channels structurally altered by AD progression causes abnormal action potentials, and (3) donepezil intake can recover the structural fitness of the ion channels as well as healthy action potentials by using anomalous release donepezil either combined or not with various drug dosages depending on the AD severity;
3. Haar piecewise-constant approximations can be used to characterize the creep and relaxation properties of the variable-order fractional Maxwell linear viscoelastic model assumed to describe the mechanical behavior of ion channels.

Figure 1 shows a schematic of this paper's main contributions (represented by blue, green and red fonts) and how they link to specific knowledge about AD and donepezil treatment that was mentioned earlier. It is hoped that this work may inspire future studies of the neuronal electromechanical properties in health and AD that could lead to improved personalized therapies for AD and other brain disorders.

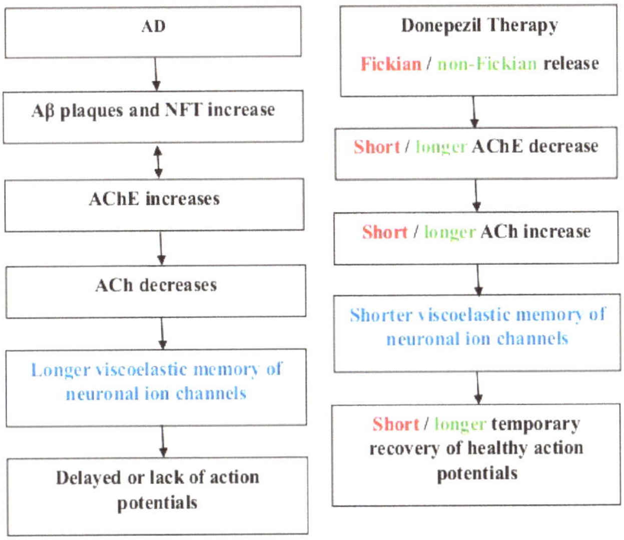

Figure 1. Schematics of simplistic linkages between AD/donepezil treatment and action potentials via the viscoelasticity of ion channels. The abnormal accumulations of $A\beta$ plaques and NFT specific to AD cause and benefit from an increase in the amount of AChE that leads to a decrease in the quantity of ACh available for binding to the nicotinic ACh receptors (and cholinergic ACh receptors, in general). The lack of ACh changes the structure of these receptors and their viscoelastic fading memory becomes longer which further causes delayed or loss of action potentials. On the other hand, donepezil treatment reduces temporarily the quantity of AChE and thus contributes to an increase in the amount of ACh. The binding between ACh and nicotinic ACh receptors changes the viscoelastic behavior of these receptors. The shorter fading memory modulated by the ACh binding leads to a temporary recovery of the healthy action potentials. Lastly, an anomalous release donepezil decays slower in the brain than a classic Fickian release, thus facilitating the propagation of healthy action potentials for a longer time. The blue, green and red fonts highlight this paper's main contributions.

The paper is organized as follows. A brief review of Caputo fractional derivatives of constant and variable orders and Haar wavelets is given in Section 2. Section 3 introduces the variable-order fractional Maxwell linear viscoelastic model, the generalized Hodgkin-Huxley and pharmacokinetic models and the coupling of these generalized models. Numerical results are shown in Section 4. Lastly, a discussion and concluding remarks are presented in Section 5.

2. Mathematical Preliminaries

The mathematical concepts needed in this paper are reviewed in this section. Readers familiar with these concepts may skip this section.

Definition 1. *1. [53] (p. 33), [54]: The* ***left-sided Riemann-Liouville fractional integral of constant order*** δ*,* $\delta > 0$*, of a function* $f \in L^1(a,b)$*,* $(a,b) \subset \mathbb{R}$*, is:*

$$I_{a+}^{\delta} f(x) = \frac{1}{\Gamma(\delta)} \int_a^x \frac{f(y)}{(x-y)^{1-\delta}} dy, \quad (1)$$

where $\Gamma(s) = \int_0^\infty t^{s-1}e^{-t}dt$ *is the gamma function. In particular,* $I_{a+}^0 f(x) = f(x)$.

2. *[54,55]: Let* $f : [a,b] \to \mathbb{R}$ *be a p-times differentiable function on* $(a,b) \subset \mathbb{R}$ *whose* p^{th}*-order derivative denoted by* $f^{(p)}$ *satisfies* $f^{(p)} \in L^1(a,b)$*, where* $p \in \{1,2,3,\dots\}$*. The* ***left-sided Caputo fractional derivative of constant order*** δ, $p-1 < \delta < p$*, of the function* f *is:*

$$D_{a+}^{\delta} f(x) = \frac{1}{\Gamma(p-\delta)} \int_a^x \frac{f^{(p)}(y)}{(x-y)^{\delta+1-p}} dy = I_{a+}^{p-\delta} f^{(p)}(x). \tag{2}$$

In particular, $D_{a+}^p f(x) = f^{(p)}(x)$.

3. *[56,57]: The* ***left-sided Caputo fractional derivative of variable order*** $\delta(x)$, $p-1 < \delta(x) < p$ *for* $p \in \{1,2,3,\dots\}$ *and* $x \in (a,b) \subset \mathbb{R}$ *of a function* $f : [a,b] \to \mathbb{R}$ *that is p-times differentiable on* (a,b) *and has* $f^{(p)} \in L^1(a,b)$ *is:*

$$D_{a+}^{\delta(x)} f(x) = \frac{1}{\Gamma(p-\delta(x))} \int_a^x \frac{f^{(p)}(y)}{(x-y)^{\delta(x)+1-p}} dy. \tag{3}$$

In [58], the variable-order Caputo fractional derivative (3) is called the variable order fractional derivative type 1.

Proposition 1. *1.* *[53] (p. 40), [54]: If* $\delta \in \mathbb{R}$ *and* $x > a$ *then:*

$$I_{a+}^{\delta}(x-a)^{\gamma-1} = \frac{\Gamma(\gamma)}{\Gamma(\delta+\gamma)} (x-a)^{\delta+\gamma-1}, \ Re(\gamma) > 0, \tag{4}$$

$$D_{a+}^{\delta}\, c = 0, \ \forall c \in \mathbb{R}, \tag{5}$$

and, if $p-1 < \delta < p$ *for* $p \in \{1,2,3\dots\}$*:*

$$D_{a+}^{\delta}(x-a)^{\gamma-1} = \begin{cases} \frac{\Gamma(\gamma)}{\Gamma(\gamma-\delta)} (x-a)^{\gamma-\delta-1}, & \gamma > p, \ \gamma \in \mathbb{R}, \\ 0, & \gamma \in \{1,2,\dots,p\}. \end{cases} \tag{6}$$

2. *[54], [59] (p. 18): If* $p-1 < \delta \le p$ *for* $p \in \{1,2,3\dots\}$ *and* $f : [a,b] \to \mathbb{R}$ *is a p-times differentiable function such that there exist* $r > \gamma \ge -1$ *and a continuous function* $g : (a,b) \to \mathbb{R}$ *such that* $f^{(p)}(x) = x^r g(x)$ *for* $x > a$ *then*

$$\begin{aligned} &D_{a+}^{\delta} I_{a+}^{\delta} f(x) = f(x), \\ &I_{a+}^{\delta} D_{a+}^{\delta} f(x) = f(x) - \sum_{j=0}^{p-1} f^{(j)}(a+)\frac{x^j}{j!}. \end{aligned} \tag{7}$$

3. *[59] (pp. 4, 19), [50] (pp. 60, 222–229): If* f, $D_{0+}^{\delta} f \in L^1(0,+\infty)$*, for* $p-1 < \delta \le p$ *and* $p \in \{1,2,3\dots\}$*, then the following identity holds:*

$$\mathcal{L}\left[D_{0+}^{\delta} f\right](s) = s^{\delta}\mathcal{L}[f](s) - \sum_{k=0}^{p-1} s^{\delta-k-1} f^{(k)}(0+), \tag{8}$$

where $\mathcal{L}[f](s) = \int_0^\infty e^{-sx} f(x)dx$, $s \ge 0$ *is the Laplace transform of the function* f*. Also, for* $0 < \delta \le 1$*:*

$$\begin{aligned} &\mathcal{L}\left[\frac{x^{-\delta}}{\Gamma(1-\delta)}\right](s) = s^{1-\delta}, \ \mathcal{L}\left[\frac{x^{\delta}}{\Gamma(1+\delta)}\right](s) = \frac{1}{s^{1+\delta}}, \\ &\mathcal{L}\left[E_{\delta}\left(-(x/a)^{\delta}\right)\right](s) = \frac{s^{\delta-1}}{s^{\delta}+(1/a)^{\delta}}, \end{aligned} \tag{9}$$

where $E_\delta(z) = \sum_{k=0}^{\infty} \frac{z^k}{\Gamma(k\delta+1)}$, $\delta > 0,\ z \in \mathbb{C}$ *is the Mittag-Leffler function. In particular,* $E_1(z) = e^z$.

4. *[58]: Let* $f : [0,T] \to \mathbb{R}$ *be a differentiable function,* $\delta : [0,T] \to (0,1]$, *and* $0 = x_0 < x_1 < \cdots < x_{N-1} < x_N = T$ *be an equally-spaced discretization of the interval* $[0,T]$ *of constant step size* Δx. *Denote by* $f^k = f(x_k)$, $\delta^k = \delta(x_k)$ *for* $x_k = k\Delta x$ *and* $k \in \{0,1,2,\ldots,N\}$. *Then a finite difference scheme for the variable-order fractional derivative (3) is:*

$$D_{0+}^{\delta(t)} f^{k+1} = \frac{\Delta x^{-\delta^{k+1}}}{\Gamma(2-\delta^{k+1})} \sum_{j=0}^{k} \left((j+1)^{1-\delta^{k+1}} - j^{1-\delta^{k+1}} \right) \left(f^{k+1-j} - f^{k-j} \right) + \mathcal{O}(\Delta x), \tag{10}$$

for $k = 0,1,2,\ldots,N-1$.

Fractional order integrals and derivatives of other special functions can be found in [53] (pp. 173–174), [60].

First introduced by Haar in [61], Haar wavelets are well-localized oscillatory functions in $L^2(\mathbb{R})$ (they have finite energy) with zero mean (derived from an admissibility condition that establishes the decay pattern of the Fourier transforms of the wavelets) [62] that have been used successfully in various applications such as signal/image processing [63] and numerically solving integro-differential equations [64–68]. In this paper, piecewise-constant approximants of variable fractional orders will be found using Haar wavelets and properties of the constant-order Caputo fractional derivative (2) will be applied to gain some insights into the behavior of the variable-order Caputo fractional derivative (3).

Definition 2 ([62] (pp. 69–82)). *1. The function* $H : \mathbb{R} \to \mathbb{R}$ *given by:*

$$H(x) = \begin{cases} 1, & 0 \le x < \frac{1}{2} \\ -1, & \frac{1}{2} \le x \le 1 \\ 0, & \text{otherwise} \end{cases}$$

is called the ***Haar mother wavelet****.*

2. *The functions* $H_{j,k} : \mathbb{R} \to \mathbb{R}$, $j,k \in \mathbb{Z}$, *given by:*

$$H_{j,k}(x) = 2^{j/2} H(2^j x - k),$$

are called the ***Haar wavelets****. They are obtained from function H by dilation with parameter* $j \in \mathbb{Z}$ *and translation with parameter* $k \in \mathbb{Z}$.

3. *The functions* $\varphi : \mathbb{R} \to \mathbb{R}$ *given by:*

$$\varphi(x) = \chi_{[0,1)}(x) = \begin{cases} 1, & 0 \le x < 1 \\ 0, & \text{otherwise} \end{cases}$$

is called the ***Haar scaling function****. Also, the dilation and translation of function* φ *generate scaling functions* $\varphi_{j,k} : \mathbb{R} \to \mathbb{R}$, $j,k \in \mathbb{Z}$, *defined as:*

$$\varphi_{j,k}(x) = 2^{j/2} \varphi(2^j x - k).$$

Proposition 2 ([62] (pp. 69–82), [69] (ch. 12)). *1. The following identity holds:*

$$H(x) = \frac{1}{\sqrt{2}} \varphi_{1,0}(x) - \frac{1}{\sqrt{2}} \varphi_{1,1}(x),\ x \in \mathbb{R}.$$

2. *The sequence* $\{\varphi_{J,k}\}_{k\in\mathbb{Z}} \cup \{H_{j,k}\}_{j\geq J, j,k\in\mathbb{Z}}$, $\forall J \in \mathbb{Z}$ *is a complete orthonormal sequence (orthonormal basis) in* $L^2(\mathbb{R})$. *Thus, every function* $f \in L^2(\mathbb{R})$ *can be represented as:*

$$f(x) = \sum_{k=-\infty}^{+\infty} c_{J,k}\varphi_{J,k}(x) + \sum_{j=J}^{+\infty}\sum_{k=-\infty}^{+\infty} d_{j,k}H_{j,k}(x),\ \forall J \in \mathbb{Z} \tag{11}$$

where the wavelet coefficients of f *are* $c_{J,k} =< f, \varphi_{J,k} >_{L^2(\mathbb{R})}$, $d_{j,k} =< f, H_{j,k} >_{L^2(\mathbb{R})}$, *and the* L^2*-inner product of two functions* $f, g \in L^2(\mathbb{R})$ *is:*

$$< f, g >_{L^2(\mathbb{R})} = \int_{\mathbb{R}} f(x)g(x)dx.$$

Representation (11) is known as the discrete Haar wavelet transform.

3. *The* ***scale spaces*** $V_J = \overline{\text{span}\{\varphi_{J,k}\}_{k\in\mathbb{Z}}}^{\|\cdot\|_{L^2(\mathbb{R})}}$, $J \in \mathbb{Z}$, *is a* ***multiresolution analysis***, *i.e.,*

$$V_J \subset V_{J+1} \subset L^2(\mathbb{R}),\ \forall J \in \mathbb{Z},\ \text{ and }\ \overline{\cup_{J\in\mathbb{Z}} V_J}^{\|\cdot\|_{L^2(\mathbb{R})}} = L^2(\mathbb{R}).$$

The ***detail spaces*** $W_J = \overline{\text{span}\{H_{J,k}\}_{k\in\mathbb{Z}}}^{\|\cdot\|_{L^2(\mathbb{R})}}$, $J \in \mathbb{Z}$, *satisfy the scale-step property:*

$$V_{J_2} = V_{J_1} \bigoplus \bigoplus_{j=J_1}^{J_2-1} W_j,\ \forall J_1, J_2 \in \mathbb{Z},\ J_1 < J_2.$$

Above, the L^2*-norm is* $\|f\|_{L^2(\mathbb{R})} = \sqrt{< f, f >_{L^2(\mathbb{R})}}$ *for* $f \in L^2(\mathbb{R})$.

Thus, for a function $f \in L^2(\mathbb{R})$ and a scale $J \in \mathbb{Z}$,

1. the scale functions act as low-pass filters providing a coarse, piecewise-constant approximation (first series of representation (11)):

$$f(x) = \sum_{k=-\infty}^{+\infty} c_{J,k}\varphi_{J,k}(x) \in V_J, \tag{12}$$

2. wavelets are band-pass filters providing the details of f as $\sum_{k=-\infty}^{+\infty} d_{J,k}H_{J,k}(x) \in W_J$ (second series of representation (11)).

Figures 2 and 3g–i show plots of Caputo fractional derivatives of the function $f(x) = x$ calculated using Formula (6). For a constant fractional order $\delta \in (0, 1]$, the fractional derivative becomes flatter as δ increases to 1 (Figure 2). Variable-order Caputo fractional derivatives of the function $f(x) = x$ are also calculated using the Haar piecewise-constant approximants of the following variable fractional orders: $\delta(x) = 0.25 + 0.15x$, $\delta(x) = 0.5 + 0.25\sin\frac{8x}{\pi}$, and $\delta(x) = 0.25 + 2e^{-x}$. Plots of the chosen variable orders and their corresponding Haar piecewise-constant approximants given by Formula (12) are shown in Figure 3a–f. The piecewise-constant approximants are found by applying the direct and inverse Haar wavelet transforms using Matlab's built-in functions **haart** and **ihaart** [70]. On each subinterval of a numerical discretization of the interval $[0, 5]$ the approximant is constant and Formula (6) can be used to calculate the corresponding constant-order Caputo fractional derivative. The plots in Figure 3g–i show not only that the variable-order Caputo fractional derivatives are bounded between the constant-order fractional derivatives for $\delta = 0.25$ and $\delta = 1$ since $\delta(x) \in (0.25, 1)$ for all considered cases, but also that certain features of the variable orders $\delta(x)$ seen in Figure 3a–c are transferred to the corresponding derivatives.

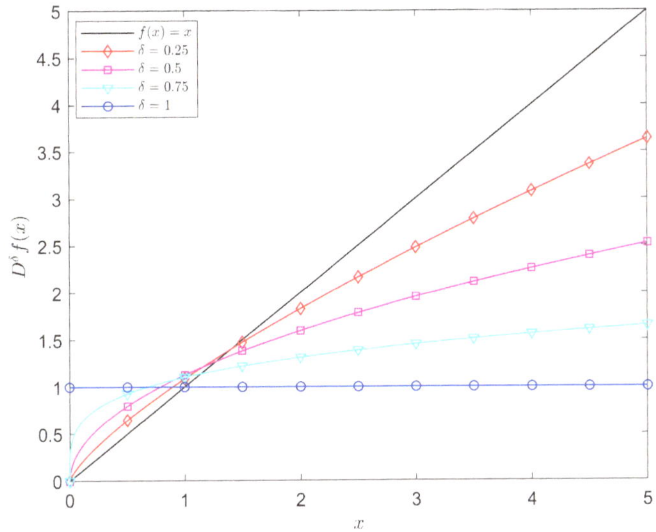

Figure 2. Plots of the function $f(x) = x$, $x \in [0,5]$, and its constant-order Caputo fractional derivatives calculated for various constant fractional orders $\delta \in (0,1]$.

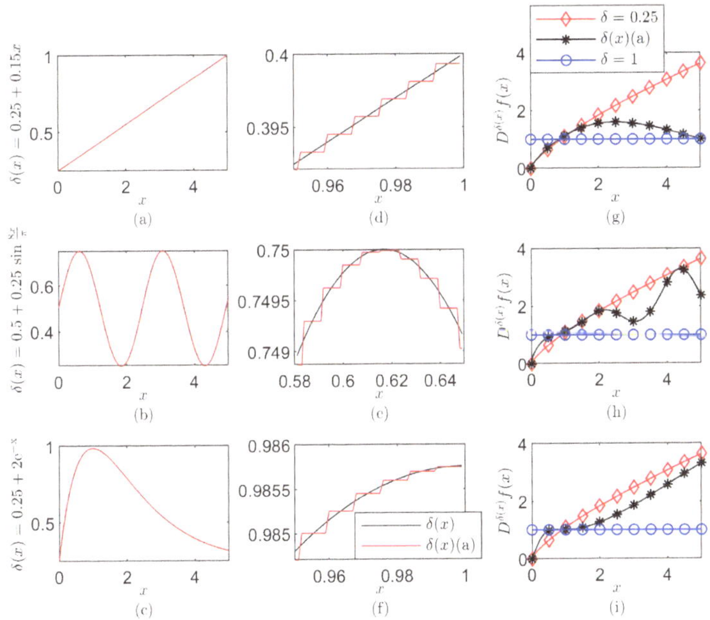

Figure 3. Plots of variable fractional orders $\delta(x)$ (black line) and their corresponding Haar piecewise-constant approximants (red line): (**a**) $\delta(x) = 0.25 + 0.15x$, (**b**) $\delta(x) = 0.5 + 0.25 \sin \frac{8x}{\pi}$, and (**c**) $\delta(x) = 0.25 + 2e^{-x}$. (**d–f**) Zoomed-in views of the plots shown in (**a–c**). (**g–i**) Plots of the variable-order Caputo fractional derivatives of the function $f(x) = x$, $x \in [0,5]$, for the Haar piecewise-constant approximants of the variable orders plotted in (**a–c**). (**a–f**) share the legend located in (**f**), while (**g–i**) share the legend in (**g**). In the legend maps, (**a**) stands for the piecewise-constant approximant of $\delta(x)$.

In this paper it is assumed that $0 < \delta,\ \delta(x) \leq 1$. Then, the constant-order and variable-order Caputo fractional derivatives (2) and respectively (3) model fading memory, and the memory gets smaller as the fractional order approaches 1. Also, the variable-order fractional derivative (3) conserves physical causality [71] for continuous functions $f(x)$ and has no memory of its past order [72]. The lack of past order memory gives the derivative (3) the ability to transfer to the output specific features (such as monotonicity or oscillatory behavior) of the fractional order $\delta(x)$ [73].

Remark 1. *Fading memory is terminology commonly used in the theory of viscoelasticity. However, anomalous diffusion processes are described using the mathematical concepts and language of probability theory. The integro-differential representation with a slowly decreasing power law kernel of the Caputo fractional derivative has been linked to the long-tailed waiting time probability density function with an asymptotic inverse power law expression specific to anomalous diffusion [49].*

3. Mathematical Modeling

The model of a variable-order fractional Maxwell linear viscoelastic material introduced in Section 3.1 is used to describe the mechanical behavior of neuronal ion channels and the corresponding generalized Hodgkin-Huxley model is presented in Section 3.2. The transport of a controlled release donepezil through various compartments of the body is described by the generalized pharmacokinetic model presented in Section 3.3. Lastly, the coupling of the proposed generalized Hodgkin-Huxley and pharmacokinetic models is given in Section 3.4. Figure 4 shows how these subsections are linked. The descriptions and physical units of the models' variables are given in Table A1.

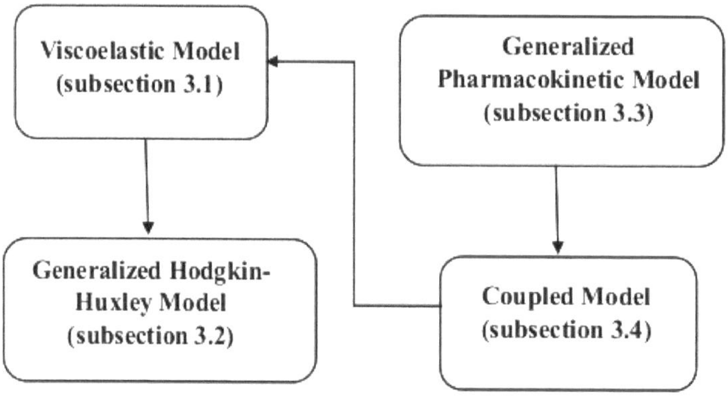

Figure 4. Linkages among Sections 3.1–3.4 of this section.

3.1. Viscoelastic Model

The constitutive equation of a variable-order fractional Maxwell linear viscoelastic model (schematically represented as an elastic spring connected in series to a viscous dashpot with long memory) is [52]:

$$D_{0+}^{\delta(t)}\sigma = E\left(D_{0+}^{\delta(t)}\epsilon - \frac{\sigma}{\mu_\delta}\right), \tag{13}$$

where $\sigma(t)$ and $\epsilon(t)$ are the stress and respectively infinitesimal strain in the one-dimensional case, and $t \geq 0$ denotes the physical time. The model has three physical parameters: E, the modulus of elasticity (of the spring element), μ_δ, and $0 < \delta(t) \leq 1$ (μ_δ, $\delta(t)$ correspond to the dashpot element). If $\delta(t) = \delta$ is a constant then Equation (13) becomes the fractional Maxwell model for $\delta \in (0,1)$ [50] (p. 62), [74], or the classic Maxwell model for $\delta = 1$ [75] (pp. 53–55), [50] (p. 32). Also, in the case $\delta(t) = 1$, $\forall t \geq 0$, parameter μ_δ is the dynamic viscosity μ and, thus, μ_δ is called the generalized viscosity. Lastly, the initial condition compatible with Equation (13) is the same as for the classic Maxwell model, namely:

$$\sigma(0+) = E\,\epsilon(0+). \tag{14}$$

If M, L, and T denote characteristic mass, length, and time, then dimensional analysis gives the dimensions of the physical quantities in Equation (13):

$$[\epsilon] = 1,\ [\delta(t)] = 1,\ [\sigma] = [E] = ML^{-1}T^{-2},\ [\mu_\delta] = ML^{-1}T^{\delta(t)-2} \tag{15}$$

since $\left[D_{0+}^{\delta(t)} f\right] = [f]T^{-\delta(t)}$ for a physical quantity f of dimension $[f]$.

When the variable fractional order is a constant $\delta \in (0,1]$, then [50] (p. 62):

1. the creep compliance $J(t)$ is the strain per unit applied constant stress σ_0 calculated from Equation (13). By replacing $\sigma = \sigma_0$ in Equation (13), using Formula (5), and then applying the Laplace transform to the updated Equation (13) combined with $\sigma_0 = E\epsilon(0+)$ (Formula (14) corresponding to this case), the following expression for J can be obtained from Formulas (8) and (9):

$$J(t) = \frac{1}{E} + \frac{1}{\mu_\delta \Gamma(1+\delta)} t^\delta. \tag{16}$$

2. the relaxation modulus $G(t)$ is the stress per unit applied constant strain ϵ_0 calculated from Equation (13). By replacing $\epsilon = \epsilon_0$ in Equation (13), using Formula (5), and then applying the Laplace transform to the updated Equation (13) combined with $\sigma(0+) = E\epsilon_0$ (Formula (14) corresponding to this case), the following expression for G can be obtained from Formulas (8) and (9):

$$G(t) = E\,E_\delta\left(-\frac{E}{\mu_\delta} t^\delta\right). \tag{17}$$

In particular, if $\delta(t) = 1, \forall t \geq 0$, then the corresponding creep compliance and relaxation modulus can also be obtained as closed-form solutions of first order linear (and separable) ordinary differential equations derived from Equation (13) [75] (pp. 53–55):

$$J(t) = \frac{1}{E} + \frac{1}{\mu} t, \tag{18}$$

$$G(t) = E\,e^{-\frac{E}{\mu} t}. \tag{19}$$

While Formula (18) is precisely Formula (16) for $\delta = 1$, Formula (19) admits a Taylor series representation given by Formula (17) for $\delta = 1$. In this case, Formula (17) approximates the function (19). Figure 5 shows plots of dimensionless functions $J(t)$ and $G(t)$ given by Formulas (16) and (17) versus a dimensionless time t for dimensionless parameters $E = 1$, $\mu_\delta = 1$ and various constant values of δ. The tail of the relaxation modulus gets longer as δ decreases to 0 (the decaying exponential behavior corresponding to $\delta = 1$ has the shortest tail; see Figure 5b).

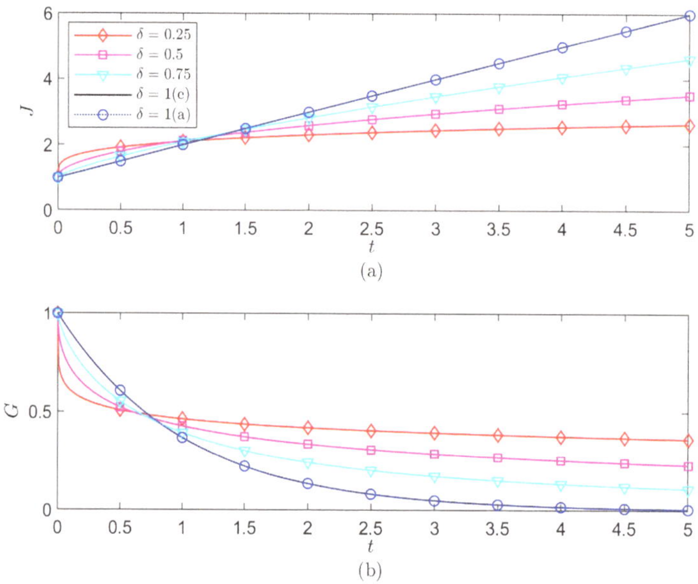

Figure 5. Plots of dimensionless (**a**) creep compliance $J(t)$ (Formula (16)) and (**b**) relaxation modulus $G(t)$ (Formula (17)) versus a dimensionless time t for dimensionless parameters $E = 1$, $\mu_\delta = 1$ and various constant fractional orders $\delta \in (0, 1]$. For $\delta = 1$, a perfect agreement is observed between the closed-form expressions (18), (19) (denoted by (e) in the legend) and the approximations (16), (17) (denoted by (a) in the legend). (**a**,**b**) share the legend in (**a**).

For a variable fractional order $\delta(t)$, the creep compliance and relaxation modulus are combinations of the creep and relaxation functions corresponding to multiple constant fractional orders. Indeed, if the variable fractional orders introduced in Section 2 and their corresponding Haar piecewise-constant approximants are used again, the creep compliance and relaxation modulus can be calculated using Formula (16) and respectively (17) on each subinterval of a numerical discretization of the chosen dimensionless time interval where the approximant is constant. The plots in Figure 6g–l, obtained for dimensionless parameters $E = 1$, $\mu_\delta = 1$, show that the approximated creep compliances and relaxation moduli corresponding to the given variable fractional orders $\delta(t)$ are bounded between the creep compliances and respectively relaxation moduli corresponding to the constant fractional orders $\delta = 0.25$ and $\delta = 1$ since $\delta(t) \in (0.25, 1)$ for all considered cases, and specific features of the variable orders $\delta(t)$ is transferred to the corresponding functions $J(t)$ and $G(t)$. For the sake of simplicity, plots of the chosen functions $\delta(t)$ and their corresponding piecewise-constant approximants are shown in Figure 6a–f. As expected from Formula (16), the approximated creep compliances corresponding to the variable fractional orders shown in Figure 6a–c resemble the plots in Figure 3g–i. For larger values of t, the relaxation modulus corresponding to the linear variable fractional order $\delta(t)$ plotted in Figure 6a decreases almost linearly (Figure 6j), while the tail of $G(t)$ plotted in Figure 6l becomes longer on the region where the variable fractional order $\delta(t)$ shown in Figure 6c experiences exponential decay. Lastly, the oscillatory decay of the relaxation modulus seen in Figure 6k is due to the oscillations of the variable fractional order $\delta(t)$ shown in Figure 6b.

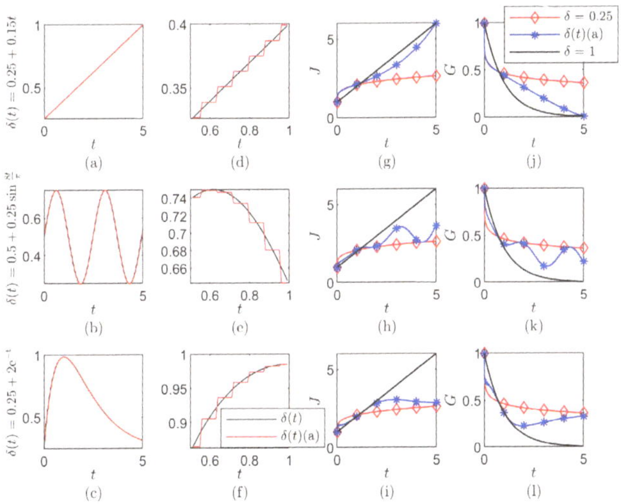

Figure 6. Plots of variable fractional orders $\delta(t)$ (black line) and their corresponding Haar piecewise-constant approximants (red line): (**a**) $\delta(t) = 0.25 + 0.15t$, (**b**) $\delta(t) = 0.5 + 0.25\sin\frac{8t}{\pi}$, and (**c**) $\delta(t) = 0.25 + 2e^{-t}$. (**d–f**) Zoomed-in views of the plots shown in (**a–c**). (**g–i**) Plots of dimensionless creep compliance $J(t)$ versus a dimensionless time t for dimensionless parameters $E = 1$, $\mu_\delta = 1$ and the Haar piecewise-constant approximants of the variable orders plotted in (**a–c**). (**j–l**) Plots of dimensionless relaxation modulus $G(t)$ versus a dimensionless time t for dimensionless parameters $E = 1$, $\mu_\delta = 1$ and and the Haar piecewise-constant approximants of the variable orders plotted in (**a–c**). (**a–f**) share the legend located in (**f**), while (**g–l**) share the legend in (**j**). In the legend maps, (**a**) stands for the piecewise-constant approximant of $\delta(t)$.

Since a variable fractional order $\delta(t)$ admits a Haar piecewise-constant approximant and the creep compliance $J(t)$ and relaxation modulus $G(t)$ corresponding to a constant fractional order are given by power law expressions (16) and respectively (17), the dimensionless physical parameter $\delta(t)$ of model (13) can be called the memory parameter.

Remark 2. *By using a Haar piecewise-constant approximation (12) of the memory parameter $\delta(t)$, definition (1) and property (7) of fractional calculus, and the initial condition (14), it can be shown that the constitutive law (13) can be written as a linear relationship between the stress $\sigma(t)$ and strain $\epsilon(t)$ involving a single integral with a power law memory kernel. For infinitesimal deformations, this one integral representation of the constitutive law can also be derived via the linearization of the Pipkin-Rogers model of non-linear viscoelasticity [76] that uses the Fréchet approximation of continuous functionals proposed in [77] to represent the stress as an infinite sum of multiple integrals of the finite strain-rate histories (also called a Volterra series, see [78]) whose so-called memory kernels are positive, at least continuous and monotonic decreasing functions of time so that they model viscoelastic fading memory. For model (13), the memory kernel is chosen to be a power law.*

3.2. Generalized Hodgkin-Huxley Model

In [51], a Lagrangian formulation and Hamilton's principle are used to derive the coupled equations describing neuronal electromechanics. The three ion gates m, n, and h whose open or close states produce action potentials in the classic Hodgkin-Huxley model [42] are assumed to behave mechanically like variable-order fractional Maxwell linear viscoelastic materials of constitutive Equation (13). Thus, in this context, m, n, and h, that in the Hodgkin-Huxley model are variables representing the activations of the Na^+ and K^+ channels and the inactivation of the Na^+ channel, respectively, become nondimensional displacements of the dashpots in the Maxwell elements modeling the material structures of the ion channels. The equations of the generalized Hodgkin-Huxley model link the electric activity of a neuronal membrane and the viscoelastic behavior of the membrane's ion channels as follows (these equations have simpler expressions than those in [51] since the neuronal mechanical behavior is neglected here; also, in [51], the focus was on the effects of traumatic mechanical events on neuronal deformation and thus only the classic Hodgkin-Huxley model was used in the numerical simulations):

$$c_m V^{(1)} = I - \left(G_{Na} m^3 h + G_{NaL}\right)(V - E_{Na}) - \left(G_K n^4 + G_{KL}\right)(V - E_K) - G_{ClL}(V - E_{Cl}), \tag{20}$$

$$D_{0+}^{\delta(t)} m = \alpha_m(1-m) - \beta_m m, \tag{21}$$

$$D_{0+}^{\delta(t)} n = \alpha_n(1-n) - \beta_n n, \tag{22}$$

$$D_{0+}^{\delta(t)} h = \alpha_h(1-h) - \beta_h h, \tag{23}$$

with initial conditions:

$$V(0) = V_{\text{rest}},\ m(0) = \frac{\alpha_m(V(0))}{\alpha_m(V(0)) + \beta_m(V(0))},$$
$$n(0) = \frac{\alpha_n(V(0))}{\alpha_n(V(0)) + \beta_n(V(0))},\ h(0) = \frac{\alpha_h(V(0))}{\alpha_h(V(0)) + \beta_h(V(0))}, \tag{24}$$

where:

$$\alpha_m = \frac{0.32(V+54)}{1 - e^{-0.25(V+54)}},\ \beta_m = \frac{0.28(V+27)}{e^{0.2(V+27)} - 1},$$
$$\alpha_n = \frac{0.032(V+52)}{1 - e^{-0.2(V+52)}},\ \beta_n = 0.5e^{-(V+57)/40},$$
$$\alpha_h = 0.128e^{-(V+50)/18},\ \beta_h = \frac{4}{1 + e^{-0.2(V+27)}}. \tag{25}$$

In Equation (20), c_m is the specific membrane capacitance, I is an externally applied current per unit area, E_{Na}, E_K, and E_{Cl} are the reverse potentials, and G_{Na}, G_K, G_{NaL}, G_{KL}, and G_{ClL} are, respectively, the voltage-gated maximal conductances of Na^+ and K^+, and the leak conductances of Na^+, K^+, and Cl^-. The membrane potential and its first order derivative are denoted by $V(t)$ and $V^{(1)}(t)$, respectively. The resting membrane potential is V_{rest}. The values of the physical parameters in Equations (20)–(23) are given in Table 1. These values are assumed to correspond to a properly functioning neuron.

Table 1. The values, physical units and descriptions of the model's parameters.

Parameters	Values and Physical Units [Reference]	Description
V_{rest}	−65 mV [43]	Normal resting potential
E_{Na}	60 mV [43]	Reversal potential of persistent sodium current
E_K	−88 mV [43]	Reversal potential of potassium current
E_{Cl}	−61 mV [43]	Reversal potential of chloride current
G_{Na}	0.3 mS/mm^2 [43]	Maximal conductance of sodium current
G_K	0.25 mS/mm^2 [43]	Maximal conductance of potassium current
G_{NaL}	0.000247 mS/mm^2 [43]	Conductance of leak sodium current
G_{KL}	0.0005 mS/mm^2 [43]	Conductance of leak potassium current
G_{ClL}	0.001 mS/mm^2 [43]	Conductance of leak chloride current
c_m	0.01 µF/mm^2 [43]	Membrane capacitance
I	0.1 µA/mm^2 [43]	Applied current

Like in the classic Hodgkin-Huxley model, Equation (20) represents Kirchhoff's current law. However, Equations (21)–(23) represent now evolution equations for the nondimensional displacements of the m, n, and h gates. In the classic Hodgkin-Huxley model,

Equations (21)–(23) were derived ad hoc from curve fitting to experiments and not from first principles. Also, given that there are no published experimental characterizations of the ion channels' viscoelasticity, it is assumed, for simplicity, that the viscoelastic fading memories of the ion channels are described by the same memory parameter $\delta(t)$. It is, however, possible to use different memory parameters in Equations (21)–(23) to model distinct patterns of viscoelastic fading memories of the channels. Furthermore, it is important to notice that this electromechanical model (20)–(23) can describe structurally intact as well as damaged neuronal membranes and thus predict various patterns of structurally dependent action potentials corresponding to both functional and dysfunctional neurons for the same externally applied current I and, for now, constant intra- and extra-cellular ion concentrations.

Lastly, the physical units of the parameters given by Formula (25) should be the same as those of the left-hand sides of the Equations (21)–(23), and thus all the constants multiplying potential V are meant to remove its physical units (mV). Furthermore, based on Formula (15) and the work in [51], these parameters have the following dimensions:

$$[\alpha_o] = [\beta_o] = \frac{[E_o]}{[\mu_{\delta_o}]} = T^{-\delta(t)}, \tag{26}$$

where $o \in \{m, n, h\}$ and E_o, μ_{δ_o} are the modulus of elasticity and, respectively, generalized viscosity of ion gate o. Formula (26) shows that the memory parameter $\delta(t)$ is intrinsic to the relaxation of the ion gates, which agrees with Formula (17) in the case of a constant fractional order δ. Thus, the relaxation time of ion gate o is given by $(\alpha_o + \beta_o)^{-1/\delta(t)}$.

3.3. Generalized Pharmacokinetic Model

In this paper the pharmacokinetic model of donepezil transport found in [32] is combined with the results on controlled release donepezil in [31]. In [32] mice received a donepezil solution orally and then the donepezil amounts were extracted from homogenized serum and brain tissue samples collected from the mice. By fitting the data to compartmental pharmacokinetic models using the software NONMEM 7.4 (NONlinear Mixed Effects Modeling), a three-compartment model that best describes the data was found. The compartments are: a depot for drug storage and distribution which for orally administered drugs is the stomach, the blood compartment, the brain, and the peripheral compartment. In [31], donepezil hydrochloride-encapsulated sodium alginate microspheres for oral administration are prepared. The controlled release of donepezil in vitro from the microspheres was fitted to a power law model. The exponent of release depended on the cross-linker concentration and time and suggested either a Fickian release pattern if its value was less than or equal to 0.5, or a non-Fickian (anomalous) diffusion pattern if the value was strictly greater than 0.5 and less than 1.

Following [49], in this paper anomalous diffusion will be modeled using a temporal Caputo fractional derivative of constant order $\varepsilon \in (0,1)$ and, thus, the generalized pharmacokinetic model describing the transport to the brain of controlled release donepezil is:

$$\begin{aligned}
D^{\varepsilon}_{0+} c_d(t) &= -k_{12}\, c_d(t), \\
D^{\varepsilon}_{0+} c_{\text{blood}}(t) &= k_{12}\, c_d(t) - (k_{20} + k_{23} + k_{24})\, c_{\text{blood}}(t) + k_{42}\, c_p(t), \\
D^{\varepsilon}_{0+} c_{\text{brain}}(t) &= \frac{V_{\text{blood}}}{V_{\text{brain}}}\, k_{23}\, c_{\text{blood}}(t) - k_{30}\, c_{\text{brain}}(t), \\
D^{\varepsilon}_{0+} c_p(t) &= k_{24}\, c_{\text{blood}}(t) - k_{42}\, c_p(t),
\end{aligned} \tag{27}$$

where $0 < \varepsilon \leq 1$ is the dimensionless exponent of drug release, and c_d, c_{blood}, c_{brain}, and c_p are the drug's (mass) concentrations in the depot, blood, brain, and peripheral compartments, respectively. The case $\varepsilon = 1$ describes classic Fickian diffusion and also corresponds to the donepezil solution available on the market that was used in [32].

Parameter k_{12} gives the drug's absorption rate through the depot to the blood, and parameters k_{20} and k_{30} are drug's elimination rates due to biochemical processes taking place within the blood and brain, respectively. Parameters k_{23}, k_{24}, and k_{42} are drug's transfer rates from the blood to the brain, and, respectively, between the blood and peripheral compartments. The drug's production rate in the brain is rescaled because the volumes of drug distribution in the brain, V_{brain}, and in the blood, V_{blood} are different [32]. The values of these parameters are assumed to be the same as those found using data fitting in [32] and are given in Table 2. A dimensional analysis of system (27) finds the dimension of the rates in system (27) to be $T^{-\varepsilon}$ which explains the physical units in Table 2.

Table 2. The values, physical units and descriptions of the model's parameters.

Parameters	Values and Physical Units [Reference]	Description
k_{12}	$0.059\,\text{min}^{-\varepsilon}$ [32]	Absorption rate through the depot
k_{20}	$0.012\,\text{min}^{-\varepsilon}$ [32]	Elimination rate in the blood
k_{23}	$0.014\,\text{min}^{-\varepsilon}$ [32]	Transfer rate from blood to brain
k_{24}	$0.034\,\text{min}^{-\varepsilon}$ [32]	Transfer rate from blood to periphery
k_{42}	$0.011\,\text{min}^{-\varepsilon}$ [32]	Transfer rate from periphery to blood
k_{30}	$0.325\,\text{min}^{-\varepsilon}$ [32]	Elimination rate in the brain
V_{blood}	279 mL [32]	Volume of distribution in blood
V_{brain}	18 mL [32]	Volume of distribution in brain
ε	$\{0.5, 0.658, 0.868, 1\}$ [31,32]	Exponent of release

Since system (27) is linear, and the eigenvalues of the system's matrix with the values in Table 2 are real numbers, the analytic solution exists and is given by [79,80]:

$$\begin{pmatrix} c_d \\ c_{\text{blood}} \\ c_{\text{brain}} \\ c_p \end{pmatrix}(t) = \sum_{i=1}^{4} c_i\, \mathbf{v}_i\, E_\varepsilon(\lambda_i\, t^\varepsilon), \tag{28}$$

where λ_i, $i \in \{1,2,3,4\}$ are the eigenvalues of the system's matrix, $\mathbf{v}_i$, $i \in \{1,2,3,4\}$ are the corresponding eigenvectors, and c_i, $i \in \{1,2,3,4\}$ are constants of integration. The constants of integration are found from the following initial condition taken from [32]:

$$\begin{pmatrix} c_d \\ c_{\text{blood}} \\ c_{\text{brain}} \\ c_p \end{pmatrix}(0) = \begin{pmatrix} 0.89\,\frac{\mu\text{g}}{\text{mL}} \\ 0 \\ 0 \\ 0 \end{pmatrix}. \tag{29}$$

which is calculated from an orally administered donepezil dose of 250 μg.

3.4. Coupled Model

Nicotinic ACh receptors are large transmembrane proteins with complex structures made of long twisted and folded chains of amino acids [47,48] that resemble the microstructures of viscoelastic materials [81] (incidentally, voltage-gated ion channels have comparable structures and, thus, also exhibit viscoelastic behavior [82]). Since nicotinic ACh receptors and collagen molecules are polymers (with different internal conformations) and collagen molecules exhibit viscoelastic behavior [83], it is assumed that nicotinic ACh receptors are also viscoelastic materials, and their electro-mechanic response is described by the generalized Hodgkin-Huxley model. The binding of ACh molecules to nicotinic ACh receptors briefly changes the internal conformations of the receptors' structures to create transmembrane pores for the transport of ions across the neuronal membrane (action potentials). This fast structural rearrangement modulated by the ACh ligand can be modeled as a short viscoelastic fading memory. Thus, it can be assumed that the ligand-gated ion channels (and, based on [82], neuronal ion channels in general) are viscoelastic materials with short viscoelastic memory such that they produce healthy action potentials that define a properly functioning neuron.

As mentioned in the introduction, the binding between AChE and donepezil increases the amount of ACh in the brain. In this paper it is assumed that the amount of donepezil in a brain with AD is proportional to the concentration of ACh and the following formula is proposed to model donepezil's effects on the viscoelastic behavior of the ligand-gated ion channels:

$$\delta(t) = \delta_{AD} + c_{\text{brain}}(t), \quad (30)$$

where $\delta_{AD} \in (0,1)$ is a constant for the AD-specific longer viscoelastic fading memory of the ligand-gated ion channels that causes abnormal action potentials (see later) and, by an abuse of notation, c_{brain} represents a nondimensional and rescaled concentration of donepezil in the brain calculated from Formula (28) such that $\delta(t) \in (0,1]$. Formula (30) that couples the generalized Hodgkin-Huxley and pharmacokinetic models presented earlier was chosen merely due to its mathematical simplicity since, currently, there are no published experimental observations supporting this formula.

Remark 3. *The extensions of the Picard-Lindeloff method proposed in [84,85] can be used to prove the existence and uniqueness of solutions to systems (20)–(23) and respectively (27) for constant fractional orders δ, $\varepsilon \in (0,1]$. Also, the extension of the method of Green's functions presented in [79] can be used to prove the existence of a unique solution of the system of linear fractional differential Equations (27). For a variable fractional order $\delta(t) \in (0,1]$, the Haar piecewise-constant approximant (12) of $\delta(t)$ can be used in combination with the extensions of the Picard-Lindeloff method in [84,85] to straightforwardly prove the existence and uniqueness of the solution of system (20)–(23).*

4. Results

This section presents numerical simulations obtained in Matlab [70]. Matlab's built-in function **ode15s** (implementation of variable-step, variable-order numerical differentiation formulas of orders 1 to 5) is used to numerically solve system (20)–(23) for $\delta = 1$, and Matlab's built-in function **ode45** (implementation of a single-step explicit Runge-Kutta (4,5) formula) is used to solve system (27) for $\varepsilon = 1$ [86]. For constant fractional orders δ, $\varepsilon \in (0,1)$, systems (20)–(23) and, respectively, (27) are solved numerically using Matlab's function **fde_pi1_ex** that implements the explicit product-integration rule of rectangular type with convergence order equal to one [87]. For a variable fractional order $\delta(t)$ given by Formula (30), system (20)–(23) is solved using the explicit Euler numerical scheme for Equation (20) and numerical scheme (10) for Equations (21)–(23) implemented in Matlab.

4.1. Action Potentials in Health and AD

Figures 7 and 8 show plots of the solutions of system (20)–(23) with initial conditions (24), parameters given by Formula (25) and in Table 1, and various constant fractional orders $\delta \in (0,1]$. Numerical solutions were obtained using the step size $\Delta t = 0.00025$ ms. The case $\delta = 1$ corresponds to the classic Hodgkin-Huxley model which is considered the healthy case (Figure 7a,e). It is assumed that the healthy action potentials are characterized by positive overshoots and ability to reach the value of the resting membrane potential given in Table 1. As δ decreases to 0, the viscoelastic memory of the ion gates gets longer slowing down the opening or closing of the m, n, and h gates (Figure 8). This causes the oscillations of the membrane's potential to become delayed (Figure 7b,c,e) and eventually cease to exist (Figure 7d,e). The amplitude of these oscillations also decreases with decreasing δ, and the positive overshoot stops to happen (Figure 7). Lastly, as δ gets smaller, the peak-to-peak amplitude of the membrane's potential gets shorter and the membrane potential does not reach the resting potential V_{rest} given in Table 1 anymore.

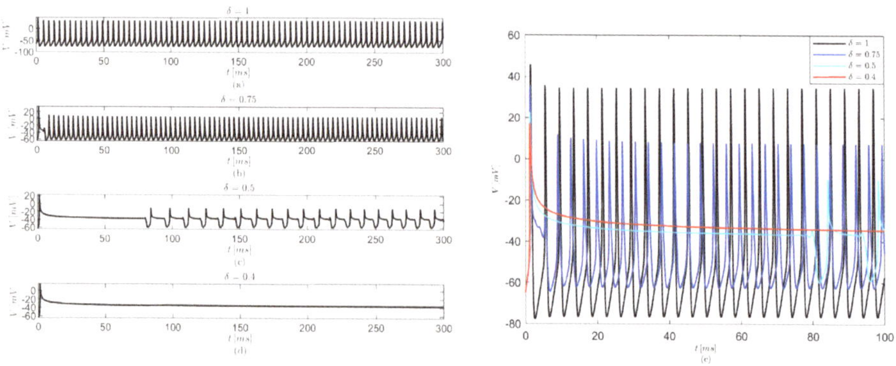

Figure 7. (**a–d**) Plots of membrane potentials for various constant fractional orders $\delta \in (0, 1]$. (**e**) A comparison of zoomed-in views of the plots shown in (**a–d**). As δ decreases to 0, the viscoelastic memory gets longer causing the peak-to-peak amplitude of the action potential V to get shorter and the resting potential to get higher. Also, plateau (delayed) potentials followed by higher frequency of firing are observed for $\delta = 0.75$, while for $\delta = 0.5$ a first long plateau region is followed by periodic shorter delayed potentials. Lastly, the action potentials cease to exist for values of δ closer to 0.

It is further assumed that a shorter viscoelastic memory (δ closer to 1) describes structurally intact ion gates that generate healthy action potentials and thus this neuron is considered functional. A longer viscoelastic memory (δ closer to 0) describes structurally altered ion gates assumed here to be caused by the decreased amount of ACh due to AD that generates abnormal action potentials and thus this is a dysfunctional neuron. It is also assumed that the abnormal action potentials for $\delta_{AD} = 0.75$ correspond to earlier stages of AD, and the abnormal action potentials for $\delta_{AD} = 0.5$ correspond to later stages of AD.

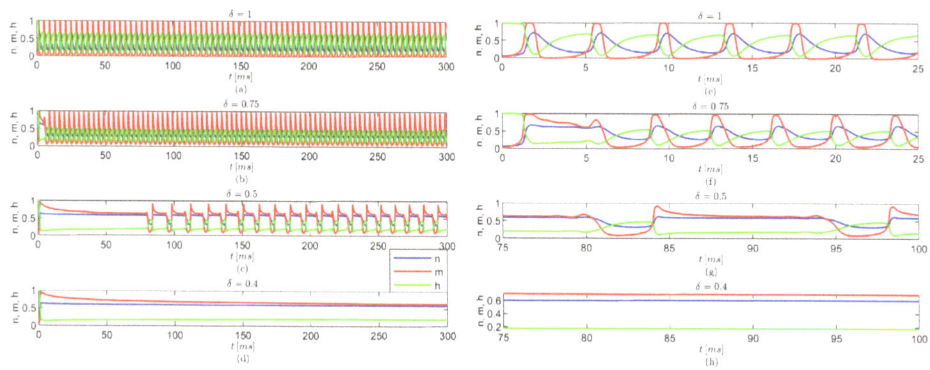

Figure 8. (**a–d**) Plots of displacements m, n, and h for various constant fractional orders $\delta \in (0, 1]$. (**e–h**) Zoomed-in views of the plots shown in (**a–d**). (**a–h**) share the legend in (**d**). As δ goes to 0, the periodic oscillations of m, n, and h gates start changing their shapes, plateau regions appear and get longer until the gates stop opening or closing.

4.2. Transport of Controlled Release Donepezil

Figures 9 and 10 show the solutions of system (27) with initial conditions (29) and parameters given in Table 2. The numerical solutions are obtained for a step size of $\Delta t = 0.005$ min. For each value of $\varepsilon \in (0, 1]$, a perfect agreement between the the analytic solution (28) and the numerical solution of system (27) is observed (Figure 9). This validates the numerical solutions obtained using Matlab's functions **fde_pi1_ex** and **ode45**. A comparison of the concentrations of donepezil for various value of ε in each of the four compartments is shown in Figure 10. As ε decreases from 1, the concentration of controlled release donepezil gets longer tails at later times (Figure 10a–c). Since the very slow decay of donepezil in the stomach for $\varepsilon = 0.5$ may elevate gastrointestinal adverse effects (Figure 10a), the value $\varepsilon = 0.658$ is chosen as the optimal controlled release based on the slow decrease of donepezil in the brain (Figure 10c).

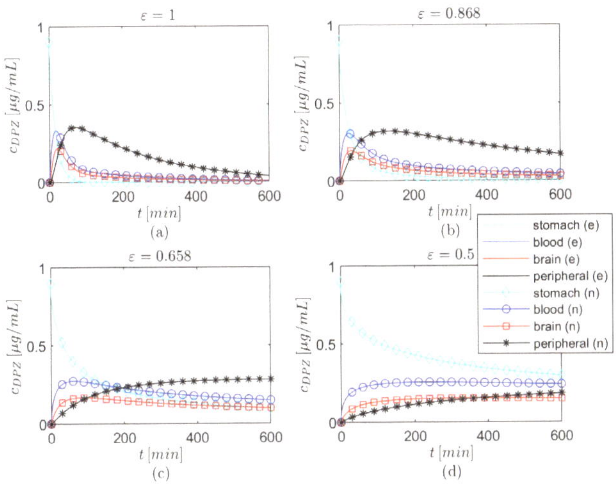

Figure 9. Plots of concentrations of donepezil in the depot, blood, brain and peripheral compartments for various constant values of ε. (**a**–**d**) share the legend in (**d**). In the legend map, (n) stands for numerical solution and (e) for exact (analytic) solution. The subscript DPZ stands for donepezil. For each value of ε, a perfect agreement is noticed between the exact and numerical solutions for each compartment.

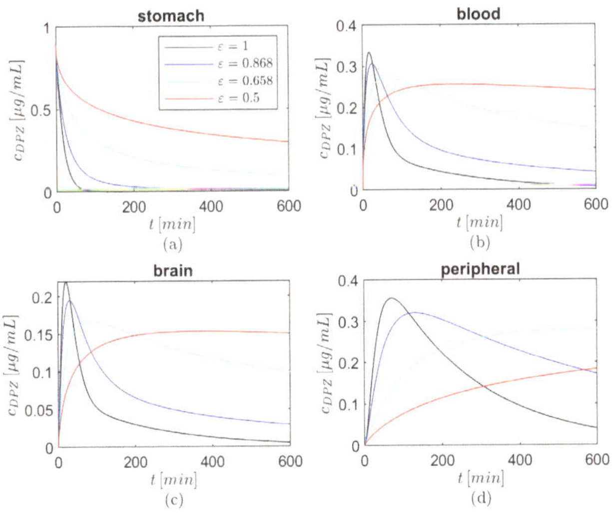

Figure 10. A comparison of concentrations of donepezil for various values of ε in each compartment: (**a**) depot, (**b**) blood, (**c**) brain, and (**d**) peripheral. The subscript DPZ stands for donepezil. The tail of the concentration in each compartment gets longer as ε decreases. Specifically, in the brain, as ε decreases, the maximum concentration of donepezil decreases and is reached at a later time, and the concentration decays slower with time.

4.3. Donepezil Effects on Action Potentials

Figure 11 shows plots of viscoelastic memories $\delta(t)$ given by Formula (30) for $\delta_{AD} \in \{0.5, 0.75\}$ and dimensionless and rescaled concentrations of controlled release donepezil in brain given by Formula (28) for $\varepsilon \in \{0.658, 1\}$. For $\delta_{AD} = 0.5$, the following cases are also considered: (1) the initially administered donepezil dose for $\varepsilon = 0.658$ is doubled to 500 μg, (2) the donepezil dosage is 250 μg every 120 min during the observed period of 600 min for $\varepsilon \in \{0.658, 1\}$. The concentrations of donepezil in brain are rescaled by interpolating using Matlab's built-in function **interp1** (Figure 11). It is further assumed that the shapes of the concentrations remain the same on a time scale of ms. This assumption allows to numerically solve system (20)–(23) for a variable fractional order $\delta(t)$ without exceeding Matlab's maximum array size and memory allowed by the computer used for calculations (HP Spectre 360× laptop with 64×-based PC and 8 GB RAM), and, thus, the patterns of the action potentials corresponding to the various shapes of $\delta(t)$ shown in Figure 11 can be visualized. The plots in Figure 11 resemble the variable fractional orders in Figure 6b,c so the creep and relaxation responses of viscoelastic materials with memories shown in Figure 11 will look like those in Figure 6h,k and respectively Figure 6i,l.

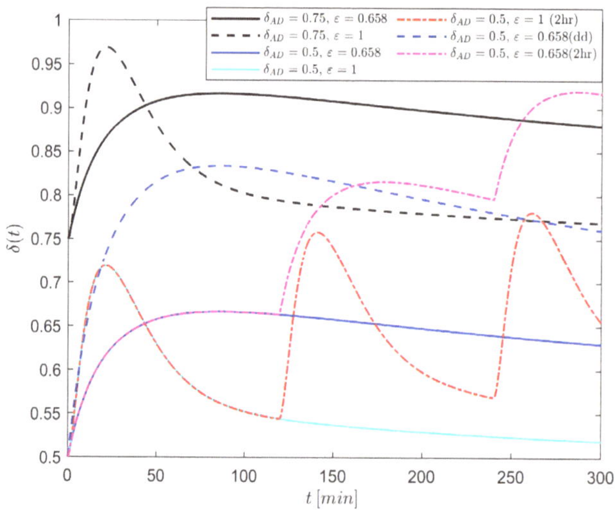

Figure 11. Plots of $\delta(t)$ given by Formula (30) for $\delta_{AD} \in \{0.5, 0.75\}$ and $\varepsilon \in \{0.658, 1\}$. In the legend map, (dd) stands for double dose, and (2 h) stands for the administration of donepezil every two hours. Due to computer memory constraints, it is assumed that the shapes of $\delta(t)$ remain unchanged on the time domain [0, 300] ms of the membrane potential.

Figures 12–17 show plots of the solutions to system (20)–(23) with initial conditions (24), parameters given by Formula (25) and in Table 1 and variable fractional orders shown in Figure 11. The solutions are obtained for a step size $\Delta t = 0.005$ ms. The plots show the effects of controlled release donepezil on action potentials.

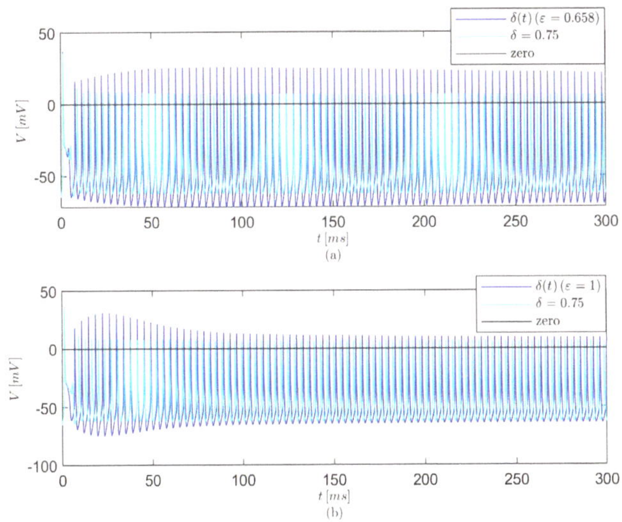

Figure 12. Action potentials for viscoelastic memories $\delta_{AD} = 0.75$ and, respectively, $\delta(t)$ given by Formula (30) with $\delta_{AD} = 0.75$ and controlled release donepezil for (**a**) $\varepsilon = 0.658$ and (**b**) $\varepsilon = 1$. Plots not to scale. Donepezil intake recovers the ability of the membrane potential to have positive overshoots and reach the normal resting potential for both values of ε.

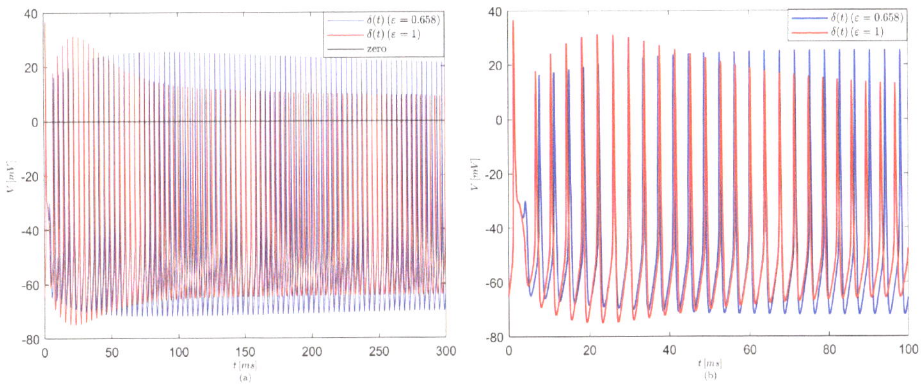

Figure 13. (**a**) Comparison of action potentials for viscoelastic memory $\delta(t)$ given by Formula (30) with $\delta_{AD} = 0.75$ and with Fickian ($\varepsilon = 1$) and non-Fickian ($\varepsilon = 0.658$) release of donepezil. (**b**) Zoomed-in view of the plots in (**a**). Plots not to scale. The membrane potential after donepezil treatment with non-Fickian control release has positive overshoots and reach the normal resting potential for a longer time and the peak-to-peak amplitude of the oscillations are larger than the membrane potential after donepezil intake with Fickian control release.

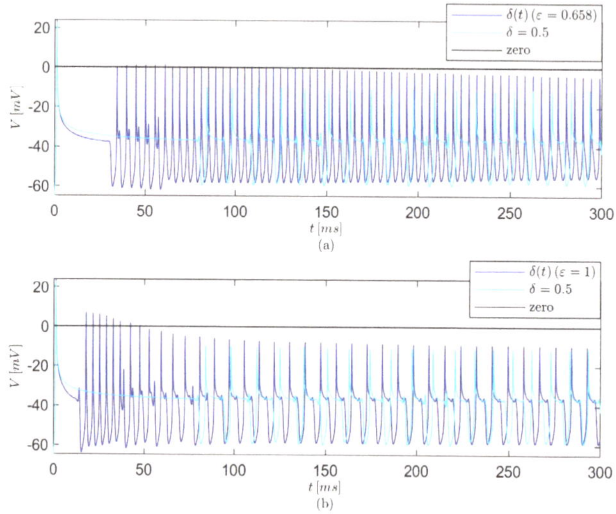

Figure 14. Action potentials for viscoelastic memories $\delta_{AD} = 0.5$ and, respectively, $\delta(t)$ given by Formula (30) with $\delta_{AD} = 0.5$ and with controlled release donepezil for (**a**) $\varepsilon = 0.658$ and (**b**) $\varepsilon = 1$. Plots not to scale. For a short time, donepezil intake recovers the ability of the membrane potential to have positive overshoots for $\varepsilon = 1$ and reach the normal resting potential for both values of ε. Also, the first plateau regions of the action potentials after donepezil treatment are shorter than the first plateau region of the potential corresponding to $\delta_{AD} = 0.5$. At later times, the action potentials corresponding to a donepezil treatment with $\varepsilon = 1$ develop shapes with small delays similar to those seen in the action potentials for the case $\delta_{AD} = 0.5$, while the action potentials associated with a donepezil intake with $\varepsilon = 0.658$ do not develop similar delays.

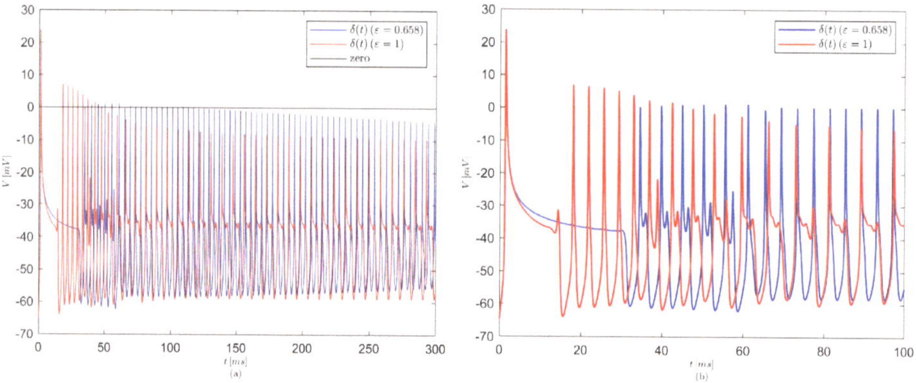

Figure 15. (**a**) Comparison of action potentials for viscoelastic memory $\delta(t)$ given by Formula (30) with $\delta_{AD} = 0.5$ and with Fickian ($\varepsilon = 1$) and non-Fickian ($\varepsilon = 0.658$) release of donepezil. (**b**) Zoomed-in view of the plots in (**a**). Plots not to scale. The action potentials after donepezil intake with Fickian release have positive overshoots and reach the normal resting potential for a short time. The first plateau region is shorter than the one corresponding to the action potentials after donepezil treatment with non-Fickian release and develop short plateau regions at later times. On the other hand, the action potentials after donepezil intake with non-Fickian drug release do not have positive overshoots, have a longer first plateau region, followed by a few shorter plateau regions, and oscillations with no delays at later times.

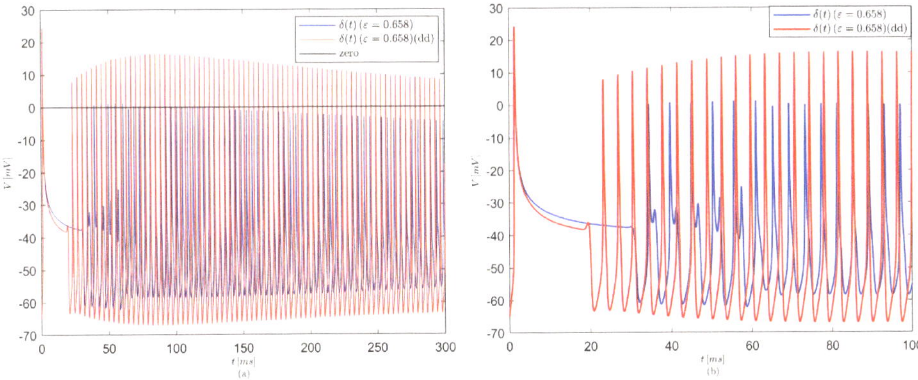

Figure 16. (**a**) Comparison of action potentials for viscoelastic memory $\delta(t)$ given by Formula (30) with $\delta_{AD} = 0.5$, non-Fickian ($\varepsilon = 0.658$) release of donepezil, and with either single (250 µg; blue line) or double (500 µg; red line) dose of orally administered donepezil. (**b**) Zoomed-in view of the plots in (**a**). In the legend maps, (dd) stands for double dose. Plots not to scale. The potential corresponding to the double dose has a shorter initial delay, reaches the normal rest potential and higher amplitudes of positive overshoots that slowly decay at later times.

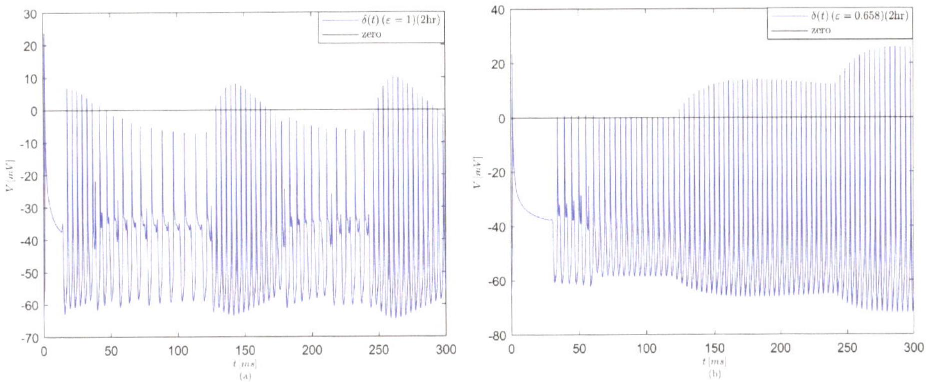

Figure 17. Action potentials for viscoelastic memory $\delta(t)$ given by Formula (30) with $\delta_{AD} = 0.5$ and with (**a**) Fickian ($\varepsilon = 1$) and (**b**) non-Fickian ($\varepsilon = 0.658$) release of donepezil. The administered dosage is 250 µg every 120 min during the observed period of 600 min. In the legend maps, (2 h) stands for the administration of donepezil every two hours. Plots not to scale. The action potentials corresponding to $\varepsilon = 1$ start sooner, the amplitudes of the positive overshoots slowly increase with each administration of donepezil, and there are regions with no positive overshoots which have smaller frequencies and small delays. The action potentials corresponding to $\varepsilon = 0.658$ start later, the amplitudes of the positive overshoots increase with each administration of donepezil, and there are no regions that lack positive overshoots.

If $\delta_{AD} = 0.75$, the intake of donepezil recovers the ability of the membrane potential to have positive overshoots and reach the resting membrane potential V_{rest} in Table 1 for a longer time (about 600 min, see Figure 10c) and both the amplitude and peak-to-peak amplitude of the oscillations are larger when the controlled release uses $\varepsilon = 0.658$ instead of $\varepsilon = 1$ (Figure 13). On the other hand, if $\delta_{AD} = 0.5$ the action potentials reach the resting potential V_{rest} and have positive overshoots for a shorter time (about 100 min, see Figure 10c) if the administered donepezil is released with $\varepsilon = 1$. The action potentials for

$\delta(t) = 0.5 + c_{\text{brain}}$ do not change significantly from those corresponding to $\delta_{AD} = 0.5$ when a controlled release donepezil with $\varepsilon = 0.658$ is administered (Figure 15).

For later stages of AD described by $\delta_{AD} = 0.5$, an increase in donepezil dosage (quantity and/or frequency) is needed to get shorter viscoelastic memory (see Figure 11) and corresponding improved action potentials. Figure 16 shows action potentials for $\delta_{AD} = 0.5$, $\varepsilon = 0.658$ when a double dose of donepezil (500 µg) is administered initially. The potential experiences a shorter initial delay than the case when the administered drug dose is 250 µg, reaches the rest potential and higher amplitudes of positive overshoots that slowly decay at later times in a manner similar to the associated $\delta(t)$ (Figure 11). If $\delta_{AD} = 0.5$ and the drug dose of 250 µg is administered every two hours then the action potentials corresponding to $\varepsilon = 1$ start sooner and although the amplitudes of the positive overshoots slowly increase with each administration of donepezil there are still regions with no positive overshoots, smaller frequencies and small delays (Figure 17a). If $\varepsilon = 0.658$, the positive overshoots are delayed, and their amplitudes increase with each administration of donepezil. Also, in this case, the positive overshoots are not interrupted by regions lacking positive overshoots (Figure 17b). These observations suggest that combining various donepezil's release modalities and dosages may provide better treatments for AD than treatments that use the same release approach with different donepezil dosages for all stages of AD.

Lastly, the convergence of the numerical scheme used to solve system (20)–(23) for a variable fractional order $\delta(t)$ was studied for the case $\delta_{AD} = 0.75$, $\varepsilon = 0.658$, and the time step sizes $\Delta t_1 = 0.005$, $\Delta t_{1/2} = \Delta t_1/2$, and $\Delta t_{1/4} = \Delta t_1/4$. Let $V(\Delta t_i)$ denote the numerical solution of Equation (20) for the step size Δt_i and $i \in \{1, 1/2, 1/4\}$. Then the following relative errors were calculated [88]:

$$E_1 = \frac{\|V(\Delta t_1) - V(\Delta t_{1/4})\|_{L^2(\mathbb{R})}}{\|V(\Delta t_{1/4})\|_{L^2(\mathbb{R})}}, E_{1/2} = \frac{\|V(\Delta t_{1/2}) - V(\Delta t_{1/4})\|_{L^2(\mathbb{R})}}{\|V(\Delta t_{1/4})\|_{L^2(\mathbb{R})}}.$$

The order of convergence was also calculated as:

$$p = \log_2\left(\frac{\|V(\Delta t_1) - V(\Delta t_{1/2})\|_{L^2(\mathbb{R})}}{\|V(\Delta t_{1/2}) - V(\Delta t_{1/4})\|_{L^2(\mathbb{R})}}\right).$$

The results of these calculations are shown in Table 3. The convergence is linear since $p \approx 1$.

Table 3. Relative errors and order of convergence for the numerical scheme used to solve the generalized Hodgkin-Huxley equations with a variable fractional order $\delta(t)$.

E_1	$E_{1/2}$	p
0.3557	0.1474	0.8662($\approx$1)

5. Discussion and Conclusion Remarks

In this paper, mathematical modeling is used to study the transport to the brain of anomalous release donepezil and the effect of donepezil on action potentials. Caputo fractional derivatives are used to model anomalous release patterns of donepezil and the viscoelastic behavior of ligand-gated ion channels. A generalized pharmacokinetic model for the transport to the brain of anomalous release donepezil is proposed and its analytic and numeric solutions are found and shown to be in perfect agreement. A generalized Hodgkin-Huxley model is also proposed that couples the electric activity of neuronal membrane and the viscoelastic behavior of the ion channels. The variable-order fractional Maxwell linear viscoelastic model is used to describe the viscoelasticity of ion channels. The variable-order Caputo fractional derivative is chosen to model viscoelastic fading memory since it has no memory of its past order and thus it can transfer to the output specific features (such as monotonicity or oscillatory behavior) of the fractional variable order. The Haar piecewise-constant approximants of specific variable fractional orders were used, in combination with properties of constant-order Caputo fractional derivatives, to characterize the corresponding creep and relaxation responses of the chosen viscoelastic model. The action potentials predicted by the generalized Hodgkin-Huxley model depend on the viscoelastic fading memory of the ion channels as follows. For shorter fading memory (fractional order closer to 1), the action potentials have shapes similar to those predicted by the classic Hodgkin-Huxley model and thus they are considered to be generated by a functional neuron with structurally intact ion channels. For longer fading memory (fractional order closer to 0), abnormal patterns of action potentials are obtained that correspond to dysfunctional neurons with structurally damaged ion channels. Based on a qualitative visual matching of the shapes of abnormal action potentials predicted by the proposed model and the shapes of action potentials in the presence of AD available in the literature, two constant values of the fractional order are chosen to correspond to earlier ($\delta_{AD} = 0.75$) and later ($\delta_{AD} = 0.5$) stages of AD. The coupling of the generalized pharmacokinetic and Hodgkin-Huxley models is based on the following assumptions: the concentration of donepezil is proportional to the concentration of ACh made available by donepezil intake, and the temporary binding between ACh and its ligand-gated ion channels confers the shorter viscoelastic fading memory to the channels needed to generate healthy action potentials. By taking the variable fractional order describing the viscoelastic memory of ion channels to be the addition of the δ_{AD} and a rescaled and dimensionless concentration of donepezil in the brain, the effects of donepezil intake on the action potentials can be observed. The lack of past order memory of the chosen variable-order fractional derivative allows for the transfer to the action potentials of the temporal variations of the amount of ACh. The results show that: (1) in the brain, an anomalous release donepezil decays slower than a donepezil solution available on the market, (2) donepezil intake temporarily recovers healthy action potentials which agrees with clinical observations [7], and (3) for later stages of AD, combinations of various release modalities and dosages of donepezil may provide better treatment outcomes.

Neuronal structural damage and dysfunction are pathological hallmarks of neurodegenerative disorders. Although these disorders have specific biochemical pathways of progression, they share certain patterns of structural and functional abnormalities. For instance, AD animal models showed that the accumulation of $A\beta$ plaques and NFT can cause myelin damage and oligodendrocyte deterioration leading to demyelienation that further contributes to the progression of AD. The AD-associated demyelienation happens years before the first symptoms of AD emerge and has neuroimaging presentations similar to the demyelienation seen in multiple sclerosis [10–13]. AD and stroke are also interrelated disorders [89]. Strokes cause, among others, an impaired cerebral blood flow, neural injury and depolarization, neuroinflammation, and damage to the blood-brain barrier, which could result in AD onset and progression years later. On the other hand, AD causes the disintegration of the blood-brain barrier that further leads to impaired diffusion and interstitial fluid flow within the cerebral extracellular space, enhanced cerebral inflammatory and immune responses, progression of AD, and, potentially, strokes [4,90,91]. Structural damage of ion channels were also associated with AD and other neurodegenerative disorders [15,47,48,82].

The abnormal patterns of action potentials in the presence of AD predicted by the proposed generalized Hodgkin-Huxley model are similar to those seen in untreated AD and other neurodegenerative disorders. Various studies have shown hyperexcitability and epileptiform activity in early stages of AD, followed by increasing lack of action potentials in later stages of AD [9,14,15,17,44,45]. Figure 7 shows that, as δ decreases from 1, the current activation happens at less negative voltages (higher resting potential), inactivation occurs at less positive voltages (smaller positive overshoots), and the action potentials experience decreasing amplitudes and smaller frequencies, which, by definition, correspond to a loss of function phenotype introduced in [15]. The case $\delta_{AD} = 0.75$ shows plateau (delayed) potentials followed by higher frequency of firing which correspond to hyperexcitability seen in earlier stages of AD. Similar patterns of action potentials were observed experimentally in transgenic Na v1.8-null knockout DRG neurons [92], abnormal Na v1.8 expression in some neurons that may be linked to multiple sclerosis [93], and mutations of some sodium channels that cause persistent sodium currents linked to epilepsy [94,95]. The case $\delta_{AD} = 0.5$ (and lower values of δ) shows an almost cessation of action potentials that may be associated with later stages of AD [17]. Similar abnormal action potentials were observed experimentally in neurons under spreading depolarization (that may happen in ischemic stroke) and a blockade of astrocytic glutamate transporters [96]. The abnormal action potentials also look similar to some of the membrane voltages predicted by the mathematical models proposed in [44,45,97,98]. The models in [44,45] were obtained by coupling the Hodgkin-Huxley model and ion concentration dynamics observed in AD animal models via the Nernst equations. The models in [97,98] are not specific to AD and are obtained by mathematical modifications of the Hodgkin-Huxley model. While a combination of Taylor's formula for multivariable functions and various external stimuli is used in [97], the concept of a two-terminal resistor with memory called memristor is introduced in [98] and the so-called memristive Hodgkin-Huxley model is then obtained by replacing the sodium and potassium conductances in the classic Hodgkin-Huxley model by flux-controlled memristors. A major difference between the proposed generalized Hodgkin-Huxley model and published mathematical models is the level of complexity. Models existing in the literature have a high level of complexity; they use many variables and parameters to predict abnormal action potentials. On the other hand, only the memory parameter of the generalized Hodgkin-Huxley model needs to be modified to predict healthy and various abnormal patterns of action potentials. Neither the recovery of the healthy patterns of action potentials after the administration of AD therapies nor the transport of anomalous release donepezil to the brain have been reported in the literature until now.

Lastly, the proposed generalized Hodgkin-Huxley and coupled models require experimental validation. For the calibration of the models, it is important to find the values of the moduli of elasticity, generalized viscosities, and memory parameters of the nicotinic ACh receptors (and neuronal ion channels in general) in properly functioning neurons and in neurons altered by the AD progression, as well as their dependencies on electric currents and, in the case of AD, the concentration of orally administered donepezil in the brain. Currently, there are no published viscoelastic characterizations of neuronal ion channels. However, optical tweezers-based microrheology [99–101] and optical coherence elastography [102] are non-invasive and contactless techniques that can be used in vivo and in vitro to study mechanical properties of biological structures at (sub)micron levels like ion channels. In optical tweezers-based microrheology, small particles are trapped and manipulated using a highly focused laser beam and measurements of the light scattered by the trapped particles moving through a material provide the viscoelastic properties of the material. Of particular interest to the work in this paper could be the active–passive calibration technique proposed in [101] that can measure viscoelastic properties of a material over a wide frequency range with high spatio-temporal resolution. In optical coherence elastography, the deformation in a material subjected to a mechanical force is measured using optical coherence tomography, and then an inverse problem is solved with the known deformation to find the viscoelastic properties of the material. Initially, these two technologies could be used with animal models of AD to find the viscoelastic properties of retinal neurons.

Future work will explore the generalized Hodgkin-Huxley model for distinct memory parameters of the ion channels, in the presence of various external stimuli and ion concentration dynamics. The derivation of semi-analytic solutions to the generalized Hodgkin-Huxley model using Haar wavelets will also be investigated.

Funding: This research received no external funding.

Data Availability Statement: The original contributions presented in the study are included in the article, further inquiries can be directed to the corresponding author.

Conflicts of Interest: The author declares no conflicts of interest. The funders had no role in the design of the study; in the collection, analyses, or interpretation of data; in the writing of the manuscript; or in the decision to publish the results.

Abbreviations

The following abbreviations are used in this manuscript:

AD	Alzheimer's disease
$A\beta$	Amyloid-β
NFT	Neurofibrillary tangles
AChE	Acetylcholinesterase
ACh	Acetylcholine

Appendix A

The descriptions and physical units of the variables of the models presented in Sections 3.2 and 3.3 are given in Table A1.

Table A1. The values, physical units, and descriptions of the models' variables.

Variables	Physical Units	Description
t	ms (Section 3.2), min (Section 3.3)	time
V	mV	Membrane potential
m		Nondimensional displacement of the activating sodium gate
h		Nondimensional displacement of the inactivating sodium gate
n		Nondimensional displacement of the activating potassium gate
c_d	μg/mL	Donepezil concentration in the depot
c_{blood}	μg/mL	Donepezil concentration in the blood
c_{brain}	μg/mL	Donepezil concentration in the brain
c_p	μg/mL	Donepezil concentration in the periphery

References

1. Alzheimer, A. Uber einen eigenartigen schweren Erkrankungsprozess der Hirninde. *Neurol. Centralblatt.* **1906**, *25*, 1134.
2. World Health Organization. Dementia Key Facts. Available online: https://www.who.int/news-room/fact-sheets/detail/dementia (accessed on 20 July 2024).
3. Carreiras, M.C.; Mendes, E.; Perry, M.J.; Francisco, A.P.; Marco-Contelles, J. The multifactorial nature of Alzheimer's disease for developing potential therapeutics. *Curr. Top. Med. Chem.* **2013**, *13*, 1745–1770. [CrossRef] [PubMed]
4. Sweeney, M.D.; Sagare, A.P.; Zlokovic, B.V. Blood–brain barrier breakdown in Alzheimer disease and other neurodegenerative disorders. *Nat. Rev. Neurol.* **2018**, *14*, 133–150. [CrossRef] [PubMed]
5. Salman, M.M.; Kitchen, P.; Halsey, A.; Wang, M.X.; Törnroth-Horsefield, S.; Conner, A.C.; Badaut, J.; Iliff, J.J.; Bill, R.M. Emerging roles for dynamic aquaporin-4 subcellular relocalization in CNS water homeostasis. *Brain* **2022**, *145*, 64–75. [CrossRef]
6. Kinney, J.W.; Bemiller, S.M.; Murtishaw, A.S.; Leisgang, A.M.; Salazar, A.M.; Lamb, B.T. Inflammation as a central mechanism in Alzheimer's disease. *Alzheimer's Dement. Transl. Res. Clin. Interv.* **2018**, *4*, 575–590. [CrossRef]
7. Marucci, G.; Buccioni, M.; Dal Ben, D.; Lambertucci, C.; Volpini, R.; Amenta, F. Efficacy of acetylcholinesterase inhibitors in Alzheimer's disease. *Neuropharmacology* **2021**, *190*, 108352. [CrossRef]
8. Garcia-Ayllon, M.-S.; Small, D.H.; Avila, J.; Saez-Valero, J. Revisiting the role of acetylcholinesterase in Alzheimer's disease: Cross-talk with P-tau and β-amyloid. *Front. Mol. Neurosci.* **2011**, *4*, 22. [CrossRef]
9. Anastacio, H.T.D.; Matosin, N.; Ooi, L. Neuronal hyperexcitability in Alzheimer's disease. what are the drivers behind this aberrant phenotype? *Transl. Psychiatry* **2022**, *12*, 257. [CrossRef]
10. Nasrabady, S.E.; Rizvi, B.; Goldman, J.E.; Brickman, A.M. White matter changes in Alzheimer's disease: A focus on myelin and oligodendrocytes. *Acta Neuropathol. Commun.* **2018**, *6*, 22. [CrossRef]
11. Depp, C.; Sun, T.; Sasmita, A.O.; Spieth, L.; Berghoff, S.A.; Nazarenko, T.; Overhoff, K.; Steixner-Kumar, A.A.; Subramanian, S.; Arinrad, S.; et al. Myelin dysfunction drives amyloid-β deposition in models of Alzheimer's disease. *Nature* **2023**, *618*, 349–357. [CrossRef]

12. Dan, L.; Zhang, Z. Alzheimer's disease: An axonal injury disease? *Front. Aging Neurosci.* **2023**, *15*, 1264448. [CrossRef] [PubMed]
13. Boukhvalova, M.S.; Kastrukoff, L.; Blanco, J.C.G. Alzheimer's disease and multiple sclerosis: A possible connection through the viral demyelinating neurodegenerative trigger (vDENT). *Front. Aging Neurosci.* **2023**, *15*, 1204852. [CrossRef] [PubMed]
14. Yavorsky, V.A.; Rozumna, N.M.; Lukyanetz, E.A. Influence of amyloid beta on impulse spiking of isolated hippocampal neurons. *Front. Cell. Neurosci.* **2023**, *17*, 1132092. [CrossRef] [PubMed]
15. Barbieri, R.; Nizzari, M.; Zanardi, I.; Pusch, M.; Gavazzo, P. Voltage-gated sodium channel dysfunctions in neurological disorders. *Life* **2023**, *13*, 1191. [CrossRef]
16. Melgosa-Ecenarro, L.; Doostdar, N.; Radulescu, C.I.; Jackson, J.S.; Barnes, S.J. Pinpointing the locus of GABAergic vulnerability in Alzheimer's disease. *Semin. Cell Dev. Biol.* **2023**, *139*, 35–54. [CrossRef]
17. Tzavellas, N.P.; Tsamis, K.I.; Katsenos, A.P.; Davri, A.S.; Simos, Y.V.; Nikas, I.P.; Bellos, S.; Lekkas, P.; Kanellos, F.S.; Konitsiotis, S.; et al. Firing alterations of neurons in Alzheimer's disease: Are they merely a consequence of pathogenesis or a pivotal component of disease progression? *Cells* **2024**, *13*, 434. [CrossRef]
18. Kumar, A.; Gupta, V.; Sharma, S. Donepezil. In *StatPearls*; StatPearls Publishing: Treasure Island, FL, USA, 2024. Available online: https://www.ncbi.nlm.nih.gov/books/NBK513257/ (accessed on 20 July 2024).
19. Sam, C.; Bordoni, B. Physiology, Acetylcholine. In *StatPearls*; StatPearls Publishing: Treasure Island, FL, USA, 2024. Available online: https://www.ncbi.nlm.nih.gov/books/NBK557825/ (accessed on 16 August 2024).
20. Carlson, A.B.; Kraus, G.P. Physiology, Cholinergic Receptors. In *StatPearls*; StatPearls Publishing: Treasure Island, FL, USA, 2024. Available online: https://www.ncbi.nlm.nih.gov/books/NBK526134/ (accessed on 8 December 2024).
21. Kim, J.; Lee, H.J.; Park, S.K.; Park, J.H.; Jeong, H.R.; Lee, S.; Lee, H.; Seol, E.; Hoe, H.S. Donepezil regulates LPS and $A\beta$-stimulated neuroinflammation through MAPK/NLRP3 inflammasome/STAT3 signaling. *Int. J. Mol. Sci.* **2021**, *22*, 10637. [CrossRef]
22. Lombardo, S.; Maskos, U. Role of the nicotinic acetylcholine receptor in Alzheimer's disease pathology and treatment. *Neuropharmacology* **2015**, *96 Pt B*, 255–262. [CrossRef]
23. Monaco, M.; Trebesova, H.; Grilli, M. Muscarinic receptors and Alzheimer's disease: New perspectives and mechanisms. *Curr. Issues Mol. Biol.* **2024**, *46*, 6820–6835. [CrossRef]
24. Jakki, S.L.; Ramesh, Y.V.; Gowthamarajan, K.; Senthil, V.; Jain, K.; Sood, S.; Pathak, D. Novel anionic polymer as a carrier for CNS delivery of anti-Alzheimer drug. *Drug Deliv.* **2016**, *23*, 3471–3479. [CrossRef]
25. Silva, R.O.; Counil, H.; Rabanel, J.-M.; Haddad, M.; Zaouter, C.; Khedher, M.R.B.; Patten, S.A.; Ramassamy, C. Donepezil-loaded nanocarriers for the treatment of Alzheimer's disease: Superior efficacy of extracellular vesicles over polymeric nanoparticles. *Int. J. Nanomed.* **2024**, *19*, 1077–1096. [CrossRef] [PubMed]
26. Topal, G.R.; Mészáros, M.; Porkoláb, G.; Szecskó, A.; Polgár, T.F.; Siklós, L.; Deli, M.A.; Veszelka, S.; Bozkir, A. ApoE-targeting increases the transfer of solid lipid nanoparticles with donepezil cargo across a culture model of the blood–brain barrier. *Pharmaceutics* **2021**, *13*, 38. [CrossRef] [PubMed]
27. Topal, G.R.; Kücüktürkmen, B.; Öz, U.C.; Özkan, E.; Bakar-Ates, F.; Bozkir, A. Investigation on formulation parameters of donepezil HCl loaded solid lipid nanoparticles. *Braz. J. Pharm. Sci.* **2023**, *59*, e22330. [CrossRef]
28. Tao, X.; Mao, S.; Zhang, Q.; Yu, H.; Li, Y.; He, X.; Yang, S.; Zhang, Z.; Yi, Z.; Song, Y.; et al. Brain-targeted polysorbate 80-emulsified donepezil drug-loaded nanoparticles for neuroprotection. *Nanoscale Res. Lett.* **2021**, *16*, 132. [CrossRef]
29. Akyol, E.; Senol, S.; Dogan, Ö. Controlled release of donepezil hydrochloride from the ternary sodium alginate based hydrogels. *Bulg. Chem. Commun.* **2017**, *49*, 57–63.
30. Ruela, A.L.; Carvalho, F.C.; Pereira, G.R. Exploring the phase behavior of monoolein/oleic acid/water Systems for enhanced donezepil administration for Alzheimer disease treatment. *J. Pharm. Sci.* **2016**, *105*, 71–77. [CrossRef]
31. Bulut, E.; Sanli, O. Delivery of Alzheimer's drug donepezil hydrochloride from ionically crosslinked alginate microspheres prepared by water-in-oil emulsion technique: Optimization of release conditions. *Asian J. Chem.* **2013**, *25*, 3993–4000. [CrossRef]
32. Kaikousidis, C.; Papakyriakopoulou, P.; Dokoumetzidis, A.; Valsami, G. Donepezil brain and blood pharmacokinetic modeling after nasal film and oral solution administration in mice. *Pharmaceutics* **2023**, *15*, 1409. [CrossRef]
33. Papakyriakopoulou, P.; Balafas, E.; Colombo, G.; Rekkas, D.M.; Kostomitsopoulos, N.; Valsami, G. Nose-to-brain delivery of donepezil hydrochloride following administration of an HPMC-Me-β-CD-PEG400 nasal film in mice. *J. Drug Deliv. Sci. Technol.* **2023**, *84*, 104463. [CrossRef]
34. Espinoza, L.C.; Guaya, D.; Calpena, A.C.; Perotti, R.M.; Halbaut, L.; Sosa, L.; Brito-Llera, A.; Mallandrich, M. Comparative study of donepezil-loaded formulations for the treatment of Alzheimer's disease by nasal administration. *Gels* **2022**, *8*, 715. [CrossRef]
35. Zhang, P.; Chen, L.; Gu, W.; Xu, Z.; Gao, Y.; Li, Y. In vitro and in vivo evaluation of donepezil-sustained release microparticles for the treatment of Alzheimer's disease. *Biomaterials* **2007**, *28*, 1882–1888. [CrossRef] [PubMed]
36. Kearney, M.-C.; Caffarel-Salvador, E.; Fallows, S.J.; McCarthy, H.O.; Donnelly, R.F. Microneedle-mediated delivery of donepezil: Potential for improved treatment options in Alzheimer's disease. *Eur. J. Pharm. Biopharm.* **2016**, *103*, 43–50. [CrossRef] [PubMed]
37. Bruschi, M.L. *Strategies to Modify the Drug Release from Pharmaceutical Systems*; Woodhead Publishing: Cambridge, UK, 2015; pp. 63–86.
38. Trucillo, P. Drug carriers: A review on the most used mathematical models for drug release. *Processes* **2022**, *10*, 1094. [CrossRef]
39. Khanday, M.A.; Rafiq, A.; Nazir, K. Mathematical models for drug diffusion through the compartments of blood and tissue medium. *Alex. J. Med.* **2017**, *53*, 245–249. [CrossRef]

40. Shyamsunder, S.; Bhatter, K.J.; Purohit, S.D. Fractionalized mathematical models for drug diffusion. *Chaos Solitons Fractals* **2022**, *165*, 112810. [CrossRef]
41. Alijani, Z.; Shiri, B.; Perfilieva, I.; Baleanu, D. Numerical solution of a new mathematical model for intravenous drug administration. *Evol. Intel.* **2024**, *17*, 559–575. [CrossRef]
42. Hodgkin, A.L.; Huxley, A.F. A quantitative description of membrane current and its application to conduction and excitation in nerve. *J. Physiol.* **1952**, *117*, 500–544. [CrossRef]
43. Wei, Y.; Ullah, G.; Schiff, S.J. Unification of neuronal spikes, seizures, and spreading depression. *J. Neurosci.* **2014**, *34*, 11733–11743. [CrossRef]
44. Perez, C.; Ziburkus, J.; Ullah, G. Analyzing and modeling the dysfunction of inhibitory neurons in Alzheimer's disease. *PLoS ONE* **2016**, *11*, e0168800. [CrossRef]
45. Adeoye, T.; Shah, S.I.; Demuro, A.; Rabson, D.A.; Ullah, G. Upregulated Ca^{2+} release from the endoplasmic reticulum leads to impaired presynaptic function in familial Alzheimer's disease. *Cells* **2022**, *11*, 2167. [CrossRef]
46. Gupta, S.; Singh, J.; Kumar, K. Ionic concentration and action potential differences between a healthy and Alzheimer's disease person. In Proceedings of the 5th International Conference on Recent Developments in Science, Engineering and Technology (REDSET 2019), Gurugram, India, 15–16 November 2019; Batra, U., Roy, N., Panda, B., Eds.; Communications in Computer and Information Science; Springer Nature: Cham, Switzerland, 2020; Volume 1229, pp. 266–277.
47. Purves, D.; Augustine, G.J.; Fitzpatrick, D.; Katz, L.C.; LaMantia, A.-S.; McNamara, J.O.; Williams, S.M. Cholinergic Receptors. In *Neuroscience*, 2nd ed.; Sinauer Associates: Sunderland, MA, USA, 2001.
48. Zoli, M.; Pucci, S.; Vilella, A.; Gotti, C. Neuronal and extraneuronal nicotinic acetylcholine receptors. *Curr. Neuropharmacol.* **2018**, *16*, 338–349. [PubMed]
49. Metzler, R.; Klafter, J. The restaurant at the end of the random walk: Recent developments in the description of anomalous transport by fractional dynamics. *J. Phys. A Math. Gen.* **2004**, *37*, R161.
50. Mainardi, F. *Fractional Calculus and Waves in Linear Viscoelasticity*; Imperial College Press: London, UK, 2010.
51. Drapaca, C.S. Fractional calculus in neuronal electromechanics. *J. Mech. Mater. Struct.* **2017**, *12*, 35–55. [CrossRef]
52. Drapaca, C.S. A nitric oxide-modulated variable-order fractional Maxwell viscoelastic model of cerebral vascular walls. *Front. Mech. Eng* **2021**, *7*, 674860.
53. Samko, S.; Kilbas, A.A.; Marichev, O.I. *Fractional Integrals and Derivatives: Theory and Applications*; Gordon and Breach Science Publishers: London, UK, 2000.
54. Gorenflo, R.; Mainardi, F. Fractional calculus: Integral and differential equations of fractional order. In *Fractals and Fractional Calculus in Continuum Mechanics*; Carpinteri, A., Mainardi, F., Eds.; Springer: Wien, Austria; New York, NY, USA, 1997; pp. 223–276.
55. Sambandham, B.; Vatsala, A.S. Basic results for sequential Caputo fractional differential equations. *Mathematics* **2015**, *3*, 76–91. [CrossRef]
56. Sun, H.G.; Chang, A.; Zhang, Y.; Chen, W. A review on variable-order fractional differential equations: Mathematical foundations, physical models, numerical methods and applications. *Fract. Calc. Appl. Anal.* **2019**, *22*, 27–59.
57. Patnaik, S.; Hollkamp, J.P.; Semperlotti, F. Applications of variable-order fractional operators: A review. *Proc. R. Soc. A* **2020**, *476*, 20190498.
58. Moghaddam, B.P.; Machado, J.A.T. Extended algorithms for approximating variable order fractional derivatives with applications. *J. Sci. Comput.* **2017**, *71*, 1351–1374.
59. Baleanu, D.; Diethelm, K.; Scalas, E.; Trujillo, J.J. *Fractional Calculus Models and Numerical Methods*; World Scientific: Hackensack, NJ, USA, 2012.
60. Atangana , A.; Secer, A. A note on fractional order derivatives and table of fractional derivatives of some special functions. *Abstr. Appl. Anal.* **2013**, *2013*, 279681.
61. Haar, A. Zur Theories der orthogonalen Funktionensystem. *Math. Annal.* **1910**, *69*, 331–371. [CrossRef]
62. Volker, M. *Lectures on Constructive Approximation: Fourier, Spline, and Wavelet Methods on the Real Line, the Sphere, and the Ball*; Birkhauser: Boston, MA, USA, 2013.
63. Addison, P.S. *The Illustrated Wavelet Transform Handbook Introductory Theory and Applications in Science, Engineering, Medicine and Finance*, 2nd ed.; CRC Press: Boca Raton, FL, USA, 2020.
64. Chen, C.F.; Hsiao, C.H. Haar wavelet method for solving lumped and distributed-parameter system. *IEEE Proc. Control Theory Appl.* **1997**, *144*, 87–94. [CrossRef]
65. Lepik, U. Numerical solution of differential equations using Haar wavelets. *Math. Comput. Simul.* **2005**, *68*, 127–143. [CrossRef]
66. Lepik, U. Solving differential and integral equations by the Haar wavelet method; revisited. *Int. J. Math. Comput.* **2008**, *1*, 43–52.
67. Aziz, I.; Amin, R. Numerical solution of a class of delay differential and delay partial differential equations via Haar wavelet. *Appl. Math. Model.* **2016**, *40*, 10286–10299. [CrossRef]
68. Gupta, A.K.; Ray, S.S. Wavelet methods for solving fractional order differential equations. *Math. Probl. Eng.* **2014**, *2014*, 140453. [CrossRef]
69. Heil, C. *A Basis Theory Primer: Expanded Edition (Applied and Numerical Harmonic Analysis)*; Birkhauser: Boston, MA, USA, 2010.
70. The MathWorks Inc. *MATLAB*, version 24.1.0.2537033 (R2024a); The MathWorks Inc.: Natick, MA, USA, 2024.
71. Ramirez, L.; Coimbra, C. On the selection and meaning of variable order operators for dynamic modeling. *Int. J. Diff. Equ.* **2010**, *2010*, 846107. [CrossRef]

72. Lorenzo, C.F.; Hartley, T.T. Variable order and distributed order fractional operators. *Nonlinear Dyn.* **2002**, *29*, 57–98. [CrossRef]
73. Sun, H.; Chen, W.; Wei, H.; Chen, Y. A comparative study of constant-order and variable-order fractional models in characterizing memory property of systems. *Eur. Phys. J. Spec. Top.* **2011**, *193*, 185–192. [CrossRef]
74. Bazhlekova, E.; Bazhlekov, I. Stokes' first problem for viscoelastic fluids with a fractional Maxwell model. *Fractal Fract.* **2017**, *1*, 7. [CrossRef]
75. Findley, W.N.; Lai, J.S.; Onaran, K. *Creep and Relaxation of Nonlinear Viscoelastic Materials—With an Introduction to Linear Viscoelasticity*; Dover Publications: Garden City, NY, USA, 1989.
76. Pipkin, A.; Rogers, T. A non-linear integral representation for viscoelastic behavior. *J. Mech. Phys. Solids* **1968**, *16*, 59–72. [CrossRef]
77. Fréchet, M. Sur les fonctionnelles continues. *Ann. De L'Ecole Norm.* **1910**, *3*, 193–216. [CrossRef]
78. Volterra, V. *Theory of Functionals and the Integral and Integro-Differential Equations*; Dover Publications: Garden City, NY, USA, 1959.
79. Bonilla, B.; Rivero, M.; Trujillo, J.J. On systems of linear fractional differential equations with constant coefficients. *Appl. Math. Comput.* **2007**, *187*, 68–78. [CrossRef]
80. Odibat, Z.M. Analytic study on linear systems of fractional differential equations. *Comput. Math. Appl.* **2010**, *59*, 1171–1183. [CrossRef]
81. Wineman, A.S.; Rajagopal, K.R. *Mechanical Response of Polymers: An Introduction*, 1st ed.; Cambridge University Press: Cambridge, UK, 2000.
82. Lehmann-Horn, F.; Jurkat-Rott, K. Voltage-gated ion channels and hereditary disease. *Physiol. Rev.* **1999**, *79*, 1317–1372. [CrossRef] [PubMed]
83. Gautieri, A.; Vesentini, S.; Redaelli, A.; Buehler, M.J. Viscoelastic properties of model segments of collagen molecules. *Matrix Biol.* **2012**, *31*, 141–149. [CrossRef] [PubMed]
84. Hayek, N.; Trujillo, J.J.; Rivero, M.; Bonilla, B.; Moreno, J.C. An extension of Picard–Lindeloff theorem to fractional differential equations. *Appl. Anal.* **1999**, *70*, 347–361. [CrossRef]
85. Almeida, R.; Malinowska, A.B.; Odzijewicz, T. On systems of fractional differential equations with the ψ-Caputo derivative and their applications. *Math. Meth. Appl. Sci.* **2021**, *44*, 8026–8041. [CrossRef]
86. Shampine, L.F.; Reichelt, M.W. The MATLAB ODE suite. *SIAM J. Sci. Comput.* **1997**, *18*, 1–22. [CrossRef]
87. Garrappa, R. Numerical solution of fractional differential equations: A survey and a software tutorial. *Mathematics* **2018**, *6*, 16. [CrossRef]
88. LeVeque, R.J. *Finite Difference Methods for Ordinary and Partial Differential Equations: Steady-State and Time-Dependent Problems*, 1st ed.; Society for Industrial and Applied Mathematics: Philadelphia, PA, USA, 2007; pp. 254–258.
89. Gupta, A.; Uthayaseelan, K.; Uthayaseelan, K.; Kadari, M.; Subhan, M.; Saji Parel, N.; Krishna, P.V.; Sange, I. Alzheimer's disease and stroke: A tangled neurological conundrum. *Cureus* **2022**, *14*, e25005. [CrossRef]
90. Montagne, A.; Zhen Zhao, Z.; Zlokovic, B.V. Alzheimer's disease: A matter of blood–brain barrier dysfunction? *J. Exp. Med.* **2017**, *214*, 3151–3169. [CrossRef] [PubMed]
91. Alkhalifa, A.E.; Al-Ghraiybah, N.F.; Odum, J.; Shunnarah, J.G.; Austin, N.; Kaddoumi, A. Blood–brain barrier breakdown in Alzheimer's disease: Mechanisms and targeted strategies. *Int. J. Mol. Sci.* **2023**, *24*, 16288. [CrossRef] [PubMed]
92. Renganathan, M.; Cummins, T.R.; Waxman, S.G. Contribution of Nav 1.8 sodium channels to action potential electrogenesis in DRG neurons. *J. Neurophysiol.* **2001**, *86*, 629–640. [CrossRef] [PubMed]
93. Waxman, S.G. Sodium channels as molecular targets in multiple sclerosis. *J. Rehabil. Res. Dev.* **2002**, *39*, 233–242. [PubMed]
94. Stafstrom, C.E. Persistent sodium current and its role in epilepsy. *Epilepsy Curr.* **2007**, *7*, 15–22. [CrossRef] [PubMed]
95. Phelan, K.; Mock, M.; Kretz, O.; Shwe, U.; Kozhemyakin, M.; Greenfield, L.; Dietrich, A.; Birnbaumer, L.; Freichel, M.; Flockerzi, V.; et al. Heteromeric canonical transient receptor potential 1 and 4 channels play a critical role in epileptiform burst firing and seizure-induced neurodegeneration. *Mol. Pharmacol.* **2011**, *81*, 384–392. [CrossRef]
96. Hubel, N.; Hosseini-Zare, M.S.; Ziburkus, J.; Ullah, G. The role of glutamate in neuronal ion homeostasis: A case study of spreading depolarization. *PLoS Comput. Biol.* **2017**, *13*, e1005804. [CrossRef]
97. Liu, Y.; Liu, S.; Zhan, F.; Zhang, X. Firing patterns of the modified Hodgkin-Huxley model subject to Taylor's formula. *Phys. A* **2020**, *547*, 124405. [CrossRef]
98. Fang, X.; Duan, S.; Wang, L. Memristive Hodgkin-Huxley spiking neuron model for reproducing neuron behaviors. *Front. Neurosci.* **2021**, *15*, 730566. [CrossRef]
99. Robertson-Anderson, R.M. Optical tweezers microrheology: From the basics to advanced techniques and applications. *ACS Macro Lett.* **2018**, *7*, 968–975. [CrossRef]
100. Lenton, I.C.D.; Scott, E.K.; Rubinsztein-Dunlop, H.; Favre-Bulle, I.A. Optical tweezers exploring neuroscience. *Front. Bioeng. Biotechnol.* **2020**, *8*, 602797. [CrossRef]
101. Kumar, R.; Valerio Vitali, V.; Wiedemann, T.; Robert Meissner, R.; Paolo Minzioni, P.; Denz, C. Multi-frequency passive and active microrheology with optical tweezers. *Sci. Rep.* **2021**, *11*, 13917. [CrossRef]
102. Kennedy, B.F. *Optical Coherence Elastography: Imaging Tissue Mechanics on the Micro-Scale*; AIP Publishing LLC: Melville, NY, USA, 2021.

Mathematical and Computational Applications

MDPI

Article

Efficient Finite-Difference Estimation of Second-Order Parametric Sensitivities for Stochastic Discrete Biochemical Systems †

Fauzia Jabeen and Silvana Ilie *

Department of Mathematics, Toronto Metropolitan University, Toronto, ON M5B 2K3, Canada; fauzia.jabeen@torontomu.ca

* Correspondence: silvana@torontomu.ca

† A preliminary version of this work was presented at the AMMCS 2023 Conference, Waterloo, ON, Canada, 14–18 August 2023.

Abstract: Biochemical reaction systems in a cell exhibit stochastic behaviour, owing to the unpredictable nature of the molecular interactions. The fluctuations at the molecular level may lead to a different behaviour than that predicted by the deterministic model of the reaction rate equations, when some reacting species have low population numbers. As a result, stochastic models are vital to accurately describe system dynamics. Sensitivity analysis is an important method for studying the influence of the variations in various parameters on the output of a biochemical model. We propose a finite-difference strategy for approximating second-order parametric sensitivities for stochastic discrete models of biochemically reacting systems. This strategy utilizes adaptive tau-leaping schemes and coupling of the perturbed and nominal processes for an efficient sensitivity estimation. The advantages of the new technique are demonstrated through its application to several biochemical system models with practical significance.

Keywords: stochastic simulation algorithm; stochastic models of biochemical kinetics; sensitivity analysis; tau-leaping method; variable time stepping

1. Introduction

Modelling and simulation of biochemical interactions among molecular species in a living cell are some important topics in Computational Biology. These molecular interactions are inherently random and discrete in nature. Thus, stochastic models are necessary to capture this intrinsic stochasticity and provide a more accurate representation of cellular processes. Such models are particularly important when some molecules involved in key biochemical reactions are present in low numbers. Stochastic models have proven crucial in studying the complex dynamics of biological processes, ranging from gene regulation to cellular signalling [1,2]. The chemical master equation is a commonly used discrete stochastic model which accurately represents the dynamics of homogeneous biochemical systems, with some reactants in small molecular amounts [3]. In this model, the stochastic law of mass action is applied to find the probabilities of reaction events taking place over time. The solution of the chemical master equation can be approximated numerically using Monte Carlo techniques. Among the most popular Monte Carlo strategies that provide statistically exact trajectories are the stochastic simulation algorithm (SSA) due to Gillespie [4,5] and the next reaction method developed by Gibson and Bruck [6]. These methods are often expensive on systems with a large number of reactive events or with some species in large molecular amounts. Then, approximate Monte Carlo methods such as τ-leaping schemes [7,8] can significantly reduce the computational cost of the simulation while maintaining an excellent accuracy. Other numerical methods for the chemical master equation include [9–14] (also see the references therein). For a detailed discussion of stochastic discrete and deterministic modelling approaches for biochemical networks, along with a comparison of the associated concepts, we direct the reader to [15,16].

Sensitivity analysis plays a prominent role in understanding the robustness of biochemical models by quantifying the influence of parameter variations on model outputs [17]. In addition, it is an important tool for parameter estimation [18] and practical identifiability analysis of these models [19,20]. Sensitivity analysis in the chemical master equation context can provide critical information about how the system's response depends on various parameters, especially in biochemically reacting systems where the estimations of parameter values are unreliable [21].

Developing sensitivity analysis techniques for stochastic discrete models is a difficult task [21]. Several approaches have been proposed for such models. Strategies for computing sensitivities using the Girsanov measure transform were developed in [22] but may be computationally infeasible for models of moderately large to large systems of biochemical reactions or models which are stiff; moreover, methods offering insights into the properties of probability distributions were proposed in [21,23,24]. One of the most popular schemes to estimate parameter sensitivities are the finite-difference strategies based on Monte Carlo simulations of the perturbed and nominal processes [12,25]. Some existing finite-difference techniques for approximating local sensitivities in stochastic biochemical reaction networks utilize exact Monte Carlo methods, including the SSA for the common random number [26], the random time change algorithm employed by the common reaction path [26], or the next reaction method used by coupled finite-difference (CFD) [25] schemes. The CFD strategy has the smallest estimator variance among these finite-difference techniques. Other finite-difference methods rely on approximate Monte Carlo strategies which are appropriate for systems with various degrees of stiffness such as the coupled tau-leaping [27] and the coupled implicit tau-leaping [28] methods.

These techniques were designed to estimate first-order sensitivities for stochastic biochemical models. While first-order sensitivities show the direct influence of parameter variations on the output of the model, second-order sensitivities capture the curvature or nonlinear effects in the system's response to parameter perturbations. Moreover, the Hessian (second-order sensitivities) plays a central role in optimization, parameter estimation and model validation. If $\mathbb{E}[\cdot]$ represents the expected value, $\mathbf{f}$ the function of interest, and $\mathbf{X}(t,c)$ the state of the system at time t, depending on the vector of M parameters c, we are interested in computing the second-order sensitivity, $\partial^2\mathbb{E}[\mathbf{f}(\mathbf{X}(t,c))]/\partial c_j \partial c_k$, where c_j and c_k are some parameters of interest, for $j,k \in \{1,\ldots,M\}$. Denote, by e_j, the vector of dimension M with the j-th component equal 1 and the others 0. Using a forward finite-difference scheme, this sensitivity may be estimated by $\{\mathbb{E}[\mathbf{f}(\mathbf{X}(t,c+(h_je_j+h_ke_k)))] - \mathbb{E}[\mathbf{f}(\mathbf{X}(t,c+h_je_j))] - \mathbb{E}[\mathbf{f}(\mathbf{X}(t,c+h_ke_k))] + \mathbb{E}[\mathbf{f}(\mathbf{X}(t,c))]\}/(h_jh_k)$, with h_j and h_k being small perturbation of the parameters of interest c_j and c_k, respectively. It is worth noting that much less research was dedicated to building finite-difference second-order sensitivity estimators for the chemical master equation. The coupled finite-difference method (CFD-2) for approximating the second-order sensitivities for stochastic biochemical systems was proposed in [29]. This method utilizes the modified next reaction method, an exact Monte Carlo strategy, to generate the four correlated processes and is working well on problems which are non-stiff.

When biochemical systems evolve on multiple scales in time, their mathematical models exhibit stiffness [16,30–33]. For stiff systems, exact simulation strategies become prohibitively costly. Another limitation of existing coupled finite-difference second-order sensitivity estimators is their lack of scalability when applied to large systems with many reacting species and parameters, which may also be stiff. The computational cost of exact simulations grows rapidly as the system size increases. An important difficulty when designing finite-difference second-order sensitivity estimators is that they are more sensitive to noise than those for first-order derivatives. Thus, strong coupling and reliable Monte Carlo simulation methods are required for an accurate approximation of the second-order derivatives. This paper proposes a novel method to estimate second-order parametric sensitivities of stochastic discrete biochemical models, which relies on an approximate Monte Carlo strategy to generate realizations of the correlated processes $\mathbf{X}(t, c + (h_j e_j + h_k e_k))$, $\mathbf{X}(t, c + h_j e_j)$, $\mathbf{X}(t, c + h_k e_k)$, and $\mathbf{X}(t, c)$. This approximate strategy is based on efficient adaptive τ-leaping schemes, expanding the techniques developed by Cao et al. [8]. We utilized a coupling of the four processes above, which ensures that the variance of the sensitivity estimator is reduced, leading to an effective sensitivity approximation. This coupling builds on the foundation of the coupled finite-difference framework [29]. Our tau-leaping based method extends the existing approach by offering improved scalability with the system size and increased efficiency for moderately stiff problems. The ability to take larger time steps and reduce the number of simulated reactions enables our method to handle realistic biochemical systems that would be computationally intensive using finite-difference methods relying on exact stochastic simulation strategies. We note that the coupled finite-difference strategy (CFD-2) considers a perturbation of a certain absolute value for both parameters ($h_j = h_k$). If there is a significant difference in the parameter values, comparing their relative importance may become challenging. To avoid this problem, our method defines the perturbations in terms of the relative value h, such that $h_j = hc_j$ and $h_k = hc_k$ are small.

The outline of this paper is given below. In Section 2, the background on stochastic discrete models of well-stirred biochemical networks and their simulation strategies is discussed. Section 3 describes various existing parametric sensitivity techniques and introduces a new finite-difference second-order sensitivity estimator for stochastic discrete biochemical networks. Section 4 illustrates the benefits of the proposed sensitivity estimator through numerical tests performed on three models arising in applications. Lastly, Section 5 provides our conclusions.

2. Background

Stochastic Modelling of Well-Mixed Biochemical Networks

This work focuses on developing tools to analyse a stochastic model that describes the evolution of the molecules' number of some biochemically reacting species confined to a constant volume at a constant temperature. More precisely, we assume that a collection of N biochemical species $S_1, \ldots, S_N$, of a homogeneous (well-mixed) biochemical system are interacting through M chemical reactions, $R_1, \ldots, R_M$. The state of the system is represented by the vector $\mathbf{X}(t) = (X_1(t), \ldots, X_N(t))^T$ at time t, wherein each $X_i(t)$ represents a non-negative integer denoting the quantity of molecules of species S_i at the given time t, where $i = 1, \ldots, N$. The state vector at the initial time t_0 is given, $\mathbf{X}(t_0) = \mathbf{x}_0$.

A reaction channel R_ℓ is defined by its stoichiometric vector ν_ℓ. Its state change vector is denoted by $^{\circ}{}_{\ell} = (\nu_{1\ell}, \nu_{2\ell}, \ldots, \nu_{N\ell})^T$, where each element $\nu_{i\ell}$ corresponds to the change in the quantity of the i-th molecular species after one reaction R_ℓ fires. Thus, $\{\nu_{i\ell}\}_{1 \le i \le N, 1 \le \ell \le M}$ is the biochemical system's stoichiometric matrix. Additionally, each reaction R_ℓ is characterized by its propensity function $a_\ell(\cdot)$. The propensity of the reaction R_ℓ is such that $a_\ell(\mathbf{x})dt$ quantifies the probability of the reaction R_ℓ firing within the time interval $[t, t + dt)$, under the assumption that $X(t) = \mathbf{x}$.

In the case of a second-order reaction, $S_k + S_j \xrightarrow{c_\ell}$ 'products', the propensity function is expressed as $a_\ell(\mathbf{x}) = c_\ell x_k x_j$, where $k \neq j$. In the case of a first-order reaction, $S_j \xrightarrow{c_\ell}$ 'products', the propensity function is calculated as $a_\ell(\mathbf{x}) = c_\ell x_j$. Lastly, for a dimerization, $S_j + S_j \xrightarrow{c_\ell}$ 'products', the propensity function is given by $a_\ell(\mathbf{x}) = c_\ell x_j (x_j - 1)/2$, since the likelihood of the reaction depends on the number of ways two S_j molecules can form unordered pairs. These expressions of the propensity functions are based on the principles of molecular physics and kinetic theory [3]. By applying further simplifying assumptions, one may obtain (in some cases) the reaction rates commonly used in deterministic chemical kinetics [16].

Define, by $P(\mathbf{x}, t|\mathbf{x}_0, t_0)$, the conditional probability of the system state being $\mathbf{X}(t) = \mathbf{x}$ at time t, provided that the initial state was $\mathbf{X}(t_0) = \mathbf{x}_0$. The evolution of this conditional probability over time is governed by the chemical master equation [3]

$$\frac{dP(\mathbf{x}, t|\mathbf{x}_0, t_0)}{dt} = \sum_{\ell=1}^{M} [a_\ell(\mathbf{x} - \nu_\ell)P(\mathbf{x} - \nu_\ell, t|\mathbf{x}_0, t_0) - a_\ell(\mathbf{x})P(\mathbf{x}, t|\mathbf{x}_0, t_0)] \, . \tag{1}$$

This is a stochastic discrete model of homogeneous biochemical networks. An alternative description of the system state $\mathbf{X}(t)$ which satisfies Equation (1) is given by the random time change (RTC) representation (see [34]):

$$\mathbf{X}(t) = \mathbf{x}_0 + \sum_{\ell=1}^{M} \nu_\ell Y_\ell \left(\int_0^t a_\ell(\mathbf{X}(s))ds \right) , \tag{2}$$

where Y_ℓ with $1 \leq \ell \leq M$ are independent Poisson processes of unit rate and $\mathbf{X}(t_0) = \mathbf{x}_0$. It should be noted that, in general, the expected value of the solution of the chemical master Equation (1) may not agree with the solution of the reaction rate equation.

For most practical applications, solving Equation (1) directly is computationally challenging due to its high dimensionality, particularly for systems with species in large molecular counts and with many reactions. To address this challenge, Monte Carlo strategies can be employed to produce sample paths consistent with the solution of the chemical master equation. An exact Monte Carlo method for generating such realizations of the stochastic process governed by the chemical master equation is the stochastic simulation algorithm (SSA) or Gillespie's direct method [4,5]. The steps of this algorithm are summarized in Algorithm 1.

Algorithm 1 Stochastic Simulation Algorithm

Initialization: $t = t_0$, $X = x_0$.
While $t < T$
Calculate $a_\ell(X)$ for each ℓ, $1 \leq \ell \leq M$, and $a_0(X) = \sum_{\ell=1}^{M} a_\ell(X)$.
Compute η_1 and η_2, two independent uniform random numbers over $[0, 1]$.
Determine the next reaction time $\tau = -\frac{1}{a_0(X)} \ln \eta_1$.
Determine the next reaction index, j, with $\sum_{\ell=1}^{j-1} a_\ell(X) < \eta_2 a_0(X) < \sum_{\ell=1}^{j} a_\ell(X)$.
Take $X = X + \nu_j$ and $t = t + \tau$.
End while.

Gillespie's algorithm and other exact Monte Carlo methods for the Chemical Master Equation can be computationally intensive for realistic biochemical networks. Simulating individual reaction events sequentially may require substantial computational resources, particularly for systems with some fast reactions.

To achieve a faster computation compared to exact methods, one could utilize approximate Monte Carlo schemes to estimate the solution of the chemical master equation. One such approximate method is the tau-leaping strategy developed by Gillespie [7]. This method may accelerate the simulation significantly for models with multiple scales in time. It does so by advancing the system in time with a predetermined step τ, which leaps over many reactions, instead of simulating individual reaction events in a sequential manner. The stepsize is chosen such that the leap condition [7] is satisfied: given that the current system state at time t is $\mathbf{X}(t) = \mathbf{x}$, the step $\tau > 0$ is small enough that each propensity varies insignificantly during $[t, t + \tau)$. With this assumption, the number of reactions R_ℓ taking place within $[t, t + \tau)$ can be approximated by a Poisson random variable with mean and variance equal to $a_\ell(\mathbf{x})\tau$, namely $P_\ell(a_\ell(\mathbf{x})\tau)$. Therefore, the system state may be estimated by

$$\mathbf{X}(t + \tau) = \mathbf{x} + \sum_{\ell=1}^{M} \nu_\ell P_\ell(a_\ell(\mathbf{x})\tau) \, , \tag{3}$$

which is known as the (explicit) tau-leaping method [7]. The Poisson random variables $\{P_\ell\}_{1\le\ell\le M}$ are independent.

Cao et al. [8] proposed a leap-selection strategy which increases the accuracy and efficiency of the simulation and which is extensively used in applications. To improve the accuracy of the numerical solution, the leap condition is adjusted to require that the relative change in the molecular count of each reactant species, rather than in each propensity, is negligible over the stepsize τ. Also, for avoiding negative population numbers, the reacting species are partitioned into critical and non-critical ones. For this purpose, a control parameter $n_c \in [2,20]$ is introduced. A reaction with a positive propensity is considered critical if n_c occurrences of this reaction drive the molecular amount of one of its reactants negative. Otherwise, the reaction is deemed non-critical. Here, J_{ncr} represents the set of indices of the non-critical reactions, I_r is the set of all reacting species indices and I_{ncr} is the set of indices of the reacting species participating in non-critical reactions. For a time t, a state $\mathbf{X}(t) = \mathbf{x}$ and a reaction R_ℓ with $a_\ell(\mathbf{x}) > 0$, the minimum number of firings of that reaction needed to consume one of its reacting species is

$$L_\ell(\mathbf{x}) = \min_{i=1,\dots,N;\nu_{i\ell}<0} \left\lfloor \frac{x_i}{|\nu_{i\ell}|} \right\rfloor \tag{4}$$

where $\lfloor\cdot\rfloor$ denotes the greatest integer that is lower than the given value. With this notation, a reaction R_ℓ with a positive propensity is classified as *critical* if $L_\ell(\mathbf{x}) < n_c$. For any index $i \in I_r$, consider r_i to be the highest order of a reaction in which the species S_i is a reactant. Then, the functions $g_i = g(\mathbf{x})$ are defined as follows [8]:

$$g_i = \begin{cases} 1, & \text{if } r_i = 1 \\ 2, & \text{if } r_i = 2 \text{ and } S_i + S_j \to \cdots, i \neq j \\ 2 + \frac{1}{x_i - 1}, & \text{if } r_i = 2 \text{ and } S_i + S_i \to \cdots \\ 3, & \text{if } r_i = 3 \text{ and } S_i + S_j + S_k \to \cdots, i \neq j \neq k \neq i \\ \frac{3}{2}\left(2 + \frac{1}{x_i - 1}\right), & \text{if } r_i = 3 \text{ and } S_i + S_i + S_j \to \cdots, i \neq j \\ \left(3 + \frac{1}{x_i - 1} + \frac{2}{x_i - 2}\right), & \text{if } r_i = 3 \text{ and } S_i + S_i + S_i \to \cdots \end{cases} \tag{5}$$

A second control parameter is the tolerance ε, which satisfies $0 < \varepsilon < 1$. The modified leap condition requires that the relative change in the molecular counts of each reactant species is (in some sense) below the tolerance ε. More precisely, if at time t the state is $\mathbf{X}(t) = \mathbf{x}$, then τ should be small enough such that

$$|X_i(t+\tau) - x_i| \le \max\{\varepsilon x_i / g_i, 1\}$$

for any $i \in I_r$. According to [8], the leap size τ is computed as

$$\tau(\mathbf{x}) = \min_i \left\{ \frac{\max\{\varepsilon x_i / g_i, 1\}}{|\hat{\mu}_i(\mathbf{x})|}, \frac{\max\{\varepsilon x_i / g_i, 1\}^2}{\hat{\sigma}_i^2(\mathbf{x})} \right\}, \tag{6}$$

with the auxiliary equations for each species S_i being

$$\hat{\mu}_i(\mathbf{x}) = \sum_{\ell \in J_{ncr}} \nu_{i\ell} a_\ell(\mathbf{x}), \quad \hat{\sigma}_i^2(\mathbf{x}) = \sum_{\ell \in J_{ncr}} \nu_{i\ell}^2 a_\ell(\mathbf{x}). \tag{7}$$

Our strategy for estimating second-order sensitivities relies on adaptive tau-leaping methods which utilize τ-selection schemes developed by Cao et al. [8].

3. Second-Order Parametric Sensitivity

Sensitivity analysis is a fundamental tool in the modelling and analysis of biochemical systems, measuring the impact of perturbations in parameter values on the behaviour of the system. This analysis aims to identify essential parameters that significantly influence system dynamics, assess the robustness of the model to parameter variations, and provide guidance for experimental design and parameter estimation. Local sensitivity analysis examines the influence of small changes in input parameters on the model outputs. In the case of biochemical systems, these parameters include initial molecular amounts of various species or kinetic parameters. A large local sensitivity in a parameter indicates that the model is sensitive with respect to that parameter; thus, an accurate estimation of its value is necessary.

For stochastic models, the first-order local parametric sensitivity can be computed by $\partial\mathbb{E}[\mathbf{f}(\mathbf{X}(t,c))]/\partial c_j$, where $\mathbb{E}[\cdot]$ symbolizes the expected value, $\mathbf{f}$ denotes a smooth function of interest, and $\mathbf{X}(t,c)$ is the state of the system at time t, corresponding to the vector of parameters c, and c_j is the parameter under investigation. Finite-difference techniques are among the most commonly utilized strategies for estimating these sensitivities.

In the case of stochastic models, designing effective methods for estimating the second-order parametric sensitivities $\partial^2\mathbb{E}[\mathbf{f}(\mathbf{X}(t,c))]/\partial c_j\partial c_k$, where c_j and c_k, with $j,k \in \{1,\ldots,M\}$ represent two model parameters, is a difficult problem. A forward finite-difference estimation of the second-order sensitivity with respect to these parameters is

$$\begin{aligned}&\frac{\partial^2}{\partial c_j\partial c_k}\mathbb{E}[\mathbf{f}(\mathbf{X}(t,c))] \approx \\ &\frac{1}{c_jc_kh^2}\left(\mathbb{E}[\mathbf{f}(\mathbf{X}(t,c+h(c_je_j+c_ke_k)))] - \mathbb{E}[\mathbf{f}(\mathbf{X}(t,c+hc_je_j))] - \mathbb{E}[\mathbf{f}(\mathbf{X}(t,c+hc_ke_k))] + \mathbb{E}[\mathbf{f}(\mathbf{X}(t,c))]\right),\end{aligned} \tag{8}$$

where h is a small relative perturbation. The goal is to estimate the expected value of the random variable $[\mathbf{f}(\mathbf{X}(t,c+h(c_je_j+c_ke_k))) - \mathbf{f}(\mathbf{X}(t,c+hc_je_j)) - \mathbf{f}(\mathbf{X}(t,c+hc_ke_k)) + \mathbf{f}(\mathbf{X}(t,c))]/(c_jc_kh^2)$. Monte Carlo methods may be utilized to generate realizations of this random variable. Then, an estimator of the second-order sensitivity with respect to c_j and c_k is the sample mean of a set of realizations

$$\Gamma = [\mathbf{f}(X(t,c+h(c_je_j+c_ke_k))) - \mathbf{f}(X(t,c+hc_je_j)) - \mathbf{f}(X(t,c+hc_ke_k)) + \mathbf{f}(X(t,c))]/(c_jc_kh^2)\,, \tag{9}$$

where we denoted, by $X(t,c)$, a sample path of the stochastic process $\mathbf{X}(t,c)$.

When the four processes $\mathbf{X}(t,c+h(c_je_j+c_ke_k))$, $\mathbf{X}(t,c+hc_je_j)$, $\mathbf{X}(t,c+hc_ke_k)$ and $\mathbf{X}(t,c)$ are correlated, the variance of the sensitivity estimator is reduced. The challenge consists of designing variance reduction schemes that result in a more effective sensitivity approximation. Such a finite-difference strategy based on correlating the four processes was proposed in [29]. For this strategy, an exact Monte Carlo scheme, the next reaction method or Gillespie's algorithm, is utilized to generate realizations of these processes. This technique estimates second-order sensitivities by ensuring that the four paths share reactions, thereby producing low-variance estimates. We shall compare our proposed strategy that shares groups of reactions to this scheme, which is the state-of-the-art method for computing second-order sensitivities for the chemical master equation model.

The four processes, $\mathbf{X}(t,c+h(c_je_j+c_ke_k))$, $\mathbf{X}(t,c+hc_je_j)$, $\mathbf{X}(t,c+hc_ke_k)$, $\mathbf{X}(t,c)$, have a unified representation:

$$\mathbf{X}(t,c+h(pc_je_j+qc_ke_k)),\ \text{where } p,q \in \{0,1\}\,, \tag{10}$$

and they share the same initial condition, $\mathbf{X}(0,c+h(pc_je_j+qc_ke_k)) = \mathbf{x}_0$.

Denote the propensity of the reaction R_ℓ, $1 \le \ell \le M$, of each process by

$$\begin{aligned} a_{\ell,11} &= a_\ell(c+h(c_je_j+c_ke_k), X(t,c+h(c_je_j+c_ke_k)), \\ a_{\ell,10} &= a_\ell(c+he_j, X(t,c+hc_je_j)), \\ a_{\ell,01} &= a_\ell(c+he_k, X(t,c+hc_ke_k)), \\ a_{\ell,00} &= a_\ell(c, X(t,c))\,, \end{aligned} \tag{11}$$

where the notation ignores the dependence on the time t and the parameters, for simplicity. Also, consider the set

$$B = \{1111,\ 1100,\ 0011,\ 1010,\ 0101,\ 1000,\ 0100,\ 0010,\ 0001\}.$$

In order to couple the four processes above, which produces a low variance sensitivity estimator, we define the following terms, $\alpha_{\ell,b}$, for $1 \leq \ell \leq M$ and $b \in B$, in accordance with [29]:

$$\begin{aligned}
\alpha_{\ell,1111} &= \min(a_{\ell,11}, a_{\ell,10}, a_{\ell,01}, a_{\ell,00}) \\
\alpha_{\ell,1100} &= \min(a_{\ell,11}, a_{\ell,10}) - \alpha_{\ell,1111} \\
\alpha_{\ell,0011} &= \min(a_{\ell,01}, a_{\ell,00}) - \alpha_{\ell,1111} \\
\alpha_{\ell,1010} &= \min(a_{\ell,11} - \min(a_{\ell,11}, a_{\ell,10}), a_{\ell,01} - \min(a_{\ell,01}, a_{\ell,00})) \\
\alpha_{\ell,0101} &= \min(a_{\ell,10} - \min(a_{\ell,11}, a_{\ell,10}), a_{\ell,00} - \min(a_{\ell,01}, a_{\ell,00})) \\
\alpha_{\ell,1000} &= (a_{\ell,11} - \min(a_{\ell,11}, a_{\ell,10})) - \alpha_{\ell,1010} \\
\alpha_{\ell,0100} &= (a_{\ell,10} - \min(a_{\ell,11}, a_{\ell,10})) - \alpha_{\ell,0101} \\
\alpha_{\ell,0010} &= (a_{\ell,01} - \min(a_{\ell,01}, a_{\ell,00})) - \alpha_{\ell,1010} \\
\alpha_{\ell,0001} &= (a_{\ell,00} - \min(a_{\ell,01}, a_{\ell,00})) - \alpha_{\ell,0101}\,.
\end{aligned} \tag{12}$$

The coupled finite-difference strategy for estimating second-order sensitivities, CFD-2 [29], relies on a strong correlation of the four processes (10). This correlation, which utilizes the RTC representation (2), is provided below.

$$\begin{aligned}
\mathbf{X}(t, c + h(c_j e_j + c_k e_k)) &= \mathbf{x}_0 + \sum_{\ell=1}^{M} \nu_\ell \sum_{b \in \{1111,1100,1010,1000\}} Y_{\ell,b}\Big(\int_0^t \alpha_{\ell,b}(s)ds\Big) \\
\mathbf{X}(t, c + hc_j e_j) &= \mathbf{x}_0 + \sum_{\ell=1}^{M} \nu_\ell \sum_{b \in \{1111,1100,0101,0100\}} Y_{\ell,b}\Big(\int_0^t \alpha_{\ell,b}(s)ds\Big) \\
\mathbf{X}(t, c + hc_k e_k) &= \mathbf{x}_0 + \sum_{\ell=1}^{M} \nu_\ell \sum_{b \in \{1111,0011,1010,0010\}} Y_{\ell,b}\Big(\int_0^t \alpha_{\ell,b}(s)ds\Big) \\
\mathbf{X}(t, c) &= \mathbf{x}_0 + \sum_{\ell=1}^{M} \nu_\ell \sum_{b \in \{1111,0011,0101,0001\}} Y_{\ell,b}\Big(\int_0^t \alpha_{\ell,b}(s)ds\Big)\,.
\end{aligned}$$

Remark that every counting process Y_ℓ in the RTC representation is split into sub-processes shared among the four continuous time Markov chains (10). Moreover, each of the correlated processes above has the same distribution as that of their uncoupled counterparts. The CFD-2 algorithm utilizes the stochastic simulation algorithm [4,5] or the next reaction method [25].

Coupled τ-Leaping Method for Estimating Second-Order Sensitivities

We propose below a strategy which relies on the novel combination of tau-leaping with a finite-difference sensitivity estimation approach that is designed for efficient and accurate second-order parametric sensitivity analysis for stochastic discrete models of homogeneous biochemical networks. We show that second-order sensitivities can be estimated accurately and efficiently by utilizing a variable time-stepping strategy for tau-leaping. This is a valuable and non-trivial extension of an existing finite-difference method, which allows sensitivity analysis to scale efficiently to larger, more complex systems. Also, this technique is effective and reliable for models of biochemical reactions that operate on multiple time scales (stiff problems). Stiff systems are commonly found in practical applications, as biochemical processes typically entail both rapid and slow reactions. For such cases, exact stochastic simulation schemes are computationally intensive, since they advance the system one reaction at a time.

Unlike the coupled finite-difference scheme CFD-2 [29], which utilizes exact methods to generate the coupled quadruple Monte Carlo trajectories, our approach employs variable

τ-leaping techniques to compute the four correlated paths. The proposed coupled τ-leaping (CTL-2) strategy is based on sharing Poisson random variables among the nominal and perturbed trajectories. Moreover, the impact of the shared terms is anticipated to be substantial, resulting in a strong coupling and thus a notable decrease in the variance of the sensitivity estimation.

In the coupled tau-leaping algorithm for approximating second-order sensitivities, the nominal and the three perturbed sample paths are correlated according to

$$\begin{aligned}
X(t+\tau, c+h(c_j e_j + c_k e_k)) &= \mathbf{x}^{c+h(c_j e_j + c_k e_k)} + \sum_{\ell=1}^{M} \nu_\ell \sum_{b\in\{1111,1100,1010,1000\}} P_{\ell,b}(\alpha_{\ell,b}\tau) \\
X(t+\tau, c+hc_j e_j) &= \mathbf{x}^{c+hc_j e_j} + \sum_{\ell=1}^{M} \nu_\ell \sum_{b\in\{1111,1100,0101,0100\}} P_{\ell,b}(\alpha_{\ell,b}\tau) \\
X(t+\tau, c+he_k) &= \mathbf{x}^{c+hc_k e_k} + \sum_{\ell=1}^{M} \nu_\ell \sum_{b\in\{1111,0011,1010,0010\}} P_{\ell,b}(\alpha_{\ell,b}\tau) \\
X(t+\tau, c) &= \mathbf{x}^{c} + \sum_{\ell=1}^{M} \nu_\ell \sum_{b\in\{1111,0011,0101,0001\}} P_{\ell,b}(\alpha_{\ell,b}\tau)
\end{aligned} \tag{13}$$

where $X(t,\tilde{c}) = \mathbf{x}^{\tilde{c}}, \tilde{c} \in \{c+h(c_j e_j + c_k e_k), c+hc_j e_j, c+hc_k e_k, c\}$. Here, $P_{\ell,b}(\alpha_{\ell,b}\tau), 1 \leq \ell \leq M$ and $b \in B$, are independent Poisson random variables of mean and variance $\alpha_{\ell,b}\tau$.

To reduce the computational time and avoid negative population numbers in the simulation, we shall extend the variable time-stepping strategy introduced by Cao et al. [8] to generate the four correlated trajectories with the explicit tau-leaping strategy (13). First, a potential stepsize is calculated separately for both critical and non-critical reactions on each of the four trajectories. Then, the smallest of these stepsizes is selected as the next τ for the tau-leaping strategy on each path. Note that for the case of double derivatives with respect to one parameter (i.e., $j = k$), the paths $X(c + hc_j e_j)$ and $X(c + hc_k e_k)$ are still generated separately. The steps of the algorithm are given below.

Coupled $\boldsymbol{\tau}$-Leaping Algorithm

1. **Set model parameters**: tolerance ϵ, critical threshold n_c, simulation final time T, and relative perturbation h.
2. **Initialize trajectories** at $t = 0$: $X(c + h(pc_j e_j + qc_k e_k)) = \mathbf{x}_0$, for all $p, q \in \{0, 1\}$.
3. **While** $t < T$ **do** steps (a)–(h):
   (a) **Calculate the propensities** from (11): for $\ell = 1, \ldots, M$ and $p, q \in \{0, 1\}$,
   $$a_{\ell,pq} = a_\ell(c + h(pc_j e_j + qc_k e_k), X(c + h(pc_j e_j + qc_k e_k))$$
   (b) **Find critical reactions** for each of the four trajectories: for each reaction R_ℓ and $p, q \in \{0, 1\}$, if $a_{\ell,pq} > 0$, then
   $$L_{\ell,pq} = L(X(c + h(pc_j e_j + qc_k e_k))) \,,$$
   with L defined by (4). Put $J_{ncr} = \{\ell \mid L_{\ell,pq} \geq n_c \text{ for all } p, q \in \{0, 1\}\}$.
   (c) **Find candidate step sizes, $\tau_{ncr,pq}$, for the non-critical reactions**; for all $p, q \in \{0, 1\}$:
      i. If $J_{ncr} = \emptyset$, then $\tau_{ncr,pq} = \infty$.
      ii. If $J_{ncr} \neq \emptyset$, then find set of indices of reactant species of non-critical reactions I_{ncr}.
      iii. For each $i \in I_{ncr}$ and each trajectory, find the highest order of reaction r_i and the function g_i defined by (5).

iv. Calculate the auxiliary values $\hat{\mu}_i(X(c+h(pc_je_j+qc_ke_k)))$ and $\hat{\sigma}_i^2(X(c+h(pc_je_j+qc_ke_k)))$ from (7) and, from (6),

$$\tau_{ncr,pq} = \tau(X(c+h(pc_je_j+qc_ke_k)))\,.$$

(d) **Find candidate step sizes, $\tau_{cr,pq}$, for critical reactions**; for all $p,q \in \{0,1\}$:

i. Calculate $a^0_{cr,pq}$ as the sum of the propensities for critical reactions corresponding to the trajectory for $c+h(pc_je_j+qc_ke_k)$.

ii. Choose samples η_{pq} from the uniform distribution $U(0,1)$ and generate the time to the first critical reaction:

$$\tau_{cr,pq} = -\frac{\ln(\eta_{pq})}{a^0_{cr,pq}}\,.$$

(e) **Calculate the step size, τ:**
$\tau_{ncr} = \min_{p,q\in\{0,1\}}\{\tau_{ncr,pq}\}$, $\tau_{cr} = \min_{p,q\in\{0,1\}}\{\tau_{cr,pq}\}$, $\tau = \min(\tau_{ncr}, \tau_{cr})$.

(f) **Calculate the number of reactions R_ℓ**, namely k_ℓ:

i. If R_ℓ is a critical reaction, then $k_{\ell,pq} = 0$, for all $p,q \in \{0,1\}$.

ii. If $\tau_{cr} \le \tau_{ncr}$, then at least one critical reaction happens. For each p,q with $\tau_{cr,pq} = \tau_{cr}$, generate ℓ^* as a sample of the integer random variable with probabilities $a_{\ell,pq}/a^0_{cr,pq}$, with ℓ running only over critical reactions. Put
$k_{\ell^*,pq} = 1$.

iii. For $\ell \in J_{ncr}$, compute $\alpha_{\ell,b}, b \in B$, from (12) and sample Poisson random variables:

$$k_{\ell,b} = \text{Poisson}(\alpha_{\ell,b}\tau)$$

iv. Put:
$$\begin{aligned}
k_{\ell,11} &= \sum_{b\in\{1111,1100,1010,1000\}} k_{\ell,b}\\
k_{\ell,10} &= \sum_{b\in\{1111,1100,0101,0100\}} k_{\ell,b}\\
k_{\ell,01} &= \sum_{b\in\{1111,0011,1010,0010\}} k_{\ell,b}\\
k_{\ell,00} &= \sum_{b\in\{1111,0011,0101,0001\}} k_{\ell,b}
\end{aligned}$$

(g) **Update the time and the states on all trajectories**:

i. $t = t+\tau$.

ii. $X(c+h(pc_je_j+qc_ke_k)) = X(c+h(pc_je_j+qc_ke_k)) + \sum_{\ell=1}^{M} k_{\ell,pq}\nu_\ell$, for all $p,q \in \{0,1\}$.

(h) **Estimate the second-order sensitivity**:

$$\Gamma = [\mathbf{f}(X(c+h(c_je_j+c_ke_k))) - \mathbf{f}(X(c+hc_je_j)) - \mathbf{f}((X(c+hc_ke_k)) + \mathbf{f}(X(c))]/(c_jc_kh^2)\,.$$

We note that the relative perturbation size h is a key parameter, which influences the variance of the sensitivity approximation. In [29], it is shown that the variance of the CFD-2 sensitivity estimator is of order $\mathcal{O}(h^{-3})$. In our numerical experiments, we observe a variance of the CTL-2 sensitivity estimator similar to that of CFD-2, both achieving a lower variance, of order $\mathcal{O}(h^{-2})$. Note that the explicit tau-leaping strategy introduces an error compared to exact stochastic simulation methods, but this error is often small and depends on the leap size [35]. If the perturbation h is too small, the variance of the estimator increases, leading to more noise affecting the sensitivity estimate. On the other hand, if the perturbation h is too large, a finite-difference approximation may become inaccurate, while the sensitivity estimates are less noisy. Choosing an appropriate h is important for balancing the approximation error of the finite-difference scheme and the variance in the sensitivity estimates, in particular for second-order ones.

The computational cost of the proposed sensitivity estimation method based on tau-leaping is generally not expected to vary significantly with h, when the absolute values of the perturbations are small. The computational cost of estimating these sensitivities depends on the tolerance ε of the adaptive tau-leaping scheme. The tolerance ε is trade-off parameter that can be used to balance computational efficiency and accuracy of the sensitivity estimation. Significant computational efficiency may be obtained by introducing a small error.

4. Numerical Experiments

In this section, we test the efficiency and accuracy of the proposed technique for estimating second-order sensitivities of the chemical master equation, on three models of biochemical networks arising in applications. For each model, we simulate 20,000 Monte Carlo paths with the new method based on tau-leaping, CTL-2, and the existing coupled finite-difference scheme for approximating the Hessian, CFD-2. In each case, we applied a forward finite-difference strategy for estimating sensitivities, with the same relative perturbation h of the parameters of interest. We report the evolution in time of the mean trajectories corresponding to the nominal value of the parameter for both methods, and of the standard deviation of these trajectories. We also compare the time dependence of the estimated second-order sensitivities for both strategies and of the standard deviation of their sensitivity estimators. In addition, we indicate the speed-up of the CTL-2 algorithm over the CFD-2 scheme for each of the three systems. We measure the efficiency gain of our algorithm over the CFD-2 one as

$$\text{Speed-up} = \frac{\text{CPU time(CFD-2)}}{\text{CPU time(CTL-2)}}.$$

A substantial computational efficiency of the new method compared to the existing scheme is obtained in our tests, which were performed on models with some degree of stiffness.

4.1. Decay-Dimerization Model

We start by testing our strategy to estimate sensitivities on the decay-dimerization system of Gillespie [7]. In this model, three molecular species participate in four reactions. The reactions, their propensities and the values of the kinetic parameters are included in Table 1. The system is studied on the time interval $[0,4]$ with the initial molecular amounts $\mathbf{X}(0) = [10^5, 10, 0]$.

In our simulations, we apply the CTL-2 method with the value of the tolerance $\varepsilon = 0.05$. The averages of the S_3 molecular counts, computed with the proposed CTL-2 and the existing CFD-2 algorithms, are compared in Figure 1a. The standard deviations of the molecular amounts of S_3, generated with the CTL-2 and CFD-2 methods, are shown in Figure 1b. The plots show an excellent match of the means and standard deviations obtained with these strategies, built on the next reaction method (CFD-2) and adaptive tau-leaping (CTL-2). This shows that the adaptive tau-leaping method utilized by our algorithm is very accurate for the chosen tolerance.

Table 1. Decay-dimerization model.

	Reaction	Propensity	Parameter Value
R_1	$S_1 \xrightarrow{c_1} \emptyset$	$a_1 = c_1 X_1$	$c_1 = 1$
R_2	$S_1 + S_1 \xrightarrow{c_2} S_2$	$a_2 = c_2 X_1 (X_1 - 1)/2$	$c_2 = 0.004$
R_3	$S_2 \xrightarrow{c_3} S_1 + S_1$	$a_3 = c_3 X_2$	$c_3 = 0.5$
R_4	$S_2 \xrightarrow{c_4} S_3$	$a_4 = c_4 X_2$	$c_4 = 0.04$

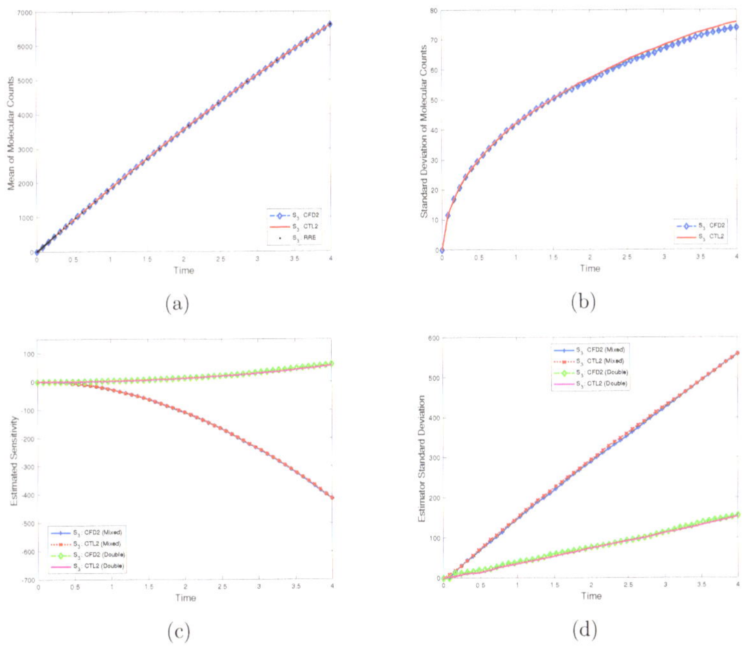

Figure 1. Decay-dimerization model: (**a**) Mean. (**b**) Standard deviation of the molecular counts of S_3 by the τ-leaping and the NRM methods. (**c**) Forward finite-difference sensitivity estimators of the second order with respect to c_1 (double) and with respect to c_1 and c_3 (mixed) of the molecular amount of S_3 computed with the CTL-2 and CFD-2 algorithms. (**d**) Standard deviation of the above CTL-2 and CFD-2 sensitivity estimators. In total, 20,000 quadruples of paths are generated with the CFD-2 and CTL-2 algorithms on the interval $[0, 4]$.

First, let us compute the *double derivative with respect to parameter* c_1. The relative value of the perturbation is $h = 0.05$, which is 5% of the nominal value of the parameter of interest. Figure 1c compares the second-order sensitivity with respect to the first parameter as a function of time, $\frac{\partial^2}{\partial c_1^2}\mathbb{E}[X_3]$, approximated with our method and the CFD-2 scheme. The agreement is excellent, validating our method. In Figure 1d, a very good match is observed for the evolution in time of the standard deviation of the CTL-2 and the CFD-2 sensitivity estimators, which verifies the accuracy of our technique. Furthermore, we investigate experimentally the dependence of the CFD-2 and CTL-2 sensitivity estimator standard deviations on the perturbation h. The findings, provided in Figure 3a, show very good agreement of the standard deviations of the two sensitivity estimators. The plots suggest an order $\mathcal{O}(h^{-2})$ variance of the CFD-2 and CTL-2 methods for this nonlinear model.

Now, let us determine the *mixed derivative with respect to parameters* c_1 *and* c_3. In our simulations, the relative value of the perturbation is $h = 0.05$ (i.e., the absolute values of the perturbations are 5% of the parameter values of c_3 and c_1, respectively). The time dependence of the estimated second-order sensitivity of the molecular amount of species S_3 with respect to parameters c_1 and c_3, $\frac{\partial^2}{\partial c_1 \partial c_3}\mathbb{E}[\mathbf{X}_3]$, approximated with the CTL-2 and the CFD-2 schemes, is given in Figure 1c. Again, the plots show that our estimator is very accurate. This is confirmed by the excellent agreement obtained for the standard deviations of the CTL-2 and CFD-2 sensitivity estimators, which are plotted versus time in Figure 1d.

When computing the mixed derivative, the computational efficiency of the new method to estimate second-order sensitivities, CTL-2, over the CFD-2 scheme is

$$\text{Speed-up} = 33.42$$

which shows substantial reductions in the computational cost of our method compared to the existing technique, for a similar accuracy of the sensitivity estimation. For instance, in this scenario, simulating the tau-leaping based algorithm (CTL-2) across 100 sets of four correlated paths for $\varepsilon = 0.05$ required 2.92 s in MATLAB, on a Macbook Pro with a 3.7 GHz Apple M2 Max Chip (12 Cores). In contrast, performing the equivalent computation for the CFD-2 algorithm took 97.61 s.

4.2. Two-Step Closed Reaction Chain Model

Consider below a two-step closed reaction chain system [36] consisting of three species that interact through four reactions. Table 2 describes the reaction channels and their propensity functions, and specifies the values of the kinetic parameters. This biochemical model is studied with the initial conditions $\mathbf{X}(0) = [2000, 1000, 100]$, on the time interval $[0, 0.1]$. The first two reactions are significantly faster compared to the last two ones and, as a consequence, the model is quite stiff. For convenience, we include below the classic deterministic model of mass action kinetics for this system, written in terms of species population numbers rather than concentrations,

$$\begin{aligned} \frac{dX_1}{dt} &= -c_1 X_1 + c_2 X_2 \\ \frac{dX_2}{dt} &= c_1 X_1 - (c_2 + c_3) X_2 + c_4 X_3 \\ \frac{dX_3}{dt} &= c_3 X_2 - c_4 X_3 \,. \end{aligned}$$

In general, the expected value of the stochastic process governed by the chemical master Equation (1) and the solution of the reaction rate equation may be different [14]. However, for this linear system, they are the same. Thus, we can use this biochemical model to validate the accuracy of the tau-leaping method employed by the CTL-2 scheme.

Table 2. Two-step closed reaction model.

	Reaction	Propensity	Parameter Value
R_1	$S_1 \xrightarrow{c_1} S_2$	$a_1 = c_1 X_1$	$c_1 = 800$
R_2	$S_2 \xrightarrow{c_2} S_1$	$a_2 = c_2 X_2$	$c_2 = 3200$
R_3	$S_2 \xrightarrow{c_3} S_3$	$a_3 = c_3 X_2$	$c_3 = 0.1$
R_4	$S_3 \xrightarrow{c_4} S_2$	$a_4 = c_4 X_3$	$c_4 = 1$

We estimate the *double derivative with respect to parameter* c_4 of the molecular abundance of species S_3, which is $\frac{\partial^2}{\partial c_4^2}\mathbb{E}[X_3]$. The finite-difference sensitivity estimators are applied with a value of the relative perturbation parameter $h = 0.05$ (the perturbation being 5% of the nominal value of the parameter of interest). We investigated the performance of our tau-leaping second-order sensitivity estimator with the tolerances $\varepsilon = 0.04$ and 0.05. Figure 2a,b show the mean and standard deviation of the number of molecules of species S_3 versus time, respectively, computed for the nominal value of the parameter, with the adaptive tau-leaping scheme utilized in the CTL-2 algorithm and the exact modified next reaction method. The agreement of the results produced by these methods is excellent, indicating that the tau-leaping strategy employed by our algorithm is accurate.

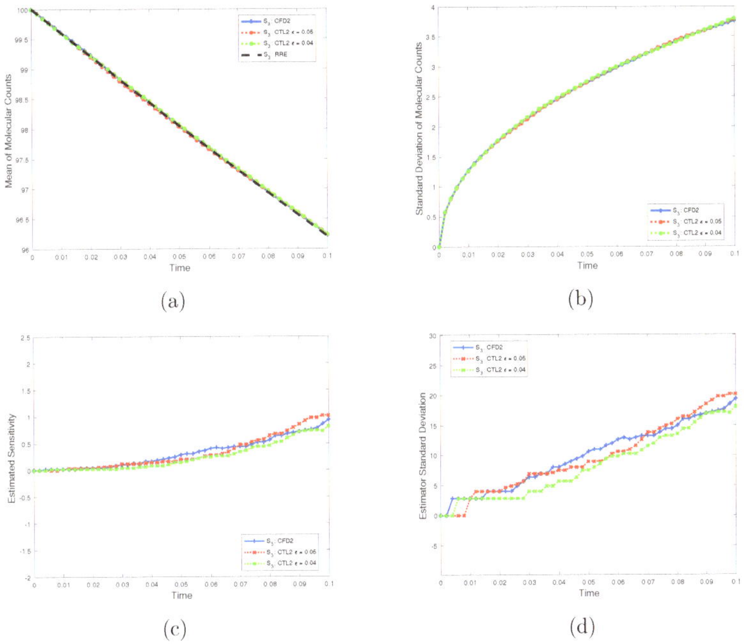

Figure 2. Two-step closed reaction chain model: (**a**) Mean. (**b**) Standard deviation of the molecular counts of S_3 by the τ-leaping and the NRM methods (the mean is also compared to the solution of the deterministic model). (**c**) Forward finite-difference sensitivity estimators of second-order with respect to c_4 of the molecular amount of S_3 computed with the CTL-2 and CFD-2 algorithms. (**d**) Standard deviation of the above CTL-2 and CFD-2 sensitivity estimators. In total, 20,000 quadruples of paths are generated with the CFD-2 and CTL-2 algorithms on the interval $[0, 0.1]$.

The second-order sensitivity of the number of S_3 molecules with respect to parameter c_4 is estimated by finite-difference schemes with the CTL-2, with tolerances 0.04 and 0.05, and the CFD-2 scheme, with their evolution being displayed in Figure 2c. The results match very well, demonstrating the accuracy of our technique for various tolerances. Furthermore, the standard deviations of the CTL-2 and CFD-2 sensitivity estimators depending on time are in agreement, as shown in Figure 2d. This further confirms the accuracy of the proposed strategy for estimating sensitivities. Furthermore, the computational efficiency of the CTL-2 algorithm over the CFD-2 is displayed in Table 3. On this model, the new technique has a considerably improved computational efficiency compared to the existing scheme, for a similar accuracy of the sensitivity estimation.

Table 3. Two-step closed reaction chain model: speed-up of CTL-2 compared to CFD-2, for estimating the second-order sensitivities with perturbation parameter $h = 0.05$. The time interval is $[0, 0.01]$.

Algorithm	Tolerance	Speed-Up
CTL-2	$\varepsilon = 0.05$	48.47
CTL-2	$\varepsilon = 0.04$	34.97
CFD-2	-	1

In addition, we numerically analysed the dependence on the perturbation parameter h of the standard deviation of the CFD-2 and CTL-2 sensitivity estimators (for a range of tolerances). The results are displayed in Figure 3b. The two algorithms give very similar results, which indicates that the new sensitivity estimator has an accuracy comparable to that of the CFD-2 scheme. Numerically, the variance of the proposed adaptive tau-leaping second-order sensitivity estimator seems to be $\mathcal{O}(h^{-2})$ for this model. As mentioned earlier, the smaller values of the perturbation parameter h yield more accurate approximations of the true second-order derivative. Yet, these small perturbation sizes increase the variance of the sensitivity estimates, resulting in larger fluctuations in the output. However, large values of the perturbation parameter can introduce considerable truncation errors when approximating the derivative. At the same time, they also lower the variance of the sensitivity estimates, thereby reducing stochastic noise. We determined satisfactory perturbation parameter values experimentally. Exploring the optimal selection of perturbation size, which provides a balance between derivative approximation accuracy and sensitivity estimator variance reduction, falls outside the scope of this paper and remains a subject for future investigation.

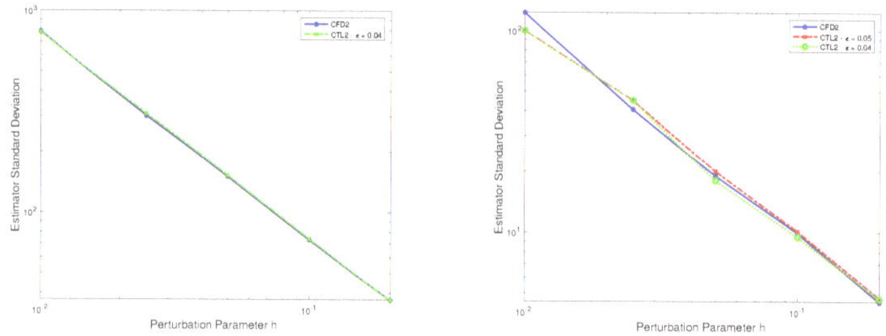

(a) (b)

Figure 3. Variation of the standard deviation of the sensitivity estimator of the S_3 molecular counts with respect to h, for the CTL-2 and CFD-2 algorithms: (**a**) the double derivative with respect to c_1 (the decay-dimerization model) and (**b**) the double derivative with respective to c_4 (the two-step model).

The speed-up of our strategy over the existing one, for the tolerance values ε tested, is recorded in Table 3. Note that, for a similar accuracy of the sensitivity estimation, the CTL-2 algorithm is significantly more efficient compared to the CFD-2 one.

4.3. Gene Regulatory Network Model

The third model represents a gene regulatory network [27,37]. This model was also studied in [13] and, in a different setting not examined here, can display bistable behaviour [38]. For this biochemical network, 8 species interact through 12 reaction channels. Table 4 includes the reactions, their propensities and the values of the reaction rate parameters. The system is analysed on the time interval $[0, 0.01]$ with initial conditions $\mathbf{X}(0) = [800, 800, 500, 500, 400, 500, 400, 500]^T$. This relatively large model is also stiff, with several orders of magnitude separating the fast and slow reactions.

The numerical simulations are carried out on 20,000 correlated trajectories with the coupled finite-difference-2 and the coupled tau-leaping-2 methods, on the time-interval $[0, 0.01]$. The CTL-2 algorithm is tested with the tolerance values $\varepsilon = 0.05$, 0.07 and 0.1. We investigated the accuracy of the tau-leaping strategy on which the CTL-2 scheme is based, by comparing the results with those of the exact Monte Carlo strategy of the next reaction method, for the nominal parameter values. The mean and standard deviations of the molecular numbers of the species S_2 and S_5 as functions of time are displayed in Figure 4a and Figure 4b, respectively. The two strategies lead to similar results and the accuracy of the tau-leaping method is improving with a smaller tolerance ε, as expected.

Table 4. Gene regulatory system model.

	Reaction	Propensity	Parameter Value
R_1	$S_3 \xrightarrow{c_1} S_3 + S_1$	$a_1 = c_1 X_3$	$c_1 = 0.16$
R_2	$S_4 \xrightarrow{c_2} S_4 + S_2$	$a_2 = c_2 X_4$	$c_2 = 0.16$
R_3	$S_3 + S_2 \xrightarrow{c_3} S_5$	$a_3 = c_3 X_3 X_2$	$c_3 = 5$
R_4	$S_5 \xrightarrow{c_4} S_3 + S_2$	$a_4 = c_4 X_5$	$c_4 = 3000$
R_5	$S_5 + S_2 \xrightarrow{c_5} S_6$	$a_5 = c_5 X_5 X_2$	$c_5 = 2.5$
R_6	$S_6 \xrightarrow{c_6} S_5 + S_2$	$a_6 = c_6 X_6$	$c_6 = 1600$
R_7	$S_1 \xrightarrow{c_7} \emptyset$	$a_7 = c_7 X_1$	$c_7 = 0.1$
R_8	$S_4 + S_1 \xrightarrow{c_8} S_7$	$a_8 = c_8 X_4 X_1$	$c_8 = 2$
R_9	$S_7 \xrightarrow{c_9} S_4 + S_1$	$a_9 = c_9 X_7$	$c_9 = 3000$
R_{10}	$S_7 + S_1 \xrightarrow{c_{10}} S_8$	$a_{10} = c_{10} X_7 X_1$	$c_{10} = 2.5$
R_{11}	$S_8 \xrightarrow{c_{11}} S_7 + S_1$	$a_{11} = c_{11} X_8$	$c_{11} = 1600$
R_{12}	$S_2 \xrightarrow{c_{12}} \emptyset$	$a_{12} = c_{12} X_2$	$c_{12} = 0.1$

We start by computing the *second-order sensitivity with respect to parameter* c_5, i.e., $\frac{\partial^2}{\partial c_5^2}\mathbb{E}[X_i]$ for the indices $i = 1, \ldots, N$. The finite-difference sensitivity estimators are employed with the relative perturbation size $h = 0.05$ (a perturbation representing 5% of the parameter value). The time-dependence of the estimated sensitivities of the S_2 and S_5 molecular counts, computed with the CFD-2 and the CTL-2 method with the tolerance values 0.05, 0.07 and 0.1, are shown in Figure 4c. The agreement is very good for each species. This observation is further verified by comparing the evolution of the standard deviation of the CTL-2 and CFD-2 sensitivity estimators for each of these species, which are plotted in Figure 4d. These results are similar, with the accuracy of the CTL-2 algorithm depending on the tolerance ε.

Finally, we estimate the *second-order sensitivity with respect to parameters* c_3 *and* c_5, namely $\frac{\partial^2}{\partial c_3 \partial c_5}\mathbb{E}[X_i]$ for each species. The perturbation parameter has the same relative perturbation size as before, $h = 0.05$ (that is, 5% of the parameter values of c_3 and c_5, respectively). The second-order sensitivity estimators of the S_2 and S_5 generated with the CTL-2 and the CFD-2 strategies are presented in Figure 4e, while the standard deviation of their sensitivity estimators are displayed in Figure 4f. Once more, the similar values of the sensitivity estimators show the accuracy of the CTL-2 method, which is also confirmed in Figure 4e. Additionally, we measured the computational cost of the two strategies, for a similar level of accuracy. Table 5 displays the speed-up of the tau-leaping sensitivity estimator over the existing one, for several values of the tolerance ε. A *significant efficiency gain* of the CTL-2 compared to the CFD-2 algorithm is observed for this model of a relatively large size, which also exhibits stiffness.

Table 5. Gene regulatory network model: speed-up of CTL-2 compared to CFD-2, for estimating the second-order sensitivities, with perturbation parameter $h = 0.05$. The time interval is $[0, 0.01]$.

Algorithm	Tolerance	Speed-Up
CTL-2	$\varepsilon = 0.05$	8.06
CTL-2	$\varepsilon = 0.07$	14.43
CTL-2	$\varepsilon = 0.1$	24.68
CFD-2	–	1

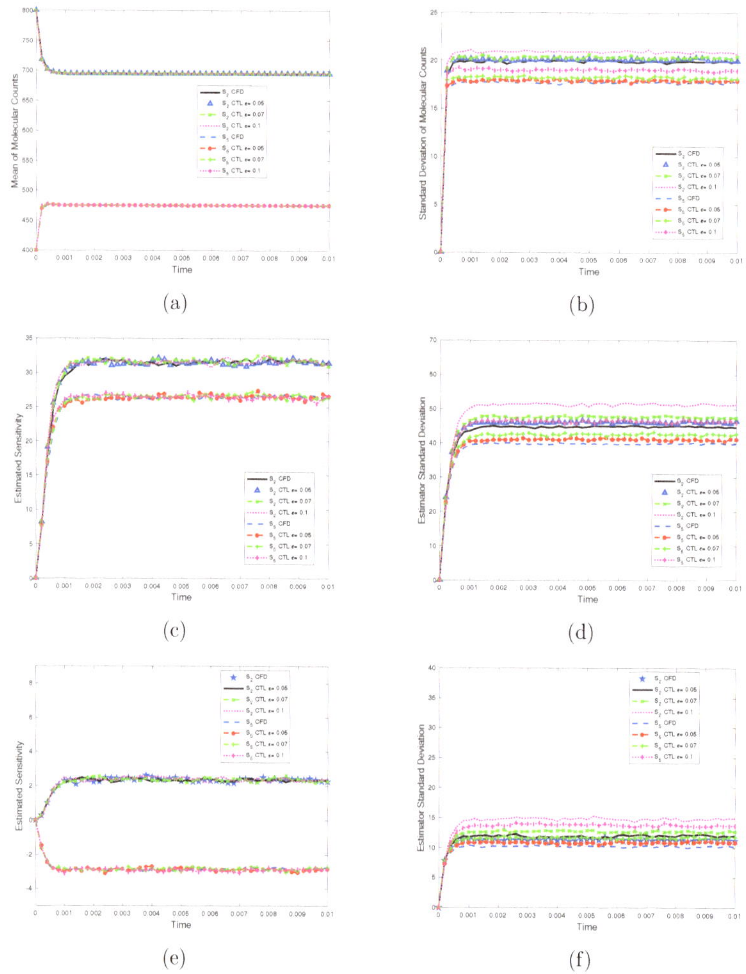

Figure 4. Gene regulatory network model: (**a**) Mean. (**b**) Standard deviation of the molecular counts of S_2 and S_5 by the τ-leaping and the NRM methods. (**c**) Forward finite-difference sensitivity estimators of a second order with respect to c_5 and (**e**) with respect to c_3 and c_5 of the molecular amount of S_2 and S_5, computed with the CTL-2 and CFD-2 algorithms. (**d**) Standard deviation sensitivity estimators of second-order with respect to c_5 and (**f**) with respect to c_3 and c_5, of the CTL-2 and CFD-2 algorithms. In total, 20,000 quadruples of paths are generated with the CFD-2 and CTL-2 algorithms on the interval $[0, 0.01]$.

5. Conclusions

Cellular processes frequently involve complex networks of molecular interactions, including enzyme–substrate reactions, molecular binding and dissociation, and biochemical signalling pathways. These systems exhibit stochastic behaviour due to the discrete and probabilistic nature of molecular interactions. The chemical master equation is a stochastic discrete model of well-stirred biochemical systems, which accurately captures the randomness and variability observed at the molecular level. Many biochemical systems arising in applications evolve on multiple scales in time, which means that their mathematical models are stiff.

In this study, we proposed a new technique for estimating second derivative sensitivities with respect to various parameters in stochastic discrete models of homogeneous biochemical systems. By accounting for curvature effects and parameter correlations, second-order sensitivity analysis enhances the accuracy and efficiency of parameter estimation, leading to improved model predictions and a better understanding of the underlying biological mechanisms. The proposed finite-difference method to estimate second-order local sensitivities is based on an adaptive tau-leaping strategy to compute correlated quadruples of trajectories.

The new sensitivity estimator utilizes a strong coupling strategy of the nominal and perturbed trajectories, thereby reducing estimator variance. The tau-leaping framework for second-order sensitivities combines variance reduction with computational effectiveness. This approach is valuable not only for its efficiency and accuracy of the sensitivity estimation but also for making second-order sensitivity analysis feasible for large, complex biochemical systems, where exact methods would be computationally prohibitive.

We successfully tested our strategy on several models of biochemical networks, from small to relatively large systems. The proposed coupled tau-leaping-2 technique is significantly more efficient than the existing coupled finite-difference-2 scheme, with a similar accuracy on biochemical models, which are moderately stiff to stiff. In the future, we plan to investigate how the accuracy of the sensitivity estimator depends on the perturbation parameter, and we aim to explore the estimation of higher-order sensitivities.

Author Contributions: Conceptualization, S.I.; methodology, S.I.; software, F.J.; validation, S.I.; investigation, F.J.; writing—original draft preparation, S.I.; writing—review and editing, S.I.; visualization, F.J.; supervision, S.I.; funding acquisition, S.I. All authors have read and agreed to the published version of the manuscript.

Funding: This research was supported by a grant from the National Sciences and Engineering Research Council of Canada (NSERC), Grant No. RGPIN-2020-05469, Toronto Metropolitan University Mathematics Department and Faculty of Science Dean's Research Fund.

Data Availability Statement: Data is contained within the article.

Acknowledgments: The authors wish to thank the anonymous reviewers for their suggestions.

Conflicts of Interest: The authors declare no conflicts of interest.

References

1. Raser, J.M.; O'Shea, E.K. Control of stochasticity in eukaryotic gene expression. *Science* **2004**, *304*, 1811–1814. [CrossRef]
2. Batada, N.N.; Shepp, L.A.; Siegmund, D.O. Stochastic model of protein-protein interaction: Why signalling proteins need to be collocalized. *Proc. Natl. Acad. Sci. USA* **2004**, *101*, 6445–6449. [CrossRef] [PubMed]
3. Gillespie, D.T. A rigorous derivation of the Chemical Master Equation. *Physica A* **1992**, *188*, 402–425. [CrossRef]
4. Gillespie, D.T. A general method for numerically simulating the stochastic time evolution of coupled chemical reactions. *J. Comp. Phys.* **1976**, *22*, 403–434. [CrossRef]
5. Gillespie, D.T. Exact stochastic simulation of coupled chemical reactions. *J. Phys. Chem.* **1977**, *81*, 2340–2361. [CrossRef]
6. Gibson, M.A.; Bruck, J. Efficient exact stochastic simulation of chemical systems with many species and many channels. *J. Phys. Chem.* **2000**, *105*, 1876–1889. [CrossRef]
7. Gillespie, D.T. Approximate accelerated stochastic simulation of chemically reacting systems. *J. Chem. Phys.* **2001**, *115*, 1716–1733. [CrossRef]
8. Cao, Y.; Gillespie, D.T.; Petzold, L.R. Efficient stepsize selection for the tau-leaping simulation method. *J. Chem. Phys.* **2006**, *124*, 044109. [CrossRef] [PubMed]
9. Simoni, G.; Reali, F.; Priami, C.; Marchetti, L. Stochastic simulation algorithms for computational Systems Biology: Exact, approximate, and hybrid methods. *Wiley Interdiscip. Rev. Syst. Biol. Med.* **2019**, *11*, e1459. [CrossRef] [PubMed]
10. Cao, Y.; Gillespie, D.T.; Petzold, L.R. The slow scale stochastic simulation algorithm. *J. Chem. Phys.* **2005**, *122*, 014116. [CrossRef] [PubMed]
11. Tian, T.; Burrage, K. Binomial leap methods for simulating stochastic chemical kinetics. *J. Chem. Phys.* **2004**, *121*, 10356–10364. [CrossRef]
12. Thanh, V.H. A critical comparison of rejection-based algorithms for simulation of large biochemical reaction networks. *Bull. Math. Biol.* **2019**, *81*, 3053–3073. [CrossRef]
13. Hu, Y.; Li, T. Highly accurate tau-leaping methods with random corrections. *J. Chem. Phys.* **2009**, *130*, 124109. [CrossRef] [PubMed]

14. Wilkinson, D.J. *Stochastic Modelling for Systems Biology*; Taylor & Francis: Boca Raton, FL, USA, 2019.
15. Higham, D.J. Modeling and simulating chemical reactions. *SIAM Rev.* **2008**, *50*, 347–368. [CrossRef]
16. Gillespie, D.T.; Hellander, A.; Petzold, L.R. Perspective: Stochastic algorithms for chemical kinetics. *J. Chem. Phys.* **2013**, *138*, 170901. [CrossRef] [PubMed]
17. Petre, I.; Mizera, A.; Hyder, C.L.; Meinander, A.; Mikhailov, A.; Morimoto, R.I.; Sistonen, L.; Eriksson, J.E.; Back, R.J. A simple mass-action model for the eukaryotic heat shock response and its mathematical validation. *Nat. Comput.* **2011**, *10*, 595–612. [CrossRef]
18. Barrows, D.; Ilie, S. Parameter estimation for the reaction-diffusion master equation. *AIP Adv.* **2023**, *13*, 065318. [CrossRef]
19. Ashyraliyev, M.; Fomekong-Nanfack, Y.; Kaandorp, J.A.; Blom, J.G. Systems biology: Parameter estimation for biochemical models. *FEBS J.* **2009**, *276*, 886–902. [CrossRef] [PubMed]
20. Gholami, S.; Ilie, S. Quantifying parameter interdependence in stochastic discrete models of biochemical systems. *Entropy* **2023**, *25*, 1168. [CrossRef]
21. Gunawan, R.; Cao, Y.; Petzold, L.; Doyle, F.J. Sensitivity analysis of discrete stochastic systems. *Biophys. J.* **2005**, *88*, 2530–2540. [CrossRef] [PubMed]
22. Plyasunov, S.; Arkin, A. Efficient stochastic sensitivity analysis of discrete event systems. *J. Comput. Phys.* **2007**, *221*, 724–738. [CrossRef]
23. Degasperi, A.; Gilmore, S. Sensitivity analysis of stochastic models of bistable biochemical reactions. In *International School on Formal Methods for the Design of Computer, Communication and Software Systems*; Springer: Berlin/Heidelberg, Germany, 2008; pp. 1–20.
24. Komorowski, M.; Costa, M.J.; Rand, D.A.; Stumpf, M.P.H. Sensitivity, robustness, and identifiability in stochastic chemical kinetics models. *Proc. Natl. Acad. Sci. USA* **2011**, *108*, 8645–8650. [CrossRef]
25. Anderson, D.F. An efficient finite difference method for parameter sensitivities of continuous time Markov chains. *SIAM J. Numer. Anal.* **2012**, *50*, 2237. [CrossRef]
26. Rathinam, M.; Sheppard, P. W.; Khammash, M. Efficient computation of parameter sensitivities of discrete stochastic chemical reaction networks. *J Chem. Phys.* **2010**, *132*, 034103. [CrossRef]
27. Morshed, M.; Ingalls, B.; Ilie, S. An efficient finite-difference strategy for sensitivity analysis of stochastic models of biochemical systems. *Biosystems* **2017**, *151*, 43–52. [CrossRef]
28. Morshed, M.; Ingalls, B.; Ilie, S. An effective implicit finite-difference method for sensitivity analysis of stiff stochastic discrete biochemical systems. *IET Syst. Biol.* **2018**, *12*, 123 – 130. [CrossRef] [PubMed]
29. Wolf, E.S.; Anderson, D.F. A finite difference method for estimating second order parameter sensitivities of discrete stochastic chemical reaction networks. *J. Chem. Phys.* **2012**, *137*, 224112. [CrossRef]
30. Strehl, R.; Ilie, S. Hybrid stochastic simulation of reaction-diffusion systems with slow and fast dynamics. *J. Chem. Phys.* **2015**, *143*, 234108. [CrossRef]
31. Chen, A.; Ren, Q.; Zhou, T.; Burrage, P.; Tian, T.; Burrage, K. Balanced implicit Patankar-Euler methods for positive solutions of stochastic differential equations of biological regulatory systems. *J. Chem. Phys.* **2024**, *160*, 064117. [CrossRef] [PubMed]
32. Padgett, J.M.A.; Ilie, S. An adaptive tau-leaping method for stochastic simulations of reaction-diffusion systems. *AIP Adv.* **2016**, *6*, 035217. [CrossRef]
33. Sayyidmousavi, A.; Ilie, S. An efficient hybrid method for stochastic reaction-diffusion biochemical systems with delay. *AIP Adv.* **2017**, *7*, 125305. [CrossRef]
34. Ethier, S.N.; Kurtz, T.G. *Markov Processes: Characterization and Convergence*; John Wiley & Sons: New York, NY, USA, 1986.
35. Li, T. Analysis of explicit tau-leaping schemes for simulating chemically reacting systems. *SIAM Multi. Model. Simul.* **2007**, *6*, 417–436. [CrossRef]
36. Ingalls, B.P. *Mathematical Modeling in Systems Biology: An Introduction*; MIT Press: Cambridge, MA, USA, 2013.
37. Lu, H.; Li, P. Stochastic projective methods for simulating stiff chemical reacting systems. *Comput. Phys. Commun.* **2012**, *183*, 1427–1442. [CrossRef]
38. Marquez-Lago, T.; Burrage, K. Binomial tau-leap spatial stochastic simulation algorithm for applications in chemical kinetics. *J. Chem. Phys.* **2007**, *127*, 104101. [CrossRef] [PubMed]

Mathematical and Computational Applications

MDPI

Article

Analysis and Optimal Control of Propagation Model for Malware in Multi-Cloud Environments with Impact of Brownian Motion Process

Othman A. M. Omar [1,*], Hamdy M. Ahmed [2], Taher A. Nofal [3], Adel Darwish [4] and A. M. Sayed Ahmed [5]

1 Department of Physics and Engineering Mathematics, Faculty of Engineering, Ain Shams University, Cairo 11517, Egypt
2 Department of Physics and Engineering Mathematics, Higher Institute of Engineering, El-Shorouk Academy, El-Shorouk City 11837, Egypt
3 Department of Mathematics, College of Science, Taif University, P.O. Box 11099, Taif 21944, Saudi Arabia
4 Department of Mathematics, Faculty of Science, Helwan University, Cairo 11795, Egypt
5 Department of Mathematics and Computer Science, Faculty of Science, Alexandria University, Alexandria 21526, Egypt
* Correspondence: osman_ahmed@eng.asu.edu.eg

Abstract: Today, cloud computing is a widely used technology that provides a wide range of services to numerous sectors around the world. This technology depends on the interaction and cooperation of virtual machines (VMs) to complete various computing tasks, propagating malware attacks quickly due to the complexity of cloud computing environments and users' interfaces. As a result of the rising demand for cloud computing from multiple perspectives for complete analysis and decision-making across a range of life disciplines, multi-cloud environments (MCEs) are established. Therefore, in this work, we discuss impacted mathematical modeling for the MCEs' network dynamics using two deterministic and stochastic approaches. In both approaches, appropriate assumptions are considered. Then, the proposed networks' VMs are classified to have six different possible states covering media, healthcare, finance, and educational servers. After that, the two developed modeling approaches' solution existence, uniqueness, equilibrium, and stability are carefully investigated. Using an optimal control strategy, both proposed models are tested for sustaining a certain level of security of the VMs' states and reducing the propagation of malware within the networks. Finally, we verify the theoretical results by employing numerical simulations to track the malware's propagation immunization. Results showed how the implemented control methods maintained the essential objectives of managing malware infections.

Keywords: stochastic dynamical modeling; malware propagation; multi-cloud environments; optimal control

MSC: 37A50; 49J15; 34D20

1. Introduction

In cloud computing, virtualization is a crucial method that enables the transcendence of temporal and physical boundaries. It accomplishes this by splitting a single physical computing resource into several virtual machines (VMs) with identical functions. This enables VM migration [1] to enable the on-demand deployment of computing resources. Regretfully, malware attacks are becoming more frequent targets for the new hidden risks that virtualization has brought about [2]. In addition to resulting in financial losses, these vulnerabilities have the potential to seriously injure people and organizations and even endanger lives [3]. Thus, investigating practical ways to protect virtual environments against malware attacks in the cloud is essential. The issue of malware spreading in cloud systems has grown in prominence due to the rapid development of the cloud computing industry [4]. Large-scale, quickly spreading, and hard-to-track-and-control malware are typical traits found in cloud systems. Furthermore, cloud providers enable users to install applications and upload items (such pictures, movies, and other documents) to the cloud. This makes it possible for hackers to infiltrate the cloud with malware like ransomware and spyware. In addition, ransomware attacks rank among the most prevalent in recent years, indicating that hackers are increasingly likely to exploit user private data for financial benefit [5]. For this purpose, numerous related efforts have been presented by researchers to control the spread of malware in cloud systems. These efforts mostly consist of malware detection, analysis of the malware propagation path, and control of the malware propagation. In cloud systems, malware detection is an essential technique for limiting the spread of malware. To detect malware, researchers currently use feature-based and machine learning-based approaches [6]. Researchers can successfully identify malware by examining its behavior and code [7]. Understanding the laws and mechanisms of malware propagation in cloud systems also requires an understanding of malware propagation paths. Furthermore, one of the most important technologies for limiting the spread of malware in cloud systems is malware propagation control. Certain models of the epidemic dynamics of malware spread have been proposed, based on comparisons between malware and biological counterparts. These models include the SEIQR model [8], the SIR1R2 model [9], the SE1E2IQR model [10], the SIWQ model [11], the SIAR model [12,13], the SEIRF model [14], the SDIQTPR model [15], and the SV1V2EIRF model [16]. These models, however, are not specifically made for cloud environments. The authors in [17] suggested a five-state dynamic model—susceptible, infected high, infected low, first recovered, and completely recovered—to enhance the security and resilience of IoT networks against malicious activities. This model offers a robust solution for mitigating cyber threats using model predictive control (MPC) to effectively manage malware propagation rates by reducing the number of infected IoT devices during periods of high infection. In [18], the authors applied a different optimal control strategy in the dynamic model in [17], considering the count of IoT devices infected during the high infection stage, costs related to executing the immediate response strategy for the restricted environment and additional security defenses, and the costs linked to the installation of the latest and most resilient cybersecurity patch during the complete recovery stage. The SPI (susceptible–protected–infected) model was presented by the authors of [19], who investigated the impact of anti-malware with infectious nodes in cloud environments. However, they made the assumption that all machines interacting with the cloud were vulnerable, which is not realistic in real-world scenarios, because the cost and efficacy of malware control are beyond the scope of this model. Moreover, the authors assert that this is the first mathematical model to investigate how anti-malware software affects malware's ability to proliferate in cloud environments. Conversely, the susceptible–infected–protected–susceptible (SIPS) model was a novel dynamical model put forth by the authors in [20]. Nonetheless, it is important to note that this model assumes that all virtual machines (VMs) join the virtual network cleanly and were developed in an initially uninfected cloud environment.

In light of all potential devices joining the presumptive dynamic cloud network, this work introduces novel six-state VMs, deterministic and stochastic dynamic models, and four popular multi-cloud environments: media, healthcare, finance, and educational servers with solution existence, uniqueness, positiveness, and stability examination. In addition, a unique integration of assumptions in the cloud network simulates the reality of defensive behavior against malware propagation. Further, the dynamic interaction between the six-state VMs and malware propagation is fully studied and visualized. Finally, a security control function is employed to control the dynamic behavior of some states for investment loss reduction against malware propagation.

The remaining material of our work is organized as follows: Section 2 introduces the suggested deterministic dynamic model of multi-cloud environments with mathematical uniqueness and existence investigation and presents the model optimality control criteria for minimizing a predefined cost function of servers' isolation, tracing, and protection. Section 3 investigates the constructed stochastic model of multi-cloud environments with full mathematical analysis and shows how model controllability can be achieved. Finally, Section 4 shows the two obtained models numerical results and conclusions.

2. Deterministic MMCE Model

With reference to the flowchart presented in Figure 1, we accumulate a mathematical model of malware in multi-cloud environments (MMCE) that consists of twelve ordinary differential equations. The compartments of the mathematical model are divided into susceptible servers $S(3)$, infected academic and educational servers $I_1(3)$, infected healthcare servers $I_2(3)$, infected financial services servers $I_3(3)$, infected media servers $I_4(3)$, isolated academic and educational servers $Q_1(3)$, isolated healthcare servers $Q_2(3)$, isolated financial services servers $Q_3(3)$, isolated media servers $Q_4(3)$, traced servers $T(3)$, semi-protected servers $P_1(3)$, and protected servers $P_2(3)$, where $S(3) + \sum_{k=1}^{4} I_k(3) + \sum_{k=1}^{4} Q_k(3) + T(3) + \sum_{k=1}^{2} P_k(3) = N(t)$.

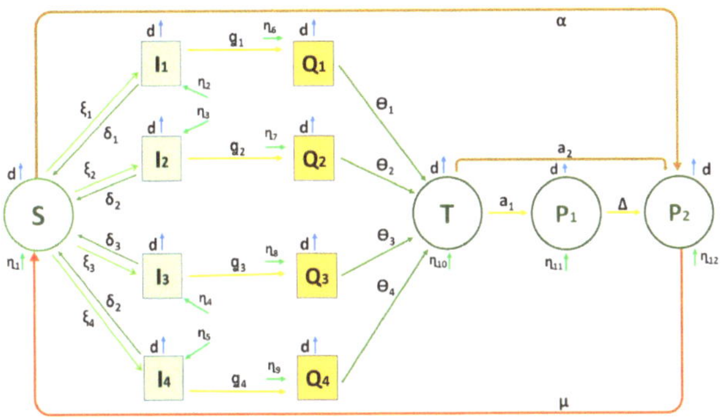

Figure 1. Proposed models' dynamic network.

Through Figure 1, the following presumptions are taken into account:

(1) All of the involved transmission rates are positive.
(2) New servers are added to the dynamic network in every possible network state.
(3) The virus can only propagate across infected servers.
(4) Susceptible servers have the ability to transition from one state to the protected one without going through further phases.
(5) Without first going through the tracing state, isolated servers are unable to transition to the semi-protected state.
(6) Traced servers have the ability to enter the protective state instantly.
(7) There is a constant rate of disconnection for every server on the network.

We are now on the model formulation. The flow diagram presented in Figure 1 illustrates how the various states of the servers nodes display behaviors that may be explained by the subsequent governing equations.

System (1) presents the suggested deterministic model equations:

$$
\begin{aligned}
\frac{dS(\mathfrak{Z})}{d\mathfrak{Z}} &= \eta_1 - \xi_1 I_1(\mathfrak{Z})S(\mathfrak{Z}) - \xi_2 I_2(\mathfrak{Z})S(\mathfrak{Z}) - \xi_3 I_3(\mathfrak{Z})S(\mathfrak{Z}) - \xi_4 I_4(\mathfrak{Z})S(\mathfrak{Z}) + \delta_1 I_1(\mathfrak{Z}) \\
&\quad + \delta_2 I_2(\mathfrak{Z}) + \delta_3 I_3(\mathfrak{Z}) + \delta_4 I_4(\mathfrak{Z}) + \mu P_2(\mathfrak{Z}) - (\alpha + d)S(\mathfrak{Z}) \\
\frac{dI_1(\mathfrak{Z})}{d\mathfrak{Z}} &= \eta_2 + \xi_1 S(\mathfrak{Z})I_1(\mathfrak{Z}) - (\delta_1 + g_1 + d)I_1(\mathfrak{Z}) \\
\frac{dI_2(\mathfrak{Z})}{d\mathfrak{Z}} &= \eta_3 + \xi_2 S(\mathfrak{Z})I_2(\mathfrak{Z}) - (\delta_2 + g_2 + d)I_2(\mathfrak{Z}) \\
\frac{dI_3(\mathfrak{Z})}{d\mathfrak{Z}} &= \eta_4 + \xi_3 S(\mathfrak{Z})I_3(\mathfrak{Z}) - (\delta_3 + g_3 + d)I_3(\mathfrak{Z}) \\
\frac{dI_4(\mathfrak{Z})}{d\mathfrak{Z}} &= \eta_5 + \xi_4 S(\mathfrak{Z})I_4(\mathfrak{Z}) - (\delta_4 + g_4 + d)I_4(\mathfrak{Z}) \\
\frac{dQ_1(\mathfrak{Z})}{d\mathfrak{Z}} &= \eta_6 + g_1 I_1(\mathfrak{Z}) - (\Theta_1 + d)Q_1(\mathfrak{Z}) \\
\frac{dQ_2(\mathfrak{Z})}{d\mathfrak{Z}} &= \eta_7 + g_2 I_2(\mathfrak{Z}) - (\Theta_2 + d)Q_2(\mathfrak{Z}) \\
\frac{dQ_3(\mathfrak{Z})}{d\mathfrak{Z}} &= \eta_8 + g_3 I_3(\mathfrak{Z}) - (\Theta_3 + d)Q_3(\mathfrak{Z}) \\
\frac{dQ_4(\mathfrak{Z})}{d\mathfrak{Z}} &= \eta_9 + g_4 I_4(\mathfrak{Z}) - (\Theta_4 + d)Q_4(\mathfrak{Z}) \\
\frac{dT(\mathfrak{Z})}{d\mathfrak{Z}} &= \eta_{10} + \Theta_1 Q_1(\mathfrak{Z}) + \Theta_2 Q_2(\mathfrak{Z}) + \Theta_3 Q_3(\mathfrak{Z}) + \Theta_4 Q_4(\mathfrak{Z}) - (a_1 + a_2 + d)T(\mathfrak{Z}) \\
\frac{dP_1(\mathfrak{Z})}{d\mathfrak{Z}} &= \eta_{11} + a_1 T(\mathfrak{Z}) - (\Delta + d)P_1(\mathfrak{Z}) \\
\frac{dP_2(\mathfrak{Z})}{d\mathfrak{Z}} &= \eta_{12} + a_2 T(\mathfrak{Z}) + \Delta P_1(\mathfrak{Z}) + \alpha S(\mathfrak{Z}) - (\mu + d)P_2(\mathfrak{Z}) \qquad (1)
\end{aligned}
$$

with initial conditions

$$
\begin{aligned}
&S(0) = \check{S}, I_1(0) = \check{I}_1, \cdots, I_4(0) = \check{I}_4, Q_1(0) = \check{Q}_1, \cdots, Q_4(0) = \check{Q}_4, \\
&T(0) = \check{T}, P_1(0) = \check{P}_1, P_2(0) = \check{P}_2 \qquad (2)
\end{aligned}
$$

where the feasible region is

$$
\begin{aligned}
\Xi &= \Big\{(S, I_1, \cdots, I_4, Q_1, \cdots, Q_4, T, P_1, P_2) \in \mathbb{R}^+_{12} : 0 \le S + \sum_{k=1}^{4} I_k + \sum_{k=1}^{4} Q_k + T + P_1 + P_2 \\
&\le \sum_{k=1}^{12} \eta_k / d\Big\}
\end{aligned}
$$

Table 1 provides a comprehensive interpretation of the parameters used in the model.

Lemma 1. *The set Ξ is positive invariant and attracts all solutions in $\mathbb{R}^+_{12}$.*

Remark 1. *Let the solution set*

$$
\mathfrak{S}(\mathfrak{Z}) \equiv \Big(S(\mathfrak{Z}), I_1(\mathfrak{Z}), \cdots, I_4(\mathfrak{Z}), Q_1(\mathfrak{Z}), \cdots, Q_4(\mathfrak{Z}), T(\mathfrak{Z}), P_1(\mathfrak{Z}), P_2(\mathfrak{Z})\Big)
$$

Lemma 2. *Let the initial data set be $\mathfrak{S}(\mathfrak{Z}) \mid_{\mathfrak{Z}=0} \ge 0 \in \Xi$; then, the solution set $\mathfrak{S}(\mathfrak{Z})$ of the model (1) is non-negative $\forall\, \mathfrak{Z} > 0$.*

Remark 2. *Since $S + \sum_{k=1}^{4} I_k + \sum_{k=1}^{4} Q_k + T + P_1 + P_2 = N$, then all functions in model (1) are bounded,*

$$0 \leq S(\mathfrak{Z}) \leq N,\ 0 \leq I_1(\mathfrak{Z}) \leq N, \cdots,\ 0 \leq I_4(\mathfrak{Z}) \leq N,\ 0 \leq Q_1(\mathfrak{Z}) \leq N, \cdots,\ 0 \leq Q_4(\mathfrak{Z}) \leq N$$
$$0 \leq T(\mathfrak{Z}) \leq N, 0 \leq P_1(\mathfrak{Z}) \leq N,\ 0 \leq P_2(\mathfrak{Z}) \leq N.$$

Table 1. Dynamic network transmission rates.

Rates	Description
η_1	Entering Rate of Susceptible Servers
η_2	Entering Rate of Infected Academic and Educational Servers
η_3	Entering Rate of Infected Healthcare Servers
η_4	Entering Rate of Infected Financial Services Servers
η_5	Entering Rate of Infected Media Servers
η_6	Entering Rate of Isolated Academic and Educational Servers
η_7	Entering Rate of Isolated Healthcare Servers
η_8	Entering Rate of Isolated Financial Services Servers
η_9	Entering Rate of Isolated Media Servers
η_{10}	Entering Rate of Traced Servers
η_{11}	Entering Rate of Semi-protected Servers
η_{12}	Entering Rate of Protected Servers
ξ_1	Infection Rate of Academic and Educational Servers
ξ_2	Infection Rate of Healthcare Servers
ξ_3	Infection Rate of Financial Services Servers
ξ_4	Infection Rate of Media Servers
δ_1	Reinstalling System Rate of Infected Academic and Educational Servers
δ_2	Reinstalling Rate of Healthcare Servers
δ_3	Reinstalling Rate of Financial Services Servers
δ_4	Reinstalling Rate of Media Servers
g_1	Isolation Rate of Infected Academic and Educational Servers
g_2	Isolation Rate of Infected Healthcare Servers
g_3	Isolation Rate of Infected Financial Services Servers
g_4	Isolation Rate of Infected Media Servers
Θ_1	Tracing Rate of Isolated Academic and Educational Servers
Θ_2	Tracing Rate of Isolated Healthcare Servers
Θ_3	Tracing Rate of Isolated Financial Services Servers
Θ_4	Tracing Rate of Isolated Media Servers
a_1	Transmission Rate from Traced into Semi-protected Servers
a_2	Transmission Rate from Traced into Protected Servers
Δ	Transmission Rate from Semi-protected into Protected Servers
μ	Antivirus Expiry Rate
α	Antivirus Installing Rate
d	Removal Rate of the Cloud Environment
κ	Stochastic Diffusion Coefficient

2.1. Existence and Uniqueness

The existence and uniqueness of the deterministic MMCE model solutions ((1) and (2)) are examined in this section. We apply the well-known Banach fixed-point theorem to demonstrate the existence of the solution to model one described in (3). To prove the existence and uniqueness of the solution, we execute the following: employing the integral operator $\int_0^{\mathfrak{Z}} (\cdot) d\mathfrak{Z}$ on the equations (Eqns.) in model (1) yields

$$
\begin{aligned}
S(\mathfrak{Z}) &= \check{S} + \int_0^{\mathfrak{Z}} \Big(\eta_1 - \xi_1 I_1(\mathfrak{G})S(\mathfrak{G}) - \xi_2 I_2(\mathfrak{Z})S(\mathfrak{G}) - \xi_3 I_3(\mathfrak{G})S(\mathfrak{G}) - \xi_4 I_4(\mathfrak{G})S(\mathfrak{G}) \\
&\quad + \delta_1 I_1(\mathfrak{G}) + \delta_2 I_2(\mathfrak{G}) + \delta_3 I_3(\mathfrak{G}) + \delta_4 I_4(\mathfrak{G}) + \mu P_2(\mathfrak{G}) - (\alpha + d)S(\mathfrak{G}) \Big) d\mathfrak{G} \\
I_1(\mathfrak{Z}) &= \check{I}_1 + \int_0^{\mathfrak{Z}} \Big(\eta_2 + \xi_1 S(\mathfrak{G}) I_1(\mathfrak{G}) - (\delta_1 + g_1 + d) I_1(\mathfrak{G}) \Big) d\mathfrak{G} \\
I_2(\mathfrak{Z}) &= \check{I}_2 + \int_0^{\mathfrak{Z}} \Big(\eta_3 + \xi_2 S(\mathfrak{G}) I_2(\mathfrak{G}) - (\delta_2 + g_2 + d) I_2(\mathfrak{G}) \Big) d\mathfrak{G} \\
I_3(\mathfrak{Z}) &= \check{I}_3 + \int_0^{\mathfrak{Z}} \Big(\eta_4 + \xi_3 S(\mathfrak{G}) I_3(\mathfrak{G}) - (\delta_3 + g_3 + d) I_3(\mathfrak{G}) \Big) d\mathfrak{G} \\
I_4(\mathfrak{Z}) &= \check{I}_4 + \int_0^{\mathfrak{Z}} \Big(\eta_5 + \xi_4 S(\mathfrak{G}) I_4(\mathfrak{G}) - (\delta_4 + g_4 + d) I_4(\mathfrak{G}) \Big) d\mathfrak{G} \\
Q_1(\mathfrak{Z}) &= \check{Q}_1 + \int_0^{\mathfrak{Z}} \Big(\eta_6 + g_1 I_1(\mathfrak{G}) - (\Theta_1 + d) Q_1(\mathfrak{G}) \Big) d\mathfrak{G} \\
Q_2(\mathfrak{Z}) &= \check{Q}_2 + \int_0^{\mathfrak{Z}} \Big(\eta_7 + g_2 I_2(\mathfrak{G}) - (\Theta_2 + d) Q_2(\mathfrak{G}) \Big) d\mathfrak{G} \\
Q_3(\mathfrak{Z}) &= \check{Q}_3 + \int_0^{\mathfrak{Z}} \Big(\eta_8 + g_3 I_3(\mathfrak{G}) - (\Theta_3 + d) Q_3(\mathfrak{G}) \Big) d\mathfrak{G} \\
Q_4(\mathfrak{Z}) &= \check{I}_2 + \int_0^{\mathfrak{Z}} \Big(\eta_9 + g_4 I_4(\mathfrak{G}) - (\Theta_4 + d) Q_4(\mathfrak{G}) \Big) d\mathfrak{G} \\
T(\mathfrak{Z}) &= \check{I}_2 + \int_0^{\mathfrak{Z}} \Big(\eta_{10} + \Theta_1 Q_1(\mathfrak{G}) + \Theta_2 Q_2(\mathfrak{G}) + \Theta_3 Q_3(\mathfrak{G}) + \Theta_4 Q_4(\mathfrak{G}) \\
&- (a_1 + a_2 + d) T(\mathfrak{G}) \Big) d\mathfrak{G} \\
P_1(\mathfrak{Z}) &= \check{I}_2 + \int_0^{\mathfrak{Z}} \Big(\eta_{11} + a_1 T(\mathfrak{G}) - (\Delta + d) P_1(\mathfrak{G}) \Big) d\mathfrak{G} \\
P_2(\mathfrak{Z}) &= \check{I}_2 + \int_0^{\mathfrak{Z}} \Big(\eta_{12} + a_2 T(\mathfrak{G}) + \Delta P_1(\mathfrak{G}) + \alpha S(\mathfrak{G}) - (\mu + d) P_2(\mathfrak{G}) \Big) d\mathfrak{G} \qquad (3)
\end{aligned}
$$

For convenience, the kernels listed below are defined:

$$
\begin{aligned}
\aleph_1(\mathfrak{Z}, S(\mathfrak{Z})) &= \eta_1 - \xi_1 I_1(\mathfrak{Z})S(\mathfrak{Z}) - \xi_2 I_2(\mathfrak{Z})S(\mathfrak{Z}) - \xi_3 I_3(\mathfrak{Z})S(\mathfrak{Z}) - \xi_4 I_4(\mathfrak{Z})S(\mathfrak{Z}) + \delta_1 I_1(\mathfrak{Z}) \\
&\quad + \delta_2 I_2(\mathfrak{Z}) + \delta_3 I_3(\mathfrak{Z}) + \delta_4 I_4(\mathfrak{Z}) + \mu P_2(\mathfrak{Z}) - (\alpha + d) S(\mathfrak{Z}) \\
\aleph_2(\mathfrak{Z}, I_1(\mathfrak{Z})) &= \eta_2 + \xi_1 S(\mathfrak{Z}) I_1(\mathfrak{Z}) - (\delta_1 + g_1 + d) I_1(\mathfrak{Z}) \\
\aleph_3(\mathfrak{Z}, I_2(\mathfrak{Z})) &= \eta_3 + \xi_2 S(\mathfrak{Z}) I_2(\mathfrak{Z}) - (\delta_2 + g_2 + d) I_2(\mathfrak{Z}) \\
\aleph_4(\mathfrak{Z}, I_3(\mathfrak{Z})) &= \eta_4 + \xi_3 S(\mathfrak{Z}) I_3(\mathfrak{Z}) - (\delta_3 + g_3 + d) I_3(\mathfrak{Z}) \\
\aleph_5(\mathfrak{Z}, I_4(\mathfrak{Z})) &= \eta_5 + \xi_4 S(\mathfrak{Z}) I_4(\mathfrak{Z}) - (\delta_4 + g_4 + d) I_4(\mathfrak{Z}) \\
\aleph_6(\mathfrak{Z}, Q_1(\mathfrak{Z})) &= \eta_6 + g_1 I_1(\mathfrak{Z}) - (\Theta_1 + d) Q_1(\mathfrak{Z}) \\
\aleph_7(\mathfrak{Z}, Q_2(\mathfrak{Z})) &= \eta_7 + g_2 I_2(\mathfrak{Z}) - (\Theta_2 + d) Q_2(\mathfrak{Z}) \\
\aleph_8(\mathfrak{Z}, Q_3(\mathfrak{Z})) &= \eta_8 + g_3 I_3(\mathfrak{Z}) - (\Theta_3 + d) Q_3(\mathfrak{Z}) \\
\aleph_9(\mathfrak{Z}, Q_4(\mathfrak{Z})) &= \eta_9 + g_4 I_4(\mathfrak{Z}) - (\Theta_4 + d) Q_4(\mathfrak{Z}) \\
\aleph_{10}(\mathfrak{Z}, T(\mathfrak{Z})) &= \eta_{10} + \Theta_1 Q_1(\mathfrak{Z}) + \Theta_2 Q_2(\mathfrak{Z}) + \Theta_3 Q_3(\mathfrak{Z}) + \Theta_4 Q_4(\mathfrak{Z}) - (a_1 + a_2 + d) T(\mathfrak{Z}) \\
\aleph_{11}(\mathfrak{Z}, P_1(\mathfrak{Z})) &= \eta_{11} + a_1 T(\mathfrak{Z}) - (\Delta + d) P_1(\mathfrak{Z}) \\
\aleph_{12}(\mathfrak{Z}, P_2(\mathfrak{Z})) &= \eta_{12} + a_2 T(\mathfrak{Z}) + \Delta P_1(\mathfrak{Z}) + \alpha S(\mathfrak{Z}) - (\mu + d) P_2(\mathfrak{Z})
\end{aligned}
$$

Since $S, I_1, \cdots, I_4, Q_1, \cdots, Q_4, T, P_1, P_2$ are non-negative bounded functions, $\exists$ positive values $\wp_k, k = \mathbb{N}_1^{12}$ such that (s.t.)

$$\begin{aligned} \|S\| &\leq \wp_1, \|I_1\| \leq \wp_2, \|I_2\| \leq \wp_3, \|I_3\| \leq \wp_4, \|I_4\| \leq \wp_5 \|Q_1\| \leq \wp_6, \|Q_2\| \leq \wp_7 \\ \|Q_3\| &\leq \wp_8, \|Q_4\| \leq \wp_9, \|T\| \leq \wp_{10}, \|P_1\| \leq \wp_{11}, \|P_2\| \leq \wp_{12}. \end{aligned}$$

In light of this, Equation (3) can be expressed as

$$\begin{aligned} S(\mathfrak{Z}) &= \check{S} + \int_0^{\mathfrak{Z}} \aleph_1(\mathfrak{G}, S(\mathfrak{G}))d\mathfrak{G}, \ I_1(\mathfrak{Z}) = \check{I}_1 + \int_0^{\mathfrak{Z}} \aleph_2(\mathfrak{G}, I_1(\mathfrak{G}))d\mathfrak{G} \\ I_2(\mathfrak{Z}) &= \check{I}_2 + \int_0^{\mathfrak{Z}} \aleph_3(\mathfrak{G}, I_2(\mathfrak{G}))d\mathfrak{G}, \ I_3(\mathfrak{Z}) = \check{I}_3 + \int_0^{\mathfrak{Z}} \aleph_4(\mathfrak{G}, I_3(\mathfrak{G}))d\mathfrak{G} \\ I_4(\mathfrak{Z}) &= \check{I}_4 + \int_0^{\mathfrak{Z}} \aleph_5(\mathfrak{G}, I_4(\mathfrak{G}))d\mathfrak{G}, \ Q_1(\mathfrak{Z}) = \check{Q}_1 + \int_0^{\mathfrak{Z}} \aleph_6(\mathfrak{G}, Q_1(\mathfrak{G}))d\mathfrak{G} \\ Q_2(\mathfrak{Z}) &= \check{Q}_2 + \int_0^{\mathfrak{Z}} \aleph_7(\mathfrak{G}, Q_2(\mathfrak{G}))d\mathfrak{G}, \ Q_3(\mathfrak{Z}) = \check{Q}_3 + \int_0^{\mathfrak{Z}} \aleph_8(\mathfrak{G}, Q_3(\mathfrak{G}))d\mathfrak{G} \\ Q_4(\mathfrak{Z}) &= \check{Q}_4 + \int_0^{\mathfrak{Z}} \aleph_9(\mathfrak{G}, Q_4(\mathfrak{G}))d\mathfrak{G}, \ T(\mathfrak{Z}) = \check{T} + \int_0^{\mathfrak{Z}} \aleph_{10}(\mathfrak{G}, T(\mathfrak{G}))d\mathfrak{G} \\ P_1(\mathfrak{Z}) &= \check{P}_1 + \int_0^{\mathfrak{Z}} \aleph_{11}(\mathfrak{G}, P_1(\mathfrak{G}))d\mathfrak{G}, \ P_2(\mathfrak{Z}) = \check{P}_2 + \int_0^{\mathfrak{Z}} \aleph_{12}(\mathfrak{G}, P_2(\mathfrak{G}))d\mathfrak{G} \end{aligned} \tag{4}$$

Theorem 1. *If $0 \leq \Upsilon < 1$, then the kernels $\aleph_k, k = \mathbb{N}_1^{12}$ are contraction mappings since they fulfill the Lipschitz condition, where $\Upsilon = \max_{1 \leq k \leq 12} \Re_k$.*

Proof. Take the kernel $\aleph_1$ into account. If $S(\mathfrak{Z})$ and $S(\bar{\mathfrak{Z}})$ are two arbitrary functions, then we obtain

$$\begin{aligned} \|\aleph_1(\mathfrak{Z}, S(\mathfrak{Z})) - \aleph_1(\mathfrak{Z}, S(\bar{\mathfrak{Z}}))\| &\leq \Big(\xi_1\wp_2 + \xi_2\wp_3 + \xi_3\wp_4 + \xi_4\wp_5 + (\alpha + d)\Big)\|S(\mathfrak{Z}) - S(\bar{\mathfrak{Z}})\| \\ &= \Re_1 \|S(\mathfrak{Z}) - S(\bar{\mathfrak{Z}})\| \end{aligned}$$

For kernels $\aleph_k, k = \mathbb{N}_2^{12}$, comparable outcomes are achievable:

$$\begin{aligned} \|\aleph_2(\mathfrak{Z}, I_1(\mathfrak{Z})) - \aleph_2(\mathfrak{Z}, I_1(\bar{\mathfrak{Z}}))\| &\leq \Big(\xi_1\wp_1 + \delta_1 + g_1 + d\Big)\|I_1(\mathfrak{Z}) - I_1(\bar{\mathfrak{Z}})\| = \Re_2\|I_1(\mathfrak{Z}) - I_1(\bar{\mathfrak{Z}})\| \\ &\vdots \\ \|\aleph_{12}(\mathfrak{Z}, P_2(\mathfrak{Z})) - \aleph_{12}(\mathfrak{Z}, P_2(\bar{\mathfrak{Z}}))\| &\leq \Big(\mu + d\Big)\|P_2(\mathfrak{Z}) - P_2(\bar{\mathfrak{Z}})\| = \Re_{12}\|P_2(\mathfrak{Z}) - P_2(\bar{\mathfrak{Z}})\| \end{aligned}$$

Consequently, $\aleph_k, k = \mathbb{N}_1^{12}$, satisfying the Lipschitz condition. Since $0 \leq \Upsilon < 1$, the kernels are contraction mappings. □

The subsequent recursive formulae are now presented using Equation (4):

$$\begin{aligned} S^n(\mathfrak{Z}) &= \int_0^{\mathfrak{Z}} \aleph_1(\mathfrak{G}, S^{n-1}(\mathfrak{G}))d\mathfrak{G}, \ Q_1^n(\mathfrak{Z}) = \int_0^{\mathfrak{Z}} \aleph_6(\mathfrak{G}, Q_1^{n-1}(\mathfrak{G}))d\mathfrak{G} \\ I_1^n(\mathfrak{Z}) &= \int_0^{\mathfrak{Z}} \aleph_2(\mathfrak{G}, I_1^{n-1}(\mathfrak{G}))d\mathfrak{G}, \ Q_2^n(\mathfrak{Z}) = \int_0^{\mathfrak{Z}} \aleph_7(\mathfrak{G}, Q_2^{n-1}(\mathfrak{G}))d\mathfrak{G} \\ &\vdots \\ I_4^n(\mathfrak{Z}) &= \int_0^{\mathfrak{Z}} \aleph_5(\mathfrak{G}, I_4^{n-1}(\mathfrak{G}))d\mathfrak{G}, \ Q_4^n(\mathfrak{Z}) = \int_0^{\mathfrak{Z}} \aleph_9(\mathfrak{G}, Q_4^{n-1}(\mathfrak{G}))d\mathfrak{G} \\ T^n(\mathfrak{Z}) &= \int_0^{\mathfrak{Z}} \aleph_{10}(\mathfrak{G}, T^{n-1}(\mathfrak{G}))d\mathfrak{G}, \ P_1^n(\mathfrak{Z}) = \int_0^{\mathfrak{Z}} \aleph_{11}(\mathfrak{G}, P_1^{n-1}(\mathfrak{G}))d\mathfrak{G}, \\ P_2^n(\mathfrak{Z}) &= \int_0^{\mathfrak{Z}} \aleph_{12}(\mathfrak{G}, P_2^{n-1}(\mathfrak{G}))d\mathfrak{G} \end{aligned}$$

Among the recursive formulae, the following are the subtraction of two successive forms:

$$\begin{aligned}
\Im_1^n(\Im) &= S^n(\Im) - S^{n-1}(\Im) = \int_0^{\Im} \Big(\aleph_1(\mathfrak{G}, S^{n-1}(\mathfrak{G})) - \aleph_1(\mathfrak{G}, S^{n-2}(\mathfrak{G}))\Big) d\mathfrak{G} \\
\Im_2^n(\Im) &= I_1^n(\Im) - I_1^{n-1}(\Im) = \int_0^{\Im} \Big(\aleph_2(\mathfrak{G}, I_1^{n-1}(\mathfrak{G})) - \aleph_2(\mathfrak{G}, I_1^{n-2}(\mathfrak{G}))\Big) d\mathfrak{G} \\
&\vdots \\
\Im_{12}^n(\Im) &= P_2^n(\Im) - P_2^{n-1}(\Im) = \int_0^{\Im} \Big(\aleph_{12}(\mathfrak{G}, P_2^{n-1}(\mathfrak{G})) - \aleph_{12}(\mathfrak{G}, P_2^{n-2}(\mathfrak{G}))\Big) d\mathfrak{G} \qquad (5)
\end{aligned}$$

Consequently, it may be claimed that

$$S^n(\Im) = \sum_{k=1}^{n} \Im_1^k(\Im),\ I_1^n(\Im) = \sum_{k=1}^{n} \Im_2^k(\Im), \cdots,\ P_2^n(\Im) = \sum_{k=1}^{n} \Im_{12}^k(\Im) \qquad (6)$$

Now,

$$\begin{aligned}
\|\Im_1^n(\Im)\| &= \left\|S^n(\Im) - S^{n-1}(\Im)\right\| = \left\|\int_0^{\Im} \Big(\aleph_1(\mathfrak{G}, S^{n-1}(\mathfrak{G})) - \aleph_1(\mathfrak{G}, S^{n-2}(\mathfrak{G}))\Big) d\mathfrak{G}\right\| \\
&\leq \int_0^{\Im} \left\|\aleph_1(\mathfrak{G}, S^{n-1}(\mathfrak{G})) - \aleph_1(\mathfrak{G}, S^{n-2}(\mathfrak{G}))\right\| d\mathfrak{G} \leq \Re_1\Im \left\|S^{n-1}(\Im) - S^{n-2}(\Im)\right\| \\
&= \Re_1\Im \left\|\Im_1^{n-1}(\Im)\right\| \qquad (7)
\end{aligned}$$

Similarly,

$$\begin{aligned}
\|\Im_2^n(\Im)\| &\leq \Re_2\Im\left\|\Im_2^{n-1}(\Im)\right\|,\ \|\Im_3^n(\Im)\| \leq \Re_3\Im\left\|\Im_3^{n-1}(\Im)\right\|,\ \|\Im_4^n(\Im)\| \leq \Re_4\Im\left\|\Im_4^{n-1}(\Im)\right\| \\
\|\Im_5^n(\Im)\| &\leq \Re_5\Im\left\|\Im_5^{n-1}(\Im)\right\|,\ \|\Im_6^n(\Im)\| \leq \Re_6\Im\left\|\Im_6^{n-1}(\Im)\right\|,\ \|\Im_7^n(\Im)\| \leq \Re_7\Im\left\|\Im_7^{n-1}(\Im)\right\| \\
\|\Im_8^n(\Im)\| &\leq \Re_8\Im\left\|\Im_8^{n-1}(\Im)\right\|,\ \|\Im_9^n(\Im)\| \leq \Re_9\Im\left\|\Im_9^{n-1}(\Im)\right\|,\ \|\Im_{10}^n(\Im)\| \leq \Re_{10}\Im\left\|\Im_{10}^{n-1}(\Im)\right\| \\
\|\Im_{11}^n(\Im)\| &\leq \Re_{11}\Im\left\|\Im_{11}^{n-1}(\Im)\right\|,\ \|\Im_{12}^n(\Im)\| \leq \Re_{12}\Im\left\|\Im_{12}^{n-1}(\Im)\right\| \qquad (8)
\end{aligned}$$

Theorem 2. *A unique solution for model (1) occurs if, for a period $\mathcal{T}_0 > 0$, the following condition holds:*

$$0 < \Re_k \mathcal{T}_0 < 1,\ \ k = \mathbb{N}_1^{12}.$$

Proof. Two items comprise the proof:

- Existence. Model (1)'s functions are bounded, and the supplied kernels meet the Lipschitz requirements. Consequently, using (6), the subsequent inequalities may be obtained:

$$\begin{aligned}
\|\Im_1^n(\Im)\| &\leq \Re_1\Im\left\|\Im_1^{n-1}(\Im)\right\| \leq (\Re_1\Im)^2\left\|\Im_1^{n-2}(\Im)\right\| \leq \cdots \leq (\Re_1\Im)^n\|\check{S}\| \\
\|\Im_2^n(\Im)\| &\leq \Re_2\Im\left\|\Im_2^{n-1}(\Im)\right\| \leq (\Re_2\Im)^2\left\|\Im_2^{n-2}(\Im)\right\| \leq \cdots \leq (\Re_2\Im)^n\|\check{I}_1\| \\
&\vdots \\
\|\Im_{12}^n(\Im)\| &\leq \Re_{12}\Im\left\|\Im_{12}^{n-1}(\Im)\right\| \leq (\Re_{12}\Im)^2\left\|\Im_{12}^{n-2}(\Im)\right\| \leq \cdots \leq (\Re_{12}\Im)^n\|\check{P}_2\| \qquad (9)
\end{aligned}$$

The smoothness and existence of the functions specified in (6) are confirmed by the inequalities (9). Define $\Psi_k^n, k = \mathbb{N}_1^{12}$ as the remaining terms after n iterations, that is,

$$\begin{aligned} S(\mathfrak{Z}) - \check{S} &= S^n(\mathfrak{Z}) + \Psi_1^n,\ I_1(\mathfrak{Z}) - \check{I}_1 = I_1^n(\mathfrak{Z}) + \Psi_2^n,\ I_2(\mathfrak{Z}) - \check{I}_2 = I_2^n(\mathfrak{Z}) + \Psi_3^n \\ I_3(\mathfrak{Z}) - \check{I}_3 &= I_3^n(\mathfrak{Z}) + \Psi_4^n,\ I_4(\mathfrak{Z}) - \check{I}_4 = I_4^n(\mathfrak{Z}) + \Psi_5^n,\ Q_1(\mathfrak{Z}) - \check{Q}_1 = Q_1^n(\mathfrak{Z}) + \Psi_6^n \\ Q_2(\mathfrak{Z}) - \check{Q}_2 &= Q_2^n(\mathfrak{Z}) + \Psi_7^n,\ Q_3(\mathfrak{Z}) - \check{Q}_3 = Q_3^n(\mathfrak{Z}) + \Psi_8^n,\ Q_4(\mathfrak{Z}) - \check{Q}_4 = Q_4^n(\mathfrak{Z}) + \Psi_9^n \\ T(\mathfrak{Z}) - \check{T} &= T^n(\mathfrak{Z}) + \Psi_{10}^n,\ P_1(\mathfrak{Z}) - \check{P}_1 = P_1^n(\mathfrak{Z}) + \Psi_{11}^n,\ P_2(\mathfrak{Z}) - \check{P}_2 = P_2^n(\mathfrak{Z}) + \Psi_{12}^n \end{aligned}$$

Using the Lipschitz condition for $\aleph_1$, we obtain

$$\begin{aligned} \|\Psi_1^n(\mathfrak{Z})\| &= \left\| \int_0^{\mathfrak{Z}} \left(\aleph_1(\mathfrak{G}, S(\mathfrak{G})) - \aleph_1(\mathfrak{G}, S^{n-1}(\mathfrak{G})) \right) d\mathfrak{G} \right\| \\ &\leq \Re_1 \mathfrak{Z} \left\| S(\mathfrak{Z}) - S^{n-1}(\mathfrak{Z}) \right\| \leq \cdots \leq (\Re_1 \mathfrak{Z})^n \wp_1 \end{aligned}$$

Setting $\mathfrak{Z} = \mathcal{T}_0$, we obtain

$$\|\Psi_1^n(\mathfrak{Z})\| \leq (\Re_1 \mathcal{T}_0)^n \wp_1 \tag{10}$$

$\|\Psi_1^n(\mathfrak{Z})\| \to 0$ is obtained by taking the limit of inequality (10) as $n \to \infty$ and applying the constraint $0 < \Re_1 \mathcal{T}_0 < 1$. This means that $\lim_{n\to\infty} S^n(\mathfrak{Z}) = S(\mathfrak{Z}) - \check{S}$. In a similar manner, the following inequalities are obtained:

$$\begin{aligned} \|\Psi_2^n(\mathfrak{Z})\| &\leq (\Re_2 \mathcal{T}_0)^n \wp_2,\ \|\Psi_3^n(\mathfrak{Z})\| \leq (\Re_3 \mathcal{T}_0)^n \wp_3,\ \|\Psi_4^n(\mathfrak{Z})\| \leq (\Re_4 \mathcal{T}_0)^n \wp_4, \\ \|\Psi_5^n(\mathfrak{Z})\| &\leq (\Re_5 \mathcal{T}_0)^n \wp_5 \\ \|\Psi_6^n(\mathfrak{Z})\| &\leq (\Re_6 \mathcal{T}_0)^n \wp_6,\ \|\Psi_7^n(\mathfrak{Z})\| \leq (\Re_7 \mathcal{T}_0)^n \wp_7,\ \|\Psi_8^n(\mathfrak{Z})\| \leq (\Re_8 \mathcal{T}_0)^n \wp_8, \\ \|\Psi_9^n(\mathfrak{Z})\| &\leq (\Re_9 \mathcal{T}_0)^n \wp_9 \\ \|\Psi_{10}^n(\mathfrak{Z})\| &\leq (\Re_{10} \mathcal{T}_0)^n \wp_{10},\ \|\Psi_{11}^n(\mathfrak{Z})\| \leq (\Re_{11} \mathcal{T}_0)^n \wp_{11},\ \|\Psi_{12}^n(\mathfrak{Z})\| \leq (\Re_{12} \mathcal{T}_0)^n \wp_{12} \end{aligned}$$

Hence, $\|\Psi_k^n(\mathfrak{Z})\| \to 0, k = \mathbb{N}_1^{12}$, as $n \to \infty$. System (1)'s existence is thus established.

- Uniqueness. Let $\mathfrak{X}$ and $\mathfrak{Y}$ be the solution sets to the model (1) s.t.

$$\begin{aligned} \mathfrak{X} &= \left(S(\mathfrak{Z}), I_1(\mathfrak{Z}), \cdots, I_4(\mathfrak{Z}), Q_1(\mathfrak{Z}), \cdots, Q_4(\mathfrak{Z}), T(\mathfrak{Z}), P_1(\mathfrak{Z}), P_2(\mathfrak{Z}) \right) \\ \mathfrak{Y} &= \left(\tilde{S}(\mathfrak{Z}), \tilde{I}_1(\mathfrak{Z}), \cdots, \tilde{I}_4(\mathfrak{Z}), \tilde{Q}_1(\mathfrak{Z}), \cdots, \tilde{Q}_4(\mathfrak{Z}), \tilde{T}(\mathfrak{Z}), \tilde{P}_1(\mathfrak{Z}), \tilde{P}_2(\mathfrak{Z}) \right) \end{aligned}$$

Then, using $0 < \Re_1 \mathcal{T}_0 < 1$, we obtain

$$\|S(\mathfrak{Z}) - \tilde{S}(\mathfrak{Z})\| \leq \Re_1 \mathfrak{Z} \|S(\mathfrak{Z}) - \tilde{S}(\mathfrak{Z})\|$$

So, $(1 - \Re_1 \mathfrak{Z}) \|S(\mathfrak{Z}) - \tilde{S}(\mathfrak{Z})\| \leq 0$. At last, one obtains $\|S(\mathfrak{Z}) - \tilde{S}(\mathfrak{Z})\| = 0$, that is, $S(\mathfrak{Z}) \equiv \tilde{S}(\mathfrak{Z})$. In the same way, we can obtain

$$I_1(\mathfrak{Z}) \equiv \tilde{I}_1(\mathfrak{Z}),\ \cdots,\ P_2(\mathfrak{Z}) \equiv \tilde{P}_2(\mathfrak{Z})$$

It is demonstrated that model (1)'s solutions are unique.

□

2.2. Equilibrium Points and Stability Analysis

To calculate the equilibrium points of the proposed deterministic model (1), set the right side of the modeling equations to zero.

$$\begin{aligned}
\frac{\mathbf{d}S(\mathfrak{Z})}{\mathbf{d}\mathfrak{Z}} &= \eta_1 - \xi_1 I_1(\mathfrak{Z})S(\mathfrak{Z}) - \xi_2 I_2(\mathfrak{Z})S(\mathfrak{Z}) - \xi_3 I_3(\mathfrak{Z})S(\mathfrak{Z}) - \xi_4 I_4(\mathfrak{Z})S(\mathfrak{Z}) + \delta_1 I_1(\mathfrak{Z}) \\
&\quad + \delta_2 I_2(\mathfrak{Z}) + \delta_3 I_3(\mathfrak{Z}) + \delta_4 I_4(\mathfrak{Z}) + \mu P_2(\mathfrak{Z}) - (\alpha + d)S(\mathfrak{Z}) = 0 \\
\frac{\mathbf{d}I_1(\mathfrak{Z})}{\mathbf{d}\mathfrak{Z}} &= \eta_2 + \xi_1 S(\mathfrak{Z})I_1(\mathfrak{Z}) - (\delta_1 + g_1 + d)I_1(\mathfrak{Z}) = 0 \\
&\vdots \\
\frac{\mathbf{d}P_2(\mathfrak{Z})}{\mathbf{d}\mathfrak{Z}} &= \eta_{12} + a_2 T(\mathfrak{Z}) + \Delta P_1(\mathfrak{Z}) + \alpha S(\mathfrak{Z}) - (\mu + d)P_2(\mathfrak{Z}) = 0
\end{aligned} \tag{11}$$

The malware-free equilibrium point is provided by

$$MF^0\Big(\frac{\eta_1(\mu + d) + \eta_{12}\mu}{d^2 + \alpha d + \mu d}, 0, 0, 0, 0, 0, 0, 0, 0, 0, \frac{\eta_1\alpha + \eta_{12}(\alpha + d)}{d^2 + \alpha d + \mu d}\Big).$$

The endemic equilibrium points are represented by

$$ME^*\Big(S^*, I_1^*, \cdots, I_4^*, Q_1^*, \cdots, Q_4^*, T^*, P_1^*, P_2^*\Big),$$

which may be found by concurrently solving the set of Equation (11). Discussions on the stability of the proposed model with regard to equilibrium points are particularly problematic because of its significantly higher dimension. As a result, we shall examine the stability analysis using $\mathfrak{R}_0$ as the basic reproduction number of model given by (1), approached through the next-generation matrix technique [21].

$$\mathfrak{R}_0 = \sum_{k=1}^{4} \frac{g_k \xi_k (\eta_1(\mu + d) + \eta_{12}\mu)}{(\Theta_k + d)(\delta_k + g_k + d)(d^2 + \alpha d + \mu d)}$$

Theorem 3. *The malware-free equilibrium (MF^0) of the deterministic model (1) is locally asymptotically stable if $\mathfrak{R}_0 < 1$ and unstable if $\mathfrak{R}_0 > 1$.*

2.3. Analyzing Optimal Control

Malware attacks are starting to have an increasing impact on global economies, resulting in significant direct losses such as labor charges, expenses for removing and fixing malware from systems, lost user productivity, income from malfunctioning or decaying systems, and other costs directly associated with a malware attack. To effectively prevent malware attacks and safeguard against malware threats, both individuals and corporations use protective barriers and regularly update anti-malware software. Nevertheless, these precautionary measures come at a substantial security cost. Our objective is to reduce total investment costs as well as direct losses from security investments, which makes it plausible to claim that the quantity of infected devices correlates with the amount of money lost due to malware. Additionally, we assume that user security awareness can result in maintaining a certain control on the isolated, traced, and semi-protected servers. So, let us define the next optimum control (O.C.) problem to maximize the ensuing objective functional (O.F.):

$$\mathcal{J}(\Delta(\mathfrak{Z})) = \int_0^{\mathcal{T}} \Big[C_1 T(\mathfrak{Z}) + C_2 \sum_{k=1}^{4} Q_k(\mathfrak{Z}) + C_3 P_1(\mathfrak{Z}) + \frac{C_4}{2}\Delta^2(\mathfrak{Z})\Big] d\mathfrak{Z} \tag{12}$$

subject to

$$\begin{aligned}
\frac{dS(\mathfrak{Z})}{d\mathfrak{Z}} &= \eta_1 - \xi_1 I_1(\mathfrak{Z})S(\mathfrak{Z}) - \xi_2 I_2(\mathfrak{Z})S(\mathfrak{Z}) - \xi_3 I_3(\mathfrak{Z})S(\mathfrak{Z}) - \xi_4 I_4(\mathfrak{Z})S(\mathfrak{Z}) + \delta_1 I_1(\mathfrak{Z}) \\
&\quad + \delta_2 I_2(\mathfrak{Z}) + \delta_3 I_3(\mathfrak{Z}) + \delta_4 I_4(\mathfrak{Z}) + \mu P_2(\mathfrak{Z}) - (\alpha + d)S(\mathfrak{Z}) \\
\frac{dI_1(\mathfrak{Z})}{d\mathfrak{Z}} &= \eta_2 + \xi_1 S(\mathfrak{Z})I_1(\mathfrak{Z}) - (\delta_1 + g_1 + d)I_1(\mathfrak{Z}) \\
\frac{dI_2(\mathfrak{Z})}{d\mathfrak{Z}} &= \eta_3 + \xi_2 S(\mathfrak{Z})I_2(\mathfrak{Z}) - (\delta_2 + g_2 + d)I_2(\mathfrak{Z}) \\
\frac{dI_3(\mathfrak{Z})}{d\mathfrak{Z}} &= \eta_4 + \xi_3 S(\mathfrak{Z})I_3(\mathfrak{Z}) - (\delta_3 + g_3 + d)I_3(\mathfrak{Z}) \\
\frac{dI_4(\mathfrak{Z})}{d\mathfrak{Z}} &= \eta_5 + \xi_4 S(\mathfrak{Z})I_4(\mathfrak{Z}) - (\delta_4 + g_4 + d)I_4(\mathfrak{Z}) \\
\frac{dQ_1(\mathfrak{Z})}{d\mathfrak{Z}} &= \eta_6 + g_1 I_1(\mathfrak{Z}) - (\Theta_1 + d)Q_1(\mathfrak{Z}) \\
\frac{dQ_2(\mathfrak{Z})}{d\mathfrak{Z}} &= \eta_7 + g_2 I_2(\mathfrak{Z}) - (\Theta_2 + d)Q_2(\mathfrak{Z}) \\
\frac{dQ_3(\mathfrak{Z})}{d\mathfrak{Z}} &= \eta_8 + g_3 I_3(\mathfrak{Z}) - (\Theta_3 + d)Q_3(\mathfrak{Z}) \\
\frac{dQ_4(\mathfrak{Z})}{d\mathfrak{Z}} &= \eta_9 + g_4 I_4(\mathfrak{Z}) - (\Theta_4 + d)Q_4(\mathfrak{Z}) \\
\frac{dT(\mathfrak{Z})}{d\mathfrak{Z}} &= \eta_{10} + \Theta_1 Q_1(\mathfrak{Z}) + \Theta_2 Q_2(\mathfrak{Z}) + \Theta_3 Q_3(\mathfrak{Z}) + \Theta_4 Q_4(\mathfrak{Z}) - (a_1 + a_2 + d)T(\mathfrak{Z}) \\
\frac{dP_1(\mathfrak{Z})}{d\mathfrak{Z}} &= \eta_{11} + a_1 T(\mathfrak{Z}) - (\Delta + d)P_1(\mathfrak{Z}) \\
\frac{dP_2(\mathfrak{Z})}{d\mathfrak{Z}} &= \eta_{12} + a_2 T(\mathfrak{Z}) + \alpha S(\mathfrak{Z}) + \Delta P_1(\mathfrak{Z}) - (\mu + d)P_2(\mathfrak{Z}) \qquad (13)
\end{aligned}$$

where C_1 is the public awareness coefficient of tracing servers per unit time, C_2 is the public awareness coefficient of isolated servers per unit time, C_3 is the public awareness coefficient of semi-protected servers per unit time, and C_4 is the awareness coefficient of semi-protected servers due to heavy infections, and the total maintained security level is symbolized by $\mathfrak{J}(\Delta)$. The controller objective framework is as indicated in Figure 2.

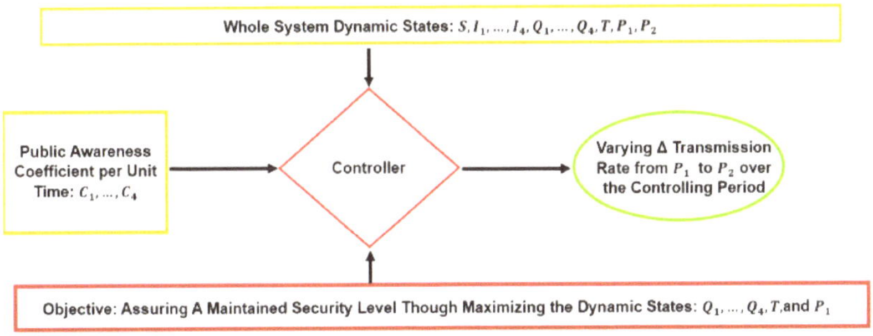

Figure 2. Proposed controller framework.

To obtain an optimum solution for the O.C. problems ((12) and (13)), we need to provide the Lagrangian and Hamiltonian. In actuality, the Lagrangian of the optimal problem is supplied by

$$\mathfrak{L}(T, Q_1, \cdots, Q_4, P_1, \Delta) = C_1 T(\mathfrak{Z}) + C_2 \sum_{k=1}^{4} Q_k(\mathfrak{Z}) + C_3 P_1(\mathfrak{Z}) + \frac{C_4}{2}\Delta^2(\mathfrak{Z})$$

Next, we need to choose a suitable $\Delta(\mathfrak{z})$ s.t. the Lagrangian integral discussed before achieves its minimum. This may be accomplished by defining the Hamiltonian $\mathfrak{H}$ of the control problem as follows:

$$\begin{aligned}
&\mathfrak{H}(S, I_1, \cdots, I_4, Q_1, \cdots, Q_4, T, P_1, P_2, \Delta, \zeta_1, \cdots, \zeta_{12}, \mathfrak{z}) \\
&\quad = \mathfrak{L}(T, Q_1, \cdots, Q_4, P_1, \Delta) \\
&\quad + \Big[\zeta_1 \frac{\mathbf{d}S(\mathfrak{z})}{\mathbf{d}\mathfrak{z}} + \zeta_2 \frac{\mathbf{d}I_1(\mathfrak{z})}{\mathbf{d}\mathfrak{z}} + \zeta_3 \frac{\mathbf{d}I_2(\mathfrak{z})}{\mathbf{d}\mathfrak{z}} + \zeta_4 \frac{\mathbf{d}I_3(\mathfrak{z})}{\mathbf{d}\mathfrak{z}} + \zeta_5 \frac{\mathbf{d}I_4(\mathfrak{z})}{\mathbf{d}\mathfrak{z}} + \zeta_6 \frac{\mathbf{d}Q_1(\mathfrak{z})}{\mathbf{d}\mathfrak{z}} \\
&\quad + \zeta_7 \frac{\mathbf{d}Q_2(\mathfrak{z})}{\mathbf{d}\mathfrak{z}} + \zeta_8 \frac{\mathbf{d}Q_3(\mathfrak{z})}{\mathbf{d}\mathfrak{z}} + \zeta_9 \frac{\mathbf{d}Q_4(\mathfrak{z})}{\mathbf{d}\mathfrak{z}} + \zeta_{10} \frac{\mathbf{d}T(\mathfrak{z})}{\mathbf{d}\mathfrak{z}} + \zeta_{11} \frac{\mathbf{d}P_1(\mathfrak{z})}{\mathbf{d}\mathfrak{z}} + \zeta_{12} \frac{\mathbf{d}P_2(\mathfrak{z})}{\mathbf{d}\mathfrak{z}}\Big]
\end{aligned} \tag{14}$$

with $\zeta_k, k = \mathbb{N}_1^{12}$ representing the adjoint variables that need to be appropriately estimated.

Theorem 4. *For the O.C. problem ((12) and (13)), let $\hat{S}, \hat{I}_1, \cdots, \hat{P}_1$ and $\hat{P}_2$ be the optimal state solutions linked to the O.C. variable $\hat{\Delta}$. Following that, adjoint variables $\zeta_k, k = \mathbb{N}_1^{12}$ exist and they all fulfill*

$$\begin{aligned}
\frac{\mathbf{d}\zeta_1(\mathfrak{z})}{\mathbf{d}\mathfrak{z}} &= \zeta_1\Big(\sum_{k=1}^{4} \xi_k \hat{I}_k + \alpha + d\Big) - \zeta_2\xi_1\hat{I}_1 - \zeta_3\xi_2\hat{I}_2 - \zeta_4\xi_3\hat{I}_3 - \zeta_5\xi_4\hat{I}_4 - \zeta_{12}\alpha \\
\frac{\mathbf{d}\zeta_2(\mathfrak{z})}{\mathbf{d}\mathfrak{z}} &= \zeta_1(\xi_1\hat{S} - \delta_1) - \zeta_2(\xi_1\hat{S} - \delta_1 - g_1 - d) - \zeta_6 g_1 \\
\frac{\mathbf{d}\zeta_3(\mathfrak{z})}{\mathbf{d}\mathfrak{z}} &= \zeta_1(\xi_2\hat{S} - \delta_2) - \zeta_3(\xi_2\hat{S} - \delta_2 - g_2 - d) - \zeta_7 g_2 \\
\frac{\mathbf{d}\zeta_4(\mathfrak{z})}{\mathbf{d}\mathfrak{z}} &= \zeta_1(\xi_3\hat{S} - \delta_3) - \zeta_4(\xi_3\hat{S} - \delta_3 - g_3 - d) - \zeta_8 g_3 \\
\frac{\mathbf{d}\zeta_5(\mathfrak{z})}{\mathbf{d}\mathfrak{z}} &= \zeta_1(\xi_4\hat{S} - \delta_4) - \zeta_5(\xi_4\hat{S} - \delta_4 - g_4 - d) - \zeta_9 g_4 \\
\frac{\mathbf{d}\zeta_6(\mathfrak{z})}{\mathbf{d}\mathfrak{z}} &= \zeta_6(\Theta_1 + d) - \zeta_{10}\Theta_1 - C_2 \\
\frac{\mathbf{d}\zeta_7(\mathfrak{z})}{\mathbf{d}\mathfrak{z}} &= \zeta_7(\Theta_2 + d) - \zeta_{10}\Theta_2 - C_2 \\
\frac{\mathbf{d}\zeta_8(\mathfrak{z})}{\mathbf{d}\mathfrak{z}} &= \zeta_8(\Theta_3 + d) - \zeta_{10}\Theta_3 - C_2 \\
\frac{\mathbf{d}\zeta_9(\mathfrak{z})}{\mathbf{d}\mathfrak{z}} &= \zeta_9(\Theta_3 + d) - \zeta_{10}\Theta_3 - C_2 \\
\frac{\mathbf{d}\zeta_{10}(\mathfrak{z})}{\mathbf{d}\mathfrak{z}} &= \zeta_{10}(a_1 + a_2 + d) - \zeta_{11}a_1 - \zeta_{12}a_2 - C_1 \\
\frac{\mathbf{d}\zeta_{11}(\mathfrak{z})}{\mathbf{d}\mathfrak{z}} &= \zeta_{11}(\Delta + d) - \zeta_{12}\Delta - C_3 \\
\frac{\mathbf{d}\zeta_{12}(\mathfrak{z})}{\mathbf{d}\mathfrak{z}} &= \zeta_{12}(\mu + d) - \zeta_1\mu
\end{aligned} \tag{15}$$

with

$$\zeta_k(\mathcal{T}) = 0, \quad k = \mathbb{N}_1^{12} \tag{16}$$

Additionally, the optimal control $\hat{\Delta}(\mathfrak{z})$ is provided by

$$\hat{\Delta}(\mathfrak{z}) = \max\left[\min\left(0, \frac{\zeta_{11} - \zeta_{12}}{C_4}\hat{P}_1\right), 1\right] \tag{17}$$

Proof. First, the Hamiltonian (14) is utilized to create the adjoint equations. Let $S(\mathfrak{Z}) = \hat{S}(\mathfrak{Z}), I_1(\mathfrak{Z}) = \hat{I}_1(\mathfrak{Z}), \cdots, P_2(\mathfrak{Z}) = \hat{P}_2(\mathfrak{Z})$, and we obtain

$$
\begin{aligned}
\frac{\mathbf{d}\zeta_1(\mathfrak{Z})}{\mathbf{d}\mathfrak{Z}} &= -\frac{\partial \mathfrak{H}}{\partial S} = \zeta_1\Big(\sum_{k=1}^{4}\xi_k\hat{I}_k + \alpha + d\Big) - \zeta_2\xi_1\hat{I}_1 - \zeta_3\xi_2\hat{I}_2 - \zeta_4\xi_3\hat{I}_3 - \zeta_5\xi_4\hat{I}_4 - \zeta_{12}\alpha \\
\frac{\mathbf{d}\zeta_2(\mathfrak{Z})}{\mathbf{d}\mathfrak{Z}} &= -\frac{\partial \mathfrak{H}}{\partial I_1} = \zeta_1(\xi_1\hat{S} - \delta_1) - \zeta_2(\xi_1\hat{S} - \delta_1 - g_1 - d) - \zeta_6 g_1 \\
\frac{\mathbf{d}\zeta_3(\mathfrak{Z})}{\mathbf{d}\mathfrak{Z}} &= -\frac{\partial \mathfrak{H}}{\partial I_2} = \zeta_1(\xi_2\hat{S} - \delta_2) - \zeta_3(\xi_2\hat{S} - \delta_2 - g_2 - d) - \zeta_7 g_2 \\
\frac{\mathbf{d}\zeta_4(\mathfrak{Z})}{\mathbf{d}\mathfrak{Z}} &= -\frac{\partial \mathfrak{H}}{\partial I_3} = \zeta_1(\xi_3\hat{S} - \delta_3) - \zeta_4(\xi_3\hat{S} - \delta_3 - g_3 - d) - \zeta_8 g_3 \\
\frac{\mathbf{d}\zeta_5(\mathfrak{Z})}{\mathbf{d}\mathfrak{Z}} &= -\frac{\partial \mathfrak{H}}{\partial I_4} = \zeta_1(\xi_4\hat{S} - \delta_4) - \zeta_5(\xi_4\hat{S} - \delta_4 - g_4 - d) - \zeta_9 g_4 \\
\frac{\mathbf{d}\zeta_6(\mathfrak{Z})}{\mathbf{d}\mathfrak{Z}} &= -\frac{\partial \mathfrak{H}}{\partial Q_1} = \zeta_6(\Theta_1 + d) - \zeta_{10}\Theta_1 - C_2 \\
\frac{\mathbf{d}\zeta_7(\mathfrak{Z})}{\mathbf{d}\mathfrak{Z}} &= -\frac{\partial \mathfrak{H}}{\partial Q_2} = \zeta_7(\Theta_2 + d) - \zeta_{10}\Theta_2 - C_2 \\
\frac{\mathbf{d}\zeta_8(\mathfrak{Z})}{\mathbf{d}\mathfrak{Z}} &= -\frac{\partial \mathfrak{H}}{\partial Q_3} = \zeta_8(\Theta_3 + d) - \zeta_{10}\Theta_3 - C_2 \\
\frac{\mathbf{d}\zeta_9(\mathfrak{Z})}{\mathbf{d}\mathfrak{Z}} &= -\frac{\partial \mathfrak{H}}{\partial Q_4} = \zeta_9(\Theta_4 + d) - \zeta_{10}\Theta_4 - C_2 \\
\frac{\mathbf{d}\zeta_{10}(\mathfrak{Z})}{\mathbf{d}\mathfrak{Z}} &= -\frac{\partial \mathfrak{H}}{\partial T} = \zeta_{10}(a_1 + a_2 + d) - \zeta_{11}a_1 - \zeta_{12}a_2 - C_1 \\
\frac{\mathbf{d}\zeta_{11}(\mathfrak{Z})}{\mathbf{d}\mathfrak{Z}} &= -\frac{\partial \mathfrak{H}}{\partial P_1} = \zeta_{11}(\Delta + d) - \zeta_{12}\Delta - C_3 \\
\frac{\mathbf{d}\zeta_{12}(\mathfrak{Z})}{\mathbf{d}\mathfrak{Z}} &= -\frac{\partial \mathfrak{H}}{\partial P_2} = \zeta_{12}(\mu + d) - \zeta_1\mu
\end{aligned}
$$

By virtue of the optimality criteria, we have

$$\frac{\partial \mathfrak{H}}{\partial \Delta}\Big|_{\Delta=\hat{\Delta}} = C_4\hat{\Delta}(\mathfrak{Z}) - \zeta_{11}\hat{P}_1(\mathfrak{Z}) + \zeta_{12}\hat{P}_1(\mathfrak{Z}) = 0$$

The identification indicated above suggests that

$$\hat{\Delta}(\mathfrak{Z}) = \frac{\zeta_{11} - \zeta_{12}}{C_4}\hat{P}_1$$

Hence,

$$\hat{\Delta}(\mathfrak{Z}) = \begin{cases} 0 & , \quad \frac{\partial \mathfrak{H}}{\partial \Delta} > 0 \\ \frac{\zeta_{11} - \zeta_{12}}{C_4}\hat{P}_1 & , \quad \frac{\partial \mathfrak{H}}{\partial \Delta} = 0 \\ 1 & , \quad \frac{\partial \mathfrak{H}}{\partial \Delta} < 0 \end{cases}$$

Consequently, $\hat{\Delta}(\mathfrak{Z})$ is the optimal control, and it may be represented in the concise notation that follows:

$$\hat{\Delta}(\mathfrak{Z}) = \max\left[\min\left(0, \frac{\zeta_{11} - \zeta_{12}}{C_4}\hat{P}_1\right), 1\right]$$

Here, the optimum control is characterized by Formula (17) for $\hat{\Delta}$. The characterization of the optimal control (17), the adjoint system (15) and (16), the state system (13) with boundary conditions, and the optimality system can be solved to yield the optimum control and states. To solve the optimality system, we incorporate the characterization of the optimum control $\hat{\Delta}$ given by (17) along with the transversality and initial conditions. By changing the $\hat{\Delta}$ values in the control system, the following system is produced for the state system (13).

$$\begin{aligned}
\frac{dS(\mathfrak{Z})}{d\mathfrak{Z}} &= \eta_1 - \xi_1 I_1(\mathfrak{Z})S(\mathfrak{Z}) - \xi_2 I_2(\mathfrak{Z})S(\mathfrak{Z}) - \xi_3 I_3(\mathfrak{Z})S(\mathfrak{Z}) - \xi_4 I_4(\mathfrak{Z})S(\mathfrak{Z}) + \delta_1 I_1(\mathfrak{Z}) \\
&\quad + \delta_2 I_2(\mathfrak{Z}) + \delta_3 I_3(\mathfrak{Z}) + \delta_4 I_4(\mathfrak{Z}) + \mu P_2(\mathfrak{Z}) - (\alpha + d)S(\mathfrak{Z}) \\
\frac{dI_1(\mathfrak{Z})}{d\mathfrak{Z}} &= \eta_2 + \xi_1 S(\mathfrak{Z})I_1(\mathfrak{Z}) - (\delta_1 + g_1 + d)I_1(\mathfrak{Z}) \\
\frac{dI_2(\mathfrak{Z})}{d\mathfrak{Z}} &= \eta_3 + \xi_2 S(\mathfrak{Z})I_2(\mathfrak{Z}) - (\delta_2 + g_2 + d)I_2(\mathfrak{Z}) \\
\frac{dI_3(\mathfrak{Z})}{d\mathfrak{Z}} &= \eta_4 + \xi_3 S(\mathfrak{Z})I_3(\mathfrak{Z}) - (\delta_3 + g_3 + d)I_3(\mathfrak{Z}) \\
\frac{dI_4(\mathfrak{Z})}{d\mathfrak{Z}} &= \eta_5 + \xi_4 S(\mathfrak{Z})I_4(\mathfrak{Z}) - (\delta_4 + g_4 + d)I_4(\mathfrak{Z}) \\
\frac{dQ_1(\mathfrak{Z})}{d\mathfrak{Z}} &= \eta_6 + g_1 I_1(\mathfrak{Z}) - (\Theta_1 + d)Q_1(\mathfrak{Z}) \\
\frac{dQ_2(\mathfrak{Z})}{d\mathfrak{Z}} &= \eta_7 + g_2 I_2(\mathfrak{Z}) - (\Theta_2 + d)Q_2(\mathfrak{Z}) \\
\frac{dQ_3(\mathfrak{Z})}{d\mathfrak{Z}} &= \eta_8 + g_3 I_3(\mathfrak{Z}) - (\Theta_3 + d)Q_3(\mathfrak{Z}) \\
\frac{dQ_4(\mathfrak{Z})}{d\mathfrak{Z}} &= \eta_9 + g_4 I_4(\mathfrak{Z}) - (\Theta_4 + d)Q_4(\mathfrak{Z}) \\
\frac{dT(\mathfrak{Z})}{d\mathfrak{Z}} &= \eta_{10} + \Theta_1 Q_1(\mathfrak{Z}) + \Theta_2 Q_2(\mathfrak{Z}) + \Theta_3 Q_3(\mathfrak{Z}) + \Theta_4 Q_4(\mathfrak{Z}) - (a_1 + a_2 + d)T(\mathfrak{Z}) \\
\frac{dP_1(\mathfrak{Z})}{d\mathfrak{Z}} &= \eta_{11} + a_1 T(\mathfrak{Z}) - \left(\max\left[\min\left(0, \frac{\zeta_{11} - \zeta_{12}}{C_4}\hat{P}_1\right), 1\right] + d \right) P_1(\mathfrak{Z}) \\
\frac{dP_2(\mathfrak{Z})}{d\mathfrak{Z}} &= \eta_{12} + a_2 T(\mathfrak{Z}) + \max\left[\min\left(0, \frac{\zeta_{11} - \zeta_{12}}{C_4}\hat{P}_1\right), 1\right] P_1(\mathfrak{Z}) \\
&+ \alpha S(\mathfrak{Z}) - (\mu + d)P_2(\mathfrak{Z})
\end{aligned} \tag{18}$$

The goal is to determine the optimal control and state system by numerically solving the aforementioned system (18). □

3. Stochastic MMCE Model

Ecologists and biologists argue that a variety of environmental disturbances, including temperature, water availability, and climate change, inevitably impact the spread of infectious diseases, human contact, human movement, and other processes [22–24]. In contrast, the deterministic model mentioned above does not account for the contributions of random variables. Furthermore, as far as continuous-time system modeling is concerned, Brownian motion, or B.M., is the preferred method for modeling random motion and noise. Based on B.M.'s strong statistical features, this decision is well founded. One may solve the B.M. issue with the aid of strong analytical methods, as well as finite moments of all orders and continuous sample-path trajectories. Even complicated systems with a high degree of unpredictability as a result of applied external noise are discussed using stochastic models. We designed the stochastic model to generate realistic curves characterizing malware in multi-cloud configurations. Further originality is brought to our article by this modeling method. One generates the stochastic model by first adding white noise aspects to each of the deterministic model equations. The terminologies that have been defined now encompass a wider spectrum of potential viral dynamics and more probable situations. Following the addition of the additional stochastic components, the resulting model equations are as described in the model ((19) and (20)).

$$
\begin{aligned}
\frac{dS(\mathfrak{z})}{d\mathfrak{z}} &= \eta_1 - \xi_1 I_1(\mathfrak{z})S(\mathfrak{z}) - \xi_2 I_2(\mathfrak{z})S(\mathfrak{z})I_2(\mathfrak{z}) - \xi_3 I_3(\mathfrak{z})S(\mathfrak{z}) - \xi_4 I_4(\mathfrak{z})S(\mathfrak{z}) + \delta_1 I_1(\mathfrak{z}) \\
&\quad + \delta_2 I_2(\mathfrak{z}) + \delta_3 I_3(\mathfrak{z}) + \delta_4 I_4(\mathfrak{z}) + \mu P_2(\mathfrak{z}) - (\alpha + d)S(\mathfrak{z}) + \kappa S(\mathfrak{z})\frac{d\mathfrak{W}}{d\mathfrak{z}} \\
\frac{dI_1(\mathfrak{z})}{d\mathfrak{z}} &= \eta_2 + \xi_1 S(\mathfrak{z})I_1(\mathfrak{z}) - (\delta_1 + g_1 + d)I_1(\mathfrak{z}) + \kappa I_1(\mathfrak{z})\frac{d\mathfrak{W}}{d\mathfrak{z}} \\
\frac{dI_2(\mathfrak{z})}{d\mathfrak{z}} &= \eta_3 + \xi_2 S(\mathfrak{z})I_2(\mathfrak{z}) - (\delta_2 + g_2 + d)I_2(\mathfrak{z}) + \kappa I_2(\mathfrak{z})\frac{d\mathfrak{W}}{d\mathfrak{z}} \\
\frac{dI_3(\mathfrak{z})}{d\mathfrak{z}} &= \eta_4 + \xi_3 S(\mathfrak{z})I_3(\mathfrak{z}) - (\delta_3 + g_3 + d)I_3(\mathfrak{z}) + \kappa I_3(\mathfrak{z})\frac{d\mathfrak{W}}{d\mathfrak{z}} \\
\frac{dI_4(\mathfrak{z})}{d\mathfrak{z}} &= \eta_5 + \xi_4 S(\mathfrak{z})I_4(\mathfrak{z}) - (\delta_4 + g_4 + d)I_4(\mathfrak{z}) + \kappa I_4(\mathfrak{z})\frac{d\mathfrak{W}}{d\mathfrak{z}} \\
\frac{dQ_1(\mathfrak{z})}{d\mathfrak{z}} &= \eta_6 + g_1 I_1(\mathfrak{z}) - (\Theta_1 + d)Q_1(\mathfrak{z}) + \kappa Q_1(\mathfrak{z})\frac{d\mathfrak{W}}{d\mathfrak{z}} \\
\frac{dQ_2(\mathfrak{z})}{d\mathfrak{z}} &= \eta_7 + g_2 I_2(\mathfrak{z}) - (\Theta_2 + d)Q_2(\mathfrak{z}) + \kappa Q_2(\mathfrak{z})\frac{d\mathfrak{W}}{d\mathfrak{z}} \\
\frac{dQ_3(\mathfrak{z})}{d\mathfrak{z}} &= \eta_8 + g_3 I_3(\mathfrak{z}) - (\Theta_3 + d)Q_3(\mathfrak{z}) + \kappa Q3(\mathfrak{z})\frac{d\mathfrak{W}}{d\mathfrak{z}} \\
\frac{dQ_4(\mathfrak{z})}{d\mathfrak{z}} &= \eta_9 + g_4 I_4(\mathfrak{z}) - (\Theta_4 + d)Q_4(\mathfrak{z}) + \kappa Q_4(\mathfrak{z})\frac{d\mathfrak{W}}{d\mathfrak{z}} \\
\frac{dT(\mathfrak{z})}{d\mathfrak{z}} &= \eta_{10} + \Theta_1 Q_1(\mathfrak{z}) + \Theta_2 Q_2(\mathfrak{z}) + \Theta_3 Q_3(\mathfrak{z}) + \Theta_4 Q_4(\mathfrak{z}) - (a_1 + a_2 + d)T(\mathfrak{z}) \\
&+ \kappa T(\mathfrak{z})\frac{d\mathfrak{W}}{d\mathfrak{z}} \\
\frac{dP_1(\mathfrak{z})}{d\mathfrak{z}} &= \eta_{11} + a_1 T(\mathfrak{z}) - (\Delta + d)P_1(\mathfrak{z}) + \kappa P_1(\mathfrak{z})\frac{d\mathfrak{W}}{d\mathfrak{z}} \\
\frac{dP_2(\mathfrak{z})}{d\mathfrak{z}} &= \eta_{12} + a_2 T(\mathfrak{z}) + \Delta P_1(\mathfrak{z}) + \alpha S(\mathfrak{z}) - (\mu + d)P_2(\mathfrak{z}) + \kappa P_2(\mathfrak{z})\frac{d\mathfrak{W}}{d\mathfrak{z}}
\end{aligned}
\tag{19}
$$

with initial conditions

$$S(0) = S_0, I_1(0) = I_{1_0}, \cdots, P_2(0) = P_{2_0} \tag{20}$$

where $\mathfrak{W}(\mathfrak{z})$ is a standard Brownian motion process and $\kappa > 0$.

3.1. Existence and Uniqueness of Positive Solution

Theorem 5. *For any initial condition $(S, I_1, \cdots, I_4, Q_1, \cdots, Q_4, T, P_1, P_2)(0) \in \mathbb{R}^+_{12}$, there is a unique global positive solution $(S, I_1, \cdots, I_4, Q_1, \cdots, Q_4, T, P_1, P_2)(\mathfrak{z})$ for stochastic model (19), which will be maintained in $\mathbb{R}^+_{12}$ with probability one.*

Proof. Since the state variables' initial conditions are known to satisfy $(S, I_1, \cdots, I_4, Q_1, \cdots, Q_4, T, P_1, P_2)(0) \in \mathbb{R}^+_{12}$, the coefficients given in Equation (19) are defined and locally Lipschitzian. So, there lies a local unique solution $(S, I_1, \cdots, I_4, Q_1, \cdots, Q_4, T, P_1, P_2)(\mathfrak{z})$ of the problem over $\mathfrak{z} \in [0, \vartheta_e)$. In order to determine the solution's global nature, we must demonstrate that $\vartheta_e = \infty$ almost surely (a.s.). Assume that we have a sufficiently large, non-negative number ϱ_0, and that all initial approximations of the state variables are given within $[\frac{1}{\varrho_0}, \varrho_0]$. For every non-negative integer $\varrho \geq \varrho_0$, let the time required to reach completion be given as

$$
\begin{aligned}
\vartheta_{\varrho} = & \left\{ \mathfrak{Z} \in [0, \vartheta_e) : \min\left\{ (S, I_1, \cdots, I_4, Q_1, \cdots, Q_4, T, P_1, P_2)(\mathfrak{Z}) \right\} \leq \frac{1}{\varrho} \right. \\
& \left. \quad or \ \max\left\{ (S, I_1, \cdots, I_4, Q_1, \cdots, Q_4, T, P_1, P_2)(\mathfrak{Z}) \right\} \right\}
\end{aligned}
$$

the choice of $\inf \chi = \infty$ is considered, where χ is for an empty set. From the result of ϑ_ϱ, we know that this grows as $\varrho \to \infty$. Set $\vartheta_\infty = \lim_{\varrho \to \infty} \vartheta_\varrho$ and $\vartheta_e \geq \vartheta_\infty$ a.s., showing $\vartheta_\infty = \infty$ a.s. Consider the subsequent non-negative C^2-function Ψ:

$$
\begin{aligned}
& \Psi(S, I_1, \cdots, I_4, Q_1, \cdots, Q_4, T, P_1, P_2) = S - 1 - \ln S + I_1 - 1 - \ln I_1 + I_2 - 1 - \ln I_2 \\
& \quad + I_3 - 1 - \ln I_3 + I_4 - 1 - \ln I_4 + Q_1 - 1 - \ln Q_1 + Q_2 - 1 - \ln Q_2 + Q_3 - 1 - \ln Q_3 \\
& \quad + Q_4 - 1 - \ln Q_4 + T - 1 - \ln T + P_1 - 1 - \ln P_1 + P_2 - 1 - \ln P_2
\end{aligned}
$$

Utilizing Itô's formula yields

$$
\begin{aligned}
\mathbf{d}\Psi = & \ \mathbb{L}\Psi \mathbf{d}\mathfrak{Z} + \kappa(S-1)\mathbf{d}\mathfrak{W} + \kappa(I_1 - 1)\mathbf{d}\mathfrak{W} + \kappa(I_2 - 1)\mathbf{d}\mathfrak{W} + \kappa(I_3 - 1)\mathbf{d}\mathfrak{W} \\
& + \kappa(I_4 - 1)\mathbf{d}\mathfrak{W} + \kappa(Q_1 - 1)\mathbf{d}\mathfrak{W} + \kappa(Q_2 - 1)\mathbf{d}\mathfrak{W} + \kappa(Q_3 - 1)\mathbf{d}\mathfrak{W} + \kappa(Q_4 - 1)\mathbf{d}\mathfrak{W} \\
& + \kappa(T - 1)\mathbf{d}\mathfrak{W} + \kappa(P_1 - 1)\mathbf{d}\mathfrak{W} + \kappa(P_2 - 1)\mathbf{d}\mathfrak{W}
\end{aligned}
$$

where

$$
\begin{aligned}
\mathbb{L}\Psi = & \left(1 - \frac{1}{S}\right)\Big(\eta_1 - \xi_1 I_1(\mathfrak{Z})S(\mathfrak{Z}) - \xi_2 I_2(\mathfrak{Z})S(\mathfrak{Z}) - \xi_3 I_3(\mathfrak{Z})S(\mathfrak{Z}) - \xi_4 I_4(\mathfrak{Z})S(\mathfrak{Z}) + \delta_1 I_1(\mathfrak{Z}) \\
& + \delta_2 I_2(\mathfrak{Z}) + \delta_3 I_3(\mathfrak{Z}) + \delta_4 I_4(\mathfrak{Z}) + \mu P_2(\mathfrak{Z}) - (\alpha + d)S(\mathfrak{Z})\Big) + \frac{\kappa^2}{2} \\
& + \left(1 - \frac{1}{I_1}\right)\Big(\eta_2 + \xi_1 S(\mathfrak{Z})I_1(\mathfrak{Z}) - (\delta_1 + g_1 + d)I_1(\mathfrak{Z})\Big) + \frac{\kappa^2}{2} \\
& + \left(1 - \frac{1}{I_2}\right)\Big(\eta_3 + \xi_2 S(\mathfrak{Z})I_2(\mathfrak{Z}) - (\delta_2 + g_2 + d)I_2(\mathfrak{Z})\Big) + \frac{\kappa^2}{2} \\
& + \left(1 - \frac{1}{I_3}\right)\Big(\eta_4 + \xi_3 S(\mathfrak{Z})I_3(\mathfrak{Z}) - (\delta_3 + g_3 + d)I_3(\mathfrak{Z})\Big) + \frac{\kappa^2}{2} \\
& + \left(1 - \frac{1}{I_4}\right)\Big(\eta_5 + \xi_4 S(\mathfrak{Z})I_4(\mathfrak{Z}) - (\delta_4 + g_4 + d)I_4(\mathfrak{Z})\Big) + \frac{\kappa^2}{2} \\
& + \left(1 - \frac{1}{Q_1}\right)\Big(\eta_6 + g_1 I_1(\mathfrak{Z}) - (\Theta_1 + d)Q_1(\mathfrak{Z})\Big) + \frac{\kappa^2}{2} \\
& + \left(1 - \frac{1}{Q_2}\right)\Big(\eta_7 + g_2 I_2(\mathfrak{Z}) - (\Theta_2 + d)Q_2(\mathfrak{Z})\Big) + \frac{\kappa^2}{2} \\
& + \left(1 - \frac{1}{Q_3}\right)\Big(\eta_8 + g_3 I_3(\mathfrak{Z}) - (\Theta_3 + d)Q_3(\mathfrak{Z})\Big) + \frac{\kappa^2}{2} + \left(1 - \frac{1}{Q_4}\right)\Big(\eta_9 + g_4 I_4(\mathfrak{Z}) \\
& - (\Theta_4 + d)Q_4(\mathfrak{Z})\Big) + \frac{\kappa^2}{2} \\
& + \left(1 - \frac{1}{T}\right)\Big(\eta_{10} + \Theta_1 Q_1(\mathfrak{Z}) + \Theta_2 Q_2(\mathfrak{Z}) + \Theta_3 Q_3(\mathfrak{Z}) + \Theta_4 Q_4(\mathfrak{Z}) - (a_1 + a_2 + d)T(\mathfrak{Z})\Big) + \frac{\kappa^2}{2} \\
& + \left(1 - \frac{1}{P_1}\right)\Big(\eta_{11} + a_1 T(\mathfrak{Z}) - (\Delta + d)P_1(\mathfrak{Z})\Big) + \frac{\kappa^2}{2} \\
& + \left(1 - \frac{1}{P_2}\right)\Big(\eta_{12} + a_2 T(\mathfrak{Z}) + \Delta P_1(\mathfrak{Z}) + \alpha S(\mathfrak{Z}) - (\mu + d)P_2(\mathfrak{Z})\Big) + \frac{\kappa^2}{2} \leq \mathfrak{K},
\end{aligned}
$$

and

$$
\mathfrak{K} := \sum_{k=1}^{12} \eta_k + \sum_{k=1}^{4} \Big(\delta_k + g_k + \Theta_k + \xi_k\Big) + \alpha + \mu + \Delta + a_1 + a_2 + 12d + 6\kappa^2
$$

where the constant $\mathfrak{K} > 0$. The approach of proof is the same as in [24], and the remaining proof is omitted. □

3.2. Extinction

Theorem 6. *The solution of model (19) with initial condition* $(S, I_1, \cdots, I_4, Q_1, \cdots, Q_4, T, P_1, P_2)(0)$ *can be represented as* $(S, I_1, \cdots, I_4, Q_1, \cdots, Q_4, T, P_1, P_2)(\mathfrak{z})$*; then*

$$\limsup_{\mathfrak{z}\to\infty}\Big(S(\mathfrak{z}) + \sum_{k=1}^{4} I_k(\mathfrak{z}) + \sum_{k=1}^{4} Q_k(\mathfrak{z}) + T(\mathfrak{z}) + P_1(\mathfrak{z}) + P_2(\mathfrak{z})\Big) < \infty, \ \ a.s.$$

$$\lim_{\mathfrak{z}\to\infty}\frac{S(\mathfrak{z})}{\mathfrak{z}} = 0, \lim_{\mathfrak{z}\to\infty}\frac{I_1(\mathfrak{z})}{\mathfrak{z}} = 0, \cdots, \lim_{\mathfrak{z}\to\infty}\frac{I_4(\mathfrak{z})}{\mathfrak{z}} = 0, \lim_{\mathfrak{z}\to\infty}\frac{Q_1(\mathfrak{z})}{\mathfrak{z}} = 0, \cdots, \lim_{\mathfrak{z}\to\infty}\frac{Q_4(\mathfrak{z})}{\mathfrak{z}} = 0,$$
$$\lim_{\mathfrak{z}\to\infty}\frac{T(\mathfrak{z})}{\mathfrak{z}} = 0, \lim_{\mathfrak{z}\to\infty}\frac{P_1(\mathfrak{z})}{\mathfrak{z}} = 0, \lim_{\mathfrak{z}\to\infty}\frac{P_2(\mathfrak{z})}{\mathfrak{z}} = 0,$$

Furthermore, when $d > \kappa^2/2$ *holds, then*

$$\lim_{\mathfrak{z}\to\infty}\frac{1}{\mathfrak{z}}\int_0^{\mathfrak{z}} S(\mathfrak{G})\mathbf{d}\mathfrak{W}(\mathfrak{z}) = 0, \lim_{\mathfrak{z}\to\infty}\frac{1}{\mathfrak{z}}\int_0^{\mathfrak{z}} I_1(\mathfrak{G})\mathbf{d}\mathfrak{W}(\mathfrak{z}) = 0, \cdots, \lim_{\mathfrak{z}\to\infty}\frac{1}{\mathfrak{z}}\int_0^{\mathfrak{z}} I_4(\mathfrak{G})\mathbf{d}\mathfrak{W}(\mathfrak{z}) = 0,$$
$$\lim_{\mathfrak{z}\to\infty}\frac{1}{\mathfrak{z}}\int_0^{\mathfrak{z}} Q_1(\mathfrak{G})\mathbf{d}\mathfrak{W}(\mathfrak{z}) = 0, \cdots, \lim_{\mathfrak{z}\to\infty}\frac{1}{\mathfrak{z}}\int_0^{\mathfrak{z}} Q_4(\mathfrak{G})\mathbf{d}\mathfrak{W}(\mathfrak{z}) = 0,$$
$$\lim_{\mathfrak{z}\to\infty}\frac{1}{\mathfrak{z}}\int_0^{\mathfrak{z}} T(\mathfrak{G})\mathbf{d}\mathfrak{W}(\mathfrak{z}) = 0, \ \lim_{\mathfrak{z}\to\infty}\frac{1}{\mathfrak{z}}\int_0^{\mathfrak{z}} P_1(\mathfrak{G})\mathbf{d}\mathfrak{W}(\mathfrak{z}) = 0, \lim_{\mathfrak{z}\to\infty}\frac{1}{\mathfrak{z}}\int_0^{\mathfrak{z}} P_2(\mathfrak{G})\mathbf{d}\mathfrak{W}(\mathfrak{z}) = 0, \ a.s.$$

Proof. The derivation of Theorem 6 is similar to [23]; thus, we skip it. □

3.3. Optimal Control Analysis

In order to avoid complexity, we take a vector of the type

$$\mathbf{X} := [S, I_1, \cdots, I_4, Q_1, \cdots, Q_4, T, P_1, P_2]^{Tr}$$

where Tr is the transpose, and

$$\mathbf{d}\mathbf{X}(\mathfrak{z}) = \ell(\mathbf{X}, \Delta)\mathbf{d}\mathfrak{z} + \mathbb{M}(\mathbf{X})\mathbf{d}\mathfrak{W}(\mathfrak{z})$$

where

$$\begin{aligned}
\ell_1(\mathbf{X}, \Delta) &= \eta_1 - \xi_1 I_1(\mathfrak{z})S(\mathfrak{z}) - \xi_2 I_2(\mathfrak{z})S(\mathfrak{z}) - \xi_3 I_3(\mathfrak{z})S(\mathfrak{z}) - \xi_4 I_4(\mathfrak{z})S(\mathfrak{z}) + \delta_1 I_1(\mathfrak{z}) \\
&\quad + \delta_2 I_2(\mathfrak{z}) + \delta_3 I_3(\mathfrak{z}) + \delta_4 I_4(\mathfrak{z}) + \mu P_2(\mathfrak{z}) - (\alpha + d)S(\mathfrak{z}) \\
\ell_2(\mathbf{X}, \Delta) &= \eta_2 + \xi_1 S(\mathfrak{z})I_1(\mathfrak{z}) - (\delta_1 + g_1 + d)I_1(\mathfrak{z}) \\
\ell_3(\mathbf{X}, \Delta) &= \eta_3 + \xi_2 S(\mathfrak{z})I_2(\mathfrak{z}) - (\delta_2 + g_2 + d)I_2(\mathfrak{z}) \\
\ell_4(\mathbf{X}, \Delta) &= \eta_4 + \xi_3 S(\mathfrak{z})I_3(\mathfrak{z}) - (\delta_3 + g_3 + d)I_3(\mathfrak{z}) \\
\ell_5(\mathbf{X}, \Delta) &= \eta_5 + \xi_4 S(\mathfrak{z})I_4(\mathfrak{z}) - (\delta_4 + g_4 + d)I_4(\mathfrak{z}) \\
\ell_6(\mathbf{X}, \Delta) &= \eta_6 + g_1 I_1(\mathfrak{z}) - (\Theta_1 + d)Q_1(\mathfrak{z}) \\
\ell_7(\mathbf{X}, \Delta) &= \eta_7 + g_2 I_2(\mathfrak{z}) - (\Theta_2 + d)Q_2(\mathfrak{z}) \\
\ell_8(\mathbf{X}, \Delta) &= \eta_8 + g_3 I_3(\mathfrak{z}) - (\Theta_3 + d)Q_3(\mathfrak{z}) \\
\ell_9(\mathbf{X}, \Delta) &= \eta_9 + g_4 I_4(\mathfrak{z}) - (\Theta_4 + d)Q_4(\mathfrak{z}) \\
\ell_{10}(\mathbf{X}, \Delta) &= \eta_{10} + \Theta_1 Q_1(\mathfrak{z}) + \Theta_2 Q_2(\mathfrak{z}) + \Theta_3 Q_3(\mathfrak{z}) + \Theta_4 Q_4(\mathfrak{z}) - (a_1 + a_2 + d)T(\mathfrak{z}) \\
\ell_{11}(\mathbf{X}, \Delta) &= \eta_{11} + a_1 T(\mathfrak{z}) - (\Delta + d)P_1(\mathfrak{z}) \\
\ell_{12}(\mathbf{X}, \Delta) &= \eta_{12} + a_2 T(\mathfrak{z}) + \Delta P_1(\mathfrak{z}) + \alpha S(\mathfrak{z}) - (\mu + d)P_2(\mathfrak{z}),
\end{aligned}$$

$$\begin{aligned}\mathbb{M}_1 &= \kappa S, \mathbb{M}_2 = \kappa I_1, \mathbb{M}_3 = \kappa I_2, \mathbb{M}_4 = \kappa I_3, \mathbb{M}_5 = \kappa I_4, \mathbb{M}_6 = \kappa Q_1,\\ \mathbb{M}_7 &= \kappa Q_2, \mathbb{M}_8 = \kappa Q_3, \mathbb{M}_9 = \kappa Q_4, \mathbb{M}_{10} = \kappa T, \mathbb{M}_{11} = \kappa P_1, \mathbb{M}_{12} = \kappa P_2\end{aligned}$$

We provide an optimum control (O.C.) problem to maximize the ensuing objective functional (O.F.)

$$\mathbb{J}(\Delta(\mathfrak{Z})) = \mathbb{E}\Big\{ \int_0^{\mathcal{T}} \Big[C_1 T(\mathfrak{Z}) + C_2 \sum_{k=1}^{4} Q_k(\mathfrak{Z}) + C_3 P_1(\mathfrak{Z}) + \frac{C_4}{2}\Delta^2(\mathfrak{Z})\Big] d\mathfrak{Z} \Big\} \tag{21}$$

subject to

$$\mathbf{dX}(\mathfrak{Z}) = \ell(\mathbf{X}, \Delta)\mathbf{d}\mathfrak{Z} + \mathbb{M}(\mathbf{X})\mathbf{d}\mathfrak{W}(\mathfrak{Z}) \tag{22}$$

where C_1 is the public awareness coefficient of tracing servers per unit time, C_2 is the public awareness coefficient of isolated servers per unit time, C_3 is the public awareness coefficient of semi-protected servers per unit time, and C_4 is the awareness coefficient of semi-protected servers due to heavy infections, and the total maintained security level is symbolized by $\mathbb{J}(\Delta)$. To obtain an optimum solution for the O.C. problems ((21)–(22)), we need to provide the Lagrangian and Hamiltonian. In actuality, the Lagrangian of the optimal problem is supplied by

$$\mathbb{L}(T, Q_1, \cdots, Q_4, P_1, \Delta) = C_1 T(\mathfrak{Z}) + C_2 \sum_{k=1}^{4} Q_k(\mathfrak{Z}) + C_3 P_1(\mathfrak{Z}) + \frac{C_4}{2}\Delta^2(\mathfrak{Z})$$

Next, we need to choose a suitable $\Delta(\mathfrak{Z})$ s.t. the Lagrangian integral discussed before achieves its maximum. This may be accomplished by defining the Hamiltonian $\mathbb{H}$ of the control problem as follows:

$$\begin{aligned}&\mathbb{H}(S, I_1, \cdots, I_4, Q_1, \cdots, Q_4, T, P_1, P_2, \Delta, \beta_1, \cdots\cdots, \beta_{12}, \gamma_1, \cdots\cdots, \gamma_{12}, \mathfrak{Z})\\ &= \mathbb{L}(T, Q_1, \cdots, Q_4, P_1, \Delta)\\ &+\Big[\beta_1 \frac{\mathbf{d}S(\mathfrak{Z})}{\mathbf{d}\mathfrak{Z}} + \beta_2 \frac{\mathbf{d}I_1(\mathfrak{Z})}{\mathbf{d}\mathfrak{Z}} + \beta_3 \frac{\mathbf{d}I_2(\mathfrak{Z})}{\mathbf{d}\mathfrak{Z}} + \beta_4 \frac{\mathbf{d}I_3(\mathfrak{Z})}{\mathbf{d}\mathfrak{Z}} + \beta_5 \frac{\mathbf{d}I_4(\mathfrak{Z})}{\mathbf{d}\mathfrak{Z}} + \beta_6 \frac{\mathbf{d}Q_1(\mathfrak{Z})}{\mathbf{d}\mathfrak{Z}}\\ &+\beta_7 \frac{\mathbf{d}Q_2(\mathfrak{Z})}{\mathbf{d}\mathfrak{Z}} + \beta_8 \frac{\mathbf{d}Q_3(\mathfrak{Z})}{\mathbf{d}\mathfrak{Z}} + \beta_9 \frac{\mathbf{d}Q_4(\mathfrak{Z})}{\mathbf{d}\mathfrak{Z}} + \beta_{10} \frac{\mathbf{d}T(\mathfrak{Z})}{\mathbf{d}\mathfrak{Z}} + \beta_{11} \frac{\mathbf{d}P_1(\mathfrak{Z})}{\mathbf{d}\mathfrak{Z}} + \beta_{12} \frac{\mathbf{d}P_2(\mathfrak{Z})}{\mathbf{d}\mathfrak{Z}}\Big]\\ &+\Big[\kappa S\gamma_1 + \kappa I_1\gamma_2 + \kappa I_2\gamma_3 + \kappa I_3\gamma_4 + \kappa I_4\gamma_5 + \kappa Q_1\gamma_6 + \kappa Q_2\gamma_7 + \kappa Q_3\gamma_8 + \kappa Q_4\gamma_9\\ &+\kappa T\gamma_{10} + \kappa P_1\gamma_{11} + \kappa P_2\gamma_{12}\Big]\end{aligned} \tag{23}$$

The adjoint variables are represented by $\beta_k, k = \mathbb{N}_1^{12}$ and $\gamma_k, k = \mathbb{N}_1^{12}$.

Theorem 7. *For the O.C. problem ((21)–(22)), let $\hat{S}, \hat{I}_1, \cdots, \hat{I}_4, \hat{Q}_1, \cdots, \hat{Q}_4, \hat{T}, \hat{P}_1$, and $\hat{P}_2$ be the optimal state solutions linked to the O.C. variable $\hat{\Delta}$. Following that, adjoint variables $\beta_k, k = \mathbb{N}_1^{12}$ exist and they all fulfill*

$$\begin{aligned}\mathbf{d}\beta_1(\mathfrak{Z}) &= \Big[\beta_1\Big(\sum_{k=1}^{4}\xi_k\hat{I}_k + \alpha + d\Big) - \beta_2\xi_1\hat{I}_1 - \beta_3\xi_2\hat{I}_2 - \beta_4\xi_3\hat{I}_3 - \beta_5\xi_4\hat{I}_4 - \beta_{12}\alpha + \kappa\gamma_1\Big]\mathbf{d}\mathfrak{Z}\\ &\quad+\gamma_1\mathbf{d}\mathfrak{W}\\ \mathbf{d}\beta_2(\mathfrak{Z}) &= \Big[\beta_1(\xi_1\hat{S} - \delta_1) - \beta_2(\xi_1\hat{S} - \delta_1 - g_1 - d) - \beta_6 g_1 + \kappa\gamma_2\Big]\mathbf{d}\mathfrak{Z} + \gamma_2\mathbf{d}\mathfrak{W}\\ \mathbf{d}\beta_3(\mathfrak{Z}) &= \Big[\beta_1(\xi_2\hat{S} - \delta_2) - \beta_3(\xi_2\hat{S} - \delta_2 - g_2 - d) - \beta_7 g_2 + \kappa\gamma_3\Big]\mathbf{d}\mathfrak{Z} + \gamma_3\mathbf{d}\mathfrak{W}\\ \mathbf{d}\beta_4(\mathfrak{Z}) &= \Big[\beta_1(\xi_3\hat{S} - \delta_3) - \beta_4(\xi_3\hat{S} - \delta_3 - g_3 - d) - \beta_8 g_3 + \kappa\gamma_4\Big]\mathbf{d}\mathfrak{Z} + \gamma_4\mathbf{d}\mathfrak{W}\end{aligned}$$

$$\begin{aligned}
\mathbf{d}\beta_5(\mathfrak{Z}) &= \Big[\beta_1(\xi_4\hat{S}-\delta_4)-\beta_5(\xi_4\hat{S}-\delta_4-g_4-d)-\beta_9 g_4+\kappa\gamma_5\Big]\mathbf{d}\mathfrak{Z}+\gamma_5\mathbf{d}\mathfrak{W} \\
\mathbf{d}\beta_6(\mathfrak{Z}) &= \Big[\beta_6(\Theta_1+d)-\beta_{10}\Theta_1-C_2+\kappa\gamma_6\Big]\mathbf{d}\mathfrak{Z}+\gamma_6\mathbf{d}\mathfrak{W} \\
\mathbf{d}\beta_7(\mathfrak{Z}) &= \Big[\beta_7(\Theta_2+d)-\beta_{10}\Theta_2-C_2+\kappa\gamma_7\Big]\mathbf{d}\mathfrak{Z}+\gamma_7\mathbf{d}\mathfrak{W} \\
\mathbf{d}\beta_8(\mathfrak{Z}) &= \Big[\beta_8(\Theta_3+d)-\beta_{10}\Theta_3-C_2+\kappa\gamma_8\Big]\mathbf{d}\mathfrak{Z}+\gamma_8\mathbf{d}\mathfrak{W} \\
\mathbf{d}\beta_9(\mathfrak{Z}) &= \Big[\beta_9(\Theta_4+d)-\beta_{10}\Theta_4-C_2+\kappa\gamma_9\Big]\mathbf{d}\mathfrak{Z}+\gamma_9\mathbf{d}\mathfrak{W} \\
\mathbf{d}\beta_{10}(\mathfrak{Z}) &= \Big[\beta_{10}(a_1+a_2+d)-\beta_{11}a_1-\beta_{12}a_2-C_1+\kappa\gamma_{10}\Big]\mathbf{d}\mathfrak{Z}+\gamma_{10}\mathbf{d}\mathfrak{W} \\
\mathbf{d}\beta_{11}(\mathfrak{Z}) &= \Big[\beta_{11}(\Delta+d)-\beta_{12}\Delta-C_3+\kappa\gamma_{11}\Big]\mathbf{d}\mathfrak{Z}+\gamma_{11}\mathbf{d}\mathfrak{W} \\
\mathbf{d}\beta_{12}(\mathfrak{Z}) &= \Big[\beta_{12}(\mu+d)-\beta_1\mu+\kappa\gamma_{12}\Big]\mathbf{d}\mathfrak{Z}+\gamma_{12}\mathbf{d}\mathfrak{W} \qquad (24)
\end{aligned}$$

with

$$\beta_k(\mathcal{T})=0, \quad k=\mathbb{N}_1^{12} \qquad (25)$$

Additionally, the optimal control $\hat{\Delta}(\mathfrak{Z})$ is provided by

$$\hat{\Delta}(\mathfrak{Z})=\max\left[\min\left(0,\frac{\beta_{11}-\beta_{12}}{C_4}\hat{P}_1\right),1\right] \qquad (26)$$

Proof. First, the Hamiltonian (23) is utilized to create the adjoint equations. Let $S(\mathfrak{Z})=\hat{S}(\mathfrak{Z}), I_1(\mathfrak{Z})=\hat{I}_1(\mathfrak{Z}),\cdots,P_2(\mathfrak{Z})=\hat{P}_2(\mathfrak{Z})$, and we obtain

$$\begin{aligned}
\mathbf{d}\beta_1(\mathfrak{Z}) &= -\frac{\partial\mathbb{H}}{\partial S}\mathbf{d}\mathfrak{Z}+\gamma_1\mathbf{d}\mathfrak{W}=\Big[\beta_1\Big(\sum_{k=1}^{4}\xi_k\hat{I}_k+\alpha+d\Big)-\beta_2\xi_1\hat{I}_1-\beta_3\xi_2\hat{I}_2-\beta_4\xi_3\hat{I}_3-\beta_5\xi_4\hat{I}_4 \\
&\quad -\beta_{12}\alpha+\kappa\gamma_1\Big]\mathbf{d}\mathfrak{Z}+\gamma_1\mathbf{d}\mathfrak{W} \\
\mathbf{d}\beta_2(\mathfrak{Z}) &= -\frac{\partial\mathbb{H}}{\partial I_1}\mathbf{d}\mathfrak{Z}+\gamma_2\mathbf{d}\mathfrak{W}=\Big[\beta_1(\xi_1\hat{S}-\delta_1)-\beta_2(\xi_1\hat{S}-\delta_1-g_1-d)-\beta_6 g_1+\kappa\gamma_2\Big]\mathbf{d}\mathfrak{Z} \\
&\quad +\gamma_2\mathbf{d}\mathfrak{W} \\
\mathbf{d}\beta_3(\mathfrak{Z}) &= -\frac{\partial\mathbb{H}}{\partial I_2}\mathbf{d}\mathfrak{Z}+\gamma_3\mathbf{d}\mathfrak{W}=\Big[\beta_1(\xi_2\hat{S}-\delta_2)-\beta_3(\xi_2\hat{S}-\delta_2-g_2-d)-\beta_7 g_2+\kappa\gamma_3\Big]\mathbf{d}\mathfrak{Z} \\
&\quad +\gamma_3\mathbf{d}\mathfrak{W} \\
\mathbf{d}\beta_4(\mathfrak{Z}) &= -\frac{\partial\mathbb{H}}{\partial I_3}\mathbf{d}\mathfrak{Z}+\gamma_4\mathbf{d}\mathfrak{W}=\Big[\beta_1(\xi_3\hat{S}-\delta_3)-\beta_4(\xi_3\hat{S}-\delta_3-g_3-d)-\beta_8 g_3+\kappa\gamma_4\Big]\mathbf{d}\mathfrak{Z} \\
&\quad +\gamma_4\mathbf{d}\mathfrak{W} \\
\mathbf{d}\beta_5(\mathfrak{Z}) &= -\frac{\partial\mathbb{H}}{\partial I_4}\mathbf{d}\mathfrak{Z}+\gamma_5\mathbf{d}\mathfrak{W}=\Big[\beta_1(\xi_4\hat{S}-\delta_4)-\beta_5(\xi_4\hat{S}-\delta_4-g_4-d)-\beta_9 g_4+\kappa\gamma_5\Big]\mathbf{d}\mathfrak{Z} \\
&\quad +\gamma_5\mathbf{d}\mathfrak{W} \\
\mathbf{d}\beta_6(\mathfrak{Z}) &= -\frac{\partial\mathbb{H}}{\partial Q_1}\mathbf{d}\mathfrak{Z}+\gamma_6\mathbf{d}\mathfrak{W}=\Big[\beta_6(\Theta_1+d)-\beta_{10}\Theta_1-C_2+\kappa\gamma_6\Big]\mathbf{d}\mathfrak{Z}+\gamma_6\mathbf{d}\mathfrak{W} \\
\mathbf{d}\beta_7(\mathfrak{Z}) &= -\frac{\partial\mathbb{H}}{\partial Q_2}\mathbf{d}\mathfrak{Z}+\gamma_7\mathbf{d}\mathfrak{W}=\Big[\beta_7(\Theta_2+d)-\beta_{10}\Theta_2-C_2+\kappa\gamma_7\Big]\mathbf{d}\mathfrak{Z}+\gamma_7\mathbf{d}\mathfrak{W}
\end{aligned}$$

$$\begin{aligned}
\mathbf{d}\beta_8(\mathfrak{Z}) &= -\frac{\partial \mathbb{H}}{\partial Q_3}\mathbf{d}\mathfrak{Z} + \gamma_8 \mathbf{d}\mathfrak{W} = \Big[\beta_8(\Theta_3 + d) - \beta_{10}\Theta_3 - C_2 + \kappa\gamma_8\Big]\mathbf{d}\mathfrak{Z} + \gamma_8\mathbf{d}\mathfrak{W} \\
\mathbf{d}\beta_9(\mathfrak{Z}) &= -\frac{\partial \mathbb{H}}{\partial Q_4}\mathbf{d}\mathfrak{Z} + \gamma_9 \mathbf{d}\mathfrak{W} = \Big[\beta_9(\Theta_3 + d) - \beta_{10}\Theta_3 - C_2 + \kappa\gamma_9\Big]\mathbf{d}\mathfrak{Z} + \gamma_9\mathbf{d}\mathfrak{W} \\
\mathbf{d}\beta_{10}(\mathfrak{Z}) &= -\frac{\partial \mathbb{H}}{\partial T}\mathbf{d}\mathfrak{Z} + \gamma_{10} \mathbf{d}\mathfrak{W} = \Big[\beta_{10}(a_1 + a_2 + d) - \beta_{11}a_1 - \beta_{12}a_2 - C_1 + \kappa\gamma_{10}\Big]\mathbf{d}\mathfrak{Z} + \gamma_{10}\mathbf{d}\mathfrak{W} \\
\mathbf{d}\beta_{11}(\mathfrak{Z}) &= -\frac{\partial \mathbb{H}}{\partial P_1}\mathbf{d}\mathfrak{Z} + \gamma_{11} \mathbf{d}\mathfrak{W} = \Big[\beta_{11}(\Delta + d) - \beta_{12}\Delta - C_3 + \kappa\gamma_{11}\Big]\mathbf{d}\mathfrak{Z} + \gamma_{11}\mathbf{d}\mathfrak{W} \\
\mathbf{d}\beta_{12}(\mathfrak{Z}) &= -\frac{\partial \mathbb{H}}{\partial P_2}\mathbf{d}\mathfrak{Z} + \gamma_{12} \mathbf{d}\mathfrak{W} = \Big[\beta_{12}(\mu + d) - \beta_1\mu + \kappa\gamma_{12}\Big]\mathbf{d}\mathfrak{Z} + \gamma_{12}\mathbf{d}\mathfrak{W}
\end{aligned}$$

As a consequence of the optimality criteria, we have

$$\frac{\partial \mathbb{H}}{\partial \Delta}\,|_{\Delta=\hat{\Delta}} = C_4\hat{\Delta}(\mathfrak{Z}) - \beta_{11}\hat{P}_1(\mathfrak{Z}) + \beta_{12}\hat{P}_1(\mathfrak{Z}) = 0$$

The identification indicated above suggests that

$$\hat{\Delta}(\mathfrak{Z}) = \frac{\beta_{11} - \beta_{12}}{C_4}\hat{P}_1$$

Hence,

$$\hat{\Delta}(\mathfrak{Z}) = \begin{cases} 0 & , \quad \frac{\partial \mathbb{H}}{\partial \Delta} > 0 \\ \frac{\beta_{11}-\beta_{12}}{C_4}\hat{P}_1 & , \quad \frac{\partial \mathbb{H}}{\partial \Delta} = 0 \\ 1 & , \quad \frac{\partial \mathbb{H}}{\partial \Delta} < 0 \end{cases}$$

Consequently, $\hat{\Delta}(\mathfrak{Z})$ is the optimal control, and it may be represented in the concise notation that follows:

$$\hat{\Delta}(\mathfrak{Z}) = \max\left[\min\left(0, \frac{\beta_{11} - \beta_{12}}{C_4}\hat{P}_1\right), 1\right]$$

Here, the optimum control is characterized by Formula (26) for $\hat{\Delta}$. The characterization of the optimal control (26), the adjoint system (24) and (25), the state system (22) with boundary conditions, and the optimality system can be solved to yield the optimum control and states. To solve the optimality system, we incorporate the characterization of the optimum control $\hat{\Delta}$ given by (26) along with the transversality and initial conditions. By changing the $\hat{\Delta}$ values in the control system, the following system is produced for the state system (22).

$$\begin{aligned}
\mathbf{d}S(\mathfrak{Z}) &= \Big[\eta_1 - \xi_1 I_1(\mathfrak{Z})S(\mathfrak{Z}) - \xi_2 I_2(\mathfrak{Z})S(\mathfrak{Z}) - \xi_3 I_3(\mathfrak{Z})S(\mathfrak{Z}) - \xi_4 I_4(\mathfrak{Z})S(\mathfrak{Z}) + \delta_1 I_1(\mathfrak{Z}) \\
&\quad + \delta_2 I_2(\mathfrak{Z}) + \delta_3 I_3(\mathfrak{Z}) + \delta_4 I_4(\mathfrak{Z}) + \mu P_2(\mathfrak{Z}) - (\alpha + d)S(\mathfrak{Z})\Big]\mathbf{d}\mathfrak{Z} + \kappa S(\mathfrak{Z})\mathbf{d}\mathfrak{W} \\
\mathbf{d}I_1(\mathfrak{Z}) &= \Big[\eta_2 + \xi_1 S(\mathfrak{Z})I_1(\mathfrak{Z}) - (\delta_1 + g_1 + d)I_1(\mathfrak{Z})\Big]\mathbf{d}\mathfrak{Z} + \kappa I_1(\mathfrak{Z})\mathbf{d}\mathfrak{W} \\
\mathbf{d}I_2(\mathfrak{Z}) &= \Big[\eta_3 + \xi_2 S(\mathfrak{Z})I_2(\mathfrak{Z}) - (\delta_2 + g_2 + d)I_2(\mathfrak{Z})\Big]\mathbf{d}\mathfrak{Z} + \kappa I_2(\mathfrak{Z})\mathbf{d}\mathfrak{W} \\
\mathbf{d}I_3(\mathfrak{Z}) &= \Big[\eta_4 + \xi_3 S(\mathfrak{Z})I_3(\mathfrak{Z}) - (\delta_3 + g_3 + d)I_3(\mathfrak{Z})\Big]\mathbf{d}\mathfrak{Z} + \kappa I_3(\mathfrak{Z})\mathbf{d}\mathfrak{W}
\end{aligned}$$

$$
\begin{aligned}
\mathbf{d}I_4(\mathfrak{Z}) &= \Big[\eta_5 + \xi_4 S(\mathfrak{Z}) I_4(\mathfrak{Z}) - (\delta_4 + g_4 + d) I_4(\mathfrak{Z})\Big]\mathbf{d}\mathfrak{Z} + \kappa I_4(\mathfrak{Z})\mathbf{d}\mathfrak{W} \\
\mathbf{d}Q_1(\mathfrak{Z}) &= \Big[\eta_6 + g_1 I_1(\mathfrak{Z}) - (\Theta_1 + d) Q_1(\mathfrak{Z})\Big]\mathbf{d}\mathfrak{Z} + \kappa Q_1(\mathfrak{Z})\mathbf{d}\mathfrak{W} \\
\mathbf{d}Q_2(\mathfrak{Z}) &= \Big[\eta_7 + g_2 I_2(\mathfrak{Z}) - (\Theta_2 + d) Q_2(\mathfrak{Z})\Big]\mathbf{d}\mathfrak{Z} + \kappa Q_2(\mathfrak{Z})\mathbf{d}\mathfrak{W} \\
\mathbf{d}Q_3(\mathfrak{Z}) &= \Big[\eta_8 + g_3 I_3(\mathfrak{Z}) - (\Theta_3 + d) Q_3(\mathfrak{Z})\Big]\mathbf{d}\mathfrak{Z} + \kappa Q3(\mathfrak{Z})\mathbf{d}\mathfrak{W} \\
\mathbf{d}Q_4(\mathfrak{Z}) &= \Big[\eta_9 + g_4 I_4(\mathfrak{Z}) - (\Theta_4 + d) Q_4(\mathfrak{Z})\Big]\mathbf{d}\mathfrak{Z} + \kappa Q_4(\mathfrak{Z})\mathbf{d}\mathfrak{W} \\
\mathbf{d}T(\mathfrak{Z}) &= \Big[\eta_{10} + \Theta_1 Q_1(\mathfrak{Z}) + \Theta_2 Q_2(\mathfrak{Z}) + \Theta_3 Q_3(\mathfrak{Z}) + \Theta_4 Q_4(\mathfrak{Z}) - (a_1 + a_2 + d) T(\mathfrak{Z})\Big]\mathbf{d}\mathfrak{Z} \\
&\quad + \kappa T(\mathfrak{Z})\mathbf{d}\mathfrak{W} \\
\mathbf{d}P_1(\mathfrak{Z}) &= \Bigg[\eta_{11} + a_1 T(\mathfrak{Z}) - \Bigg(\max\Bigg[\min\Bigg(0, \frac{\beta_{11} - \beta_{12}}{C_4}\hat{P}_1\Bigg), 1\Bigg] + d\Bigg) P_1(\mathfrak{Z})\Bigg]\mathbf{d}\mathfrak{Z} + \kappa P_1(\mathfrak{Z})\mathbf{d}\mathfrak{W} \\
\mathbf{d}P_2(\mathfrak{Z}) &= \Bigg[\eta_{12} + a_2 T(\mathfrak{Z}) + \max\Bigg[\min\Bigg(0, \frac{\beta_{11} - \beta_{12}}{C_4}\hat{P}_1\Bigg), 1\Bigg] P_1(\mathfrak{Z}) + \alpha S(\mathfrak{Z}) - (\mu + d) P_2(\mathfrak{Z})\Bigg]\mathbf{d}\mathfrak{Z} \\
&\quad + \kappa P_2(\mathfrak{Z})\mathbf{d}\mathfrak{W}
\end{aligned} \tag{27}
$$

The goal is to determine the optimal control and state system by numerically solving the aforementioned system (27). □

4. Results and Conclusions

4.1. Results Analysis

The deterministic MMCE model and Stochastic MMCE are numerically simulated using Euler and Euler–Maruyama schemes, respectively. The used time step takes the value 0.02, and the initial assumed values of the twelve dynamic servers states take the values 1000, 20, 25, 27, 35, 30, 18, 15, 23, 60, 40, and 80 for both suggested models, while the transmission rates η_1, η_2, η_3, η_4, η_5, η_6, η_7, η_8, η_9, η_{10}, η_{11}, η_{12}, ξ_1, ξ_2, ξ_3, ξ_4, δ_1, δ_2, δ_3, δ_4, g_1, g_2, g_3 , g_4, Θ_1, Θ_2, Θ_3, Θ_4, a_1, a_2, Δ, μ, α, d, and κ take the values 15, 5, 4, 3, 6, 7, 8, 10, 9, 3, 5, 8, 0.00013, 0.00012, 0.0001, 0.0001, 0.05, 0.05, 0.05, 0.05, 0.1, 0.15, 0.08, 0.06, 0.0244, 0.03, 0.02, 0.02,0.1,0.05, 0.07, 0.0365, 0.015, 0.025, and 0.035. These parameters values are obtained through trial-and-error techniques to visualize the models' behaviors. MATLAB software version 2021a is used to run all simulations.

The twelve dynamic states of the network that were produced by first simulating the deterministic MMCE model without taking any control into account are shown in Figure 3. It is clear that susceptible servers are presenting the higher core of total servers with a settling value 789, which means that a lot of servers will be at risk of exposure to infection. Following that, the protected servers have a settling value of about 700 servers. The four infected servers' categories—healthcare, educational, financial, and media servers—had settling values of 69, 30, 40, and 107, respectively, while the isolated servers of the same four categories had settling values of 240, 229, 292, and 341, respectively. The traced, semi-protected, and protected servers of all categories settled at 162, 222, and 706, respectively. The deterministic controlled MMCE model results are as shown in Figure 4. The susceptible server numbers decreased to a settling value 648 with a reduction of 18%. The four infected servers' categories settling values reduced to 55, 27, 33, and 85, with a total reduction of 20%, while the isolated servers of the same four categories had reduced settling values of

254, 220, 281, 315, with a total reduction of 3%. The traced servers had a value of 159, with a reduction of 2%. The semi-protected servers of the four categories increased significantly, by 263%, reaching a settling value of 807 servers. So, the controller action succeeded in increasing the total number of servers in the traced, isolated, and semi-protected states and decreased the total number of servers in the susceptible and infected states.

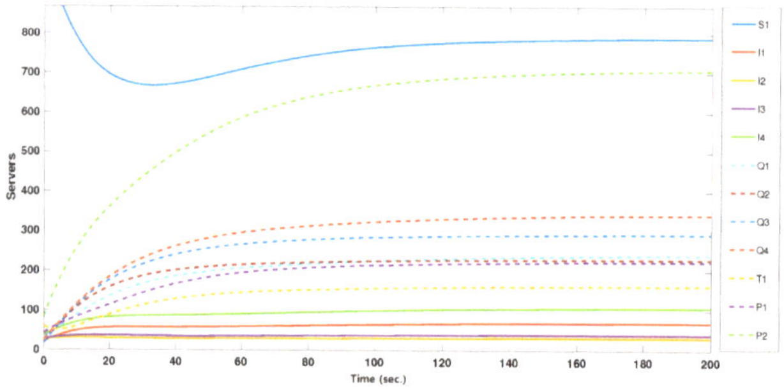

Figure 3. Deterministic MMCE model dynamics.

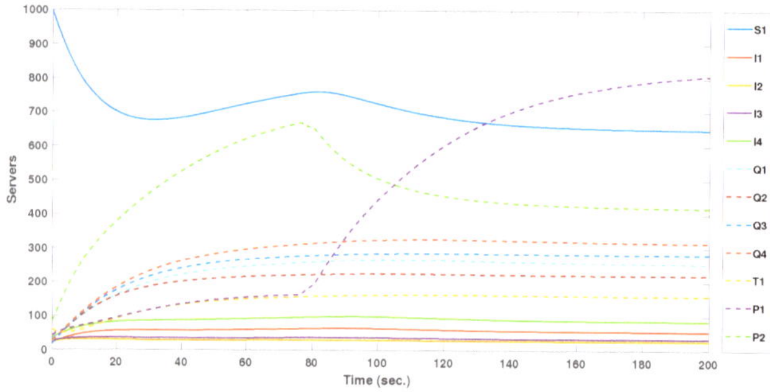

Figure 4. Controlled deterministic MMCE model dynamics.

The stochastic and stochastic controlled MMCE models results are as shown in Figures 5 and 6, respectively. The stochastic modeling process provides a more realistic representation of the servers' state dynamics. Starting by analyzing Figure 5, the susceptible and protected servers settled at a value of 978, while the four infected servers' categories settled at 101, 41, 54, and 153. And after control, as shown in Figure 6, the susceptible reduced to 717 servers with a reduction of 27%. In addition, the infected servers reduced to the settle values of 62, 35, 37, and 97. The semi-protected servers of the four categories increased significantly, by 166%, from a settle value of 325, as shown in Figure 5, to a controlled value of 863, as shown in Figure 6. So, the controller action also succeeded in maintaining a higher level of security in the stochastic MMCE by increasing the total number of servers on the traced, isolated, and semi-protected states and decreasing the total number of servers in the susceptible and infected states. Figures 7–13 show a visualization comparison of the susceptible, isolated, traced, and semi-protected server dynamics of all supposed models under study with and without control.

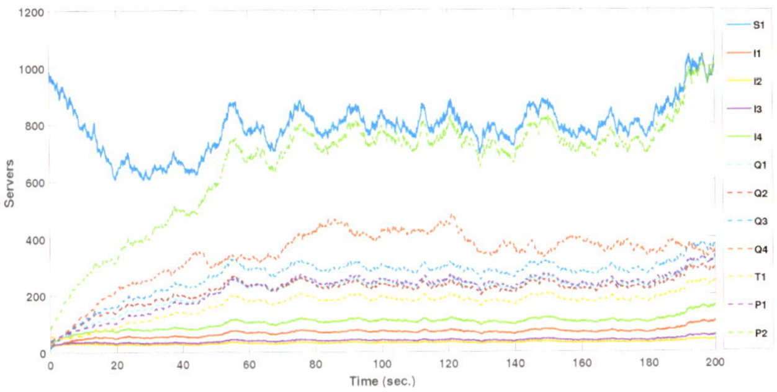

Figure 5. Stochastic MMCE model dynamics.

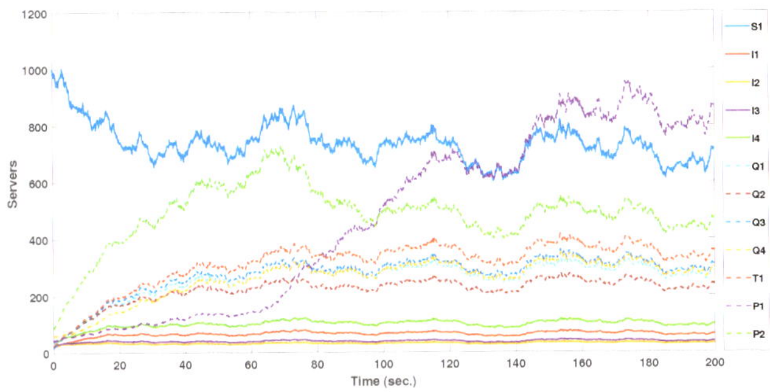

Figure 6. Controlled stochastic MMCE model dynamics.

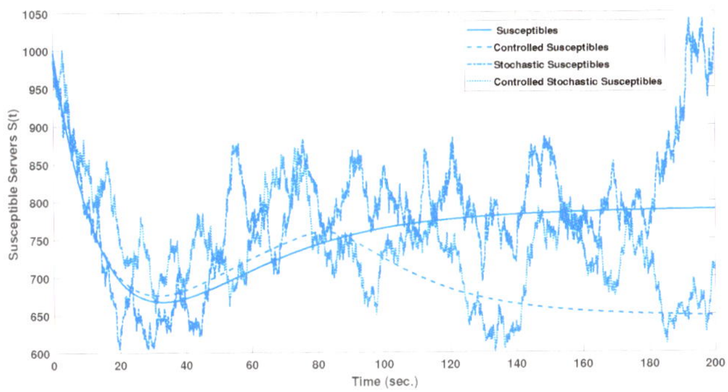

Figure 7. Susceptible server dynamics.

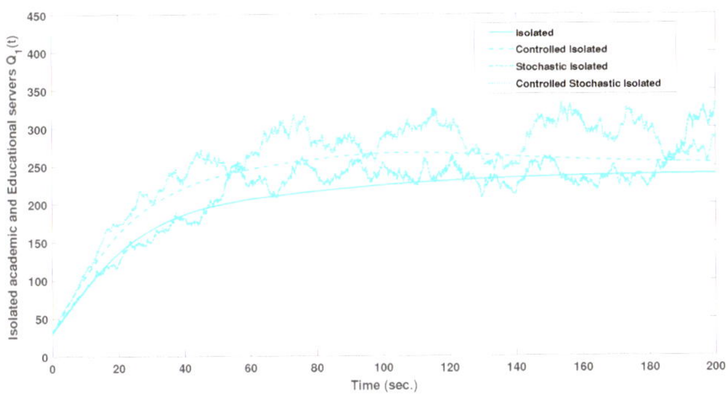

Figure 8. Isolated academic and educational server dynamics.

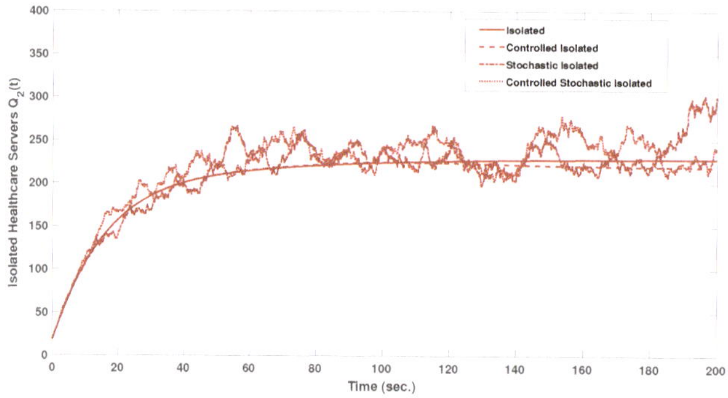

Figure 9. Isolated healthcare server dynamics.

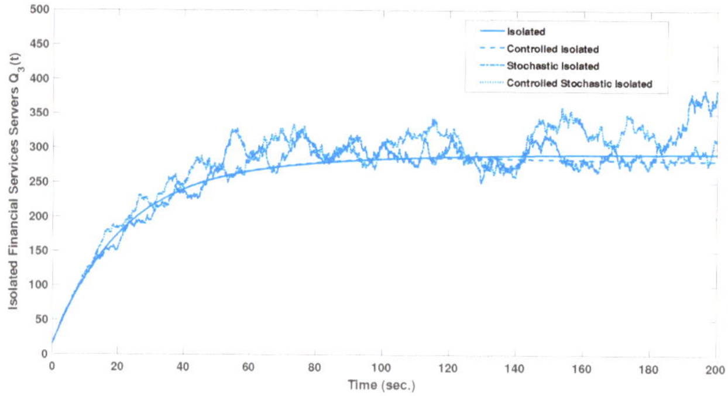

Figure 10. Isolated financial service server dynamics.

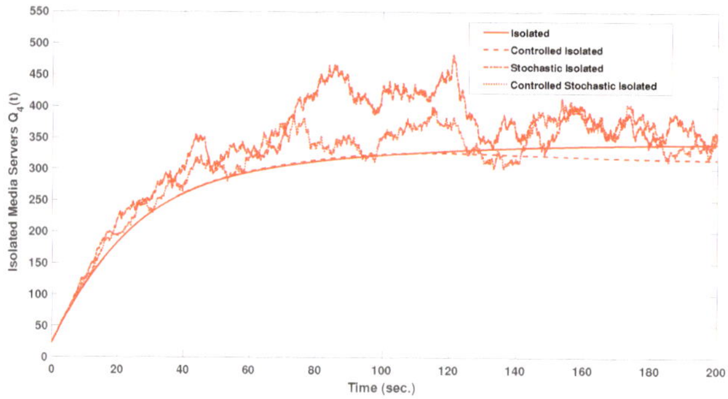

Figure 11. Isolated media server dynamics.

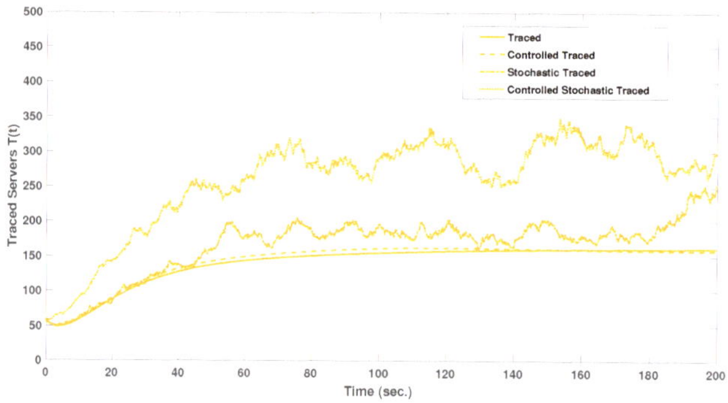

Figure 12. Traced server dynamics.

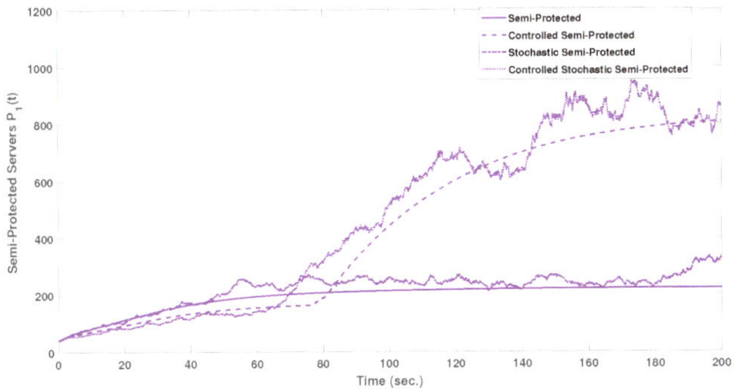

Figure 13. Semi-protected server dynamics.

Both of the controlled dynamic models in this study have the limitation that all transmission rates between the six different types of virtual machines (VMs) and their variety among media, financial, health, and educational servers have fixed values. The four public awareness coefficients were also taken into consideration as constants inside the maximizing objective functional under investigation when the control framework was imposed. Therefore, when applying these models over longer time periods, both transmission rates and awareness coefficients should be time-dependent to make the description more realistic. And for more generalization, if the controller has a multi-objective framework, modern numerical optimization algorithms can be applied.

4.2. Conclusions

To fully describe multi-cloud environments under malware propagation, this work obtained two state-of-the-art dynamical models: the deterministic MMCE model and the stochastic MMCE model. The dynamical network of system servers' states is categorized into six different states. Two of these states, isolated and infected, are scattered into four types of servers according to their terms of use: healthcare servers, academic and educational servers, financial services servers, and media servers. Then, a control function is recommended for application in both models to maintain a higher level of security in isolated, traced, and semi-protected states. After running both models with initial state values and using the numerical simulations, the obtained results showed the effectiveness of the recommended control strategy in decreasing the level of susceptible and infected servers, as well as increasing the level of security of the three mentioned states to a significant degree. The semi-protected servers increased by 263% for the deterministic MMCE and 166% for the stochastic MMCE model in terms of their settling values. In our future studies, we will work on modifying the proposed models considering fractional-order stochastic frameworks and then applying them to actual mutli-cloud environments for modeling validations.

Author Contributions: Conceptualization, O.A.M.O.; methodology, O.A.M.O.; software, O.A.M.O.; validation, H.M.A. and A.D.; formal analysis, A.M.S.A. and H.M.A.; investigation, A.M.S.A.; resources, A.D.; writing—original draft preparation, A.M.S.A.; writing—review and editing, T.A.N.; supervision, T.A.N. All authors have read and agreed to the published version of the manuscript.

Funding: This work was partially funded by Taif University, Saudi Arabia, Project No. (TU-DSPP-2024-46).

Data Availability Statement: Data sharing is not applicable to this article.

Conflicts of Interest: The authors declare no conflicts of interest.

References

1. Zhu, X.; Wang, J.; Guo, H.; Zhu, D.; Yang, L.T.; Liu, L. Fault-Tolerant Scheduling for Real-Time Scientific Workflows with Elastic Resource Provisioning in Virtualized Clouds. *IEEE Trans. Parallel Distrib. Syst.* **2016**, *27*, 3501–3517. [CrossRef]
2. Pluzhnik, E.; Nikulchev, E. Virtual laboratories in cloud infrastructure of educational institutions. In Proceedings of the 2nd International Conference on Emission Electronics (ICEE), Saint-Petersburg, Russia, 30 June–4 July 2014; pp. 1–3.
3. Ali, M.; Khan, S.U.; Vasilakos, A.V. Security in cloud computing: Opportunities and challenges. *Inform. Sci.* **2015**, *305*, 357–383. [CrossRef]
4. Ezhilchelvan, P.D.; Mitrani, I. Evaluating the probability of malicious coresidency in public clouds. *IEEE Trans. Cloud Comput.* **2015**, *5*, 420–427. [CrossRef]
5. Jansen, P.W.A. Cloud hooks: Security and privacy issues in cloud computing. In Proceedings of the 44th Hawaii International Conference on System Sciences (HICSS), Kauai, HI, USA, 4–7 January 2011; pp. 1–10.
6. Merabet, H.E.; Hajraoui, A. A survey of malware detection techniques based on machine learning. *Int. J. Adv. Comput. Sci. Appl.* **2019**, *10*. [CrossRef]
7. Lu, K.; Cheng, J.; Yan, A. Malware detection based on the feature selection of a correlation information decision matrix. *Mathematics* **2023**, *11*, 961. [CrossRef]
8. Zhu, Q.; Liu, Y.; Luo, X.; Cheng, K. A malware propagation model considering conformity psychology in social networks. *Axioms* **2022**, *11*, 632. [CrossRef]
9. Ye, X.; Xie, S.; Shen, S. SIR_1R_2: Characterizing malware propagation in wsns with second immunization. *IEEE Access* **2021**, *9*, 82083–82093. [CrossRef]
10. Dong, N.P.; Long, H.V.; Son, N.T.K. The dynamical behaviors of fractional-order SE_1E_2IQR epidemic model for malware propagation on wireless sensor network. *Commun. Nonlinear Sci. Numer. Simul.* **2022**, *111*, 106428. [CrossRef]
11. Al-Tuwairqi, S.M.; Bahashwan, W.S. The impact of quarantine strategies on malware dynamics in a network with heterogeneous immunity. *Math. Model. Anal.* **2022**, *27*, 282–302. [CrossRef]
12. Martin del Rey, A.; Hernandez, G.; Bustos Tabernero, A.; Queiruga Dios, A. Advanced malware propagation on random complex networks. *Neurocomputing* **2021**, *423*, 689–696. [CrossRef]
13. Piqueira, J.R.C.; Cabrera, M.A.; Batistela, C.M. Malware propagation in clustered computer networks. *Phys. A Stat. Mech. Appl.* **2021**, *573*, 125958. [CrossRef]
14. Omar, O.A.; Elbarkouky, R.A.; Ahmed, H.M. Fractional stochastic models for COVID-19: Case study of Egypt. *Results Phys.* **2021**, *23*, 104018. [CrossRef] [PubMed]
15. Ahmed, A.M.S.; Ahmed, H.M.; Nofal, T.A.; Darwish, A.; Omar, O.A.M. Hilfer-Katugampola fractional epidemic model for malware propagation with optimal control. *Ain Shams Eng. J.* **2024**, *15*, 102945. [CrossRef]
16. Omar, O.A.M.; Alnafisah, Y.; Elbarkouky, R.A.; Ahmed, H.M. COVID-19 deterministic and stochastic modelling with optimized daily vaccinations in Saudi Arabia. *Results Phys.* **2021**, *28*, 104629. [CrossRef] [PubMed]
17. Jafar, M.T.; Yang, L.-X.; Li, G.; Zhu, Q.; Gan, C. Minimizing Malware Propagation in Internet of Things Networks: An Optimal Control Using Feedback Loop Approach. *IEEE Trans. Inf. Forensics Secur.* **2024**, *19*, 9682–9697. [CrossRef]
18. Jafar, M.T.; Yang, L.X.; Li, G.; Zhu, Q.; Gan, C.; Yang, X. Malware containment with immediate response in IoT networks: An optimal control approach. *Comput. Commun.* **2024**, *228*, 107951. [CrossRef]
19. Abazari, F.; Analoui, M.; Takabi, H. Dynamical propagation model of malware for cloud computing security. *Comput. Secur.* **2016**, *58*, 139–148. [CrossRef]
20. Gan, C.; Feng, Q.; Zhang, X.; Zhang, Z.; Zhu, Q. Effect of anti-malware software on infectious nodes in cloud environment. *IEEE Access* **2020**, *8*, 20325–20333. [CrossRef]
21. Ndaïrou, F.; Area, I.; Nieto, J.J.; Torres, D.F.M. Mathematical modeling of COVID-19 transmission dynamics with a case study of Wuhan. *Chaos Solit. Fract.* **2020**, *135*, 10846. [CrossRef]
22. Mao, X.; Marion, G.; Renshaw, E. Environmental Brownian noise suppresses explosions in population dynamics. *Stoch. Process. Appl.* **2002**, *97*, 95–110. [CrossRef]
23. May, R. *Stability and Complexity in Model Ecosystems*; Princeton University: Princeton, NJ, USA, 1973.
24. Li, R.; Guo, X. Dynamics of a stochastic SEIR epidemic model with vertical transmission and standard incidence. *Mathematics* **2024**, *12*, 359. [CrossRef]

Mathematical and Computational Applications

MDPI

Article

Reduced-Order Model for Cell Volume Homeostasis: Application to Aqueous Humor Production

Riccardo Sacco [1,2,*], Greta Chiaravalli [3], Giovanna Guidoboni [4], Anita Layton [5], Gal Antman [6,7], Keren Wood Shalem [1], Alice Verticchio [1], Brent Siesky [1] and Alon Harris [1]

1 Department of Ophthalmology, Icahn School of Medicine at Mount Sinai, 1468 Madison Avenue, Annenberg 22-86, New York, NY 10029, USA; keren.woodshalem@mssm.edu (K.W.S.); alice.verticchio@mssm.edu (A.V.); brent.siesky@mssm.edu (B.S.); alon.harris@mssm.edu (A.H.)

2 Dipartimento di Matematica, Politecnico di Milano, Piazza Leonardo da Vinci 32, 20133 Milano, Italy

3 Dipartimento di Fisica, Politecnico di Milano, Piazza Leonardo da Vinci 32, 20133 Milano, Italy; greta.chiaravalli@polimi.it

4 Maine College of Engineering and Computing, University of Maine, 168 College Avenue, Orono, ME 04469, USA; giovanna.guidoboni@maine.edu

5 Department of Applied Mathematics, University of Waterloo, 200 University Ave W, Waterloo, ON N2L 3G1, Canada; anita.layton@uwaterloo.ca

6 Department of Ophthalmology, Rabin Medical Center, Zeev Jabotinsky St 39, Petah Tikva 4941492, Israel; antmangal@gmail.com

7 Faculty of Medicine, Tel Aviv University, Klachkin 35 Street, P.O. Box 39040, Tel Aviv 6997801, Israel

* Correspondence: riccardo.sacco@mssm.edu

Abstract: The ability of a cell to keep its volume constant irrespective of intra- and extracellular conditions is essential for cellular homeostasis and survival. The purpose of this study is to elaborate a theoretical model of cell volume homeostasis and to apply it to a simulation of human aqueous humor (AH) production. The model assumes a cell with a spherical shape and only radial deformation satisfying the property that the cell volume in rest conditions equals that of the cell couplets constituting the ciliary epithelium of the human eye. The cytoplasm is described as a homogeneous mixture containing fluid, ions, and neutral solutes whose evolution is determined by net production mechanisms occurring in the intracellular volume and by water and solute exchange across the membrane. Averaging the balance equations over the cell volume leads to a coupled system of nonlinear ordinary differential equations (ODEs) which are solved using the θ-method and the Matlab function `ode15s`. Simulation tests are conducted to characterize the set of parameters corresponding to baseline conditions in AH production. The model is subsequently used to investigate the relative importance of (a) impermeant charged proteins; (b) sodium–potassium (Na+/K+) pumps; (c) carbonic anhydrase (CA) in the AH production process; and (d) intraocular pressure. Results suggest that (a) and (b) play a role; (c) lacks significant weight, at least for low carbon dioxide values; and (d) plays a role for the elevated values of intraocular pressure. Model results describe a higher impact from charged proteins and Na+/K+ ATPase than CA on AH production and cellular volume. The computational virtual laboratory provides a method to further test in vivo experiments and machine learning-based data analysis toward the prevention and cure of ocular diseases such as glaucoma.

Keywords: homeostasis; cell volume; homogeneous mixtures; mathematical modeling; aqueous humor production

1. Introduction

The mechanical integrity of a cell is essential to its function and the tissues and organ systems that it supports [1,2]. Cell volume stabilization is a fundamental function to preserve the mechanical integrity of a cell [3]. Cell volume stabilization over time is based on the balance among forces acting on the cell membrane from both intra- and extracellular sides, which emanate from electrochemical and fluid–mechanical mechanisms [4]. The experimental characterization of force equilibrium at the nanoscale level is a highly complex task, on the one hand, because of the difficulty of performing measurements in vivo (see [5]) and, on the other hand, the presence of two regimes of mechanical behavior in living cells, on short and long time scales (see [6]). In fact, a combination of experimentation and theoretical analysis shows that living mammalian cytoplasm behaves as an equilibrium material on short time scales whereas it behaves as an out-of-equilibrium material on long time scales [6].

Based on these considerations, in this article we address the theoretical study of cell volume dynamics and apply the formulation to the process of the production of aqueous humor (AH) in the eye. AH is a slowly moving fluid that is continuously produced by the ciliary body of the eye, whose main function is to keep the anterior segment of the eye clean from metabolic wastes and to preserve the spherical shape of the eyeball by establishing the intraocular pressure (IOP) (see [7–10]).

The clinical motivation of our work is the fact that an elevated value of intraocular pressure (IOP) is an established risk factor for glaucoma. Therefore, keeping IOP within normal levels (10 to 21 mmHg) is central to prevent the occurrence and progression of neurodegenerative diseases of the retina such as glaucoma [11–13]. IOP is the result of the balance between the production and drainage of aqueous humor (AH) in the anterior segment of the eye. AH is produced by the ciliary epithelium (CE) double cell layer, which lines the ciliary processes of the ciliary body, and is drained out from the anterior chamber of the eye throughout the trabecular meshwork and uveoscleral outflow pathways. In this article, we focus our attention on the process of AH production with the goal of characterizing the fluid–mechanical and electrochemical conditions under which the cells constituting the CE maintain their volume, allowing at the same time stationary water flow from the ciliary body toward the posterior and anterior chamber of the eye. Understanding these conditions endows clinicians with a supporting tool to identify the causes of a pathological increase (or decrease) in AH production which may give rise to a potentially pathological increase in IOP, the final aim being to devise effective therapies for its reduction [14]. In this respect, the development of a mathematical model and of a computational virtual laboratory (CVL) for the simulation of AH production has been the object of investigation in recent years (see [15–18]).

In this article, we propose a theoretical formulation based on homogeneous mixtures including neutral and charged solutes (see [19] (Chapter 13)) and utilizing a model reduction procedure from three spatial dimensions to zero spatial dimensions. The model resulting from the reduction is a coupled system of highly stiff, nonlinear, ordinary differential equations (ODEs) which are solved using the θ-method and the `Matlab` function `ode15s`. Simulation tests are run to characterize the values of model parameters in baseline conditions in AH production. The model is subsequently used to investigate the relative importance of (a) impermeant charged proteins; (b) the sodium–potassium pump; (c) carbonic anhydrase; and (d) intraocular pressure in the AH production process. Results suggest that (a) and (b) play a role; (c) does not have a significant weight, at least for low CO_2 values; and (d) plays a role for elevated values of intraocular pressure, as in the case of hypertensive patients.

2. Materials and Methods

In Section 2.1, we describe the cellular mechanisms of AH production. In Section 2.2, we describe the geometrical representation of the cell volume and surface; in Section 2.3, we describe the heterogeneous structure of the cell membrane; and in Section 2.4, we provide a conceptual scheme of the transport of water and solute across the cell membrane. In Section 2.5, we introduce a continuum model of the cell based on the theory of mixtures to describe the spatial and temporal response of the cell volume to the concomitant action of fluid and electrochemical forces. In Section 2.6, we derive a reduced-order model of the cell volume that comprises ordinary differential and algebraic equations. Finally, in Sections 2.7 and 2.8, we describe the compact form of the reduced-order model and its numerical approximation, respectively. In Appendices A and B, we provide the expressions of the fluid velocity, molar flux densities, and net production rates that are involved in cellular metabolism.

2.1. Cellular Mechanisms of Aqueous Humor Production

Aqueous humor (AH) is a slowly moving fluid that is (1) continuously produced by the ciliary body epithelium of the eye (CE); (2) flows from the posterior to the anterior chamber of the eye; and (3) is drained out from the eye throughout two main outflow pathways (see [7–10]). The sequence (1) AH production, (2) AH flow, and (3) AH outflow is schematically represented in Figure 1. The main function of AH is to keep the anterior segment of the eye clean from the metabolic waste of the surrounding tissues and to preserve the spherical shape of the eyeball by establishing the intraocular pressure (IOP) as the value of the AH fluid pressure. This corresponds to the volumetric flow rate of AH that is produced by the ciliary body and equals the volumetric flow rate of AH that is drained out by the outflow pathways in the anterior segment of the eye (see [20]).

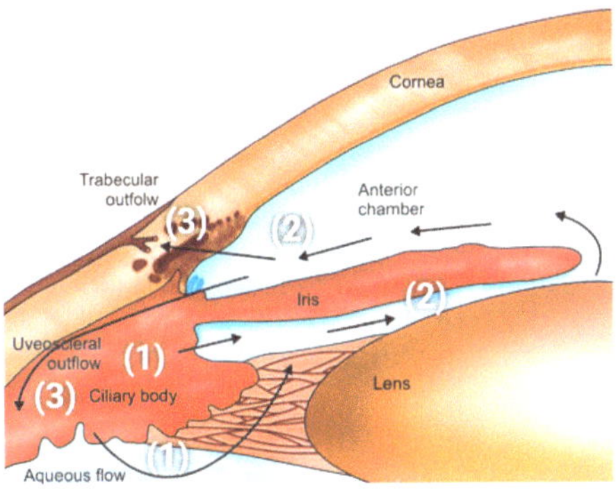

Figure 1. A schematic representation of the processes involved in AH dynamics. (1): AH production; (2): AH flow; (3): AH outflow. Reprinted from R. Ramakrishnan et al., *Diagnosis & Management of Glaucoma*, Chapter 9 Aqueous Humor Dynamics, 10.5005/jp/books/11801_9, (2013) [21]; used in accordance with the Creative Commons Attribution (CC BY) license.

The left panel of Figure 2 is a micrograph photo of a histology of the CE. The CE is a double layer of apex-to-apex connected cells (pigmented cell, PE, and nonpigmented cell, NPE) which have cuboidal and columnar shapes, respectively (see [22]). The CE cell couplet is schematically illustrated in the compartment-based representation shown in the right panel of Figure 2. This scheme may be used to characterize the various cellular mechanisms that concur in the process of AH production (see [9,20]). The first mechanism is the transport of the "metabolic fuel", a mixture constituted by water, solutes, ions, and proteins, by blood flow in the ciliary capillary (CC). The second mechanism is the ultrafiltration of the metabolic fuel across the fenestrations in the CC wall. The third mechanism is the motion of the metabolic fuel across the stromal tissue. Then, the metabolic fuel divides across two different pathways. The first pathway is the intracellular transport throughout the CE cell couplet; the second pathway is the paracellular transport throughout the lateral interstitial space separating two neighbouring CE cell couplets. In its motion across the CE, the osmolarity gradient that is established between the cytoplasm of the NPE cell and the interstitial space, on the one hand, and between the NPE cell and the posterior chamber (PC), on the other hand, gives rise to a water efflux into the PC which eventually drains out through the trabecular meshwork and uveoscleral outflow pathways.

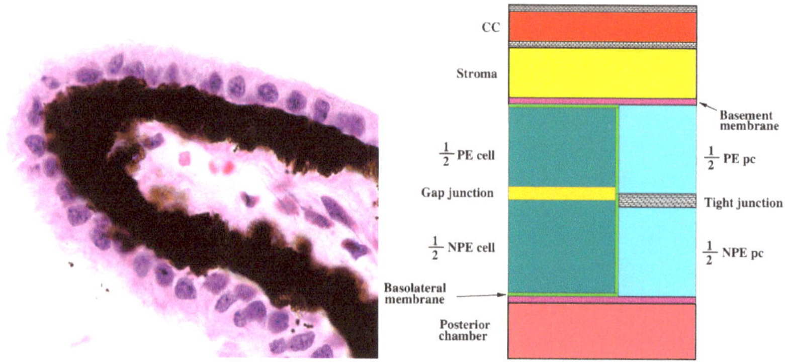

Figure 2. Left panel: a light micrograph of the ciliary body epithelium. It consists of two epithelial cell layers: a non-pigmented inner layer and an outer pigmented layer. Under the epithelium, there is a highly vascularized stroma. Reprinted from [23]; used in accordance with the Creative Commons Attribution (CC BY) license. **Right** panel: a compartmental representation of the CE cell couplet.

2.2. *Geometrical Description of the Cell*

Figure 3 illustrates the "equivalent cell", representing a simplified model of the CE cell couplet illustrated in Section 2.1. The "equivalent cell" has the property that its volume is equal to the volume of the PE/NPE cell couplet, and the reason supporting the choice of a spherical shape is that the PE/NPE cell couplet acts as a functional syncytium [8] so that the process of AH production can be mathematically described as the collective contribution of all the cell couplets that constitute the CE. Despite its over-simplified geometry, we believe that our model may serve as a basis for extension to future closer representations of the morphology and physiology of the CE.

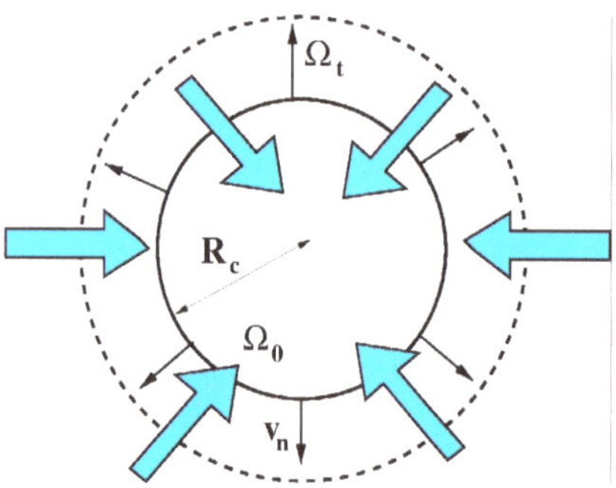

Figure 3. The "equivalent cell". Solid line: initial cell configuration. Dashed line: deformed cell configuration. The cyan arrows indicate water flow. The cell is increasing its volume (cell swelling).

Let $t \geq 0$ denote the time variable (units: s). We denote by $\Omega_t := \Omega(t)$ the three-dimensional (3D) body representing the cell at the time t and $\mathcal{V}_t := \mathcal{V}(t)$, the volume of the cell at the time t, such that $\mathcal{V}_0 = 4\pi R_c^3/3$ is the value of the cell volume at the time $t = 0$, with R_c being the cell radius in the initial (undeformed) configuration (units: m). From the point of view of Continuum Mechanics, the body is a deformable, electrically charged, homogeneous mixture comprising a fluid constituent, with N_α moving charged solute constituents (ions) and N_β moving neutral solute constituents (see [19] (Chapter 13)). The cell volume also contains impermeant charged proteins whose chemical valency and molar density are such that electroneutrality holds in ionic homeostasis conditions (see [3,24]). The boundary of Ω_t is denoted by $\partial\Omega_t$ and $\mathbf{n}$ is the outward unit normal vector for $\partial\Omega_t$. The surface of the cell at the time t is denoted by $\mathcal{S}_t := \mathcal{S}(t)$, such that $\mathcal{S}_0 = 4\pi R_c^2$ is the value of the cell surface at the time $t = 0$. In the remainder of this article, we denote by $\mathbf{x}$ the spatial coordinate vector of any point in the cell with respect to a fixed system of reference. We also refer to the interior of Ω_t as the intracellular region (shortly, *in*), to the exterior of Ω_t as the extracellular region (shortly, *out*), and to the surface of Ω_t as the membrane (shortly, *m*).

Assumption 1. *We assume that cell deformation occurs only in the radial direction and that it does not depend on the position on the cell surface.*

Assumption 2. *We assume that the electric potential, solute concentrations, and fluid pressure in the extracellular region are given functions of space and time.*

The solid line and dashed line in Figure 3 illustrate the initial and deformed configurations of the cell, respectively. In agreement with Assumption 1, both configurations are spherical.

2.3. Porous and Lipid Structure of Cell Membrane

The intra- and extracellular regions are separated by a 3D membrane whose thickness, t_m, is such that the ratio $\delta := t_m/R_c$ is $\ll 1$. In the limit $\delta \to 0$, the geometrical representation of the 3D membrane degenerates into the two-dimensional (2D) manifold $\partial\Omega_t$. The membrane surface can be represented as a heterogeneous porous medium whose solid constituent is lipid material. The lipid surface is impermeable to water and solutes, whereas the pore surface allows the exchange of fluid and solutes between intra- and extracellular regions. The pore surface contains a variety of membrane proteins, denoted by mp. Each membrane protein, mp, is characterized by a dimensionless function, Φ^{mp}, henceforth referred to as the surface fraction, defined as

$$\Phi^{mp}(t) := \frac{\mathcal{S}^{mp}(t)}{\mathcal{S}(t)} \qquad \mathcal{S}(t) > 0, \quad t \geq 0, \tag{1a}$$

where $\mathcal{S}^{mp}(t)$ is the total area occupied by the protein mp at the time t.

Assumption 3. *Let $\mathcal{S}_t > 0$. We assume that*

$$\frac{d\Phi^{mp}(t)}{dt} = 0 \qquad t > 0. \tag{1b}$$

Equation (1b) *implies that*

$$\Phi^{mp}(t) = \Phi^{mp}(0) = \frac{\mathcal{S}^{mp}(0)}{\mathcal{S}(0)} \qquad t \geq 0. \tag{1c}$$

The first type of membrane protein that we introduce in this article is an aquaporin (AQP), which is a specialized protein for rapid transmembrane water exchange between intracellular and extracellular regions (see [25–27]). The quantity Φ^{AQP} represents the AQP surface fraction. Other membrane proteins on the cell surface comprise carrier proteins, which permit neutral solute exchange through the cell membrane by the mechanism of facilitated diffusion (see [28,29]), and ion channels, ion exchangers, and ion pumps, which permit charged solute exchange through the cell membrane (see [30]). The total surface fractions for each considered membrane protein are defined as

$$\Phi^{carr} = \sum_{\beta \in S_\beta} \Phi_\beta^{carr}, \tag{2a}$$

$$\Phi^{ch} = \sum_{\alpha \in S_\alpha} \Phi_\alpha^{ch}, \tag{2b}$$

$$\Phi^{exch} = \sum_{\alpha \in S_\alpha} \Phi_\alpha^{exch}, \tag{2c}$$

$$\Phi^{pump} = \sum_{\alpha \in S_\alpha} \Phi_\alpha^{pump}, \tag{2d}$$

in such a way that the total pore surface fraction is

$$\Phi_{tot}^{p} = \Phi^{carr} + \Phi^{ch} + \Phi^{exch} + \Phi^{pump} + \Phi^{AQP}. \tag{3}$$

The remainder of the cell surface is occupied by the lipid constituent of the membrane, whose surface fraction is

$$\Phi^{lip} = 1 - \Phi_{tot}^{p}. \tag{4}$$

Assumption 3 implies that the membrane protein surface fractions in (2) do not depend on the time and also that the lipid surface fraction does not depend on the time because of (4).

2.4. Water and Solute Transport Across the Cell Membrane

In this section, we provide a conceptual scheme of the transport of water and solute across the cell membrane and we refer to [26,31] for more details about the subject.

Figure 4 illustrates a schematic representation of a zoomed view of the cell membrane and of the intra- and extracellular regions. The scheme shows water molecules and solutes (neutral and charged) in motion across the membrane.

Assumption 4. *Based on the scheme in Figure 4, we make the following assumptions:*

A1. *Water is transported across the membrane, in a selective manner, via aquaporins;*

A2. *Water and solutes are co-transported across the membrane via ionic channels and carrier proteins;*

A3. *Water velocity inside ionic channels and carrier proteins is the same as water velocity inside aquaporins.*

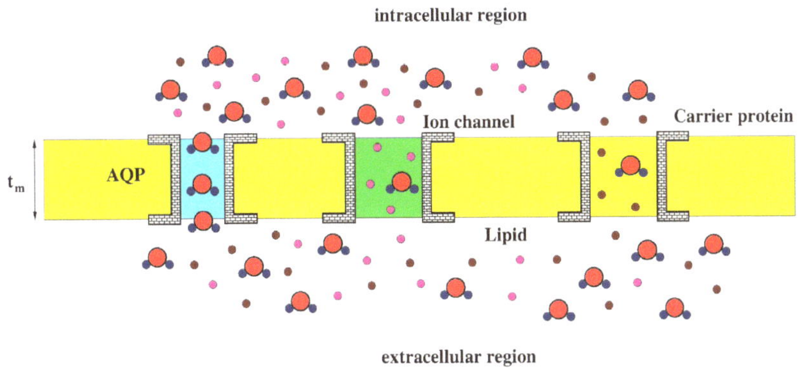

Figure 4. A schematic representation of the structure of the cell membrane. AQP: aquaporin (cyan). The ion channel is drawn in green. Water molecules (red and dark blue), charged solutes (magenta), and neutral solutes (brown) are illustrated. The lipid constituent is drawn in yellow. The AQP is selective to water molecules whereas the ion channel permits the co-transport of ions and water.

2.5. Continuum-Based Model of Cell

This section is structured with four subsections devoted to fluid motion, neutral and charged solutes, and electric potential.

2.5.1. Fluid Motion

Assumption 5. *We assume that water flow is slow and inertial forces can be neglected with respect to viscous effects.*

According to Assumption 5, the motion of water flow across the cell membrane can be described by the linear Stokes equation system (see [19] (Section 10.5.1)):

$$\nabla \cdot \mathbf{v} = 0, \tag{5a}$$

$$\nabla \cdot \mathbb{T} + \mathbf{b} = \mathbf{0}, \tag{5b}$$

$$\mathbb{T} = -p\mathbb{I} + 2\mu_f \nabla_S \mathbf{v}, \tag{5c}$$

where $\mathbf{v}$ (units: $\mathrm{m\,s^{-1}}$) and p (units: $\mathrm{N\,m^{-2}}$) are the fluid velocity and pressure inside the aquaporin, respectively, $\mathbb{T}$ is the fluid stress tensor (units: $\mathrm{N\,m^{-2}}$), $\mathbf{b}$ is the force density acting on the fluid (units: $\mathrm{N\,m^{-3}}$), μ_f is the fluid dynamic viscosity (units: $\mathrm{N\,m^{-2}\,s} = \mathrm{Pa\,s}$), and ∇_S is the symmetric gradient operator. Equation (5a) expresses mass conservation in a local form; Equation (5b) expresses linear momentum balance in a local form; and Equation (5c) is the constitutive law for a Newtonian fluid.

2.5.2. Neutral Solutes' Motion

We denote by S_β the set of neutral solutes, with $\mathrm{card}(S_\beta) = N_\beta$. Each solute, $\beta \in S_\beta$, is described by its number density, n_β, and molar density, c_β (units: $\mathrm{mol\,m^{-3}} = \mathrm{mM}$). The advection–diffusion model is adopted to represent the motion of neutral solutes across the cell membrane (see [19] (Chapter 12)):

$$\frac{\partial c_\beta}{\partial t} + \nabla \cdot \mathbf{j}_\beta = P_\beta - C_\beta \qquad \beta = 1, \ldots, N_\beta, \tag{6a}$$

$$\mathbf{j}_\beta = c_\beta \mathbf{v} - D_\beta \nabla c_\beta \qquad \beta = 1, \ldots, N_\beta. \tag{6b}$$

Equation (6a) expresses mass balance in a local form for each solute, $\beta \in S_\beta$. The vector $\mathbf{j}_\beta$ is the molar flux density of the solute β (units: $\mathrm{mol\,m^{-2}\,s^{-1}} = \mathrm{mM\,m\,s^{-1}}$), accounting for a passive advective contribution due to water motion inside the carrier protein channel (see Assumption 4-A2) and a diffusive contribution expressed by Fick's law in which D_β is the diffusion coefficient of the solute β (units: $\mathrm{m^2\,s^{-1}}$). The scalar functions P_β and C_β represent the production and consumption rates of c_β (units: $\mathrm{mol\,m^{-3}\,s^{-1}} = \mathrm{mM\,s^{-1}}$).

2.5.3. Charged Solutes' Motion

We denote by S_α the set of charged solutes (ions) with $\mathrm{card}(S_\alpha) = N_\alpha$. Each ion, $\alpha \in S_\alpha$, is described by its number density, n_α, molar density, c_α, charge number, z_α, diffusion coefficient, D_α, and electric mobility, μ_α^{el} (units: $\mathrm{m^2\,V^{-1}\,s^{-1}}$). The number and molar densities are related by the equation $n_\alpha = N_{\mathrm{Av}} c_\alpha$ (units: $\mathrm{m^{-3}}$), where $N_{\mathrm{Av}} = 6.02214076 \cdot 10^{23}\,\mathrm{mol^{-1}}$ is Avogadro's constant. The diffusion coefficient and the electric mobility are proportional through Einstein's relation:

$$D_\alpha = \frac{\mu_\alpha^{el} V_{th}}{|z_\alpha|} \qquad \alpha = 1, \ldots, N_\alpha, \tag{7}$$

where $V_{th} = (K_B T)/q$ is the thermal voltage (units: V) and K_B, T, and q denote Boltzmann's constant, absolute temperature, and electron charge, respectively. The Nernst–Planck (NP) model is adopted to represent ion transport and exchange across the cell membrane (see [19] (Chapter 13)):

$$\frac{\partial c_\alpha}{\partial t} + \nabla \cdot \mathbf{j}_\alpha = P_\alpha - C_\alpha \qquad \alpha = 1, \ldots, N_\alpha, \tag{8a}$$

$$\mathbf{j}_\alpha = \mathbf{j}_\alpha^p + \mathbf{j}_\alpha^a \qquad \alpha = 1, \ldots, N_\alpha, \tag{8b}$$

$$\mathbf{j}_\alpha^p = \mathbf{j}_\alpha^{p,edw} + \mathbf{j}_\alpha^{p,exch} \qquad \alpha = 1, \ldots, N_\alpha, \tag{8c}$$

$$\mathbf{j}_\alpha^{p,edw} = c_\alpha \mathbf{v}_\alpha - D_\alpha \nabla c_\alpha \qquad \alpha = 1, \ldots, N_\alpha, \tag{8d}$$

$$\mathbf{v}_\alpha = \mathbf{v} + \mu_\alpha^{el} \frac{|z_\alpha|}{z_\alpha} \mathbf{E}. \tag{8e}$$

Equation (8a) expresses mass balance in a local form for each ion $\alpha \in S_\alpha$. Equation (8b) is the constitutive law to describe ion transport and contains two main contributions to the vector $\mathbf{j}_\alpha$ which is the molar flux density of the ion α. The first contribution accounts for passive ion transport and is represented by the ion molar flux density $\mathbf{j}_\alpha^p$. The second contribution, which is a novel aspect introduced by the model proposed in the present work, accounts for active ion transport and is represented by the ion molar flux density $\mathbf{j}_\alpha^a$. The vector $\mathbf{j}_\alpha^{p,edw}$ is the velocity-extended NP equation for ion electrodiffusion and accounts for a passive advective contribution and a diffusive contribution expressed by Fick's law (see [19] (Chapter 13)). The vector field $\mathbf{v}_\alpha$ is the generalized drift velocity of the ion α due to water motion inside the ion channel (see Assumption 4-A2) and Faraday's law under quasi-static approximation (see [19] (Chapter 14))

$$\mathbf{E} = -\nabla \psi, \tag{8f}$$

where $\mathbf{E}$ and ψ are the electric field (units: $\mathrm{V\,m^{-1}}$) and electric potential (units: V), respectively. The vector $\mathbf{j}_\alpha^{p,exch}$ expresses passive ion exchange mediated by transporters and co-transporters. The scalar functions P_α and C_α represent the production and consumption rates of c_α.

2.5.4. Electric Potential

In the classical VE-PNP equation system, the Poisson equation,

$$\nabla \cdot (-\varepsilon_m \nabla \psi) = \rho^{el}, \tag{9}$$

is used to determine the spatial distribution of the electric potential, with ε_m denoting the dielectric permittivity of the medium (units: $\mathrm{F\,m^{-1}}$) and ρ^{el} denoting the electric charge density (units: $\mathrm{C\,m^{-3}}$). Since, in the present article, we are interested in the study of the spatial and temporal distribution of the electric potential ψ across the cell membrane, we replace (9) (which, for a given ρ^{el}, is an elliptic equation) with the following differential system:

$$\nabla \cdot \mathbf{J}_{tot} = 0, \tag{10a}$$

$$\mathbf{J}_{tot} = \mathbf{J}^{disp} + \mathbf{J}^{cond}_{tot}, \tag{10b}$$

$$\mathbf{J}^{disp} = \frac{\partial}{\partial t}(\varepsilon_m \mathbf{E}) = \frac{\partial}{\partial t}(-\varepsilon_m \nabla \psi), \tag{10c}$$

$$\mathbf{J}^{cond}_{tot} = F \sum_{\alpha \in S_\alpha} z_\alpha \mathbf{j}_\alpha, \tag{10d}$$

where F is Faraday's constant (units: $\mathrm{C\,mol^{-1}}$). Equation (10a) is a consequence of Maxwell's equations (see [19,32] (Chapter 4)) and expresses the continuity of the total electric current density $\mathbf{J}_{tot}$ (units: $\mathrm{C\,m^{-2}\,s^{-1}} = \mathrm{A\,m^{-2}}$), given by the sum of the total conduction current density $\mathbf{J}^{cond}_{tot}$ and the displacement current density $\mathbf{J}^{disp}$, in which the electric field $\mathbf{E}$ is expressed as a function of the electric potential ψ via Equation (8f). The adoption of System (10) to determine the electric potential ψ is a novel aspect of this article and extends to the time-dependent case of the approach originally proposed in [33] to enforce electroneutrality in the mathematical study of cellular electric activity.

2.6. Reduced-Order Cell Model

This section is structured with five subsections devoted to the derivation of the reduced-order model for the cell surface normal velocity, cell volume, neutral and charged solute molar densities, and electric potential.

2.6.1. Time Evolution of Normal Velocity Across Single AQP

We make the following assumption about the geometry of the AQP shown in Figure 4.

Assumption 6. *The AQP is geometrically represented as a cylinder, ω_p, with radius r_p and axial length t_m, as depicted in Figure 5.*

Moreover, we make the following assumption:

Assumption 7. *We assume the following:*

1. *The fluid velocity only has the axial component V_s;*
2. *$V_s = v_{p,n}(t)\eta(r)$, with $t \geq 0$ and $r \in [0, r_p]$;*
3. *$\eta(r) = 2\left[1 - (r/r_p)^2\right]$ (Poiseuille flow);*
4. *The force density only has the axial component b_s;*
5. *$b_s = b_{p,s}(s,t)$, with $s \in [0, t_m]$ and $t \geq 0$.*

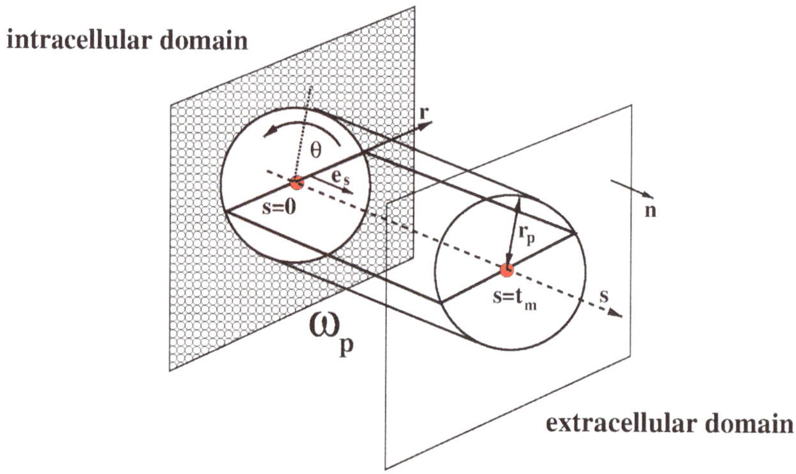

Figure 5. A three-dimensional schematic representation of an aquaporin. The cylindrical domain ω_p is the pore channel, t_m is the membrane thickness, and r_p is the aquaporin radius.

Inserting Assumption 7 into Equation (5), we obtain the following expression of the normal fluid velocity inside the pore channel of the AQP:

$$v_{p,n}(t) = \frac{r_p^2}{8\mu_f}\left[-\frac{\partial p(s,t)}{\partial s} + b_{p,s}(s,t)\right] \qquad s \in [0, t_m], \ t \geq 0. \tag{11a}$$

Integrating (11a), we obtain

$$p(s,t) = p(0,t) - \frac{8\mu_f v_{p,n}(t)}{r_p^2} s + \int_0^s b_{p,s}(\xi,t)\, d\xi \qquad s \in [0, t_m], \ t \geq 0, \tag{11b}$$

For any function, $U = U(t)$, we define the transmembrane difference operator:

$$\Delta U(t) := U_{in}(t) - U_{out}(t) \qquad t \geq 0. \tag{11c}$$

Setting $s = t_m$ in (11b) and applying the definition (11c) to the variables p and Π, we obtain

$$v_{p,n}(t) = \mathcal{L}_p[\Delta p(t) - \Delta\Pi(t)] \qquad t \geq 0, \tag{11d}$$

where

$$\mathcal{L}_p := \frac{r_p^2}{8\mu_f t_m} \tag{11e}$$

is the hydraulic conductivity of the membrane (units: $\mathrm{m(Pa\,s)^{-1}}$) and

$$\Pi(t) := \int b_{p,s}(\xi,t)\, d\xi + C \tag{11f}$$

is the total osmo-oncotic pressure, with C being an arbitrary integration constant. Relation (11d) is Starling's equation [34] and represents the mathematical expression of the normal fluid velocity across a single aquaporin.

2.6.2. Time Evolution of Cell Surface Normal Velocity

In order to determine the motion of the whole cell surface, we need to define an average normal fluid velocity to represent the collective contribution of the AQPs. For this purpose, we denote by σ_{AQP} the surface density of aquaporins, defined as the number, N_{AQP}, of AQPs per square meter, and by $\mathcal{S}^{tot}_{AQP}(t)$ the total surface area occupied by the AQPs. Using Assumption 3, we obtain

$$\Phi^{AQP} = \frac{\mathcal{S}^{tot}_{AQP}(0)}{\mathcal{S}_0} = \frac{\sigma_{AQP}\,\mathcal{S}_0\,\pi r_p^2}{\mathcal{S}_0} = \sigma_{AQP}\pi r_p^2. \tag{12a}$$

Let us introduce the total water volumetric flow rate across the cell surface (units: $\mathrm{m^3\,s^{-1}}$),

$$Q_w(t) := \int_{\partial\Omega_t} \mathbf{v}(\mathbf{x},t)\cdot\mathbf{n}\,d(\partial\Omega_t) = v_{p,n}(t)\mathcal{S}^{tot}_{AQP}(t) \qquad t \geq 0, \tag{12b}$$

and the average normal fluid velocity across the cell surface,

$$\langle \mathbf{v}(\mathbf{x},t)\cdot\mathbf{n}\rangle(t) := \frac{Q_w(t)}{\mathcal{S}_t} = \frac{\mathcal{S}^{tot}_{AQP}(t)}{\mathcal{S}_t} v_{p,n}(t) = \Phi^{AQP} v_{p,n}(t) \qquad t \geq 0, \tag{12c}$$

where Φ^{AQP} is given by (12a) and $v_{p,n}$ is given by (11d).

The pore fluid velocity $v_{p,n}$ is typically quite large (even thousands of $\mathrm{\mu m\,s^{-1}}$; see [35]) to favor a fast transmembrane exchange of water molecules. The average normal fluid velocity $\langle \mathbf{v}\cdot\mathbf{n}\rangle$, instead, is considerably smaller (less that 1 $\mathrm{\mu m\,s^{-1}}$; see [36]). For this reason, in the remainder of this article, we introduce the following definition.

Definition 1. *Let us denote by $v_{cell,n}$ the cell surface normal velocity. We set*

$$v_{cell,n}(t) := \langle \mathbf{v}(\mathbf{x},t)\cdot\mathbf{n}\rangle = \Phi^{AQP} v_{p,n}(t) \qquad \mathbf{x} \in \partial\Omega_t,\ t \geq 0, \tag{13}$$

where $v_{p,n}$ is defined in (11d).

2.6.3. Time Evolution of Cell Volume

The time evolution of $\mathcal{V}_t$ is described by the cell mass balance equation in an integral form (see [19] (Chapter 6)):

$$\frac{dM(t)}{dt} = -\int_{\partial\Omega_t} \rho_w \mathbf{v}(\mathbf{x},t)\cdot\mathbf{n}\,d(\partial\Omega_t) + \int_{\Omega_t} R_w(\mathbf{x},t)\,d\Omega_t \qquad t \geq 0, \tag{14}$$

where ρ_w is the mass density of water (units: $\mathrm{kg\,m^{-3}}$), $M(t) = \int_{\Omega_t} \rho_w\,d\Omega_t$ is the mass of the cell at the time t (units: kg), and

$$R_w(\mathbf{x},t) := P_w(\mathbf{x},t) - C_w(\mathbf{x},t) \qquad \mathbf{x} \in \Omega_t,\ t \geq 0 \tag{15}$$

is the intracellular water mass density net production rate, with P_w and C_w being the water mass density production and consumption rates, respectively (units: $\mathrm{kg\,m^{-3}\,s^{-1}}$).

Assumption 8. *We assume that*

$$P_w(\mathbf{x},t) = \mathcal{P}_w(t)\Phi(\mathbf{x}) \qquad \mathbf{x} \in \Omega_t,\ t \geq 0, \tag{16a}$$

$$C_w(\mathbf{x},t) = \mathcal{C}_w(t)\Phi(\mathbf{x}) \qquad \mathbf{x} \in \Omega_t,\ t \geq 0, \tag{16b}$$

where $\mathcal{P}_w$ and $\mathcal{C}_w$ represent the time-dependent intracellular water mass density production and consumption rates, and the shape function Φ is such that

$$\int_{\Omega_t} \Phi(\mathbf{x})\, d\Omega_t = \mathcal{V}_t. \tag{16c}$$

Using Assumption 4-A1, using Definitions 1 and 8 in the right-hand side of (14), and noting that $\mathcal{S}(t) = \gamma(\mathcal{V}(t))^{2/3}$, with $\gamma := (36\pi)^{1/3}$, we obtain the following reduced-order model of cell volume motion:

$$\frac{d\mathcal{V}(t)}{dt} = -\gamma\, j_{v,n}(t)(\mathcal{V}(t))^{2/3} + \mathcal{R}_w(t)\mathcal{V}(t) \qquad t > 0, \tag{17a}$$

$$j_{v,n}(t) := v_{cell,n}(t) = \Phi^{AQP} v_{p,n}(t) \qquad t > 0, \tag{17b}$$

$$\mathcal{V}(0) = \mathcal{V}_0, \tag{17c}$$

where $v_{p,n}$ is given by (11d) and

$$\mathcal{R}_w(t) := k_{w,prod}(t) - k_{w,cons}(t) \qquad t \geq 0 \tag{17d}$$

is the water volume net production rate (units: s^{-1}), with $k_{w,prod}(t) := \mathcal{P}_w(t)/\rho_w$ and $k_{w,cons}(t) := \mathcal{C}_w(t)/\rho_w$ set for every $t \geq 0$.

2.6.4. Time Evolution of Neutral Solutes

The time evolution of neutral solute molar density is described by the mass balance equation in an integral form:

$$\int_{\Omega_t} \frac{\partial c_\beta}{\partial t}\, d\Omega_t = -\int_{\partial\Omega_t} j_{\beta,n}\, d(\partial\Omega_t) + \int_{\Omega_t} R_\beta\, d\Omega_t \qquad \beta \in S_\beta, \tag{18}$$

where $j_{\beta,n}$ is the normal molar flux density of c_β over the cell surface $\mathcal{S}_t$, and

$$R_\beta(\mathbf{x},t) := P_\beta(\mathbf{x},t) - C_\beta(\mathbf{x},t) \qquad \mathbf{x} \in \Omega_t,\ t \geq 0 \tag{19}$$

is the net production rate of the intracellular solute β, with P_β and C_β being the production and consumption rates, respectively (units: $\mathrm{mM\,s^{-1}}$).

Assumption 9. *We assume that*

$$c_\beta(\mathbf{x},t) = c^\beta(t)\Phi(\mathbf{x}) \qquad \mathbf{x} \in \Omega_t,\ t \geq 0, \tag{20a}$$

$$j_{\beta,n}(\mathbf{x},t) = j_n^\beta(t)\eta_\beta(\mathbf{x}) \qquad \mathbf{x} \in \partial\Omega_t,\ t \geq 0, \tag{20b}$$

where c^β and j_n^β are time-dependent intracellular neutral solute molar densities and normal molar flux densities on the cell surface, respectively, and the shape function η_β is such that

$$\int_{\partial\Omega_t} \eta_\beta(\mathbf{x})\, d(\partial\Omega_t) = \Phi_\beta^{carr}\mathcal{S}_t, \tag{20c}$$

where Φ_β^{carr} is the surface fraction of the carrier protein that allows the transmembrane exchange of the neutral solute β. We also assume that equations similar to (16) *apply to P_β and C_β.*

Using Assumption 9 in (18), we obtain the following reduced-order model for each neutral solute molar density:

$$\frac{dc^\beta(t)}{dt} = -\gamma\, j^{tot}_{\beta,n}(t)(\mathcal{V}(t))^{-1/3} + \mathcal{R}_\beta(t) \qquad \beta \in S_\beta,\ t > 0, \tag{21a}$$

$$j^{tot}_{\beta,n}(t) := \Phi^{carr}_\beta j^\beta_n(t) \qquad \beta \in S_\beta,\ t > 0, \tag{21b}$$

$$c^\beta(0) = c^\beta_0, \tag{21c}$$

where c^β_0 is the initial value of the intracellular neutral solute molar density, $\beta \in S_\beta$, and

$$\mathcal{R}_\beta(t) := \mathcal{P}_\beta(t) - \mathcal{C}_\beta(t) \qquad \beta \in S_\beta,\ t > 0, \tag{21d}$$

is the net production rate of the solute β (units: $\mathrm{mM\,s^{-1}}$).

2.6.5. Time Evolution of Charged Solutes

The time evolution of the charged solute molar density is described by the mass balance equation in an integral form:

$$\int_{\Omega_t} \frac{\partial c_\alpha}{\partial t}\, d\Omega_t = -\int_{\partial\Omega_t} j_{\alpha,n}\, d(\partial\Omega_t) + \int_{\Omega_t} R_\alpha\, d\Omega_t \qquad \alpha \in S_\alpha, \tag{22}$$

where $j_{\alpha,n}$ is the normal molar flux density of c_α on the cell surface $\mathcal{S}_t$, and

$$R_\alpha(\mathbf{x},t) := P_\alpha(\mathbf{x},t) - C_\alpha(\mathbf{x},t) \qquad \mathbf{x} \in \Omega_t,\ t \geq 0 \tag{23}$$

is the net production rate of the intracellular ion α, with P_α and C_α being the production and consumption rates, respectively (units: $\mathrm{mM\,s^{-1}}$).

Assumption 10. *We assume that*

$$c_\alpha(\mathbf{x},t) = c^\alpha(t)\Phi(\mathbf{x}) \qquad \mathbf{x} \in \Omega_t,\ t \geq 0, \tag{24a}$$

$$j_{\alpha,n}(\mathbf{x},t) = j^{ch}_{\alpha,n}(t)\eta^{ch}_\alpha(\mathbf{x}) + j^{exch}_{\alpha,n}(t)\eta^{exch}_\alpha(\mathbf{x}) + j^{pump}_{\alpha,n}(t)\eta^{pump}_\alpha(\mathbf{x}) \qquad \mathbf{x} \in \partial\Omega_t,\ t \geq 0, \tag{24b}$$

where c^α is the time-dependent intracellular molar density of the ion $\alpha \in S_\alpha$; the functions $j^{ch}_{\alpha,n}$ and $j^{exch}_{\alpha,n}$ are time-dependent normal molar flux densities on the cell surface representing passive ion transport and exchange, respectively; the function $j^{pump}_{\alpha,n}$ is a time-dependent normal molar flux density on the cell surface representing ion exchange through active pumps; and the shape functions η^{ch}_α, η^{exch}_α, and η^{pump}_α are such that

$$\int_{\partial\Omega_t} \eta^{ch}_\alpha(\mathbf{x})\, d(\partial\Omega_t) = \Phi^{ch}_\alpha \mathcal{S}_t, \tag{24c}$$

$$\int_{\partial\Omega_t} \eta^{exch}_\alpha(\mathbf{x})\, d(\partial\Omega_t) = \Phi^{exch}_\alpha \mathcal{S}_t, \tag{24d}$$

$$\int_{\partial\Omega_t} \eta^{pump}_\alpha(\mathbf{x})\, d(\partial\Omega_t) = \Phi^{pump}_\alpha \mathcal{S}_t, \tag{24e}$$

where Φ^{ch}_α, Φ^{exch}_α, and Φ^{pump}_α are the surface fractions of the membrane protein associated with the passive transport, passive exchange, and active pump-mediated exchange of the ion α, respectively. We also assume that equations similar to (8) apply to P_α and C_α.

Using Assumption 10 in (22), we obtain the following reduced-order model for each charged solute molar density:

$$\frac{dc^{\alpha}(t)}{dt} = -\gamma\, j^{tot}_{\alpha,n}(t)(\mathcal{V}(t))^{-1/3} + \mathcal{R}_{\alpha}(t) \qquad \alpha \in S_{\alpha},\, t > 0, \tag{25a}$$

$$j^{tot}_{\alpha,n}(t) := \left[\Phi^{ch}_{\alpha} j^{ch}_{\alpha,n}(t) + \Phi^{exch}_{\alpha} j^{exch}_{\alpha,n}(t) + \Phi^{pump}_{\alpha} j^{pump}_{\alpha,n}(t)\right] \qquad \alpha \in S_{\alpha},\, t > 0, \tag{25b}$$

$$c^{\alpha}(0) = c^{\alpha}_{0}, \tag{25c}$$

where c^{α}_{0} is the initial value of the intracellular charged solute molar density, $\alpha \in S_{\alpha}$, and

$$\mathcal{R}_{\alpha}(t) := \mathcal{P}_{\alpha}(t) - \mathcal{C}_{\alpha}(t) \qquad \alpha \in S_{\alpha},\, t > 0, \tag{25d}$$

is the net production rate of the ion α (units: $\mathrm{mM\,s^{-1}}$).

2.6.6. Time Evolution of Membrane Potential

For any $t \geq 0$, the membrane potential is defined as

$$\psi_m(t) := \psi^{in}(t) - \psi^{out}(t). \tag{26}$$

The time evolution of the membrane potential is described by the charge balance equation in an integral form:

$$\int_{\partial\Omega_t} -\varepsilon_m \frac{\partial}{\partial t}\left(\frac{\partial\psi}{\partial n}\right) d(\partial\Omega_t) = -\int_{\partial\Omega_t} J_{cond,n}\, d(\partial\Omega_t) \tag{27}$$

where $\frac{\partial\psi}{\partial n}$ and $J_{cond,n}$ are the normal derivative of the electric potential and the normal total conduction current density on the cell surface $\mathcal{S}_t$, respectively.

Assumption 11. *We assume that the electric potential is a piecewise linear continuous function across the cell membrane thickness (see Figure 6).*

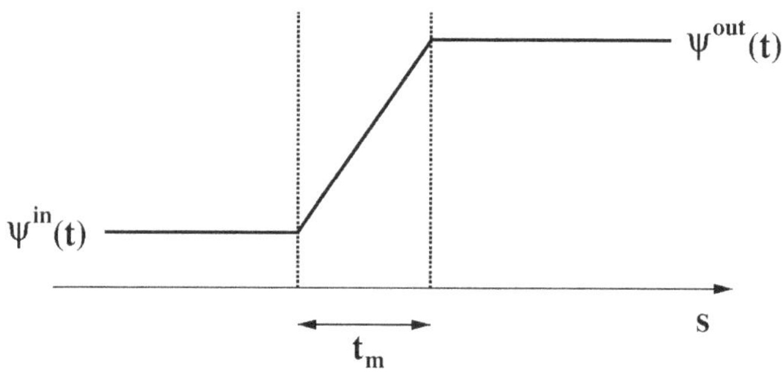

Figure 6. Transmembrane electric potential.

Using Assumption 11 in the left-hand side of (27), we obtain

$$\int_{\partial\Omega_t} -\varepsilon_m \frac{\partial}{\partial t}\left(\frac{\partial\psi}{\partial n}\right) d(\partial\Omega_t) = c_m \frac{d\psi_m(t)}{dt} \Phi^{lip} \mathcal{S}(t) \qquad t \geq 0, \tag{28}$$

where

$$c_m := \frac{\varepsilon_m}{t_m} \tag{29}$$

is the cell specific capacitance (units: $\mathrm{F\,m^{-2}}$). The right-hand side of (27) is given by the following expression:

$$-\int_{\partial\Omega_t} J_{cond,n}\, d(\partial\Omega_t) = -I^{tot}_{cond}(t) \qquad t \geq 0, \tag{30}$$

where I^{tot}_{cond} is the total conduction current flowing across the cell membrane at the time t (units: A), defined as

$$I^{tot}_{cond}(t) = \left[\sum_{\alpha\in S_\alpha} F\, z_\alpha j^{tot}_{\alpha,n}(t)\right] \mathcal{S}(t) \qquad t \geq 0. \tag{31}$$

Replacing (28) and (30) in (27), we obtain the following reduced-order model for the membrane potential:

$$\frac{d\psi_m(t)}{dt} = -\left(\Phi^{lip}\, c_m\right)^{-1} j^{\psi}_n(t) \qquad t > 0, \tag{32a}$$

$$j^{\psi}_n(t) := j^{tot}_{cond,n}(t) \qquad t > 0, \tag{32b}$$

$$\psi_m(0) = \psi_{m,0}, \tag{32c}$$

where

$$j^{tot}_{cond,n}(t) := \sum_{\alpha\in S_\alpha} F z_\alpha j^{tot}_{\alpha,n}(t) \qquad t \geq 0 \tag{32d}$$

is the normal total conduction current density over the cell surface defined in (25b), and $\psi_{m,0}$ is the initial value of the membrane potential.

2.7. Compact Form of Reduced-Order Cell Model

Let us introduce the vector of dependent variables:

$$\mathbf{Y}(t) := \begin{bmatrix} \mathcal{V}(t) \\ \mathbf{c}_\beta(t) \\ \mathbf{c}_\alpha(t) \\ \psi_m(t) \end{bmatrix} \qquad t \geq 0, \tag{33}$$

where

$$\mathbf{c}_\beta(t) := \left[c_{\beta,1}(t), \ldots, c_{\beta,N_\beta}(t)\right]^T \qquad t \geq 0, \tag{34}$$

$$\mathbf{c}_\alpha(t) := \left[c_{\alpha,1}(t), \ldots, c_{\alpha,N_\alpha}(t)\right]^T \qquad t \geq 0, \tag{35}$$

are the column vectors containing the time values of the neutral and charged solute molar densities.

Let us define the vector of source terms:

$$\mathbf{F}(t, \mathbf{Y}(t)) := \begin{bmatrix} -\gamma\, j_{v,n}(t)(\mathcal{V}(t))^{2/3} + \left(k_{w,prod}(t) - k_{w,cons}(t)\right)\mathcal{V}(t) \\ -\gamma\, \mathbf{j}^{tot}_{\beta,n}(t)(\mathcal{V}(t))^{-1/3} + (\mathbf{P}_\beta(t) - \mathbf{C}_\beta(t)) \\ -\gamma\, \mathbf{j}^{tot}_{\alpha,n}(t)(\mathcal{V}(t))^{-1/3} + (\mathbf{P}_\alpha(t) - \mathbf{C}_\alpha(t)) \\ -\left(c_m\, \Phi^{lip}\right)^{-1} j^{\psi}_n(t) \end{bmatrix} \qquad t \geq 0, \tag{36}$$

where $\mathbf{j}^{tot}_{\beta,n}$ and $\mathbf{j}^{tot}_{\alpha,n}$ are column vectors of the sizes N_β and N_α, respectively, containing the time values of the neutral and charged solute normal molar flux densities, whereas

$\mathbf{P}_\beta$ (and $\mathbf{C}_\beta$) and $\mathbf{P}_\alpha$ (and $\mathbf{C}_\alpha$) are column vectors of the sizes N_β and N_α, respectively, containing the time values of the intracellular neutral and charged solute production (and consumption) terms.

Finally, let us define the column vector containing the initial condition of the model:

$$\mathbf{Y}_0 := \begin{bmatrix} \mathcal{V}_0 \\ \mathbf{c}_0^\beta \\ \mathbf{c}_0^\alpha \\ \psi_{m,0} \end{bmatrix}, \tag{37}$$

where $\mathbf{c}_0^\beta$ and $\mathbf{c}_0^\alpha$ are column vectors of the sizes N_β and N_α, respectively, which contain the values of the initial conditions for the intracellular neutral and charged solute molar densities.

The equation system constituting the model of cell volume motion can be written in a compact form as

$$\frac{d\mathbf{Y}(t)}{dt} = \mathbf{F}(t, \mathbf{Y}(t)) \qquad t > 0, \tag{38a}$$

$$\mathbf{Y}(0) = \mathbf{Y}_0. \tag{38b}$$

System (38) comprises differential equations and algebraic equations, represented by the constitutive laws for the normal molar flux densities of neutral and charged solutes and for the production and consumption rates of water and solutes. The expressions of the normal fluid velocity on the cell surface and normal molar flux densities are provided in Appendix A whereas the expressions of the production and consumption rates are provided in Appendix B.

2.8. Numerical Approximation

The mathematical model proposed in this article was implemented within a computational virtual laboratory (CVL) for the simulation of cell volume motion.

The θ-method (see [19] (Chapeter 3)) and the `Matlab` tool `ode suite` were used for the numerical approximation of (38) (see [37]). In the case of the θ-method, the values $\theta = 1$ and $\theta = 0.5$ were utilized, with $\theta = 1$ corresponding to the Backward Euler (BE) method and $\theta = 0.5$ corresponding to the Crank–Nicolson (CN) method. In the case of the `Matlab` tool `ode suite`, the functions `ode15s` or `ode23tb` were utilized because they are specifically tailored to solve stiff problems like the object of the present article (see [38] (Chapter 11)). The function `ode15s` is endowed with a variable-order Backward Differentiation Formulae (BDF) method, whereas the function `ode23tb` uses the trapezoidal rule coupled with a BDF of the order 3.

The CVL allows the user to consider physical situations of increasing complexity, starting from the solution of the sole Cauchy problem (17) in which the fluid velocity is a given function of time and the production and consumption terms are functions of the time and cell volume. Simulation complexity can be increased by adding charged and neutral solutes, intracellular reactions, and transmembrane ion exchange mechanisms, as well as the presence of impermeant protein charges in both intra- and extracellular regions. This hierarchy of scenarios is investigated in Section 3 where the accuracy and reliability of the model predictions is verified against analytical solutions and available data.

3. Results

In this section, we validate the proposed CVL through the solution of case studies characterized by increasing complexity. In Section 3.1, we consider a reduced version of the model described in the previous pages, solely accounting for the cell volume equation. In Section 3.2, we use the full model and the concept of the "equivalent cell" introduced in Section 2.2 to simulate the process of AH production.

3.1. The Basic Configuration

This case study is mathematically represented by the sole model for cell volume motion (17).

Assumption 12. *We make the following assumptions:*

1. $v_{p,n}(t) = \overline{v}$ *for all* $t \geq 0$, *with* $\overline{v}$ *being a given constant;*
2. $k_{w,prod}(t) = \kappa_a$ *for* $t \geq 0$, *with* κ_a *being a given positive constant (units:* s^{-1}*);*
3. $k_{w,cons}(t) = \kappa_d \frac{\mathcal{V}(t)}{\mathcal{V}_{ref}}$ *for* $t \geq 0$, *with* κ_d *(units:* s^{-1}*) and* $\mathcal{V}_{ref}$ *(units:* m^3*) being given positive constants.*

Replacing the assumptions about the water volume production and consumption rates in (17d), we obtain

$$\mathcal{R}_w(t) = \kappa_a - \kappa_d \frac{\mathcal{V}(t)}{\mathcal{V}_{ref}} \qquad t \geq 0. \tag{39}$$

Replacing Assumption 12 and (39) in (17a), we obtain the following Cauchy problem:

$$\frac{d\mathcal{V}(t)}{dt} = -\gamma\,\Phi^{AQP}\overline{v}(\mathcal{V}(t))^{2/3} + \left(\kappa_a - \kappa_d \frac{\mathcal{V}(t)}{\mathcal{V}_{ref}}\right)\mathcal{V}(t) \qquad t > 0, \tag{40a}$$

$$\mathcal{V}(0) = \mathcal{V}_{ref}, \tag{40b}$$

where $\mathcal{V}_{ref} = (4\pi/3)R_{cell}^3$ is the cell volume in resting conditions.

The equilibrium points of System (40) are the solution of the following nonlinear algebraic equation:

$$f(x) = -\gamma\,\Phi^{AQP}\overline{v}x^{2/3} + \mathcal{V}_{ref}^{1/3}x(\kappa_a - \kappa_d x) = 0, \tag{41}$$

where $x := \mathcal{V}_\infty/\mathcal{V}_{ref}$ is the dependent variable and $\mathcal{V}_\infty$ is the stationary value of the cell volume.

Figure 7 (left panel) shows a graph of f, corresponding to the following values of the input data: $R_{cell} = 10 \cdot 10^{-6}$ m; $\Phi^{AQP} = 0.5$; $\overline{v} = -30 \cdot 10^{-6}\ \mathrm{m\,s^{-1}}$; $\kappa_a = 1\ \mathrm{s}^{-1}$; and $k_d = 3\ \mathrm{s}^{-1}$. System (40) admits two equilibrium points: $\mathcal{V}_{\infty,1} = 0\ \mathrm{m}^3$ and $\mathcal{V}_{\infty,2} = 1.6124\mathcal{V}_{ref} = 6.754 \cdot 10^{-15}\ \mathrm{m}^3$. We have

$$\left.\frac{df}{dx}\right|_{x=0} = +\infty, \qquad \left.\frac{df}{dx}\right|_{x=1.6124} = -9.86 \cdot 10^{-5}\ \mathrm{m\,s^{-1}},$$

from which we can conclude that $\mathcal{V}_{\infty,2}$ is the only stable equilibrium point of System (40). The theoretical conclusions of the stability analysis are confirmed by Figure 7 (left panel) which shows the normalized cell volume computed by solving System (40) in the time interval [0, 10] s with the following values of the model parameters: $R_{cell} = 10 \cdot 10^{-6}$ m; $\Phi^{AQP} = 0.5$; $\overline{v} = -30 \cdot 10^{-6}\ \mathrm{m\,s^{-1}}$; $\kappa_a = 1\ \mathrm{s}^{-1}$; and $k_d = 3\ \mathrm{s}^{-1}$. Cell volume dynamics were studied using the BE method with a uniform time partition made of 10^5 elements. Consistent with the physical configuration illustrated in Figure 3, the cell tends to increase its volume by more than 60% because of the inflow of water from the extracellular region.

The cell volume reaches a stationary limit because of intracellular water consumption dominating over intracellular water production.

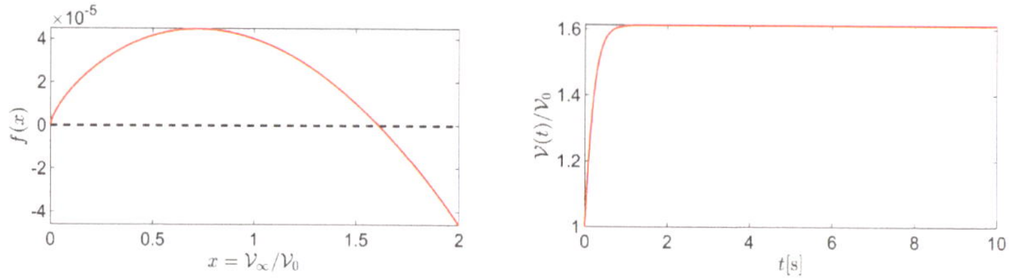

Figure 7. Left panel: a plot of $f(x)$. **Right** panel: a plot of $\mathcal{V}(t)/\mathcal{V}_{ref}$ in the time interval $[0, 10]$ s. The values of the input data are as follows: $R_{cell} = 10 \times 10^{-6}$ m, $\Phi^{AQP} = 0.5$; $\overline{v} = -30 \times 10^{-6}$ m s^{-1}; $\kappa_a = 1$ s^{-1}; and $k_d = 3$ s^{-1}.

Figure 8 (left panel) illustrates the temporal evolution of cell volume in the case where the values of $\overline{v}$ are $[-30 - 18, -6, +6, +18, +30] \times 10^{-6}$ m s^{-1}. The results suggest that the cell switches between swelling and shrinking as fluid velocity changes its sign from negative to positive. Interestingly, the BE method gives rise to positive cell volumes for each $t \geq 0$ and for each considered value of $\overline{v}$. This outcome is the consequence of the positivity principle enjoyed by the BE method when applied to the linear equation $y'(t) = -\lambda y(t)$ for $t \geq 0$, with λ being a given positive constant. The difference between using the BE method ($\theta = 1$) and another θ-method with $\theta \in [0, 1)$ is illustrated in Figure 8 (right panel) which shows a comparison between the BE method (red line) and the Crank–Nicolson method (CN, blue line) in the solution of (40) when the time interval is $[0, 1]$ s, the fluid velocity is $\overline{v} = 100 \times 10^{-6}$ m s^{-1}, and the number of time elements is 10, corresponding to the time step $\Delta t = 0.1$ s. The values of the remaining model parameters are the same as in the previous example. The results indicate that the solution computed by the CN method is affected by spurious unphysical oscillations, unlike that computed by the BE method. Such oscillations can be removed by increasing the number of discretization time elements, at the price of increasing the computational effort.

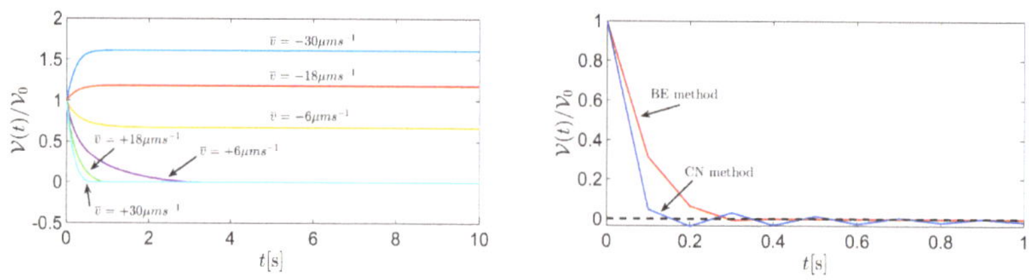

Figure 8. Left panel: a plot of $\mathcal{V}(t)/\mathcal{V}_{ref}$ in the time interval $[0, 10]$ s. The value of fluid velocity (expressed in μms^{-1}) is indicated for each computed normalized cell volume. **Right** panel: a plot of $\mathcal{V}(t)/\mathcal{V}_{ref}$ in the time interval $[0, 1]$ s.

Figure 9 (left panel) illustrates the temporal evolution of the water volume net production rate $\mathcal{R}_w$ in the time interval $[0, 10]$ s with the fluid velocity varying in the range $[-30 + 30] \cdot 10^{-6}$ m s^{-1}. We see that the stationary value of $\mathcal{R}_w$, for each considered normal fluid velocity in the range $[-30 + 30] \cdot 10^{-6}$ m s^{-1}, changes from negative to positive, with the magnitude for $v_n = -30 \cdot 10^{-6}$ m s^{-1} being three times larger than for $v_n = +30 \cdot 10^{-6}$ m s^{-1}. Figure 9 (right panel) illustrates the temporal evolution of the cell normalized volume in the time interval $[0, 1]$ s (ten times smaller than before) in the case where $\mathcal{R}(t) = 0$ for $t \geq 0$ and the fluid velocity varies in the range $[-30 + 30] \cdot 10^{-6}$ m s^{-1}. We see that in the absence of intracellular regulatory mechanisms, the model System (40) predicts an abnormal increase in cell volume for a highly negative value of normal fluid velocity.

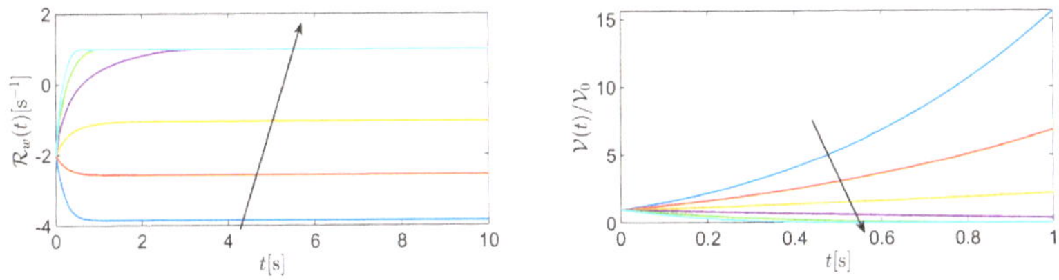

Figure 9. Left panel: a plot of the water volume net production rate $\mathcal{R}_w(t)$ in the time interval [0, 10] s. **Right** panel: a plot of $\mathcal{V}(t)/\mathcal{V}_{ref}$ in the time interval [0, 1] s in the case where $\mathcal{R}(t) = 0$ for every $t \in [0, 1]$ s. Fluid velocity varies in the range $[-30\ +30] \times 10^{-6}$ m s^{-1}. The arrow indicates velocity increase from negative to positive values.

3.2. Cell Homeostasis in the Ciliary Epithelium of the Eye

This case study is mathematically represented by the Cauchy problem (38) in which the cell represents one of the pigmented/nonpigmented couplets in the ciliary epithelium (CE) of the ciliary body of the eye (see [7,8,39]). The aim of this study is to apply the model proposed in this article to characterize the homeostatic configuration of the cell under the physiological conditions of the system. According to the data reported in [7], such conditions correspond to the following:

C1. The volume of the CE, equal to 8 µL;

C2. The number of cell couplets, $N_{cells,CE}$, constituting the CE, equal to 4 million;

C3. The intraocular pressure, equal to 15 mmHg;

C4. The AH volumetric flow rate, equal to 2.75 µL min^{-1}.

Assumption 13. *We make the following assumptions:*

1. *The considered sets of the neutral and charged solutes are*

$$S_\beta = \{CO_2, H_2CO_3\}, \tag{42}$$

$$S_\alpha = \{Na^+, K^+, H^+, Cl^-, HCO_3^-\}; \tag{43}$$

2. *The molar densities of the neutral and charged solutes are given constants denoted by $\bar{c}_{\beta,ex}$, $\beta \in S_\beta$, and $\bar{c}_{\alpha,ex}$, $\alpha \in S_\alpha$;*
3. *The hydraulic pressure difference is a given constant denoted by $\Delta\bar{p} = \bar{p}_{in} - \bar{p}_{ex}$, where $\bar{p}_{in}$ and $\bar{p}_{ex}$ are the values of the intracellular and extracellular fluid pressure, respectively;*
4. *No transmembrane ion exchangers are considered so that $\Phi^{exch} = 0$;*
5. *The model of carrier membrane proteins is described in Appendix A.3;*
6. *The model of ion channels is described in Appendix A.4;*
7. *The model of the net production rates for the neutral and charged solutes is described in Appendix B.1;*
8. *The model of the net production rate in cell volume regulation is described in Appendix B.2.*

The values of all model parameters that were used in the computer simulations illustrated in this section are listed in Table 1.

The vector containing the initial condition data is

$$\mathbf{Y}_0 = \left[2 \times 10^{-15}, 0, 0, 10, 140, 5.0119 \times 10^{-5}, 7, 10^{-4}, -8.0097 \times 10^{-2}\right]^T$$
$$\text{units: } \left[\mathrm{m}^3, \mathrm{mM}, \mathrm{mM}, \mathrm{mM}, \mathrm{mM}, \mathrm{mM}, \mathrm{mM}, \mathrm{mM}, \mathrm{V}\right]^T.$$

The vectors containing the extracellular values of the solute molar density are

$$\overline{\mathbf{c}}_\beta = \left[0.5, 1.3 \cdot 10^{-3}\right]^T \qquad \text{units: } [\mathrm{mM}, \mathrm{mM}]^T,$$
$$\overline{\mathbf{c}}_\alpha = \left[130, 5, 3.1623 \cdot 10^{-5}, 150, 9.9\right]^T \qquad \text{units: } [\mathrm{mM}, \mathrm{mM}, \mathrm{mM}, \mathrm{mM}, \mathrm{mM}]^T.$$

The values of the remaining model parameters can be found in Appendices B.1, B.2 and B.3. Computations were performed using the `Matlab` solver `ode15s`, setting both relative and absolute tolerances equal to 10^{-12}.

Table 1. Model parameters: symbol, value, and units.

Symbol	Value	Units
T	298.15	K
R_{cell}	7.816×10^{-6}	m
t_M	7.5×10^{-9}	m
c_M	5.9×10^{-3}	$\mathrm{F\,m^{-2}}$
$\overline{p}_{in}$	20	mmHg
$\overline{p}_{ex}$	15	mmHg
$\Delta\overline{p}$	5	mmHg
σ_β	$[0.1, 0.1]^T$	$[\cdot]$
σ_α	$[0.3, 0.3, 0.3, 0.3, 0.3]^T$	$[\cdot]$
σ_X	1	$[\cdot]$
P_β	$[0.228, 0.1467]^T$	$\mathrm{m\,s^{-1}}$
P_α	$[0.0013, 0.2613, 1.1587, 0.5227, 0.1467]^T$	$\mathrm{m\,s^{-1}}$
Φ_β^{carr}	$10^{-3}[0.1352, 0.1352]^T$	$[\cdot]$
Φ_α^{ch}	$10^{-4}[0.0126, 0.0785, 0.0031, 0.0196, 0.1539]^T$	$[\cdot]$
Φ^{AQP}	1.3515×10^{-4}	$[\cdot]$
Φ_α^{pump}	$[0.0011, 0.0011, 0, 0, 0]^T$	$[\cdot]$
Φ^{lip}	0.9974	$[\cdot]$

3.2.1. Electroneutrality and Impermeant Charged Proteins

Electroneutrality is essential in the maintainance of cell volume as a function of time and of phenomena occurring in the intracellular and extracellular regions (see [3] and references cited therein). The computation of the total electric charge densities $\rho_{mob,in}$ and $\rho_{mob,ex}$ due to mobile ions inside and outside the cell in resting conditions yields the following:

$$\rho_{mob,in} = F \sum_{\alpha \in S_\alpha} z_\alpha c_{\alpha,in}(0) = 1.3797 \times 10^7 \, \mathrm{C\,m^{-3}},$$
$$\rho_{mob,ex} = F \sum_{\alpha \in S_\alpha} z_\alpha \overline{c}_\alpha = -9.5519 \times 10^5 \, \mathrm{C\,m^{-3}}.$$

These results indicate that the intracellular region (at $t = 0$) is characterized by a high excess of positive charge whereas the extracellular region (at $t = 0$) is characterized by a high excess of negative charge. This charge difference gives rise to a large osmotic pressure difference across the cell membrane which may eventually lead to the disruption of cell integrity. To neutralize the excess of positive and negative charge, we need the presence of internally sequestered impermeant charges inside and outside the cell. Denoting by c_X and z_X the molar density and charge number of the impermeant charge, we have

$$c_{X,in} = 143 \text{ mM} \qquad z_{X,in} = -1, \tag{44a}$$
$$c_{X,ex} = 9.9 \text{ mM} \qquad z_{X,ex} = +1. \tag{44b}$$

These results indicate that a high molar density of fixed anions is sequestered inside the cell cytoplasm whereas a much smaller molar density of cations is required to make the extracellular solution electroneutral.

Assumption 14. *We assume that the number of moles of the intracellular impermeant charge n_X is constant during the time evolution of the cell.*

Let $c_X(0)$ denote the intracellular molar density of the impermeant charge at the time $t = 0$ (units: mM). By the definition of molar density, we have

$$c_X(0) = \frac{n_X}{\mathcal{V}_0}. \tag{45}$$

Using Assumption 14 and (45), we can express the intracellular impermeant charge molar density for any time, $t \geq 0$, as

$$c_{X,in}(t) = \frac{n_X}{\mathcal{V}(t)} = c_X(0)\frac{\mathcal{V}_0}{\mathcal{V}(t)} \qquad t \geq 0. \tag{46}$$

The simulations illustrated in the next sections were conducted using the values of z_X and c_X in (44) and the constitutive Equation (46).

3.2.2. Fast-Time-Scale Cell Evolution

We investigated the evolution of the cell in the time interval $[t_0, t_{end}]$, where $t_0 = 0$ s and $t_{end} = 50 \times 10^{-12}$ s.

Figure 10 (left panel) illustrates the time evolution of the intracellular protonated hydrogen (blue curve) and the corresponding intracellular pH (red curve). Hydrogen fast diffusion from the intracellular side into the extracellular side determines a sharp decrease in $c_{H^+,in}$ so that the cytoplasm solution turns into a very basic condition (the maximum value of the intracellular pH is 12.5). Figure 10 (right panel) illustrates the time evolution of the average cell normal surface velocity (blue curve) and the corresponding percentage variation, $\Delta\mathcal{V}_{\%}$, in the cell volume with respect to the initial condition (red curve). The cell surface velocity is positive and very small in magnitude (less than 0.02 $\mu\mathrm{m\,s^{-1}}$) so the volume of the cell experiences a very little decrease (maximum magnitude equal to $2 \times 10^{-11}\%$) with respect to resting conditions. Figure 10 (middle panel, bottom) illustrates the time evolution of the total AH volumetric flow rate Q_{AH} throughout the CE (units: $\mu L\,min^{-1}$). The quantity $Q_{AH}(t)$ is computed for every $t \in [t_0, t_{end}]$ using the following relation

$$Q_{AH}(t) = (N_{cells,CE} v_{cell,n}(t) S(t)) 60 \times 10^9 \qquad \text{units: } \mu L\,min^{-1} \qquad t \in [t_0, t_{end}]. \tag{47}$$

In less than 50 ps, Q_{AH} reaches almost 72% of the volumetric flow rate that is physiologically expected in a normal-tension individual (see condition C4).

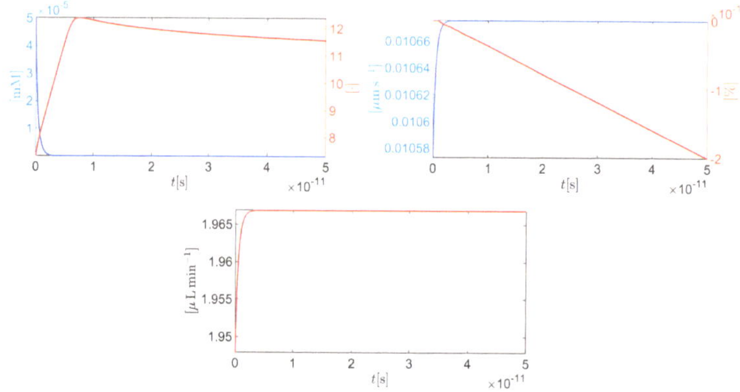

Figure 10. Left panel: blue curve, $c_{H^+,in}(t)$; red curve, $pH_{,in}(t)$, $t \in [0, 50 \cdot 10^{-12}]$ s. **Right** panel: blue curve, $v_{cell,n}(t)$; red curve, $\Delta\mathcal{V}_{\%}(t)$, $t \in [0, 50 \cdot 10^{-12}]$ s. Middle panel (bottom): total AH volumetric flow rate $Q_{AH}(t)$ for $t \in [0, 50 \cdot 10^{-12}]$ s.

3.2.3. Medium-Time-Scale Cell Evolution

We investigated the evolution of the cell in the time interval $[t_0, t_{end}]$, where $t_0 = 0$ s and $t_{end} = 5$ s.

Figure 11 (top left panel) illustrates the time evolution of the intracellular CO_2 (blue curve) and H_2CO_3 (red curve) molar densities. In less than 0.5 s, the hydration process gives rise to a significant production of carbon dioxide and carbonic acid, which is eventually followed by a stationary condition corresponding to the dynamic equilibrium of the reaction (A8a). The consequence of the CO_2 hydration can also be seen from Figure 11 (top right panel) which illustrates the time evolution of the intracellular H^+ molar density and the corresponding pH (blue and red curves, respectively). Protonated hydrogen concentration increases significantly until almost $t = 0.125$ s because of carbonic acid dissociation (forward reaction in (A8b)). Then, the association reaction (backward reaction in (A8b)) with bicarbonate gives rise to a decrease in $c_{H^+,in}$ until a stationary condition is reached. In such a condition, the value of the intracellular pH (almost 6) indicates that the cytoplasm solution is acidic. Figure 11 (bottom left panel) illustrates the time evolution of the average cell normal surface velocity (blue curve) and the corresponding percentage variation $\Delta\mathcal{V}_{\%}$ of the cell volume with respect to the initial condition (red curve). As in the fast-scale cell evolution, the cell surface velocity is positive and very small in magnitude (less than 0.02 $\mu m\,s^{-1}$). In this case, however, the much larger time duration of the analysis (5 s instead of 50 ps) allows the cell to decrease by a larger percentage amount with respect to resting conditions (less than 0.5% instead of 2×10^{-11}%). Figure 11 (bottom right panel) illustrates the time evolution of the total AH volumetric flow rate Q_{AH} throughout the CE (units: $\mu L\,min^{-1}$). In 5 s, Q_{AH} reaches more than 78% of the volumetric flow rate that is physiologically expected in a normal-tension individual.

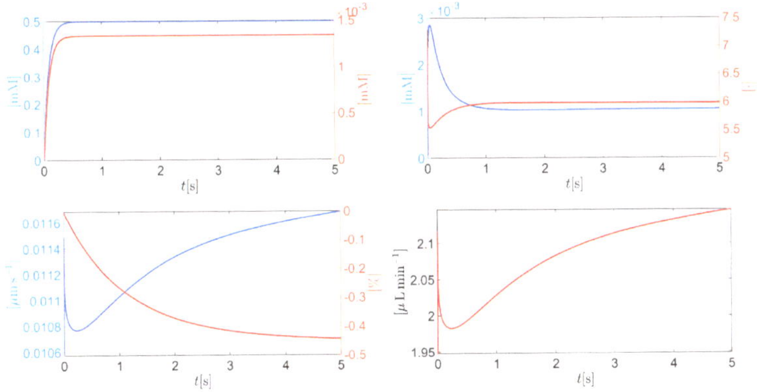

Figure 11. Top left panel: blue curve, $c_{CO_2,in}(t)$; red curve, $H_2CO_{3,in}(t)$, $t \in [0, 5]$ s. **Top right** panel: blue curve, $H^+_{in}(t)$; red curve, $pH_{in}(t)$, $t \in [0, 5]$ s. **Bottom left** panel: blue curve, average cell normal surface velocity, $v_{cell,n}(t)$; red curve, percentage volume variation, $\Delta\mathcal{V}_{\%}(t)$, $t \in [0, 5]$ s. **Bottom right** panel: total AH volumetric flow rate $Q_{AH}(t)$ for $t \in [0, 5]$ s.

3.2.4. Long-Time-Scale Cell Evolution

We investigated the evolution of the cell in the time interval $[t_0, t_{end}]$, where $t_0 = 0$ s and $t_{end} = 5400$ s. The value of t_{end} corresponds to the time that is needed by the eye to completely replace the AH content of the anterior chamber (see [7,9]).

Figure 12 (top left panel) illustrates a zoomed view of the time evolution of the membrane potential in the time interval $t \in [0, 10]$ s. After an initial ultra-fast transient due to the mismatch between the intracellularly applied initial conditions and the conditions in the extracellular bath, the cell reaches a stationary state of -85.9 mV, corresponding to an after-hyperpolarization of -5.9 mV. Figure 12 (top right panel) illustrates a zoomed view of the time evolution of the intracellular molar densities of sodium (blue curve), potassium (red curve), and chlorine (green curve) for $t \in [0, 120]$ s. The concentration of sodium experiences a significant depletion (from 10 to 5.4 mM) because of the NaK ATPase pump activity. Similarly, the concentration of potassium increases from 140 mM up to a stationary value of almost 144 mM. The green curve in Figure 12 (top right panel) indicates that the concentration of intracellular chlorine experiences a depletion from 7 mM to 5.3 mM. This can be explained by Figure 12 (middle left panel) which illustrates a zoom of the chlorine molar flux density $j_{Cl^-,n}(t)$ (units: $\mathrm{mM\,m\,s^{-1}}$) for $t \in [0, 10]$ s. The positive value of $j_{Cl^-,n}$ indicates that chlorine is swept out of the cell cytoplasm with a progressively reducing magnitude over time. Figure 12 (middle right panel) illustrates a zoom of the time evolution of the average cell normal surface velocity (blue curve) and the corresponding percentage variation $\Delta\mathcal{V}_{\%}$ of the cell volume with respect to the initial condition (red curve) for $t \in [0, 120]$ s. As in the previous conditions, the cell surface velocity is positive and very small in magnitude (less than 0.015 $\mathrm{\mu m\,s^{-1}}$). The stationary cell volume percentage decrease with respect to resting conditions is less than 0.57 %. Figure 12 (bottom center panel) illustrates a zoom of the time evolution of the total AH volumetric flow rate Q_{AH} throughout the CE (units: $\mathrm{\mu L\,min^{-1}}$) for $t \in [0, 300]$ s. The stationary value of Q_{AH} is 2.75017 $\mathrm{\mu L\,min^{-1}}$, with a percentage error of -0.0062% with respect to the physiological value of 2.75 $\mathrm{\mu L\,min^{-1}}$.

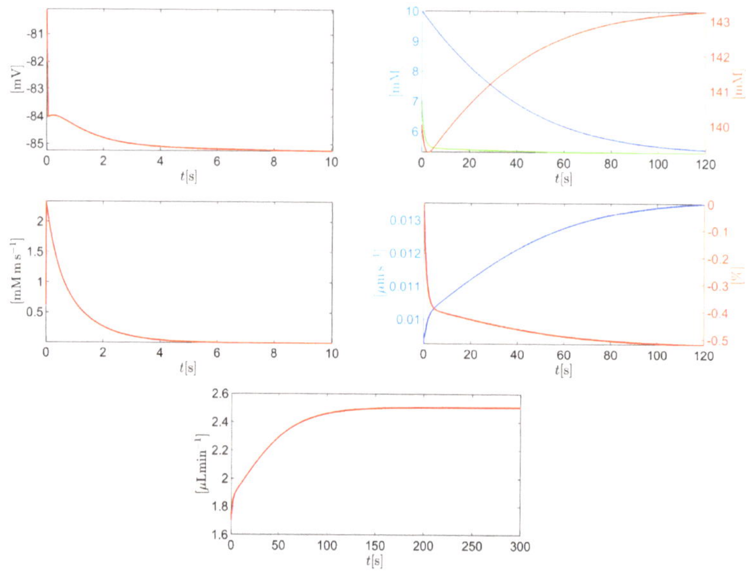

Figure 12. Top left panel: a zoom of the membrane potential $\psi_m(t)$ (units: mV) in the time interval $t \in [0, 10]$ s. **Top right** panel: a zoom of $c_{\mathrm{Na}^+,in}(t)$ (blue curve), $c_{\mathrm{K}^+,in}(t)$ (red curve), and $c_{\mathrm{Cl}^-,in}(t)$ (green curve) (units: mM) in the time interval $t \in [0, 120]$ s. **Middle left** panel: a zoom of the chlorine molar flux density $j^{ecw}_{\mathrm{Cl}^-}(t)$ (units: $\mathrm{mM\,m\,s^{-1}}$) in the time interval $t \in [0, 10]$ s. **Middle right** panel: a zoom of $v_{cell,n}(t)$ (blue curve) and $\Delta\mathcal{V}_{\%}(t)$ (red curve) in the time interval $t \in [0, 120]$ s. **Bottom center** panel: a zoom of the total AH volumetric flow rate in the time interval $t \in [0, 300]$ s.

4. Discussion

The development of a mathematical model and of a CVL for the simulation of the process of the production, flow, and outflow of AH has been subject of investigation in recent years (see [15–18]). Our proposed formulation is characterized by the following features: (F.1) it is based on the use of homogeneous mixtures including neutral and charged solutes (see [19] (Chapter 13)); (F.2) it is defined at the level of the single cell; and (F.3) it utilizes a model reduction procedure from three spatial dimensions to zero spatial dimensions. Feature (F.1) confers a solid theoretical foundation to the proposed model. Features (F.2) and (F.3) make the model structure simple and the computational schemes fast and suitable for adoption in a clinical environment. The model is a consistent generalization of previous approaches [3,4] as it shares with them the same conceptual, simplifying assumption of working at the level of the single cell. This assumption is applied here to evaluate the collective behaviour of the cells in the CE of the eye in the process of the production of AH. The model that we propose in these pages also has limitations: (L.1) the dependence of all the variables from the spatial coordinate is neglected; (L.2) several important transmembrane mechanisms regulating solute exchange are not considered in the simulations; (L.3) the statistical variability of the parameters is not accounted for. Limitation (L.1) is a consequence of the use of the 3D-to-0D reduction procedure. Limitation (L.2) is a choice to prevent the proliferation of model parameters, thereby rendering the analysis of simulation predictions more easily. Limitation (L.3) is a consequence of the choice of using a mechanistic (continuum-based) approach. We intend to remove all these limitations in future extensions of the formulation considered in the present article.

In the next sections, we use the CVL developed in the present article to address specific questions of clinical importance in the study of AH production and its relation to ocular diseases. In Section 4.1, we assess the impact of oncotic pressure due to impermeant charge.

In Section 4.2, we assess the impact of Na^+/K^+ ATPase. In Section 4.3, we assess the impact of the carbonic anhydrase enzyme. In Section 4.4, we assess the impact of IOP.

4.1. The Impact of Oncotic Pressure Due to Impermeant Charge

In this simulation, we solve System (38) in the time interval $[0, 600]$ s with the same set of parameters as in Section 3.2, except the reflection coefficient σ_X which is set equal to [0:0.2:1] (`Matlab` vector notation). By doing so, the weight of the contribution of the oncotic pressure difference (A5h) to the total osmo-oncotic pressure difference (A5k) increases progressively from 0% ($\sigma_X = 0$) to 100 % ($\sigma_X = 1$), thereby allowing us to investigate the impact of oncotic pressure difference on the AH simulation.

Figure 13 (left panel) illustrates the time evolution of the total volumetric flow rate $Q_{\text{AH}}(t)$ as a function of the time t in the interval $t \in [0, 600]$ s. The results indicate that the smaller σ_X is, the larger the predicted total volumetric flow rate of AH is. In particular, the value of Q_{AH} predicted by the model which is compatible with the given intracellular and extracellular fluid pressures ($\overline{p}_{in} = 20$ mmHg and $\overline{p}_{ex} = 15$ mmHg, respectively) is obtained for $\sigma_X = 1$. To better understand the effect of properly including impermeant charge in AH modeling, we illustrate in Figure 13 (right panel) the time evolution of the total osmo-oncotic pressure difference $\Delta\Pi(t)$ in the interval $t \in [0, 600]$ s. We see that the magnitude of $\Delta\Pi$ largely exceeds the contribution from hydrostatic pressure difference $\Delta\overline{p} = 5$ mmHg for every value of $\sigma_X \in$ [0:0.2:1]. Moreover, for every $\sigma_X \in$ [0:0.2:1], the oncotic pressure difference $\Delta\Pi_X$ is always positive whereas the total osmotic pressure difference $\Delta\Pi_{osm}$ is always negative, so, as σ_X increases, the magnitude of the total osmo-oncotic pressure difference decreases to reach the value of -1500 mmHg.

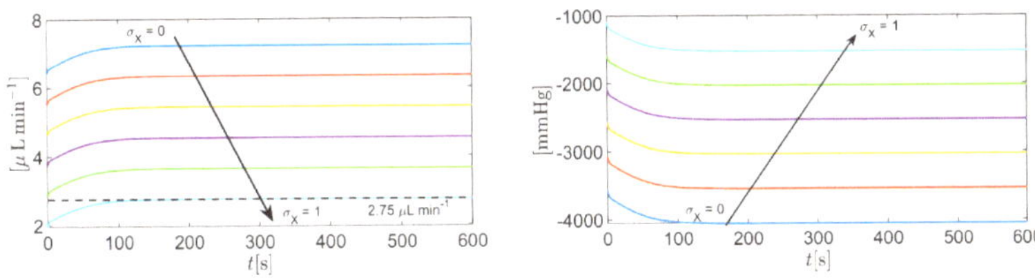

Figure 13. Left panel: total AH volumetric flow rate Q_{AH} for $t \in [0, 600]$ s as a function of σ_X. The black dashed line indicates the physiological value of Q_{AH}, equal to 2.75 μL min^{-1}, when IOP = 15 mmHg. **Right** panel: the total osmo-oncotic pressure difference $\Delta\Pi(t)$ for $t \in [0, 600]$ s as a function of σ_X.

4.2. The Impact of Na^+/K^+ ATPase

In this simulation, we solve System (38) in the time interval $[0, 5400]$ s with the same set of parameters as in Section 3.2, except the amplification coefficient M_{ATP} in (A12e) which is set equal to [0:0.5:2] (`Matlab` vector notation). By doing so, the weight of the contribution of the sodium–potassium pump to the electrochemical balance of the cell increases progressively from 0% ($M_{ATP} = 0$) to 100 % ($M_{ATP} = 2$), with respect to the working conditions of Section 3.2, thereby allowing us to investigate the impact of Na^+/K^+ ATPase on the AH simulation.

Figure 14 illustrates the time evolution of the total volumetric flow rate $Q_{\text{AH}}(t)$ as a function of the time t in the interval $t \in [0, 600]$ s. The results indicate that for $M_{ATP} < 1$ (corresponding to values of $c_{ATP} < c_{ATP,ref}$), the predicted total volumetric flow rate of AH is negative. This means that the pump does not have enough energy to move sodium out from the cell and potassium into the cell, so the accumulation of sodium in the cytoplasm attracts chlorine from the extracellular region, eventually leading to the inversion of fluid flow. The predicted total volumetric flow rate of AH becomes positive for $M_{ATP} = 1$.

This means that the pump does have enough energy to move sodium out from the cell and potassium into the cell, preventing chlorine accumulation in the cell cytoplasm. For increasing values in ATP concentration ($M_{ATP} > 1$), water flux is favored, reaching a physiological level when $M_{ATP} = 2$.

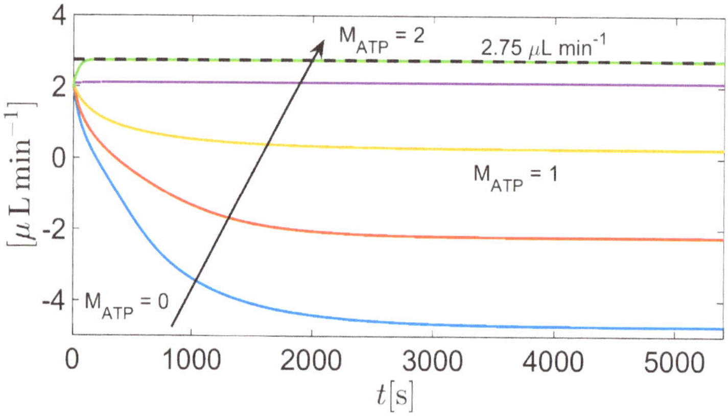

Figure 14. Total AH volumetric flow rate Q_{AH} for $t \in [0, 5400]$ s as a function of M_{ATP}. The black dashed line indicates the physiological value of Q_{AH}, equal to 2.75 µL min^{-1}, when IOP = 15 mmHg.

4.3. The Impact of Carbonic Anhydrase

In this simulation, we solve System (38) in the time interval $[0, 5400]$ s with the same set of parameters as in Section 3.2, except the amplification coefficient A_{CA} in Equations (A9c) and (A9d) which is set equal to [0:1:5] (`Matlab` vector notation). By doing so, the weight of the contribution of the CA enzyme to improve the reaction rate of CO_2 hydration increases progressively from 0% ($A_{CA} = 0$) to 100 % ($A_{CA} = 5$), with respect to the working conditions of Section 3.2, thereby allowing us to investigate the impact of the CA enzyme on the AH simulation. The six computed total AH volumetric flow rates over the considered time interval do not show any visible difference with respect to the change in A_{CA}. Therefore, to investigate the impact of CA on AH production, we define the reference AH volumetric flow rate $q_{ref} = q_{ref}(t)$ as the function of time predicted by the model for $A_{CA} = 0$. Then, we evaluate the maximum percentage variation between each of the other five AH volumetric flow rates predicted by the model and q_{ref}. The results show that the maximum percentage variation ranges between 6.2963×10^{-3}% and 6.3129×10^{-3}%. At the same time, the value of the predicted total AH volumetric flow rate for $t \in [500, 5400]$ s varies between 2.7501731 µL min^{-1} and 2.7501736 µL min^{-1}. Correspondingly, the percentage difference between these values and the physiological value of 2.75 µL min^{-1} (for IOP = 15 mmHg) ranges between 6.2945×10^{-3}% and 6.3127×10^{-3}%. All the above obtained results indicate that $Q_{AH}(t)$ is practically independent of the concentration of the CA enzyme. Further tests with higher values of the amplification parameter A_{CA} do not show significant variation of the predicted value of $Q_{AH}(t)$; this probably to be ascribed to the low value of the intracellular carbon dioxide molar density (cf. Figure 11).

4.4. The Impact of IOP

In this simulation, we solve System (38) in the time interval $[0, 5400]$ s with the same set of parameters as in Section 3.2, except the value of IOP which is taken in the range $[15, 150]$ mmHg. By doing so, we investigate the response of the model in the case of highly hypertensive patients. In the remainder of this section, we denote by $Q_{AH,b} = 2.75$ µLmin^{-1} the value of the AH volumetric flow rate at the baseline condition IOP = 15 mmHg and by

$Q_{AH}(\text{IOP})$ the value of the AH volumetric flow rate predicted by the model corresponding to a given value of IOP in the range [15, 150] mmHg. We define by

$$\Delta Q_{\%}(\text{IOP}) = \frac{Q_{AH,b} - Q_{AH}(\text{IOP})}{Q_{AH,b}} \times 100 \qquad \text{IOP} \in [15, 150]\ \text{mmHg}$$

the percentage difference between the AH volumetric flow rate in baseline conditions and the AH volumetric flow rate predicted by the model corresponding to a given value of IOP in the range [15, 150] mmHg.

Figure 15 is the graph of $\Delta Q_{\%}(\text{IOP})$ for IOP in the interval [15, 150] mmHg. The trend is linearly increasing with IOP and indicates that the AH volumetric flow rate predicted by the model decreases with respect to baseline conditions as IOP increases. This agrees with observations in patients affected by Graves' disease (also known as Thyroid Eye Disease, TED). TED is a chronic, autoimmune, inflammatory orbital disease, causing an increase in the volume and swelling of the soft orbital tissue behind the globe, for which an increase in IOP, as a consequence of an increase in the episcleral vein pressure, determines a reduction in the aqueous humor outflow facility (see [40,41]). The results in Figure 15 also agree with the following expression for fluid velocity:

$$v_f = L_p(p_{\text{in}}(t) - p_{\text{ex}}(t) - \Delta\Pi(t)) \qquad t \geq 0, \tag{48}$$

where $p_{\text{in}}(t) = \overline{p}_{in} = 20$ mmHg for $t \geq 0$ is the intracellular fluid pressure (corresponding to the estimated fluid pressure in the CE upon assuming 25 mmHg in the ciliary capillaries), $p_{\text{ex}}(t)$ is the considered value of IOP in the interval [15, 150] mmHg for $t \geq 0$, and $\Delta\Pi(t)$ is the total osmo-oncotic difference for $t \geq 0$. Since $\Delta\Pi(t)$ turns out to be negative with respect to t, as IOP increases, we see from (48) that v_f diminishes, which explains the fact that $Q_{AH}(\text{IOP})$ also diminishes as IOP increases. For IOP = 30 mmHg, which is a typical value of intraocular pressure in hypertensive individuals (see [42]), we have $\Delta Q_{\%}(30) = 1.09\%$, and for IOP = 50 mmHg, we have $\Delta Q_{\%}(30) = 2.54\%$.

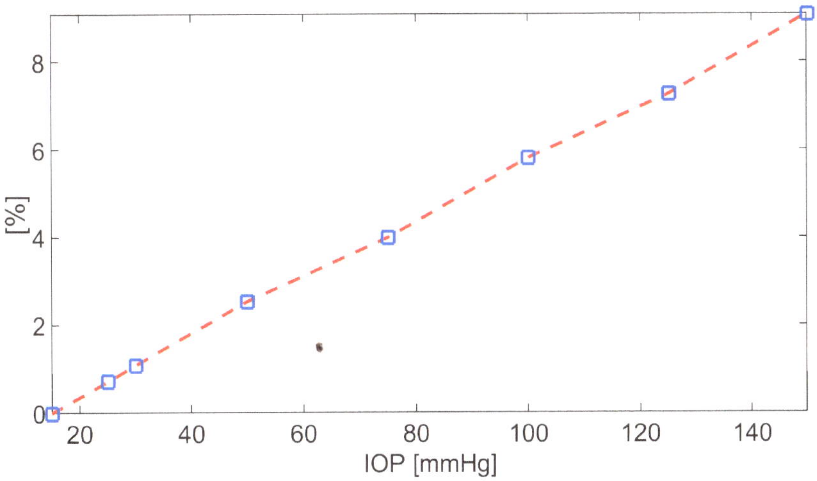

Figure 15. Plot of $\Delta Q_{\%}(\text{IOP})$ for IOP in interval [15, 150] mmHg, with $t \in [0, 5400]$ s.

5. Conclusions

In this article, we proposed, analyzed, and numerically investigated a reduced-order mathematical description of cellular volume homeostasis. The model accounts for intracellular reactions and transmembrane mechanisms for neutral and charged solute exchange. Hydrostatic and osmo-oncotic pressure differences were used in conjunction with Starling's Law to compute the velocity of the fluid which drives the motion and radial deformation of the cell volume.

The model was implemented within the context of a CVL that was applied to the study of the process of AH production in the human eye. The scope of the simulations was to test the potential of the CVL to assess the relative quantitative importance of the biophysical mechanisms that underlie AH production, from the perspective of their use as supporting tools to integrate and complement in vivo experiments and artificial intelligence-based methodologies for the analysis of data with a statistically significant number of patients.

This study identified for the first time three novel sources of influence on AH production that contribute understanding to previously established models. The first source of influence is that impermeant charged proteins and Na^+/K^+ ATPase are important on the level of AH production. The second source of influence is that AH production is independent of CA concentration, at least for low values of CO_2 concentration. The third source of influence is that AH production decreases with an increase in IOP.

Author Contributions: Conceptualization, R.S. and G.C.; methodology, R.S.; software, R.S.; validation, R.S., B.S. and G.A.; formal analysis, R.S., A.L. and G.G.; investigation, R.S. and G.A.; resources, A.H. and G.G.; data curation, R.S. and K.W.S.; writing—original draft preparation, R.S.; writing—review and editing, R.S., B.S., A.L., K.W.S. and A.H.; visualization, R.S.; supervision, A.H.; project administration, A.H. and G.G.; funding acquisition, A.H., A.V., B.S. and G.G. All authors have read and agreed to the published version of the manuscript.

Funding: This work was partially supported by the NIH grants R01EY030851 and R01EY034718, the NSF grants 2108711/2327640 and 2412130, the NYEE Foundation grants, The Glaucoma Foundation, and in part by a Challenge Grant award from Research to Prevent Blindness, NY. Professor Alon Harris was supported by the NIH grants (R01EY030851 and R01EY034718), the NYEE Foundation grants, The Glaucoma Foundation, and in part by the Challenge Grant award from Research to Prevent Blindness, NY. Dr. Alice Verticchio was supported by the NYEE Foundation grant.

Data Availability Statement: The original contributions presented in this study are included in the article. Further inquiries can be directed to the corresponding author.

Conflicts of Interest: The funders had no role in the design of this study; in the collection, analyses, or interpretation of data; in the writing of the manuscript; or in the decision to publish the results.

Abbreviations

The following abbreviations are used in this manuscript:

AH	Aqueous humor
CA	Carbonic anhydrase
ATP	Adenosinetriphosphate
IOP	Intraocular pressure
CVL	Computational virtual laboratory
AQP	Aquaporin
BE	Backward Euler
CN	Crank–Nicolson
CE	Ciliary epithelium

Appendix A. Normal Fluid Velocity and Solute Molar Flux Densities

In this section, we provide the expressions of the normal fluid velocity on the cell surface and solute molar flux densities in (36).

Appendix A.1. Normal Fluid Velocity Inside the Pore Channel of the AQP

Let us proceed by proving Equation (11a). The expression of the fluid stress tensor in cylindrical coordinates is

$$\mathbb{T} = \begin{bmatrix} -p(r,s,\theta,t) & 0 & \mu_f \frac{\partial V_s(r,t)}{\partial r} \\ 0 & -p(r,s,\theta,t) & 0 \\ \mu_f \frac{\partial V_s(r,t)}{\partial r} & 0 & -p(r,s,\theta,t) \end{bmatrix}, \tag{A1a}$$

so its divergence is the following vector:

$$\nabla \cdot \mathbb{T} = \begin{bmatrix} \frac{1}{r}\frac{\partial}{\partial r}(-rp(r,s,\theta,t)) \\ \frac{1}{r}\frac{\partial}{\partial \theta}(-p(r,s,\theta,t)) \\ \frac{1}{r}\frac{\partial}{\partial r}\left(r\mu_f \frac{\partial V_s(r,t)}{\partial r}\right) - \frac{\partial p(r,s,\theta,t)}{\partial s} \end{bmatrix}. \tag{A1b}$$

The linear momentum Equation (5b) becomes

$$\frac{1}{r}\frac{\partial}{\partial r}(-rp(r,s,\theta,t)) = 0, \tag{A1c}$$

$$\frac{1}{r}\frac{\partial}{\partial \theta}(-p(r,s,\theta,t)) = 0, \tag{A1d}$$

$$\frac{1}{r}\frac{\partial}{\partial r}\left(r\mu_f \frac{\partial V_s(r,t)}{\partial r}\right) - \frac{\partial p(r,s,\theta,t)}{\partial s} + b_{p,s}(s,t) = 0. \tag{A1e}$$

Equations (A1c) and (A1d) imply that p does not depend on r and θ, i.e., $p = p(s,t)$. Equation (A1e) becomes

$$\frac{1}{r}\mu_f v_{p,n}(t)\frac{\partial}{\partial r}\left(r\frac{d\eta(r)}{dr}\right) - \frac{\partial p(s,t)}{\partial s} + b_{p,s}(s,t) = 0. \tag{A1f}$$

Replacing the expression of η in (A1f) yields

$$\frac{1}{r}\mu_f v_{p,n}(t)\frac{\partial}{\partial r}\left(-\frac{4r^2}{r_p^2}\right) - \frac{\partial p(s,t)}{\partial s} + b_{p,s}(s,t) = 0,$$

from which we obtain

$$-\frac{8\mu_f}{r_p^2}v_{p,n}(t) - \frac{\partial p(s,t)}{\partial s} + b_{p,s}(s,t) = 0.$$

Rearranging the above equation, we obtain the following expression for the normal fluid velocity in the AQP:

$$v_{p,n}(t) = \frac{r_p^2}{8\mu_f}\left[-\frac{\partial p(s,t)}{\partial s} + b_{p,s}(s,t)\right],$$

which is Equation (11a).

Appendix A.2. Normal Fluid Velocity on the Cell Surface

To determine the normal fluid velocity on the cell surface, we consider the following two cases:

(B.C.1) The hydraulic pressure difference $\Delta\overline{p}(t)$ is given (units: Pa).

(B.C.2) The total water volumetric flow rate across the cell surface $\overline{Q}_{w,cell}(t)$ is given (units: $m^3\,s^{-1}$).

In case (B.C.2), we recall the definition of the water volumetric flow rate across the cell surface:

$$Q_{w,cell}(t) = \int_{\mathcal{S}_t} \mathbf{v}_{fl}(\mathbf{x},t) \cdot \mathbf{n}\, d(\mathcal{S}_t) \qquad t \geq 0. \tag{A2}$$

Since $Q_{w,cell}(t)$ is given and equal to $\overline{Q}_{w,cell}(t)$, we obtain

$$v_{cell,n}(t) = \frac{\overline{Q}_{w,cell}(t)}{\mathcal{S}(t)} = \frac{\overline{Q}_{w,cell}(t)}{\gamma(\mathcal{V}(t))^{2/3}} \qquad t \geq 0, \tag{A3}$$

where we use the fact that $\mathcal{S}(t) = \gamma(\mathcal{V}(t))^{2/3}$.

In case (B.C.1), we use (11d) to obtain

$$v_{cell,n}(t) = \Phi^{AQP} v_{p,n}(t) = \Phi^{AQP} \mathcal{L}_p[\Delta\overline{p}(t) - \Delta\Pi(t)] \qquad t \geq 0. \tag{A4}$$

To determine $v_{cell,n}$, we need to provide a mathematical model for the total osmo-oncotic pressure difference $\Delta\Pi$. Let us define the electrochemical potential of the ion α:

$$\varphi_\alpha^{ec} = \psi + \frac{V_{th}}{z_\alpha} \ln\left(\frac{c_\alpha}{c_{ref}}\right) \qquad \alpha \in S_\alpha, \tag{A5a}$$

where c_{ref} is a reference molar density. Let us introduce the pressure differences:

$$\Delta\Pi_\alpha(t) := \int_0^{t_m} -\sigma_\alpha F z_\alpha c_\alpha(s,t) \frac{\partial \varphi_\alpha^{ec}(s,t)}{\partial s}\, ds \qquad \alpha \in S_\alpha, \quad t \geq 0, \tag{A5b}$$

$$\Delta\Pi_\beta(t) := \int_0^{t_m} -\sigma_\beta RT \frac{\partial c_\beta(s,t)}{\partial s}\, ds \qquad \beta \in S_\beta \quad t \geq 0, \tag{A5c}$$

$$\Delta\Pi_X(t) := \int_0^{t_m} -\sigma_X RT \frac{\partial c_X(s,t)}{\partial s}\, ds \qquad t \geq 0. \tag{A5d}$$

The quantities $\Delta\Pi_\alpha$ and $\Delta\Pi_\beta$ are the osmotic pressure differences associated with the charged solute $\alpha \in S_\alpha$ and neutral solute $\beta \in S_\beta$, respectively, whereas $\Delta\Pi_X$ is the oncotic pressure difference associated with the impermeant charge c_X. The quantity R is the gas constant (units: $J\,mol^{-1}\,K^{-1}$), and σ_α, σ_β, and σ_X are the membrane reflection coefficients associated with the ion α, solute β, and impermeant fixed charge c_X, respectively. We notice that each coefficient is in the range [0, 1], with the value 0 corresponding to a completely permeable membrane and the value 1 corresponding to a completely impermeable membrane. To evaluate the integral on the right-hand side of (A5b), we replace $c_\alpha(s,t)$ with its spatial average:

$$\langle c_\alpha \rangle(t) := \frac{\int_0^{t_m} c_\alpha(s,t)\, ds}{t_m} \qquad t \geq 0. \tag{A5e}$$

Postponing to Appendix A.4 the computation of (A5e) and replacing it in (A5b), we obtain

$$\Delta\Pi_\alpha(t) = \sigma_\alpha F z_\alpha \langle c_\alpha \rangle(t) \Delta\varphi_\alpha^{ec}(t) \qquad \alpha \in S_\alpha \quad t \geq 0, \tag{A5f}$$

$$\Delta\Pi_\beta(t) = \sigma_\beta RT \Delta c_\beta(t) \qquad \beta \in S_\beta \quad t \geq 0, \tag{A5g}$$

$$\Delta\Pi_X(t) = \sigma_X RT \Delta c_X(t) \qquad t \geq 0. \tag{A5h}$$

We define the total osmotic pressure difference as

$$\Delta\Pi_{osm}(t) = \sum_{\alpha\in S_\alpha} \Delta\Pi_\alpha(t) + \sum_{\beta\in S_\beta} \Delta\Pi_\beta(t) \qquad t \geq 0, \tag{A5i}$$

and the oncotic pressure difference as

$$\Delta\Pi_{onc}(t) = \Delta\Pi_X(t) \qquad t \geq 0. \tag{A5j}$$

Then, we define the total osmo-oncotic pressure difference as

$$\begin{aligned}\Delta\Pi(t) &= \Delta\Pi_{osm}(t) + \Delta\Pi_{onc}(t)\\ &= \sum_{\alpha\in S_\alpha} \sigma_\alpha \left[F z_\alpha \langle c_\alpha\rangle(t)\psi_m(t) + RT\langle c_\alpha\rangle(t)\ln\left(\frac{c_{\alpha,in}(t)}{c_{\alpha,ex}(t)}\right)\right]\\ &+ RT\sum_{\beta\in S_\beta} \sigma_\beta \Delta c_\beta(t) + \sigma_X RT\Delta c_X(t) \qquad t \geq 0.\end{aligned} \tag{A5k}$$

Relation (A5k) is the generalization of Equation (31) in [4].

Appendix A.3. Neutral Solutes

Let us consider a carrier protein whose geometrical structure is the same as that of the pore ω_p, illustrated in Figure 5, in the case of an AQP. To determine the normal molar flux density of the solute β inside ω_p, we make the following assumptions.

Assumption 15. *We assume the following:*

1. *The molar flux density of the neutral solute β only has the axial component $j_{\beta,s}$;*
2. *For any $t \geq 0$, the quantity $j_{\beta,s}$ is spatially constant inside ω_p, i.e., $j_{\beta,s} = j_{\beta,n}(t)$, with $t \geq 0$ and $s \in [0, t_m]$.*

Using Assumption 15, Assumption 4-A3, and the fact that $v_{p,n}$ is spatially constant in the constitutive Equation (6b) gives the following second-order equation for the solute molar density inside the carrier protein channel:

$$v_{p,n}(t)\frac{\partial c_\beta(s,t)}{\partial s} - D_\beta \frac{\partial^2 c_\beta(s,t)}{\partial s^2} = 0 \qquad t \geq 0,\ s \in [0, t_m], \tag{A6a}$$

whose solution is

$$c_\beta(s,t) = A(t) + B(t)\exp\left(\frac{v_{p,n}(t)}{D_\beta}s\right) \qquad t \geq 0,\ s \in [0, t_m], \tag{A6b}$$

with $A = A(t)$ and $B = B(t)$ being time-dependent arbitrary functions. Inserting (A6b) into the constitutive Equation (6b) yields

$$j_{\beta,n}(t) = v_{p,n}(t)A(t) \qquad t \geq 0. \tag{A6c}$$

To determine $A(t)$, we need to enforce the following boundary conditions:

$$c_\beta(0,t) = c_{\beta,in}(t), \quad c_\beta(t_m,t) = c_{\beta,ex}(t) \qquad t \geq 0, \tag{A6d}$$

where $c_{\beta,in}(t)$ and $c_{\beta,ex}(t)$ are the intra- and extracellular values of the neutral solute molar density for any $t \geq 0$. Using (A6d) in (A6b), we obtain the following expression of the normal molar flux density of β inside the carrier protein channel:

$$j_{\beta,n}(t) = -P_\beta \left[c_{\beta,ex}(t)\mathrm{Be}(X_\beta(t)) - c_{\beta,in}(t)\mathrm{Be}(-X_\beta(t)) \right] \qquad t \geq 0, \tag{A6e}$$

where $P_\beta := D_\beta / t_m$ is the membrane permeability of the neutral solute β (units: $\mathrm{m\,s^{-1}}$),

$$\mathrm{Be}(W) := \frac{W}{e^W - 1} \qquad W \in \mathbb{R}$$

is the inverse of the Bernoulli function, and

$$X_\beta(t) := \frac{v_{p,n}(t)}{P_\beta} \qquad t \geq 0. \tag{A6f}$$

Appendix A.4. Charged Solutes

We can proceed in the same way as in Appendix A.3 to obtain the following expression for the normal molar flux density of the ion $\alpha \in S_\alpha$ inside an ionic channel.

$$j^{edw}_{\alpha,n}(t) = -P_\alpha \left[c_{\alpha,ex}(t)\mathrm{Be}(X_\alpha(t)) - c_{\alpha,in}(t)\mathrm{Be}(-X_\alpha(t)) \right] \qquad t \geq 0, \tag{A7a}$$

where $P_\alpha := D_\alpha / t_m$ is the membrane permeability of the charged solute α and

$$X_\alpha(t) := \frac{v_{p,n}(t)}{P_\alpha} + z_\alpha \frac{\psi_m(t)}{V_{th}} \qquad t \geq 0. \tag{A7b}$$

Using the definition (8e) of the generalized drift velocity of the ion α in (A7b), we find the spatial distribution of c_α inside the ionic channel:

$$c_\alpha(s,t) = \widetilde{A}(t) + \widetilde{B}(t) \exp\left(\frac{v_{\alpha,n}(t)}{D_\alpha} s \right) \qquad t \geq 0,\ s \in [0,\ t_m], \tag{A7c}$$

where $\widetilde{A} = \widetilde{A}(t)$ and $\widetilde{B} = \widetilde{B}(t)$ are time-dependent arbitrary functions, and $v_{\alpha,n}(t)$ is the average normal component of $\mathbf{v}_\alpha = \mathbf{v}_\alpha(\mathbf{x}, t)$ on the cell surface, with $\mathbf{x} \in \partial\Omega_t$ and $t \geq 0$. Replacing (A7c) in (A5e), we obtain

$$\langle c_\alpha \rangle(t) = c_{\alpha,in}(t)\xi_{in}(t) + c_{\alpha,ex}(t)\xi_{ex}(t) \qquad t \geq 0, \tag{A7d}$$

where

$$\xi_{in}(t) = \frac{\mathrm{Be}(-X_\alpha(t)) - 1}{X_\alpha(t)} \qquad t \geq 0, \tag{A7e}$$

$$\xi_{ex}(t) = \frac{1 - \mathrm{Be}(X_\alpha(t))}{X_\alpha(t)} \qquad t \geq 0. \tag{A7f}$$

We illustrate in Figure A1 a graphical representation of $\langle c_\alpha \rangle(\tau)$ (red solid line) compared to the arithmetic value $c^{arith}_\alpha(\tau) = (c_{\alpha,in}(\tau) + c_{\alpha,ex}(\tau))/2$ for $\tau = [0 : 20]$ (`Matlab` vector notation); $X_\alpha(\tau) = \tau - 10$, $c_{\alpha,in}(\tau) = 1$ mM; and $c_{\alpha,ex}(\tau) = 5$ mM. We see that $\langle c_\alpha \rangle$ and c^{arith}_α are comparably close only if $|X_\alpha| \simeq 0$, which corresponds to a diffusive regime of transport inside the channel. Conversely, their distance increases for larger values of $|X_\alpha|$, which corresponds to an advection-dominated regime of transport inside the channel. On the basis of these considerations, the adoption of (A7d) instead of the customary choice

c_α^{arith} (as performed, for example, in [4]) is expected to warrant an accurate and robust model simulation.

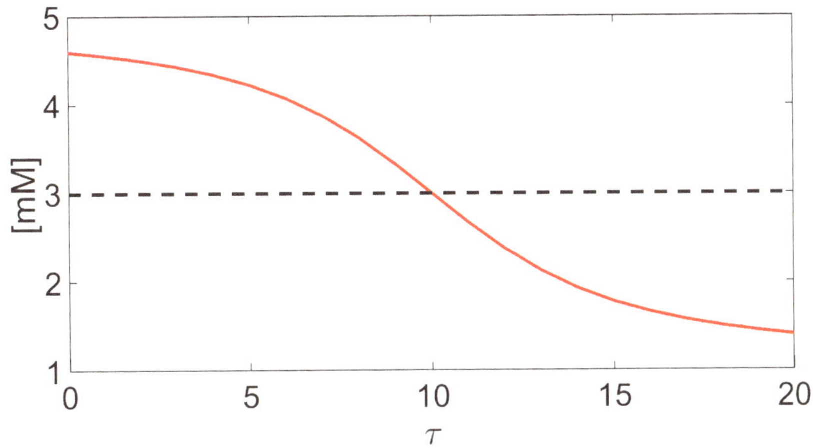

Figure A1. Red solid line: a plot of $\langle c_\alpha \rangle$ for $\tau = [0 : 20]$ and $X_\alpha(\tau) = \tau - 10$, with τ being a dimensionless time. The endpoint values of c_α are $c_{\alpha,in}(\tau) = 1$ mM and $c_{\alpha,ex}(t) = 5$ mM for every $\tau \in [0 : 20]$. Black dashed line: the arithmetic average of c_α.

Appendix B. Mathematical Modeling of Cellular Metabolism

The following conceptual scheme of cellular metabolism is considered in this article:

1. Glucose is absorbed by mitochondria to produce ATP and CO_2.
2. ATP provides the energy needed by the Na^+/K^+ pump to export three sodium ions and import two potassium ions.
3. The carbonic anhydrase enzyme (CA) catalizes the hydrolysis of CO_2, which is a waste product of mitochondrial metabolism.
4. Specialized exchangers supervise the transmembrane transport of the proton (H^+) and bicarbonate (HCO_3^-), which are the products of CO_2 hydrolysis.

Figure A2 illustrates the intracellular reactions and transmembrane transport mechanisms that are involved in cellular metabolism.

In the following, we provide the expressions of:

1. The production and consumption rates for the neutral and charged solutes involved in the CA enzyme-mediated carbon dioxide conversion into carbonic acid and its subsequent dissociation into protonated hydrogen and bicarbonate;
2. The production and consumption rates in cellular volume regulation;
3. The molar flux densities representing the mathematical model of transmembrane sodium and potassium solute exchange throughout Na^+/K^+ ATPase.

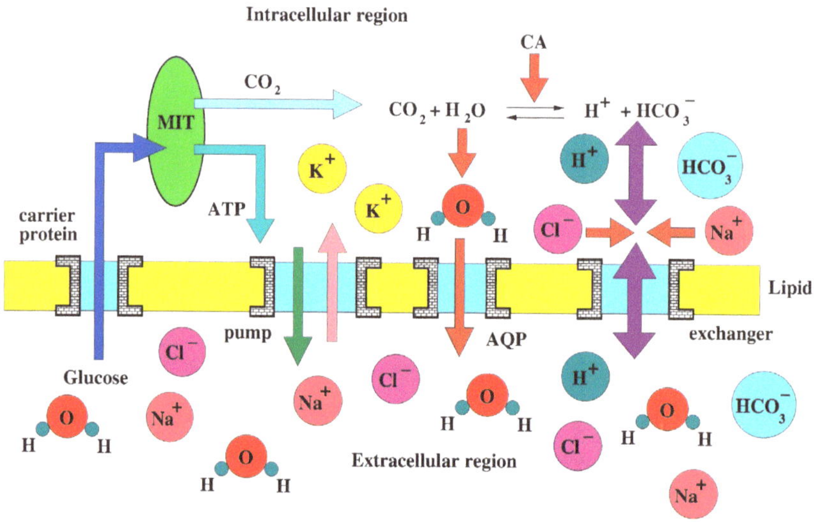

Figure A2. Schematic representation of intracellular reactions and transmembrane transport mechanisms. MIT: mitochondrium. ATP: adenosinetriphosphate. CA: carbonic anhydrase. Exchangers perform multiple ion transport across membrane.

Appendix B.1. Mathematical Model of CA-Mediated CO_2 Hydrolysis

In this section, we illustrate the mathematical modeling of the CA enzyme-mediated CO_2 hydrolysis. This process can be conveniently represented as the following two-step chemical reaction (see [43,44]):

$$\text{STEP 1:} \qquad CO_2 + H_2O \rightleftharpoons H_2CO_3. \tag{A8a}$$

$$\text{STEP 2:} \qquad H_2CO_3 \rightleftharpoons H^+ + HCO_3^-. \tag{A8b}$$

STEP 1 is the conversion of intracellular carbon dioxide into carbonic acid under the mediation of the CA enzyme, which is a very fast catalyzer of the CO_2 hydration process (see [45,46]). STEP 1 is characterized by the quantities k_{hydr} and k_{dehydr} (units: s^{-1}) depending on the molar density of the carbonic anhydrase enzyme (CA) and representing the rate constants of the forward and backward reactions in STEP 1, respectively.

Assumption 16. *Let $c_{CO_{2(aq)},in}$ denote the molar density of intracellular CO_2 that is hydrated by water molecules. We assume that each CO_2 molecule that is produced by mitochondria's respiration is hydrated by a corresponding water molecule. This allows us to denote by $c_{CO_{2,in}}$ the molar density of the intracellular $c_{CO_{2(aq)}}$.*

Using the Law of Conservation of Mass and Assumption 16, the net production rates in the mass balance equation for $\beta = CO_2$ and $\beta = H_2CO_3$ have the following expressions:

$$\mathcal{R}_{CO_2}(t) = k_{dehydr} c_{H_2CO_3,in}(t) - k_{hydr} c_{CO_{2,in}}(t) \qquad t \geq 0, \tag{A9a}$$

$$\mathcal{R}_{H_2CO_3}(t) = -\mathcal{R}_{CO_2}(t) \qquad t \geq 0. \tag{A9b}$$

The hydration and dehydration rate constants experimentally depend on the amount of CA that is present in the compartment where the hydration reaction takes place (see [47]). Following [43,44], in this article, we use the following model for the hydration and dehydration rate constants:

$$k_{hydr} = A_{\text{CA}} k_{hydr,ref}, \tag{A9c}$$

$$k_{dehydr} = A_{\text{CA}} k_{dehydr,ref}, \tag{A9d}$$

where A_{CA} is a nonnegative given constant, whereas $k_{hydr,ref} = 0.037\ \text{s}^{-1}$ and $k_{dehydr,ref} = 13.7\ \text{s}^{-1}$ are experimental reference values measured at $T = 25\ ^\circ\text{C}$, reported in [48]. Taking $A_{\text{CA}} > 1$ is a way to represent the catalyzing effect of CA compared to the uncatalyzed reaction corresponding to $A_{\text{CA}} = 1$. In the numerical simulations illustrated in Section 3.2, we set $A_{\text{CA}} = 5$.

STEP 2 is the dissociation of intracellular carbonic acid into bicarbonate and protonated hydrogen and is characterized by the quantities k_{diss} (units: $\text{mM}^{-1}\ \text{s}^{-1}$) and k_{assoc} (units: $\text{mM}^{-2}\ \text{s}^{-1}$), the rate constants of the forward and backward reactions in STEP 2, respectively. The dissociation reaction of H_2CO_3 is extremely rapid, so the values of the forward and backward rate constants in STEP 2 are very large. In the numerical simulations illustrated in Section 3.2, we use the data of [43,44] and set $k_{diss} = 10^{16}\ \text{s}^{-1}$ and $k_{assoc} = k_{diss}/K_{eq}$, where $K_{eq} = 0.2804$ mM is the equilibrium constant of STEP 2. Using the Law of Conservation of Mass, we obtain the following expressions for the net production rates in the mass balance equations for $\alpha = \text{H}^+$ and $\alpha = \text{HCO}_3^-$:

$$\mathcal{R}_\alpha(t) = k_{diss} c_{\text{H}_2\text{CO}_3}(t) - k_{assoc} c_{\text{H}^+}(t) c_{\text{HCO}_3^-}(t) \qquad \alpha = \{\text{H}^+, \text{HCO}_3^-\}, \quad t \geq 0. \tag{A10}$$

Assumption 17. *We assume that the ions* $\alpha \in \{Na^+, K^+, Cl^-\}$ *are non-reacting (see [49] (Section 8.3.3)). Therefore, we have* $\mathcal{R}_\alpha(t) = 0$, *with* $t \geq 0$.

Appendix B.2. Net Production Rate in Cell Volume Regulation

The mechanisms which govern intracellular water production/consumption are the object of considerable debate and investigation because of their importance in cell life and survival. One example is provided by the process of normotonic cell shrinkage which is the major hallmark of cellular apoptosis [50]. Another example is provided by the TCA cycle (Krebs cycle) in cellular respiration, whose products of metabolism of fuels are ATP, CO_2, and so-called "metabolic" water [51].

The difficulty of accessing and sampling the contents of intact cells makes the study of the intracellular fluid environment, and, more generally, of body fluid content a challenging problem. Specific approaches to accurately detect metabolic water content as a result of intracellular metabolic activity have been proposed in [52,53].

In this article, we focus on the role of the net production and consumption of intracellular water, $\mathcal{R}_w$, in driving the motion of the cell surface. Our proposed model is

$$\mathcal{R}_w(t) = k_{w,p} - k_{w,c} \frac{\mathcal{V}(t)}{\mathcal{V}_0} \qquad t \geq 0, \tag{A11}$$

where $k_{w,p}$ and $k_{w,c}$ are given constants (units: s^{-1}) and $\mathcal{V}_0$ is the value of cell volume in resting conditions. In the simulations illustrated in Section 3.2, we set $k_{w,p} = k_{w,c} = 1\ \text{s}^{-1}$.

Appendix B.3. Na^+/K^+ ATPase

The mathematical model of the sodium–potassium pump (Na^+/K^+ ATPase) adopted in this article follows the idea proposed in [3]. The molar flux densities for sodium and potassium are

$$j^a_{\mathrm{Na},n}(t) = j_{pump}(t) \qquad t \geq 0, \tag{A12a}$$

$$j^a_{\mathrm{K},n}(t) = -\frac{2}{3} j_{pump}(t) \qquad t \geq 0, \tag{A12b}$$

where

$$j_{pump}(t) = j^{MAX}_{pump} \left(\frac{c_{\mathrm{Na}_{in}}(t)}{c_{\mathrm{Na}_{in}}(t) + c_{\mathrm{Na},\,1/2}} \right)^3 \left(\frac{c_{\mathrm{K}_{ex}}(t)}{c_{\mathrm{K}_{ex}}(t) + c_{\mathrm{K},\,1/2}} \right)^2 \qquad t \geq 0, \tag{A12c}$$

$$j^{MAX}_{pump} := c_{ATP} v_{pump}. \tag{A12d}$$

The quantity j^{MAX}_{pump} is the maximum pump molar flux density (units: $\mathrm{mM\,m\,s^{-1}}$), c_{ATP} is the intracellular molar density of ATP (units: mM), and v_{pump} is the ion transfer velocity of the pump (units: $\mathrm{m\,s^{-1}}$). The quantities c_{ATP} and v_{pump} are defined as

$$c_{ATP} = M_{ATP} c_{ATP,ref}, \tag{A12e}$$

$$v_{pump} = r_{turn} t_M, \tag{A12f}$$

where $c_{ATP,ref}$ is the reference value of the ATP molar density (units: mM), r_{turn} is pump turnover rate (units: $\mathrm{s^{-1}}$), and M_{ATP} is a nonnegative given constant. The quantities $c_{\mathrm{Na},\,1/2}$ and $c_{\mathrm{K},\,1/2}$ are the Michaelis constants of the pump model (units: mM). In the simulations illustrated in Section 3.2, we set $c_{ATP,ref} = 10^{-2}$ mM, $M_{ATP} = 2$, $r_{turn} = 10^7\ \mathrm{s^{-1}}$, $c_{\mathrm{Na},\,1/2} = 1.3$ mM and $c_{\mathrm{K},\,1/2} = 0.14$ mM.

References

1. Cid, V.J.; Durán, A.; del Rey, F.; Snyder, M.P.; Nombela, C.; Sánchez, M. Molecular basis of cell integrity and morphogenesis in Saccharomyces cerevisiae. *Microbiol. Rev.* **1995**, *59*, 345–386. [CrossRef] [PubMed]
2. Ammendolia, D.A.; Bement, W.M.; Brumell, J.H. Plasma membrane integrity: Implications for health and disease. *BMC Biol.* **2021**, *19*, 71. [CrossRef] [PubMed]
3. Armstrong, C.M. The Na/K pump, Cl ion, and osmotic stabilization of cells. *Proc. Natl. Acad. Sci. USA* **2003**, *100*, 6257–6262. [CrossRef]
4. Cheng, X.; Pinsky, P.M. The Balance of Fluid and Osmotic Pressures across Active Biological Membranes with Application to the Corneal Endothelium. *PLoS ONE* **2016**, *10*, e0145422. [CrossRef]
5. Pan, J.; Kmieciak, T.; Liu, Y.T.; Wildenradt, M.; Chen, Y.S.; Zhao, Y. Quantifying molecular- to cellular-level forces in living cells. *J. Phys. Appl. Phys.* **2021**, *54*, 483001. [CrossRef]
6. Gupta, S.K.; Guo, M. Equilibrium and out-of-equilibrium mechanics of living mammalian cytoplasm. *J. Mech. Phys. Solids* **2017**, *107*, 284–293. [CrossRef]
7. Brubaker, R.F. Flow of Aqueous Humor in Humans. *Invest. Ophthalmol. Vis. Sci.* **1991**, *32*, 3145–3166.
8. Delamare, N.A. Ciliary Body and Ciliary Epithelium. *Adv. Organ. Biol.* **2005**, *1*, 127–148.
9. Kiel, J.W.; Hollingsworth, M.; Rao, R.; Chen, M.; Reitsamer, H.A. Ciliary blood flow and aqueous humor production. *Prog. Retin. Eye Res.* **2011**, *30*, 1–17. [CrossRef]
10. Toris, C.B.; Tye, G.; Pattabiraman, P. Changes in Parameters of Aqueous Humor Dynamics Throughout Life. In *Ocular Fluid Dynamics: Anatomy, Physiology, Imaging Techniques, and Mathematical Modeling*; Guidoboni, G., Harris, A., Sacco, R., Eds.; Springer International Publishing: Cham, Switzerland, 2019; pp. 161–190.
11. Heijl, A.; Leske, M.C.; Bengtsson, B.; Hyman, L.; Bengtsson, B.; Hussein, M.; Group, E.M.G.T. Reduction of Intraocular Pressure and Glaucoma Progression: Results From the Early Manifest Glaucoma Trial. *Arch. Ophthalmol.* **2002**, *120*, 1268–1279. [CrossRef]
12. Noecker, R.J. The management of glaucoma and intraocular hypertension: Current approaches and recent advances. *Ther. Clin. Risk Manag.* **2006**, *2*, 193–206. [CrossRef] [PubMed]

13. Harris, A.; Guidoboni, G.; Siesky, B.; Mathew, S.; Verticchio Vercellin, A.C.; Rowe, L.; Arciero, J. Ocular blood flow as a clinical observation: Value, limitations and data analysis. *Prog. Retin. Eye Res.* **2020**, *78*, 100841. [CrossRef] [PubMed]
14. Killer, H.E.; Pircher, A. What is the optimal glaucoma treatment: Reducing aqueous humour production or facilitating its outflow? *Eye* **2020**, *34*, 1719–1721. [CrossRef] [PubMed]
15. Dvoriashyna, M.; Pralits, J.O.; Tweedy, J.H.; Repetto, R. Mathematical Models of Aqueous Production, Flow and Drainage. In *Ocular Fluid Dynamics: Anatomy, Physiology, Imaging Techniques, and Mathematical Modeling*; Guidoboni, G., Harris, A., Sacco, R., Eds.; Springer International Publishing: Cham, Switzerland, 2019; pp. 227–263.
16. Sala, L.; Mauri, A.G.; Sacco, R.; Messenio, D.; Guidoboni, G.; Harris, A. A theoretical study of aqueous humor secretion based on a continuum model coupling electrochemical and fluid-dynamical transmembrane mechanisms. *Comm. App. Math. Comp. Sci.* **2019**, *14*, 65–103. [CrossRef]
17. Dvoriashyna, M.; Foss, A.J.E.; Gaffney, E.A.; Repetto, R. A Mathematical Model of Aqueous Humor Production and Composition. *Invest. Ophthalmol. Vis. Sci.* **2022**, *63*, 1. [CrossRef]
18. Sacco, R.; Chiaravalli, G.; Antman, G.; Guidoboni, G.; Verticchio, A.; Siesky, B.; Harris, A. The role of conventional and unconventional adaptive routes in lowering of intraocular pressure: Theoretical model and simulation. *Phys. Fluids* **2023**, *35*, 061902. [CrossRef]
19. Sacco, R.; Guidoboni, G.; Mauri, A.G. *A Comprehensive Physically Based Approach to Modeling in Bioengineering and Life Sciences*; Elsevier: Amsterdam, The Netherlands; Academic Press: Cambridge, MA, USA, 2019.
20. Kiel, J.W. Physiology of the intraocular pressure. In *Pathophysiology of the Eye: Glaucoma*; Feher, J., Ed.; Akadémiai Kiadó: Budapest, Hungary, 1998; Volume 4, pp. 109–144.
21. Ramakrishnan, R.; Krishnadas, S.R.; Khurana, M.; Robin, A.L. Aqueous Humor Dynamics. In *Diagnosis & Management of Glaucoma*; Jaypee Brothers Medical Publishers (P) Ltd.: Rijeka, Croatia, 2006; Chapter 9.
22. Raviola, G.; Raviola, E. Intercellular junctions in the ciliary epithelium. *Invest. Ophthalmol. Vis. Sci.* **1978**, *17*, 958–981.
23. Calvo, J. Science Photo Gallery. Available online: https://www.sciencephoto.com/media/1112469/view/ciliary-body-epithelium-light-micrograph (accessed on 11 December 2024).
24. Dmitriev, A.V.; Dmitriev, A.A.; Linsenmeier, R.A. The logic of ionic homeostasis: Cations are for voltage, but not for volume. *PLoS Comput. Biol.* **2019**, *15*, 1–45. [CrossRef]
25. Hamann, S.; Zeuthen, T.; Cour, M.L.; Nagelhus, E.A.; Ottersen, O.P.; Agre, P.; Nielsen, S. Aquaporins in complex tissues: Distribution of aquaporins 1–5 in human and rat eye. *Am. J. -Physiol.-Cell Physiol.* **1998**, *274*, C1332–C1345. [CrossRef] [PubMed]
26. Pohl, P. Combined transport of water and ions through membrane channels. *Biol. Chem.* **2004**, *385*, 921–926. [CrossRef]
27. Verkman, A.S. Aquaporins. *Curr. Biol.* **2013**, *23*, R52–R55. [CrossRef] [PubMed]
28. Keener, J.; Sneyd, J. *Mathematical Physiology: I: Cellular Physiology*; Springer: New York, NY, USA, 2009.
29. Pizzagalli, M.D.; Bensimon, A.; Superti-Furga, G. A guide to plasma membrane solute carrier proteins. *FEBS J.* **2021**, *288*, 2784–2835. [CrossRef]
30. Hille, B. *Ion Channels of Excitable Membranes*, 3rd ed.; Sinauer Associates Inc.: Sunderland, MA, USA, 2001.
31. Wegner, L.H. Cotransport of water and solutes in plant membranes: The molecular basis, and physiological functions. *AIMS Biophys.* **2017**, *4*, 192–209. [CrossRef]
32. Jackson, J. *Classical Electrodynamics*; Wiley: Hoboken, NJ, USA, 2012.
33. Mori, Y. A Three Dimensional Model of Cellular Electrical Activity. Ph.D. Thesis, Department of Mathematics, New York University, New York, NY, USA, 2006.
34. Starling, E.S. On the Absorption of Fluids from the Connective Tissue Spaces. *J. Physiol.* **1896**, *19*, 312–326. [CrossRef]
35. Omir, A.; Satayeva, A.; Chinakulova, A.; Kamal, A.; Kim, J.; Inglezakis, V.J.; Arkhangelsky, E. Behaviour of Aquaporin Forward Osmosis Flat Sheet Membranes during the Concentration of Calcium-Containing Liquids. *Membranes* **2020**, *10*, 108. [CrossRef]
36. Li, Y.; He, L.; Gonzalez, N.A.; Graham, J.; Wolgemuth, C.; Wirtz, D.; Sun, S.X. Going with the Flow: Water Flux and Cell Shape during Cytokinesis. *Biophys. J.* **2017**, *113*, 2487–2495. [CrossRef]
37. Shampine, L.F.; Reichelt, M.W. The MATLAB ODE Suite. *SIAM J. Sci. Comput.* **1997**, *18*, 1–22. [CrossRef]
38. Quarteroni, A.; Sacco, R.; Saleri, F. *Numerical Mathematics*, 2nd ed.; Texts in Applied Mathematics; Springer: Berlin/Heidelberg, Germany, 2007; Volume 37.
39. Goel, M.; Picciani, R.G.; Lee, R.K.; Bhattacharya, S.K. Aqueous humor dynamics: A review. *Open Ophthalmol. J.* **2010**, *4*, 52–59. [CrossRef]
40. Klysik, A.; Kozakiewicz, M. Blood–aqueous barrier integrity in patients with Graves' ophthalmopathy (GO), before and after rehabilitative surgery. *Eye* **2015**, *29*, 542–551. [CrossRef]
41. Haefliger, I.O.; von Arx, G.; Pimentel, A.R. Pathophysiology of intraocular pressure increase and glaucoma prevalence in thyroid eye disease: A mini-review. *Klin. Monbl. Augenheilkd.* **2010**, *227*, 292–293. [CrossRef]

42. Kass, M.A.; Heuer, D.K.; Higginbotham, E.J.; Johnson, C.A.; Keltner, J.L.; Miller, J.P.; Parrish, R.K.I.; Wilson, M.R.; Gordon, M.O.; for the Ocular Hypertension Treatment Study Group. The Ocular Hypertension Treatment Study: A Randomized Trial Determines That Topical Ocular Hypotensive Medication Delays or Prevents the Onset of Primary Open-Angle Glaucoma. *Arch. Ophthalmol.* **2002**, *120*, 701–713. [CrossRef] [PubMed]
43. Somersalo, E.; Occhipinti, R.; Boron, W.F.; Calvetti, D. A reaction–diffusion model of CO2 influx into an oocyte. *J. Theor. Biol.* **2012**, *309*, 185–203. [CrossRef] [PubMed]
44. Bocchinfuso, A.; Calvetti, D.; Somersalo, E. Modeling surface pH measurements of oocytes. *Biomed. Phys. Eng. Express* **2022**, *8*, 045006. [CrossRef] [PubMed]
45. Larachi, F. Kinetic Model for the Reversible Hydration of Carbon Dioxide Catalyzed by Human Carbonic Anhydrase II. *Ind. Eng. Chem. Res.* **2010**, *49*, 9095–9104. [CrossRef]
46. Uchikawa, J.; Zeebe, R.E. The effect of carbonic anhydrase on the kinetics and equilibrium of the oxygen isotope exchange in the CO2–H2O system: Implications for δ^{18} vital effects in biogenic carbonates. *Geochim. Cosmochim. Acta* **2012**, *95*, 15–34. [CrossRef]
47. Donaldson, T.L.; Quinn, J.A. Kinetic Constants Determined from Membrane Transport Measurements: Carbonic Anhydrase Activity at High Concentrations. *Proc. Natl. Acad. Sci. USA* **1974**, *71*, 4995–4999. [CrossRef]
48. Gibbons, B.H.; Edsall, J.T. Rate of Hydration of Carbon Dioxide and Dehydration of Carbonic Acid at 25°. *J. Biol. Chem.* **1963**, *238*, 3502–3507. [CrossRef]
49. Layton, A.T.; Edwards, A. *Mathematical Modeling in Renal Physiology*; Lecture Notes on Mathematical Modelling in the Life Sciences; Springer: Berlin/Heidelberg, Germany, 2014.
50. Maeno, E.; Ishizaki, Y.; Kanaseki, T.; Hazama, A.; Okada, Y. Normotonic cell shrinkage because of disordered volume regulation is an early prerequisite to apoptosis. *Proc. Natl. Acad. Sci. USA* **2000**, *97*, 9487–9492. [CrossRef]
51. Litwack, G. Chapter 3—Introductory Discussion on Water, pH, Buffers and General Features of Receptors, Channels and Pumps. In *Human Biochemistry*; Litwack, G., Ed.; Academic Press: Boston, MA, USA, 2018; pp. 39–61. [CrossRef]
52. Kreuzer, H.W.; Quaroni, L.; Podlesak, D.W.; Zlateva, T.; Bollinger, N.; McAllister, A. Detection of Metabolic Fluxes of O and H Atoms into Intracellular Water in Mammalian Cells. *PLoS ONE* **2012**, *7*, e39685. [CrossRef]
53. Li, H.; Yu, C.; Wang, F.; Chang, S.J.; Yao, J.; Blake, R.E. Probing the metabolic water contribution to intracellular water using oxygen isotope ratios of PO_4. *Proc. Natl. Acad. Sci. USA* **2016**, *113*, 5862–5867. [CrossRef]

Mathematical and Computational Applications

MDPI

Article

Modeling of Nonlinear Systems: Method of Optimal Injections

Anatoli Torokhti [1] **and Pablo Soto-Quiros** [2,*]

1 STEM Discipline, University of South Australia, Mawson Lakes, GPO Box 2471, Adelaide, SA 5001, Australia; anatoli.torokhti@gmail.com
2 Escuela de Matemática, Instituto Tecnológico de Costa Rica, Cartago 30101, Costa Rica
* Correspondence: jusoto@tec.ac.cr

Abstract: In this paper, a nonlinear system is interpreted as an operator F transforming random vectors. It is assumed that the operator is unknown and the random vectors are available. It is required to find a model of the system represented by a best constructive operator F approximation. While the theory of operator approximation with any given accuracy has been well elaborated, the theory of best constrained constructive operator approximation is not so well developed. Despite increasing demands from various applications, this subject is minimally tractable because of intrinsic difficulties with associated approximation techniques. This paper concerns the best constrained approximation of a nonlinear operator in probability spaces. The main conceptual novelty of the proposed approach is that, unlike the known techniques, it targets a constructive optimal determination of all $3p + 2$ ingredients of the approximating operator where p is a nonnegative integer. The solution to the associated problem is represented by a combination of new best approximation techniques with a special iterative procedure. The proposed approximating model of the system has several degrees of freedom to minimize the associated error. In particular, one of the specific features of the developed approximating technique is special random vectors called injections. It is shown that the desired injection is determined from the solution of a special Fredholm integral equation of the second kind. Its solution is called the optimal injection. The determination of optimal injections in this way allows us to further minimize the associated error.

Keywords: approximation of nonlinear mappings; error minimization; optimization

1. Introduction

1.1. Motivation

Over the last few decades, the problem of constructive approximation of nonlinear operators has been a topic of profound research. A number of fundamental papers have appeared, which have established significant advances in this research area. Some relevant references can be found, in particular, in [1–14].

The known related results mainly concern proving the existence and uniqueness of operators approximating a given map, and justifying the bounds of errors arising from the approximation methods. The assumptions are that preimages and images are deterministic and can be represented in an analytical form, that is, by equations. At the same time, in many applications, the sets of preimages and images are stochastic and cannot be described by equations. Nevertheless, it is possible to represent these sets in terms of their numerical characteristics, such as the expectation and covariance matrices. Typical examples include stochastic signal processing [15–19], statistics [20–24], engineering [25–28], and image processing [29,30]; in the latter case, a digitized image, presented by a matrix, is often interpreted as the sample of a stochastic signal.

While the theory of operator approximation with any given accuracy is well elaborated (see, e.g., [1–14]), the theory of *best constrained* constructive operator approximation is not particularly well developed, although this is an area of intensive recent research (see, e.g., [31–36]). Despite increasing demands from applications [17–19,21–23,25–28,30–34,36–46], this subject is minimally tractable because of intrinsic difficulties in best approximation techniques, especially when the approximating operator should have a specific structure implied by the underlying problem. Recent studies related to this topic can be found in [47–55]. In particular, in [47], an approach to modeling of nonlinear systems was proposed based on observed input-output data. The proposed method constructs an approximate model by decomposing the unknown function into multiple components, aiming to minimize the associated error. Special random vectors, called injections, are used to refine the approximation while specific transformations simplify the optimization process. The approach provides flexibility in adjusting parameters to improve accuracy.

We wish to extend the known results in this area to the case when the sets of preimages and images of map $\mathcal{F}$ are stochastic, and the approximating operator we search is constructive in the sense it can numerically be realized and, therefore, is applicable to problems in applications.

1.2. Short Description of the Method

Let Ω be a set of outcomes in the probability space (Ω, Σ, μ) for which Σ is a σ–field of measurable subsets of Ω and $\mu : \Sigma \to [0,1]$ is an associated probability measure. Let $\mathbf{x} \in L^2(\Omega, \mathbb{R}^m)$ and $\mathbf{y} \in L^2(\Omega, \mathbb{R}^n)$ be random vectors, $\mathcal{F} : L^2(\Omega, \mathbb{R}^n) \to L^2(\Omega, \mathbb{R}^m)$ be a nonlinear map, and $\mathbf{x} = \mathcal{F}(\mathbf{y})$.

A nonlinear system is interpreted as the map $\mathcal{F}$ where $\mathbf{x}$ and $\mathbf{y}$ are random output and input signals, respectively. It is assumed that $\mathcal{F}$ is unknown, and $\mathbf{x}$ and $\mathbf{y}$ are available. We propose and justify a new method for the constructive $\mathcal{F}$ approximation such that first, an associated error is minimized and second, a structure of the approximating operator satisfies the special constraint related to the dimensionality reduction of vector $\mathbf{y}$. More specifically, we develop a new approach to the best constructive approximation of the map $\mathcal{F}$ in probability spaces subject to a specialized criterion associated with the dimensionality reduction of random preimages. The latter constraint follows from the requirements in applications such as those considered in [20–24,26,41,42]. In particular, in signal processing and system theory, a dimensionality reduction of random signals is used to optimize the cost of signal transmission. It is assumed that the only available information on $\mathcal{F}$ is given by certain covariance matrices formed from the preimages and images. This is a typical assumption used in applications such as those considered, e.g., in [15–23,37,38,44–46]. Here, we adopt that assumption. As mentioned, in particular, in [56–61], a priori knowledge of the covariances can come either from specific data models, or after sample estimation during a training phase.

The problem we consider (see (7) below) concerns finding the best approximating operator that depends on $2p+2$ unknown matrices G_j and H_j, for $j = 0, \ldots, p$ and p more unknown random vectors $\mathbf{v}_1, \ldots, \mathbf{v}_p$. We call $\mathbf{v}_1, \ldots, \mathbf{v}_p$ the *injections*. Here, p is a non-negative integer p. The injections $\mathbf{v}_1, \ldots, \mathbf{v}_p$ are aimed to further diminish the associated error. The difficulty is that $3p+2$ unknowns should be determined from a minimization of the single cost function given in (7).

The main difference between the approach in [47] and the method proposed in this paper is in their approach to modeling of nonlinear systems. The method in [47] constructs the model $\mathcal{T}_p$ as a sum of $p+1$ specific components, and the original optimization problem is decomposed into $p+1$ simpler sub-problems. The method empirically determines the special random vectors, called injections, to reduce the associated error. In contrast, the method proposed here solves the problem of a best constrained approximation of a nonlinear operator in probability spaces. It introduces the concept of an *optimal injection*, which is determined by solving a Fredholm integral equation of the second kind. While the method in [47] focuses on a structured numerical implementation, the proposed method provides a more general theoretical framework that optimally determines all $3p+2$ parameters through a combination of best approximation techniques and an iterative procedure.

The solution is represented in Sections 3 and 4, and is based on the following observation. Methods for best approximation are aimed at obtaining the best solution within a certain class; the accuracy of the solution is limited by the extent to which the class is satisfactory. By contrast, iterative methods are normally convergent, but the convergence can be quite slow. Moreover, in practice, only a finite number of iteration loops can be carried out, and therefore, the final approximate solution is often unsatisfactorily inaccurate. A natural idea is to combine the methods for best approximation and iterative techniques to exploit their advantageous features.

In Sections 3 and 4, we present an approach which realizes this. First, a special iterative procedure is proposed which aims to improve the accuracy of approximation with each consequent iteration loop. Secondly, the best approximation problem is solved providing the smallest associated error within the chosen class of approximants for each iteration loop. In Section 4, we show that the combination of these techniques allows us to build a computationally efficient and flexible method. In particular, we prove that the error in approximating $\mathcal{F}$ by the proposed method decreases with an increase in the number of iterations. An application is made to the optimal filtering of stochastic signals.

1.3. Novelty and Advantages

The novelty of the proposed method is in its approach to the best constructive approximation of nonlinear operators in probability spaces. Unlike methods that primarily focus on the existence and uniqueness of approximating operators, the proposed method aims to achieve an optimal constructive approximation by targeting a specific structure for the approximating operator, which is constrained by dimensionality reduction of random preimages. The focus on dimensionality reduction and the optimization of random vectors, known as injections, distinguishes the proposed method from existing approaches.

The advantages of the proposed method are twofold. First, it allows us to determine optimal injections which are derived from solving a Fredholm integral equation of the second kind, thereby providing a more robust and theoretically sound approach to approximating nonlinear systems. Second, the method combines best approximation techniques with a special iterative procedure, ensuring that the accuracy of the approximation improves with each iteration. This combination not only enhances the computational efficiency of the model but also provides a flexible framework that can be adapted to various applications, particularly those associated with signal processing and system theory, where dimensionality reduction plays a crucial role in optimizing signal transmission costs. By approximating $\mathcal{F}$ using the best constrained constructive operator approximation, one can design efficient algorithms for signal reconstruction or denoising. In system identification, this approach can be used to model and estimate dynamic systems with stochastic inputs, enabling more accurate predictions and adaptive control strategies.

2. The Proposed Approach

2.1. Some Special Notation

Let us write $\mathbf{x} = [\mathbf{x}_{(1)}, \ldots, \mathbf{x}_{(m)}]^T$ and $\mathbf{y} = [\mathbf{y}_{(1)}, \ldots, \mathbf{y}_{(n)}]^T$ where $\mathbf{x}_{(i)}, \mathbf{y}_{(j)} \in L^2(\Omega, \mathbb{R})$, for $i = 1, \ldots, m$ and $j = 1, \ldots, n$, and $\mathbf{x}(\omega) \in \mathbb{R}^m$ and $\mathbf{y}(\omega) \in \mathbb{R}^n$ for all $\omega \in \Omega$. Each matrix $A \in \mathbb{R}^{m\times n}$ defines a bounded linear transformation $\mathcal{A} : L^2(\Omega, \mathbb{R}^n) \to L^2(\Omega, \mathbb{R}^m)$. It is customary to write A rather then $\mathcal{A}$ since $[\mathcal{A}(\mathbf{x})](\omega) = A[\mathbf{x}(\omega)]$, for each $\omega \in \Omega$. Let us also write

$$\|\mathbf{x}\|_\Omega^2 = \int_\Omega \sum_{j=1}^{m} [\mathbf{x}_j(\omega)]^2 d\mu(\omega) < \infty. \tag{1}$$

The covariance matrix formed from $\mathbf{x}$ and $\mathbf{y}$ is denoted by E_{xy} such that

$$E_{xy} = \int_\Omega \mathbf{x}(\omega)[\mathbf{y}(\omega)]^T d\mu(\omega) = \left\{ \int_\Omega \mathbf{x}_{(i)}(\omega)\mathbf{y}_{(j)}(\omega) d\mu(\omega) \right\}_{i,j=1}^{m,n}. \tag{2}$$

The Moore–Penrose pseudo-inverse [62] of matrix M is denoted by $M^\dagger$.

2.2. Generic Structure of Approximating Operator

Let $\mathbf{v}_1, \ldots, \mathbf{v}_p$ be random vectors such that $\mathbf{v}_j \in L^2(\Omega, \mathbb{R}^{q_j})$, for $j = 1, \ldots, p$. We write $\mathbf{y} = \mathbf{v}_0$ and $q_0 = n$. As mentioned before, we call $\mathbf{v}_1, \ldots, \mathbf{v}_p$ the injections. This is because $\mathbf{v}_1, \ldots, \mathbf{v}_p$ contribute to the decrease of the associated error, as shown in Section 4.4 below. The choice of $\mathbf{v}_1, \ldots, \mathbf{v}_p$ is considered in Section 4.2, where each $\mathbf{v}_j$, for $j = 1, \ldots, p$ is defined by a nonlinear transformation φ_j of $\mathbf{y}$, i.e., $\mathbf{v}_j = \varphi_j(\mathbf{y})$. To facilitate the numerical implementation of the approximating technique introduced below, each vector $\mathbf{v}_j$, for $j = 1, \ldots, p$, is transformed to vector $\mathbf{z}_j \in L^2(\Omega, \mathbb{R}^{q_j})$ by transformation Q_j so that

$$\mathbf{z}_j = Q_j(\mathbf{v}_j, Z_{j-1}), \tag{3}$$

where $Z_{j-1} = \{\mathbf{z}_0, \ldots, \mathbf{z}_{j-1}\}$. The choice of Q_j is considered in Section 3.1.

Further, for $i = 0, 1, \ldots, p$, let $G_i \in \mathbb{R}^{m\times r_i}$, $H_i \in \mathbb{R}^{r_i \times q_i}$ where r_i is given, $0 < r_i < r$, and

$$r = r_0 + \ldots + r_p. \tag{4}$$

Here, r is a positive integer such that $r \le \min\{m, n\}$.

It is convenient to set $Q_0 = I$ and $\mathbf{z}_0 = \mathbf{v}_0 = \mathbf{y}$. To approximate $\mathcal{F}$ for a given reduction ratio

$$c = r/\min\{m, n\}, \tag{5}$$

we consider operator $\mathcal{T}_p : L^2(\Omega, \mathbb{R}^{q_0}) \times \ldots \times L^2(\Omega, \mathbb{R}^{q_p}) \to L^2(\Omega, \mathbb{R}^m)$ represented by

$$\mathcal{T}_p(\mathbf{v}_0, \ldots \mathbf{v}_p) = G_0 H_0 \mathbf{z}_0 + \ldots + G_p H_p \mathbf{z}_p, \tag{6}$$

where $G_j : L^2(\Omega, \mathbb{R}^{r_j}) \to L^2(\Omega, \mathbb{R}^m)$ and $H_j : L^2(\Omega, \mathbb{R}^{q_j}) \to L^2(\Omega, \mathbb{R}^{r_j})$, for $j = 0, 1, \ldots, p$, are linear operators (i.e., G_j and H_j are represented by $m \times r_j$ and $r_j \times q_j$ matrices, respectively. Recall that we use the same symbol to define a matrix and the associated linear operator).

Importantly, operators $H_0, \ldots, H_p$ imply the dimensionality reduction of vector $\mathbf{y}$. This is because $H_i \mathbf{z}_i \in L^2(\Omega, \mathbb{R}^{r_i})$ where $0 < r_i < r \le \min\{m, n\}$, for $i = 0, \ldots, p$.

We call p the degree of $\mathcal{T}_p$. It is shown below that $\mathcal{T}_p$ approximates an operator of interest $\mathcal{F} : L^2(\Omega, \mathbb{R}^n) \to L^2(\Omega, \mathbb{R}^m)$, with the accuracy represented by theorems in Sections 3.3, 3.4 and 4.4.

2.3. Statement of the Problem

Let $\mathcal{F} : L^2(\Omega, \mathbb{R}^n) \to L^2(\Omega, \mathbb{R}^m)$ be a continuous operator. We consider the problem as follows: Given $\mathbf{x}, \mathbf{y}$ and $r_0, \ldots, r_p$, find matrices $G_0, H_0, \ldots, G_p, H_p$ and vectors $\mathbf{v}_1, \ldots, \mathbf{v}_p$ that solve

$$\min_{\mathbf{v}_1,\ldots,\mathbf{v}_p} \min_{G_0,H_0,\ldots,G_p,H_p} \left\| \mathcal{F}(\mathbf{y}) - \sum_{j=0}^{p} G_j H_j \mathbf{z}_j \right\|_{\Omega}^2 \tag{7}$$

subject to

$$G_j \in \mathbb{R}^{m \times r_j} \quad \text{and} \quad H_j \in \mathbb{R}^{r_j \times q_j}, \tag{8}$$

and

$$E_{z_i z_j} = \mathbb{O}, \quad \text{for } i \neq j, \tag{9}$$

where $i, j = 0, \ldots, p$ and $\mathbb{O}$ denotes the zero matrix (and the zero vector).

It will be shown in Section 3 below that the solution of problem (7)–(9) is determined under a special condition imposed on vectors $\mathbf{v}_1, \ldots, \mathbf{v}_p$.

2.4. Related Work

2.4.1. Low-Rank Approximations

For $p = 0$ and the assumption that matrix E_{xy} is invertible, the particular case of the problem in (7)–(9),

$$\min_{G_0,H_0} \|\mathbf{x} - G_0 H_0 \mathbf{y}\|_{\Omega}^2, \tag{10}$$

has been solved, in particular, in [20,21,63–65]. Under the assumption E_{xy} is singular, it has been solved in [24,66–69]. Note it is a quite simplified case of problem in (7)–(9).

For $p = 1$, $G : L^2(\Omega, \mathbb{R}^{r_0}) \to L^2(\Omega, \mathbb{R}^m)$, $H_j : L^2(\Omega, \mathbb{R}^n) \to L^2(\Omega, \mathbb{R}^{r_0})$ where $j = 0, 1$, a particular case of the problem in (7)–(9),

$$\min_{G,H_0,H_1} \|\mathbf{x} - G(H_0 \mathbf{y} + H_1 \mathbf{v}_1)\|_{\Omega}^2, \tag{11}$$

was solved in [20]. The choice of $\mathbf{v}_1$ considered in [20] is not optimal.

2.4.2. Tensor Methods

In [70–72], the problem in (10) is generalized and studied in terms of tensors. Together with methods formulated in terms of matrices and mentioned in Sections 1 and 2.4.1, the tensor methods represent an important research subject both in the theoretical and applied sense. In this paper, the problems in (10) and (11) are generalized in a different way, which was described in Sections 2.2 and 2.3, and will be justified in Sections 3 and 4.4 below.

2.4.3. System Identification and Modeling

The problem we consider can also be represented as a black-box problem [73,74] where $\mathbf{y}$ and $\mathbf{x}$ are an available random input and output, respectively, and $\mathcal{F}$ is an unknown system. Then, the approximating operator $\mathcal{T}_p$ identifies a model of the system. Its particular features are detailed in Section 4.

2.5. Contribution

2.5.1. Challenges of High-Dimensionality

The proposed approach achieves the dimensionality reduction of preimages in the following way. In (6), let us write $\mathcal{T}_p(\mathbf{v}_0, \ldots \mathbf{v}_p)$ as

$$\mathcal{T}_p(\mathbf{v}_0, \ldots \mathbf{v}_p) = [G_0, \ldots, G_p] \begin{bmatrix} H_0\mathbf{z}_0 \\ \vdots \\ H_p\mathbf{z}_p \end{bmatrix} = GH\mathbf{z},$$

where $G = [G_0, \ldots, G_p]$, $H = \text{diag}[H_0, \ldots, H_p]$, $\mathbf{z} = [\mathbf{z}_0^T, \ldots, \mathbf{z}_p^T]^T$. Here, H is the block-diagonal matrix with $H_0, \ldots, H_p$ on the main diagonal and the dimensionality of vector $\mathbf{u} = H\mathbf{z}$ is r, which is defined by (4). Therefore, H realizes the dimensionality reduction in vector $\mathbf{y}$ with the reduction ratio defined by (5). Unlike the known techniques [20–24,26,41,42] for dimensionality reduction, the proposed approach achieves this by using $p+1$ terms, $H_0, \ldots, H_p$. This allows us to increase the accuracy associated with optimal determination of approximating operator $\mathcal{T}_p$. More details can be found in Sections 2.5.3 and 4.

2.5.2. Challenges of Accuracy

It is shown in Theorems 3, 4, and 5 below that, for the same reduction ratio, accuracy associated with the approximating operator $\mathcal{T}_p$ improves if the degree p or dimensions of injections $\mathbf{v}_1, \ldots, \mathbf{v}_p$ increase. For the case of the optimal determination of injections $\mathbf{v}_1, \ldots, \mathbf{v}_p$, the associated error is further improved. This is established in Theorems 10 and 11 in Section 4.4.

2.5.3. Novelties and Relation to Existing Concepts

Commonly, an approximating operator is represented by $\sum_{j=0}^{p} P_j\mathbf{v}_j$ where $\mathbf{v}_0, \ldots, \mathbf{v}_p$ are known basis functions and $P_0, \ldots, P_j$ are scalars or matrices which should be determined from (desirably) an error minimization.

The main conceptual novelty of the proposed approach is that, unlike the known techniques, it targets a *constructive* optimal determination of all $3p+2$ ingredients of the approximating operator in (6). They are $\mathbf{v}_1, \ldots, \mathbf{v}_p, H_0, G_0, \ldots, H_p, G_p$. The solution of the associated problem (7)–(9) is provided in Sections 3.1, 3.2, 4.1, and 4.2, and is represented by a combination of a new best approximation technique with a special iterative procedure.

A basic idea of the solution is to reduce the original problem (7)–(9) with $3p+2$ unknowns to $p+1$ simpler problems in (25) and (63) so that each of them, for $j = 1, \ldots, p$, contains only three unknowns, $\mathbf{v}_j$, H_j, and G_j. For $j = 0$, there are only two unknowns, H_0 and G_0. This is achieved due to exploiting operators $Q_0, \ldots, Q_p$ determined in Theorem 1.

The iteration procedure represented in Section 4.1 is based on the idea of the maximum block improvement method [75] which is an efficient technique for solving the spherically constrained homogeneous polynomial optimization problems. The associated novelties are in the techniques for the determination of matrices $H_0, G_0, \ldots, H_p, G_p$ given in Section 3.2, injections $\mathbf{v}_1, \ldots, \mathbf{v}_p$ represented in Section 4.2, and for the error analysis given in Sections 3.3 and 4.4. In particular, it is shown in Theorem 8 in Section 4.2 that the desired vector $\mathbf{v}_j$ is determined from the solution of a special Fredholm integral equation of the second kind (71). Its solution is called the *optimal injection*.

Further, unlike the known techniques, the proposed approximating operator $\mathcal{T}_p$ has several degrees of freedom to minimize the associated error. They are:

- 'degree' p of $\mathcal{T}_p$,
- matrices $G_0, H_0, \ldots, G_p, H_p$,
- optimal injections $\mathbf{v}_1, \ldots, \mathbf{v}_p$,
- values of $r_0, \ldots, r_p$ in (4), and

- dimensions $q_1, \ldots, q_p$ of injections $\mathbf{v}_1, \ldots, \mathbf{v}_p$.

It is shown in Sections 3.3 and 4.4 below that both the optimal choice of $G_0, H_0, \ldots, G_p, H_p$ and injections $\mathbf{v}_1, \ldots, \mathbf{v}_p$, and the increase in $r_0, \ldots, r_p$ and $q_1, \ldots, q_p$ lead to the decrease in the error associated with approximating operator $\mathcal{T}_p$. Injections $\mathbf{v}_1, \ldots, \mathbf{v}_p$ represent a new special feature of the proposed technique.

In terms of multilayer networks, the novelties are as follows. First, the associated network consists of four hidden layers, i.e., more than in the known networks. Second, it exploits the interaction between the hidden layers (see Figure 1 where a particular case of $\mathcal{T}_p(\mathbf{v}_0, \ldots, \mathbf{v}_p)$ for $p = 3$ is illustrated). These particular features imply the improvement in the associated accuracy (see Sections 3.3 and 4.4). In Figure 1, $\widetilde{\mathbf{x}}$ denotes the approximation of $\mathbf{x} = \mathcal{F}(\mathbf{y})$ determined by the proposed technique, i.e., $\widetilde{\mathbf{x}} = \mathcal{T}_p(\mathbf{v}_0, \ldots, \mathbf{v}_p)$. Details are provided in Section 4.

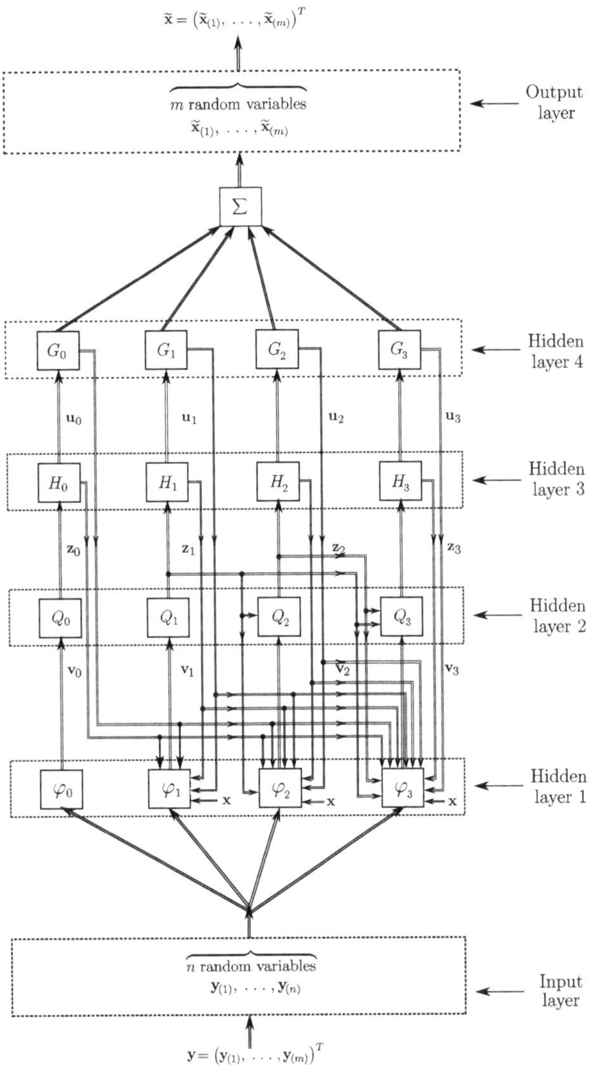

Figure 1. Diagrammatical representation of the proposed technique.

In terms of system identification, side by side with the black-box problem as in [73,74], the problem in (7)–(9) can be interpreted as a quite wide generalization of the blind system identification problem considered, in particular, in [17,76–78]. This is because vectors $\mathbf{v}_1, \ldots, \mathbf{v}_p$ are assumed to be unknown. Unlike the existing work on blind system identification, the problem in (7)–(9) is stated in terms of random vectors. Other novelties associated with system identification are similar to those considered above, i.e., the proposed model $\mathcal{T}_p(\mathbf{v}_0, \ldots, \mathbf{v}_p)$ of the system contains $2p+1$ unknowns, the method of their determination is different from the existing techniques, and the associated accuracy is improved by a variation in more degrees of freedom.

3. Preliminary Results

Here, we consider the determination of vectors $\mathbf{z}_1, \ldots, \mathbf{z}_p$ (in Definition 1 that follows, they are called pairwise uncorrelated vectors) and the solution of a particular case of the problem in (7), (8), (9) where the minimization with respect to $\mathbf{v}_1, \ldots, \mathbf{v}_p$ is not included. These preliminary results will be used in Section 4, where the solution to the original problem represented by (7), (8), (9) is provided.

Definition 1. *Random vectors $\mathbf{z}_0, \ldots, \mathbf{z}_p$ are called pairwise uncorrelated if the condition in (9) holds for any pair of vectors $\mathbf{z}_i$ and $\mathbf{z}_j$, for $i \neq j$, where $i, j = 0, \ldots, p$. Two vectors $\mathbf{z}_i$ and $\mathbf{z}_j$ belonging to the set of the pairwise uncorrelated vectors are called uncorrelated.*

For $j = 0, \ldots, p$, let $\mathcal{N}(M^{(j)})$ be a null space of matrix $M^{(j)} \in \mathbb{R}^{q_j \times q_j}$.

Definition 2. *Random vectors $\mathbf{v}_0, \ldots, \mathbf{v}_p$ are called linearly independent in the generalized sense if, for every collection of matrices $M^{(0)}, \ldots, M^{(p)}$,*

$$M^{(0)}\mathbf{v}_0(\omega) + \ldots + M^{(p)}\mathbf{v}_p(\omega) = \mathbb{O},$$

for almost all $\omega \in \Omega$, implies that $\mathbf{v}_j(\omega) \in \mathcal{N}(M^{(j)})$, for each $j = 0, \ldots, p$, and almost all $\omega \in \Omega$.

Definition 3. *Random vector $\mathbf{v}_j$, for $j = 0, \ldots, p$, is called the well-defined injection if*

$$\Gamma_{z_j} = E_{xz_j} E^{\dagger}_{z_j z_j} E_{z_j x} \neq \mathbb{O}, \tag{12}$$

where $\mathbf{z}_j$ is defined by (3). Otherwise, injection $\mathbf{v}_j$ is called ill-defined.

An explanation for introducing Definition 3 is provided by Remark 1 at the end of Section 3.2 below.

3.1. Determination of Pairwise Uncorrelated Vectors

Theorem 1. *Let random vectors $\mathbf{v}_0, \ldots, \mathbf{v}_p$ be linearly independent in the generalized sense. Then they are transformed to the pairwise uncorrelated vectors $\mathbf{z}_0, \ldots, \mathbf{z}_p$ by transformations $\mathcal{Q}_0, \ldots, \mathcal{Q}_p$ as follows:*

$$\mathbf{z}_0 = \mathcal{Q}_0(\mathbf{v}_0) = \mathbf{v}_0 \quad \textit{and, for } j = 1, \ldots, p, \tag{13}$$

$$\mathbf{z}_j = \mathcal{Q}_j(\mathbf{v}_j, Z_{j-1}) = \mathbf{v}_j - \sum_{k=0}^{j-1} E_{v_j z_k} E^{\dagger}_{z_k z_k} \mathbf{z}_k. \tag{14}$$

Proof. Suppose that the condition in (9) holds for $\mathbf{z}_0, \ldots, \mathbf{z}_{i-1}$. Then, for $\ell = 0, \ldots, i-1$,

$$\begin{aligned} E_{z_i z_\ell} &= E[(\mathbf{v}_i - \sum_{l=0}^{i-1} E_{v_i z_l} E^{\dagger}_{z_l z_l} \mathbf{z}_l)\mathbf{z}_\ell^T] \\ &= E_{v_i z_\ell} - \sum_{l=0}^{i-1} E_{v_i z_l} E^{\dagger}_{z_l z_l} E_{z_l z_\ell} \\ &= E_{v_i z_\ell} - E_{v_i z_\ell} E^{\dagger}_{z_\ell z_\ell} E_{z_\ell z_\ell} = \mathbb{O}. \end{aligned} \tag{15}$$

The latter is true because by Lemma 1 in [79],

$$E_{v_i z_\ell} E^{\dagger}_{z_\ell z_\ell} E_{z_\ell z_\ell} = E_{v_i z_\ell}.$$

Thus, by induction, (9) holds for any $i = 0, \ldots, p$. □

In terms of a multi layer network, this procedure is illustrated diagrammatically in Figure 1.

The solution device of the problem in (7)–(9) is based, in particular, on the solution of the problem

$$\min_{G_0, H_0, \ldots, G_p, H_p} \left\| \mathcal{F}(\mathbf{y}) - \sum_{j=0}^{p} G_j H_j \mathbf{z}_j \right\|_{\Omega}^2 \tag{16}$$

subject to (8) and (9). In the following Section 3.2, matrices $G_0, H_0, \ldots, G_p, H_p$ that solve this problem are given.

3.2. Determination of Matrices $G_0, H_0, \ldots, G_p, H_p$ That Solve the Problem in (16), (8) and (9)

First, recall the definition of a truncated singular value decomposition (SVD). Let the SVD of matrix $A \in \mathbb{R}^{m\times n}$ be given by $A = U_A \Sigma_A V_A^T$, where $U_A = [u_1\ u_2\ \ldots u_m] \in \mathbb{R}^{m\times m}$, $V_A = [v_1\ v_2\ \ldots v_n] \in \mathbb{R}^{n\times n}$ are unitary matrices, and $\Sigma_A = \text{diag}(\sigma_1(A), \ldots, \sigma_{\min(m,n)}(A)) \in \mathbb{R}^{m\times n}$ is a generalized diagonal matrix, with the singular values $\sigma_1(A) \geq \sigma_2(A) \geq \ldots \geq 0$ on the main diagonal. For $k < m$, $j < n$ and $\ell < \min(m,n)$, we denote

$$U_{A,k} = [u_1\ u_2\ \ldots u_k], \quad V_{A,j} = [v_1\ v_2\ \ldots v_j], \quad \Sigma_{A,\ell} = \text{diag}(\sigma_1(A), \ldots, \sigma_\ell(A)),$$

and write

$$\Pi_{A,L} = \sum_{k=1}^{\text{rank}\,(A)} u_k u_k^T \quad \text{and} \quad \Pi_{A,R} = \sum_{j=1}^{\text{rank}\,(A)} v_j v_j^T.$$

For $r = 1, \ldots, \text{rank}\,(A)$,

$$[A]_r = \sum_{i=1}^{r} \sigma_i(A) u_i v_i^T \in \mathbb{R}^{m\times n}, \tag{17}$$

is the truncated SVD of A. For $r \geq \text{rank}\,(A)$, we write $[A]_r = A\ (= A_{\text{rank}\,(A)})$.

Proposition 1. *For any random vector $\mathbf{x} \in L^2(\Omega, \mathbb{R}^m)$,*

$$\|\mathbf{x}\|_{\Omega}^2 = \text{tr}\,\{E_{xx}\}. \tag{18}$$

Proof. We have

$$\|\mathbf{x}\|_\Omega^2 = \int_\Omega \|\mathbf{x}(\omega)\|^2 d\mu(\omega) = \int_\Omega \text{tr}\,\{x(\omega)[\mathbf{x}(\omega)]^T\} d\mu(\omega)$$
$$= \sum_{j=1}^{m} \int_\Omega [\mathbf{x}_{(j)}(\omega)]^2 d\mu(\omega) = \text{tr}\,\{E_{xx}\}.$$

Thus, (18) is true. □

Theorem 2. *Let $\mathbf{v}_0, \dots, \mathbf{v}_p$ be well-defined injections and vectors $\mathbf{z}_0, \dots, \mathbf{z}_p$ be pairwise uncorrelated. Then the minimal Frobenius norm solution to the problem in (16) is given, for $j = 0, \dots, p$, by*

$$G_j = U_{\Gamma_{z_j}, r_j} \quad \text{and} \quad H_j = U^T_{\Gamma_{z_j}, r_j} E_{xz_j} E^\dagger_{z_j z_j}, \tag{19}$$

Proof. Let $P_j = G_j H_j$, $P = [P_0, \dots, P_p]$ and $\mathbf{w} = [\mathbf{z}_0^T, \dots, \mathbf{z}_p^T]^T$. Then, for $\mathbf{x} = \mathcal{F}(\mathbf{y})$,

$$\|\mathcal{F}(\mathbf{y}) - \sum_{j=0}^{p} P_j \mathbf{z}_j\|_\Omega^2 = \text{tr}\Big\{ E_{xx} - E_{xw}P^T - PE_{wx} + PE_{ww}P^T \Big\}, \tag{20}$$

where $E_{xw} = [E_{xz_0}, \dots, E_{xz_p}]$ and by Theorem 1, matrix E_{ww} is block-diagonal, $E_{ww} = diag[E_{z_0 z_0}, \dots, E_{z_p z_p}]$. Thus,

$$PE_{ww}P^T = \sum_{j=0}^{p} P_j E_{z_j z_j} P_j^T \quad \text{and} \quad PE_{wx} = \sum_{j=0}^{p} P_j E_{z_j x}$$

and then

$$\|\mathcal{F}(\mathbf{y}) - \sum_{j=0}^{p} P_j \mathbf{z}_j\|_\Omega^2 = \text{tr}\left\{ E_{xx} - \sum_{j=0}^{p} E_{xz_j} P_j^T - \sum_{j=0}^{p} P_j E_{z_j x} + \sum_{j=0}^{p} P_j E_{z_j z_j} P_j^T \right\}.$$

At the same time,

$$\|\mathcal{F}(\mathbf{y}) - P_j \mathbf{z_j}\|_\Omega^2 = \text{tr}\{E_{xx} - E_{xz_j}P_j^T - P_j E_{z_j x} + P_j E_{z_j z_j} P_j^T\}. \tag{21}$$

Therefore, (20) and (21) imply

$$\|\mathcal{F}(\mathbf{y}) - \sum_{j=0}^{p} P_j \mathbf{z}_j\|_\Omega^2 = \sum_{j=0}^{p} \|\mathcal{F}(\mathbf{y}) - P_j \mathbf{z_j}\|_\Omega^2 - \text{tr}\{pE_{xx}\}. \tag{22}$$

The RHS in (22) is nonnegative. Indeed, for $R = -\sum_{j=0}^{p} E_{xz_j}P_j^T - \sum_{j=0}^{p} P_j E_{z_j x} + \sum_{j=0}^{p} P_j E_{z_j z_j} P_j^T$, we have $\|\mathcal{F}(\mathbf{y}) - \sum_{j=0}^{p} P_j \mathbf{z}_j\|_\Omega^2 = \text{tr}\{E_{xx} + R\} \geq 0$, i.e., $\text{tr}\{E_{xx} + (pE_{xx} - pE_{xx}) + R\} \geq 0$. Here, $\text{tr}\{E_{xx} + (pE_{xx} - pE_{xx}) + R\} = \sum_{j=0}^{p} \|\mathcal{F}(\mathbf{y}) - P_j \mathbf{z_j}\|_\Omega^2 - \text{tr}\{pE_{xx}\}$. Further, the case $\mathcal{F}(\mathbf{y}) = P_j \mathbf{z_j}$ is not possible since matrix P_j is singular. Further,

$$\begin{aligned} \|\mathcal{F}(\mathbf{y}) - P_j \mathbf{z_j}\|_\Omega^2 &= \|E_{xx}^{1/2}\|^2 - \|E_{xz_j}(E_{z_j z_j}^{1/2})^\dagger\|^2 + \|(P_j - E_{xz_j}E^\dagger_{z_j z_j})E^{1/2}_{z_j z_j}\|^2 \\ &= \|E_{xx}^{1/2}\|^2 - \|E_{xz_j}(E_{z_j z_j}^{1/2})^\dagger\|^2 + \|P_j E^{1/2}_{z_j z_j} - E_{xz_j}(E^{1/2}_{z_j z_j})^\dagger\|^2 \end{aligned} \tag{23}$$

because $E^\dagger_{z_j z_j} E^{1/2}_{z_j z_j} = (E^{1/2}_{z_j z_j})^\dagger$ and

$$E_{xz_j} E^\dagger_{z_j z_j} E_{z_j z_j} = E_{xz_j} \tag{24}$$

(see [80]).

Let us now denote by $\mathbb{R}_{r_j}^{m\times n}$ the set of all $m \times n$ matrices of rank at most r_j. In the RHS of (23), only the last term depends on P_j. Therefore, on the basis of [24,68,81–84], the minimal Frobenius norm solution to the problem

$$\min_{P_j \in \mathbb{R}_{r_j}^{m\times n}} \|\mathcal{F}(\mathbf{y}) - P_j \mathbf{z}_j\|_\Omega^2, \tag{25}$$

for $j = 0, \ldots, p$, is given by

$$P_j = G_j H_j = U_{\Gamma_{z_j}, r_j} U_{\Gamma_{z_j}, r_j}^T E_{xz_j} E_{z_j z_j}^\dagger. \tag{26}$$

Then, (19) follows from (26). □

Remark 1. *Definition 3 of the well-defined injections is motivated by the following observation. It follows from (19) that if, for all $j = 0, \ldots, p$, vector $\mathbf{v}_j$ is such that $\Gamma_{z_j} = \mathbb{O}$, then $G_j = \mathbb{O}$ and $H_j = \mathbb{O}$. In other words, then approximating operator $\mathcal{T}_p = \mathbb{O}$.*

Therefore, in Theorem 2 above and in the theorems below, vectors $\mathbf{v}_0, \ldots, \mathbf{v}_p$ are assumed well-defined.

3.3. Error Analysis Associated with the Solution of Problem in (16), (8) and (9)

In Theorem 3 of this section, we obtain the constructive representation of the error associated with the solution of problem in (16), (8), and (9). In Theorem 4, we show that the error can be improved by the increase in the dimensions of injections $\mathbf{v}_1, \ldots, \mathbf{v}_p$. Further, Theorems 3, 4, and 5 establish that the error is also diminished by the increase in the degree of approximating operator $\mathcal{T}_p$.

The error associated with $\mathcal{T}_p(\mathbf{v}_0, \ldots, \mathbf{v}_p) = \sum_{j=0}^{p} G_j H_j \mathbf{z}_j$ is denoted by

$$\varepsilon_{GH}^{(p)} = \min_{G_0, H_0, \ldots, G_p, H_p} \|\mathcal{F}(\mathbf{y}) - \sum_{j=0}^{p} G_j H_j \mathbf{z}_j\|_\Omega^2.$$

Let us denote the Frobenius norm by $\|\cdot\|$.

Theorem 3. *For $j = 0, \ldots, p$, let $A_j = E_{xz_j}(E_{z_j z_j}^{1/2})^\dagger$, $\text{rank}\,(A_j) = s_j$ and $s_j \geq r_j + 1$. For $k = 1, \ldots, s_j$, let $\sigma_k(A_j)$ be a singular value of A_j. Let $G_0, H_0, \ldots, G_p, H_p$ be determined by (19). Then*

$$\varepsilon_{GH}^{(p)} = tr\{E_{xx}\} - \sum_{j=0}^{p} \sum_{k=1}^{r_j} \sigma_k^2(A_j) \tag{27}$$

In particular, the error decreases as p increases.

Proof. In the notation introduced in (17), matrix $G_j H_j$ in (19) is represented as $G_j H_j = [E_{xz_j}(E_{z_j z_j}^{1/2})^\dagger]_{r_j} (E_{z_j z_j}^{1/2})^\dagger$. Therefore, in (23),

$$\begin{aligned}
\|G_j H_j E_{z_j z_j}^{1/2} - E_{xz_j}(E_{z_j z_j}^{1/2})^\dagger\|^2 &= \|[E_{xz_j}(E_{z_j z_j}^{1/2})^\dagger]_{r_j} (E_{z_j z_j}^{1/2})^\dagger E_{z_j z_j}^{1/2} - E_{xz_j}(E_{z_j z_j}^{1/2})^\dagger\|^2 \\
&= \|[E_{xz_j}(E_{z_j z_j}^{1/2})^\dagger]_{r_j} - E_{xz_j}(E_{z_j z_j}^{1/2})^\dagger\|^2 \\
&= \|[A_j]_{r_j} - A_j\|^2 \\
&= \sum_{k=r_j+1}^{s_j} \sigma_k^2(A_j).
\end{aligned} \tag{28}$$

because $[E_{xz_j}(E_{z_jz_j}^{1/2})^{\dagger}]_{r_j}(E_{z_jz_j}^{1/2})^{\dagger}E_{z_jz_j}^{1/2} = [E_{xz_j}(E_{z_jz_j}^{1/2})^{\dagger}]_{r_j}$. Further, since

$$\|E_{xz_j}(E_{z_jz_j}^{1/2})^{\dagger}\|^2 = \|A_j\|^2 = \sum_{k=1}^{s_j} \sigma_k^2(A_j), \tag{29}$$

then (22), (23), (28), and (29) imply (27). □

Let us write

$$A_j = \{a_{ki(j)}\}_{k,i=1}^{m,q_j} \quad \text{and} \quad A_j - [A_j]_{r_j} = \{b_{ki(j)}\}_{k,i=1}^{m,q_j}$$

where $a_{ki(j)}$ and $a_{ki(j)}$ are entries of matrices A_j and $A_j - [A_j]_{r_j}$, respectively. Let us also denote

$$\gamma_{k,(j)} = \sum_{i=1}^{m}\left(a_{ki(j)}^2 - b_{ki(j)}^2\right), \quad \gamma_{(j)} = \max\{\gamma_{1,(j)}, \ldots, \gamma_{q_j,(j)}\},$$

$$\gamma = \max_{j=1,\ldots,p} \gamma_{(j)}, \alpha_0 = \text{tr}\{E_{xx} - \sum_{j=0}^{p} A_j A_j^T\} \text{ and } q = q_1 + \ldots, +q_p.$$

In the following theorem we show that injections $\mathbf{v}_1, \ldots, \mathbf{v}_p$ are useful in the sense that their dimensions increase, so the error associated with the solution to the problem in (16), (8), and (9) is diminished.

Theorem 4. *Let $\mathbf{v}_1, \ldots, \mathbf{v}_p$ be well-defined injections and let matrices $G_0, H_0, \ldots, G_p, H_p$ be defined by Theorem 2. Then the associated error decreases as the sum q of dimensions of injections $\mathbf{v}_1, \ldots, \mathbf{v}_p$ increases. In particular, there is $\beta \in (0, \gamma]$ such that, given $\alpha \geq \alpha_0$, then*

$$\alpha_0 \leq \varepsilon_{GH}^{(p)} \leq \alpha \tag{30}$$

if

$$q \geq \frac{tr\{E_{xx}\} - \sum_{k=1}^{r_0} \sigma_k^2(A_0)}{\beta}. \tag{31}$$

Proof. It follows from (22), (27), (23), and (28) that

$$\begin{aligned} \varepsilon_{GH}^{(p)} &= \|\mathcal{F}(\mathbf{y}) - G_j H_0 \mathbf{z_0}\|_\Omega^2 + \sum_{j=1}^{p} \|\mathcal{F}(\mathbf{y}) - G_j H_j \mathbf{z_j}\|_\Omega^2 - \text{tr}\{p E_{xx}\} \\ &= \text{tr}\{E_{xx}\} - \sum_{k=1}^{r_0} \sigma_k^2(A_0) + \sum_{j=1}^{p} \|\mathcal{F}(\mathbf{y}) - G_j H_j \mathbf{z_j}\|_\Omega^2 - \text{tr}\{p E_{xx}\} \end{aligned} \tag{32}$$

where

$$\begin{aligned} \|\mathcal{F}(\mathbf{y}) - G_j H_j \mathbf{z_j}\|_\Omega^2 &= \text{tr}\{E_{xx}\} - \left(\|A_j\|^2 - \left\|[A_j]_{r_j} - A_j\right\|^2\right) \\ &= \text{tr}\{E_{xx}\} - \sum_{k=1}^{q_j}\sum_{i=1}^{m}\left(a_{ki(j)}^2 - b_{ki(j)}^2\right). \end{aligned}$$

Therefore,

$$\varepsilon_{GH}^{(p)} = \text{tr}\{E_{xx}\} - \sum_{k=1}^{r_0} \sigma_k^2(A_0) - \sum_{j=1}^{p}\sum_{k=1}^{q_j}\sum_{i=1}^{m}\left(a_{ki(j)}^2 - b_{ki(j)}^2\right). \tag{33}$$

Here, $\sum_{k=1}^{q_j}\sum_{i=1}^{m}\left(a^2_{ki(j)} - b^2_{ki(j)}\right) > 0$, since by (28),

$$\|A_j\|^2 - \left\|[A_j]_{r_j} - A_j\right\|^2 = \sum_{k=1}^{r_j}\sigma_k^2(A_j) > 0. \tag{34}$$

Thus, (32)–(34) imply that $\varepsilon_{GH}^{(p)}$ decreases as q_j increases, for $j = 1, \ldots, p$, and p increases.

Further, (27) implies

$$\varepsilon_{GH}^{(p)} \geq \text{tr}\{E_{xx}\} - \sum_{j=0}^{p}\sum_{k=1}^{s_j}\sigma_k^2(A_j) = \text{tr}\left\{E_{xx} - \sum_{j=0}^{p} A_j A_j^T\right\} = \alpha_0. \tag{35}$$

Since

$$0 < \sum_{j=1}^{p}\sum_{k=1}^{q_j}\sum_{i=1}^{m}\left(a^2_{ki(j)} - b^2_{ki(j)}\right) \leq \gamma\sum_{j=1}^{p} q_j = \gamma q,$$

then

$$\sum_{j=1}^{p}\sum_{k=1}^{q_j}\sum_{i=1}^{m}\left(a^2_{ki(j)} - b^2_{ki(j)}\right) = q\beta. \tag{36}$$

Therefore, (33), (35), and (36) imply

$$\alpha_0 \leq \varepsilon_{GH}^{(p)} = \text{tr}\{E_{xx}\} - \sum_{k=1}^{r_0}\sigma_k^2(A_0) - q\beta.$$

thus, if $\varepsilon_{GH}^{(p)} \leq \alpha$, then (31) is true. Conversely, if the latter is true, then $\varepsilon_{GH}^{(p)} \leq \alpha$. □

Remark 2. *An empirical explanation of Theorem 4 is that the increase in q implies the increase in the dimensions of matrices $H_1, \ldots, H_p$ in (6) and (19). Hence, it implies the increase in the number of parameters to optimize. As a result, for a fixed parameter r given by (4), the accuracy associated with approximating operator $\mathcal{T}_p$ improves. Further, it follows from (31) that, as q increases, $\varepsilon_{GH}^{(p)}$ tends to α_0, which is the error associated with the full rank approximating operator $\mathcal{S}_h$ (see (40) and (46) below).*

Remark 3. *By Theorem 3, the error associated with solution of problem (16) decreases as the degree p of the approximating operator $\mathcal{T}_p$ increases. At the same time, the increase in degree p of the approximating operator $\mathcal{T}_p$ may involve an increase in parameter r (see (4)). However, by a condition of some applied problem in hand, r must be fixed. In the following Theorem 5, under the condition of fixed r, the case of decreasing the error as the degree p of the approximating operator increases is detailed.*

Theorem 5. *Let r and r_j, for $j = 0, \ldots, p$, be given. Let g be a nonnegative integer such that $g < p$ and let $\ell_g = r_g + r_{g+1} + \ldots + r_p$. If*

$$\sum_{k=r_g+1}^{r_{g+1}+\ldots+r_p}\sigma_k^2(A_g) < \sum_{j=g+1}^{p}\sum_{k=1}^{r_j}\sigma_k^2(A_j), \tag{37}$$

where $\sum_{j=g+1}^{p}\sum_{k=1}^{r_j}\sigma_k^2(A_j) = \sum_{j=g+1}^{p}\sum_{k=1}^{q_j}\sum_{i=1}^{m}\left(a_{ki(j)}^2 - b_{ki(j)}^2\right)$, *then*

$$\varepsilon_{GH}^{(x)} < \varepsilon_{GH}^{(c)}, \tag{38}$$

i.e., for the same r, the error associated with the approximating operator of higher degree p is less than the error associated with the approximating operator of lower degree g.

Proof. We write $r = r_0 + \ldots + r_{g-1} + \ell_g$. Then,

$$\begin{aligned} \varepsilon_{GH}^{(g)} &= \operatorname{tr}\{E_{xx}\} - \sum_{j=0}^{g-1}\sum_{k=1}^{r_j}\sigma_k^2(A_j) - \sum_{k=1}^{\ell_g}\sigma_k^2(A_g) \\ &= \operatorname{tr}\{E_{xx}\} - \sum_{j=0}^{g-1}\sum_{k=1}^{r_j}\sigma_k^2(A_j) - \sum_{k=1}^{r_g}\sigma_k^2(A_g) - \sum_{k=r_g+1}^{r_{g+1}+\ldots+r_p}\sigma_k^2(A_g). \end{aligned} \tag{39}$$

Thus, (27) and (39) imply (37) and (38). □

Remark 4. *The RHS in (37) increases as the dimension q_j of at least single injection $\mathbf{v}_j$, for $j = g+1, \ldots, p$, increases while the LHS does not depend on q_j. In other words, one can always find q_j, for $j = g+1, \ldots, p$, such that the inequality in (37) is true.*

Example 1. *Here, we wish to numerically illustrate Theorems 3, 4, and 5. To this end, we assume that $\mathbf{x} \in L^2(\Omega, \mathbb{R}^m)$ and $\mathbf{y} \in L^2(\Omega, \mathbb{R}^m)$ are uniformly and normally distributed random vectors, respectively. Injections $\mathbf{v}_1 \in L^2(\Omega, \mathbb{R}^{q_1})$ and $\mathbf{v}_2 \in L^2(\Omega, \mathbb{R}^{q_2})$ are here chosen as uniformly distributed random vectors. Covariance matrices E_{xv_j}, $E_{v_iv_j}$, for $i, j = 0, 1, 2$, are represented by $E_{xv_j} = \frac{1}{s}XV_j^T$ and $E_{v_iv_j} = \frac{1}{s}V_iV_j^T$ where $X \in \mathbb{R}^{m\times s}$ and $V_j \in \mathbb{R}^{q_j\times s}$ are sample matrices of $\mathbf{x}$ and $\mathbf{v}_j$, respectively, for $j = 0, 1, 2$.*

We choose $m = 100$ and $r = 50$. It follows from Theorems 3, 4, and 5 that the error associated with approximating operator $\mathcal{T}_p(\mathbf{v}_0, \ldots, \mathbf{v}_p)$ varies if values of p, $q = q_0 + \ldots + q_p$ and $r = r_0 + \ldots + r_p$ vary. We wish to illustrate the decrease in error when, for the same r, values of p and q increase. To this end, we provide Tables 1 and 2, where values of the errors are given for different values of q_j, r_j, for $j = 0, 1, 2$ and $p = 0, 1, 2$. In Tables 1 and 2, the abbreviation MSE (mean square error) is used to unite notations $\varepsilon_{GH}^{(p)}$ and $\varepsilon_{GH}^{(g)}$ used in Theorems 3, 4, and 5. In the tables, Cases 1 and 2 for specific values of q_j, r_j are considered.

In Figure 2 the MSE values are illustrated diagrammatically.

It follows from Tables 1 and 2 and Figure 2 that, for the same r, the error associated with the proposed system model decreases if degree p or the sum q of the injection dimensions increases. This is because the increase in the number of parameters to optimize in the operator $\mathcal{T}$ results in a decrease in the error, as stated in Theorem 5.

Table 1. Numerical characterizations of approximating operators $\mathcal{T}_0(\mathbf{v}_0)$, $\mathcal{T}_1(\mathbf{v}_0, \mathbf{v}_1)$, and $\mathcal{T}_2(\mathbf{v}_0, \mathbf{v}_1, \mathbf{v}_2)$ in Case 1.

Approximating Operator	q_0	q_1	q_2	r_0	r_1	r_2	MSE
$\mathcal{T}_0(\mathbf{v}_0)$	100	N/A	N/A	50	N/A	N/A	8.30
$\mathcal{T}_1(\mathbf{v}_0, \mathbf{v}_1)$	100	25	N/A	25	25	N/A	7.93
$\mathcal{T}_2(\mathbf{v}_0, \mathbf{v}_1, \mathbf{v}_2)$	100	25	500	17	17	16	7.03

Table 2. Numerical characterizations of operators $\mathcal{T}_0(\mathbf{v}_0)$, $\mathcal{T}_1(\mathbf{v}_0, \mathbf{v}_1)$, and $\mathcal{T}_2(\mathbf{v}_0, \mathbf{v}_1, \mathbf{v}_2)$ in Case 2.

Approximating Operator	q_0	q_1	q_2	r_0	r_1	r_2	MSE
$\mathcal{T}_0(\mathbf{v}_0)$	100	N/A	N/A	50	N/A	N/A	8.30
$\mathcal{T}_1(\mathbf{v}_0, \mathbf{v}_1)$	100	200	N/A	25	25	N/A	7.61
$\mathcal{T}_2(\mathbf{v}_0, \mathbf{v}_1, \mathbf{v}_2)$	100	200	500	17	17	16	6.28

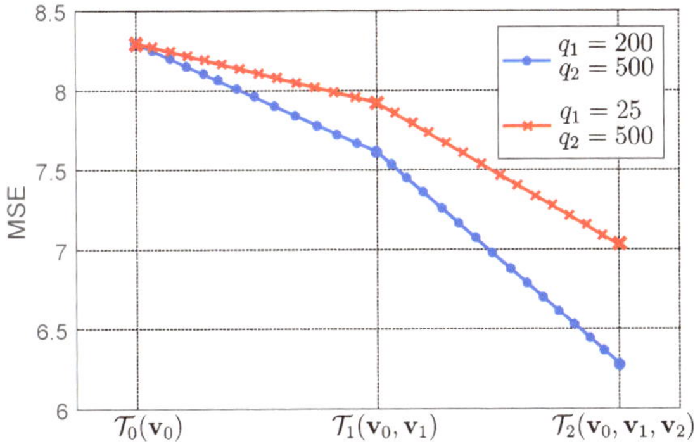

Figure 2. Example 1: Diagrams of the errors associated with $\mathcal{T}_0(\mathbf{v}_0)$, $\mathcal{T}_1(\mathbf{v}_0, \mathbf{v}_1)$, and $\mathcal{T}_2(\mathbf{v}_0, \mathbf{v}_1, \mathbf{v}_2)$.

3.4. Particular Case: No Reduction of Vector Dimensionality

An important particular case of the problem in (16) and (9) is when matrix $G_j H_j$ is replaced with a full rank matrix $M_j \in \mathbb{R}^{m \times q_j}$, for $j = 0, \ldots, h$. Operator $\mathcal{S}_h$ given by

$$\mathcal{S}_h(\mathbf{v}_0, \ldots \mathbf{v}_p) = \sum_{k=0}^{h} M_k \mathbf{z}_k \tag{40}$$

is called *the full rank* approximating operator. The problem then is to find matrices $M_0, \ldots, M_h$ that solve

$$\min_{M_0, \ldots, M_h} \|\mathcal{F}(\mathbf{y}) - \sum_{k=0}^{h} M_k \mathbf{z}_k\|_{\Omega}^2, \tag{41}$$

subject to condition (9).

In particular, searched matrices $M_0, \ldots, M_h$ can be found, for $\omega \in \Omega$, from equations

$$M_j \mathbf{z}_j(\omega) = [\mathcal{F}(\mathbf{y})](\omega), \quad \text{for } j = 0, \ldots, h, \tag{42}$$

and

$$\sum_{k=0}^{h} M_k \mathbf{z}_k(\omega) = [\mathcal{F}(\mathbf{y})](\omega), \tag{43}$$

respectively. Those equations may have an infinite number of solutions or have no solutions. Therefore, instead, the following theorem provides the solution to the problem in (41) where cases (42) and (43) are excluded.

Theorem 6. *Let $\mathbf{v}_0, \ldots, \mathbf{v}_p$ be well-defined injections and vectors $\mathbf{z}_0, \ldots, \mathbf{z}_p$ be pairwise uncorrelated. Let $\mathcal{F}(\mathbf{y}) \neq M_j\mathbf{z_j}$, for $j = 0, \ldots, h$, and $\mathcal{F}(\mathbf{y}) \neq \sum_{k=0}^{h} M_k\mathbf{z}_k$. Then, the minimal Frobenius norm solution to the problem (41) is given, for $k = 0, \ldots, h$, by*

$$M_k = E_{xz_k}E^{\dagger}_{z_kz_k}. \tag{44}$$

Proof. Similar to (22),

$$\|\mathcal{F}(\mathbf{y}) - \sum_{j=0}^{h} M_j\mathbf{z}_j\|^2_{\Omega} = \sum_{j=0}^{h} \|\mathbf{x} - M_j\mathbf{z_j}\|^2_{\Omega} - \text{tr}\{hE_{xx}\}. \tag{45}$$

It is known (see, for example, [16,20,21]) that the minimal Frobenius norm solution to the problem

$$\min_{M_j} \|\mathbf{x} - M_j\mathbf{z_j}\|^2_{\Omega}$$

is given by (44). □

Theorem 7. *Let $A_j = E_{xz_j}(E^{1/2}_{z_jz_j})^{\dagger}$. The error associated with the minimal Frobenius norm solution to the problem in (41) is represented by*

$$\min_{M_0,\ldots,M_h} \|\mathcal{F}(\mathbf{y}) - \sum_{j=0}^{h} M_j\mathbf{z}_j\|^2_{\Omega} = tr\{E_{xx}\} - \sum_{j=0}^{h} \|A_j\|^2. \tag{46}$$

Proof. For M_j determined by (44),

$$\|\mathcal{F}(\mathbf{y}) - M_j\mathbf{z}_j\|^2_{\Omega} = \text{tr}\{E_{xx}\} - \|E_{xz_j}(E^{1/2}_{z_jz_j})^{\dagger}\|^2 + \|M_jE^{1/2}_{z_jz_j} - E_{xz_j}(E^{1/2}_{z_jz_j})^{\dagger}\|^2. \tag{47}$$

Here,

$$\begin{aligned} \|M_jE^{1/2}_{z_jz_j} - E_{xz_j}(E^{1/2}_{z_jz_j})^{\dagger}\|^2 &= \|E_{xz_j}E^{\dagger}_{z_jz_j}E^{1/2}_{z_jz_j} - E_{xz_j}(E^{1/2}_{z_jz_j})^{\dagger}\|^2 \\ &= \|E_{xz_j}(E^{1/2}_{z_jz_j})^{\dagger} - E_{xz_j}(E^{1/2}_{z_jz_j})^{\dagger}\|^2 = 0. \end{aligned} \tag{48}$$

That is,

$$\|\mathcal{F}(\mathbf{y}) - M_j\mathbf{z}_j\|^2_{\Omega} = \text{tr}\{E_{xx}\} - \|E_{xz_j}(E^{1/2}_{z_jz_j})^{\dagger}\|^2. \tag{49}$$

Then, (46) follows from (22) and (48). □

4. Solution of Problem Given by (7), (8), (9)

Now we are in the position to consider a solution to the original problem in (7), (8), (9).

4.1. Device of Solution

In comparison with the problem in (16), (8), (9), a specific difficulty of the original problem in (7), (8), (9) is a determination of p additional unknowns, injections $\mathbf{v}_1, \ldots, \mathbf{v}_p$.

The device of the proposed solution is as follows. First, in (13) and (14), arbitrary linearly independent in the generalized sense vectors $\mathbf{v}_0, \ldots, \mathbf{v}_p$ are denoted by $\mathbf{v}_0^{(0)}, \ldots, \mathbf{v}_p^{(0)}$ and pairwise uncorrelated random vectors $\mathbf{z}_0, \ldots, \mathbf{z}_p$ are denoted by $\mathbf{z}_0(\mathbf{v}_0^{(0)}), \ldots, \mathbf{z}_p(\mathbf{v}_p^{(0)})$.

In (19), matrices G_j and H_j, for $j = 0, \ldots, p$, are denoted by $G_j^{(0)}$ and $H_j^{(0)}$, respectively. The associated error is still represented by (27). We also write

$$\varepsilon^{(0)} = \|\mathcal{F}(\mathbf{y}) - \sum_{k=0}^{p} G_k^{(0)} H_k^{(0)} \mathbf{z}_k(\mathbf{v}_k^{(0)})\|_\Omega^2. \tag{50}$$

Then, for $i = 0, 1, \ldots$, searched injections $\mathbf{v}_1^{(i+1)}, \ldots, \mathbf{v}_p^{(i+1)}$ and matrices $G_1^{(i+1)}$, $H_1^{(i+1)}, \ldots G_p^{(i+1)}, H_p^{(i+1)}$ are determined by the iterative procedure represented bellow. It will be shown in Theorem 10 below that $\mathbf{v}_1^{(i+1)}, \ldots, \mathbf{v}_p^{(i+1)}$ and $G_1^{(i+1)}$, $H_1^{(i+1)}, \ldots G_p^{(i+1)}, H_p^{(i+1)}$ further minimize the associated error. The i-th iterative loop of the iterative procedure consists of the steps as follows.

The i-th iterative loop, for $i = 0, 1, \ldots$.

Step 1. Given $\mathbf{v}_0 = \mathbf{y}$, $G_0^{(i)}, H_0^{(i)}, \ldots, G_p^{(i)}, H_p^{(i)}$, find $\mathbf{v}_1, \ldots, \mathbf{v}_p$ that solve

$$\min_{\mathbf{v}_1, \ldots, \mathbf{v}_p} \|\mathcal{F}(\mathbf{y}) - \sum_{k=1}^{p} G_k^{(i)} H_k^{(i)} \mathbf{z}_k(\mathbf{v}_k)\|_\Omega^2, \tag{51}$$

where

$$\mathbf{z}_j(\mathbf{v}_j) = \mathbf{v}_j - \sum_{k=0}^{j-1} E_{v_j z_k(v_k)} E^\dagger_{z_k(v_k) z_k(v_k)} \mathbf{z}_k(\mathbf{v}_k). \tag{52}$$

The solution is denoted by $\mathbf{v}_1^{(i+1)}, \ldots, \mathbf{v}_p^{(i+1)}$. We call $\mathbf{v}_1^{(i+1)}, \ldots, \mathbf{v}_p^{(i+1)}$ the *optimal injections*.

Step 2. Given $\mathbf{v}_1^{(i+1)}, \ldots, \mathbf{v}_p^{(i+1)}$, find pairwise uncorrelated random vectors $\mathbf{z}_1(\mathbf{v}_1^{(i+1)}), \ldots, \mathbf{z}_p(\mathbf{v}_p^{(i+1)})$ and denote

$$\varepsilon_v^{(i+1)} = \|\mathcal{F}(\mathbf{y}) - \sum_{k=0}^{p} G_k^{(i)} H_k^{(i)} \mathbf{z}_k(\mathbf{v}_k^{(i+1)})\|_\Omega^2, \tag{53}$$

where, for $k = 0$, we set $G_0^{(i)} = G_0^{(0)}$ and $H_0^{(i)} = H_0^{(0)}$, for all $i = 1, 2, \ldots$ and we write

$$\mathbf{z}_j(\mathbf{v}_j^{(i)}) = \mathbf{v}_j^{(i)} - \sum_{k=0}^{j-1} E_{v_j^{(i)} z_k(v_k^{(i)})} E^\dagger_{z_k(v_k^{(i)}) z_k(v_k^{(i)})} \mathbf{z}_k(\mathbf{v}_k^{(i)}). \tag{54}$$

Step 3. Given $\mathbf{z}_1(\mathbf{v}_1^{(i)}), \ldots, \mathbf{z}_p(\mathbf{v}_p^{(i)})$, find $G_1, H_1, \ldots, G_p, H_p$ that solve

$$\min_{G_1, H_1, \ldots, G_p, H_p} \left\| \mathcal{F}(\mathbf{y}) - \sum_{k=1}^{p} G_k H_k \mathbf{z}_k(\mathbf{v}_k^{(i)}) \right\|_\Omega^2. \tag{55}$$

The solution of problem (55) is denoted by $G_1^{(i+1)}, H_1^{(i+1)}, \ldots, G_p^{(i+1)}, H_p^{(i+1)}$. Further, denote

$$\varepsilon_{GH}^{(i+1)} = \|\mathcal{F}(\mathbf{y}) - \sum_{k=0}^{p} G_k^{(i+1)} H_k^{(i+1)} \mathbf{z}_k(\mathbf{v}_k^{(i)})\|_\Omega^2, \tag{56}$$

where, as before, for $k = 0$, we set $G_0^{(i+1)} = G_0^{(0)}$ and $H_0^{(i+1)} = H_0^{(0)}$, for all $i = 0, 1, \ldots$.

Step 4. Denote

$$\varepsilon^{(i+1)} = \min\{\varepsilon_v^{(i+1)}, \varepsilon_{GH}^{(i+1)}\}. \tag{57}$$

Step 5. If, for a given tolerance δ,

$$\|\varepsilon^{(i+1)} - \varepsilon^{(i)}\|_{\Omega}^{2} \leq \delta, \tag{58}$$

the iterations are stopped. If not, then Steps 1–4 are repeated to form the next iterative loop.

The above steps of the solution device are consummated as follows.

4.2. Determination of $\mathbf{v}_j^{(i+1)}$ in Step 1

Let us denote

$$\mathbf{z}_k^{(i+1)} = \mathbf{z}_k(\mathbf{v}_k^{(i+1)}), \quad P_j^{(i)} = G_j^{(i)} H_j^{(i)}, \quad B_{jk}^{(i+1)} = (P_j^{(i)})^{\dagger} E_{x z_k^{(i+1)}},$$
$$\gamma_\ell^{(i+1)} = E^{\dagger}_{z_\ell^{(i+1)} z_\ell^{(i+1)}} \mathbf{z}_\ell^{(i+1)} \quad \text{and} \quad A_{\ell k}^{(i+1)} = E^{\dagger}_{z_\ell^{(i+1)} z_\ell^{(i+1)}} E_{z_\ell^{(i+1)} z_k^{(i+1)}}.$$

Here, $\gamma_\ell^{(i+1)}(\omega) \in \mathbb{R}^{q_\ell \times q_\ell}$, $B_{jk}^{(i+1)} \in \mathbb{R}^{q_j \times q_k}$ and $A_{\ell k}^{(i+1)} \in \mathbb{R}^{q_\ell \times q_k}$.

Theorem 8. *Let*

$$B_j^{(i)} = [B_{j0}^{(i)}, \ldots, B_{j,j-1}^{(i)}] \quad \text{and} \quad A_j^{(i)} = \{A_{ks}^{(i)}\}_{k,s=0}^{j-1}, \tag{59}$$

where $B_j^{(i)} \in \mathbb{R}^{q_j \times q}$, $A_j^{(i)} \in \mathbb{R}^{q \times q}$ and $q = q_0 + \ldots + q_{j-1}$. Then the optimal injection $\mathbf{v}_j^{(i+1)}$, for $j = 1, \ldots, p$, that solves the problem in (51) is determined by

$$\mathbf{v}_j^{(i+1)}(\omega) = (G_j^{(i)} H_j^{(i)})^{\dagger}[\mathcal{F}(\mathbf{y})](\omega) + \sum_{\ell=0}^{j-1} C_{j\ell}^{(i)} \gamma_\ell^{(i)}(\omega), \tag{60}$$

where matrices $C_{j0}^{(i)}, \ldots, C_{j,j-1}^{(i)}$ are defined by

$$[C_{j0}^{(i)}, \ldots, C_{j,j-1}^{(i)}] = B_j^{(i)}(I - A_j^{(i)})^{\dagger}. \tag{61}$$

Here, $C_{j\ell}^{(i)} \in \mathbb{R}^{q_j \times q_\ell}$, for $\ell = 0, \ldots, j-1$.

Proof. Based on (22), the problem in (51) is represented as follows:

$$\begin{aligned}\min_{\mathbf{v}_1,\ldots,\mathbf{v}_p} \left\| \mathcal{F}(\mathbf{y}) - \sum_{j=0}^{p} P_j^{(i)} \mathbf{z}_j(\mathbf{v}_j) \right\|_{\Omega}^{2} &= \left\| \mathcal{F}(\mathbf{y}) - P_0^{(0)} \mathbf{z}_0(\mathbf{v}_0) \right\|_{\Omega}^{2} \\ &+ \sum_{j=1}^{p} \min_{\mathbf{v}_j} \left\| \mathcal{F}(\mathbf{y}) - P_j^{(i)} \mathbf{z}_j(\mathbf{v}_j) \right\|_{\Omega}^{2} - \mathrm{tr}\{p E_{xx}\}. \end{aligned} \tag{62}$$

Then

$$\min_{\mathbf{v}_j} \left\| \mathcal{F}(\mathbf{y}) - P_j^{(i)} \mathbf{z}_j(\mathbf{v}_j) \right\|_{\Omega}^{2} = \min_{\mathbf{v}_j} \left\| \mathcal{F}(\mathbf{y}) - P_j^{(i)} \left(\mathbf{v}_j - \sum_{k=0}^{j-1} E_{v_j z_k} E^{\dagger}_{z_k z_k} \mathbf{z}_k(\mathbf{v}_k) \right) \right\|_{\Omega}^{2}. \tag{63}$$

Here,

$$
\begin{aligned}
&\left\|\mathcal{F}(\mathbf{y}) - P_j^{(i)}\left(\mathbf{v}_j - \sum_{k=0}^{j-1} E_{v_j z_k} E_{z_k z_k}^{\dagger} \mathbf{z}_k(\mathbf{v}_k)\right)\right\|_{\Omega}^2 \\
&= \int_{\Omega}\left\|[\mathcal{F}(\mathbf{y})](\omega) - P_j^{(i)}\left(\mathbf{v}_j(\omega) - \sum_{k=0}^{j-1} E_{v_j z_k} E_{z_k z_k}^{\dagger} \mathbf{z}_k(\mathbf{v}_k(\omega))\right)\right\|^2 d\mu(\omega) \\
&= \int_{\Omega}\left\|[\mathcal{F}(\mathbf{y})](\omega) - P_j^{(i)} \mathbf{z}_j(\mathbf{v}_j(\omega))\right\|^2 d\mu(\omega). \qquad (64)
\end{aligned}
$$

For $\mathbf{u} = \mathbf{z}_j((\mathbf{v}_j)$, let us denote $\mathcal{U}_v = \{\mathbf{u} \in L^2(\Omega, \mathbb{R}^{q_j}) \mid \mathbf{u} = \mathbf{z}_j((\mathbf{v}_j)\}$. For any $\mathbf{u}$, let us write $\mathcal{U} = \{\mathbf{u} \in L^2(\Omega, \mathbb{R}^{q_j})\}$. For $j = 1, \ldots, p$, the vector $\tilde{\mathbf{u}}(\omega) \in \mathbb{R}^{q_j}$ of the smallest Euclidean norm of all minimizers that solves

$$
\min_{\mathbf{u}(\omega)} \|[\mathcal{F}(\mathbf{y})](\omega) - P_j^{(i)} \mathbf{u}(\omega)\|^2
$$

is given by (see [85], p. 257)

$$
\tilde{\mathbf{u}}(\omega) = (P_j^{(i)})^{\dagger}[\mathcal{F}(\mathbf{y})](\omega).
$$

Let $\mathbf{v}_j^{(i+1)}$ be such that

$$
\min_{\mathbf{v}_j(\omega)} \|[\mathcal{F}(\mathbf{y})](\omega) - P_j^{(i)} \mathbf{z}_j(\mathbf{v}_j(\omega))\|^2 = \|[\mathcal{F}(\mathbf{y})](\omega) - P_j^{(i)} \mathbf{z}_j(\mathbf{v}_j^{(i+1)}(\omega))\|^2.
$$

Since $\mathcal{U}_v \subset \mathcal{U}$ then

$$
\min_{\mathbf{v}_j(\omega)} \|[\mathcal{F}(\mathbf{y})](\omega) - P_j^{(i)} \mathbf{z}_j(\mathbf{v}_j(\omega))\|^2 \geq \min_{\mathbf{u}(\omega)} \|[\mathcal{F}(\mathbf{y})](\omega) - P_j^{(i)} \mathbf{u}(\omega)\|^2, \qquad (65)
$$

i.e.,

$$
\begin{aligned}
\min_{\mathbf{v}_j(\omega)} \|[\mathcal{F}(\mathbf{y})](\omega) - P_j^{(i)} \mathbf{z}_j(\mathbf{v}_j(\omega))\|^2 &\geq \|[\mathcal{F}(\mathbf{y})](\omega) - P_j^{(i)} \tilde{\mathbf{u}}(\omega)\|^2 \\
&= \|[\mathcal{F}(\mathbf{y})](\omega) - P_j^{(i)} (P_j^{(i)})^{\dagger}[\mathcal{F}(\mathbf{y})](\omega)\|^2 \qquad (66)
\end{aligned}
$$

Then (66) implies that, for all $\mathbf{v}_j$,

$$
\|[\mathcal{F}(\mathbf{y})](\omega) - P_j^{(i)} \mathbf{z}_j(\mathbf{v}_j(\omega))\|^2 \geq \|[\mathcal{F}(\mathbf{y})](\omega) - P_j^{(i)} (P_j^{(i)})^{\dagger}[\mathcal{F}(\mathbf{y})](\omega)\|^2. \qquad (67)
$$

Since it is true for all $\mathbf{v}_j$, then

$$
\|[\mathcal{F}(\mathbf{y})](\omega) - P_j^{(i)} (P_j^{(i)})^{\dagger}[\mathcal{F}(\mathbf{y})](\omega)\|^2 = \min_{\mathbf{v}_j(\omega)} \|[\mathcal{F}(\mathbf{y})](\omega) - P_j^{(i)} \mathbf{z}_j(\mathbf{v}_j(\omega))\|^2, \qquad (68)
$$

i.e.,

$$
\|[\mathcal{F}(\mathbf{y})](\omega) - P_j^{(i)} (P_j^{(i)})^{\dagger}[\mathcal{F}(\mathbf{y})](\omega)\|^2 = \|[\mathcal{F}(\mathbf{y})](\omega) - P_j^{(i)} \mathbf{z}_j(\mathbf{v}_j^{(i+1)}(\omega))\|^2.
$$

Therefore,

$$
\mathbf{z}_j(\mathbf{v}_j^{(i+1)}(\omega)) = (P_j^{(i)})^{\dagger}[\mathcal{F}(\mathbf{y})](\omega). \qquad (69)
$$

Taking into account (14), Equation (69) is written as

$$\mathbf{v}_j^{(i+1)}(\omega) = (P_j^{(i)})^\dagger[\mathcal{F}(\mathbf{y})](\omega) + \sum_{k=0}^{j-1} E_{v_j^{(i+1)} z_k^{(i+1)}} E^\dagger_{z_k^{(i+1)} z_k^{(i+1)}} \mathbf{z}_k(\mathbf{v}_k^{(i+1)}(\omega)),$$

where we denote

$$E_{v_j^{(i+1)} z_k^{(i+1)}} = \int_\Omega \mathbf{v}_j^{(i+1)}(\xi)\mathbf{z}_k[\mathbf{v}_k^{(i+1)}(\xi)]^T d\mu(\xi),$$

and

$$E_{z_k^{(i+1)} z_k^{(i+1)}} = \int_\Omega \mathbf{z}_k[\mathbf{v}_k^{(i+1)}(\xi)]\mathbf{z}_k[\mathbf{v}_k^{(i+1)}(\xi)]^T d\mu(\xi).$$

Thus,

$$\mathbf{v}_j^{(i+1)}(\omega) = (P_j^{(i)})^\dagger[\mathcal{F}(\mathbf{y})](\omega) + \sum_{\ell=0}^{j-1}\left[\int_\Omega \mathbf{v}_j^{(i+1)}(\xi)[\mathbf{z}_\ell(\mathbf{v}_\ell^{(i+1)}(\xi))]^T d\mu(\xi)\right]\gamma_\ell^{(i+1)}(\omega). \quad (70)$$

Let us now write Equation (70) as follows:

$$\mathbf{v}_j^{(i+1)}(\omega) = \boldsymbol{\alpha}_j^{(i)}(\omega) + \int_\Omega K_j^{(i+1)}(\omega,\xi)\mathbf{v}_j^{(i+1)}(\xi)d\mu(\xi), \quad (71)$$

where

$$\boldsymbol{\alpha}_j^{(i)}(\omega) = (P_j^{(i)})^\dagger[\mathcal{F}(\mathbf{y})](\omega) \text{ and } K_j^{(i+1)}(\omega,\xi) = \sum_{\ell=0}^{j-1}[\mathbf{z}_\ell(\mathbf{v}_\ell^{(i+1)}(\xi))]^T\gamma_\ell^{(i+1)}(\omega). \quad (72)$$

Recall that in (72), matrix $P_j^{(i)} = G_j^{(i)}H_j^{(i)}$ depends on $\mathbf{v}_j^{(i)}(\omega)$, not on $\mathbf{v}_j^{(i+1)}(\omega)$. Therefore, Equation (71) is a vector version of the Fredholm integral equation of the second kind [86] with respect to $\mathbf{v}_j^{(i+1)}(\omega)$. Its solution is provided as follows. Write Equation (71) as

$$\mathbf{v}_j^{(i+1)}(\omega) = \boldsymbol{\alpha}_j^{(i)}(\omega) + \sum_{\ell=0}^{j-1} C_{j\ell}^{(i+1)}\gamma_\ell^{(i+1)}(\omega), \quad (73)$$

where

$$C_{j\ell}^{(i+1)} = \int_\Omega \mathbf{v}_j^{(i+1)}(\xi)[\mathbf{z}_\ell(\mathbf{v}_\ell^{(i+1)}(\xi))]^T d\mu(\xi).$$

Let us now multiply both sides of (73) by $[\mathbf{z}_k(\mathbf{v}_k^{(i+1)}(\omega))]^T$, for $k = 0, \ldots, j-1$, and integrate. It implies

$$\begin{aligned}\int_\Omega \mathbf{v}_j^{(i+1)}(\omega)[\mathbf{z}_k(\mathbf{v}_k^{(i+1)}(\omega))]^T d\mu(\omega) &= \int_\Omega \boldsymbol{\alpha}_j^{(i)}(\omega)[\mathbf{z}_k(\mathbf{v}_k^{(i+1)}(\omega))]^T d\mu(\omega)\\ &\quad + \sum_{\ell=0}^{j-1} C_{j\ell}^{(i+1)}\int_\Omega \gamma_\ell^{(i+1)}(\omega)[\mathbf{z}_k(\mathbf{v}_k^{(i+1)}(\omega))]^T d\mu(\omega)\end{aligned}$$

and

$$C_{jk}^{(i+1)} = B_{jk}^{(i+1)} + \sum_{\ell=0}^{j-1} C_{j\ell}^{(i+1)} A_{\ell k}^{(i+1)}, \quad (74)$$

where, for $k = 0, \ldots, j-1$,

$$\begin{aligned} B_{jk}^{(i+1)} &= \int_\Omega \boldsymbol{\alpha}_j^{(i+1)}(\omega)[\mathbf{z}_k(\mathbf{v}_k^{(i+1)}(\omega))]^T d\mu(\omega) \\ &= (P_j^{(i)})^\dagger \int_\Omega [\mathcal{F}(\mathbf{y})](\omega)[\mathbf{z}_k(\mathbf{v}_k^{(i+1)}(\omega))]^T d\mu(\omega) = (P_j^{(i)})^\dagger E_{xz_k^{(i+1)}} \in \mathbb{R}^{q_j \times q_k} \end{aligned}$$

and

$$A_{\ell k}^{(i+1)} = \int_\Omega \gamma_\ell^{(i+1)}(\omega)[\mathbf{z}_k(\mathbf{v}_k^{(i+1)}(\omega))]^T d\mu(\omega) = E^\dagger_{z_\ell^{(i+1)} z_\ell^{(i+1)}} E_{z_\ell^{(i+1)} z_k^{(i+1)}} \in \mathbb{R}^{q_\ell \times q_k}.$$

Let $C_j^{(i+1)} = [C_{j0}^{(i+1)}, \ldots, C_{j,j-1}^{(i+1)}] \in \mathbb{R}^{q_j \times q}$. Then the set of matrix equations in (74) can be written as a single equation

$$C_j^{(i+1)} = B_j^{(i+1)} + C_j^{(i+1)} A_j^{(i+1)} \quad \text{or} \quad B_j^{(i+1)} = C_j^{(i+1)}(I - A_j^{(i+1)}). \tag{75}$$

If matrix $I - A_j^{(i+1)}$ is invertible, then (75) implies

$$C_j^{(i+1)} = B_j^{(i+1)}(I - A_j^{(i+1)})^{-1}. \tag{76}$$

If matrix $I - A_j^{(i+1)}$ is singular, then instead of Equation (75), we consider the problem

$$\min_{C_j^{(i+1)}} \|B_j^{(i+1)} - C_j^{(i+1)}(I - A_j^{(i+1)})\|^2. \tag{77}$$

Its minimal Frobenius norm solution is given by [62]:

$$C_j^{(i+1)} = B_j^{(i+1)}(I - A_j^{(i+1)})^\dagger. \tag{78}$$

Thus, injection $\mathbf{v}_j^{(i+1)}$ follows from (73), (76), and (78). □

In terms of a multilayer network, this procedure is illustrated diagrammatically in Figure 1. To simplify the notation, superscripts (i) and $(i+1)$ are omitted in Figure 1.

4.3. Determination of Matrices $G_1^{(i+1)}, H_1^{(i+1)}, \ldots, G_p^{(i+1)}, H_p^{(i+1)}$ in Step 3

In Step 3, the matrices $G_1^{(i+1)}, H_1^{(i+1)}, \ldots, G_p^{(i+1)}, H_p^{(i+1)}$ solve problem (55). At the same time, the problems in (16) and (55) differ in notation only. Therefore, in Step 3, matrices $G_1^{(i+1)}, H_1^{(i+1)}, \ldots, G_p^{(i+1)}, H_p^{(i+1)}$ are determined, in fact, by Theorem 2, where only the notation should be changed.

Nevertheless, to avoid any confusion, we below provide Theorem 9, where matrices $G_1^{(i+1)}, H_1^{(i+1)}, \ldots, G_p^{(i+1)}, H_p^{(i+1)}$ are represented. To this end, we denote $\Gamma_{z_j^{(i)}} = E_{xz_j^{(i)}} E^\dagger_{z_j^{(i)} z_j^{(i)}} E_{z_j^{(i)} x}$. To simplify the notation, we set $\Gamma_{z_j} := \Gamma_{z_j^{(i)}}$.

Theorem 9. *Let $\mathbf{v}_0, \mathbf{v}_1^{(i)}, \ldots, \mathbf{v}_p^{(i)}$ be well-defined injections and let vectors $\mathbf{z}_1^{(i)}, \ldots, \mathbf{z}_p^{(i)}$ be pairwise uncorrelated. Then, the minimal Frobenius norm solution to the problem in (55) is given, for $j = 1, \ldots, p$, by*

$$G_j^{(i+1)} = U_{\Gamma_{z_j}, r_j} \quad \text{and} \quad H_j^{(i+1)} = U^T_{\Gamma_{z_j}, r_j} E_{xz_j^{(i)}} E^\dagger_{z_j^{(i)} z_j^{(i)}}. \tag{79}$$

Proof. The proof follows from the proof of Theorem 2. □

4.4. Error Analysis of the Solution of Problem in (7), (8), (9)

Theorem 10. *Let $G_0^{(0)}$, $H_0^{(0)}$, $G_k^{(i)}$, $H_k^{(i)}$, and $\mathbf{v}_k^{(i)}$, for $k = 1, \dots, p$, $i = 0, 1, \dots$, be determined by Theorems 2, 8, and 9. Let $\varepsilon^{(i)}$ be the associated error defined by (50) and (57). Then, the increase in the number of iterations i of the procedure represented in Sections 4.1–4.3 implies the decrease in the associated error, i.e.,*

$$\varepsilon^{(i+1)} \le \varepsilon^{(i)}. \tag{80}$$

Proof. Let us consider the initial case of the proposed technique when $i = 0$.

- **The case** $i = 0$. For $i = 0$, the i-th iteration loop represented in Section 4.1 implies

$$\varepsilon_v^{(1)} = \min_{\mathbf{v}_1,\dots,\mathbf{v}_p} \|\mathcal{F}(\mathbf{y}) - \sum_{k=1}^{p} G_k^{(0)} H_k^{(0)} \mathbf{z}_k(\mathbf{v}_k)\|_\Omega^2.$$

Here, for $j = 1, \dots, p$, $G_j^{(1)} = G_j^{(0)}$ and $H_j^{(1)} = H_j^{(0)}$. Therefore,

$$\varepsilon_v^{(1)} = \min_{\mathbf{v}_1,\dots,\mathbf{v}_p} \|\mathcal{F}(\mathbf{y}) - \sum_{k=1}^{p} G_k^{(1)} H_k^{(1)} \mathbf{z}_k(\mathbf{v}_k)\|_\Omega^2,$$

i.e., for any $\mathbf{v}_j$, in particular, for $\mathbf{v}_j = \mathbf{v}_j^{(0)}$,

$$\varepsilon_v^{(1)} \le \|\mathcal{F}(\mathbf{y}) - \sum_{k=1}^{p} G_k^{(1)} H_k^{(1)} \mathbf{z}_k(\mathbf{v}_k^{(0)})\|_\Omega^2 = \varepsilon_{GH}^{(1)} \tag{81}$$

and

$$\varepsilon_v^{(1)} \le \|\mathcal{F}(\mathbf{y}) - \sum_{k=1}^{p} G_k^{(0)} H_k^{(0)} \mathbf{z}_k(\mathbf{v}_k^{(0)})\|_\Omega^2 = \varepsilon^{(0)}. \tag{82}$$

Taking into account (81), we denote

$$\varepsilon^{(1)} = \varepsilon_v^{(1)}. \tag{83}$$

Then, by (82),

$$\varepsilon^{(1)} \le \varepsilon^{(0)}. \tag{84}$$

For $i = 1, 2, \dots$, let us prove inequality (80) by induction. To this end, we first consider the basis step of the induction, which consists of cases $i = 1$ and $i = 2$.

- *The basis step: Case* $i = 1$. If $i = 1$ then the i-th iteration loop (see Section 4.1) implies

$$\begin{aligned}\varepsilon_{GH}^{(2)} &= \min_{G_1,H_1,\dots,G_p,H_p} \left\| \mathcal{F}(\mathbf{y}) - \sum_{k=1}^{p} G_k H_k \mathbf{z}_k(\mathbf{v}_k^{(1)}) \right\|_\Omega^2 \\ &= \left\| \mathcal{F}(\mathbf{y}) - \sum_{k=1}^{p} G_k^{(2)} H_k^{(2)} \mathbf{z}_k(\mathbf{v}_k^{(1)}) \right\|_\Omega^2,\end{aligned}$$

i.e., for all G_k and H_k with $k = 1, \dots, p$,

$$\varepsilon_{GH}^{(2)} \le \left\| \mathcal{F}(\mathbf{y}) - \sum_{k=1}^{p} G_k H_k \mathbf{z}_k(\mathbf{v}_k^{(1)}) \right\|_\Omega^2.$$

In particular, for $G_k = G_k^{(1)}$ and $H_k = H_k^{(1)}$ with $k = 1, \ldots, p$,

$$\varepsilon_{GH}^{(2)} \leq \left\| \mathcal{F}(\mathbf{y}) - \sum_{k=1}^{p} G_k^{(1)} H_k^{(1)} \mathbf{z}_k(\mathbf{v}_k^{(1)}) \right\|_{\Omega}^{2}. \tag{85}$$

Further, because for $k = 1, \ldots, p$, $G_k^{(1)} = G_k^{(0)}$ and $H_k^{(1)} = H_k^{(0)}$, then

$$\min_{\mathbf{v}_1,\ldots,\mathbf{v}_p} \|\mathcal{F}(\mathbf{y}) - \sum_{k=1}^{p} G_k^{(1)} H_k^{(1)} \mathbf{z}_k(\mathbf{v}_k)\|_{\Omega}^{2} = \min_{\mathbf{v}_1,\ldots,\mathbf{v}_p} \|\mathcal{F}(\mathbf{y}) - \sum_{k=1}^{p} G_k^{(0)} H_k^{(0)} \mathbf{z}_k(\mathbf{v}_k)\|_{\Omega}^{2}. \tag{86}$$

Therefore, for $k = 1, \ldots, p$,

$$\mathbf{v}_k^{(2)} = \mathbf{v}_k^{(1)} \quad \text{and} \quad \mathbf{z}_k(\mathbf{v}_k^{(2)}) = \mathbf{z}_k(\mathbf{v}_k^{(1)}) \tag{87}$$

(see (51)). Thus, (85) and (87) imply

$$\varepsilon_{GH}^{(2)} \leq \left\| \mathcal{F}(\mathbf{y}) - \sum_{k=1}^{p} G_k^{(1)} H_k^{(1)} \mathbf{z}_k(\mathbf{v}_k^{(2)}) \right\|_{\Omega}^{2} = \varepsilon_v^{(2)}. \tag{88}$$

But since $G_k^{(1)} = G_k^{(0)}$, $H_k^{(1)} = H_k^{(0)}$ and $\mathbf{z}_k(\mathbf{v}_k^{(2)}) = \mathbf{z}_k(\mathbf{v}_k^{(1)})$, for $k = 1, \ldots, p$, then

$$\varepsilon_v^{(2)} = \varepsilon_v^{(1)}. \tag{89}$$

Then by (83), (88), and (89),

$$\varepsilon_{GH}^{(2)} \leq \varepsilon_v^{(1)} = \varepsilon^{(1)}. \tag{90}$$

Taking into account (88), we denote

$$\varepsilon^{(2)} = \varepsilon_{GH}^{(2)}. \tag{91}$$

Then, (90) and (91) imply

$$\varepsilon^{(2)} \leq \varepsilon^{(1)}. \tag{92}$$

- *The basis step: Case* $i = 2$. In this case, $\varepsilon_v^{(i+1)}$ in (53) is written as

$$\varepsilon_v^{(3)} = \min_{\mathbf{v}_1,\ldots,\mathbf{v}_p} \|\mathcal{F}(\mathbf{y}) - \sum_{k=0}^{p} G_k^{(2)} H_k^{(2)} \mathbf{z}_k(\mathbf{v}_k)\|_{\Omega}^{2} = \|\mathcal{F}(\mathbf{y}) - \sum_{k=1}^{p} G_k^{(2)} H_k^{(2)} \mathbf{z}_k(\mathbf{v}_k^{(3)})\|_{\Omega}^{2}.$$

That is, for any $\mathbf{v}_1, \ldots, \mathbf{v}_p$,

$$\varepsilon_v^{(3)} \leq \|\mathcal{F}(\mathbf{y}) - \sum_{k=0}^{p} G_k^{(2)} H_k^{(2)} \mathbf{z}_k(\mathbf{v}_k)\|_{\Omega}^{2}. \tag{93}$$

At the same time,

$$\begin{aligned} \varepsilon_{GH}^{(3)} &= \min_{G_1,H_1,\ldots,G_p,H_p} \left\| \mathcal{F}(\mathbf{y}) - \sum_{k=1}^{p} G_k H_k \mathbf{z}_k(\mathbf{v}_k^{(2)}) \right\|_{\Omega}^{2} \\ &= \left\| \mathcal{F}(\mathbf{y}) - \sum_{k=1}^{p} G_k^{(3)} H_k^{(3)} \mathbf{z}_k(\mathbf{v}_k^{(2)}) \right\|_{\Omega}^{2}. \end{aligned} \tag{94}$$

By (87), $\mathbf{z}_k(\mathbf{v}_k^{(2)}) = \mathbf{z}_k(\mathbf{v}_k^{(1)})$, therefore, $G_k^{(3)}, H_k^{(3)}$ solve, in fact,

$$\min_{G_1,H_1,\ldots,G_p,H_p} \left\| \mathcal{F}(\mathbf{y}) - \sum_{k=1}^{p} G_k H_k \mathbf{z}_k(\mathbf{v}_k^{(1)}) \right\|_\Omega^2,$$

i.e., $G_k^{(3)} = G_k^{(2)}, H_k^{(3)} = H_k^{(2)}$. Therefore, (94) implies

$$\varepsilon_{GH}^{(3)} = \left\| \mathcal{F}(\mathbf{y}) - \sum_{k=1}^{p} G_k^{(2)} H_k^{(2)} \mathbf{z}_k(\mathbf{v}_k^{(1)}) \right\|_\Omega^2 = \varepsilon_{GH}^{(2)}. \tag{95}$$

Further, (93) is true for all $\mathbf{v}_k$ and, in particular, for $\mathbf{v}_k^{(1)}$. Therefore, (93) and (95) imply

$$\varepsilon_v^{(3)} \le \varepsilon_{GH}^{(3)} = \varepsilon_{GH}^{(2)}. \tag{96}$$

Denote $\varepsilon^{(3)} = \varepsilon_v^{(3)}$. By (91), $\varepsilon^{(2)} = \varepsilon_{GH}^{(2)}$. Thus,

$$\varepsilon^{(3)} \le \varepsilon^{(2)}. \tag{97}$$

- *The inductive step.* For $s = 1,2,\ldots$, let us suppose that if $\mathbf{z}_k(\mathbf{v}_k^{(2s)}) = \mathbf{z}_k(\mathbf{v}_k^{(2s-1)})$, and $\varepsilon^{(2s)} = \varepsilon_{GH}^{(2s)}$ and $\varepsilon^{(2s-1)} = \varepsilon_v^{(2s-1)}$ then

$$\varepsilon^{(2s)} \le \varepsilon^{(2s-1)}. \tag{98}$$

Below, we show that then $\varepsilon^{(2s+1)} \le \varepsilon^{(2s)}$ and $\varepsilon^{(2s+2)} \le \varepsilon^{(2s+1)}$, i.e., that (80) is true. To this end, for the i-th iterative loop, let us consider case $i = 2s$ where $s = 1,2,\ldots$. We have

$$\begin{aligned} \varepsilon_v^{(2s+1)} &= \min_{\mathbf{v}_1,\ldots,\mathbf{v}_p} \|\mathcal{F}(\mathbf{y}) - \sum_{k=1}^{p} G_k^{(2s)} H_k^{(2s)} \mathbf{z}_k(\mathbf{v}_k)\|_\Omega^2 \\ &= \|\mathcal{F}(\mathbf{y}) - \sum_{k=1}^{p} G_k^{(2s)} H_k^{(2s)} \mathbf{z}_k(\mathbf{v}_k^{(2s+1)})\|_\Omega^2. \end{aligned}$$

That is, for any $\mathbf{v}_1,\ldots,\mathbf{v}_p$,

$$\varepsilon_v^{(2s+1)} \le \|\mathcal{F}(\mathbf{y}) - \sum_{k=1}^{p} G_k^{(2s)} H_k^{(2s)} \mathbf{z}_k(\mathbf{v}_k)\|_\Omega^2. \tag{99}$$

At the same time,

$$\begin{aligned} \varepsilon_{GH}^{(2s+1)} &= \min_{G_1,H_1,\ldots,G_p,H_p} \left\| \mathcal{F}(\mathbf{y}) - \sum_{k=1}^{p} G_k H_k \mathbf{z}_k(\mathbf{v}_k^{(2s)}) \right\|_\Omega^2 \\ &= \left\| \mathcal{F}(\mathbf{y}) - \sum_{k=1}^{p} G_k^{(2s+1)} H_k^{(2s+1)} \mathbf{z}_k(\mathbf{v}_k^{(2s)}) \right\|_\Omega^2. \end{aligned} \tag{100}$$

By the assumption $\mathbf{z}_k(\mathbf{v}_k^{(2s)}) = \mathbf{z}_k(\mathbf{v}_k^{(2s-1)})$, $G_k^{(2s+1)}, H_k^{(2s+1)}$ solve, in fact,

$$\min_{G_1,H_1,\ldots,G_p,H_p} \left\| \mathcal{F}(\mathbf{y}) - \sum_{k=1}^{p} G_k H_k \mathbf{z}_k(\mathbf{v}_k^{(2s-1)}) \right\|_\Omega^2,$$

i.e., $G_k^{(2s+1)} = G_k^{(2s)}, H_k^{(2s+1)} = H_k^{(2s)}$. Therefore, (100) implies

$$\varepsilon_{GH}^{(2s+1)} = \left\| \mathcal{F}(\mathbf{y}) - \sum_{k=1}^{p} G_k^{(2s)} H_k^{(2s)} \mathbf{z}_k(\mathbf{v}_k^{(2s-1)}) \right\|_{\Omega}^{2} = \varepsilon_{GH}^{(2s)}. \tag{101}$$

Further, (99) is true for all $\mathbf{v}_k$ and, in particular, for $\mathbf{v}_k^{(2s-1)}$. Therefore, (99) and (101) imply

$$\varepsilon_v^{(2s+1)} \leq \varepsilon_{GH}^{(2s+1)} = \varepsilon_{GH}^{(2s)}. \tag{102}$$

Denote $\varepsilon^{(2s+1)} = \varepsilon_v^{(2s+1)}$. By the assumption $\varepsilon^{(2s)} = \varepsilon_{GH}^{(2s)}$, then

$$\varepsilon^{(2s+1)} \leq \varepsilon^{(2s)}. \tag{103}$$

Further, for $s = 1, 2, \ldots$, let us now consider the case $i = 2s+1$. Then

$$\begin{aligned} \varepsilon_{GH}^{(2s+2)} &= \min_{G_1, H_1, \ldots, G_p, H_p} \left\| \mathcal{F}(\mathbf{y}) - \sum_{k=1}^{p} G_k H_k \mathbf{z}_k(\mathbf{v}_k^{(2s+1)}) \right\|_{\Omega}^{2} \\ &= \left\| \mathcal{F}(\mathbf{y}) - \sum_{k=1}^{p} G_k^{(2s+2)} H_k^{(2s+2)} \mathbf{z}_k(\mathbf{v}_k^{(2s+1)}) \right\|_{\Omega}^{2}. \end{aligned}$$

Therefore, for any G_k and H_k,

$$\varepsilon_{GH}^{(2s+2)} \leq \left\| \mathcal{F}(\mathbf{y}) - \sum_{k=1}^{p} G_k H_k \mathbf{z}_k(\mathbf{v}_k^{(2s+1)}) \right\|_{\Omega}^{2}. \tag{104}$$

We also need the following. By (55), $G_k^{(2s+1)}$ and $H_k^{(2s+1)}$ solve

$$\min_{G_1, H_1, \ldots, G_p, H_p} \left\| \mathcal{F}(\mathbf{y}) - \sum_{k=1}^{p} G_k H_k \mathbf{z}_k(\mathbf{v}_k^{(2s)}) \right\|_{\Omega}^{2}. \tag{105}$$

Since by the assumption, $\mathbf{z}_k(\mathbf{v}_k^{(2s)}) = \mathbf{z}_k(\mathbf{v}_k^{(2s-1)})$, then (105) is equivalent to

$$\min_{G_1, H_1, \ldots, G_p, H_p} \left\| \mathcal{F}(\mathbf{y}) - \sum_{k=1}^{p} G_k H_k \mathbf{z}_k(\mathbf{v}_k^{(2s-1)}) \right\|_{\Omega}^{2}. \tag{106}$$

Thus,

$$G_k^{(2s+1)} = G_k^{(2s)} \quad \text{and} \quad H_k^{(2s+1)} = H_k^{(2s)}. \tag{107}$$

For $G_k = G_k^{(2s+1)}$ and $H_k = H_k^{(2s+1)}$, (104) implies

$$\varepsilon_{GH}^{(2s+2)} \leq \left\| \mathcal{F}(\mathbf{y}) - \sum_{k=1}^{p} G_k^{(2s+1)} H_k^{(2s+1)} \mathbf{z}_k(\mathbf{v}_k^{(2s+1)}) \right\|_{\Omega}^{2} \tag{108}$$

and then by (107),

$$\varepsilon_{GH}^{(2s+2)} \leq \left\| \mathcal{F}(\mathbf{y}) - \sum_{k=1}^{p} G_k^{(2s)} H_k^{(2s)} \mathbf{z}_k(\mathbf{v}_k^{(2s+1)}) \right\|_{\Omega}^{2} = \varepsilon_v^{(2s+1)}. \tag{109}$$

Because of (109), denote $\varepsilon^{(2s+2)} = \varepsilon_{GH}^{(2s+2)}$. Recall that according to the above, $\varepsilon^{(2s+1)} = \varepsilon_{v}^{(2s+1)}$. Therefore, (109) implies

$$\varepsilon^{(2s+2)} \leq \varepsilon^{(2s+1)}. \tag{110}$$

Thus, (80) is true.

□

Further, we wish to show that the proposed procedure for the solution of problem in (7), (8), (9) converges to a so-called coordinate-wise minimum, which is defined in Theorem 11 below. To this end, let us denote

$$f(P, \mathbf{v}) = \|\mathcal{F}(\mathbf{y}) - \sum_{j=0}^{p} P_j \mathbf{z}_j(\mathbf{v}_j)\|_{\Omega}^2, \tag{111}$$

where, for $j = 0, 1, \ldots, p$, as in (20), $P_j = G_j H_j$, $P = [P_0, \ldots, P_p]$, $\mathbf{v} = [\mathbf{v}_0, \ldots, \mathbf{v}_p]^T$ and $\mathbf{v}_0 = \mathbf{y}$.

For $i = 1, 2, \ldots$, we also write $P_j^{(i)} = G_j^{(i)} H_j^{(i)}$, $P^{(i)} = [P_0^{(0)}, P_1^{(i)}, \ldots, P_p^{(i)}]$ and $\mathbf{v}^{(i)} = [\mathbf{v}_0^{(i)}, \ldots, \mathbf{v}_p^{(i)}]^T$. As before, $\mathbf{v}_0^{(i)} = \mathbf{y}$.

Let us now define compact sets K_1 and K_2 such that

$$K_1 = \{P : 0 \leq f(P, \mathbf{v}^{(0)}) \leq f(P^{(0)}, \mathbf{v}^{(0)})\} \tag{112}$$

and

$$K_2 = \{\mathbf{v} : 0 \leq f(P^{(0)}, \mathbf{v}) \leq f(P^{(0)}, \mathbf{v}^{(1)})\}. \tag{113}$$

Theorem 11. *Let $P \in K_1$ and $\mathbf{v} \in K_2$. Let $\mathbf{v}^{(i)} = [\mathbf{v}_0^{(i)}, \ldots, \mathbf{v}_p^{(i)}]^T$ and $P^{(i)}$ be determined by Theorems 2, 8, and 9, respectively, and let $S^{(i)} = (P^{(i)}, \mathbf{v}^{(i)})$. Then, any cluster point of sequence $\{S^{(i)}\}$, say $S^* = (P^*, \mathbf{v}^*)$ is a coordinate-wise minimum point of $f(P, \mathbf{v})$, i.e.,*

$$P^* \in \arg\min_{P \in K_1} f(P, \mathbf{v}^*) \quad \text{and} \quad \mathbf{v}^* \in \arg\min_{\mathbf{v} \in K_2} f(P^*, \mathbf{v}).$$

Proof. Since each K_j, for $j = 1, 2$, is compact, then there is a subsequence $\{S^{(j_t)}\} = \{(P^{(j_t)}, \mathbf{v}^{(j_t)})\}$ such that $S^{(j_t)} \to S^*$ when $t \to \infty$. Let $\overline{P}^*$ and $\overline{P}^{(j_t)}$ be defined by

$$\overline{P}^* \in \arg\min_{P \in K_1} f(P, \mathbf{v}^*) \quad \text{and} \quad \overline{P}^{(j_t)} \in \arg\min_{P \in K_1} f(P, \mathbf{v}^{(j_t)}).$$

Similar to [75], $\overline{P}^*$ and $\overline{P}^{(j_t)}$ are called the best responses to P associated with $\mathbf{v}^*$ and $\mathbf{v}^{(j_t)}$, respectively. Consider entry $P^{(j_t)}$ of $S^{(j_t)}$. Then, on the basis of the proof of Theorem 3.1 in [75], we have

$$f(\overline{P}^*, \mathbf{v}^{(j_t)}) \geq f(\overline{P}^{(j_t)}, \mathbf{v}^{(j_t)}) \geq f(P^{(j_t+1)}, \mathbf{v}^{(j_t+1)}) \geq f(P^{(j_{t+1})}, \mathbf{v}^{(j_{t+1})}).$$

By continuity, as $t \to \infty$,

$$f(\overline{P}^*, \mathbf{v}^*) \geq f(P^*, \mathbf{v}^*). \tag{114}$$

This implies that (114) should hold as an equality, since the inequality is false by the definition of the best response $\overline{P}^*$. Thus, P^* is the best response for $\mathbf{v}^*$, or equivalently, P^* is the solution of the problem $\arg\min_{P \in K_1} f(P, \mathbf{v}^*)$.

The proof is similar if we consider entry $\mathbf{v}^{(j_t)}$ of $S^{(j_t)}$. □

Example 2. *In the above Example 1, we illustrated the decrease in the error associated with approximating operator $\mathcal{T}_p(\mathbf{v}_0, \ldots, \mathbf{v}_p)$ when parameters p and q increase. In Example 1, injections were not optimal.*

Here, for the case of optimal injections, we wish to numerically illustrate Theorem 10, i.e., that the error $\varepsilon^{(i)}$ decreases if the number of iterations i increases.

For $i = 0, 1, \ldots$, we denote

$$\mathcal{T}_p^{(i,i+1)}(\mathbf{v}_0, \mathbf{v}_1^{(i+1)}, \ldots, \mathbf{v}_p^{(i+1)}) = \sum_{k=0}^{p} G_k^{(i)} H_k^{(i)} \mathbf{z}_k(\mathbf{v}_k^{(i+1)})$$

and

$$\mathcal{T}_p^{(i+1,i)}(\mathbf{v}_0, \mathbf{v}_1^{(i)}, \ldots, \mathbf{v}_p^{(i)}) = \sum_{k=0}^{p} G_k^{(i+1)} H_k^{(i+1)} \mathbf{z}_k(\mathbf{v}_k^{(i)}).$$

According to the procedure described in Section 4, for each i-th iteration where $i = 0, 1, \ldots$, the proposed method is represented by either

$$\mathcal{T}_p^{(i,i+1)}(\mathbf{v}_0, \mathbf{v}_1^{(i+1)}, \ldots, \mathbf{v}_p^{(i+1)})$$

or

$$\mathcal{T}_p^{(i+1,i)}(\mathbf{v}_0, \mathbf{v}_1^{(i)}, \ldots, \mathbf{v}_p^{(i)}).$$

To simplify the notation, both $\mathcal{T}_p^{(i,i+1)}(\mathbf{v}_0, \mathbf{v}_1^{(i+1)}, \ldots, \mathbf{v}_p^{(i+1)})$ and $\mathcal{T}_p^{(i+1,i)}(\mathbf{v}_0, \mathbf{v}_1^{(i)}, \ldots, \mathbf{v}_p^{(i)})$ are denoted by $\mathcal{T}_p^{(i)}(\mathbf{v}^{(i)})$.

We assume that $\mathbf{x} \in L^2(\Omega, \mathbb{R}^m)$ and $\mathbf{y} \in L^2(\Omega, \mathbb{R}^m)$ are uniformly and normally distributed random vectors, respectively. Initial injections $\mathbf{v}_1^{(0)} \in L^2(\Omega, \mathbb{R}^{q_1}), \ldots, \mathbf{v}_p^{(0)} \in L^2(\Omega, \mathbb{R}^{q_p})$ are here chosen as uniformly distributed random vectors. It is assumed that covariance matrices E_{xv_j}, $E_{v_iv_j}$, for $i, j = 0, 1, 2$, are given by $E_{xv_j} = \frac{1}{s} X V_j^T$ and $E_{v_iv_j} = \frac{1}{s} V_i V_j^T$ where $X \in \mathbb{R}^{m \times s}$ and $V_j \in \mathbb{R}^{q_j \times s}$ are samples of $\mathbf{x}$ and $\mathbf{v}_j$, respectively, for $j = 0, 1, 2$.

We choose $m = 400$ and $r_0 = 200$. In Table 3, values of error $\varepsilon^{(i)}$ associated with operators $\mathcal{T}_0^{(i)}(\mathbf{v}^{(i)})$, $\mathcal{T}_1^{(i)}(\mathbf{v}^{(i)})$ and $\mathcal{T}_2^{(i)}(\mathbf{v}^{(i)})$, for $i = 14$ are represented.

In Figure 3, diagrams of the errors are represented. It follows from Table 3 and Figure 3 that the error decreases if the number of iterations i increases. In particular, the error remains the same after $i = 14$.

Similar to what is explained in Example 1, the proposed method is more effective because the increase in the number of parameters to optimize in the operator $\mathcal{T}$ leads to the reduction in the error, as stated in Theorem 10.

Table 3. Numerical characterizations of approximating operators $\mathcal{T}_0^{(i)}(\mathbf{v}^{(i)})$, $\mathcal{T}_1^{(i)}(\mathbf{v}^{(i)})$ and $\mathcal{T}_2^{(i)}(\mathbf{v}^{(i)})$.

$\mathcal{T}_p^{(i)}(\mathbf{v}^{(i)})$	q_0	q_1	q_2	r_0	r_1	r_2	i	$\varepsilon^{(i)}$
$\mathcal{T}_0^{(i)}(\mathbf{v}^{(i)})$	400	N/A	N/A	200	N/A	N/A	N/A	30.9
$\mathcal{T}_1^{(i)}(\mathbf{v}^{(i)})$	400	400	400	100	100	N/A	14	21.8
$\mathcal{T}_2^{(i)}(\mathbf{v}^{(i)})$	400	400	400	67	67	66	14	18.9

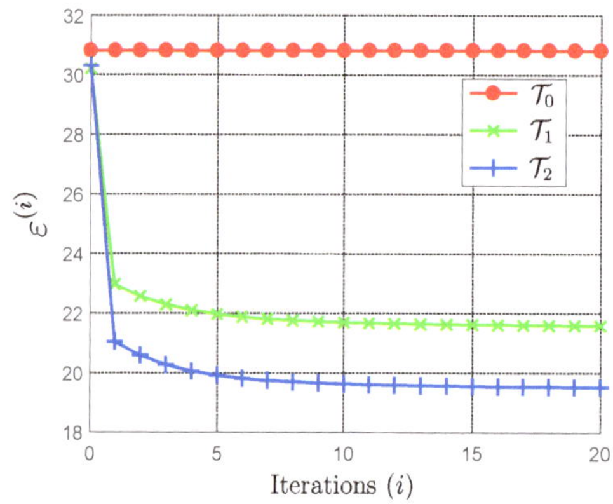

Figure 3. Example 2: Diagrams of the errors associated with $\mathcal{T}_0 := \mathcal{T}_0^{(i)}(\mathbf{v}^{(i)})$, $\mathcal{T}_1 := \mathcal{T}_1^{(i)}(\mathbf{v}^{(i)})$ and $\mathcal{T}_2 := \mathcal{T}_2^{(i)}(\mathbf{v}^{(i)})$.

Author Contributions: Conceptualization, P.S.-Q. and A.T.; methodology, A.T.; software, P.S.-Q.; validation, A.T.; writing—original draft, A.T.; visualization, P.S.-Q. All authors have read and agreed to the published version of the manuscript.

Funding: This research received no external funding.

Data Availability Statement: The raw data supporting the conclusions of this article will be made available by the authors on request.

Acknowledgments: This work was financially supported by *Vicerrectoría de Investigación y Extensión* from *Instituto Tecnológico de Costa Rica* (Research #1440054).

Conflicts of Interest: The authors declare no conflicts of interest.

References

1. Amat, S.; Busquier, S.; Negra, M. Adaptive Approximation of Nonlinear Operators. *Numer. Funct. Anal. Optim.* **2004**, *25*, 397–405. [CrossRef]
2. Bruno, V.I. An approximate Weierstrass theorem in topological vector space. *J. Approx. Theory* **1984**, *42*, 1–3. [CrossRef]
3. Chen, T.; Chen, H. Universal approximation to nonlinear operators by neural networks with arbitrary activation functions and its application to dynamical systems. *IEEE Trans. Neural Netw.* **1995**, *6*, 911–917. [CrossRef] [PubMed]
4. Dingle, K.; Camargo, C.Q.; Louis, A.A. Input–output maps are strongly biased towards simple outputs. *Nat. Commun.* **2018**, *9*, 761. [CrossRef]
5. Gallman, P.G.; Narendra, K. Representations of nonlinear systems via the Stone-Weierstrass theorem. *Automatica* **1976**, *12*, 619–622. [CrossRef]
6. Howlett, P.G.; Torokhti, A.P.; Pearce, C.E.M. A Philosophy for the Modelling of Realistic Non-linear Systems. *Proc. Am. Math. Soc.* **2003**, *131*, 353–363. [CrossRef]
7. Istrăţescu, V.I. A Weierstrass theorem for real Banach spaces. *J. Approx. Theory* **1977**, *19*, 118–122. [CrossRef]
8. Prenter, P.M. A Weierstrass Theorem for Real, Separable Hilbert Spaces. *J. Approx. Theory* **1970**, *4*, 341–357. [CrossRef]
9. Prolla, J.B.; Machado, S. Weierstrass-Stone theorems for set-valued mappings. *J. Approx. Theory* **1982**, *36*, 1–15. [CrossRef]
10. Rao, N.V. Stone-Weierstrass theorem revisited. *Am. Math. Mon.* **2005**, *112*, 726–729. [CrossRef]
11. Sandberg, I.W. Notes on uniform approximation of time–varying systems on finite time intervals. *IEEE Trans. Circuits Syst. I Fundam. Theory Appl.* **1998**, *45*, 863–864. [CrossRef]
12. Sandberg, I.W. Time delay polynomial networks and quality of approximation. *IEEE Trans. Circuits Syst. I Fundam. Theory Appl.* **2000**, *47*, 40–49. [CrossRef]
13. Sandberg, I.W. R+ Fading Memory and Extensions of Input-Output Maps. *IEEE Trans. Circuits Syst. I Fundam. Theory Appl.* **2002**, *49*, 1586–1591. [CrossRef]
14. Timofte, V. Stone-Weierstrass theorem revisited. *J. Approx. Theory* **2005**, *536*, 45–59. [CrossRef]
15. Piotrowski, T.; Cavalcante, R.; Yamada, I. Stochastic MV-PURE Estimator-Robust Reduced-Rank Estimator for Stochastic Linear Model. *IEEE Trans. Signal Process.* **2009**, *57*, 1293–1303. [CrossRef]

16. Hua, Y.; Nikpour, M.; Stoica, P. Optimal reduced-rank estimation and filtering. *IEEE Trans. Signal Process.* **2001**, *49*, 457–469.
17. Zhu, X.L.; Zhang, X.D.; Ding, Z.Z.; Jia, Y. Adaptive nonlinear PCA algorithms for blind source separation without prewhitening. *IEEE Trans. Circuits Syst. I Regul. Pap.* **2006**, *53*, 745–753.
18. Wang, Q.; Jing, Y. New rank detection methods for retuced-rank MIMO systems. *EURASIP J. Wirel. Comm. Netw.* **2015**, *2015*, 1–16. [CrossRef]
19. Bai, L.; Dou, S.; Xiao, Z.; Choi, J. Doubly iterative multiple-input-multiple-output-bit-interleaved coded modulation receiver with joint channel estimation and randomised sampling detection. *IET Signal Process.* **2016**, *10*, 335–341. [CrossRef]
20. Brillinger, D.R. *Time Series: Data Analysis and Theory*; SIAM: San Francisco, CA, USA, 2001.
21. Jolliffe, I. *Principal Component Analysis*, 2nd ed.; Springer: New York, NY, USA, 2002.
22. Bühlmann, P.; van de Geer, S. *Statistics for High-Dimensional Data*; Springer: New York, NY, USA, 2011.
23. She, Y.; Li, S.; Wu, D. Robust Orthogonal Complement Principal Component Analysis. *J. Am. Stat. Assoc.* **2016**, *111*, 763–771. [CrossRef]
24. Torokhti, A.; Friedland, S. Towards theory of generic Principal Component Analysis. *J. Multivar. Anal.* **2009**, *100*, 661–669. [CrossRef]
25. Saghri, J.A.; Schroeder, S.; Tescher, A.G. Adaptive two-stage Karhunen-Loeve-transform scheme for spectral decorrelation in hyperspectral bandwidth compression. *Opt. Eng.* **2010**, *49*, 057001. [CrossRef]
26. Kmuntchav, R.; Kountcheva, R. Hierarchical Adaptive KL-Based Transform: Algorithms and Applications. In *Computer Vision in Control Systems-1*; Favorskaya, M.N., Jain, L.C., Eds.; Springer: Cham, Switzerland, 2015; Volume 73, pp. 91–136.
27. de Moura, E.P.; de Abreu Melo Junior, F.; Damasceno, F.; Figueiredo, L.; de Andrade, C.; de Almeida, M. Classification of imbalance levels in a scaled wind turbine through detrended fluctuation analysis of vibration signals. *Renew. Energy* **2016**, *96 Pt A*, 993–1002. [CrossRef]
28. Azimia, R.; Ghayekhlooa, M.; Ghofranib, M.; Sajedic, H. A novel clustering algorithm based on data transformation approaches. *Expert Syst. Appl.* **2017**, *76*, 59–70. [CrossRef]
29. Burge, M.J.; Burger, W. *Digital Image Processing—An Algorithmic Introduction Using Java*; Springer: London, UK, 2006.
30. Phophalia, A.; Rajwade, A.; Mitra, S.K. Rough set based image denoising for brain MR images. *Signal Process.* **2014**, *103*, 24–35. [CrossRef]
31. Chenn, S.; Billings, S.A.; Grant, P. Non-linear system identification using neural networks. *Int. J. Control* **1990**, *51*, 1191–1214. [CrossRef]
32. Chenn, S.; Billings, S.A. Neural networks for nonlinear dynamic system modelling and identification. *Int. J. Control* **1992**, *56*, 319–346. [CrossRef]
33. Fomin, V.N.; Ruzhansky, M.V. Abstract Optimal Linear Filtering. *SIAM J. Control Optim.* **2000**, *38*, 1334–1352. [CrossRef]
34. Temlyakov, V.N. Nonlinear Methods of Approximation. *Found. Comput. Math.* **2003**, *3*, 33–107. [CrossRef]
35. Torokhti, A.; Howlett, P. Best approximation of identity mapping: The case of variable memory. *J. Approx. Theory* **2006**, *143*, 111–123. [CrossRef]
36. Liu, W.; Zhang, H.; Yu, K.; Tan, X. Optimal linear filtering for networked systems with communication constraints, fading measurements, and multiplicative noises. *Int. J. Adapt. Control Signal Process.* **2016**. [CrossRef]
37. Aminzare, Z.; Sontag, E.D. Synchronization of Diffusively-Connected Nonlinear Systems: Results Based on Contractions with Respect to General Norms. *IEEE Trans. Netw. Sci. Eng.* **2014**, *1*, 91–106. [CrossRef]
38. Schneider, M.K.; Willsky, A.S. A Krylov Subspace Method for Covariance Approximation and Simulation of Random Processes and Fields. *Multidimens. Syst. Signal Process.* **2003**, *14*, 295–318. [CrossRef]
39. Chenn, S.; Billings, S.A. Representations of non-linear systems: The NARMAX model. *Int. J. Control* **1989**, *49*, 1013–1032. [CrossRef]
40. Piroddi, L.; Spinelli, W. An identification algorithm for polynomial NARX models based on simulation error minimization. *Int. J. Control* **2003**, *76*, 1767–1781. [CrossRef]
41. Alter, O.; Golub, G.H. Singular value decomposition of genome-scale mRNA lengths distribution reveals asymmetry in RNA gel electrophoresis band broadening. *Process. Natl. Acad. Sci. USA* **2006**, *103*, 11828–11833. [CrossRef]
42. Gianfelici, F.; Turchetti, C.; Crippa, P. A non-probabilistic recognizer of stochastic signals based on KLT. *Signal Process.* **2009**, *89*, 422–437. [CrossRef]
43. Formaggia, L.; Quarteroni, A.; Veneziani, A. (Eds.) *Cardiovascular Mathematics—Modeling and Simulation of the Circulatory System*; Springer: Milano, Italy, 2009; Volume 1.
44. Ambrosi, D.; Quarteroni, A.; Rozza, G. (Eds.) *Modelling of Physiological Flows*; Springer: Milano, Italy, 2011; Volume V.
45. Piotrowski, T.; Yamada, I. Performance of the stochastic MV-PURE estimator in highly noisy settings. *J. Frankl. Insitute* **2014**, *351*, 3339–3350. [CrossRef]
46. Poor, H.V. *An Introduction to Signal Processing and Estimation*, 2nd ed.; Springer: New York, NY, USA, 2001.

47. Torokhti, A.; Soto-Quiros, P. Optimal modeling of nonlinear systems: Method of variable injections. *Proyecciones* **2024**, *43*, 189–224. [CrossRef]
48. Yang, S.; Xing, T.; Ke, C.; Liang, J.; Ke, X. Effect of wavefront distortion on the performance of coherent detection systems: Theoretical analysis and experimental research. *Photonics* **2023**, *10*, 493. [CrossRef]
49. Stanković, L.; Brajović, M.; Stanković, I.; Lerga, J.; Daković, M. RANSAC-based signal denoising using compressive sensing. *Circuits Syst. Signal Process.* **2021**, *40*, 3907–3928. [CrossRef]
50. Wang, Y.; Cheng, K.; Zhao, S.; Xu, E. Human ear image recognition method using PCA and Fisherface complementary double feature extraction. *J. Artif. Intell. Technol.* **2023**, *3*, 18–24. [CrossRef]
51. Al-Saffar, N.F.H.; Al-Saiq, I.R. Symmetric text encryption scheme based Karhunen Loeve transform. *J. Discret. Math. Sci. Cryptogr.* **2022**, *25*, 2773–2781. [CrossRef]
52. Soto-Quiros, P.; Torokhti, A. Fast random vector transforms in terms of pseudo-inverse within the Wiener filtering paradigm. *J. Comput. Appl. Math.* **2024**, *448*, 115927. [CrossRef]
53. Howlett, P.; Torokhti, A. Optimal approximation of a large matrix by a sum of projected linear mappings on prescribed subspaces. *Electron. J. Linear Algebra* **2024**, *40*, 585–605. [CrossRef]
54. Torokhti, A.; Howlett, P. Optimal estimation of distributed highly noisy signals within KLT-Wiener archetype. *Digit. Signal Process.* **2023**, *143*, 104225. [CrossRef]
55. Howlett, P.; Torokhti, A. An optimal linear filter for estimation of random functions in Hilbert space. *ANZIAM J.* **2020**, *62*, 274–301. [CrossRef]
56. Ledoit, O.; Wolf, M. A well-conditioned estimator for large-dimensional covariance matrices. *J. Multivar. Anal.* **2004**, *88*, 365–411. [CrossRef]
57. Ledoit, O.; Wolf, M. Nonlinear shrinkage estimation of large-dimensional covariance matrices. *Ann. Stat.* **2012**, *40*, 1024–1060. [CrossRef]
58. Adamczak, R.; Litvak, A.E.; Pajor, A.; Tomczak-Jaegermann, N. Quantitative estimates of the convergence of the empirical covariance matrix in log-concave ensembles. *J. Am. Math. Soc.* **2009**, *2*, 535–561. [CrossRef]
59. Vershynin, R. How Close is the Sample Covariance Matrix to the Actual Covariance Matrix? *J. Theor. Probab.* **2012**, *25*, 655–686. [CrossRef]
60. Won, J.-H.; Lim, J.; Kim, S.-J.; Rajaratnam, B. Condition-number-regularized covariance estimation. *J. R. Stat. Soc. Ser. B* **2013**, *75*, 427–450. [CrossRef] [PubMed]
61. Yang, R.; Berger, J.O. Estimation of a covariance matrix using the reference prior. *Ann. Statist.* **1994**, *22*, 1195–1211. [CrossRef]
62. Ben-Israel, A.; Greville, T.N.E. *Generalized Inverses: Theory and Applications*, 2nd ed.; Springer: New York, NY, USA, 2003.
63. Hotelling, H. Analysis of a Complex of Statistical Variables into Principal Components. *J. Educ. Psychol.* **1933**, *6*, 417–441 and 498–520. [CrossRef]
64. Karhunen, K. Über Lineare Methoden in der Wahrscheinlichkeitsrechnung. *Ann. Acad. Sci. Fenn. Ser. A I. Math.-Phys.* **1947**, *1*, 1–79.
65. Loève, M. Fonctions aléatoires de second ordre. In *Processus Stochastiques et Mouvement Brownien*; Lévy, P., Ed.; Hermann: Paris, France, 1948.
66. Scharf, L. The SVD and reduced rank signal processing. *Signal Process.* **1991**, *25*, 113–133. [CrossRef]
67. Hua, Y.; Liu, W. Generalized Karhunen-Loeve transform. *IEEE Signal Process. Lett.* **1998**, *5*, 141–142.
68. Torokhti, A.; Soto-Quiros, P. Generalized Brillinger-Like Transforms. *IEEE Signal Process. Lett.* **2016**, *23*, 843–847. [CrossRef]
69. Torokhti, A.; Miklavcic, S. Data compression under constraints of causality and variable finite memory. *Signal Process.* **2010**, *90*, 2822–2834. [CrossRef]
70. Lathauwer, L.D.; Moor, B.D.; Vandewalle, J. On the Best Rank-1 and Rank-(R1,R2, ..., RN) Approximation of Higher-Order Tensors. *SIAM J. Matrix Anal. Appl.* **2006**, *21*, 1324–1342. [CrossRef]
71. Grasedyck, L.; Kressner, D.; Tobler, C. A literature survey of low-rank tensor approximation techniques. *GAMM-Mitteilungen* **2013**, *36*, 53–78. [CrossRef]
72. Friedland, S.; Tammali, V. Low-Rank Approximation of Tensors. In *Numerical Algebra, Matrix Theory, Differential-Algebraic Equations and Control Theory*; Benner, P., Bollhöfer, M., Kressner, D., Mehl, C., Stykel, T., Eds.; Springer: Cham, Switzerland, 2015; pp. 377–411.
73. Billings, S.A. *Nonlinear System Identification-Narmax Methods in the Time, Frequency, and Spatio-Temporal Domains*; John Wiley and Sons, Ltd.: Hoboken, NJ, USA, 2013.
74. Schoukens, M.; Tiels, K. Identification of Block-oriented Nonlinear Systems Starting from Linear Approximations: A Survey. *Automatica* **2017**, *85*, 272–292. [CrossRef]
75. Chen, B.; He, S.; Li, Z.; Zhang, S. Maximum Block Improvement and Polynomial Optimization. *SIAM J. Optim.* **2012**, *22*, 87–107. [CrossRef]
76. Abed-Meraim, K.; Qui, W.; Hua, Y. Blind system identification. *Proc. IEEE* **1997**, *85*, 1310–1322. [CrossRef]

77. Hua, Y. Blind methods of system identification. *Circuits Syst. Signal Process.* **2002**, *21*, 91–108. [CrossRef]
78. Shi, X. Mathematical Description of Blind Signal Processing. In *Blind Signal Processing: Theory and Practice*; Springer: Berlin/Heidelberg, Germany, 2011; pp. 27–59.
79. Torokhti, A.; Howlett, P. Optimal Fixed Rank Transform of the Second Degree. *IEEE Trans. Circuits Syst. II Analog. Digit. Signal Process.* **2001**, *48*, 309–315. [CrossRef]
80. Torokhti, A.; Howlett, P. *Computational Methods for Modelling of Nonlinear Systems*; Elsevier: Amsterdam, The Netherlands, 2007.
81. Friedland, S.; Torokhti, A. Generalized Rank-Constrained Matrix Approximations. *SIAM J. Matrix Anal. Appl.* **2007**, *29*, 656–659. [CrossRef]
82. Liu, X.; Li, W.; Wang, H. Rank constrained matrix best approximation problem with respect to (skew) Hermitian matrices. *J. Comput. Appl. Math.* **2017**, *319*, 77–86. [CrossRef]
83. Boutsidis, C.; Woodruff, D.P. Optimal CUR Matrix Decompositions. *SIAM J. Comput.* **2017**, *46*, 543–589. [CrossRef]
84. Wang, H. Rank constrained matrix best approximation problem. *Appl. Math. Lett.* **2015**, *50*, 98–104. [CrossRef]
85. Golub, G.; Loan, C.F.V. *Matrix Computations*; Jons Hopkins University Press: Baltimore, MD, USA, 1996.
86. Kanwal, R. *Linear Integral Equations*; Birkhäuser Boston: Boston, MA, USA, 1996.

MDPI AG
Grosspeteranlage 5
4052 Basel
Switzerland
Tel.: +41 61 683 77 34

Mathematical and Computational Applications Editorial Office
E-mail: mca@mdpi.com
www.mdpi.com/journal/mca

www.ingramcontent.com/pod-product-compliance
Lightning Source LLC
Chambersburg PA
CBHW041427210926
51850CB00013B/3371